Residential Construction Academy

House Wiring

Residential Construction Academy

House Wiring

Greg Fletcher

THOMSON

DELMAR LEARNING

Australia Canada Mexico Singapore Spain United Kingdom United States

DELMAR

THOMSON LEARNING

Residential Construction Academy: House Wiring
Greg Fletcher

Vice President, Technology and Trades SBU:
Alar Elken

Editorial Director:
Sandy Clark

Acquisitions Editor:
Alison Weintraub

Development Editor:
Monica Ohlinger

Marketing Director:
Cynthia Eichelman

Channel Manager:
Fair Huntoon

Marketing Coordinator:
Erin Coffin

Production Director:
Mary Ellen Black

Production Manager:
Andrew Crouth

Senior Project Editor:
Christopher Chien

Art/Design Specialist:
Mary Beth Vought

Editorial Assistant:
Jennifer Luck

Full Production Services:
Carlisle Publishers Services

Cover Photo:
Randall Perry

Library of Congress Cataloging-in-Publication Data
Fletcher, Greg.
   Residential construction academy : house wiring / Greg Fletcher.—1st ed.
      p. cm.
Includes index.
   ISBN 1-4018-1371-2 (alk. paper)
   1. Electric wiring, Interior. 2. Dwellings—Electric equipment. I. Title.
   TK3285.F58 2003
   621.319'24—dc21

Card Number: 2003012990

ISBN: 1-4018-1371-2

NOTICE TO THE READER

Table of Contents

SECTION **4 Residential Electrical System Trim-Out. 465**

SECTION 5 Maintaining and Troubleshooting a Residential Electrical Wiring System 539

Preface

Home Builders Institute Residential Construction Academy: House Wiring

About the Residential Construction Academy Series

One of the most pressing problems confronting the building industry today is the shortage of skilled labor. It is estimated that the construction industry must recruit 200,000 to 250,000 new craft workers each year to meet future needs. This shortage is expected to continue well into the next decade because of projected job growth and a decline in the number of available workers. At the same time, the training of available labor is becoming an increasing concern throughout the country. This lack of training opportunities has resulted in a shortage of 65,000 to 80,000 skilled workers per year. The crisis is affecting all construction trades and is threatening the ability of builders to build quality homes.

These are the reasons for the creation of the innovative *Residential Construction Academy Series*. The *Residential Construction Academy Series* is the perfect way to introduce people of all ages to the building trades while guiding them in the development of essential workplace skills including carpentry, electrical, HVAC, plumbing, and facilities maintenance. The products and services offered through the *Residential Construction Academy* are the result of cooperative planning and rigorous joint efforts between industry and education. The program was originally conceived by the National Association of Home Builders—the premier association of over 200,000 member groups in the residential construction industry—and its work-force development arm, the Home Builders Institute.

For the first time, Construction professionals and educators created National Skill Standards for the Construction trades. In the summer of 2001, the National Association of Home Builders (NAHB), through the Home Builders Institute (HBI), began the process of developing residential craft standards in five trades: carpentry, electrical wiring, HVAC, plumbing, and facilities maintenance. Groups of electrical employers from across the country met with an independent research and measurement organization to begin the development of new craft training Skill Standards. The guidelines from the National Skills Standards Board were followed in developing the new standards. In addition, the process met or exceeded the American Psychological Association standards for occupational credentialing.

Then, through a partnership between HBI and Delmar Learning, learning materials—textbooks, videos, and instructor's curriculum and teaching tools—were created to effectively teach these standards. A foundational tenant of this series is that students *learn by doing*. A constant focus of the *Residential Construction Academy* is teaching the skills needed to be successful in the Construction industry and constantly applying the learning to real world applications.

Perhaps most exciting to learners and industry is the creation of a National Registry of students who have successfully completed courses in the *Residential Construction Academy Series*. This registry, like a transcript service, provides an opportunity for easy access for verification of skills and competencies achieved. The Registry links construction industry employers and qualified potential employees together in an online database facilitating student job search and the employment of skilled workers.

About This Book

A home is an essential part of life. It provides protection, security, and privacy to the occupants. It is often viewed as the single most important thing a family can own. This book is written for the students who want to learn how to wire a home.

House Wiring covers the basic electrical wiring principles and practices, with *National Electrical Code®* references, used in the installation of residential electrical wiring systems. Wiring practices that are commonly used in today's residential electrical market are discussed in detail and presented in a way that not only tells what needs to be done, but also shows how to do it. Both general safety and electrical safety are stressed throughout the textbook.

This textbook provides a resource for the areas in residential wiring that are required of an entry-level electrician, including the basic "hands-on" skills, as well as more advanced theoretical knowledge needed to gain job proficiency. In addition to topics such as calculating conductor size, calculating voltage drop, determining appliance circuit requirements, sizing service entrance conductors, grounding services and equipment and many other aspects of residential electrical installation, this text focuses on "hands-on" wiring skills, such as the proper usage of hand and power tools, splicing wires together properly, attaching electrical boxes to a wood or metal stud or fishing a cable in an existing wall. The format is intended to be easy to learn and easy to teach.

Organization

This textbook is organized in the same way that a typical residential wiring project unfolds. The five major sections cover the installation of a residential wiring system from start to finish:

- **Section 1: Preparing and Planning a Residential Wiring Job** is designed to show students how to apply common safety practices, how to use materials, tools and testing instruments, and how to read and understand building plans.
- **Section 2: Residential Service Entrances and Equipment** shows how to install the necessary equipment to get electrical power from the electric utility to the dwelling unit before permanent power is established.
- **Section 3: Residential Electrical System Rough-In** demonstrates how to install electrical boxes and run cable or raceway according to the electrical circuit requirements.
- **Section 4: Residential Electrical System Trim-Out** involves installing all of the switches, receptacles, and luminaires (lighting fixtures) throughout the house.
- **Section 5: Maintaining and Troubleshooting a Residential Electrical Wiring System** explains how to test each circuit to make sure they are to code and in proper working order. It also shows how to troubleshoot and correct problems to ensure a satisfied customer.

Features

This innovative series was designed with input from educators and industry and informed by the curriculum and training objectives established by the Standards Committee. The following features aid learning:

Learning Features such as the **Introduction, Objectives,** and **Glossary** set the stage for the coming body of knowledge and help the learner identify key concepts and information. These learning features serve as a road map for continuing through the chapter. The learner also may use them as a reference later.

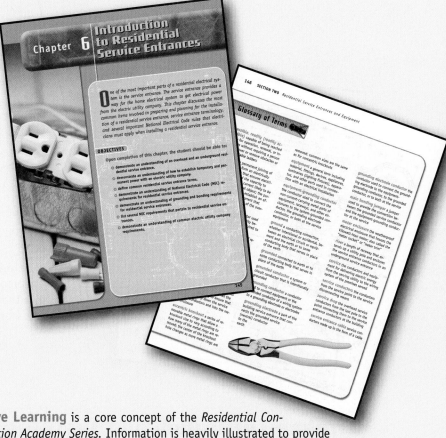

Active Learning is a core concept of the *Residential Construction Academy Series*. Information is heavily illustrated to provide a visual of new tools and tasks encountered by the learner. Chapters also contain a **Procedures** section that takes the information and applies it so that learning is accomplished through doing. In the **Procedures**, various tasks used in home construction are grouped in a step-by-step approach. The overall effect is a clear view of the task, making learning easier.

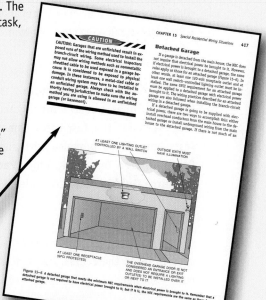

Safety is featured throughout the text to instill safety as an "attitude" among learners. Safe jobsite practices by all workers is essential; if one person acts in an unsafe manner all workers on the job are at risk of being injured too. Learners will come to appreciate that safety is a blend of ability, skill, and knowledge that should be continuously applied to all they do in the Construction industry.

Caution features highlight safety issues and urgent safety reminders for the trade.

From Experience provides tricks of the trade and mentoring wisdom that make a particular task a little easier for the novice to accomplish.

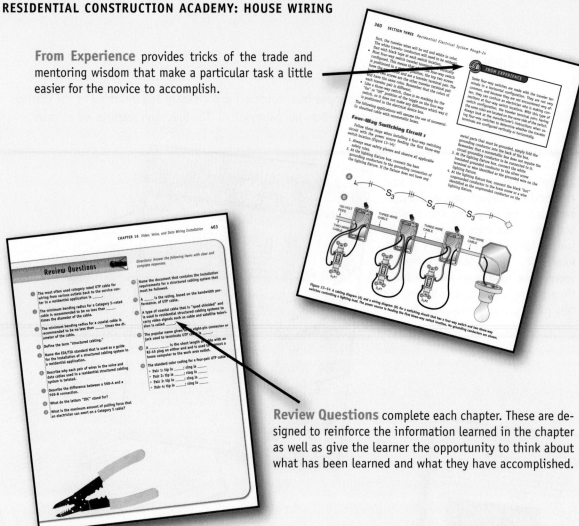

Review Questions complete each chapter. These are designed to reinforce the information learned in the chapter as well as give the learner the opportunity to think about what has been learned and what they have accomplished.

Turnkey Curriculum and Teaching Material Package

We understand that a text is only one part of a complete, turnkey educational system. We also understand that Instructors want to spend their time on teaching, not preparing to teach. The *Residential Construction Academy Series* is committed to providing thorough curriculum and prepatory materials to aid Instructors and alleviate some of their heavy preparation commitments. An integrated teaching solution is ensured with the text, Instructor's e.resource ™, print Instructor's Resource Guide, Student Videos, and CD Courseware.

e.resource™

Delmar Learning's **e.resource**™ is a complete guide to classroom management. The CD-ROM contains lecture outlines, notes to instructors with teaching hints, cautions, and answers to review questions, and other aids for the Instructor using this *Series*. Designed as a complete and integrated package, the Instructor is also provided with suggestions for when and how to use the accompanying **PowerPoint, Computerized Test Bank, Video Package,** and **CD Courseware** package components. A print **Instructor's Resource Guide** is also available.

PowerPoint

The series includes a complete set of PowerPoint Presentations providing lecture outlines that can be used to teach the course. Instructors may teach from this outline or can make changes to suit individual classroom needs.

Computerized Testbank

The Computerized Testbank contains hundreds of questions that can be used for in-class assignments, homework, quizzes, or tests. Instructors can edit the questions in the testbank, or create and save new questions.

Videos

The *House Wiring Video Series* is an integrated part of the *Residential Construction Academy House Wiring* package. The series contains a set of eight, 20-minute videos that provide step-by-step instruction for wiring a house. All the essential information is covered in this series, beginning with the important process of reviewing the plans and following through to the final phase of testing and troubleshooting. Need to know *NEC®* articles are highlighted, and Electrician's Tips and Safety Tips offer practical advice from the experts.

The complete set includes the following: Video #1: Safety and Safe Practices, Video #2: Hardware, Video #3: Tools, Video #4: Initial Review of Plans, Video #5: Rough-In, Video #6: Service Entrance, Video #7: Trim-Out, Video #8: Testing & Troubleshooting.

CD Courseware

This package also includes computer-based training that uses video, animation, and testing to introduce, teach, or remediate the concepts covered in the videos. Students will be pre-tested on the material and then, if needed, provided with the suitable remediation to ensure understanding of the concepts. Post-tests can be administered to ensure that students have gained mastery of all material.

Online Companion

The Online Companion is an excellent supplement for students. It features many useful resources to support the *House Wiring* book, videos, and CDs. Linked from the Student Materials section of www.residentialacademy.com, the Online Companion includes chapter quizzes, an online glossary, product updates, related links, and more.

About the Author

The author of this textbook, Gregory W. Fletcher, has over 25 years of experience in the electrical field. He has taught electrical wiring practices at both the secondary level and the post-secondary level and has been licensed, first as a Journeyman and then a Master Electrician, since 1976. He has taught apprenticeship electrical courses and has facilitated workshops ranging from Fiber Optics for Electricians to Understanding Electrical Calculations. The knowledge gained over those years, specifically on what works and what does not work to effectively teach electrical wiring practices, was used as a guide to help determine the focus of this text.

Since 1988 he has been Department Chairman of the Trades and Technology Department and an Electrical Instructor at Kennebec Valley Community College in Fairfield, Maine. He holds an Associate of Applied Science Degree in Electrical Construction and Maintenance, a Bachelor of Science Degree in Applied Technical Education and a Master of Science Degree in Industrial Education. Mr. Fletcher is a member of the International Association of Electrical Inspectors, The National Fire Protection Association, and the Instrument Society of America. At present, he lives in Waterville, Maine with his wife and daughter.

Acknowledgments

House Wiring National Skill Standards

The NAHB and the HBI would like to thank the many individual members and companies that participated in the creation of the House Wiring National Skills Standards. Special thanks are extended to the following individuals and companies:

John Gaddis, Home Builders Institute Electrical Instructor
Stephen L. Herman, Lee College
Roy Hogue, TruRoy Electrical
Fred Humphreys, Home Builders Institute
Mark Huth, Delmar Learning
Ray Mullin, Wisconsin Schools of Vocational, Technical and Adult Education
Ron Rodgers, Wasdyke Associates/Employment Research
Jack Sanders, Home Builders Institute
Clarence Tibbs, STE Electrical Systems, Inc.
Ray Wasdyke, Wasdyke Associates

In addition to the standards committee, many other people contributed their time and expertise to the project. They have spent hours attending focus groups, reviewing and contributing to the work. Delmar Learning and the author extend our sincere gratitude to:

DeWain Belote, Pinellas Technical Education Centers
Mike Brumbach, York Technical College
Mark Caskey, Lancaster Vocational School
Gary Reiman, Dunwoody Institute

Finally, the author would like to express a special thanks to David Gehlauf of Tri-County Vocational School in Nelsonville, Ohio, for lending his time and expertise to this writing project. Many of David's ideas and observations found their way into this textbook. David's contributions, especially in Chapters 17, 18, and 19, helped the author meet his goal of writing a residential wiring textbook that was up-to-date, easy-to-use, and technically accurate.

Preparing and Planning a Residential Wiring Job

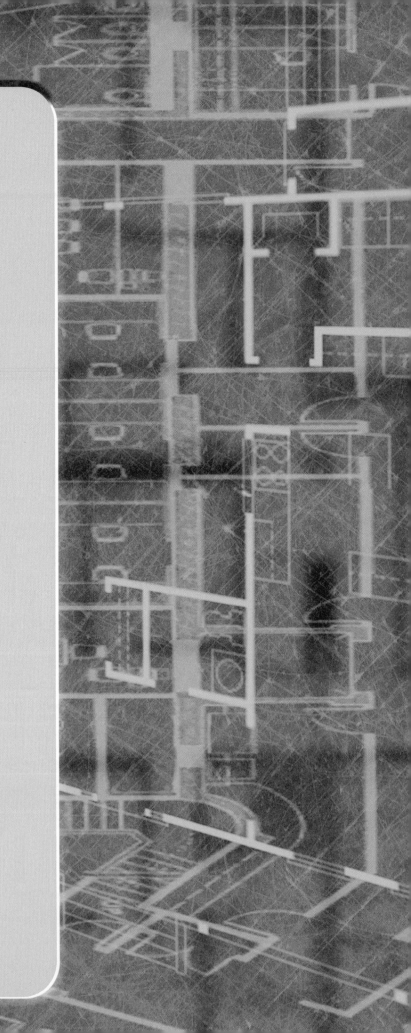

Chapter 1 | Residential Workplace Safety

afety should be the main concern of every worker. Each person on the job should work in a safe manner, no matter what the occupation. Too often, failure on the part of workers to follow recommended safe practices results not only in serious injury to themselves and fellow workers but also in costly damage to equipment and property. The electrical trades, perhaps more than most other occupations, require constant awareness of the hazards associated with the occupation. The difference between life and death is a very fine line. There is no room for mistakes or mental lapses. Trial-and-error practices are not acceptable! Electricity plays a big part in our lives and serves us well. Being able to control electricity allows us to make it do the things we want it to do. Control comes with a thorough understanding of how electricity works and with an equally thorough appreciation of the hazards and consequences involved when this control is not present. A good residential electrician will, in addition to being proficient in the technology of the trade, possess and display respect for the hazards associated with the occupation. Residential electrical workers must realize from the beginning of their training that if they do not observe safe practices when installing, maintaining, and troubleshooting an electrical system, there will be a good chance that they could be injured on the job. Both general and electrical safety is serious business. It is very important that you make workplace safety a part of your everyday life.

OBJECTIVES

Upon completion of this chapter, the student should be able to:

- demonstrate an understanding of the shock hazard associated with electrical work.
- demonstrate an understanding of the purpose of the *National Electrical Code®*.
- demonstrate an understanding of the arrangement of the *National Electrical Code®*.
- cite examples of rules from the *National Electrical Code®* pertaining to common residential electrical safety hazards.
- identify common electrical hazards and how to avoid them on the job.
- demonstrate an understanding of the purpose of OSHA.
- cite specific OSHA provisions pertaining to various general and electrical safety hazards associated with residential wiring.
- demonstrate an understanding of the personal protective equipment used by residential electricians.
- list several safety practices pertaining to general and electrical safety.
- demonstrate an understanding of material safety data sheets.
- demonstrate an understanding of various classes of fires and the types of extinguishers used on them.

Glossary of Terms

ampere the unit of measure for electrical current flow

arc the flow of a high amount of current across an insulating medium, like air

arc blast a violent electrical condition that causes molten metal to be thrown through the air

circuit (electrical) an arrangement consisting of a power source, conductors, and a load

conductor a material that allows electrical current to flow through it; examples are copper, aluminum, and silver

current the intensity of electron flow in a conductor

double insulated an electrical power tool type constructed so the case is isolated from electrical energy and is made of a nonconductive material

electrical shock the sudden stimulation of nerves and muscle caused by electricity flowing through the body

grounding an electrical connection to an object that conducts electrical current to the earth

ground fault circuit interrupter (GFCI) a device that protects people from dangerous levels of electrical current by measuring the current difference between two conductors of an electrical circuit and tripping to an open position if the measured value exceeds 6 milliamperes

hazard a potential source of danger

insulator a material that does not allow electrical current to flow through it; examples are rubber, plastic, and glass

load (electrical) a part of an electrical circuit that uses electrical current to perform some function; an example would be a lightbulb (produces light) or electric motor (produces mechanical energy)

Material Safety Data Sheet (MSDS) a form that lists and explains each of the hazardous materials that electricians may work with so they can safely use the material and respond to an emergency situation

National Electrical Code® (NEC®) a document that establishes minimum safety rules for an electrician to follow when performing electrical installations; it is published by the National Fire Protection Association (NFPA)

Occupational Safety and Health Administration (OSHA) since 1971, OSHA's job has been to establish and enforce workplace safety rules

ohm the unit of measure for electrical resistance

Ohm's law the mathematical relationship between current, voltage, and resistance in an electrical circuit

polarized plug a two-prong plug that distinguishes between the grounded conductor and the "hot" conductor by having the grounded conductor prong wider than the hot conductor prong; this plug will fit into a receptacle only one way

power source a part of an electrical circuit that produces the voltage required by the circuit

resistance the opposition to current flow

scaffolding also referred to as staging; a piece of equipment that provides a platform for working in high places; the parts are put together at the job site and then taken apart and reconstructed when needed at another location

shall a term used in the *National Electrical Code®* that means that the rule *must* be followed

ventricular fibrillation very rapid irregular contractions of the heart that result in the heartbeat and pulse going out of rhythm with each other

volt the unit of measure for voltage

voltage the force that causes electrons to move from atom to atom in a conductor

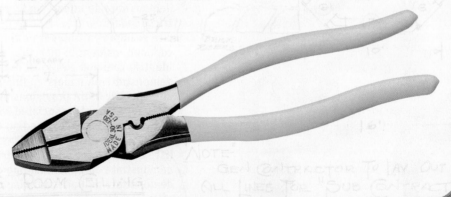

Understanding the Shock Hazard

Electrical shock is considered the biggest safety **hazard** associated with doing electrical work. Many residential electricians think that the voltages encountered in residential work will not really hurt them. Others think that residential wiring just does not present the same opportunities for an electrical shock that commercial or industrial electrical work does. They are wrong. The shock hazard exists in residential wiring to the same degree that it exists in other wiring areas. To understand and appreciate the shock hazard in residential wiring, a review of basic electrical theory is provided in the following paragraphs.

Electricity refers to the flow of electrons through a material. The force that drives the electrons and makes this electron flow possible is known as the **voltage**. Any material or substance through which electricity flows is called a **conductor**. Examples of conductors used in electrical work include copper and aluminum. These substances offer very little resistance to electron flow. Some materials offer very high resistance to electron flow and are classified as **insulators**. Examples are plastic, rubber, and porcelain. Electricity flows along a path or **circuit**. Typically, this path begins with a **power source** and follows through a conductor to a **load**. The path then flows back along another conductor to the power source (Figure 1–1).

A very important point to consider at this time is that the human body can, under certain conditions, readily become a conductor and a part of the electrical circuit (Figure 1–2). When this happens, the result is often fatal. Electrons flowing in the circuit have no way of detecting the difference between human beings and electrical equipment.

There is a certain relationship between the **current**, voltage, and **resistance** in an electrical circuit. Georg Simon Ohm discovered this relationship many years ago. His discovery resulted in a mathematical formula that became known as **Ohm's law**. According to Ohm's law,

Current = Electrical Force/Resistance

Current flow is measured in **amperes**, electrical force is measured in **volts**, and electrical resistance is measured in **ohms**. The lower the circuit resistance, the greater the amperage, or quantity of current, that a voltage can push through a circuit.

Table 1–1 shows some common resistance values for the human body. Dry skin offers much more opposition to current flow through the human body than does a perspiring or wet body. Remember that the higher the ohm value, the more opposition to current flow there is.

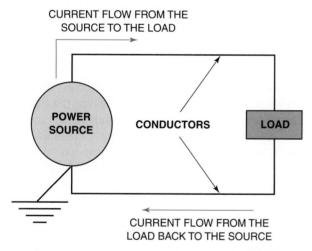

CURRENT FLOW FROM THE
SOURCE TO THE LOAD

POWER SOURCE   CONDUCTORS   LOAD

CURRENT FLOW FROM THE
LOAD BACK TO THE SOURCE

Figure 1–1 A basic electrical circuit showing the relationship of the electrical source, the conductors, and the load.

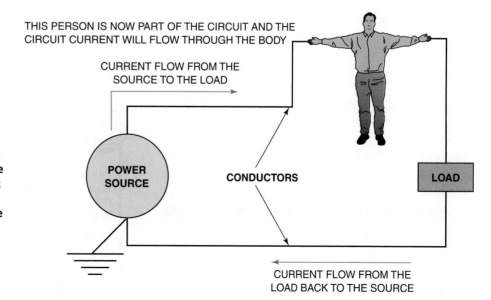

THIS PERSON IS NOW PART OF THE CIRCUIT AND THE
CIRCUIT CURRENT WILL FLOW THROUGH THE BODY

CURRENT FLOW FROM THE
SOURCE TO THE LOAD

POWER SOURCE   CONDUCTORS   LOAD

CURRENT FLOW FROM THE
LOAD BACK TO THE SOURCE

Figure 1–2 An electrical circuit that includes a human body as part of the circuit. Because of the way the body is in the circuit, current will flow into the right hand and travel through the right arm, through the heart and lungs, through the left arm, and back into the circuit through the left hand. The amount of current that flows through the body will depend on the body's resistance and the voltage of the power source.

Table 1–1   Human Body Resistance under Wet and Dry Conditions

Skin Condition	Resistance
Dry skin	100,000 to 500,000 ohms
Perspiring (sweaty hands)	1,000 ohms
In water (completely wet)	150 ohms

Table 1–2   Effect of Current on the Human Body

Current Flow	Effect on the Human Body
Less than 1 milliampere	No sensation
1 to 20 milliamperes	Can feel the shock, it may be painful, and you could experience muscular contraction (which could cause a person to hold on)
20 to 50 milliamperes	Painful shock, inability to let go
50 to 200 milliamperes	Heart convulsions (**ventricular fibrillation**), usually means death
200 or more milliamperes	Severe burns and paralysis of breathing

The human body reacts differently to the level of current flowing through it. Table 1–2 shows some typical reactions when a body is subjected to various amounts of current. Remember that a milliampere (mA) is 1/1,000 (0.001) of an ampere. It is a very small amount of current.

Consider a situation where a residential electrician is operating an electric-powered drill to bore holes through 2 × 6-inch wall studs. It is a hot and humid day in the middle of the summer. Like many of us, the electrician will perspire heavily and end up using the drill with wet, sweaty hands. According to Table 1–2, the electrician's body resistance would be reduced to about 1,000 ohms. Assuming that the voltage is 120 volts and applying Ohm's law,

$$\text{Current in Amperes} = 120 \text{ volts}/1{,}000 \text{ ohms}$$
$$= 0.12 \text{ ampere, or } 120 \text{ mA}$$

CAUTION: High voltage (over 600 volts) is more likely to cause death than low voltage. However, more electrocution deaths occur from low voltage because electricians are exposed more often to low voltages and do not usually use the same caution around low voltage that they do around high voltage.

With this amount of current, the electrician is going to receive a shock that is sufficient to paralyze breathing. The electrician is in big trouble and may not survive without proper medical treatment.

CAUTION: The longer the body is in the circuit, the greater will be the severity of the injury.

Let us look at another situation, this time in the home. After a hard day working in the heat, an electrician goes home and decides to cool off by taking a dip in his new aboveground pool. The house the electrician lives in is an older home and does not have the electrical safety features that newer homes have, such as ground fault circuit interrupter receptacles. As the electrician cools off in the pool, he reaches to change the channel on a radio he is listening to. The radio is plugged into an outdoor receptacle located close to the pool and is accidentally knocked into the pool. Unfortunately, the outcome is probably death or, at the very least, a serious shock. Here is why. Body resistance in this case is about 150 ohms, and the current amount is well in excess of the shock required to cause paralysis of breathing (see Tables 1–1 and 1–2):

$$\text{Current in Amperes} = 120 \text{ volts}/150 \text{ ohms}$$
$$= 0.800 \text{ ampere, or } 800 \text{ mA}$$

Do not become a shock victim like the electricians described in the preceding paragraphs. By understanding the shock hazard and following electrical safety procedures at all times, your chance of becoming another victim is greatly reduced.

Burns caused by electricity are another hazard encountered by electricians. Usually the severity of the burn depends on the voltage of the circuit. While the chance of a burn in residential electrical work is certainly present, the chances for burn injuries tend to be greater on commercial and industrial electrical job sites, where the voltages are typically higher.

An electrical burn is sometimes a result of getting an electrical shock. Burns occur whenever electrical current flows through bone or tissue. This is a very serious type of injury since it happens inside the body. It may not look that severe from the outside, but you need to be aware that severe tissue damage could have taken place under the skin where you cannot see it. Seek medical help whenever you get an electrical shock, especially if it could result in internal burning.

FROM EXPERIENCE

Current that flows from one finger to another on the same hand will not pass through vital organs like the heart. For this reason, it is recommended that electricians try to use only one hand when taking measurements or working on "live" circuits. Current flowing from one hand to the other would pass through the heart or lungs.

Another type of burn that electricians are exposed to in residential work is referred to as an **arc** burn. This type of burn is not a result of electrical shock but rather a result of electrical equipment malfunctioning and causing an extremely high temperature area around the arc. If you or any part of your body is in that area, you will get burned.

Another hazard associated with an arc is what is called the **arc blast**. When an arc occurs, there is a blast that causes molten metal to be thrown through the air and onto the skin or into the eyes of an electrical worker. The arc itself and the pieces of molten metal that it produces can reach temperatures as high as 35,000 degrees Fahrenheit. This temperature is many times hotter than the surface of the sun, so it is no wonder that workers can be severely burned by an arc blast. Poor electrical connections or insulation that has failed are the usual causes of arcing that can result in personal injury. Proper personal protection equipment must be used where the possibility of arcing exists. This type of protection is covered later in this chapter.

CAUTION

CAUTION: Always wear the proper personal protective equipment and test the equipment for voltage before working on electrical equipment. Never work on energized equipment unless it is absolutely necessary and you have permission from your supervisor!

National Electrical Code®

The *National Electrical Code®* (*NEC®*) is the guide for safe practices and procedures in the electrical field and is published by the National Fire Protection Association (NFPA). Every three years, the *NEC®* is brought up to date to reflect the latest changes and trends in the electrical industry. It contains specific rules to help safeguard people and property from the hazards arising from the use of electricity. Its content should become very familiar to all residential electricians since all electrical work done in a dwelling must conform to the *NEC®*. Many electricians refer to the *NEC®* as "the electrician's bible," which demonstrates their idea of the importance of knowing and applying the *NEC®*. It should be noted that only those *NEC®* areas that pertain to residential wiring practices are presented in this book, although many of the residential *NEC®* rules can also apply to commercial and industrial situations.

NEC® Arrangement

New electricians who first take a look at the *NEC®* are often intimidated. It is not the easiest book to read, and it

even tells us in the introduction that it is not intended as a how-to book or a book to use for someone who has no electrical training. The best way for a new electrician to learn how to use the *NEC®* is to use it in conjunction with a textbook. However, the first step in learning how to use the *NEC®* is to understand how it is arranged. Figure 1–3 shows the layout of the *NEC®*.

The *NEC®* contains nine chapters that cover all the major areas of the electrical field. Chapters 1 through 4 apply generally to all electrical wiring applications unless modified in some way by Chapters 5 through 8. Chapter 9 contains tables of information necessary to apply the *NEC®*. Each chapter is broken down into "articles." Each article covers a specific aspect of the electrical field, and each starts with the number of the chapter in which it is located. For example, Article 250 covers grounding and is located in Chapter 2; Article 450 covers transformers and is located in Chapter 4. The articles are broken down into "parts" that cover a specific topic area of that article. The parts are numbered using roman numerals. For example, Part VI of Article 250 covers equipment grounding and equipment grounding conductors. The parts are further broken down into "sections." A section will cover a specific item of that part. For example, to find the section that contains the requirements for sizing the motor circuit conductors to a well pump motor in a residential application, you

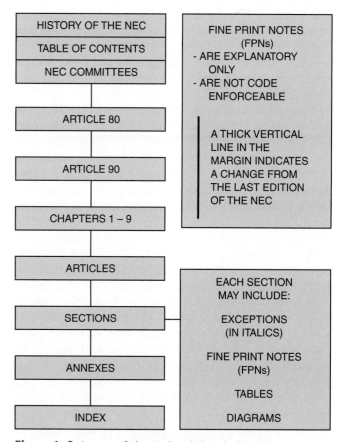

Figure 1–3 Layout of the *National Electrical Code®*.

would go to Chapter 4: Motors, Motor Circuits, and Controllers; look in Part II, which covers motor circuit conductors; and then look in Section 430.22, which covers sizing conductors for a single electric motor. Sections can be broken down further into parts that are labeled alphabetically, for example, 220.3(A). Many sections also contain exceptions, which are always printed using *italics*. The exceptions apply only to the section they are in and typically give the electrician an alternative to the general rule stated in the section. Tables are also found in many sections. They provide important information that is used to help apply the code to your particular wiring application. For example, Table 314.16(A) is where you would look to find out what size the electrical boxes that you need to install for a residential wiring job should be.

You will notice in the margins of the *NEC*® certain lengths of solid black line. The lines indicate that there is a change in that section from the previous *NEC*®. This is a quick way for electricians to see what areas of the *NEC*® have been updated from the previous version. Remember that the *NEC*® is updated every three years.

Two articles are not located in chapters. They are Articles 80 and 90. Article 80 is new to the 2002 *NEC*® and is simply a suggested way for a state or local electrical code enforcement agency to be organized. It is not enforceable unless adopted by the state or local agency. Article 90, Introduction, is there to tell us what is and what is not covered by the *NEC*®, how the *NEC*® is arranged, who has the authority to enforce the *NEC*®, and other introductory information. Article 90 clearly points out that it is the authority having jurisdiction (AHJ) who will inspect the building to make sure that all electrical materials used and wiring techniques used are acceptable. It is usually the local or state electrical inspector who inspects the electrical work and who is usually considered to be the AHJ.

There are also several annexes in the back of the *NEC*® that contain additional information to the nine main chapters. For example, Annex C contains information for sizing various conduit types when they contain a certain number of conductors.

An index located at the very back of the *NEC*® is broken down into various topics and is very useful for finding a particular section in the *NEC*®. For example, say you are looking for information in the *NEC*® on the location of ground fault circuit interrupter receptacles around swimming pools. Look in the index under "Swimming Pools" and then scan down until you find "Ground Fault Circuit Interrupters." The index tells us that information is located in Section 680.5.

There is also an excellent table of contents that gives the breakdown of all nine chapters and a page number for each article. It should be noted that the *NEC*® is NFPA brochure 70 and that each page in the *NEC*® starts out with number 70. For example, Article 230, Services, starts on page 70-75 in the 2002 edition of the *NEC*®. At this time, it would be a good idea for you to familiarize yourself with the table of contents in the *NEC*®.

(See page 21 for a suggested procedure to find specific information in the NEC®*.)*

NEC® Applications

The following paragraphs cover some examples of common residential wiring applications and the *NEC*® rules that should be followed. The purpose of this section is simply to show you, the beginning electrician, how the *NEC*® can be applied to residential wiring situations.

Working Spaces and Clearances

Article 110 of the *NEC*® provides specifications for working space around electrical equipment. Specifications are outlined for two situations: (1) 600 V or less (110.26) and (2) over 600 V (110.30). In residential work, we will be concerned only with the requirements for 600 V or less.

In your copy of the *NEC*®, locate Table 110.26(A). This table specifies the space requirements for working space around electrical equipment. Since residential electrical systems never exceed 150 volts to ground, the 0–150-volt row is referenced in the table. We see that for each "condition," the minimum free space that must be maintained in front of residential electrical equipment is 3 feet. So, when you install the electrical panel for a residential wiring job, there must be at least 3 feet of free space in front of it (see Figure 1–4). Notice that "Conditions" 1, 2, and 3 are explained under the Table but for residential wiring it really doesn't make any difference which Condition we are under since they all have the same clearance for the 0–150-volt locations.

Ground-Fault Protection of Personnel

The *NEC*® requires **ground fault circuit interrupter (GFCI)** protection at various locations. This protection is for personnel and is not designed specifically to protect equipment. Look up Section 210.8(A) in your *NEC*®, and let us look at a few examples.

1. All 125V, 15A, and 20A receptacles installed in bathrooms shall have GFCI protection.
2. All 125V, 15A, and 20A receptacles installed outside of a dwelling unit shall have GFCI protection.
3. All 125V, 15A, and 20A receptacles located to serve the kitchen countertop shall have GFCI protection.

GFCI protection in residential situations is covered in detail later in this book.

In addition, ground-fault protection for personnel on construction site wiring installations must be provided that complies with Section 527.6(A) and (B) of the *NEC*®. This section applies only to temporary wiring installations used to supply temporary power to equipment used by personnel during construction or remodeling of buildings. All 125-volt, single-phase, 15-, 20-, and 30-ampere receptacle outlets that are used by electricians to plug in portable power tools or lighting equipment on a construction site must have GFCI protection for personnel. One way to meet the intent of this *NEC*® requirement is for a residential electrician to use cord sets with built-in GFCI protection and to plug their portable power equipment and lighting into the GFCI-protected cord sets.

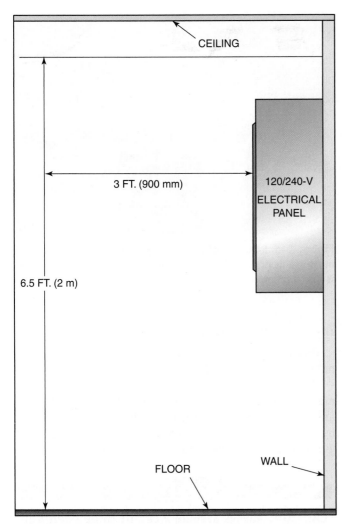

Figure 1–4 Working space in front of an electrical panel. According to Section 110.26 of the *NEC®*, the working space must extend out at least 3 feet in front of the electrical panel in a residential application. The work space must also be at least 30 inches wide and extend from the floor to a height of 6.5 feet. Nothing can be allowed to exist in this space.

Figure 1–5 A three-prong electrical plug. This device is considered to be a grounding type.

Grounding

Grounding is covered in Article 250 of the *NEC®*. Grounding of electrical tools and equipment is one of the most important methods of controlling the hazards of electricity. If the insulation in electrical equipment deteriorates or if a wire works loose and comes into contact with the frame or some other part that does not normally carry current, these parts can become energized. The electricity is no longer controlled. It is ready to follow a path to ground—any path! The unprotected human body will work just fine! As we discussed earlier in this chapter, the consequences could be fatal.

It is the green or bare ground wire of an electrical circuit that is connected to the non-current-carrying metal parts of electrical equipment. This wire provides a low-resistance path to ground for more precise control of the electrical current flow. This results in fuses and circuit breakers working

faster in the event of an electrical problem, causing the circuit to open and stopping the flow of electricity until the problem can be found and corrected. If the metal parts of electrical equipment were not grounded and a loose energized wire came in contact with the metal, the metal would now be energized. If a human touches the equipment, a fatal shock could be delivered. Grounding the equipment properly can all but eliminate the shock hazard. This discussion should alert you to the importance of the third prong on an approved cord with a three-prong plug (Figure 1–5).

CAUTION

CAUTION: Never cut off the grounding prong on an electrical plug!

These are just a few of the many *NEC®* articles and sections that have a direct bearing on the safe installation of electrical wiring. Ongoing study of the *NEC®* is a must for every electrician.

Occupational Safety and Health Administration (OSHA)

Another set of safety rules and regulations that apply to both general and electrical safety are those specified by the Occupational Safety and Health Act of 1970. The Code of Federal Regulations (CFR) is published and administered by the **Occupational Safety and Health Administration (OSHA)**. The purpose of OSHA is to ensure safe and healthy working conditions for working men and women on the job site by authorizing enforcement of the standards developed under the act; by assisting and encouraging the states in their efforts to ensure safe and healthful working conditions;

and by providing for research, information, education, and training in the field of occupational safety and health and for other purposes. CFR Part 1910 covers the OSHA regulations for general industry, and CFR Part 1926 covers the regulations for the construction industry. Additionally, regulations on general electrical safety are covered in CFR 1910, Subpart S—Electrical, and electrical safety on the construction site is covered in CFR 1926, Subpart K—Electrical.

The National Fire Protection Association's Standard 70-E, Electrical Safety for Employee Workplaces, also contains safety regulations that an electrician should be aware of. The NFPA 70-E standards are incorporated into the OSHA regulations.

It is not the intent of this book to cover in detail all the regulations on safety outlined in OSHA CFR 1910 and CFR 1926 or NFPA 70-E, but a residential electrician should be familiar with some of the more important regulations. The following paragraphs cover some specific OSHA provisions that have a bearing on proper safety practices in residential electrical work.

Safety Training

OSHA 1910.332 requires employees to be trained in safety-related work practices and procedures that are necessary for safety from electrical hazards. Employers are required to provide safety training to those employees who face the greatest risk of injury from electrical hazards. This certainly includes electricians. OSHA is fairly flexible and allows this training to consist of classroom and on-the-job training.

Lockout/Tagout

This rule is one of the more important safety regulations that you need to follow. It is covered in OSHA 1910.333 and 1926.417 and outlines the requirements for locking out and then tagging out the electrical source for the circuit or equipment you may be working on. OSHA provides only the minimum requirements for lockout/tagout. Each company is required to establish its own lockout/tagout procedure, so make sure to follow the procedure of the company you are working for.

Only devices that are designed specifically for locking out and tagging are to be used (Figure 1-6). When using padlocks they must be numbered and then assigned to only one person. No duplicate or master keys to the lockout devices are to be available, except possibly the site supervisor. The tags are supposed to be of white, red, and black and need to include the employee's name, the date and company you work for. The information you put on the tag should be in permanent marker.

CAUTION: Always follow the lockout/tagout procedure that is used by the company you work for.

(See page 21 for a suggested procedure for lockout/tagout.)

Figure 1–6 Examples of lockout/tagout devices. An electrician must use devices of this type when a lockout/tagout procedure is being followed.

Initial Inspections, Tests, or Determinations

OSHA 1926.950 requires that existing conditions be determined by inspection or testing with a test instrument before work is started (test and measurement instruments are covered in Chapter 4). All electrical equipment and conductors are to be considered energized until determined deenergized by testing methods. Work is not to proceed on or near energized parts until the operating voltage is determined.

CAUTION: Always assume a circuit to be energized until you verify otherwise.

(See page 21 for a suggested procedure for verifying that circuits are deenergized.)

Power Tools

OSHA 1926.951 contains some important regulations for power tools. The requirements for electrical power tools stipulate that they be equipped with a three-wire cord. The ground wire (green wire) must be permanently connected to the tool frame, and a three-prong plug used as a means for grounding must be at the other end. If a tool is constructed and labeled as **double insulated**, a two-prong **polarized plug** can be used (Figure 1-7).

Trenching and Excavating

OSHA Section 1926.651 sets out specific requirements for trenching and excavating that are of interest to the residential electrical worker who may be involved in the installation of underground wiring. Of special interest is the requirement for sloping, or shoring up, the sides of trenches. Without shoring, the threat of an excavated hole caving in while you

Figure 1–7 A two-prong polarized electrical plug. This device is considered to be a nongrounding type of plug. Notice that one prong is wider than the other. This will enable the plug to be inserted into a receptacle only in the proper way.

are in the hole is very real and also very sobering. There have been many workers who have been buried alive. Do not put yourself in a position where this could happen to you.

As you can see from the preceding paragraphs, OSHA provides many regulations for safe work practices that all electricians need to follow. A more in-depth study of OSHA regulations is a good idea for anyone doing electrical work. Understanding and following these regulations can mean the difference between being hurt seriously on the job and enjoying a long, injury-free career as a residential electrician.

Personal Protective Equipment

The topic of personal protection equipment is so important that it is being covered in a separate section. OSHA CFR 1910.335 and 1926.951 outline the regulations concerning personal protective equipment. OSHA requires that electricians working where there is a potential electrical hazard wear personal protective equipment. It is not a choice for the electrician—it is a requirement! The protective equipment must also be maintained and inspected according to CFR 1910.137 and 1926.951, to ensure that it is always in good working condition. In addition:

- All employees who are exposed to hazardous conditions, such as falling objects, electrical shocks, or burns, are required to wear hard hats on the job site. Do not use a metal hard hat, as it could conduct electricity.
- All employees who are exposed to electrical hazards that could produce electric arcs or an arc blast that could throw objects into an electrician's face or eyes must wear eye and/or face protection.

- Wear fiberglass-toe safety shoes. The metal of steel-toe shoes can become exposed and present a safety hazard around energized equipment. Keep your shoes in good condition so that your footing will always be solid and secure.
- OSHA 1910.132 and 1910.335 require that electricians wear low-voltage-rated gloves with leather protectors when working on energized 120- or 240-volt residential circuits and equipment. The leather protectors worn over the rubber gloves should not be worn for any other purpose.
- Wear long-sleeved cotton shirts and pants made with a sturdy, comfortable material like denim. Do not wear clothing with exposed zippers, buttons, or other metal fasteners, and do not wear loose or flapping clothing.
- Remove your rings, wristwatch, and any other metal jewelry before beginning work near equipment with exposed current-carrying parts.
- Any tools or equipment used on or around energized equipment must be of the nonconductive type.
- Fuse-handling equipment must be insulated at or higher than the system voltage and must be used by electricians to remove fuses when the fuse holder is energized.

CAUTION

CAUTION: Electrical tools such as screwdrivers and pliers with rubber grips supplied by the manufacturer are _not_ considered nonconductive. The grips are for comfort only. You must use tools that are specifically made to be nonconductive when working on energized circuits.

Fall Protection

Fall protection is an area that many residential electricians do not think should concern them. The fact is that fall protection must be used whenever workers are on a walking or work surface that is 6 feet or more high and has unprotected sides. These areas include finished or unfinished floors and roof areas in a residential application. If you are working above unguarded dangerous equipment, fall protection must be used at all times, no matter what the height. The type of fall protection used is typically chosen by your supervisor and can include items such as guardrails, personal fall-arrest systems, or safety nets.

Ground Rules for General and Electrical Safety

Residential electricians must follow industry-suggested practices for both general safety and electrical safety on the job site. The rules for each area of safety are similar; however, general safety rules should be followed at all times, while electrical safety rules are followed when the hazard of electricity is present.

General Safety Rules

General safety in residential electrical work takes into account the proper planning of your work and includes choosing the proper materials and the right tools to do the wiring job. It also includes how an electrician should behave on the job. Be serious about your work and never fool around on the job. This type of behavior can result in serious injury to you or a coworker. Intoxicating drugs or alcohol must never be used on the job. They impair judgment, and the consequences could be fatal. Remember, you are working with electricity, and it takes a very small quantity to kill you. Do not talk unnecessarily when working. A lapse in concentration could be fatal when engaged in electrical work.

During the installation, an electrician must practice safe work methods and exercise good judgment in both giving and receiving instructions. The following paragraphs cover some of the more important rules for general safety.

Material Handling

Handling the materials that you use to install a residential wiring system requires practicing the following safety rules:

- Learn the correct way to lift: Get solid footing, stand close to the load, bend your knees, and lift with your legs, not your back (see Figure 1–8).
- Never get on or off moving equipment like delivery trucks or company vans.

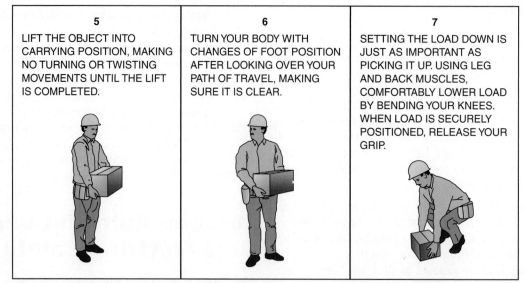

Figure 1–8 How to lift safely. As part of their jobs, electricians are required to lift objects that weigh a fair amount and are sometimes of a shape that make them hard to handle. Following proper lifting techniques will keep your body injury free.

- When two or more persons are carrying long objects like a ladder, the object should be carried on the same shoulder (right or left) of each person. One person should give the signal for raising or lowering (Figure 1–9).

- Use hand lines to raise or lower bulky materials and heavy tools from a scaffold or staging. Never drop down or throw up an item to another worker (Figure 1–10).

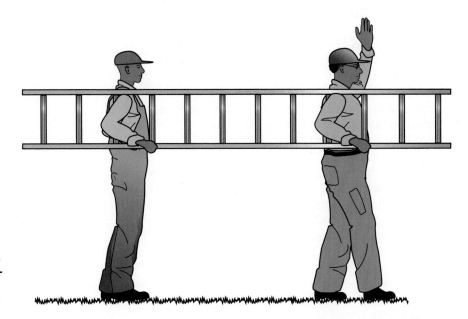

Figure 1–9 The proper technique for two people carrying a long object, like a ladder, is to carry it on the same side and to designate one person to signal when the object is to be lifted or lowered.

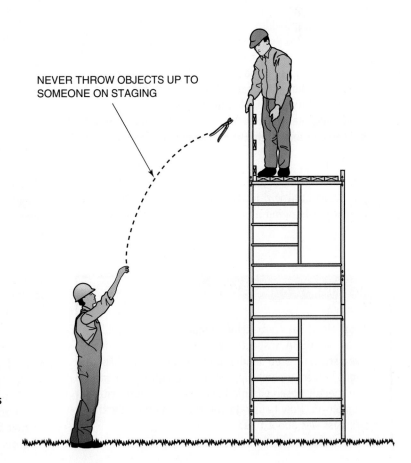

NEVER THROW OBJECTS UP TO SOMEONE ON STAGING

Figure 1–10 Never throw objects up to a person standing on staging. Hand lines should be used to haul up the tools and materials that are needed. Electricians can get injured from being hit by the thrown object or, even worse, fall off the staging while reaching for the thrown object.

Ladders

Improper use of step ladders and straight or extension ladders cause many workplace injuries. The following rules should be followed when using ladders:

- OSHA stipulates that portable metal ladders should not be used near energized lines or equipment. Use a nonconductive ladder made of a material such as fiberglass.
- Inspect the ladders before use. Look for any missing, loose, or cracked parts. If the ladder is not in good shape, do not use it!
- Always place a straight or extension ladder at the proper angle. It is strongly suggested that you place the ladder so that the bottom of the ladder is about one-fourth the vertical height from the structure it is up against (Figure 1–11).
- The ladder should extend at least 3 feet above the top support when placed against a structure that is not as tall as the ladder (Figure 1–12).
- Set ladders on firm footing and tie them off where possible. If there is a danger of a ladder moving, station someone to hold on to it as you climb.
- Use both hands and face the ladder when going up or down.
- Keep hands free of tools; materials should be hoisted or lowered with a hand line, never carried up or down.
- Make sure to open a step ladder all the way and lock the spreaders in place.
- Never use a step ladder as a straight ladder. If an OSHA inspector catches you doing this, your company is in violation and subject to a fine (Figure 1–13).
- The top two rungs are not for standing on a step ladder. Standing on them may cause the ladder to fall, resulting in serious injury to you (Figure 1–14).
- Do not leave materials or tools on the top of a step ladder. They can fall off and injure someone.

Scaffolding

Scaffolding, or staging as it is often called, is used regularly in residential wiring. For example, it may be used to stand on when hanging a ceiling paddle fan in a residence

USE BLOCKING TO LEVEL THE LADDER WHEN NECESSARY.

16 FT.

4 FT.

CORRECT BASE POSITION IS ¼ THE VERTICAL HEIGHT.

Figure 1–11 The safe ladder angle results from having the base position no more or less than one-fourth of the vertical height. In this example, the vertical height is 16 feet, and one-fourth of 16 feet is 4 feet. Therefore, the location of the base position should be 4 feet out from the building.

FROM EXPERIENCE

A local electrical contracting company was assessed a fine by an OSHA inspector because a new electrical worker stood a step ladder up against a wall like a straight ladder to climb up and hang a lighting fixture. The inspector observed this behavior and quickly cited the employer. How happy do you think the new electrician's boss was about this?

with a high cathedral ceiling. The following rules should be followed when using staging (note that some of the following information is derived from OSHA regulations):

- When appropriate, wear an approved harness and lanyard to secure yourself to the staging so you cannot fall.
- Scaffold planks must be secured and must extend over their end supports at least 6 inches but not more than 12 inches.
- Guardrails and toe boards must be installed when the actual standing platform is higher than 6 feet off the ground.

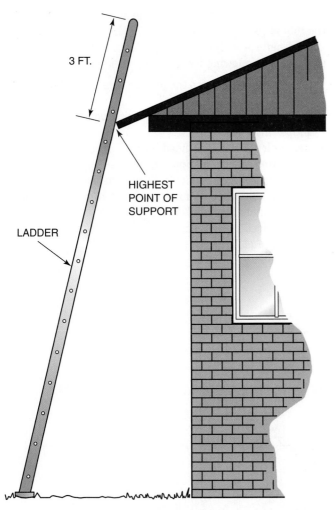

Figure 1–12 When using a straight or extension ladder, place it so that the top of the ladder extends 3 feet above the highest point of support. This will allow you to mount or dismount the ladder in a safe manner.

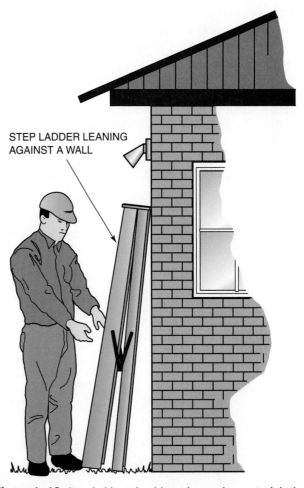

Figure 1–13 Step ladders should not be used as a straight ladder and placed up against a wall. OSHA can cite electrical contractors for using step ladders this way and assess a fine.

SAFETY PRACTICES
FOR STEP LADDERS

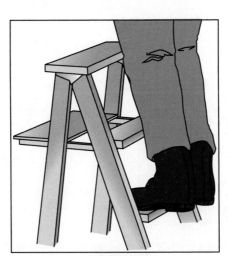

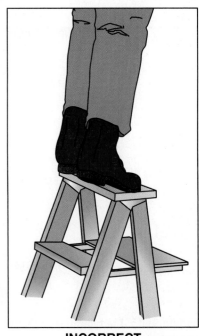

Figure 1–14 Safe practices for using step ladders include not standing on the very top step. Two steps down from the top is the highest point on the step ladder that is suitable for standing.

CORRECT

INCORRECT

- If there are people who have to work or pass under the scaffolding, a screen of no more than half-inch openings must be installed from the top of the toe board to the midpoint on the side rails.
- All workers must be trained in the proper methods for construction of the scaffolding (Figure 1–15).
- Keep scaffold platforms clear of unnecessary materials, tools, and scrap; they may become a tripping hazard or be knocked off, endangering people.
- The platform must completely cover the staging work area. There must be no holes for people to possibly fall through.

Tools

The safe and proper use of the tools used to install residential electrical systems is very important. If a tool comes with an owner's manual, read it! The place where you bought the tool is also an excellent source to find out how to safely and properly use the tool. A residential electrician should consider the following items:

- Use the right tools for the job. Be sure that they are in good working order.
- Use only those power tools for which you have received proper instruction.
- Use tools for their intended purpose (e.g., pliers are not to be used for hammering) (Figure 1–16).
- Portable power tools should be provided with a three-prong grounding plug for attachment to a ground unless it is a double-insulated power tool.
- Sharp-edged or pointed tools should be carried in tool pouches and not in clothing pockets.

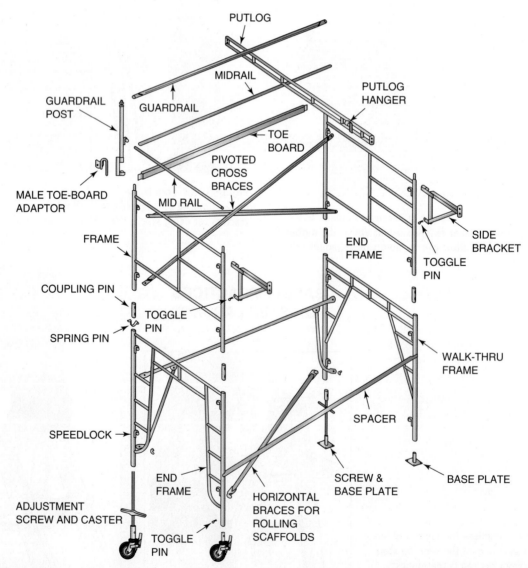

Figure 1–15 The typical parts that make up a scaffold. Electricians should be given training on how to construct a scaffolding or staging system.

Figure 1–16 Use tools properly. Do not hammer with a pair of pliers like this electrician is doing. Personal injury can result from not using a tool properly.

 FROM EXPERIENCE

A residential electrician was mounting electrical boxes on the side of studs and using nails to attach them. He was using lineman pliers to hammer the nails. Since the side of the pliers is not designed to take the tremendous force that hammering nails produces, a piece of metal broke off from the pliers and embedded itself in the electrician's arm. He was fortunate that the metal did not go into his eye. As it was, a visit to the doctor was required to have the metal piece removed from his arm. The electrician learned the hard way that you must always use a tool for what is was designed to do. If you do not, serious injury could result. By the way, the electrician who learned this lesson is me, the author of this book!

 **CAUTION**

CAUTION: Don't carry tools like screwdrivers or other sharp tools in your pants pockets. One slip or stumble could result in the tool doing serious injury to you.

- Tools should not be left on any overhead workspace or structure unless held in suitable containers that will prevent them from falling.

Material Safety Data Sheets

An electrician may have to use or may come in contact with hazardous materials on the job site. Solvents used to clean tools and other equipment are examples of materials that are considered hazardous. Electricians must follow specific procedures and methods for using, storing, and disposing of most solvents and other chemicals. **Material Safety Data Sheets (MSDS)** must be made available to a worker using any hazardous material.

Material Safety Data Sheets (see Figure 1–17) are designed to inform the electrician of a certain material's physical properties as well as the effects on health that make the material

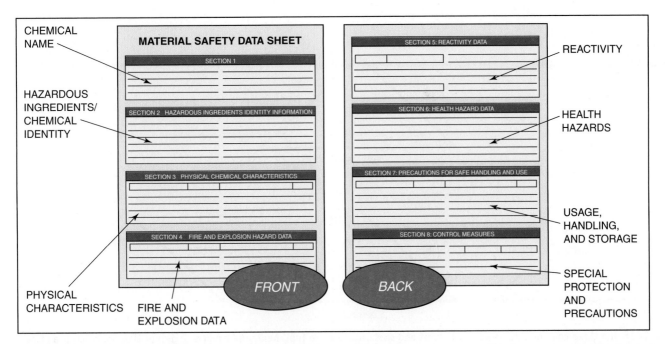

Figure 1–7 Material Safety Data Sheets (MSDS) contain information about hazardous materials that you may need to know. A typical MSDS will have eight parts. The different parts are shown in this illustration.

dangerous to handle. They instruct you in the proper type and style of protective equipment needed when using the material and in proper first aid treatment if you are exposed to the material or its hazards. Other information given includes proper storage methods, how to safely handle spills of material, and how to properly dispose of the material.

Each company is required by law to develop a hazard communications (HazCom) program that must contain, at a minimum, warning labels on containers of hazardous material, employee training on the safe use and handling of hazardous material, and MSDS. The law requires the employer to keep the MSDS up to date and to keep them readily accessible so that employees can access them quickly when they are needed. Check with your supervisor on the job site to find out where your company's MSDS are located.

Electrical Safety Rules

As we discussed earlier in this chapter, when electricity is present, a whole new set of hazards are present for an electrical worker. To minimize the chance of an accident occurring, a residential electrician should practice the following electrical safety rules:

- Install all electrical wiring according to the *NEC®*.
- Whenever possible, work with a buddy. Avoid working alone.
- Always turn off the power and lock it out before working on any electrical circuits or equipment. If possible, when working on electrical equipment, stand on a rubber mat or a wooden floor or wear rubber-soled shoes.
- Never cut off the grounding prong from a three-prong plug on any power extension cord or from a power cord on any piece of equipment.

- Assume all electrical equipment to be "live" and treat them as such.
- Do not defeat the purpose of any safety devices, such as fuses or circuit breakers. Shorting across these devices could cause serious damage to equipment and property in addition to serious personal injury.
- Do not open and close switches under load unless absolutely necessary. This practice could result in severe arcing.
- Clean up all wiring debris at the end of each workday—or more often if it presents a safety hazard. Keeping the job site free of debris will help eliminate tripping hazards.

Classes of Fires and Types of Extinguishers

Fire is an ever present danger when working with electricity. The following three components must be present for a fire to start and sustain itself:

- Fuel: Any material that can burn
- Heat: Raises the fuel to its ignition temperature
- Oxygen: Is required to sustain combustion

These three components are often referred to as the sides of the "fire triangle" (Figure 1–18). If any one of the three is missing, a fire cannot be started. With the removal of any one of them, the fire will be extinguished.

There are different types of fires that an electrician could encounter. Recognizing the different fire classifications and knowing the proper type of extinguisher used to combat the fire is very important.

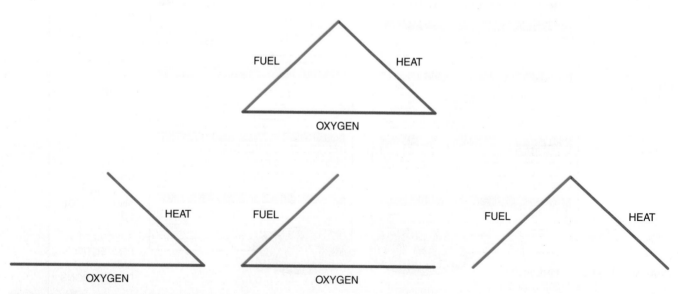

Figure 1–18 The three components necessary to start and sustain a fire are fuel, heat, and oxygen. Together they make up the "fire triangle," shown here. By removing any one of the components, a fire can be extinguished.

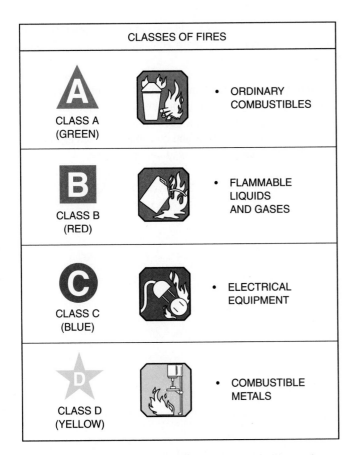

Figure 1–19 There are four classes of fires. This illustration shows what they are and the symbols that are used to represent them.

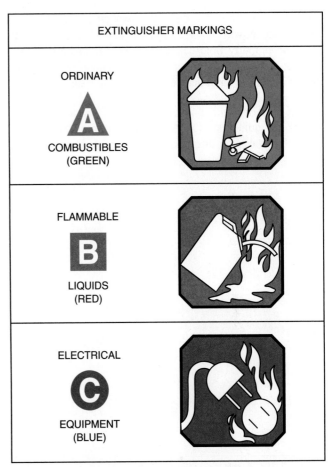

Figure 1–20 Fire extinguishers that you will encounter should have symbols on them like those illustrated here. These markings will let you know what type of fire you can use the extinguisher on.

Fires are classified as follows (see Figure 1–19):

- Class A: Fires that occur in ordinary combustible materials, such as wood, rags, and paper
- Class B: Fires that occur with flammable liquids, such as gasoline, oil, grease, paints, and thinners
- Class C: Fires that occur in or near electrical equipment, such as motors, switchboards, and electrical wiring
- Class D: Fires that occur with combustible metals, such as powdered aluminum and magnesium

There are many types of fire extinguishers. Each one is used to put out a specific class of fire (see Figure 1–20). You should become familiar with the fire extinguisher type that is available at the job site and learn how to properly use it. The extinguishers can be broken down as follows:

- Pressurized water: Operates usually by squeezing a handle or a trigger and spraying a stream of water onto the fire; used on Class A fires.
- Carbon dioxide (CO_2): Operates usually by squeezing a handle or trigger and spraying first at the base of the flames and then moving upward; used on Class B and C fires.
- Dry chemical: Operates usually by squeezing a handle, trigger, or lever and spraying at the base of the fire and then on top of any remaining materials that are burning; used on Class A, B, and C fires. This is the type of extinguisher found on most job sites and in existing buildings.
- Foam: Typically operates by turning the extinguisher upside down and spraying out the foam so that it lands on top of the fire; used on Class A and B fires.

Summary

There is a lot to know when it comes to workplace safety. You now have some knowledge about the concerns of the *NEC*® and OSHA for safety. But these rules and regulations are of little effect by themselves. You must give meaning to them by performing competently, by being knowledgeable, and most of all by being a safe, mentally alert, and responsible electrical worker. You must be able to do your work so that no injury is inflicted on yourself or your fellow workers. Following safe wiring practices will help keep you safe and will prevent equipment and property from damage. Knowledge of both general safety and electrical safety is the most important safeguard against serious or fatal accidents on the job site.

The *NEC*®, OSHA, and common safe electrical trade practices provide the guidelines for residential electrical construction safety. The stakes for unsafe practice are high. Death or injury to you or your colleagues and damage to or destruction of equipment are the unfortunate consequences of unsafe electrical installation practice.

The safety rules outlined in this chapter should serve as a guide to you as you seek to become a knowledgeable and skilled residential electrician. Learn them and the values that they imply. Practice them as you continue your career as a residential electrician. Your life may depend on it.

Procedures

Suggested Procedure to Find Information in the *NEC*®

As an example, let's find the ratings of the standard sizes of fuses recognized by the NEC®.

- Go to the index at the back of the *NEC*®.

- Locate the specific topic in the index. *In this example, you will find the topic "Fuses." Because you are looking for the rating of fuses, scan down until you find the word "Rating." The index tells us to look in Section 240.6. (Note: The index is set up alphabetically and gives only the article and the section where the information is found. There are no page numbers.)*

- Now that you have a specific section to find, go to the table of contents and find the page number of the article that the section is in. *You know that Section 240.6 is in Article 240 and that Article 240 is in Chapter 2. The table of contents tells you that Article 240 starts on page 70-85 in the 2002* NEC®.

- Turn to the article and scan through it until you find the referenced section. *Go to page 70-85 and scan down through the article until you come to Section 240.6.*

- Read the related areas and exceptions. *Section 240.6 clearly shows what the standard ratings of fuses are.*

Procedures

Suggested Lockout/Tagout Procedure

- Let all supervisors and affected workers know that you are performing a lockout/tagout on a particular electrical circuit.

- Deenergize the equipment or circuit in the normal manner.

- Lock out the electrical source to the equipment or circuit in the Off position with an approved lockout device.

- Check that after the lockout device is attached, the electrical switch cannot be placed in the On position.

- Place a tag with your name, date, and whom you work for on the switch.

- Use a proper test instrument to verify that all parts of the equipment you are working on are deenergized.

- Once you have completed the necessary work and made sure that all tools and meters are removed from the equipment you were working on, get ready to reenergize the circuit.

- Let everyone affected by the lockout know that you are reenergizing the circuit and equipment.

- Remove the lockout device and tags from the energy source.

- Turn on the electrical source to restore power.

Procedures

Suggested Procedure for Verifying That Circuits Are Deenergized

- Follow the lockout/tagout procedure.

- Make sure your test instrument is working by trying it on a known energized circuit.

- Test the circuit. There should be no voltage present.

Review Questions

Directions: For each of the following questions, choose the best answer. Indicate your choice by circling the letter in front of the answer.

1 In a work situation where an electrician's hands get wet while operating a portable drill, which of the following would be true?

a. Body resistance increases, and any shock would be mild.

b. Body resistance remains the same, and there is no danger as long as rubber boots are worn.

c. Body resistance is substantially decreased, and severe shock could occur.

d. There is no danger as long as a three-prong plug is used.

2 The amount of current that it takes to cause ventricular fibrillation is

a. 5 milliamperes

b. 100 milliamperes

c. 20 milliamperes

d. 0.25 milliampere

3 Which of the following is considered an insulator?

a. Glass

b. Plastic

c. Rubber

d. All of the above

4 Who publishes the *National Electrical Code®* (*NEC®*)?

a. OSHA

b. The National Electrical Contractors Association

c. The National Fire Protection Association

d. The Department of Labor

5 The unit of measure for electrical force is the

a. Ampere

b. Volt

c. Ohm

d. Watt

6 Energized, or "live," electrical equipment is dangerous. How does OSHA suggest that the worker deal with this potential hazard?

a. Work only on equipment that is marked or tagged "dead."

b. Inspect and test all equipment, assuming all to be energized until proven otherwise.

c. Work only on equipment that the foreman says is dead.

d. Ask someone nearby what the condition is.

7 Which of the following is considered to be the most important safeguard against serious or fatal accidents?

a. Knowledge of safety

b. Experience

c. Horse sense

d. Bravery

8 Which of the following types of extinguishers should not be used on electrical fires?

a. Carbon dioxide

b. Foam

c. Dry chemical

d. Pressurized water

9 When setting up scaffolding, all the following procedures should be followed except what?

a. When appropriate, wear an approved harness and lanyard to secure yourself to the staging so that you cannot fall.

b. Guardrails and toe boards must be installed when the actual standing platform is higher than 6 feet off the ground.

c. Set up the platform on the scaffolding so that it does not completely cover the staging work area. There must be a hole for a worker to climb up through to get on the platform.

d. Keep scaffold platforms clear of unnecessary materials, tools, and scrap; they may become a tripping hazard or be knocked off, endangering people.

10 **Fires that occur in or near electrical equipment, such as motors, switchboards, and electrical wiring, are classified as**

   a. Class A fires

   b. Class B fires

   c. Class C fires

   d. Class D fires

Directions: Circle T to indicate true statements or F to indicate false statements.

T F   **11**   The higher the resistance in a circuit, the lower the amount of current that can flow on the circuit.

T F   **12**   The resistance of the human body is fixed regardless of the conditions.

T F   **13**   OSHA requires voluntary compliance, and employers may comply at their pleasure.

T F   **14**   An energized line is the same as a "live" or "hot" line.

T F   **15**   Lower voltages (120–240 V) cannot kill.

T F   **16**   Copper is not a conductor of electricity since only materials that are magnetic can conduct electricity.

T F   **17**   The third prong (grounding prong) on a three-prong plug is optional and may be removed.

T F   **18**   Safety eyeglasses and goggles should only be worn when working on "live" circuits or equipment.

T F   **19**   The *NEC®* gives minimum safety standards for electrical work and is not a how-to manual.

T F   **20**   Electricians should wear metal helmets on job sites to avoid injury from falling objects.

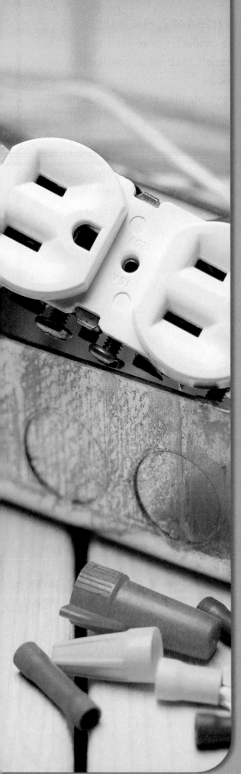

Chapter 2

Hardware and Materials Used in Residential Wiring

One of the things that a new residential electrician finds difficult about the job is recognizing and knowing when to use the large and varied amount of hardware and material used to install a residential electrical system. There is a lot to learn, and it is no wonder that some electricians find it overwhelming. However, when it comes to common hardware and materials, recognizing what they are and where to use them is a very important skill for a new electrician. This chapter introduces you to many of the common hardware and materials used to install a residential wiring system. Hardware and materials used in special applications are covered as those situations come up in future chapters.

OBJECTIVES

Upon completion of this chapter, the student should be able to:

- list several nationally recognized testing laboratories and demonstrate an understanding of the purpose of these labs.
- identify common box and enclosure types used in residential wiring.
- identify common conductor and cable types used in residential wiring.
- identify types of cable connectors, terminals, and lugs.
- identify common raceway types used in residential wiring.
- identify common devices used in residential wiring.
- identify common box covers and plates used in residential wiring.
- identify common types of fuses and circuit breakers used in residential wiring.
- describe the operation of a fuse and a circuit breaker.
- identify common panelboards, load centers, and safety switches used in residential wiring.
- identify common types of fasteners and fittings used in residential wiring.

Glossary of Terms

American Wire Gauge (AWG) a scale of specified diameters and cross sections for wire sizing that is the standard wire-sizing scale in the United States

ampacity the current in amperes that a conductor can carry continuously under the conditions of use without exceeding its temperature rating

approved when a piece of electrical equipment is approved, it means that it is acceptable to the authority having jurisdiction (AHJ)

bimetallic strip a part of a circuit breaker that is made from two different metals with unequal thermal expansion rates; as the strip heats up, it will tend to bend

cabinet an enclosure for a panel board that is designed for either flush or surface mounting; a swinging door is provided

cable a factory assembly of two or more insulated conductors that have an outer sheathing that holds everything together; the outside sheathing can be metallic or nonmetallic

circuit breaker a device designed to open and close a circuit manually and to open the circuit automatically on a predetermined overcurrent without damage to itself when properly applied within its rating

circular mils the diameter of a conductor in mils (thousandths of inches) times itself; the number of circular mils is the cross-sectional area of a conductor

connector a fitting that is designed to secure a cable or length of conduit to an electrical box

copper-clad aluminum an aluminum conductor with an outer coating of copper that is bonded to the aluminum core

device a piece of electrical equipment that is intended to carry but not use electrical energy; examples include switches, lamp holders, and receptacles

device box an electrical device that is designed to hold devices such as switches and receptacles

disconnecting means a term used to describe a switch that is able to deenergize an electrical circuit or piece of electrical equipment; sometimes referred to as the "disconnect"

fitting an electrical accessory, like a locknut, that is used to perform a mechanical rather than an electrical function

fuse an overcurrent protection device that opens a circuit when the fusible link is melted away by the extreme heat caused by an overcurrent

ganging the joining together of two or more device boxes for the purpose of holding more than one device

ground fault an accidental connection of a "hot" electrical conductor and a grounded piece of equipment or the grounded circuit conductor

handy box a type of metal device box used to hold only one device; it is surface mounted

insulated refers to a conductor that is covered by a material that is recognized by the *National Electrical Code®* as electrical insulation

junction box a box whose purpose is to provide a protected place for splicing electrical conductors

knockout (KO) a part of an electrical box that is designed to be removed, or "knocked out," so that a cable or raceway can be connected to the box

loadcenter a type of panelboard normally located at the service entrance in a residential installation and usually containing the main service disconnect switch

mil 1 mil is equal to .001 inches; this is the unit of measure for the diameter of a conductor

National Electrical Manufacturers Association (NEMA) an organization that establishes certain construction standards for the manufacture of electrical equipment; for example, a NEMA Type 1 box purchased from Company X will meet the same construction standards as a NEMA Type 1 box from Company Y

new work box an electrical box without mounting ears; this style of electrical box is used to install electrical wiring in a new installation

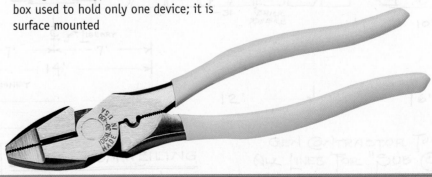

Glossary of Terms

old work box an electrical box with mounting ears; this style of electrical box is used to install electrical wiring in existing installations

outlet box a box that is designed for the mounting of a receptacle or a lighting fixture

overcurrent any current in excess of the rated current of equipment or the ampacity of a conductor; it may result from an overload, a short circuit, or a ground fault

overload a larger-than-normal current amount flowing in the normal current path

panelboard a panel designed to accept fuses or circuit breakers used for the protection and control of lighting, heating, and power circuits; it is designed to be placed in a cabinet and placed in or on a wall; it is accessible only from the front

pryout (PO) small parts of electrical boxes that can be "pried" open with a screwdriver and twisted off so that a cable can be secured to the box

Romex a trade name for nonmetallic sheathed cable (NMSC); this is the term most electricians use to refer to NMSC

safety switch a term used sometimes to refer to a disconnect switch; a safety switch may use fuses or a circuit breaker to provide overcurrent protection

service entrance the part of the wiring system where electrical power is supplied to the residential wiring system from the electric utility; it includes the main panelboard, the electric meter, overcurrent protection devices, and service conductors

sheath the outer covering of a cable that is used to provide protection and to hold everything together as a single unit

short circuit a low-resistance path which results in a high value of current that flows around rather than through a circuit; usually, short circuits are unintended and happen when two "hot" conductors touch each other by mistake

spliced connecting two or more conductors with a piece of approved equipment like a wirenut; splices must be done in approved electrical boxes

switch box a name used to refer to a box that just contains switches

torque the turning force applied to a fastener

utility box a name used to refer to a metal single gang, surface mounted device box; also called a handy box

wirenut a piece of electrical equipment used to mechanically connect two or more conductors together

Nationally Recognized Testing Laboratories

Residential electricians must use hardware and materials that are approved. The *National Electrical Code®* (*NEC®*) tells us in Article 90 that it is the authority having jurisdiction (AHJ) who has the job of approving the electrical materials used in an electrical system. The AHJ is usually the local or state electrical inspector in your area. The *NEC®* in Section 110.3(B) also requires electricians to use and install equipment and materials according to the instructions that come with their listing or labeling. Many AHJs base their approval of electrical equipment on whether it is listed and labeled by a nationally recognized testing laboratory (NRTL).

So what does it mean when a product is listed or labeled? When an electrical product is listed, it means that the product has been put on a list published by a NRTL that is acceptable to the authority having jurisdiction. The independent testing laboratory evaluates the product, and if the product meets a series of designated standards, it is found suitable for its intended purpose and is placed on the list. A label is put on the equipment (or on the box that it comes in if the item is too small to put a label on) to serve as the identifying mark of a specific testing laboratory. The label allows an electrician or, more important, the AHJ to see that the product complies with appropriate standards. If the product is installed and used according to the installation instructions, it will work in a safe manner.

Underwriters Laboratories (UL) is the most recognizable of the NRTLs. Most manufacturers of electrical products submit their products to UL, where the equipment will be subjected to several types of tests. The tests will determine if the product can be used safely under a variety of conditions. Once UL determines that the product complies with the specific standards it was tested for, the manufacturer is allowed to put the UL label on the product (Figure 2–1). The product is then listed in a UL directory.

By listing a product in its directory, UL is not saying that it approves the product, only that when if tested the product to nationally recognized safety standards, the product performed adequately in terms of fire, electric shock, and other related safety hazards.

For residential wiring, there are three UL directories that the electrician should consult: the *Electrical Construction Equipment Directory* (Green Book), the *Electrical Appliance and Utilization Equipment Directory* (Orange Book), and the *General Information for Electrical Equipment Directory* (White Book) (Figure 2–2). The White Book is the directory that will be used the most. It gives information on specific requirements, permitted uses, and limitations of the product.

Two more NRTLs that you should be aware of are CSA International and Intertek Testing Services. CSA International used to be known as the Canadian Standards Association (CSA). This lab tests, evaluates, and lists electrical equipment in Canada. Many electrical products that you may use will have the CSA International label or the older CSA label (Figure 2–3) on them. This label and listing, like the UL label, is a basis for the AHJ approval of equipment required by the *NEC®*. Intertek Testing Services (ITS), formerly known as

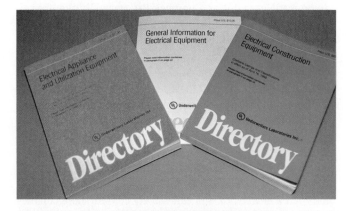

Figure 2–2 The UL Green, White, and Orange Books. Electricians should consult these books to determine specific information, such as permitted uses and installation instructions, for a listed electrical product.

Figure 2–1 The Underwriters Laboratories (UL) label. If a piece of electrical equipment has this label on it, it means that the item has met the specific standards it was tested for and has been listed and labeled by UL.

Figure 2–3 This CSA International label is found on many electrical products and indicates that the product has been tested and meets the specific testing standards of CSA International.

Electrical Testing Laboratories (ETL), also tests and evaluates electrical products according to nationally recognized safety standards. Again, the AHJ will often use the ITS labeling and listing as a basis for approval.

While the **National Electrical Manufacturers Association (NEMA)** is not actually a testing laboratory, it is an organization you should also be aware of. It develops electrical equipment standards that all member manufacturers follow when designing and producing electrical equipment. Over 500 manufacturers of electrical equipment belong to NEMA. As an example of how NEMA affects an electrician, consider that a residential electrician in Maine may use a NEMA 5-20R receptacle made by Company X that will look like, install like, and work like a NEMA 5-20R receptacle that an electrician in California is using that was manufactured by Company Y.

Electrical Boxes

The residential electrical system wiring connections must be contained in approved electrical boxes. The design of these boxes helps prevent electrical sparks and arcs from causing fires within the walls, floors, and ceilings of a house. Boxes can be made of metal or a nonmetallic material, like plastic. A **device box** is designed to contain switches and receptacles and is the *NEC's®* term used for this type of box. Many electricians refer to this type of box as an outlet box, or a "wall case" box. The term **outlet box** is used to indicate a box that may contain switches, receptacles, or a lighting fixture. Boxes that contain just switches are sometimes referred to as simply a **switch box**.

Some boxes are designed to contain only **spliced** conductors. This type of box is referred to as a **junction box**. Junction boxes, or J-boxes as they are sometimes called, are either octagonal or square in shape. They are also larger in total volume than a device box and can usually contain more conductors. Sizing electrical boxes is covered in detail in Chapter 12. Outlet boxes can also serve as a junction box. This occurs when splices are made in the same box that also is used to mount a lighting fixture or hold a receptacle or a switch.

Metal Device Boxes

Metal device boxes are used widely in residential wiring, even though the use of nonmetallic boxes has become more popular across the country. The standard metal device box has a 3 × 2-inch opening and can vary in depth from 1½ to 3½ inches. The most common metal device box used is the 3 × 2 × 3½-inch box because it can contain several wires and a device (Figure 2–4).

A metal device box has many features for the electrician to use. One common feature involves taking the sides off the box and **ganging** the boxes together to make a box that can accommodate more than one device (Figure 2–5). For example, two switches may be required beside the front door of a house. One switch is required to control an outside light, and the other

Figure 2–4 A 3 x 2 x 3½ inch metal device box with a side-mounting bracket and internal cable clamps.

Figure 2–5 Two 3 x 2 x 2¾–inch metal device boxes ganged together. A side-mounting bracket is left on one box for mounting the ganged boxes to a building framing member.

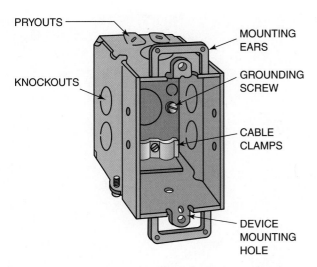

PRYOUTS

KNOCKOUTS

MOUNTING
EARS

GROUNDING
SCREW

CABLE
CLAMPS

DEVICE
MOUNTING
HOLE

Figure 2–6 **Parts of a typical metal device box. Mounting ears on a device box allow the box to be used in remodel or "old work." The mounting ears are not used on metal boxes used in "new work."**

Figure 2–7 **A handy box is used normally for surface mount applications. Another name sometimes used for this box style is "utility box."**

switch is necessary to control the lighting in the room. For this situation, the electrician would need to gang together two single-gang metal device boxes to make a two-gang device box.

Figure 2–6 shows the parts of a typical metal device box. Boxes with mounting ears are typically called old work boxes because the ears allow them to be installed in an existing wall. On the other hand, boxes without mounting ears are referred to as new work boxes and are the kind used when installing an electrical system in a new home. The device mounting holes are designed to accommodate the screws of a switch or receptacle and are used to secure the device to the box. The holes are tapped for a 6-32 screw size. In the back of the box, you will find a tapped hole that accommodates a 10-32 grounding screw. This screw will be hex headed and green in color, and its purpose is to attach the circuit electrical grounding conductor to the metal box. Knockouts (KO) are included on the sides and the top and bottom of the box. When a KO is removed, an opening exists for a connector to be used to secure a wiring method to the box. The pryouts (PO) located at the rear top and bottom of the box are "pried" off and provide an opening through which a cable can be secured to the box. Most of these boxes come with internal cable clamps that are tightened onto the cable after it has been inserted through the pryout opening. The cable clamps secure the cable to the box.

CAUTION

CAUTION: The *NEC*® requires wiring methods, like cables and raceways, to always be secured to an electrical box with a connector designed for the purpose. The design may include an internal or an external clamping mechanism.

Another type of metallic device box recognized by the *NEC*® is the handy box, sometimes also referred to as a utility box. This style of box is used primarily for surface mounting and can accommodate one device, such as a receptacle or a switch. The size of the opening for this type of box is $4 \times 2\frac{1}{8}$ inches and the depth of the box can range from $1\frac{1}{2}$ to $2\frac{1}{8}$ inches (Figure 2–7).

Nonmetallic Device Boxes

The most popular style of device box used in residential wiring is the nonmetallic device box. Typically, these boxes are made of plastic or fiberglass. They are very lightweight, strong, and very easy to install. Another feature that makes them so popular is that, when compared to metal device boxes, they are very inexpensive. In residential wiring, this box type is used only with a nonmetallic sheathed cable wiring method.

Nonmetallic device boxes are available in a variety of shapes and sizes (Figure 2–8). Common sizes include single-, two-, and three-gang boxes. The boxes come from the manufacturer with nails already attached. This makes this box style easy to mount on a wooden framing member, like wall studs.

The single-gang nonmetallic box is made so that no cable clamp is needed. This feature contributes to the popularity of this box type. The electrician simply uses a tool like a screwdriver to open a KO on the top or bottom of the box, and the cable is put into the box through the resulting hole. The *NEC*® in Section 314.17(C) requires the electrician to secure (staple) the cable within 8 inches of the single-gang nonmetallic box. Two or more ganged nonmetallic device boxes have built-in cable clamps and, when used, require securing the cable within 12 inches of the box.

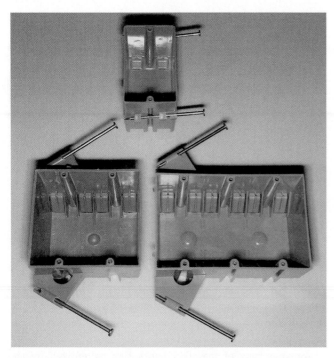

Figure 2–8 Single-gang, two-gang, and three-gang nonmetallic device boxes. All three box styles are secured to wood framing members with nails.

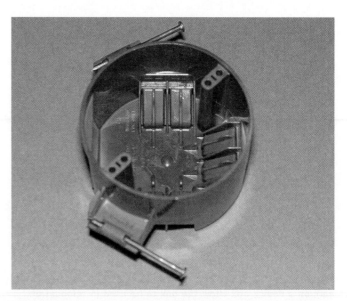

Figure 2–9 A nonmetallic round outlet box. This box is used in walls and ceilings and is designed to support luminaires (lighting fixtures). Nails are used to secure this box to a wood framing member.

Outlet and Junction Boxes

When installing lighting fixtures in a ceiling or on a wall, an outlet box is typically used. This box style can also be used when connecting small or large appliances. They are larger than a device box and provide more room for different wiring situations than do device boxes. In residential wiring, you will find these boxes in round, octagon, or square shapes. The round boxes are typically nonmetallic (Figure 2–9), while the octagon and square boxes are typically metal (Figure 2–10). Devices can be installed in this type of box but only when special mounting covers called "raised plaster rings" or "raised covers" are used (Figure 2–11).

When several conductors are spliced together at a point on the wiring system, the *NEC®* requires a junction box to be used. The *NEC®* also requires junction boxes to be accessible after installation without having to alter the finish of a building. Attics, garages, crawl spaces, and basements are typical areas where electricians locate junction boxes. Like outlet boxes, junction boxes are found in round, octagon and square shapes. Junction boxes must always be covered, and several different types of covers are available (Figure 2–12).

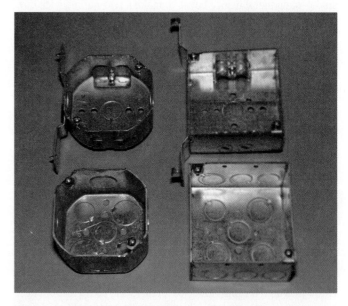

Figure 2–10 Metal octagon and square boxes are used in two common sizes for residential work: 4 × 1½ inches and 4 × 2⅛ inches. They may have side-mounting brackets or may be attached directly to framing members with screws.

CAUTION: Never bury a junction box in a wall or ceiling. Junction boxes must be accessible after they are installed.

Special Boxes for Heavy Loads

When installing heavy items like ceiling paddle fans, special boxes are required. Boxes used to support a ceiling paddle fan must be designed for that application and listed as such by an organization like UL. The boxes can be made of metal or nonmetal and are specifically designed and tested to support heavy loads (Figure 2–13).

Figure 2–11 Raised covers must be used with square outlet boxes when you want to install a switch or receptacle in them. Raised plaster rings are used when the square box is used in a wall or ceiling location. Raised covers are used in a surface-mount situation.

Figure 2–12 When conductors are spliced in a junction box, a cover must be used. Flat blank covers are available in an octagon or a square shape.

Figure 2–13 A box that is specifically designed to support ceiling-suspended paddle fans.

Conductors and Cable Types

When residential electricians install a wiring system in a dwelling, the electrical conductors required for the circuits are most often installed as part of a cable assembly. A **cable** is described as a factory assembly of two or more conductors that has an overall outer covering. The covering is referred to as a **sheath** and can be made of plastic, PVC (polyvinyl chloride), or metal. The sheathing holds the cable assembly together and provides protection to the conductors. This section of the chapter introduces you to the types of cables and conductors used in residential wiring.

Conductors

Conductors in residential wiring are usually installed in a cable assembly. A few situations in residential wiring call for conductors to be installed in raceways. (Conductors installed in raceways are discussed later in this chapter.) As you already know, conductors conduct the current that is delivered to various loads in a residential wiring system. The installed conductors are usually made of copper but may also be aluminum or copper-clad aluminum. Copper is the preferred material because of its great ability to conduct electricity, its strength, and its proven record of having very few, if any, problems over the long term when correctly installed.

Aluminum conductors are typically used in larger conductor sizes because of their lower cost and lighter weight, which makes them easier to handle. Because aluminum has a higher resistance value and does not conduct electricity as well as copper, a larger-size aluminum conductor must be used to carry the same amount of current as a copper conductor. Aluminum also tends to oxidize much quicker than copper, and electricians must use an antioxidant compound on exposed aluminum at all terminations. Without the antioxidant, the exposed aluminum will oxidize, and a white powder (aluminum oxide) will develop at the conductor termination. This will cause high resistance at the termination and result in excessive heating and possible damage to the conductor and electrical equipment.

(See page 56 for a procedure for properly preparing aluminum conductors for termination.)

A conductor that has an aluminum core and an outer coating of copper is called **copper-clad aluminum**. Copper-clad aluminum is not used very often in residential wiring since it can be used only with wiring devices that are listed and labeled for use with this type of conductor. Most of the devices are rated for copper only, but there are listed devices that can be used with copper or copper-clad aluminum. Check the device for a marking that indicates what conductor type it is listed for. Table 2–1 shows typical markings for determining the type of conductors allowed with devices and connectors.

Table 2-1 **Terminal Identification Markings**

Type of Device	Marking on Terminal or Conductor	Connector Permitted
15- or 20-ampere receptacles and switches	CO/ALR	Aluminum, copper, copper-clad aluminum
15- and 20-ampere receptacles and switches	NONE	Copper, copper-clad aluminum
30-ampere and greater receptacles and switches	AL/CU	Aluminum, copper, copper-clad aluminum
30-ampere and greater receptacles and switches	NONE	Copper only
Screwless pressure terminal connectors of the push-in type	NONE	Copper or copper-clad aluminum
Wire connectors	AL	Aluminum
Wire connectors	AL/CU or CU/AL	Aluminum, copper, copper-clad aluminum
Wire connectors	CC	Copper-clad aluminum only
Wire connectors	CC/CU or CU/CC	Copper or copper-clad aluminum
Wire connectors	CU or CC/CU	Copper only
Any of the above devices	COPPER OR CU ONLY	Copper only

Table 2-2 **Conductor Applications in Residential Wiring**

Conductor Size	Overcurrent Protection	Typical Applications (Check wattage and/or ampere rating of load to select the correct size conductors based on Table 310.16 in the *NEC*®.)
20 AWG	Class 2 circuit transformers provide overcurrent protection.	Telephone wiring is usually 20 or 22 AWG.
18 AWG	7 amperes. Class 2 circuit transformers provide overcurrent protection.	Low-voltage wiring for thermostats, chimes, security, remote control, home automation systems, etc. For these types of installations, 18 or 20 AWG conductors can be used, depending on the connected load and length of circuit.
16 AWG	10 amperes. Class 2 circuit transformers provide overcurrent protection.	Same applications as above. Good for long runs to minimize voltage drop.
14 AWG	15 amperes	Typical lighting branch circuits.
12 AWG	20 amperes	Small appliance branch circuits for the receptacles in kitchens and dining rooms. Also laundry receptacles and workshop receptacles. Often used as the "home run" for lighting branch circuits. Some water heaters.
10 AWG	30 amperes	Most clothes dryers, built-in ovens, cooktops, central air conditioners, some water heaters, heat pumps.
8 AWG	40 amperes	Ranges, ovens, heat pumps, some large clothes dryers, large central air conditioners, heat pumps.
6 AWG	50 amperes	Electric furnaces, heat pumps.
4 AWG	70 amperes	Electric furnaces, feeders to subpanels.
3 AWG and larger	100 amperes	Main service entrance conductors, feeders to subpanels, electric furnaces.

The **American Wire Gauge (AWG)** is the system used to size the conductors used in residential wiring. In this system, the smaller the number, the larger the conductor, and the larger the number, the smaller the conductor. For example, an 8 AWG conductor is larger than a 10 AWG conductor, and a 12 AWG conductor is smaller than a 10 AWG conductor. The *NEC*® recognizes building wire from 18 AWG up to 4/0. 1/0, 2/0, 3/0, and 4/0 are sometimes shown as 0, 00, 000, and 0000, respectively. Electricians pronounce these wire sizes as 1 "aught," 2 "aught," 3 "aught," and 4 "aught."

Conductor sizes larger than 4/0 are measured in **circular mils**. A **mil** is equal to .001 inch. If the diameter of a conductor is measured in mils (thousandths of inches) and then squared, you get circular mils. The next size up from a 4/0 is a 250,000-circular-mil conductor. The *NEC*® refers to conductor sizes over 4/0 as having a certain number of kcmils. A kcmil is 1,000 circular mils; therefore, a 250,000-circular-mil conductor is referred to as 250 kcmil. (It is the size of the cross-sectional area of the conductor that is actually being described when its size in kcmils is given.) The largest wire size recognized by the *NEC*® is 2,000 kcmil. Table 2–2 shows some common conductor sizes and applications in residential wiring.

In residential wiring, 14, 12, and 10 AWG conductors are solid when used in a cable assembly and are usually stranded when installed in a raceway. Conductor sizes that are 8 AWG and larger will almost always be stranded. Stranding makes the conductors easier to work with, especially in the larger sizes.

The **ampacity** of a conductor refers to the ability of a conductor to carry current. The current in amperes that a conductor can carry continuously under the conditions of use without exceeding its temperature rating is how the term "ampacity" is defined in the *NEC*®. The larger the conductor size, the higher the ampacity of the conductor. A residential electrician must be able to choose the correct conductor size depending on the ampacity needed for each circuit that is being installed. The ampacity of a conductor depends not only on the size of the conductor but also on what insulation type the conductor has. An insulated conductor has an outer covering that is approved and recognized as electrical insulation by the *NEC*®. Letters in certain combinations identify the type of conductor insulation and where the conductor with that particular insulation can be used. The letters are written on the insulation of the conductor or on the cable itself. Table 310.13 of the *NEC*® lists the recognized insulation types. Table 2–3 shows some typical insulation types used in residential wiring.

Table 2–3 **Typical Conductor Insulations Used in Residential Wiring**

Trade Name	Type Letter	Operating Temperature	Maximum Application Provisions	Insulation	AWG	Outer Covering
Heat-resistant thermoplastic	THHN	194° F (90° C)	Dry and damp locations	Flame-retardant, heat-resistant thermoplastic	14–1,000 kcmil	Nylon jacket or equivalent
Moisture- and heat-resistant thermoplastic	THHW	167° F (75° C) 194° F (90° C)	Wet location Dry location	Flame-retardant, moisture- and heat-resistant thermoplastic	14–1,000 kcmil	None
Moisture- and heat-resistant thermoplastic	THWN Note: If marked THWN-2, okay for 194° F (90° C) in dry or wet locations	167° F (75° C)	Dry and wet locations	Flame-retardant, moisture- and heat-resistant thermoplastic	14–1,000 kcmil	Nylon jacket or equivalent
Moisture- and heat-resistant thermoplastic	THW Note: If marked THW-2, okay for 194° F (90° C) in dry or wet locations	167° F (75° C)	Dry and wet locations	Flame-retardant, moisture- and heat-resistant thermoplastic	14–2,000 kcmil	None
Underground feeder and branch-circuit single conductor or multiconductor cable (see *Article 339 of the* NEC®)	UF	140° F (60° C) 167° F (75° C)	See *Article 339 of the* NEC® See *Article 339 of the* NEC®	Moisture resistant Moisture resistant	14–4/0	Integral with insulation

Table 310.16 of the *NEC®* is where residential electricians go to determine the ampacity of the conductors they will be using. This table shows the ampacities of conductors from size 18 AWG to 2,000 kcmil. It is broken down into copper wire sizes and aluminum/copper-clad aluminum wire sizes. The ampacity columns are headed by the insulation temperature rating. The ratings are 60, 75, and 90 degrees Celsius. Choosing the correct conductor size for a specific application is covered later in this book.

Conductor functions are identified with a certain color coding of their insulation. The *NEC®* requires that each conductor be color coded to indicate the function that it performs in a circuit. The following color coding applies to residential wiring applications:

- Black: Used as an ungrounded, or "hot," conductor and carries the current to the load in 120-volt circuits.
- Red: Also used as an ungrounded, or "hot," conductor and carries current to the load in 120/240-volt circuits like a dryer circuit.
- White: Used as the grounded circuit conductor and returns current from the load back to the source. It is sometimes referred to as the "neutral" conductor but is truly neutral only when used with a black and a red wire in a multiwire circuit.
- Bare: Used as an equipment-grounding conductor that bonds all non-current-carrying metal parts of a circuit together; it never carries current.
- Green: Used as an insulated equipment-grounding conductor; could be green with yellow stripes; never carries current.

Conductors are terminated to electrical equipment and each other with a variety of connector types, terminals, and lugs. A **wirenut** is the most common wire connector used by electricians to splice together two or more conductors. Figure 2–14 shows several types of wire connectors used in residential wiring.

Cable Types

The conductors installed for circuits in residential wiring are usually installed as part of a cable. Nonmetallic sheathed cable, underground feeder cable, armored-clad cable, metal-clad cable, and service entrance cable are the types of cable commonly used. Each has certain advantages and uses in residential wiring. Residential electricians need to recognize these cable types and know when and where to use them. This section introduces you to these cable types, and later chapters discuss their installation procedures.

Nonmetallic Sheathed Cable

Residential electricians refer to nonmetallic sheathed cable (NMSC) as **Romex.** This is the name first given to NMSC by the Rome Wire and Cable Company. Even though other companies manufacture NMSC, electricians call it "Romex." It is the least expensive wiring method to purchase and in-

stall, which is why it is used more than other wiring methods in residential applications.

Article 334 of the *NEC®* covers NMSC, which is defined as a factory assembly of two or more insulated conductors having an outer sheathing of moisture-resistant, flame-retardant, nonmetallic material (Figure 2–15). It is available with two or three conductors in sizes from 14 through 2 AWG copper and 12 through 2 AWG aluminum. Two-wire cable has a black and a white insulated conductor along with a bare grounding conductor. The three-wire cable will have a black, red, and a white conductor along with a bare equipment-grounding conductor. It is used in residential applications for general-purpose branch circuits, small-appliance branch circuits, and individual branch circuits. The ampacity of NMSC is found in the "60 degrees Celsius" column of Table 310.16 in the *NEC®*.

CAUTION: Even though the ampacity for 14, 12, and 10 AWG NMSC is found in the "60 degrees Celsius" column of Table 310.16, Section 240.4(D) requires a maximum fuse or circuit breaker size of 15 amps for 14 AWG, 20 amps for 12 AWG, and 30 amps for 10 AWG conductors used in residential circuits.

Underwriters Laboratories lists three types of NMSC. The type will be written on the outer jacket of the NMSC:

- Type NM-B is by far the most common type used and has a flame-retardant, moisture-resistant, nonmetallic outer jacket. It can be used in dry locations only. The conductor insulation is rated at 194 degrees Fahrenheit (90 degrees Celsius). However, the ampacity of the NM-B is based on 140 degrees Fahrenheit (60 degrees Celsius).
- Type NMC-B is not used often in residential work. It has a flame-retardant and moisture-, fungus-, and corrosion-resistant nonmetallic outer jacket and can be used in dry or damp locations. The conductor insulation and ampacity are rated the same as the Type NM-B cable.
- Type NMS-B is used in new homes that have home automation systems using the latest technology. The cable contains the power conductors, telephone wires, coaxial cable for video, and other data conductors all in the same cable. It has a moisture-resistant, flame-retardant, nonmetallic outer jacket.

Underground Feeder Cable

Type UF cable is covered in Article 340 of the *NEC®* (Figure 2–16). It is used for underground installations of branch circuits and feeder circuits. It can also be used in interior installations but must be installed following the installation requirements for NMSC. It is available in wire sizes

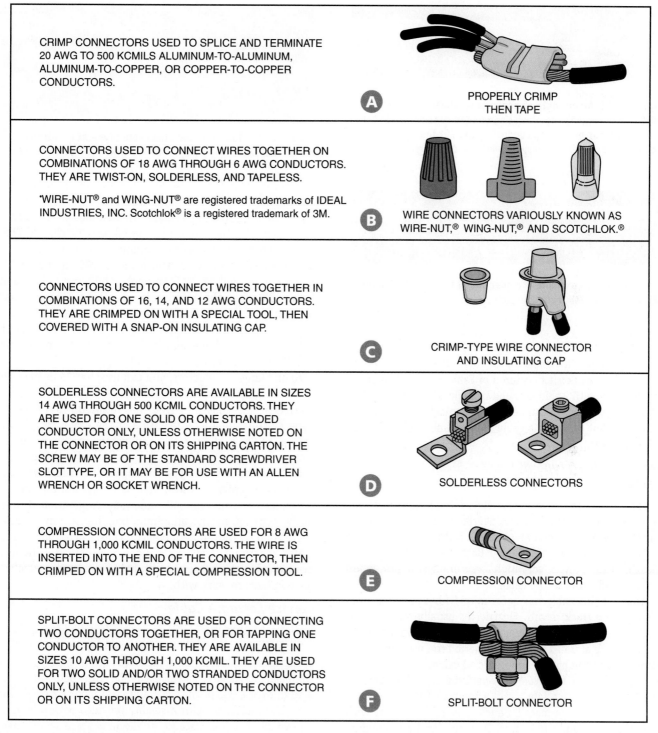

CRIMP CONNECTORS USED TO SPLICE AND TERMINATE 20 AWG TO 500 KCMILS ALUMINUM-TO-ALUMINUM, ALUMINUM-TO-COPPER, OR COPPER-TO-COPPER CONDUCTORS.

A

PROPERLY CRIMP THEN TAPE

CONNECTORS USED TO CONNECT WIRES TOGETHER ON COMBINATIONS OF 18 AWG THROUGH 6 AWG CONDUCTORS. THEY ARE TWIST-ON, SOLDERLESS, AND TAPELESS.

*WIRE-NUT® and WING-NUT® are registered trademarks of IDEAL INDUSTRIES, INC. Scotchlok® is a registered trademark of 3M.

B

WIRE CONNECTORS VARIOUSLY KNOWN AS WIRE-NUT,® WING-NUT,® AND SCOTCHLOK.®

CONNECTORS USED TO CONNECT WIRES TOGETHER IN COMBINATIONS OF 16, 14, AND 12 AWG CONDUCTORS. THEY ARE CRIMPED ON WITH A SPECIAL TOOL, THEN COVERED WITH A SNAP-ON INSULATING CAP.

C

CRIMP-TYPE WIRE CONNECTOR AND INSULATING CAP

SOLDERLESS CONNECTORS ARE AVAILABLE IN SIZES 14 AWG THROUGH 500 KCMIL CONDUCTORS. THEY ARE USED FOR ONE SOLID OR ONE STRANDED CONDUCTOR ONLY, UNLESS OTHERWISE NOTED ON THE CONNECTOR OR ON ITS SHIPPING CARTON. THE SCREW MAY BE OF THE STANDARD SCREWDRIVER SLOT TYPE, OR IT MAY BE FOR USE WITH AN ALLEN WRENCH OR SOCKET WRENCH.

D

SOLDERLESS CONNECTORS

COMPRESSION CONNECTORS ARE USED FOR 8 AWG THROUGH 1,000 KCMIL CONDUCTORS. THE WIRE IS INSERTED INTO THE END OF THE CONNECTOR, THEN CRIMPED ON WITH A SPECIAL COMPRESSION TOOL.

E

COMPRESSION CONNECTOR

SPLIT-BOLT CONNECTORS ARE USED FOR CONNECTING TWO CONDUCTORS TOGETHER, OR FOR TAPPING ONE CONDUCTOR TO ANOTHER. THEY ARE AVAILABLE IN SIZES 10 AWG THROUGH 1,000 KCMIL. THEY ARE USED FOR TWO SOLID AND/OR TWO STRANDED CONDUCTORS ONLY, UNLESS OTHERWISE NOTED ON THE CONNECTOR OR ON ITS SHIPPING CARTON.

F

SPLIT-BOLT CONNECTOR

Figure 2–14 **Examples of wire connectors.**

Figure 2–15 An example of nonmetallic sheathed cable (NMSC) showing (A) the black ungrounded conductor, (B) the bare equipment-grounding conductor, and (C) the white grounded conductor. *Courtesy of Southwire Company.*

Figure 2–16 An example of underground feeder cable (Type UF). *Courtesy of Southwire Company.*

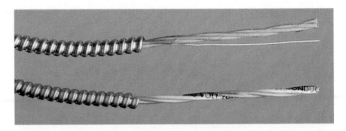

Figure 2–17 Top: Armored-clad cable (Type AC); bottom: metal-clad cable (Type MC). *Courtesy of AFC Cable Systems, Inc.*

from 14 through 4/0 AWG copper and from 12 through 4/0 AWG aluminum. The ampacity for this cable type is found in the "60 degrees Celsius" column of Table 310.16. The burial depth for Type UF cable is found in Table 300.5 of the *NEC®*. Additionally, Type UF can be used for outside residential installations but only if the cable is listed and marked as sunlight resistant.

Armored-Clad (Type AC) and Metal-Clad (Type MC) Cable

There are certain locations in the United States where the AHJ may not allow NMSC in residential construction. The alternative wiring method most often used is armored-clad or metal-clad cable (Figure 2–17). Both have a metal outer sheathing and provide very high levels of physical protection for the conductors in the cable.

Armored-clad cable has been around for a long time and has a proven track record. Electricians usually refer to it as "BX" cable. "BX" was a trademark owned by the General Electric Company and has become a generic term used as a trade name for any company's armored-clad cable.

Article 320 of the *NEC®* covers Type AC cable. Section 320.1 defines it as a fabricated assembly of insulated conductors in a flexible metallic enclosure. It is available with two, three, or four conductors. A small aluminum bonding wire is included and is used to ensure electrical continuity of the outside flexible metal sheathing. Because of the bonding wire, the outside metal sheathing can be used as the grounding conductor. Type AC is also available with a green insulated grounding conductor. This wiring method is available in wire sizes 14 through 3 AWG copper and 12 through 1 AWG aluminum. Like Romex, Type AC ampacity is found in the "60 degrees Celsius" column of Table 310.16.

Metal-clad cable looks very much like armored-clad cable. However, there are many differences. Article 330 covers Type

MC and defines it as a factory assembly of one or more insulated circuit conductors enclosed in an armor of interlocking metal tape or a smooth or corrugated metallic sheath. There is no limit to the number of conductors found in Type MC, and it is available in wire sizes from 18 AWG through 2,000-kcmil copper and 12 AWG through 2,000-kcmil aluminum. Unlike BX cable, the outer metal sheathing cannot be used for a grounding conductor unless it is specifically listed as such. A green insulated grounding conductor will always be included in the cable assembly. The ampacity is found using the "60 degrees Celsius" column of Table 310.16, but for 1/0 AWG and larger conductors, you can use the "75 degrees Celsius" column.

Electricians sometimes find that it is hard to distinguish Type AC from Type MC cable. One of the easiest ways to do this is to look at how the conductors in the cables are wrapped. Type AC cable has a light brown paper covering each conductor, while Type MC has no individual wrapping on the conductors. However, there is a polyester tape over all the conductors in Type MC cable, and as mentioned previously, Type MC has no aluminum bonding wire in it. Since there is no writing to help identify these cable types on the metal sheathing itself, knowing a few of the differences mentioned in this paragraph will help you identify which cable you have to work with.

Service Entrance Cable

Service entrance (SE) cables are installed to supply the electrical power from the utility company to the building's electrical system. It is also sometimes used to supply power to a large appliance, like an electric range. This cable type is covered in Article 338 of the *NEC®* and is defined as a single conductor or multiconductor assembly provided with or without an overall covering. Because of the larger size requirements and expense, most residential installations use aluminum conductors instead of copper in service entrance cables.

Electricians should be familiar with three types of service entrance cable. The most commonly used type is SEU. Type SEU cable is used for service entrance installations (Figure 2–18). It contains three conductors enclosed in a flame-retardant, moisture-resistant outer covering. The service neutral conductor is made up of several bare conductors wrapped around the insulated ungrounded conductors. The electrician needs to twist these bare conductors together for termination.

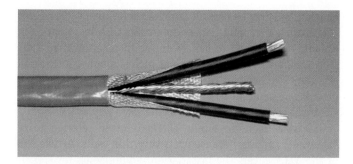

Figure 2–18 Service entrance cable (Type SEU) with two black insulated ungrounded conductors and one wraparound bare neutral grounded conductor.

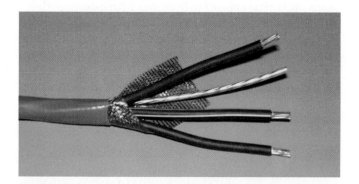

Figure 2–19 Service entrance cable (Type SER) with two black insulated ungrounded conductors, one white identified insulated grounded conductor, and one bare grounding conductor.

Type SER cable is similar to Type SEU but has four conductors wrapped in a round configuration (Figure 2–19). The neutral conductor is insulated, and the cable also contains a bare grounding conductor. SER cable is what you would use to wire a new electric range or electric dryer if you chose service entrance cable as the wiring method. Type SER is also used as a feeder from the main service panel to a subpanel.

Type USE cable is a three-conductor cable used for underground service entrance installations (Figure 2–20). It has a moisture-resistant covering but, because it is buried, is not flame retardant. It can also be used for underground feeder or branch circuits.

Cable Fittings and Supports

There are a large variety of fittings used with the cables described in the preceding paragraphs. **Fittings** allow cables to be installed in compliance with the *NEC®*. When a cable is terminated at an electrical box, a connector must be used to secure it to the box. Figure 2–21 shows some examples of cable connectors commonly used in residential wiring.

The *NEC®* also requires cables to be properly supported during their installation along their entire length. Figure 2–22 shows some common items used to support cables.

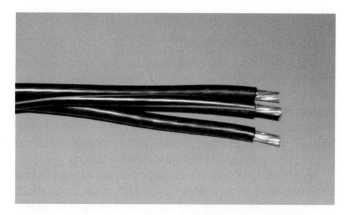

Figure 2–20 Service entrance cable (Type USE) is used to install underground service entrances. Unlike Type SEU and Type SER service entrance cable, the insulation on Type USE cable is suitable for direct burial. The conductors are simply wrapped together and do not have an overall outer sheathing.

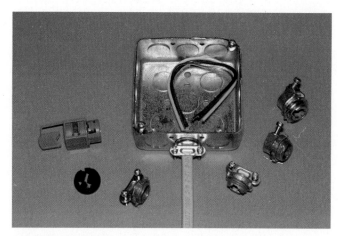

Figure 2–21 Examples of common cable connectors used in residential wiring.

Figure 2–22 Examples of common items used to support and secure cables in residential wiring.

Raceways

Article 100 of the *NEC®* defines a raceway as an enclosed channel of metal or nonmetallic materials designed expressly for holding wires or cables. Raceways that may be used in residential wiring include rigid metal conduit, rigid nonmetallic conduit, intermediate metal conduit, liquidtight flexible conduit, flexible metal conduit, electrical nonmetallic tubing, and electrical metallic tubing.

Many residential wiring installations do not have a need for the installation of any raceway. However, some installations may use a raceway for parts of a service entrance or to protect conductors going to an outside air-conditioning unit. Because raceways are used in some installations, residential electricians need to recognize common raceway types.

Rigid metal conduit (RMC) is typically made of steel and is galvanized to enable it to resist rusting. It is sometimes referred to as "heavywall" conduit. RMC is often used as a mast for a service entrance. A mast will allow the service conductors from the electric utility to be located high enough from the ground that the danger to pedestrians is minimized. This conduit is threaded on each end and is available in trade sizes of 1/2 inch through 6 inches. It is available in a standard length of 10 feet, and a coupling is included on one end. Article 344 of the *NEC®* provides installation requirements for RMC. RMC is connected to an enclosure by threading it directly into a threaded hole in the enclosure or by using two locknuts, one inside and one outside the box, to secure it to a nonthreaded KO hole. The *NEC®* requires RMC to be supported at regular intervals. The general support rule is to support it within 3 feet of each box and then no more than every 10 feet thereafter. Figure 2–23 shows an example of RMC with some associated fittings.

Intermediate metal conduit (IMC) (Figure 2–24) is a lighter version of RMC but can be used in all the locations that the heavier RMC may be used. It is threaded on each end and also comes with a coupling on one end. Because it is lighter in weight, IMC is easier to handle during installation than RMC. It is available in trade sizes of 1/2 inch through 4 inches and also comes in standard 10-foot lengths. Article 342 of the *NEC®* provides installation requirements for IMC. IMC is attached to electrical boxes and supported in the same way as RMC.

Electrical metallic tubing (EMT) is often referred to as "thinwall" conduit because of its very thin walls. Article 358 of the *NEC®* covers the installation requirements for EMT. It is much lighter and easier to install than both RMC and IMC. EMT cannot be threaded. Connectors and couplings used with EMT utilize setscrew or compression tightening systems. EMT is available in trade sizes of 1/2 inch through 4 inches and comes in standard 10-foot lengths. Figure 2–25 shows an example of EMT and various fittings used with it.

Rigid nonmetallic conduit (RNC) is a PVC type of raceway. It is very inexpensive and can be used in a variety of residential wiring applications. Article 352 of the *NEC®* covers the installation requirements of RNC. It is available in trade sizes of 1/2 inch through 6 inches and, like the other race-

Figure 2–24 Intermediate metal conduit (IMC) and associated fittings.

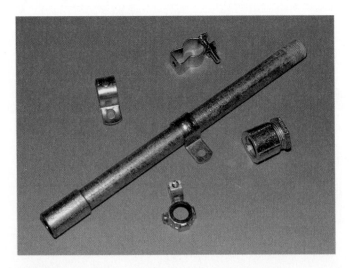

Figure 2–23 Rigid metal conduit (RMC) and associated fittings.

Figure 2–25 Electrical metallic tubing (EMT) and associated fittings.

ways mentioned in this section, comes in standard 10-foot lengths. There are two types of RNC that electricians typically will use. Schedule 40 has a heavy wall thickness, and Schedule 80 has an extra-heavy-duty wall thickness. The outside diameters of these two types are the same, but the inside diameter of Schedule 40 is thinner than Schedule 80. Schedule 40 is used where it is not subject to physical damage, while Schedule 80 is used where it is subject to physical damage. Connectors and couplings are attached to the RNC with a PVC cement in the same manner that plumbers use. However, there is a difference between plumbing PVC pipe and electrical PVC pipe. The white plumbing pipe is designed to withstand water pressure from the inside, while the gray electrical PVC is designed to withstand forces from the outside. Figure 2–26 shows an example of RNC and some associated fittings.

Flexible metal conduit (FMC) is a raceway of circular cross section made of a helically wound, formed, interlocked metal strip. Article 348 of the *NEC®* covers the use and installation requirements of this raceway type. The trade name for this raceway type is "Greenfield." It looks very similar to BX cable but does not have the conductors already installed. It is up to the electrician to install the conductors in FMC. It is designed to be very flexible and is often used to connect appliances and other equipment in residential applications. For example, a built-in oven will come from the factory with a short length of FMC enclosing the oven conductors. The electrician secures the FMC to a junction box and makes the necessary connections. FMC is available in trade sizes of 1/2 inch through 4 inches. An exception allows a trade size of 3/8 inch when lengths of not more than 6 feet are used. Also, if listed for grounding, used in lengths not longer than 6 feet, and used with connectors that are listed for grounding, the outside metal sheathing can be used as an equipment-grounding conductor. Otherwise, a green equipment-grounding conductor must be installed. Figure 2–27 shows an example of FMC and some associated fittings.

Electrical nonmetallic tubing (ENT) is a flexible nonmetallic raceway that is being used more and more in residential wiring. Electricians often refer to it as "Smurf Tube." The name comes from the blue color that most ENT has, which reminded some electricians of the blue-colored Smurf

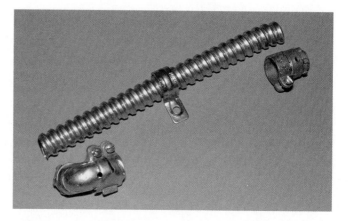

Figure 2–27 Flexible metal conduit (FMC) and associated fittings.

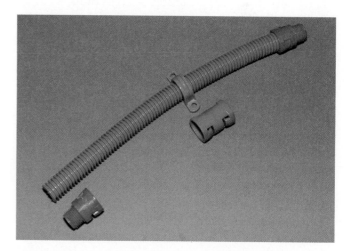

Figure 2–28 Electrical nonmetallic tubing (ENT) and associated fittings.

cartoon characters. Article 368 of the *NEC®* provides the installation requirements for ENT and defines it as a nonmetallic pliable corrugated raceway of circular cross section with integral or associated couplings, connectors, and fittings for the installation of electric conductors. ENT is composed of a material that is resistant to moisture and chemical atmospheres and is flame retardant. It is available in trade sizes of 1/2 inch through 2 inches. Figure 2–28 shows an example of ENT and some associated fittings.

Liquidtight flexible metal conduit (LFMC) and liquidtight flexible nonmetallic conduit (LFNC) are raceway types that are used where flexibility is desired in wet locations, such as outdoors. While they are not used often in residential wiring, it is important for an electrician to be able to recognize these raceway types for those times when they are used.

LFMC is defined as a raceway of circular cross section having an outer liquidtight, nonmetallic, sunlight-resistant jacket over an inner flexible metal core with associated

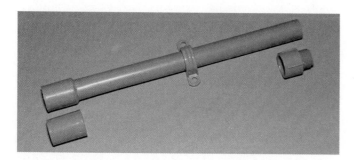

Figure 2–26 Rigid nonmetallic conduit (RNC) and associated fittings.

couplings, connectors, and fittings for the installation of electric conductors (Figure 2–29). Article 350 of the *NEC®* covers the installation requirements for LFMC. When combined with proper connectors, an installation that does not allow liquid into the raceway and around the conductors is accomplished. It is available in trade sizes of 1/2 inch through 4 inches. A 3/8-inch trade size is available for special situations outlined in Section 348.20(A). A common application for this wiring method is the connection to a central air-conditioning unit located outside a dwelling.

Article 356 of the *NEC®* covers the installation requirements for liquidtight flexible nonmetallic conduit (LFNC). It is defined as a raceway of circular cross section of various types as follows:

1. A smooth seamless inner core and cover bonded together and having one or more reinforcement layers between the core and covers, designated as Type LFNC-A
2. A smooth inner surface with integral reinforcement within the conduit wall, designated as Type LFNC-B
3. A corrugated internal and external surface without integral reinforcement within the conduit wall, designated as LFNC-C

LFNC is flame resistant and with proper fittings is approved for the installation of electrical conductors (Figure 2–30). LFNC-B is the most often used type. LFNC is used in the same applications as LFMC. It is also available in trade sizes of 1/2 inch through 4 inches, and a 3/8-inch trade size is available for the following:

1. Enclosing the leads of motors
2. Lengths not exceeding 6 feet when used as the wiring method for tap connections to luminaires (lighting fixtures)
3. Electric sign conductors

Figure 2–29 Liquidtight flexible metal conduit (LFMC) and associated fittings.

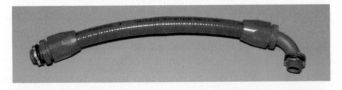

Figure 2–30 Liquidtight flexible nonmetallic conduit (LFNC) and associated fittings.

Devices

The *NEC®* defines a **device** as a unit of an electrical system that is intended to carry but not utilize electric energy. Components such as switches, receptacles, attachment plugs, and lamp holders are considered devices because they distribute or control but do not consume electricity.

Receptacles

Receptacles are probably the most recognizable parts of a residential electrical system. They provide ready access to the electrical system and are defined as a contact device installed at the outlet for the connection of an attachment plug. Even though many people, including electricians, refer to a receptacle as an "outlet," it is the wrong term to use. An outlet is the point on the wiring system at which current is taken to supply equipment. A receptacle is the device that allows the electrician to access current from the wiring system and deliver it through a cord and attachment plug to a piece of equipment (Figure 2–31).

A single receptacle is a single contact device with no other contact device on the same yoke. This type of receptacle is sometimes used in residential wiring. An example would be the receptacle installed for a washing machine, which is usually a single receptacle with a 20-ampere, 125-volt rating. However, the most common type of receptacle used in residential wiring is a duplex receptacle rated for 15 amperes at 125 volts. It consists of two single receptacles on the same mounting strap. The short contact slot on the receptacle receives the "hot" conductor from the attached cord. The long contact slot receives the grounded conductor from the attached cord. There is also a U-shaped grounding contact that receives the grounding conductor. Silver screws are located on the side with the long contact slot and are used to terminate the white, grounded circuit conductor. Brass- or bronze-colored screws are located on the same side as the short contact slot and are used to terminate the "hot," or ungrounded, circuit conductor. A green screw is located on the duplex receptacle for terminating the circuit bare or green grounding conductor (Figure 2–32).

There are several other features on a duplex receptacle that you should be familiar with. The following features are usually found on the front of a receptacle:

- Connecting tabs: These are used to connect the top half and the bottom half of the duplex receptacle. They can be taken off to provide different wiring configurations with "split" receptacles. Wiring split receptacles is covered in a later chapter.
- Mounting straps: These are used to attach the receptacle to a device box. New receptacles will have 6–32 screws held in place by small pieces of cardboard or plastic in the mounting straps.
- Ratings: Both the amperage and the voltage rating of the receptacle are written on the receptacle.

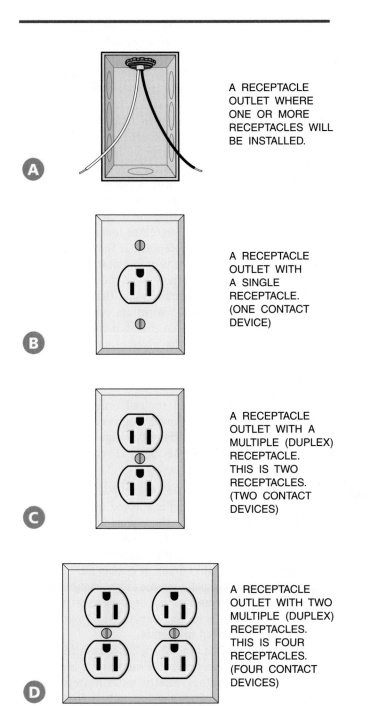

A RECEPTACLE OUTLET WHERE ONE OR MORE RECEPTACLES WILL BE INSTALLED.

A RECEPTACLE OUTLET WITH A SINGLE RECEPTACLE. (ONE CONTACT DEVICE)

A RECEPTACLE OUTLET WITH A MULTIPLE (DUPLEX) RECEPTACLE. THIS IS TWO RECEPTACLES. (TWO CONTACT DEVICES)

A RECEPTACLE OUTLET WITH TWO MULTIPLE (DUPLEX) RECEPTACLES. THIS IS FOUR RECEPTACLES. (FOUR CONTACT DEVICES)

Figure 2–31 Electricians sometimes confuse the term "outlet" with the term "receptacle." Article 100 of the *NEC*® defines a receptacle "outlet" as the branch-circuit wiring and the box where one or more receptacle devices are to be installed. A strap (yoke) with one, two, or three contact devices is defined as a "receptacle."

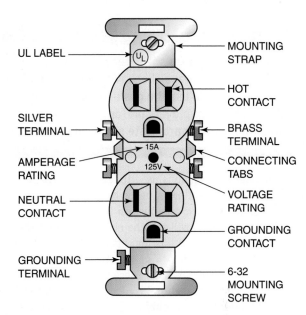

Figure 2–32 **The parts of a duplex receptacle.**

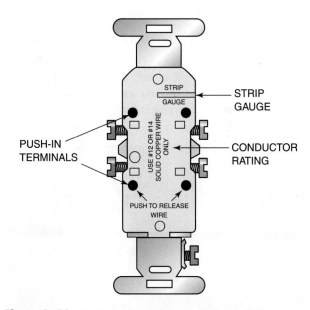

Figure 2–33 **The back of a receptacle device contains information that should be checked before installation. There are also some receptacle features located on the back of the device.**

- NRTL label: A label from an NRTL like UL will be on the receptacle. This is used as a basis for product approval by the AHJ.

  The following features are usually found on the back of a duplex receptacle (Figure 2–33):

- Push-in terminals: Sometimes called "back-stabs," these are used when electricians strip a conductor and push it into the hole rather than terminating on the screws.

◣◣◣◣◣◣▰◀ CAUTION ▶▰◢◢◢◢◢◢

CAUTION: Using the push-in terminals may cause bad connections. Always check with your supervisor to get permission to use the push-in terminations on any device. Most employers prefer that wires be terminated under the appropriate screw on the device.

- Strip gauge: This gauge is used to let the electrician know how much insulation needs to be stripped off the conductor when using the push-in fittings.
- Conductor size: This will tell the electrician what the maximum conductor size is for this device. Most duplex receptacles are rated for 14 or 12 AWG conductors.
- Conductor material markings: These markings will indicate what conductor material is okay to use with the device. "CU" indicates that only copper conductors can be used, "Cu-Clad Only" indicates that only copper-clad aluminum can be used, and "CO/ALR" indicates that either copper or copper-clad aluminum can be used. Aluminum is not allowed to be used with receptacles of this type.

Ground Fault Circuit Interrupter Receptacles

Duplex receptacles installed in residential wiring also are available in a ground fault circuit interrupter (GFCI) type (Figure 2–34). In addition to their use as duplex receptacles, GFCI receptacles also protect people from electrical shock. Sev-

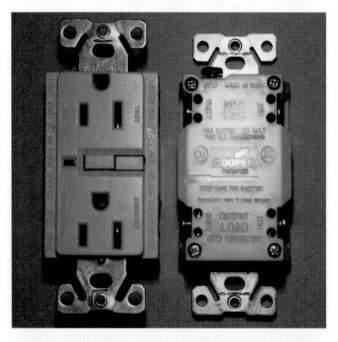

Figure 2–34 A GFCI receptacle is equipped with test and reset buttons. The back of the device contains information that is useful when installing the device in a circuit, such as which terminals are used to connect the incoming (LINE) wires and the outgoing (LOAD) wires.

eral locations in residential wiring are required by the *NEC®* to have GFCI protection. Section 210.8 (A) outlines those areas that require GFCI protection for all 15- and 20-ampere-, 125-volt-rated receptacles. Some of these locations include kitchens, bathrooms, and basements. Installation of GFCI receptacles in these locations is covered in later chapters.

The GFCI receptacle can be used to provide protection only at its location or may provide GFCI protection to other "regular" duplex receptacles connected "downstream" from the GFCI receptacle. GFCI receptacles have the same features as a regular duplex receptacle with a few exceptions. There is a "reset" and a "test" button on the front of the GFCI receptacle. They are used for testing the GFCI on a regular basis and then resetting it. An important feature to know for new electricians on the GFCI receptacle is the location of the words "LOAD" and "LINE." It is very important to make sure that the incoming electrical power, or the "line," is connected to the proper LINE terminals and that the outgoing electrical power, or the "load," wires are connected to the proper LOAD terminals. The color coding is the same as for a regular receptacle: silver for the white grounded conductor and bronze or brass for the "hot" ungrounded conductor.

Special Receptacle Types

Some circuits are intended to feed only one piece of equipment. Usually, this equipment requires a special receptacle that is larger and has a different configuration than the single or duplex receptacles previously covered. Appliances can be classified into three groups:

1. Portable appliance: This is a small appliance, like a toaster or coffeemaker, and is plugged into 15- and 20-ampere receptacles like those previously discussed. They do not require a special kind of receptacle.
2. Stationary appliance: This is an appliance like an electric range, a clothes dryer, or a room air conditioner. They usually require large amounts of current and are connected to receptacles that are designed specifically for the amperage and voltage that these appliances need to operate.
3. Fixed appliance: This type of appliance is fastened in place and is not easily moved. Examples are built-in ovens, built-in cooktops, electric water heaters, and furnaces. This type of appliance usually is "hardwired" and is not cord and plug connected.

Special receptacles are available in a flush-mount style and a surface-mount style. Flush mount is used in new construction and requires first mounting an electrical box and then installing the flush-mount receptacle in the box. It gets its name because it is "flush," or even, with the finished wall when the installation is complete. A surface-mount receptacle is not attached to an electrical box. It is a self-contained piece of equipment and is installed by attaching a back plate to the floor or wall surface, connecting the wiring to the proper terminals, and securing a plastic cover over the installation. It can be used in new construction but is most often used in remodel work when an electrical box is not easily installed in a wall.

In residential wiring, there are two appliances that typically require a special receptacle installation: the electric range and the electric clothes dryer (Figure 2–35). The range requires a heavy-duty, 50-ampere, 250-volt-rated receptacle and attachment plug for its installation. The dryer requires a heavy-duty, 30-ampere-, 250-volt-rated receptacle and attachment plug.

The range and dryer receptacles and plugs have special letter designations that you should know. The letter "G" indicates the location of the equipment-grounding conductor. The letter "W" indicates the location of the white, or grounded, conductor. The letters "X" and "Y" indicate the location of the "hot" ungrounded conductors.

The NEMA has compiled a chart that shows all the general-purpose receptacle and plug configurations (see Table 2–4).

Switches

Devices called switches are used to control the various lighting outlets installed in residential wiring. Article 404 of the *NEC®* provides the installation requirements for switches. This device type can be called many different names, such as "toggle switch," "snap switch," or "light switch," but in this section we refer to these devices simply as "switches." Single-pole, double-pole, three-way, and four-way switches are the switch types commonly used in residential wiring. This section shows you how to recognize these different switch types and how each switch works. Chapter 13 covers how to install switches.

The most common type of switch used in residential wiring is called a single-pole switch. This switch type is used in 120-volt circuits to control a lighting outlet or outlets from only one location. An installation example would be in a bedroom where the single-pole switch is located next to the door and allows a person to turn on a lighting fixture when the person enters the room. Figure 2–36 shows the parts of a basic single-pole switch. The main parts include the following:

- Switch toggle: Used to place the switch in the ON or OFF position when moved up or down.
- Screw terminals: Used to attach the lighting circuit wiring to the switch. On a single-pole switch, the two terminal screws are the same color, usually bronze.
- Grounding screw terminal: Used to attach the circuit-grounding conductor to the switch. It is green in color.
- Mounting ears: Used to secure the switch in a device box with two 6-32 size screws.

CAUTION

CAUTION: When installing single-pole switches, always mount them so that when the toggle is in the up position, the writing on the toggle says "ON." You can tell when you have installed a single-pole switch upside down because when the switch is in the ON position, it will read as "NO."

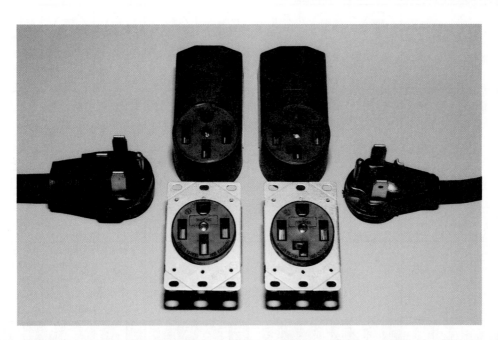

Figure 2–35 A four-wire 50-amp-rated range receptacle and a four-wire 30-amp-rated dryer receptacle. Both are available in a surface- or flush-mount design. The attachment plug used with each receptacle is also shown. Prior to the 1996 *NEC®*, three-wire receptacles and cords were permitted for the connection of ranges and dryers. Three-wire circuits are no longer permitted for ranges and dryers.

Table 2–4 **NEMA General-Purpose Nonlocking Plugs and Receptacles**

NEMA RECEPTACLE AND PLUG CHART CONFIGURATIONS

NEMA No.	15 AMPERE RECEPTACLE	15 AMPERE PLUG	20 AMPERE RECEPTACLE	20 AMPERE PLUG	30 AMPERE RECEPTACLE	30 AMPERE PLUG	50 AMPERE RECEPTACLE	50 AMPERE PLUG	60 AMPERE RECEPTACLE	60 AMPERE PLUG
TWO-POLE TWO-WIRE										
1 125V	1-15R	1-15P								
2 250V		2-15P	2-20R	2-20P	2-30R	2-30P				
TWO-POLE THREE-WIRE GROUNDING										
5 125V	5-15R	5-15P	5-20R	5-20P	5-30R	5-30P	5-50R	5-50P		
6 250V	6-15R	6-15P	6-20R	6-20P	6-30R	6-30P	6-50R	6-50P		
THREE-POLE TWO-WIRE										
7 277V AC	7-15R	7-15P	7-20R	7-20P	7-30R	7-30P	7-50R	7-50P		
THREE-POLE FOUR-WIRE GROUNDING										
10 125/ 250V			10-20R	10-20P	10-30R	10-30P	10-50R	10-50P		
11 3Ø 250V	11-15R	11-15P	11-20R	11-20P	11-30R	11-30P	11-50R	11-50P		
FOUR-POLE FOUR-WIRE										
14 125/ 250V	14-15R	14-15P	14-20R	14-20P	14-30R	14-30P	14-50R	14-50P	14-60R	14-60P
15 3Ø 250V	15-15R	15-15P	15-20R	15-20P	15-30R	15-30P	15-50R	15-50P	15-60R	15-60P
18 3ØY 120/ 208V	18-15R	18-15P	18-20R	18-20P	18-30R	18-30P	18-50R	18-50P	18-60R	18-60P

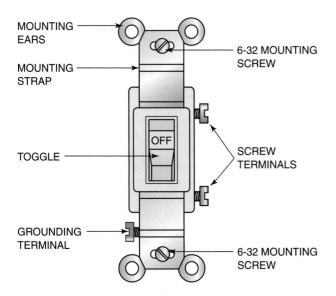

Figure 2–36 The parts of a single-pole switch.

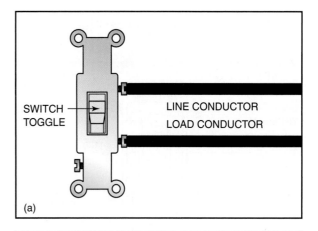

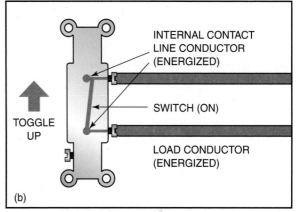

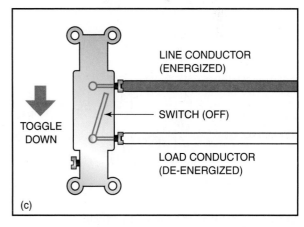

Figure 2–37 How a single-pole switch works. (A) Terminal connections. (B) The terminals are energized when the switch toggle is moved to the ON position. (C) The switch terminals are deenergized when the switch toggle is moved to the OFF position.

Single-pole switches, as well as all other switch types, will have their amperage and voltage ratings written on them. They, like receptacles, may also have push-in fittings on the back.

Figure 2–37 shows how a single-pole switch works. When the toggle is in the ON position, the internal contacts are closed, and current can flow through the switch and energize the lighting outlet. When the toggle is in the OFF position, the internal contacts are open, and current cannot flow, resulting in a deenergized lighting outlet.

A switch that is used on 240-volt circuits to control a load from one location is called a double-pole switch (Figure 2–38). It is similar in construction to a single-pole switch but has four terminal screws instead of two. The top two screws are usually labeled as the "line" terminals and have the same color. The bottom two screws are labeled as the "load" terminals and are colored the same but in a color that is different from the top two screw terminations. An installation example of this switch is as a **disconnecting means** for a 240-volt electric water heater. The double-pole switch works like a single-pole switch except that there are two sets of internal contacts that are connected in such a way that when the toggle is in the ON position, both sets of contacts are closed, and when the toggle is in the OFF position, both sets of contacts are open.

Three-way switches are used to control a lighting outlet or outlets from two locations. An installation example would be in a living room that has two doorways. One three-way switch would be located next to one doorway and another three-way switch next to the other doorway. A person entering or leaving the living room can turn the lighting outlet(s) in that room on or off from either doorway. Three-way switches get their name from the fact that they have three screw terminals on them. Figure 2–39 shows the common parts of a three-way switch. There are some characteristics about three-way switches that make them different from single-pole switches. On characteristic, mentioned earlier, is that three-way switches

have three terminals. Two of the terminals, called "traveler terminals," have the same brass color and are located directly across from each other on opposite sides of the switch. The other screw terminal is black in color and is called the "common terminal." Some electricians may refer to this as the "point" or "hinge" terminal. Identifying the common and the traveler terminals will enable you to correctly connect three-way switches in a lighting circuit.

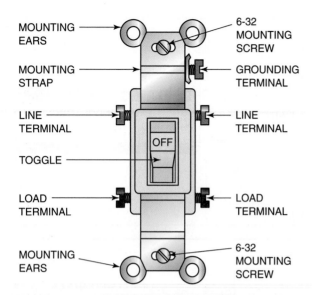

MOUNTING EARS

MOUNTING STRAP

LINE TERMINAL

TOGGLE

LOAD TERMINAL

MOUNTING EARS

6-32 MOUNTING SCREW

GROUNDING TERMINAL

LINE TERMINAL

LOAD TERMINAL

6-32 MOUNTING SCREW

OFF

Figure 2–38 The parts of a double-pole switch.

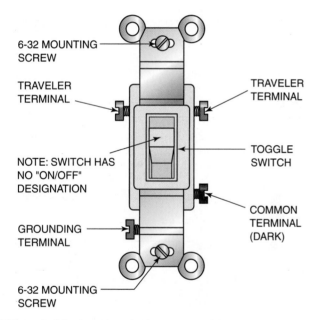

6-32 MOUNTING SCREW

TRAVELER TERMINAL

NOTE: SWITCH HAS NO "ON/OFF" DESIGNATION

GROUNDING TERMINAL

6-32 MOUNTING SCREW

TRAVELER TERMINAL

TOGGLE SWITCH

COMMON TERMINAL (DARK)

Figure 2–39 The parts of a three-way switch.

Another distinguishing characteristic of a three-way switch is that the toggle, unlike a single-pole switch, does not have any ON or OFF position written on it. In other words, there is no up or down on a three-way switch, and it does not make any difference which way the switch is mounted in a device box.

Three-way switches must always be installed in pairs, and a three-wire cable or three wires in a raceway must always be run between the two three-way switches for proper connection. Again, Chapter 13 covers installing switches.

Three-way switches work a little differently than single-pole switches. Single-pole switches either let current flow through the switch and to the load or do not. The internal configuration of a three-way switch always allows current to flow from the common terminal to either of the two traveler terminals (Figure 2–40). The three-way switch is sometimes called a single-pole double-throw switch because of this characteristic. When the three-way switch is mounted in a device box and the toggle is placed in the "down" position, contact is made between one of the traveler terminals and the common terminal. When the toggle is moved to the "up" position, contact is made between the common and the other traveler terminal. There is always one "set" of contacts in a three-way switch that are closed.

Four-way switches are used in conjunction with three-way switches to allow control of a lighting outlet or outlets from more than two locations. An installation example would be in a room with three doorways. A three-way switch would be located at two of the doorways and a four-way switch at the third doorway. When such switches are wired correctly, a person could turn on or off the room lighting outlet(s) from any of the three locations. The four-way gets its name from the fact that there are two sets or four total screw terminals on the switch (Figure 2–41). Most manufacturers distinguish the two sets of terminals with different colors. One set will typically be black in color, and the other set will be a lighter color, like brass. The two screws in a set are located opposite each other on most four-way switches. Four-way switch terminals are also called "traveler terminals." There is no common terminal as found on a three-way switch. However, like three-way switches, there is no ON or

CAUTION

CAUTION: The terminals may be located differently on three-way switches made by different manufacturers, but the color coding will be the same. Always look for the two brass terminals, which will be the traveler terminals and the black terminal, which will always be the common terminal. It does not matter where they are located on the switch.

 FROM EXPERIENCE

Four-way switches are always installed in combination with three-way switches and are wired between them. There is no limit to the number of four-way switches that could be wired between two three-way switches. If a large room had five doorways and switch control of the room lighting outlets was required from each of the doorways, the residential electrician would use two three-way switches and three four-way switches.

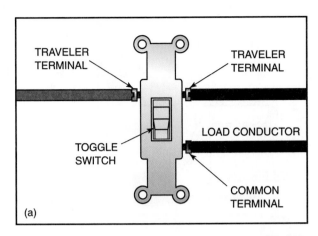

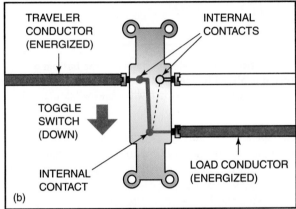

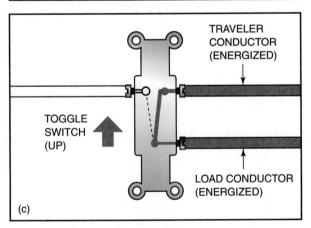

Figure 2–40 How a three-way switch works. (A) Three-way switches have two traveler terminals and one common terminal. (B) When the switch toggle is moved, a connection is made from the common terminal to one of the traveler terminals. (C) When the switch toggle is moved in the opposite direction, a connection is made from the common terminal to the other traveler terminal.

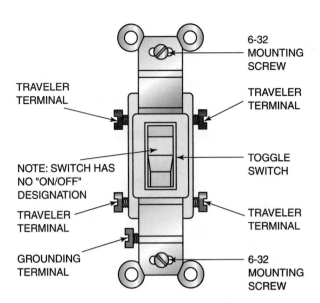

Figure 2–41 The parts of a four-way switch.

are closed or opened in a sequence determined by the position of the toggle. Figure 2–42 shows the internal configuration of the contacts when the toggle is in the up or down position.

Dimmer switches are used to dim or brighten the light output of a lighting fixture. They come in both single-pole and three-way configurations. A single-pole dimmer might be used, for example, to control a large lighting fixture located over a dining room table. Figure 2–43 shows some common styles of dimmer switches. Today's dimmers used in residential applications use electronic circuitry to provide dimming capabilities. They are usually rated at 125 volts and 600 watts, although larger wattage ratings are available.

CAUTION

CAUTION: Never load a dimmer switch higher than its wattage rating. It will malfunction and produce dangerous temperature levels that could cause fires.

OFF designation written on the toggle of a four-way switch. Therefore, there is no up or down mounting orientation when a four-way is installed in a device box.

Like single-pole and three-way switches, four-way switches work by having their internal contacts open or close. Four-ways have four sets of internal contacts. The contacts

FROM EXPERIENCE

The four-way switches made by most of today's manufacturers have vertical configurations as described in this section. However, some manufacturers make four-ways with a horizontal configuration. The residential electrician should always check the color coding and read the instructions that come with the switch to determine the proper switch configuration.

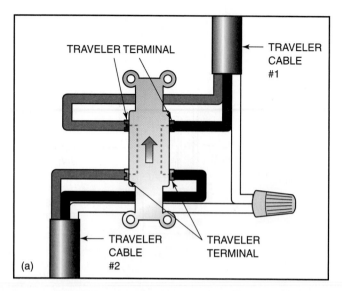
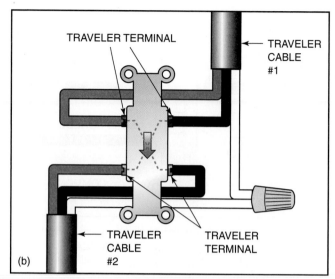

Figure 2–42 How a four-way switch works. (A) Connections are made vertically from the traveler terminals on the bottom of the switch to the traveler terminals on the top of the switch when the switch toggle is placed in one position. (B) Connections are made diagonally from the traveler terminals on the bottom to the traveler terminals on the top of the switch when the switch toggle is placed in the opposite position.

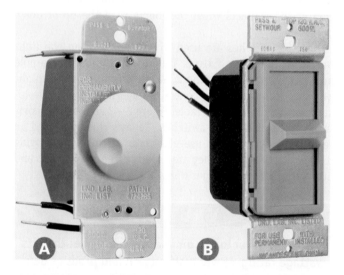

Figure 2–43 Examples of dimmer switches. (A) A rotating knob–style dimmer. (B) A sliding knob-style dimmer. Dimmer switches are available as single-pole or three-way models. Both dim or brighten a lamp by varying the applied voltage to the lamp. *Courtesy of Pass & Seymour/Legrand.*

Examples of low-voltage switches used in residential signaling systems are shown in Figure 2–44. A residential electrician would install this switch type next to a front and a rear doorway so that a signal can be sent to a chime that lets people inside the house know that there is someone at the door. This switch type has a push button rather than a toggle and is referred to as a momentary-contact switch. Momentary contact means that the internal contacts are closed and current can flow only as long as someone is pushing the button. Because the button is spring loaded, when a person stops pushing the button, the contacts automatically revert back to their normal open position.

Combination Devices

Combination devices are used when more than one device is needed at one location. This type of device has a combination of two devices, both of which are mounted on the same strap. It may be a combination of two single-pole switches, a single-pole and a three-way switch, a single-pole switch and a receptacle, or even a switch and a pilot light. Figure 2–45 shows some examples of combination devices.

Figure 2–44 Examples of low-voltage momentary-contact switches. These switches are used as push buttons to activate a door chime. *Courtesy of Nutone, Inc.*

Figure 2–45 Examples of combination switches. This switch type has two devices on one strap. Typical combinations include two single-pole switches, a single-pole and a three-way switch, or two three-way switches.

 FROM EXPERIENCE

Sometimes there are locations in residential wiring where two devices are needed but there is not enough room between the studs to mount a two-gang device box to hold the two devices. An example might be in a bathroom where one switch is needed to control a lighting outlet and another switch is needed to control the ventilation fan. There is not enough room at the desired location for a two-gang electrical box to be installed. A common practice to accomplish this installation is to mount a single-gang device box in the narrow space between the two studs and install a combination device with two single-pole switches. This installation will meet the two-device requirement and can be installed in the space available.

Overcurrent Protection Devices

In residential wiring, **overcurrent** protection devices consist of **fuses** or **circuit breakers**. The *NEC*® states that overcurrent protection for conductors and equipment is provided to open the circuit if the current reaches a value that will cause

an excessive or dangerous temperature in conductors or conductor insulation. Recognizing common fuse and circuit breaker types is important for anyone doing residential wiring.

Circuit breakers are the most often used type of overcurrent protection device in residential wiring. A circuit breaker is an automatic overcurrent device that trips into an open position and stops the current flow in an electrical circuit. An **overload**, **short circuit**, or **ground fault** can cause the circuit breaker to trip. Circuit breakers used in residential wiring are of the thermal/magnetic type. Thermal tripping is caused by an overload. When a larger current flow than the breaker's rating occurs, a **bimetallic strip** in the breaker gets hot and tends to bend. If the larger-than-normal current flow continues, the temperature of the bimetallic strip increases, and the strip continues to bend. If it bends enough, a latching mechanism is tripped, and the breaker contacts open, causing current flow to stop. When a short circuit or ground fault occurs, circuit resistance is drastically reduced, and large amounts of current flow through the breaker. This high current causes the magnetic part of the breaker to react. A strong magnetic field is created from the high current, which causes a metal bar attached to the latching mechanism to trip, opening the circuit breaker contacts and stopping current flow.

Circuit breakers used in residential wiring are available in single-pole for use on 120-volt circuits and two-pole for use on 240-volt circuits (Figure 2–46). They are rated in both amperes and voltage. Common residential amperages are 15, 20, 30, 40, and 50 amperes. Larger sizes, such as a 100 or 200 amp used as the main service entrance disconnecting means, are available. The voltage rating on residential circuit breakers is usually 120/240 volts. The slash (/) between the lower

Figure 2–46 Typical molded-case circuit breakers. The circuit breaker on the left is a single-pole breaker and is used to provide overcurrent protection for 120-volt residential circuits. The circuit breaker on the right is a two-pole breaker and is used to provide overcurrent protection for 240-volt residential circuits.

and higher voltage ratings in the marking indicates that the circuit breaker has been tested for use on a circuit with the higher voltage between "hot" conductors (240 V) and with the lower voltage from a "hot" conductor to ground (120 V).

Circuit breakers can also be found in a GFCI model. This breaker will provide regular overcurrent protection to circuits and equipment and will also provide GFCI protection to the whole circuit. It is easily distinguishable from regular circuit breakers by the test button on the front of the breaker and the length of white wire that is attached to it (Figure 2–47).

Fuses are not often used in residential wiring, but electricians will certainly encounter them from time to time on new installations or when doing remodel work. A fuse is an overcurrent protection device that opens a circuit when the fusible link is melted away by the extreme heat caused by an overcurrent. There are two styles of fuses that you will see on a residential installation: plug fuses and cartridge fuses. Both of these fuse types are available in a time-delay or non-time-delay configuration. Electrical circuits that have electric motors in them require a time-delay fuse. The reason for this is that motors draw a lot of current to get started, but once started and turning their load, they draw their normal current amount. Time-delay fuses allow the motors to start

and get up to running speed. If non-time-delay fuses were used on electric motor circuits, the fuse would blow every time the motor tried to start. Circuit breakers used in residential wiring have time-delay capabilities built in.

Plug fuses are broken down into two types: the Edison base and the Type S. The Edison-base plug fuse (Figure 2–48) can be used to replace only existing Edison-base fuses. New installations cannot use Edison-base fuses. This is because all sizes of Edison-base plug fuses have the same base size and will all fit in the same fuse holder. Overfusing can occur with this fuse type when a person puts in a 30-ampere Edison-base fuse in the fuse holder that should have a 15-ampere fuse. Type S, or safety, fuses (Figure 2–49) are used in all new installations that require plug fuses. They are considered non-tamperable and can fit and work only in a fuse holder with the same amperage rating as the fuse. In other words, a 30-ampere Type S fuse cannot be put in a 15-ampere Type S adapter and work. This feature is designed to prevent overfusing of circuits when using plug fuses.

Cartridge fuses are used sometimes in residential wiring. They come in two styles: the ferrule style and the blade-type style. The ferrule style (Figure 2–50) is basically a cylindrical tube of insulating material, like cardboard, with metal caps on each end. Standard ampacities for ferrule-type cartridge fuses

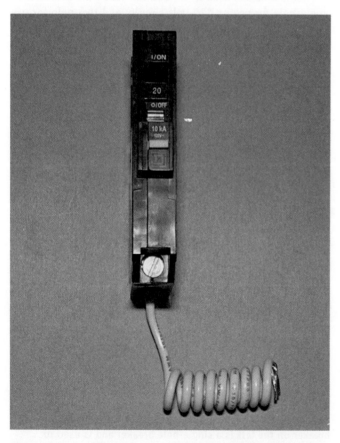

Figure 2–47 A GFCI circuit breaker. It is easily recognizable from a regular circuit breaker by the push-to-test button on the front of the breaker and the length of white insulated conductor attached to it. *Courtesy of Square D Company.*

Figure 2–48 A 15-amp Type W Edison-base plug fuse is shown on the left. Notice the hexagonal shape of the area around the window. The hexagonal shape indicates a 15-amp-rated fuse. The fuse on the right is a 20-amp-rated Type T time-delay Edison-base plug fuse. *Courtesy of Cooper Bussman, Inc.*

Figure 2–49 A 15-amp-rated (blue in color) Type S plug fuse and adapter. A 20-amp (orange in color) or a 30-amp (green in color) Type S fuse will not work in the 15-amp-rated adapter. This type of fuse will help prevent overfusing. *Courtesy of Cooper Bussman, Inc.*

Figure 2–50 Examples of Class G ferrule-type cartridge fuses. This fuse style is found in fuse sizes that are rated 60 amps and less. *Courtesy of Cooper Bussman, Inc.*

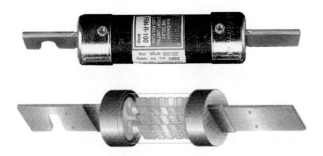

Figure 2–51 A blade-type cartridge fuse. This cartridge fuse style is found in fuses that are rated larger than 60 amps. *Courtesy of Cooper Bussman, Inc.*

are 15, 20, 30, 40, 50, and 60 amps. Cartridge fuses with amperage ratings over 60 amps are found in the blade-type style (Figure 2–51). This fuse has a cylindrical body of insulating material with protruding metal blades on each end.

Panelboards, Loadcenters, and Safety Switches

A **panelboard** is defined as a single panel that includes automatic overcurrent devices used for the protection of light, heat, or power circuits. It is designed to be placed in a **cabinet** located in or on a wall, partition, or other support. It is accessible only from the front. A **loadcenter** is a type of electrical panel that contains the main disconnecting means

for the residential service entrance as well as the fuses or circuit breakers used to protect circuits and equipment like electric water heaters, ranges, dryers, and lighting (Figure 2–52). Most load centers are placed inside a dwelling unit and are typically a NEMA Type 1 enclosure, which is the most commonly used type of enclosure in residential wiring, being designed for use indoors under usual service conditions. Sometimes an electrical enclosure, such as a meter enclosure or a loadcenter, must be installed outside. A NEMA Type 3, which is a weatherproof enclosure designed to give protection from falling dirt, rain, snow, sleet, and windblown dust, will need to be used outside. In residential work, a NEMA Type 3R is often used. The "R" stands for "rain-tight" and means that the enclosure will provide the same protection as a NEMA Type 3, except for windblown dust.

Sometimes a residential wiring system has a large concentration of electrical circuits in one area of the house, like a kitchen area. If the kitchen is located some distance from the main loadcenter, the circuits that are run from the main loadcenter to the kitchen can be quite long. This makes them more expensive to install and may result in some low-voltage problems for the kitchen equipment. It is common practice to run a set of larger-current-rated conductors from the main loadcenter to a smaller loadcenter located closer to the kitchen area. This type of loadcenter is called a "subpanel" (Figure 2–53) and allows shorter runs for the circuits going to the kitchen area. Since a subpanel does not have a main circuit breaker or main set of fuses, it is often referred to as a "main-lug-only" or simply "MLO" panel.

A **safety switch** is used as a disconnecting means for larger electrical equipment found in residential wiring. It is typically mounted on the surface and is operated with an

Figure 2–52 A typical 120/240-volt, single-phase, three-wire main breaker loadcenter. A cover is installed on this loadcenter after all the connections have been made. *Courtesy of Square D Company.*

Figure 2–53 **A typical 120/240-volt main-lug-only (MLO) loadcenter. This panel does not have a main circuit breaker.** *Courtesy of Square D Company, Group Schneider.*

Figure 2–54 **Typical 120/240-volt-rated safety switches. The fusible safety switch on the left requires cartridge fuses for 60-amp-rated and larger switches. Smaller-rated safety switches will require plug fuses. The nonfusible safety switch on the right is used as a disconnecting means only and does not provide any overcurrent protection.**

external handle or with a pullout fusible device (Figure 2–54). It can be classified as a fusible type or a nonfusible type. The fusible type contains plug fuses in the smaller sizes and cartridge fuses in the larger sizes. The nonfusible safety switch is strictly a disconnecting means and contains no fuses for overcurrent protection. Disconnect switch enclosures can also contain a circuit breaker (Figure 2–55). Safety switch enclosures are available in a variety of NEMA classifications, including NEMA Type 1 and NEMA Types 3 and 3R.

Figure 2–55 **A disconnect switch that uses a circuit breaker to provide overcurrent protection.**

Fasteners

There are several different types of fasteners used in residential wiring. They are used to assemble and install electrical equipment. Fasteners used in residential wiring include items such as anchors, screws, bolts, nuts, and tie wraps. This section covers the most common fastener types.

Anchors

Anchors are used to mount electrical boxes and enclosures on solid surfaces like concrete as well as hollow wall surfaces like wallboard. Some common anchors used in residential wiring to mount items on wallboard are toggle bolts, wallboard anchors, and molly bolts (also called a sleeve-type anchor) (Figure 2–56). Plastic anchors are sometimes used, but care must be taken not to mount very heavy items with them since they pull out of a sheetrock wall quite easily. Make sure to follow the manufacturer's recommendations when installing anchors in hollow walls and ceilings.

(See page 57 for the procedure for installing toggle bolts in a hollow wall or ceiling.)

Plastic anchors, lead (caulking) anchors, and drive studs (Figure 2–57) are used to mount electrical enclosures and equipment to masonry surfaces in residential applications. Plastic and lead anchor installation requires drilling a hole with a masonry bit and inserting the anchor. Drive studs are installed using a powder-actuated fastening system that uses

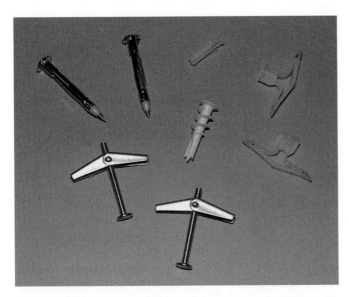

Figure 2–56 Types of wallboard anchors, including toggle bolts, wallboard screws, and molly bolts.

Figure 2–57 Types of masonry anchors, including plastic anchors, caulking anchors, and drive studs.

a tool specifically designed to use the force of a gunpowder load to drive a fastener into concrete or steel. The depth to which the stud or pin is driven depends on the strength of the gunpowder load and the density of the material you are installing on. The different powder loads are color coded, and it depends on the specific manufacturer as to what color code is being used. Consult the manufacturer's instructions.

(See page 59 for the procedure for installing plastic anchors.)

CAUTION

CAUTION: OSHA CFR 1926.302(e) governs the use of powder-actuated fastening systems and requires special training and certification before an electrician is allowed to use this type of fastening system. Usually, a manufacturer of this type of system will provide training and certification.

Many electricians use masonry screw anchors to mount items on concrete, block, and brick walls. This is a system that uses a steel fastener that screws directly into the masonry and eliminates the need for caulking anchors (Figure 2–58). The masonry screws are often installed with a special tool. The tool is used with an electric drill motor and has a masonry bit that drills a proper size hole in the concrete. The masonry bit is then retracted back into the tool, and the masonry screw is inserted in the tool's hexagonal socket head and driven into the previously drilled hole. It is an easy two-step installation, and this quick-and-easy system makes this a very popular fastening system for mounting items on masonry. Electricians commonly refer to these masonry screws as "TapCons." TapCon is actually a registered trademark for the masonry screws made by Greenlee Textron.

(See page 58 for the procedure for the installation of a lead [caulking] anchor in a concrete wall.)

Screws

Screws used to mount enclosures to wood surfaces include wood screws, sheet metal screws, and sheetrock screws (Figure 2–59). Sheet metal screws work very well because the thread extends for the full length of the screw. Sheetrock screws work well for the same reason, and they are also very easy to use with a screw gun, which saves a lot of time and energy versus having to screw them in by hand. Mounting items on metal studs or on other thin metal surfaces is accomplished using sheet metal screws or self-drilling screws (sometimes called "Tek-Screws") (Figure 2–60).

Bolts and machine screws are types of threaded fasteners used in residential wiring. Bolts are rarely used, but residential

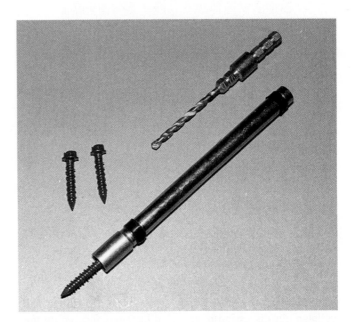

Figure 2–58 A popular anchor type used for securing electrical equipment to a masonry wall is a masonry screw. Electricians refer to these screws as "TapCons."

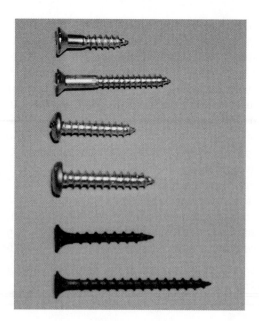

Figure 2–59 Examples of various fasteners for securing electrical equipment to a wood surface include wood screws, sheet metal screws, and sheetrock screws.

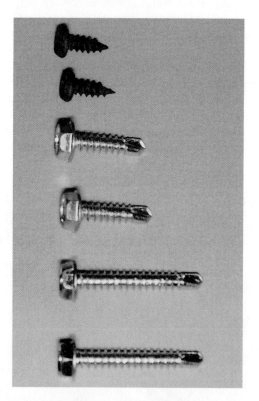

Figure 2–60 Self-drilling screws, or "Tek-Screws" as they are called by some electricians, are used to secure electrical equipment to metal studs or other sheet metal surfaces.

electricians use machine screws all the time. For this reason, we keep our discussion of threaded fasteners focused on machine screws. Machine screws are used to attach devices to electrical boxes, to secure covers to devices, and to secure covers to junction boxes. The most common sizes are 6-32, 8-32, and

10-32. The 6-32 screw size is used to attach switches and receptacles to electrical device boxes, and 8-32 screws are used to attach covers to octagonal and square electrical boxes; a 10-32 machine screw is the size of the green grounding screw used to secure an equipment-grounding conductor to an electrical box. Machine screws are available in either a slotted or a Phillips head and come in different lengths. Other common screwhead shapes used in residential wiring applications are shown in Figure 2–61.

Nuts and Washers

Many applications require the use of nuts and washers with threaded fasteners. Nuts are found in either a square or hexagonal shape and are threaded onto a bolt or machine screw to help secure an electrical item in place. Washers fit over a bolt or screw and distribute the pressure of the bolt or screw over a larger area. They also provide a larger area for the bolt head or nut to tighten against when the fastener is being used in a hole that is larger than the diameter of the fastener. Lock washers are used to help keep bolts or nuts from working loose. Figure 2–62 shows several of the most common nuts and washers used in residential work.

▰▰▰▰▰ **CAUTION** ▰▰▰▰▰

CAUTION: Section 110.3(B) in the *NEC*® states that electrical equipment must be installed according to any instructions included in the listing or labeling of the item. This includes making sure to torque a threaded fastener to the listed and labeled specification. Torque is a measure of how much the fastener is being tightened. It is normally expressed in inch pounds (in. lb.) or foot pounds (ft. lb.). A torque wrench is used to apply the proper turning force to a fastener.

(See page 60 for the procedure to be followed for installing threaded fasteners.)

Tie Wraps

Tie wraps are not often thought of as being in the fastener category, but it is one of the most often used fastening systems in residential wiring. A tie wrap is made of nylon and is a one-piece, self-locking cable tie used to fasten a bundle of wires or cables together. For example, an electrician may use tie wraps in a load center to fasten conductors together to make a neater wiring installation, or, when several Romex cables are bundled together, tie wraps can be used to hold them together. Tie wraps designed for outdoor use are often black in color and are made to resist the harmful effects of sunlight and outdoor atmospheric conditions. Indoor-use tie wraps come in a variety of colors and styles. Figure 2–63 shows some common cable tie examples.

Figure 2–61 Examples of the different heads found on machine screws.

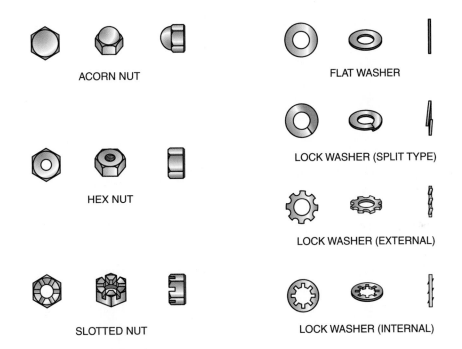

Figure 2–62 Examples of common nuts and washers.

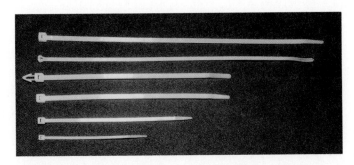

Figure 2–63 Cable ties, or tie wraps as they are sometimes called, are used often to secure wires and cables. They are available in many styles and sizes.

Summary

It is very important to be familiar with the correct names and applications for electrical hardware and materials. Using the proper name and selecting the right part for the job helps avoid confusion and will help ensure that the electrical system is properly installed.

There is a wide variety of hardware and materials used in residential electricity. This chapter covered only the most common items. You will run into other hardware and material types as you continue your residential electrical career. However, they will be similar to the items discussed in this chapter, and by following the manufacturer's instructions and using your newly acquired hardware and material knowledge, you should have no problem when using them.

Procedures   Installing Aluminum Conductors

- Observe proper safety procedures.

- Strip off the required amount of conductor insulation on the end you wish to terminate.

- Apply a liberal amount of antioxidant to the newly exposed aluminum.

- Scrape the conductor end with a wire brush. This breaks down the oxidation layer, and the antioxidant keeps the air from contacting the conductor, preventing further oxidation.

- Attach the conductor to the termination and tighten to the proper torque requirements.

- Clean up any excess antioxidant.

Procedures

Installing Toggle Bolts in a Hollow Wall or Ceiling

- Observe proper safety procedures.

- Mark the location on the wall or ceiling where the toggle bolt will be installed.

A Using the proper size drill bit or punch tool, drill or punch a hole completely through the wall or ceiling in the desired location.

- Insert the toggle bolt through the opening of the item to be mounted.

- Screw the toggle wing onto the end of the bolt, making sure that the flat part of the wing is facing the bolt head.

B While folding the wings back, insert the toggle bolt into the hole until the spring-loaded wing opens up.

C Tighten the toggle bolt with a screwdriver until the mounted item is secure. Pulling back on the item to be mounted in a careful but firm manner will help hold the toggle wing in the wall so that the toggle screw can be tightened easily.

A
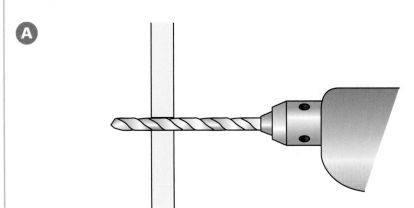

B
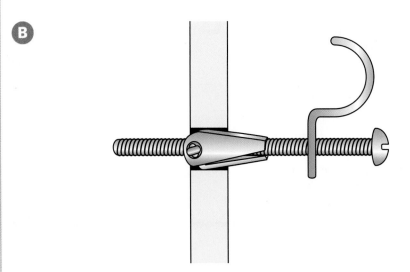

C
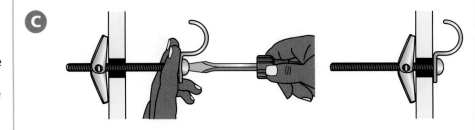

Procedures

Installing a Lead (Caulking) Anchor in a Concrete Wall Using a Setting Tool

- Observe proper safety procedures.

- Mark the location on the wall where the caulking anchor will be installed.

(A) Using the proper size masonry drill bit as found in the manufacturer's instructions, drill a hole to the required depth (usually the depth of the anchor itself for a flush installation) in the desired location.

(B) Insert the anchor, with the conical threaded end first, and tap it in with a hammer until it is flush with the surface.

(C) Using a setting tool, hit the lead outer sleeve with sharp blows.

(D) Position the item to be mounted and insert a screw or bolt into the caulking sleeve and tighten.

Note: For installing caulking anchors in hollow masonry walls like cinder block or for setting all anchors at the same depth no matter how deep the initial hole was drilled, a screw anchor expander tool is used.

(A)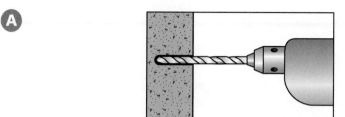

(B)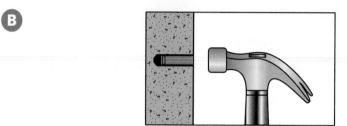

(C)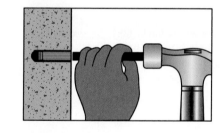

(D)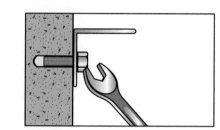

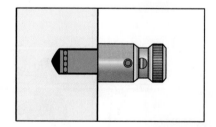 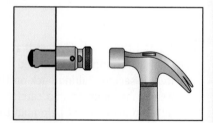

Courtesy of Greenlee Based Textron.

Procedures

Installing Plastic Anchors

- Observe all safety procedures.

 Using the proper masonry bit size, drill a hole to the depth of the anchor or up to 1/8-inch deeper.

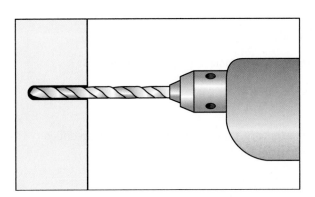

 With a hammer or other appropriate setting tool, tap the plastic anchor into the hole until it is flush with the surface.

- Insert a screw of the size designed to be used with the anchor through the item to be mounted and into the plastic anchor.

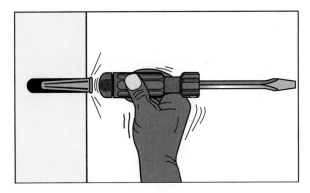

 Tighten snug, being careful not to overtighten.

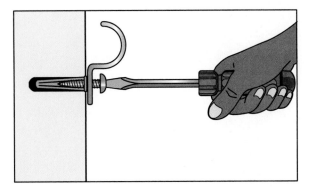

Procedures

General Procedure for Installing Threaded Fasteners

- Observe all safety procedures.

- Choose the correct bolt or screw for the job.

- Check the fastener for damage and make sure it is clean.

- If the fastener is a bolt, insert it through the predrilled hole and hand tighten the nut.

- If the fastener is a screw, insert the screw into the hole and start hand tightening.

- Tighten the bolt or screw to the proper torque requirement using a torque wrench or torque screwdriver.

Review Questions

Directions: Answer the following items with clear and complete responses.

1. Name the size of machine screw that is used to secure switches and receptacles to device boxes.

2. List three nationally recognized testing laboratories.

3. The approval of electrical equipment by the authority having jurisdiction is often based on the listing and labeling of the product. Describe what "listing" and "labeling" electrical equipment means.

4. Name the size of the most common metal electrical device box used in residential wiring.

5. Describe what "ganging" electrical device boxes is and what purpose ganging electrical device boxes serves.

6. Discuss when an electrician would use a "new work" box and an "old work" box.

7. An electrician installs a 4-inch square box in a location that requires a duplex receptacle. Describe what must be done so that the square box can accommodate the receptacle.

8. Name and describe three cable types that can be used in residential electrical installations.

9. Conductors used in residential wiring are selected for a specific circuit according to their ampacities. Define the term "ampacity."

10. Name the system that is used to size the conductors used in electrical work.

11. Name the colors of the wires found in a three-wire nonmetallic sheathed cable.

12. Type UF cable is used in _____ installations for branch circuits and feeders. Article _____ of the *NEC®* covers Type UF cable.

13. "BX" cable is the trade name used for Type _____ cable. Article _____ of the *NEC®* covers this cable type.

14. Describe how an electrician can tell BX and Type MC cable apart.

15. Describe the difference between Type SEU and Type SER cable in terms of physical characteristics.

16. List four raceway types described in this chapter that could be used in residential wiring.

17. Name the trade name used for flexible metal conduit.

18. Define the term "receptacle."

19. Describe when three-way switches would be used in a residential lighting circuit.

20. Describe a combination device.

21. List the two types of plug fuses.

22. A single-pole circuit breaker is used on a circuit voltage of _____ volts, and a two-pole circuit breaker is used on a circuit voltage of _____ volts in residential wiring.

23. Describe a load center.

24. An electrician needs to mount a chime on a sheetrock wall. Name three anchor types described in this chapter that could be used.

25. Describe a powder-actuated fastening system.

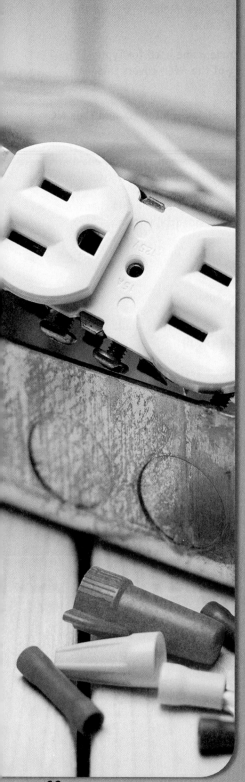

Chapter 3

Tools Used in Residential Wiring

Residential electricians must become skilled in selecting and using the right tool for the job. There are many common hand tools and specialty tools that are used to install electrical systems. Power tools come in various styles and are used often by residential electricians. In this chapter, we take a look at the common hand tools, specialty tools, and power tools that residential electricians use on a regular basis and need to become familiar with.

OBJECTIVES

Upon completion of this chapter, the student should be able to:

- identify common electrical hand tools and their uses in the residential electrical trade.
- identify common specialty tools and their uses in the residential electrical trade.
- identify common power tools and their uses in the residential electrical trade.
- list several guidelines for the care and safe use of electrical hand tools, specialty tools, and power tools.
- demonstrate an understanding of the procedures for using several common hand tools, specialty tools, and power tools.

Glossary of Terms

auger a drill bit type with a spiral cutting edge used to bore holes in wood

bender a tool used to make various bends in electrical conduit raceway

chuck a part of a power drill used for holding a drill bit in a rigid position

chuck key a small wrench, usually in a T shape, used to open or close a chuck on a power drill

crimp a process used to squeeze a solderless connector with a tool so that it will stay on a conductor

cutter a hardened steel device used to cut holes in metal electrical boxes

die the component of a knockout punch that works in conjunction with the cutter and is placed on the opposite side of the metal box or enclosure

hydraulic a term used to describe tools that use a pressurized fluid, like oil, to accomplish work

knockout punch a tool used to cut holes in electrical boxes for the attachment of cables and conduits

level perfectly horizontal; completely flat; a tool used to determine if an object is level

plumb perfectly vertical; the surface of the item you are leveling is at a right angle to the floor or platform you are working from

reciprocating to move back and forth

strip to damage the threads of the head of a bolt or screw

tempered treated with heat to maximize the metal hardness

torque the turning or twisting force applied to an object when using a torque tool; it is measured in inch-pounds or foot-pounds

wallboard a thin board formed from gypsum and layers of paper that is used often as the interior wall sheathing in residential applications; commonly called "sheetrock"

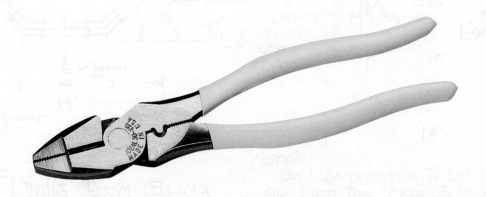

Common Hand Tools

Using hand tools to install a residential electrical system can be hazardous. Injuries to yourself and others can result if you are not careful. It is very important for electricians to be aware of some guidelines for the care and safe use of electrical hand tools. The following points should be considered whenever you are using hand tools:

- Use the correct tool for the job. Using a tool that is not designed for the job you are trying to do may result in serious damage to you as well as to the equipment you are working on.
- Keep all cutting tools sharp. It is a fact that more injuries result from dull cutting tools than with sharp ones. The reason is simple: Dull cutting tools require much more force to do their job. With the extra force being applied, a tool can more easily slip and cause serious damage to you or the equipment you are working on.
- Keep tools clean and dry. Hand tools that are kept clean simply work better. When they are free of rust and dirt, they open and close better and are easier to grip.
- Lubricate tools when necessary. This goes along with keeping the tools clean and dry. The lubrication will allow the tools with a hinged joint to work easily.
- Inspect tools frequently to make sure they are in good condition. Do not use tools that are not in good working condition. A tool could malfunction and cause serious damage. An example would be using a hammer with a loose head. The head could fly off, striking someone and causing serious injury.
- Repair damaged tools promptly and dispose of broken or damaged tools that cannot be repaired. Some hand tools are relatively easy to repair. For example, a new wooden handle can easily replace a broken wooden handle in a hammer. However, it has been this author's experience that most electrical hand tools cannot be easily repaired and need to be disposed of. Only after buying a new tool will you be assured of the tool working in a safe and reliable manner.
- Store tools properly when not in use. Cutting tools especially need to be protected so that their cutting edges will not be dulled and, more important, so that their cutting edges cannot injure someone.
- Pay attention when using tools. Using electrical hand tools is serious business. Always pay strict attention to what you are doing.
- Wear eye protection when necessary. This is a no-brainer. Using hand tools may cause pieces of wire to fly through the air. If you are not wearing proper eye protection, the pieces can easily fly into your eye.

There are many different types of hand tools used in the residential electrical trade. Hand tools are commonly referred to as "pouch" tools because of their ability to fit easily into an electrical tool pouch. The most common hand tools used by a residential electrician to install electrical systems are outlined in the following paragraphs.

Screwdrivers

Screwdrivers are available in a variety of styles, depending on the type of screw that it is designed to fit. The styles used most often in residential electrical work include the slotted (standard) and Phillips-head screwdrivers. The slotted-blade (standard) screwdriver (Figure 3–1) is used to remove and install slot-head screws and to tighten and loosen slot-head lugs. Phillips-head screwdrivers (Figure 3–2) are used to remove and install Phillips-head screws and to tighten and loosen Phillips-head lugs. Both styles are available in a short version called a "stubby" screwdriver (Figure 3–3). Stubby screwdrivers are used to remove and install screws and to tighten and loosen lugs in a limited working space.

Figure 3–1 A flat-blade screwdriver with a comfort grip. *Courtesy of Ideal Industries, Inc.*

Figure 3–2 A Phillips screwdriver with a number 2 head. *Courtesy of Ideal Industries, Inc.*

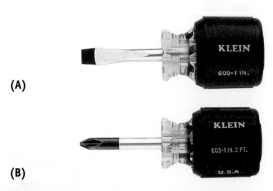

(A)

(B)

Figure 3–3 Flat-blade (A) and Phillips-head (B) stubby screwdrivers. *Courtesy of © Klein Tools, Inc.*

Another style that is popular with residential electricians is the screw-holding screwdriver (Figure 3–4). It is designed to hold on to screws for easier starting and setting of slotted screws. This style works well when you have to install a grounding screw in the bottom of a 3 × 2 × 3 ½-inch device box. The depth of the box makes it very difficult to reach to the bottom and start the screw with your fingers. Using a holding screwdriver lets you start the screw in the bottom of the box very easily. Remember that this screwdriver is designed just to start screws and that you need to use a regular screwdriver to finish tightening the screw.

Each screwdriver type has three parts: the handle, the shank, and the tip (Figure 3–5). The handle is designed to give you a firm grip. The shank is made of hardened steel and is designed to withstand a high twisting force. The tip is the end that fits into a specific screw head. A good screwdriver tip is made of **tempered** steel to resist wear and prevent bending and breaking.

> **⚠ CAUTION**
>
> **CAUTION: The comfort-grip handles on screwdrivers used in electrical work are designed for comfort and to provide a secure grip. They are not designed to act as an insulator from electrical shock and are not to be used on "live" electrical equipment.**

It is important to have the screwdriver tip fit snugly in the screw head, as using the wrong size screwdriver can cause the head of the screw to **strip.** Screwdrivers can be dangerous to use. They are very sharp, so do not point the tip toward yourself or anyone else and do not carry them in your back pocket, especially with the tip sticking up toward your back. To make sure that a screwdriver does not become damaged from misuse, do not use it as a chisel, punch, or pry bar.

(See page 80 for the correct procedure for using a screwdriver.)

Pliers

A residential electrician uses many different plier styles. These are considered pouch tools, and the most common styles include lineman pliers, long-nose pliers, diagonal cutting pliers, and pump pliers.

Lineman pliers, sometimes referred to as "sidecutter" pliers, are used to cut cables, conductors, and small screws; to form large conductors; and to pull and hold conductors. It is important to use a plier that is large enough for the job. Typically, the handles should be around 9 inches in length so that a minimum of hand pressure is required to cut the conductor or cable. In most cases, one hand is all that is needed to make the cut, although at times, when the conductor or cable is fairly large, two hands may be required to provide enough pressure to make the cut. Lineman pliers are available in a variety of styles and sizes, but the 9-inch handle with a New England nose is probably the best style for residential electrical work (Figure 3–6). Like all of the pliers used in electrical work, the handles are coated with vinyl for better comfort and to provide a secure grip.

> **⚠ CAUTION**
>
> **CAUTION: The vinyl covering on electrical pliers is designed for comfort and to provide a secure grip. It is not designed to function as an insulator from electrical shock, so do not use pliers with regular grips on "live" circuits.**

Long-nose pliers, sometimes called "needle-nose" pliers, are used to form small conductors, to cut conductors, and to hold and pull conductors (Figure 3–7). The narrow head allows working in tight areas. Residential electricians should use a long-nose plier with at least an 8-inch handle size.

Figure 3–4 A ³⁄₁₆ × 8-inch screw-holding screwdriver. *Courtesy of Ideal Industries, Inc.*

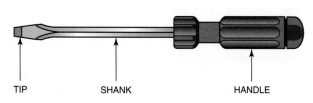

TIP   SHANK   HANDLE

Figure 3–5 The three parts of a screwdriver are the tip, the shank, and the handle.

Figure 3–6 A lineman plier with 9-inch handles and a New England nose with a crimping die is a good choice for a residential electrician. *Courtesy of Ideal Industries, Inc.*

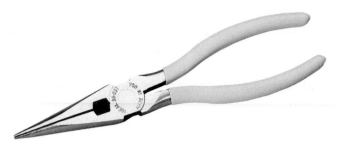

Figure 3–7 Long-nose pliers with 8½-inch handles. *Courtesy of Ideal Industries, Inc.*

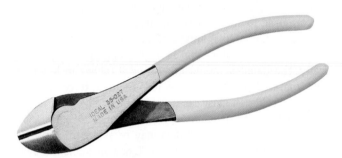

Figure 3–8 Diagonal cutting pliers with 8-inch high-leverage handles and an angled head. *Courtesy of Ideal Industries, Inc.*

Figure 3–9 A pump plier with 10-inch handles. This tool is sometimes called a "tongue-and-grooved plier." *Courtesy of © Klein Tools, Inc.*

Diagonal pliers, sometimes called "dikes," are used to cut cables and conductors in limited spaces (Figure 3–8). Electricians can cut conductors off much closer to the work with this plier type than with other cutting plier types. Diagonal cutting pliers are available in a straight-head or an angled-head design.

Pump pliers, often called "channel-lock pliers," are used to hold and tighten raceway couplings and connectors and to hold and turn conduit and tubing (Figure 3–9). They can also be used to tighten locknuts and cable connectors. Because of their unique design, the gripping heads can be adjusted to several different sizes.

Wire and Cable Strippers

Wire strippers are used to strip insulation from conductors and to cut and form conductors. They are available in several different styles. Probably the most popular style of wire strip-

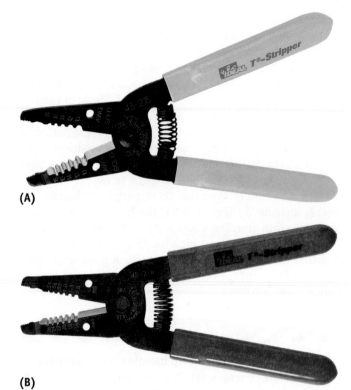

(A)

(B)

Figure 3–10 T-stripper-style wire strippers. (A) The yellow-handled model strips 10- to18-AWG solid wire. (B) The red-handled model strips 16- to 26-AWG stranded wire. *Courtesy of Ideal Industries, Inc.*

per used in residential electrical work is the nonadjustable type, commonly called a "T-stripper" (Figure 3–10). The T-stripper is designed to strip the insulation from several different wire sizes without having to be adjusted for each size. A good choice for residential electricians is a wire stripper designed to strip the insulation from 10 through 18 AWG, which are the wire sizes used most often in residential electricity.

(See page 81 for the correct procedure for stripping the insulation from a conductor using a wire stripper.)

FROM EXPERIENCE

Electrical conductors are available in either a solid or a stranded configuration. Stranded wire has a slightly larger diameter than does solid wire. For this reason, wire strippers that are designed to strip solid wire will "catch" on stranded wire and not strip the insulation off smoothly. Manufacturers make wire strippers for both solid and stranded wire. Usually, the yellow-handled strippers are for solid wire and the red-handled strippers are for stranded wire.

Figure 3–11 Cable rippers are used to slit the insulation for stripping on Romex-type cables. The standard model requires inserting the cable through the back, while the side-entry model allows the electrician to place the ripper onto the cable at the point to be stripped. *Courtesy of Ideal Industries, Inc.*

Figure 3–12 This multipurpose tool is able to do seven things: strip 6- to 16-AWG solid wire and 8- to 18-AWG stranded wire, crimp solderless connectors, cut wire, cut small bolts and screws, re-form threads on small bolts and screws, ream and de-burr conduit, and gauge wire. *Courtesy of Ideal Industries, Inc.*

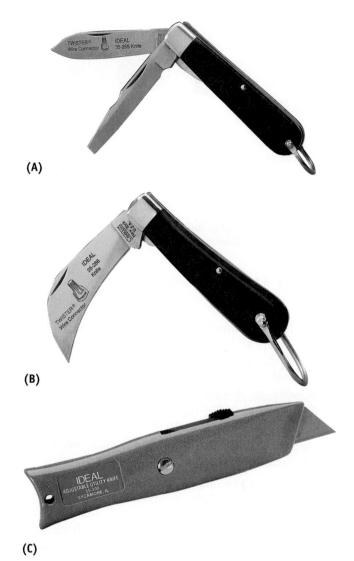

(A)

(B)

(C)

Figure 3–13 (A) An electrician's knife, (B) a hawkbill knife, and (C) a utility knife. *Courtesy of Ideal Industries, Inc.*

Cable rippers are used to more easily remove the outside sheathing from a nonmetallic sheathed cable (Figure 3–11). They are designed to slit the cable sheathing. The electrician then uses a knife or cutting pliers to cut off the sheathing from the cable.

A multipurpose tool, sometimes called a "six-in-one tool," is a versatile tool (Figure 3–12). As its name implies, it is designed to do several different things such as strip insulation, cut conductors, **crimp** smaller solderless connectors, cut small bolts and screws, thread small bolts and screws, and can be used as a wire gauge.

An electrician's knife is a valuable tool and is used often by residential electricians. It is used to open cardboard boxes containing electrical equipment and to strip large conductors and cables. Some models of an electrician's knife include both a cutting blade and a screwdriver blade that can be used to tighten and loosen small screws. Many electricians prefer a knife with a curved blade, called a "hawkbill" knife. Others prefer using a utility knife with a retractable blade. Figure 3–13 shows some examples of knives used in the residential electrical trade.

(See page 82 for the procedure for stripping insulation from large conductors using a knife.)

Other Common Hand Tools

In addition to screwdrivers, pliers, and wire strippers, there are a few other important hand tools that a residential electrician should be familiar with.

A tap tool is used to tap holes for securing equipment to metal, enlarging existing holes by tapping for larger screws, retapping damaged threads, and determining screw sizes. The most common tap tool is called a "triple tap" tool (Figure 3–14) and typically has the capability to tap 10/32, 8/32, and 6/32 hole sizes. This tool comes in handy when the 6/32 screw holes on a device box are stripped and you cannot secure a switch or receptacle to the box. By using the tap tool to redo the stripped 6/32 threads, you are now able to secure the device to the box.

Figure 3–14 **This triple-tap tool can tap or retap 6/32, 8/32, and 10/32 sizes.** *Courtesy of Ideal Industries, Inc.*

Figure 3–15 **A 10-inch adjustable wrench.** *Courtesy of © Klein Tools, Inc.*

An adjustable wrench, sometimes called a "crescent wrench" (Figure 3–15), is used to tighten couplings and connectors, tighten pressure-type wire connectors, and remove and hold nuts and bolts. This wrench has one fixed jaw and one movable jaw. A worm gear is turned to adjust the jaws to any number of head sizes. Common sizes in residential electrical work include an 8- and a 10-inch size.

(See page 83 for the correct procedure to follow when using an adjustable wrench.)

An awl, sometimes called a "scratch awl," is used to start screw holes, make pilot holes for drilling, and mark on metal (Figure 3–16). Starting wood screws or sheet metal screws in wood for the purpose of mounting electrical equipment can be difficult. Using an awl to make a small pilot hole will allow the screws to be started easily.

> ⚠️ **CAUTION** ⚠️
>
> **CAUTION: The end of an awl is very sharp. Care must be used when using and transporting an awl so the sharp end will not cause personal injury.**

A very important hand tool for the residential electrician is an electrician's hammer (Figure 3–17). It is used to drive and pull nails, pry boxes loose, break wallboard, and strike awls and chisels. A hammer used by an electrician should have long, straight claws to simplify the removal of electrical equipment. A strong, shock-absorbent fiberglass handle is recommended for electrical work. An 18- or 20-ounce hammer size is very common and works well in residential work.

A residential electrician has to take measurements for the correct location of electrical equipment. A folding rule or tape measure (Figure 3–18) can be used to check measurements on prints, determine box location on prints, and determine the depth and setout of electrical boxes. A folding

Figure 3–16 **An awl with a cushioned rubber grip.** *Courtesy of Ideal Industries, Inc.*

Figure 3–17 **An 18-ounce electrician's hammer with a fiberglass handle and a comfortable neoprene grip.** *Courtesy of Ideal Industries, Inc.*

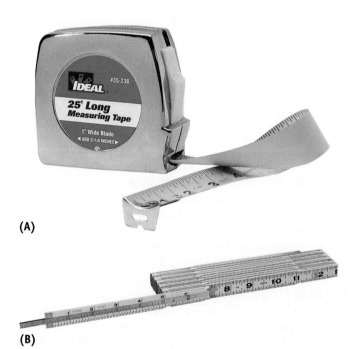

(A)

(B)

Figure 3–18 **(A) A 25-foot tape measure. (B) A 6-foot folding rule.** *Courtesy of Ideal Industries, Inc.*

rule usually opens up to a maximum length of 6 feet and works well when measuring shorter distances. Electricians use tape measures more often than folding rules. The tape measures come in standard 12-, 16-, 20-, and 25-foot lengths. The tape should be at least ¾ inch wide. If the tape

(A)

(B)

Figure 3–19 Two common tool pouch styles. *(A) courtesy of Ideal Industries, Inc. and (B) © Klein Tools, Inc.*

![From Experience icon] **FROM EXPERIENCE**

Many electricians carry their tools in a canvas or polyester tote bag. Some also use an empty 5-gallon drywall mud bucket. A good idea is to carry only those tools you are using for a specific job in your tool pouch and to leave the tools you do not need in your tote or bucket. This makes the weight of the tool pouch much lighter and will allow you to save a lot of energy over the course of a workday by not having to carry all that unnecessary weight of a full tool pouch.

is too narrow, it has very little strength and will be more likely to "break down" when it is extended.

A tool pouch is used to hold, organize, and carry an electrician's hand tools (Figure 3–19). It is made from leather or a strong fabric like nylon polyester or heavyweight canvas. A good-quality belt should be used with the pouch. Because the tool pouch is quite heavy when all of your tools are in it, a wide belt works better than a narrow one to distribute the weight around your waist.

Specialty Tools

Specialty tools are tools that are not typically used on a regular basis and are not usually found in a tool pouch. They are used to install specific parts of an electrical system. Using specialty tools when installing a residential electrical system can be just as hazardous as using common hand tools. The guidelines for the care and safe use of specialty electrical tools are the same as for the guidelines for common hand tools that were discussed earlier in this chapter. The following paragraphs describe the different types of specialty tools and their common trade uses.

A set of nut drivers (Figure 3–20) is used to tighten and loosen various sizes of nuts and bolts. They look like a screwdriver but have a head that is designed to fit on various sizes of nuts and bolts with hexagonal heads. One special feature of a nut driver is that the shank is hollow so that the tool can fit down over long bolts to loosen or tighten the nut on the bolt. A typical set of nut drivers comes with seven sizes ranging from ³/₁₆- through ½-inch trade size.

Figure 3–20 A seven-piece nut driver set with ³/₁₆-, ¼-, ⁵/₁₆-, ¹¹/₃₂-, ³/₈-, ⁷/₁₆-, and ½-inch sizes. *Courtesy of Ideal Industries, Inc.*

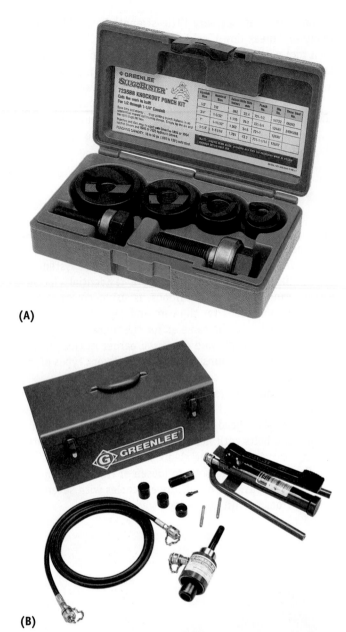

(A)

(B)

Figure 3–21 (A) A ½- through 1¼-inch manual knockout set and (B) a hydraulic knockout set. *Courtesy of Greenlee Textron.*

A **knockout punch** (Figure 3–21) is used to cut holes for installing cable or conduit in metal boxes, equipment, and appliances. This job usually requires the electrician to drill a hole through the enclosure at the spot where the hole needs to be made. The drilled hole should be just large enough for the threaded stud of the knockout set to fit easily through. The knockout set is then used to make a hole that will match the trade size of the conduit or cable connector that is going to be used. Typically a manual knockout set is used to make trade-size holes from ½ inch through 1¼ inches. For these sizes and for sizes up to 6 inches, there are also hydraulic

knockout sets. An electrician should be familiar with both the manual and the **hydraulic** types of knockout tools.

(See page 84 for the procedure for cutting a hole in a metal box with a manual knockout punch.)

A keyhole saw, sometimes called a "compass saw," is used in remodel work to cut holes in wallboard for installing electrical boxes (Figure 3–22). This saw can also be used to cut exposed lath for "old work" box installation in an older lath and plaster installation. The standard blade is 6 to 12 inches long and has seven to eight teeth per inch.

A hacksaw (Figure 3–23) is used to cut some conduit types and to cut larger conductors and cables. Today's hacksaws use rugged frames that are lightweight but provide ample rigidity for exceptional control when cutting. Hacksaw blades for electrical work are available in configurations of 18, 24, and 32 teeth per inch. The best all-around blade has 24 teeth per inch.

(See page 86 for the correct procedure for setting up and using a hacksaw.)

A fish tape and reel (Figure 3–24) is used to pull wires or cables through electrical conduit and to pull or push cables in wall or ceiling cavities. The tape itself can be made of stainless steel, standard blued steel, nonconductive fiberglass, nonconductive nylon, or multistranded steel for greater flexibility. Modern fish tapes are housed in a metal or nonmetallic case called the reel, which provides a convenient way to play out the fish tape to the length you need and to then "reel" it back in.

Figure 3–22 A keyhole saw. *Courtesy of © Klein Tools, Inc.*

Figure 3–23 A heavy-duty hacksaw that uses 12-inch blades. *Courtesy of Ideal Industries, Inc.*

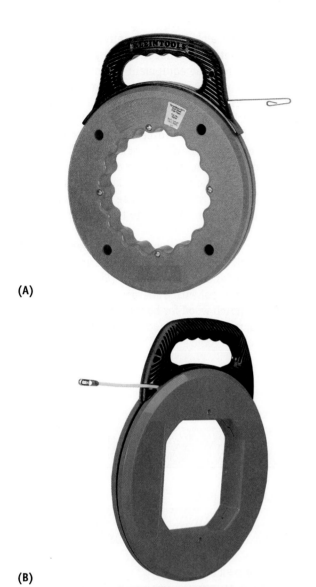

(A)

(B)

Figure 3–24 (A) Steel fish tape. *Courtesy of © Klein Tools, Inc..* **(B) Fiberglass fish tape. The cases are made from impact-resistant plastic.** *Courtesy Ideal Industries, Inc.*

Figure 3–25 This conduit hand bender has many markings to assist the electrician in making accurate bends in conduit. It is used with a 38-inch-long handle. *Courtesy of Ideal Industries, Inc.*

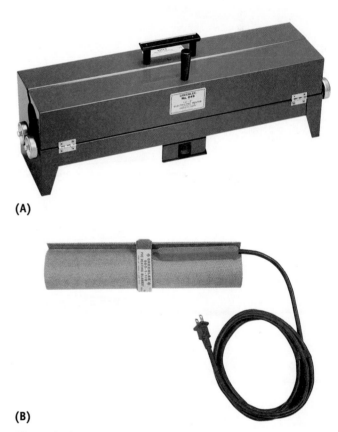

(A)

(B)

Figure 3–26 (A) A PVC conduit heating box. (B) A heating blanket. *Courtesy of Greenlee Textron.*

A conduit **bender** (Figure 3–25) is used to bend electrical metallic tubing (up to 1.25 inches) or rigid metal conduit (up to 1 inch). The bender consists of a bending head and a handle. The heads can be made of iron for durability or of aluminum for less weight. Special markings on the bender head help the electrician make the desired bend accurately. An introduction to basic conduit bending with a hand bender is given in Chapter 12 of this text.

When bending rigid nonmetallic conduit (PVC), a PVC heater box or heating blanket is used (Figure 3–26). The box uses electricity to produce a high temperature that causes the inserted length of PVC pipe to soften. After the pipe has "cooked" for the manufacturer's recommended length of time, the electrician pulls it out and bends the pipe by hand

into the desired shape. A bending blanket is wrapped around the pipe and softens it in the same manner. Again, after the blanket has heated the pipe for a certain length of time, it is taken off, and the pipe is bent into the desired shape.

Levels are used to level or **plumb** conduit, equipment, and appliances. Electricians usually use a small level called a "torpedo level" (Figure 3–27). A torpedo level is approximately

Figure 3–27 A 9-inch torpedo level with an aluminum frame and a magnetic edge on one side. *Courtesy of Ideal Industries, Inc.*

Figure 3–28 A plumb bob.

9 inches long and has a plastic or aluminum frame. One side of the level is magnetized so that it will stay on metal electrical equipment while the equipment is being installed.

A plumb bob (Figure 3–28) is a rather unusual tool for an electrician to use, but it can help make an electrical installation go much easier. It can transfer location points from ceiling to floor or floor to ceiling. For example, say you are installing a light fixture in the middle of a bedroom ceiling. You can easily find the center of the room by taking measurements on the floor and making a mark at that spot. Then, while on a step ladder, use the plumb bob to transfer the location from the floor to the ceiling and mark the location.

Metal files are used to deburr conduit, sharpen tools, and cut and form metal. Wood files or rasps can be used to enlarge drilled holes in wooden framing members (Figure 3–29).

A chisel is a metal tool with a sharpened and angled edge. It can be used to notch wood for boxes and cables and to cut and shape masonry or metal. Wood chisels often have to be

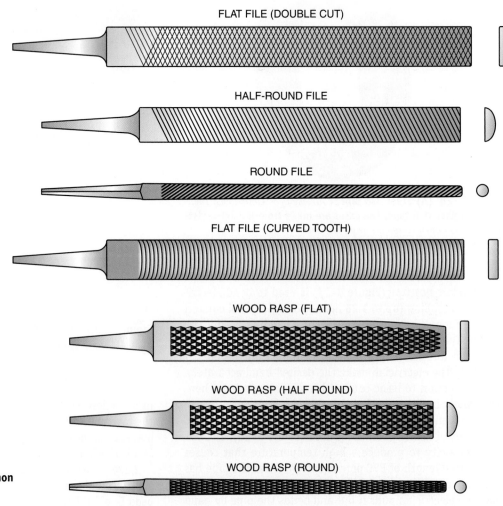

FLAT FILE (DOUBLE CUT)

HALF-ROUND FILE

ROUND FILE

FLAT FILE (CURVED TOOTH)

WOOD RASP (FLAT)

WOOD RASP (HALF ROUND)

WOOD RASP (ROUND)

Figure 3–29 Examples of common file types.

used to "fine-tune" an opening for an electrical box in a re-model job (Figure 3–30).

A cable cutter is used to cut larger size cables and con-ductors (Figure 3–31). Most cables and conductors used in residential wiring can be cut with a pair of lineman pliers, but cables that contain conductors larger than 10 gauge and single conductors larger than 2 AWG are more easily cut to size with a cable cutter.

Figure 3–31 This cable cutter can cut conductors up to 350 kcmil in size. It is only 18 inches long and is very easy to store. *Courtesy of Greenlee Textron.*

(A)

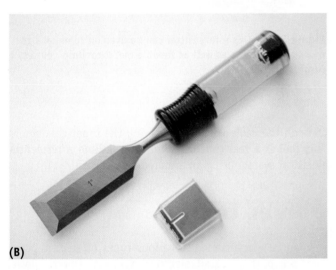

(B)

Figure 3–30 **(A) A metal chisel (sometimes called a "cold chisel").** *Courtesy of © Klein Tools, Inc.* **(B) A wood chisel.**

Figure 3–32 A 3-pound sledgehammer. *Courtesy of © Klein Tools, Inc.*

A sledgehammer is used to drive posts or other large stakes into the ground. Residential electricians may use them to install 8-foot ground rods. The *NEC*® requires that ground rods be 8 feet in the ground. Depending on the soil conditions, driving a ground rod can be a tough job. Sledge-hammers may be found in sizes from 2 through 20 pounds. A 3- to 5-pound sledgehammer is recommended for driving a ground rod (Figure 3–32).

A hex key set, often called Allen wrenches, is used to tighten and loosen countersunk hexagonal setscrews. Larger electrical panel boards will require an Allen wrench to tighten the main conductor lugs (Figure 3–33).

A fuse puller is used to remove cartridge-type fuses from electrical enclosures. They are constructed of a strong non-conductive material, such as glass-filled polypropylene (Fig-ure 3–34). Some models are hinged so that one end accommodates a certain range of fuse sizes and the other end accommodates another range of fuse sizes.

Torque screwdrivers are designed to tighten smaller screws and lugs to the manufacturer's recommended torque requirements. Torque wrenches are used to tighten Allen-head and bolt-type lugs to the manufacturer's torque recom-mendations. Section 110.3(B) of the *NEC*® requires that all equipment be installed according to the instructions that come with the equipment. Most electrical equipment will have instructions that list the torque specifications in inch-pounds or foot-pounds. It is mandatory that an electrician uses the proper torque tools to tighten the equipment to the stated requirements (Figure 3–35).

FROM EXPERIENCE

In certain parts of the United States, log cabin homes are very popular. When installing an electrical system in this type of construction, care must be taken to hide electrical cable and boxes as much as possible. One way to do this is to drill the logs as they are being installed up to the height of the outlet or switch box. At the appropriate height, holes for the electrical boxes are cut into the logs. The ca-ble is pulled into the box hole through the holes that were previously drilled. The cable is then secured to the box, and the box is secured in the box hole. Usually, the box hole needs to be fine-tuned with wood chisels and wood rasps so that the electrical box can fit properly.

Figure 3–33 An Allen wrench set provides several different hex key sizes in one tool. *Courtesy of Ideal Industries, Inc.*

Figure 3–34 Fuse pullers with notched handles for a safe grip. These pullers can handle cartridge fuse sizes from ½ to 1 inch in diameter, 0 to 100 amperes. *Courtesy of Ideal Industries, Inc.*

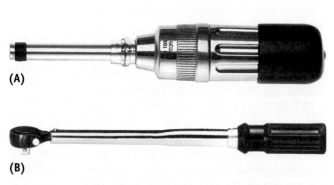

(A)

(B)

Figure 3–35 (A) This torque screwdriver has a capacity to torque from 2 to 36 inch-pounds and can be used on both slotted and Phillips-type heads. (B) The torque wrench can torque from 40 to 200 inch-pounds. *Courtesy of © Klein Tools, Inc.*

(See page 87 for the procedure to follow when using torque screwdrivers or wrenches.)

A rotary BX cutter is a tool that is used to strip the outside armor sheathing from Type AC cable, Type MC cable, and flexible metal conduit (Figure 3–36). It uses a rotating handle to turn a small cutting wheel that is designed to cut

Figure 3–36 This rotary cutter can be used on common sizes of BX and MC cable as well as flexible metal conduit. *Courtesy of Ideal Industries, Inc.*

through the cable without damaging the inner conductors. This tool is a real time-saver when installing a residential electrical system using an armored-clad cable.

Power Tools

Residential electricians use various types of electric power tools when they install electrical systems. Electric power tools include those powered by 120-volt AC and those powered by low-voltage DC electricity. The power tools that use 120-volt AC electricity have a cord and an attachment plug that is plugged into a wall receptacle or extension cord. A "double-insulated" power tool will have a two-prong attachment plug. All others will have a three-prong grounding attachment plug. A battery supplies the DC voltage necessary to power cordless power tools. This section introduces you to the common electrical power tools used by residential electricians. Guidelines for the safe use of power tools are also discussed.

CAUTION

CAUTION: This section provides an overview of common power tools used by residential electricians and should not be considered as the only instruction and training you need before you use a power tool. Never use a power tool before you have received the proper training on how to use the power tool and have read the owner's manual associated with the power tool.

Power Tool Safety

The most important thing to remember when using any power tool is to read the operator's manual that comes with every power tool. The manual will contain information about the tool's applications and limitations as well as hazards associated with that tool. The following safety guidelines should be followed when using power tools:

- Always use proper personal protection equipment when using power tools. Safety glasses or goggles need to be worn at all times. A dust mask, nonskid safety shoes, a hard hat, or hearing protection must be worn when appropriate.
- Keep the work area clean and well illuminated.
- Do not operate electric power tools in explosive atmospheres such as in the presence of flammable liquids like gasoline. Power tools create sparks that could cause the gasoline fumes to ignite.
- Make sure that people not working on the job are kept away from the work area when you are using a power tool. Flying debris could hurt them.
- Make sure that grounded tools are plugged into a properly installed grounded receptacle outlet. Never remove the grounding prong from a grounding-type attachment plug.
- Double-insulated power tools use a polarized attachment plug. Make sure that it is plugged into a correctly installed polarized receptacle.
- Do not use electric power tools in wet conditions. Water entering a power tool will increase the risk of electric shock.
- All 125-volt power tools with cords must be plugged into ground fault circuit interrupter (GFCI)-protected receptacles on a residential construction site.
- Do not abuse the cord on a power tool. Never carry the tool by hanging onto the cord.
- If using a power tool outside, be sure to use an extension cord that is designed for outdoor use. The cord will be marked with "W-A" or "W" if it is rated for outdoor use.
- Stay alert when using a power tool. Never operate a power tool when you are tired or under the influence of drugs or alcohol.
- Do not wear loose clothing or wear jewelry. Longer hair needs to be put up inside a hat. Loose clothes, jewelry, or long hair can get caught in a power tool, and serious injury can result.
- Be sure that the power tool switch is off before plugging in the tool.
- Make sure that all **chuck keys** or other tightening wrenches are removed before turning a power tool on. A wrench or key can be thrown from the tool at high speed and cause personal injury.
- Make sure you have firm footing when using a power tool. Do not overreach. Slipping or tripping when using a power tool can result in serious injury.

- Always try to keep your hands and other body parts as far away as possible from all cutting edges and moving parts of power tools.
- Always secure the material that you are using a power tool on. If the part is unstable, you may lose control of the power tool.
- Always use the correct power tool for the job.
- Do not force a power tool. If it is working too hard to do the job, it either is malfunctioning or is the wrong tool for the job.
- Always unplug a power tool or take out the batteries before you make any adjustments, change accessories, or store the tool.
- Always store power tools in a dry and clean location away from children and other untrained people.
- Maintain tools with care. Properly maintained tools are less likely to malfunction and are easier to control.
- Do not use a damaged power tool. You should tag a damaged tool with a "Do Not Use" message until it is repaired.

Power Drills

A power drill is used with the appropriate bit to bore holes for the installation of cables, conduits, and other electrical equipment in wood, metal, plastic, or other material. The power drill is the most often used power tool by residential electricians, and, like other power tools, models are available with a cord and plug or are cordless. Power drills can be broken down into pistol grip drills, right-angle drills, hammer drills, and cordless drills.

Pistol Grip Drills

Pistol grip electric drills (Figure 3–37) are very popular because they are small, relatively lightweight, and easy to use. This style gets its name from the fact that it looks like, and is held in your hand like, a pistol. They are available in three common **chuck** sizes: ¼, ⅜, and ½ inch. A chuck is the part of the drill that holds the drill bit securely in place. Most drills have chucks that are tightened and loosened with a chuck key or wrench. Some newer models are using a keyless chuck.

Electricians usually use the ⅜- or ½-inch size. The ⅜-inch size is used for small hole boring at higher speed. The ½-inch size is used for larger hole boring at lower speeds. The speed in revolutions per minute (rpm) of this drill type is inversely proportional to the chuck size. In other words, the larger the chuck size, the slower the speed of the drill. The turning force, or the **torque,** of the drill is proportional to the chuck size. That is, the larger the chuck, the more torque is available. A trigger switch controls the speed of this drill type. The harder you squeeze the trigger, the faster the drill turns. The drill can also be reversed.

The pistol grip drill can be used with a wide variety of drill bits. Drill bits are the tools that are attached to the drill and actually do the hole boring. Bits used with a pistol grip drill should be designed for use at higher speeds. A twist bit is

Figure 3–37 **This pistol grip drill has a ½-inch chuck. It reverses easily and has a trigger switch for speed control of 0 to 850 rpm.** *Courtesy of Milwaukee Electric Tool Corporation.*

(A)

(B)

(C)

(D)

Figure 3–38 **Examples of common bit types and accessories used with a pistol grip drill: (A) twist bit, (B) flat-blade spade bit, (C) masonry bit, and (D) bit extension.** *Courtesy of Milwaukee Electric Tool Corporation.*

designed to drill wood or plastic at high speed and metal at a lower speed. A flat-bladed spade bit, sometimes called a "speed-bore bit," is used to drill holes in wood at high speed. A masonry bit is used to drill holes in concrete, brick, and other masonry surfaces. An **auger** bit is used for drilling wood at a relatively slow speed. Figure 3–38 shows examples of drill bit types used with a pistol grip drill.

(See page 88 for the correct procedure on using a pistol grip power drill.)

Right-Angle Drills

In residential wiring, electricians need to drill holes for the installation of cables through wooden framing members. The tight space between wall studs can make drilling the required holes with a pistol grip drill a slow and awkward process. To allow easier drilling of wood framing members in tight spaces, the right-angle drill (Figure 3–39) was introduced by the Milwaukee Electric Tool Company in 1949. The head of the drill is at a right angle (90 degrees) in relation to the rest of the drill, which allows the drill body to be located away from the material being drilled. Right-angle drills are usually used with a ½-inch chuck and work well with auger bits, forstner bits, or hole saws to drill holes in wood framing members. Figure 3–40 shows examples of drilling bits used with a right-angle drill.

(See page 89 for the correct procedure for drilling a hole in a wood framing member with an auger bit and a right-angle drill.)

(See page 90 for the correct procedure for cutting a hole in a wood framing member with a hole saw and a right-angle drill.)

Figure 3–39 **This right-angle drill is designed for use in electrical system installation. The drill has a ½-inch chuck that is attached to a 360-degree swivel head that allows drilling in virtually any direction. The drill is reversible and has speed control from 0 to 600 rpm.** *Courtesy of Milwaukee Electric Tool Corporation.*

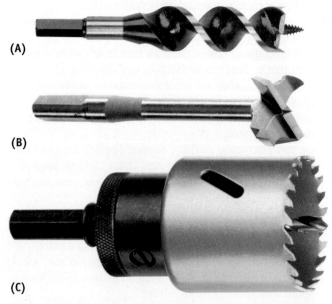

(A)

(B)

(C)

Figure 3–40 **Examples of common bit types and accessories used with a right-angle drill: (A) auger bit, (B) forstner bit, and (C) hole saw.** *Courtesy of Milwaukee Electric Tool Corporation.*

Hammer Drills

Hammer drills (Figure 3–41) are used to drill holes in masonry or concrete walls and floors. When installing anchors to hold electrical equipment on a masonry wall, hammer drills are used to drill the anchor holes. Hammer drills are used with special masonry bits to bore holes in masonry. The bits used are of the percussion carbide-tip type. While the drill is turning the masonry bit, it is also moving the bit in a reciprocating, or hammering, motion. Some hammer drills can switch back and forth from a "drill only" mode to a "hammer drill" mode. With this feature, you actually have two drill styles in one package. The drill-only mode turns off the hammering action so that you can drill wood, metal, or plastic. The hammer-drill mode restores the hammering action and allows you to drill holes in masonry. Pistol-style hammer drills come with ⅜- or ½-inch chucks.

Most hammer-drill mechanisms are designed so the more pressure you exert on the drill, the more hammering action you get. You will know that the right amount of pressure is being used when the hammering is even and smooth.

(See page 91 for the correct procedure for drilling a hole in masonry with a hammer drill.)

When larger holes need to be drilled in masonry, a rotary hammer drill is used (Figure 3–42). The rotary hammer drill bores holes by pulverizing the work with a steady rhythm of heavy blows. It does not have the same style of chuck as a regular hammer drill. The bits are still carbide tipped but have specially shaped shanks that fit into the rotary hammer drill's nose assembly and are loosely held in place with a movable collar.

Residential electricians need to decide whether to buy and use a regular hammer drill or a rotary hammer drill. The deciding factor is usually frequency of use. Electricians who

Figure 3–41 A hammer drill used for percussion carbide-bit drilling in concrete and masonry and drilling without hammering in wood or metal. *Courtesy of Milwaukee Electric Tool Corporation.*

Figure 3–42 A rotary hammer drill. *Courtesy of Milwaukee Electric Tool Corporation.*

drill only an occasional hole in masonry can get by very nicely with a regular hammer drill. Electricians who plan on drilling small or large holes in concrete on a daily basis should lean toward a rotary hammer drill.

Cordless Drills

Pistol grip drills, right-angle drills, and hammer drills are all now available in cordless models (Figure 3–43). The source of electrical power for these drills usually is the Ni-Cad battery. This battery type is rechargeable, long lasting, virtually maintenance free, and relatively cost effective and can provide the power needed to operate any of the power tools that an electrician may use. Ni-Cad is short for nickel-cadmium, the materials responsible for the electrochemical reaction in the battery that produces a voltage. The voltage of today's cordless power tools usually will be 12, 14.4, 18, or 24 volts. The larger voltage will result in more power available for the power tool. A cordless drill kit usually comes with the drill, a battery charger, and at least one battery.

Many companies are making their cordless drills with a "keyless" chuck. No wrench is needed to tighten the bit into the chuck. To secure a drill bit in a cordless drill with a keyless chuck, make sure the battery is removed and then open the chuck with your hand until it is open enough to receive the drill bit. Insert the bit and tighten the chuck by hand until the shank of the bit is firmly gripped. The last thing you need to do is a little tricky. Reinstall the battery and, with the drill set to turn in the clockwise direction, firmly grip the chuck with one hand and lightly squeeze the drill trigger with the other hand. The drill will start to turn, but the resistance that you are applying by holding onto the chuck will allow the bit shank to be locked tightly into the chuck.

Most cordless drills also have an adjustable clutch feature that allows the amount of turning force to be increased or decreased. This allows the cordless drills to act like a power screwdriver. Electricians often use this feature for installing receptacles and switches in device boxes and for other tasks that normally involve using a manual screwdriver. Using the

Figure 3–43 A ½-inch cordless drill. This drill is rated for 18 volts and comes with a case, a battery charger, and an extra battery. *Courtesy of Milwaukee Electric Tool Corporation.*

Figure 3–44 A circular saw with a 7¼-inch blade. *Courtesy of Milwaukee Electric Tool Corporation.*

 FROM EXPERIENCE

Electricians who use cordless drills need to have a battery charger and at least one or two spare batteries. You should be charging one battery while you are using the drill and using up the installed battery. Employers are not happy when you cannot do your job because of a discharged battery and you do not have a charged battery as a backup.

cordless drill as a power screwdriver is much faster than using a manual screwdriver. It also is physically easier on the installing electrician.

Power Saws

Residential electricians are required to use power saws on the job for cutting to size such things as plywood backboards for mounting electrical equipment and for cutting building framing members to facilitate the installation of the electrical wiring. There are two styles of power saws you should be familiar with: the circular saw and the reciprocating saw.

Circular Saws

Many electricians call this saw type a "skilsaw." This name comes from the first portable electric power saw, which was made by a company named Skil. A circular saw (Figure 3–44) is designed to cut wood products, such as studs and plywood, to a certain size. Most circular saws still use a cord and plug connection, but several manufacturers have recently introduced powerful cordless models. Electricians do not use this

saw type very often but should be familiar with it for those times when it is used.

The size of this saw is measured by the diameter of the saw blade. The most common size is 7¼ inches. These saws are fairly heavy and can weigh as much as 12 or 15 pounds. The handle of the saw has a trigger switch that starts the blade turning. The harder you squeeze the trigger, the faster the blade turns. It is important to make sure that the blade's teeth are pointing in the direction of rotation. Blades typically have an arrow on them that indicates their required direction of rotation. A two-part guard protects the blade. The top half is stationary, and the bottom half is hinged and spring-loaded. As you push the saw forward when cutting, the lower half of the guard is hinged up and under the top guard, allowing the saw to continue to cut.

Using a circular saw is very dangerous. You should always receive proper instruction on its use before you attempt to cut anything with it. Try to follow these guidelines:

- Always wear the proper personal protection equipment.
- Always check to see that the blade guard is working properly.
- Make sure you know what is behind or below the piece of wood you are cutting.
- Try to always keep two hands on the saw at all times.
- Do not force the saw through the work. Apply a steady but firm pressure and let the saw do its job.
- Always secure the work you are cutting.
- If the saw has a cord, be aware of where it is when you are sawing. Do not cut off the cord.
- Do not reach under the work while the saw is operating. You may lose your fingers.
- Try to position yourself so you are standing to one side of the work as you cut, not directly behind it.

(See page 92 for the correct procedure for using a circular saw.)

Figure 3–45 A Sawzall reciprocating saw. *Courtesy of Milwaukee Electric Tool Corporation.*

Reciprocating Saws

A reciprocating saw has a blade that moves back and forth to make a cut rather than using a rotating blade (Figure 3–45). It can make a straight or curved cut in many materials, including wood and metal. When cutting wood, a wood blade must be used; when cutting metal, a metal blade must be used. Electricians usually refer to this saw type as a "Sawzall." The name Sawzall is a trademark of the Milwaukee Electric Tool Corporation. Reciprocating saws are usually found with a cord and plug, but, like circular saws, manufacturers are now producing well-engineered cordless models.

Residential electricians use the reciprocating saw much more often than a circular saw. It can be used to cut heavy lumber and to cut off conduit, wood, and fasteners flush to the surface. It works well when doing new or remodel work for such things as cutting box holes in plywood or fine-tuning framing members so that an electrical box can be properly mounted. It is great for sawing in tight locations where other saw types cannot reach.

When using a reciprocating saw, the following cutting tips should be followed:

- Always have at least three teeth in a cut. Less than three teeth will result in the blade getting hung up on the material being cut. More than three teeth slow down the cutting process. The solution is to use the coarsest blade possible while still having three teeth in the cut at all times.

- Turning the blade 180 degrees makes it easier to make a flush cut. This way, the head of the saw does not get in the way as much.
- When cutting wood, splintering can be limited by putting the finished side down when cutting. You can also cover the cutting line with masking tape and then make the cut.
- If a blade breaks when cutting a thin material, switch to a blade with more and finer teeth.
- Drilling a starter hole larger than the widest part of the blade is the best way to get a cut started that is away from the edge of a material. Plunge cutting can be done in wood only by starting the saw at a shallow angle and then increasing the angle of the cut until the blade goes through the material.
- When cutting metal, cut at a lower speed and use a good-quality cutting oil.

(See page 94 for the correct procedure for using a reciprocating saw.)

Summary

This chapter has covered the common electrical hand tools, specialty tools, and power tools used in residential electricity. It has looked at how to identify the many tool types used in installing a residential wiring system as well as how to safely use them. You may run across hand tools, specialty tools, and power tools that are not covered in this chapter. Don't worry. Simply look over the owner's manual that comes with the tool and ask an experienced electrician to show you the proper way to use the tool. Attend trade shows and vendor days at your local electrical distributor to stay up to date on the newest tools for electricians. Remember, a good residential electrician knows which tool is right for the job and how to use that tool in a safe and proper manner.

Procedures

Using a Screwdriver

- Observe proper safety procedures.

- Choose the proper blade type for the screw head encountered and make sure that the screwdriver tip fits the screw correctly.

 If you are installing a screw in a new location, make a pilot hole at the spot where you are installing the screw. An awl can be used in softwood. Drill a pilot hole for the screw in sheet metal or hardwood. Insert the tip of the screw in the pilot hole.

 Insert the screwdriver tip in the head of the screw. Hold the screwdriver tip steady and position the shank perpendicular to the screw head.

 While using one hand to keep the screwdriver steady, use the other hand to turn the handle of the screwdriver. Apply firm, steady pressure to the screw head and turn the screw clockwise to tighten or counterclockwise to loosen. (A good way to remember which way a screw is turned to tighten or loosen it is to use either of the following sayings: Right Is Tight, Left Is Loose, or Righty Tighty, Lefty Loosey.)

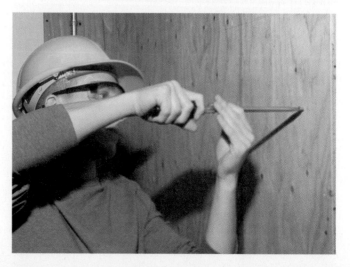

Procedures

Using a Wire Stripper

- Observe proper safety procedures.

A Insert the conductor to be stripped in the proper stripping slot.

- Close the jaws until you feel that you have reached the conductor and then open the jaws slightly.

B With an even pressure, pull back the strippers and remove the conductor insulation.

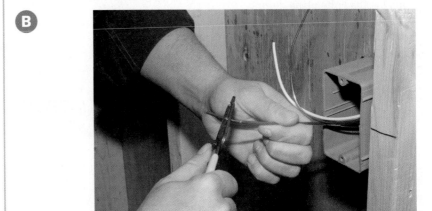

C Check the conductor for a ring or nick. If a nick occurs, restrip until the insulation is removed without any conductor damage.

Procedures

Using a Knife to Strip Insulation from Large Conductors

- Observe proper safety procedures.

- Mark the conductor to indicate the amount of insulation to be stripped.

(A) Place the knife blade on the insulation at the point marked and carefully cut around the conductor to a depth that is just shy of touching the actual conductor.

(A)

(B) Place the blade across the conductor and lay the knife blade down to a position that is almost parallel with the conductor at the cut made in the previous step.

(B)

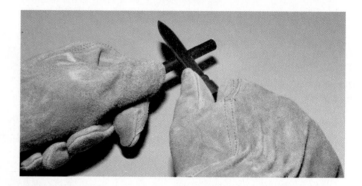

(C) Using a pushing motion away from you, cut the insulation to a depth that is as close as possible to the conductor, all the way to the end. Be careful to not nick the conductor in this process.

(C)

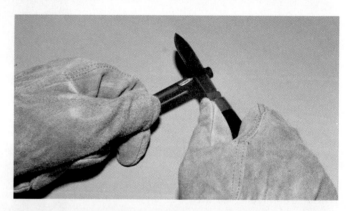

(D) Using your fingers (or possibly your long-nose pliers), peel off the remaining insulation.

(D)

Procedures

Using an Adjustable Wrench

- Observe proper safety procedures.

- Determine the size of the nut or bolt and set the jaws to the correct size.

- Verify that the jaws are not loose on the work when the wrench is in position.

 Using the proper turning force, turn the wrench in the direction required for tightening or loosening. Pressure must be put on the stationary jaw, not the movable jaw.

- Turn the wrench by pulling it toward you rather than pushing it away from you.

- Continue turning the wrench until the part is completely tightened or loosened.

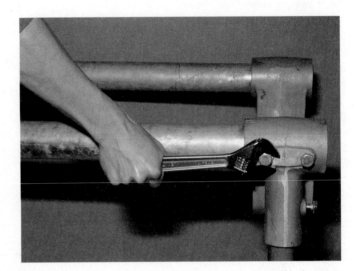

Procedures

Using a Manual Knockout Punch to Cut a Hole in a Metal Box

- Observe proper safety procedures.

- Set up a portable power drill for use. Remember to do a complete check of the power tool before using it. Choose a metal drill bit that is slightly larger than the threaded stud of the knockout tool and install the bit into the drill. Make sure it is tightened securely in the chuck using the chuck key. You may want to drill a pilot hole with a small drill bit and then use the larger bit to enlarge the hole to the size required for easy insertion of the knockout set's threaded stud.

 Drill a hole slightly larger than the knockout set's threaded stud in the center of the space you are going to punch. A center punch can be used to make an indentation for your drill to start in. This will reduce the tendency of the drill bit to move off line when you start to drill. Hold the drill firmly while drilling. A loose grip could cause an accident. Remember that the drill bit will be hot, so use caution around it until it cools.

 Insert the threaded stud through the drilled hole and put the cutting **die** back on the stud. Make sure that the cutting die is aligned so that the cutting edge is toward the metal enclosure.

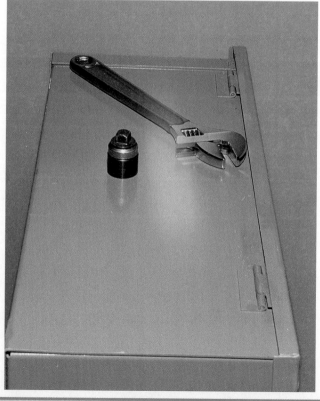

Tighten the drive nut with a properly sized wrench. An adjustable wrench works well.

- Continue to tighten the drive nut until the **cutter** is pulled all the way through the metal. Remove the knockout punch when the cutter is finally pulled through.

 Remove the cutter from the threaded stud and shake out the punched metal.

- Replace the cutter onto the threaded stud and place the knockout punch in its proper storage area.

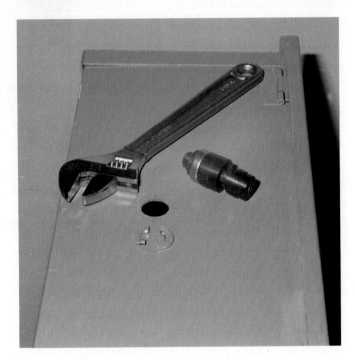

Procedures

Setting Up and Using a Hacksaw

- Observe proper safety procedures.

 Insert the blade in the frame of the hacksaw and tighten it securely in place. Be sure the teeth angles are pointed toward the front of the saw. There is usually an arrow on the blade that indicates the proper direction for the blade to be installed.

- Mark the point on the material where you want to cut and secure the item being sawed in a vise.

- Set the blade of the hacksaw on the point to be cut.

 Push gently forward until the cut is started. Do not exert too much pressure on the saw.

- Make reciprocal strokes until the cut is finished. Remember that the cutting stroke is the forward stroke. Your cut should be straight and relatively smooth. Be aware that excessive speed while cutting can ruin blades.

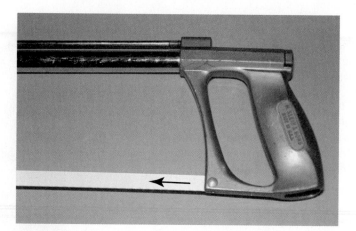

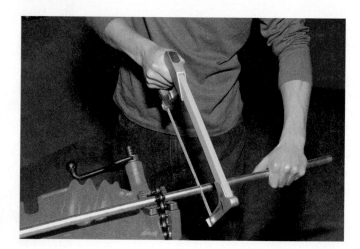

Procedures

Using Torque Screwdrivers or Torque Wrenches

- Observe proper safety procedures.

A Look at the equipment instructions and determine the required torque specifications.

 A

Panel Board Lug Torque Data		
Line Lugs and Main Breaker		
	Wire Range (awg)	**Torque (in. lbs.)**
Line Neutral Lug	4 – 2/0 CU or AL	50
Main Lugs	6 – 2/0 CU or AL	50
Main Breaker	4 – 2/0 CU or AL	(see circuit breaker)
Branch Circuit Neutral and Equipment Ground Bar		
Wire Range (awg)		**Torque (in. lbs.)**
3 – 1/0 CU or AL		50
4 CU or AL		45
6 CU or AL		40
8 CU or AL		35
14 – 10 CU; 12 – 10 AL		30
Branch Circuit Breakers		

See the individual circuit breakers for wire-binding screw torque.

B Adjust the tool to the desire number of inch-pounds (or foot-pounds).

- Set the tool on the object to be tightened.

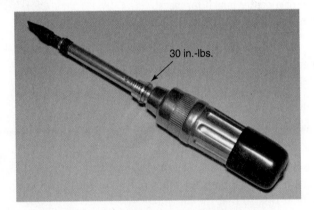 **B**

30 in.-lbs.

C Keeping everything properly aligned, turn the tool in the direction for tightening.

- Continue to tighten until you hear a click. The click means that you have tightened the fastener to the required torque specification. Some torque tools have a dial indicator that you must watch to see when it indicates that the proper torque has been applied.

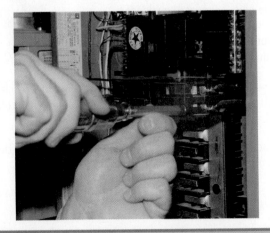 **C**

Procedures

Using a Pistol Grip Power Drill

- Observe proper safety procedures.

- Choose the proper bit type for the job and place it in the drill chuck in the following manner:

  - Make sure electrical power is off and open the chuck by turning it with your hand in a counterclockwise direction until it can accommodate the shank of the drill bit.

  - Insert the bit shank into the chuck and tighten the chuck by hand as much as you can.

  A Using the chuck key, tighten the bit securely in the chuck. If a chuck has more than one tightening hole, use the key to tighten at each hole.

  - Remove the chuck key.

- Make a mark at the location you want to drill and, using an awl for wood or a center punch for metal, make a small indent exactly where you want to drill.

- If necessary, securely clamp the material that is being drilled.

- Firmly grip the drill and place the tip of the bit at the indent.

  B Hold the drill perpendicular to the work so that the bit will not make a hole at an angle. Start the drill slowly at first so that the bit will not stray from the spot where you want the hole. Squeeze the trigger harder and speed the drill up while applying moderate pressure to the drill.

A

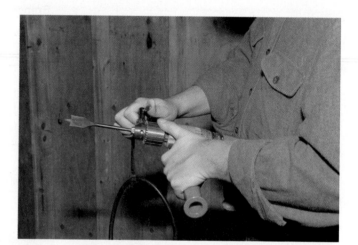

B

- When you sense that the bit is about to go through the material you are drilling, reduce the amount of pressure you are exerting on the drill and let the drill bit complete boring out the hole.

- While the drill is still turning the bit in a clockwise direction, slowly pull the drill and bit out and then release the trigger.

- Unplug the cord and take the bit out of the chuck. Place everything in its proper storage area.

Procedures

Drilling a Hole in a Wooden Framing Member with an Auger Bit and a Right-Angle Drill

- Observe proper safety procedures.

- Make a mark at the location where you want to bore the hole.

A Set up the right-angle drill with the auger bit. Make sure that the bit is secure in the drill chuck by tightening the chuck evenly with the chuck key. Remember that the cord must be unplugged while working on the drill .

- Plug the drill in and place the tip of the bit at the spot that was previously marked.

B Keeping an even force on the drill, start the drill and allow the auger bit to feed itself completely through the wooden framing member. Make sure the drill is turning in a clockwise direction.

- Once the auger bit has gone through the wood, stop the drill; reverse it and, while pulling back with an even pressure, back out the bit.

- Unplug the cord and take the auger bit out of the drill chuck. Place everything in its proper storage area.

A

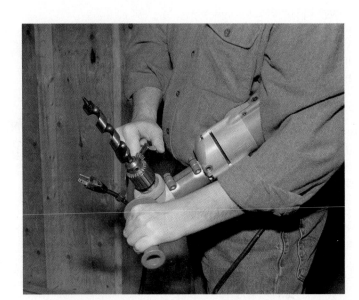

B

Procedures

Cutting a Hole in a Wooden Framing Member with a Hole Saw and a Right-Angle Drill

- Observe all proper safety procedures.

Ⓐ Choose the proper size hole saw and arbor for the size hole you wish to cut and assemble it. Set up the hole saw so that the pilot bit on the hole saw extends past the end of the hole saw by approximately 1 inch.

Ⓑ Make sure the drill is unplugged and secure the hole saw in the drill chuck by tightening the chuck evenly with the chuck key.

- Mark the location on the material where you want the hole to be made.

- Plug the drill in and place the tip of the hole saw pilot bit at the marked spot.

Ⓒ Keeping an even force on the drill, start the drill and allow the hole saw to feed itself completely through the wooden framing member. Make sure the drill is turning in a clockwise direction.

- Once the hole saw has gone through the wood, stop the drill; reverse it and, pulling back with an even pressure, back out the hole saw.

Ⓓ Unplug the drill and remove the wood from the hole saw by using a screwdriver or some other narrow tool to dislodge the wood.

- Disassemble the hole saw and put everything back in the proper storage area.

Ⓐ

Ⓑ
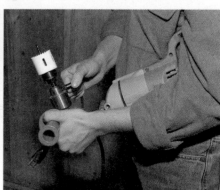

Ⓒ

Ⓓ

Procedures

Drilling a Hole in Masonry with a Hammer Drill

- Observe proper safety procedures.

 Select the proper-size carbide-tip masonry drill bit for the hole you are boring and secure it into the chuck. Make sure the drill is un-plugged.

- Check to be sure that the hammer drill is in the hammer-drill mode.

- Make a mark at the location where you want to drill the hole.

- Grip the drill firmly and place the drill bit at the mark.

 Apply some pressure to the drill and slowly squeeze the trigger to increase the speed.

- Continue to apply pressure until the hammering is smooth and even and drill out the hole to the desired depth.

 Discontinue putting pressure on the drill, and the hammering will stop. While the drill is still turning, slowly pull the bit from the hole. This will help clean the hole of masonry debris.

- Unplug the drill, remove the bit, and put everything away.

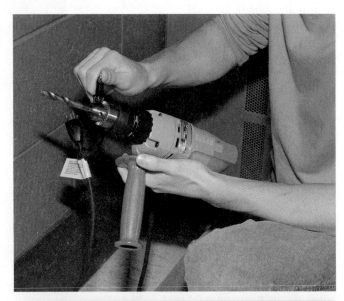

Procedures

Using a Circular Saw

- Follow proper safety procedures and wear appropriate personal protective equipment.

A Secure the material to be cut. Use clamps whenever possible.

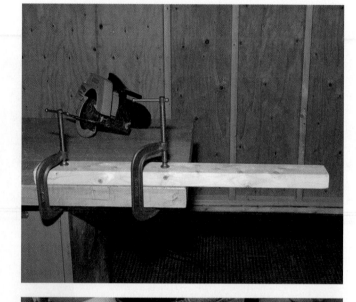

B Mark along the track you want to cut with a pencil.

C Adjust the saw's blade depth to a value that is slightly greater than the thickness of the material being cut.

- With the front edge of the saw base plate resting on the work, line up the guide slot with your cutting mark.

Procedures

Using a Circular Saw (continued)

D Grip the saw firmly with both hands and slowly exert pressure on the trigger switch. Bring the blade speed up to full speed and slowly push the saw forward to begin the cut.

- Continue to use firm and steady force to cut along the marked line, still using two hands on the saw.

D

E While maintaining the saw speed, push the saw completely through the end of the work. Make sure the blade continues to saw along the cutting line since at the end of the work the guide slot is off the work and cannot be used as a guide.

F Release the trigger switch and pull the saw up and away from the work. The blade should stop, and the lower guard should now be covering the bottom of the blade.

- Unplug the saw and store it in an appropriate location.

E

F

Procedures

Using a Reciprocating Saw

- Put on safety glasses and observe all safety rules.

A Set up the saw with a cutting blade suitable for cutting metal or wood. Make sure that the blade is secure in the saw head by tightening the setscrew with the Allen wrench provided with the saw. The cord must be unplugged while working on the saw.

- Mark the location where you want to cut on the work material.

- If possible, secure the work with a vise or clamp to reduce vibration while cutting.

B Plug in the saw and set the blade on the work at the point to be cut as indicated by your mark. Always try to position the tool so that you are cutting in a downward direction and away from your body.

C Pull gently back on the trigger switch to start the saw and begin the cut. Do not exert too much pressure on the saw. It will cut through very easily by allowing the weight of the tool itself to provide the necessary pressure.

- Let the saw make reciprocal strokes until the cut is finished. Your cut should be straight and relatively smooth.

- Unplug the tool, remove the blade, and store the tool in an appropriate location.

A

B

C
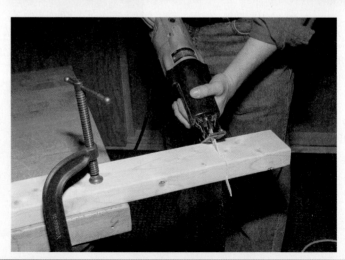

Review Questions

① State five guidelines for care and safe use of electrical hand tools.

② Match the following hand tools to their common residential electrical trades use. Write the correct number for the tool in the blanks.

1. Lineman pliers
2. Adjustable wrench
3. Wire strippers
4. Electrician's knife
5. Phillips screwdriver
6. Hammer
7. Triple-tap tool
8. Long-nose pliers
9. Multipurpose tool
10. Flat-blade screwdriver
11. Awl
12. Diagonal pliers

_____ a. Used to remove and install slot-head screws and to tighten and loosen slot-head pressure connectors

_____ b. Used to cut cables and conductors, form large conductors, and pull and hold conductors

_____ c. Used to form small conductors, cut conductors, and to hold and pull conductors

_____ d. Used to cut cables and conductors in a limited space

_____ e. Used to strip insulation from conductors

_____ f. Used to strip insulation, crimp solderless connectors, cut and size conductors, and cut and thread small bolts

_____ g. Used to strip large conductors and cables

_____ h. Used to tap $6/32$, $8/32$, and $10/32$ holes for securing equipment to metal, retap damaged threads, and determine screw sizes

_____ i. Used to start screw holes, make pilot holes for drilling, and mark metal

_____ j. Used to drive and pull nails, pry boxes loose, and strike chisels

③ Match the following specialty tools and power tools to their common residential electrical trades use. Write the correct number for the tool in the blanks.

1. Portable drill
2. Keyhole saw
3. Plumb bob
4. Hacksaw
5. Cable cutter
6. Torpedo Level
7. Fish tape
8. Knockout punch
9. Reciprocating saw
10. File
11. Chisel
12. Auger bit

_____ a. Used to cut conduit, large conductors, and cables

_____ b. Used with appropriate bits to bore holes for cables and conduits

_____ c. Used in a drill to bore holes through larger wooden framing members

_____ d. Used to cut holes for cable or conduit in metal boxes, equipment, and appliances

_____ e. Used to cut holes in wallboard for boxes and to cut lath and plaster for box installation in remodel work

_____ f. Used to cut lumber and to cut off pipe, wood, and fasteners flush to the surface

_____ g. Used to pull wires or cables through conduit and to pull cables in insulated walls

_____ h. Used to level or plumb conduit, equipment, and appliances

_____ i. Used to transfer location points from ceiling to floor or floor to ceiling and to plumb conduit and equipment

_____ j. Used to cut large cables and conductors

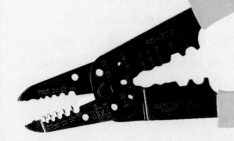

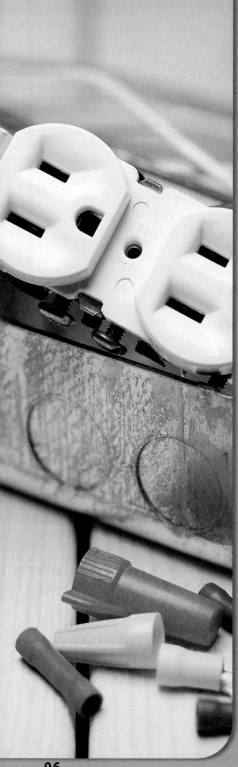

Chapter 4

Test and Measurement Instruments Used in Residential Wiring

Residential electricians must be familiar with a variety of test and measurement instruments. These instruments, usually referred to as "testers" or "meters," can provide a wealth of information to the residential electrician. For example, an electrician can use a meter to measure the current draw of an air-conditioning unit or measure the voltage value at a certain point on the wiring system. The residential electrician needs to know how to use test and measurement instruments in a safe and proper manner for a variety of residential wiring applications. This chapter looks at how to properly use several of the most common types of meters that are encountered in residential wiring. Safe meter practices, as well as common meter care and maintenance techniques, are discussed in detail.

OBJECTIVES

Upon completion of this chapter, the student should be able to:

- ⊗ demonstrate an understanding of continuity testers and how to properly use them.
- ⊗ demonstrate an understanding of the differences between a voltage tester and a voltmeter.
- ⊗ connect and properly use a voltage tester and a voltmeter.
- ⊗ demonstrate an understanding of the differences between an in-line ammeter and a clamp-on ammeter.
- ⊗ connect and properly use an in-line ammeter and a clamp-on ammeter.
- ⊗ demonstrate an understanding of ohmmeters and megohmmeters.
- ⊗ connect and properly use an ohmmeter and megohmmeter.
- ⊗ demonstrate an understanding of how to use a multimeter.
- ⊗ demonstrate an understanding of the uses for a true RMS meter.
- ⊗ demonstrate an understanding of how to read a kilowatt-hour meter.
- ⊗ demonstrate an understanding of safe practices to follow when using test and measurement instruments.
- ⊗ demonstrate an understanding of the proper care and maintenance of test and measurement instruments.

Glossary of Terms

ammeter (clamp-on) a measuring instrument that has a movable jaw that is opened and then clamped around a current-carrying conductor to measure current flow

ammeter (in-line) a measuring instrument that is connected in series with a load and measures the amount of current flow in the circuit

analog meter a meter that uses a moving pointer (needle) to indicate a value on a scale

auto-ranging meter a meter feature that automatically selects the range with the best resolution and accuracy

continuity tester a testing device used to indicate whether there is a continuous path for current flow through an electrical circuit or circuit component

digital meter a meter where the indication of the measured value will be given as an actual number in a liquid crystal display (LCD)

DMM a series of letters that stands for "digital multimeter"

harmonics a frequency that is a multiple of the 60 Hz fundamental; harmonics cause distortion of the voltage and current AC waveforms

kilowatt-hour meter an instrument that measures the amount of

electrical energy supplied by the electric utility company to a dwelling unit

manual ranging meter a meter feature that requires the user to manually select the proper range

megohmmeter a measuring instrument that measures large amounts of resistance and is used to test electrical conductor insulation

multimeter a measuring instrument that is capable of measuring many different electrical values, such as voltage, current, resistance, and frequency, all in one meter

multiwire circuit a circuit in residential wiring that consists of two ungrounded conductors that have 240 volts between them and a grounded conductor that has 120 volts between it and each ungrounded conductor

noncontact voltage tester a tester that indicates if a voltage is present by lighting up, making a noise, or vibrating; the tester is not actually connected into the electrical circuit but is simply brought into close proximity of the energized conductors or other system parts

nonlinear loads a load where the load impedance is not constant, resulting in harmonics being present on the electrical circuit

ohmmeter a measuring instrument that measures values of resistance

open circuit a break in an electrical conductor or cable

polarity the positive or negative direction of DC voltage or current

short circuit a connection in an electrical circuit or device that results in a very low resistance; this is normally not a desired condition

true RMS meter a type of meter that allows accurate measurement of AC values in harmonic environments

voltage tester a device designed to indicate approximate values of voltage or to simply indicate if a voltage is present

voltmeter a measuring instrument that measures a precise amount of voltage

VOM a name sometimes used in reference to a "multimeter"; the letters stand for "volt-ohm milliammeter"

Wiggy a trade name for a solenoid type of voltage tester

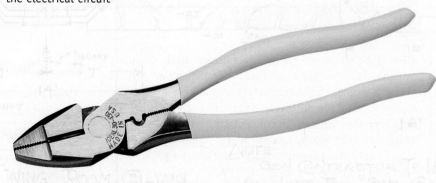

Continuity Testers

A continuity tester is a device used to indicate whether there is a continuous path for current flow in an electrical circuit or electrical device. Continuity testers are the least complicated of all the instruments that a residential electrician may use, but their use in residential wiring is extensive. For example, they can be used to test continuity in electrical conductors, to test for faulty fuses, to test for malfunctioning switches, to identify individual wires in a cable, and many other applications. Residential electricians should have some type of continuity tester as part of their complement of tools. Figure 4-1 shows a commercially available continuity tester that is inexpensive and very easy for an electrician to use. This particular model has a light that comes on to indicate continuity.

Figure 4-1 A basic continuity tester. This tester has a lamp at the end of the tester that lights up to indicate continuity. *Courtesy of Ideal Industries, Inc.*

CAUTION

CAUTION: Never attach a continuity tester to a circuit that is energized.

(See page 110 for step-by-step instructions on how to use a continuity tester.)

Voltage Testers and Voltmeters

Measuring for a certain amount of voltage or simply testing to see if a voltage is present are two of the most often done test and measurement tasks performed by a residential electrician. The instruments used to test or measure for voltage are similar in how they are connected into an electrical circuit but are quite different in other areas, such as the accuracy of the meter reading. This section explains the differences between a voltage tester and a voltmeter. It also explains how to use each of them in a variety of residential electrical applications.

Voltage Testers

The **voltage tester** is a very useful instrument for the residential electrician. This instrument is designed to indicate approximate values of voltage for either direct current (DC) or alternating current (AC) applications. It is not designed to be a precision measuring instrument. The more common types indicate the following voltages: 120, 240, 480, and 600 volts AC and, 125, 250, and 600 volts DC. Many of these instruments will also indicate the **polarity** of DC, which can come in handy when working with low-voltage DC circuits and you are not sure which conductor is the "positive" or which is the "negative."

A voltage tester can be used for a variety of applications, such as identifying the grounded conductor of a circuit, checking for blown fuses, and distinguishing between AC and DC. However, the voltage tester's main job is to test for whether a voltage is present and to provide an approximate value of what that voltage is. The voltage tester is small and rugged, making it easy to carry and store. It fits easily in a tool pouch or may have its own carrying case that can clip onto your belt for easy access.

CAUTION

CAUTION: Remember that a voltage tester gives an approximate voltage amount.

The most common style of voltage tester (Figure 4-2), or **Wiggy** as it is often called in the electrical trades, operates by having the circuit voltage that is present applied to a so-

FROM EXPERIENCE

After the electrical circuit conductors have been "roughed in," the ceiling and wall coverings go on and hide the wiring. If the conductors terminating in each electrical box have not been identified, the electrician installing the switches and receptacles will not know which wire goes where. An easy way to identify them is to use a continuity tester to track down where each wire originates and where it ends.

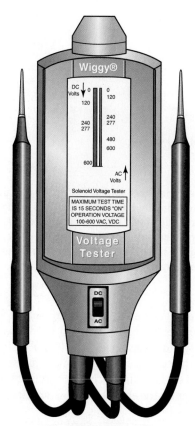

Figure 4–2 **A very common style of voltage tester. It is most often referred to as a "Wiggy."**

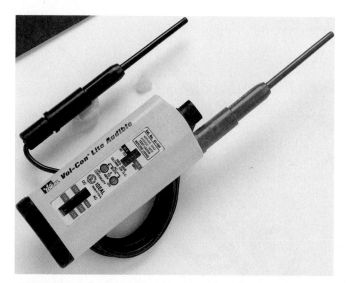

Figure 4–3 **An example of a voltage tester that also can be used as a continuity tester. This tester is an excellent choice for residential electricians.** *Courtesy of Ideal Industries, Inc.*

lenoid coil. A magnetic field is developed that is proportional to the amount of voltage present, and a movable core in the solenoid moves a certain distance, depending upon the strength of the magnetic field. There is an indicator, usually a colored band on the movable core, that will indicate the approximate value of voltage on a scale printed on the face of the tester. This solenoid action also causes the tester to vibrate in your hand. The more voltage present, the harder the tester vibrates. Small lamps may also light to indicate a voltage is present on some models.

Figure 4–3 shows a common solenoid type of voltage tester that is an excellent choice for residential wiring applications. It can indicate AC voltage up to 480 volts and DC voltage up to 240 volts, and as an added feature, it can also be used as a continuity tester. This combination voltage tester and continuity tester will allow an electrician to carry one meter for these two jobs instead of two separate meters.

CAUTION: Make sure to never use a voltage tester on circuits that could exceed the tester voltage rating. You could be seriously injured or even killed.

FROM EXPERIENCE

Many electricians use just the feeling of the vibration to verify that there is a voltage present. They are able to do this because they know that in residential wiring you will encounter either 120-volt or 240-volt AC. The electrician only has to recognize three different vibration levels to know what is going on. No vibration means that no voltage is present, a medium amount of vibration means that 120 volts is present, and a high level of vibration means that 240 volts is present. It is still always best for the electrician to look at the tester to see what approximate value of voltage is indicated.

Some newer styles of voltage testers are classified as digital voltage testers (see Figure 4–4). They tend to be more accurate since they use digital circuitry rather than the electromechanical mechanism found in solenoid testers. Typically, they use light-emitting diodes (LEDs) that light up to indicate a particular measured voltage level. Some manufacturers are even equipping their voltage testers with a liquid crystal display (LCD) digital readout. Of course, the fancier the voltage tester gets, the more it typically costs. A good-quality solenoid voltage tester is an instrument that a residential electrician can use over and over again for many years.

(See page 111 for step-by-step instructions on how to use a voltage tester.)

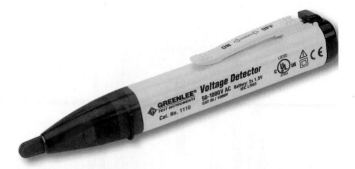

Figure 4–5 **A noncontact voltage tester like this does not need to be connected directly into the circuit. It is brought close to an electrical conductor and will light up and/or make a sound if the conductor is energized. It can be used to detect the presence of voltage at receptacle outlets, lighting fixtures, wires, and cables.** *Courtesy of Greenlee Textron.*

Figure 4–4 **An example of a digital voltage tester. This tester uses digital circuitry for more accurate voltage measurements. This model can also test for continuity.** *Courtesy of Ideal Industries, Inc.*

CAUTION: Always read and follow the instructions that are supplied with the voltage tester.

Another type of voltage tester that has become very popular with residential electricians is the **noncontact voltage tester** (see Figure 4–5). As the name implies, this tester does not need to actually be connected to the electrical circuit to get a reading. All you have to do is bring the tester in close proximity of a conductor, and it will indicate whether it is energized. The indication can occur with the tester lighting up, vibrating in your hand, making an audible tone, or some combination of all three. This type of tester is inexpensive, easy to use, and fits easily into your pocket or tool pouch. It is really no wonder that it has become so popular with residential electricians.

(See page 112 for step-by-step instructions on how to use a noncontact voltage tester.)

Voltmeters

Voltmeters can be used for the same applications as voltage testers. However, voltmeters are much more accurate than voltage testers, so more precise voltage information can be obtained. For example, if the supply voltage to a building is slightly below normal, the voltmeter can indicate this problem where the voltage tester could not. The voltmeter can also be used to determine the exact amount of voltage drop on feeder- and branch-circuit conductors. Again, the voltage tester could not be used.

Because voltmeters are connected in parallel with the circuit or the component being tested, it is necessary that they have relatively high resistance. The internal resistance of the meter keeps the current flowing through the meter to a minimum. The lower the value of current through the meter, the less effect it has on the electrical characteristics of the circuit.

CAUTION: Remember to always connect a voltmeter across (in parallel with) the load.

A meter that uses a moving pointer or needle to indicate a value on a scale is called an **analog meter** (see Figure 4–6). Analog meters used to be the only type of meter available. Now they are pretty much a thing of the past. It is virtually impossible to purchase an analog meter from any of the major meter manufacturers. **Digital meters**, a meter where the indication of the measured value will be given as an actual number in a liquid crystal display (LCD), are the style of meter most often made by meter manufacturers and are the style of choice for today's electrician. However, there are still analog meters being used in the electrical field, and if you ever have to use an analog meter, there are a few things you need to remember. For example, an accurate meter reading is obtained only by standing directly in front of the meter face and looking directly at it. An error (called "parallax error") can occur in your meter reading if you view the meter face from the side. Some higher-quality analog meters have a mirror behind the scale to help compensate for parallax error. Simply adjust the angle of your sight until there is no reflection of the indicating needle in the mirror. This means that you are looking at the meter face "dead on" and that the reading you see is accurate.

The average analog voltmeter is generally between 95 and 98 percent accurate. This range of accuracy is satisfactory for most applications. It is very important, however, that the

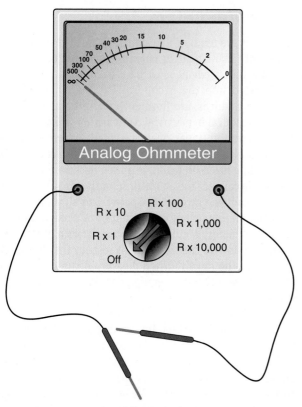

Figure 4–6 An example of an analog meter. The needle points to the value of the quantity being measured.

Figure 4–7 This digital voltmeter uses a large, easy to read digital display to show the measured voltage value for best accuracy. This model has the capability to also be a noncontact voltage tester and a continuity tester. *Courtesy of Greenlee Textron.*

electrician strives to obtain the most accurate reading possible. Today's digital voltmeters (see Figure 4–7) are extremely accurate, and an **auto-ranging** feature always gives the electrician the best accuracy possible in response to the electrical application being tested.

⚠️ **CAUTION**

CAUTION: Always read and follow the instructions that are supplied with the voltmeter.

Voltmeters often have more than one scale. Auto-ranging voltmeters will pick the proper scale for you, but many older voltmeters have a range selector switch that requires you to select the scale manually. This type of meter is called a **manual ranging meter**. It is very important to select the scale that will provide the most accurate measurement. The range selector switch is provided for this purpose. It is advisable to begin with a high scale and work down to the lowest scale so that you do not exceed the range limit of any scale. Setting the range selector switch on the lowest usable scale possible will provide the most accurate reading. It is rare for a digital voltmeter to not be auto-ranging. The auto-ranging feature is a big reason why digital meters are so popular in electrical work.

If you end up using an analog meter of any type, always check to be sure that the indicating needle is pointing to zero when you start to take a measurement. A screw is provided to make the adjustment and is located just below the face of the meter. A very slight turn will cause the needle to move. The needle can be aligned with the zero line on the scale by turning the screw one way or the other. This adjustment may have to be done so that an accurate reading can be taken.

Another point to consider is that when using analog voltmeters on DC electricity, it is very important to maintain proper polarity. Most DC power supplies and meters are color coded to indicate the polarity. Red indicates the positive terminal, and black indicates the negative terminal. If the polarity of the circuit or component is unknown, touch the leads to the terminals while observing the indicating needle. If the indicating needle attempts to move backward, the meter lead connections must be reversed. If the leads are not reversed, damage to the meter, particularly the needle, can occur. If the polarity is reversed when using a digital voltmeter, usually a (−) minus sign is shown in the display and indicates reversed polarity. However, if you do not reverse the leads, no harm will come to the digital meter.

⚠️ **CAUTION**

CAUTION: Do not leave an analog meter connected with the polarity reversed.

(See pages 113–114 for step-by-step instructions on how to use both an analog and a digital voltmeter.)

Ammeters

Ammeters are designed to measure the amount of current flowing in a circuit. Ammeters can be used to locate overloads and open circuits. They can also be used to balance the loads on multiwire circuits and to locate electrical component malfunctions. There are two ammeter types: the in-line and the clamp-on (see Figure 4–8). Clamp-on ammeters are by far the most often used type of ammeter used in residential electrical work and are covered in detail later in this chapter. In-line ammeters are seldom used because they do not have high enough amperage ratings to make them practical in electrical construction work and are not as easy to use as a clamp-on ammeter.

CAUTION

CAUTION: Always read and follow the instructions that are supplied with the ammeter.

In-Line Ammeters

When they are used, in-line ammeters are always connected in series with the circuit or circuit component being tested. This means that a circuit must be disconnected and the ammeter inserted into the circuit so that the current flowing through the circuit is also flowing through the ammeter. It is not always practical in electrical construction work to turn the electrical power off, break open the circuit, insert the ammeter in series with the load, and then reenergize the circuit. After the reading has been taken, it is then necessary to turn the power off, disconnect the ammeter from the circuit, reconnect the circuit, and reapply electrical power. It is all very time consuming and presents many opportunities to make a mistake.

In-line ammeters require that the resistance of the meter be extremely low so that it does not restrict the flow of current through the circuit. When measuring the current flowing through very sensitive equipment, even a slight change in current caused by the ammeter may cause the equipment to malfunction.

Analog in-line ammeters, like analog voltmeters, have an adjustment screw to set the indicating needle to zero. Each time a reading needs to be taken, make sure that the needle starts at zero. Adjust if necessary. Many analog in-line ammeters have mirrors to assist the user in obtaining an accurate reading by overcoming parallax error.

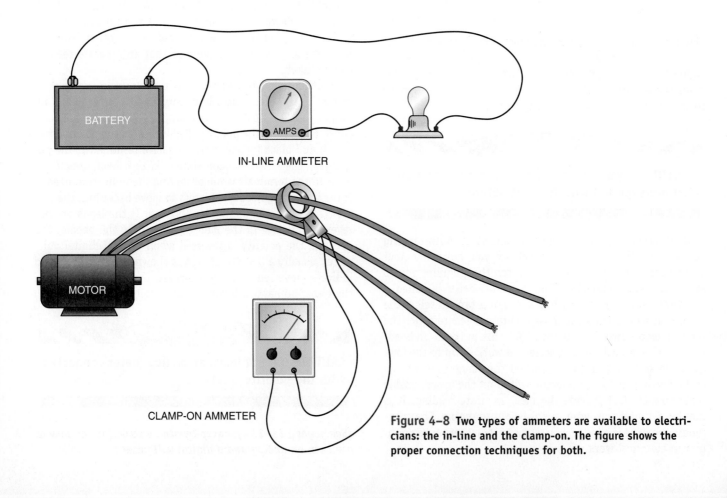

Figure 4–8 Two types of ammeters are available to electricians: the in-line and the clamp-on. The figure shows the proper connection techniques for both.

CAUTION

CAUTION: In-line ammeters should always be connected in series with the circuit or component being tested. If DC is being measured, always check the polarity.

A common application for a clamp-on ammeter is for a residential electrician to measure each "hot" conductor of the service entrance when all loads are connected to see how much current is flowing on each conductor. It is good wiring practice to balance the current flowing on each service conductor. Knowing what the current is on each conductor will allow the electrician to adjust the loads accordingly for proper balancing.

Clamp-On Ammeters

Clamp-on ammeters are much easier to use than in-line ammeters. They are designed with a movable jaw that can be opened, which allows the meter to be clamped around a current-carrying conductor to measure current flow. When current is flowing through a conductor, a magnetic field is set up around that conductor. The more current flowing through the conductor, the stronger the magnetic field will be. The clamp-on ammeter picks up the strength of the magnetic field and converts the magnetic field's strength into a proportional value of current. The display can be analog or digital (see Figure 4–9).

A clamp-on ammeter is a valuable tool for any residential electrician. Meters are now available that provide clamp-on ammeter capabilities as well as being able to measure voltage and resistance. Many of these meters have other capabilities, such as capacitor testing and frequency measurement. These meters are most often referred to as "multimeters" because they can measure several different quantities. Multimeters are covered in greater detail later in this chapter.

(See page 115 for step-by-step instructions on how to use a clamp-on ammeter.)

Ohmmeters and Megohmmeters

Ohmmeters

An **ohmmeter** is used to measure the resistance of a circuit or circuit component (see Figure 4–10). It is not often that a residential electrician will need to take a resistance measurement, but for those times when it is necessary, an electrician should be familiar with the following information.

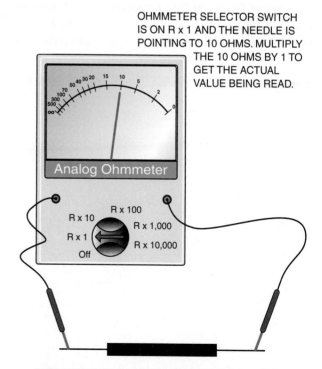

Figure 4–10 An analog ohmmeter being used to measure the resistance of a 10-ohm resistor. The range selector switch is on R × 1, and the needle is pointing to 10 ohms. Multiply the indicated amount (10 ohms) by the range switch value (1) to get the actual value being shown by the meter. 10 ohms × 1 = 10 ohms.

Figure 4–9 Examples of (A) an analog clamp-on ammeter. *Courtesy of Ideal Industries, Inc.* and (B) a digital clamp-on ammeter. *Courtesy of Greenlee Textron.*

CAUTION: Always read and follow the instructions that are supplied with the ohmmeter.

Batteries located in the meter case furnish the power for the operation of an ohmmeter, and just like the other meters discussed in this chapter, there is an analog and a digital model. In the analog model, the ohmmeter scale is designed to be read in the direction opposite other meters—in other words, from right to left, not left to right (see Figure 4–11). When the electrical circuit or component being measured is "open," the indicating needle should point to infinity (indicated with a symbol that looks like the number "8" lying on its side: ∞). The infinity symbol is located on the far left side of the scale. The needle can be aligned with the infinity mark by turning the adjustment screw located in the face of the meter in the same manner as adjusting the needle to point to zero on analog voltmeters and analog ammeters.

CAUTION: It is very important to be sure that the circuit or component is disconnected from its regular power source before connecting an ohmmeter. Connecting an ohmmeter to a circuit that has not been deenergized can result in damage to the meter and possible injury to the user.

Analog ohmmeters have several ranges. The range selector switch must be set on the scale that will provide the most accurate measurement (see Figure 4–12). The ranges are generally indicated as follows: R × 1, R × 10, R × 100, and R × 10,000. If the selector switch is set on R × 1, the value indicated on the scale is the actual value. If the selector switch is set on R × 10, the value indicated on the scale is multiplied by 10. For R × 100, the value indicated on the scale must be multiplied by 100. For R × 10,000, the value indicated on the scale must be multiplied by 10,000. For example, an ohmmeter selector switch is set on the R × 100 setting, and the indicating needle is pointing to the number "10" on the scale. The correct reading is 10 × 100, or 1,000 ohms.

CAUTION: Remember that the analog ohmmeter scale is read from right to left.

The analog ohmmeter is also an excellent continuity tester. Connect the ohmmeter across the circuit or component to be tested. If there is continuity through the circuit or component, the needle will move. A reading of zero generally indicates a **short circuit**. A reading of infinity indicates an **open circuit**.

(See page 116 for step-by-step instructions on how to use an analog ohmmeter.)

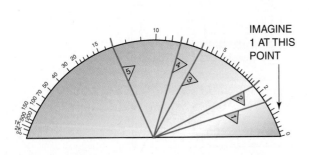

Needle Position	Value
1	1.75
2	2.5
3	7
4	8.5
5	13

Figure 4–11 An example of an analog ohmmeter scale. Notice that the scale reads from the right to the left. Take a look at the five needle positions shown and their values as indicated in the chart. Make sure you understand why each needle position has the value it does.

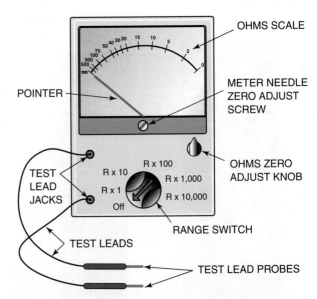

Figure 4–12 The common parts of an analog ohmmeter. Notice that the range selector switch has an "Off" position. Make sure that when the meter is not in use, the switch is set to the "Off" position. This will make sure that the batteries do not lose their charge.

Megohmmeter

A megohmmeter, commonly known by the trade name "Megger," is an instrument used to measure very high values of resistance (see Figure 4–13). Residential electricians will seldom be asked to use a megohmmeter, but they should still be familiar with them. For this reason, the following information is offered.

▰▰▰◣◣◣◣ ⬭ CAUTION ⬭ ◢◢◢◣ ▰▰▰

CAUTION: Always read and follow the instructions that are supplied with the megohmmeter.

Megohmmeters may be used to test the resistance of the insulation on circuit conductors, transformer windings, and motor windings. A megohmmeter is designed to measure the resistance in megohms. One megohm (MΩ) is equal to 1 million ohms.

Visual inspection of insulation and leakage tests with voltmeters are not always reliable. A megohmmeter test is one of the most reliable tests available to an electrician. Insulation tests should be made at the time of the installation and periodically thereafter. For residential circuits and equipment rated at 600 volts or less, the 1,000-volt setting can be used.

A small generator called a "magneto" is contained within the megohmmeter housing. The magneto furnishes the power for the instrument just as batteries do for an ohmmeter. The magneto can be hand powered or driven by batteries. Some megohmmeters use batteries with electronic circuitry to produce the high voltage levels required to measure high values of resistance. Megohmmeters have many different voltage ratings. Some of the most common are designed to operate on one of the following values: 250, 500, and 1,000 volts.

▰▰▰◣◣◣◣ ⬭ CAUTION ⬭ ◢◢◢◣ ▰▰▰

CAUTION: Never touch the test leads of a megohmmeter while a test is being conducted. Also, isolate whatever it is that you are conducting the test on. High voltage is present and could injure you and/or the item you are testing.

▰▰▰◣◣◣◣ ⬭ CAUTION ⬭ ◢◢◢◣ ▰▰▰

CAUTION: Before a megohmmeter is connected to a conductor or a circuit, the circuit must be deenergized. When testing the circuit insulation, the testing is generally done between each conductor and ground. A good ground is a vital part of the testing procedure. The ground connection should be checked with the megohmmeter and with a low-range ohmmeter to ensure good continuity.

(See page 117 for step-by-step instructions on how to use a megohmmeter.)

Multimeters

Multimeters are designed to measure more than one electrical value. For example, the basic volt-ohm milliammeter (VOM) can measure DC and AC voltages, DC and AC current, and resistance. The major advantage of this type of meter is that several different types of test and measurement can be

(A)

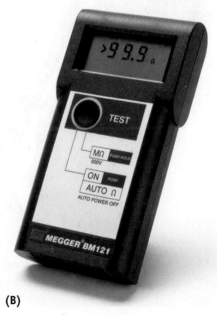

Figure 4–13 A crank-type (A) and a battery-type (B) Megohmmeter. Many electricians refer to any megohmmeter as a "Megger," but "Megger" is a registered trademark of Megger (formerly AVO International), and only their megohmmeters actually bear the Megger name. *Courtesy of Megger.*

(B)

taken with only one meter. This means that electricians can get by with one meter rather than having separate meters for each value that they want to measure. Various accessories such as clamp-on ammeter adapters and temperature probes are available for most multimeters. Analog (Figure 4–14) and digital (Figure 4–15) models are available. However, most major meter manufacturers offer only the digital version. These manufacturers refer to their meters as **DMMs**, or digital multimeters.

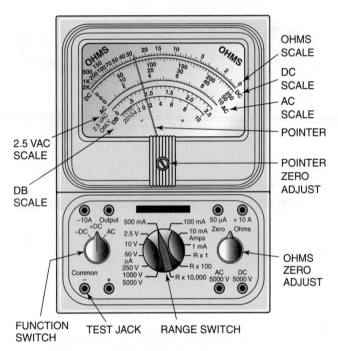

Figure 4–14 An analog multimeter showing the major parts and their locations on the meter.

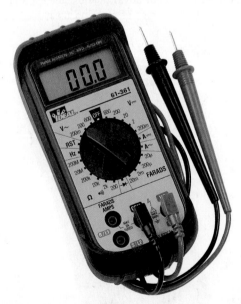

Figure 4–15 A digital multimeter. *Courtesy of Ideal Industries, Inc.*

CAUTION: Always read and follow the instructions that are supplied with the multimeter.

Measurements taken with an analog multimeter are basically done the same way as with an analog voltmeter, analog ammeter, or analog ohmmeter. These operations were all discussed earlier in this chapter. Remember that with the analog multimeter, several electrical quantities can be measured with the meter. You must specify which quantity you are measuring by setting a selector switch on the meter to the desired electrical quantity.

Measurements taken with a digital multimeter are quite easy to do. A selector switch must be used to indicate which electrical quantity you wish to measure. The selector switch is set to point at an "icon" (Figure 4–16) on the meter face that represents the particular electrical quantity you are measuring.

(See page 118 for step-by-step instructions on how to use a digital multimeter.)

Digital multimeters are used extensively in electrical construction work. They are very rugged and take up less room to store than analog multimeters. The DMMs have advanced features, such as auto-zeroing, auto-polarity, auto-ranging, and an automatic shut off. These features help make DMMs easier to use than analog multimeters. They are also much easier to read since the electrical value is displayed directly in digits. An analog multimeter is considered a good "bench meter" and is often found on a repair bench in the shop, but residential electricians will find that a DMM will be the multimeter of choice in the field.

Nonlinear loads, harmonics, and true RMS meters are discussed next. **Nonlinear loads** are load types where the load impedance is not constant. This type of load can cause harmonics to occur on electrical circuits. Examples of these loads include computers and fluorescent lighting fixtures with electronic ballasts. **Harmonics** are frequencies that

ICON	MEASUREMENT
V ⎓	DC VOLTAGE
A ⎓	DC AMPERAGE
V ∿	AC VOLTAGE
A ∿	AC AMPERAGE
Ω	OHMS
➔‖)))	CONTINUITY

Figure 4–16 Common digital multimeter icons.

Figure 4–17 An example of a true RMS multimeter. This type of meter must be used to take accurate measurements on electrical circuits and equipment that have harmonics present. *Courtesy of Ideal Industries, Inc.*

Figure 4–18 A typical single-phase kilowatt-hour meter with dials. This is the most common style of watt-hour meter found in residential applications. *Courtesy of Landis+Gyr Energy Management Inc.*

are multiples of the fundamental frequency (60 hertz). For example, a third harmonic would be at 180 hertz (3 × 60 Hz). Harmonics cause distortion of the basic alternating current waveforms. Regular measuring instruments, called "average" reading instruments, do not respond fast enough to accurately read values caused by harmonic distortion. **True RMS meters** (Figure 4–17) provide accurate measurement of AC values in environments with harmonics, but true RMS meters can still be used when harmonics are not present. True RMS meters tend to be more expensive than "average" reading meters. Residential electrical systems rarely have the types of loads that produce harmonics. However, as computers and other loads that produce harmonics are used more often in residential settings, electricians should be aware of what harmonics are and the type of meter that must be used to accurately measure electrical values when harmonics are present.

Watt-Hour Meters

Residential electricians install the meter enclosure as part of the service entrance that connects the dwelling unit electrical system to the electric utility. (Service entrances and equipment are covered in Section II of this book.) The meter enclosure contains a **kilowatt-hour meter** (Figure 4–18). The local electric utility meter department usually installs the meter into the meter enclosure once the service entrance is done and the dwelling is ready to receive electrical power. This meter measures the amount of electrical energy used by

the dwelling's electrical system. Residential electricians should have a basic understanding of how these meters work and how to read them.

Electrical energy is the product of power and time. Electrical power is measured in watts. Larger amounts of electrical power are measured in kilowatts (1,000 watts is equal to 1 kilowatt). The watt-hour meter measures the amount of power consumed over a specific amount of time. It is a meter that registers the amount of watt-hours delivered by the electric utility to the customer. Because residential customers require a large amount of electrical energy, the standard meter is designed to indicate kilowatt-hours (1 kilowatt-hour is equal to 1,000 watt-hours.)

The kilowatt-hour meter works on the principle of magnetic induction. Moving magnetic fields cause currents to flow in an aluminum disk. These currents, called "eddy currents," produce magnetic fields that interact with the moving magnetic fields, causing the disk to rotate like a small electric motor. The rotating disk drives a gear, which in turn drives a series of smaller gears, which ultimately position an indicating needle on a dial. Kilowatt-hour meters have either four or five dials. Each dial has an indicating needle and a scale from 0 to 9. Many electric utility companies are now using kilowatt-hour meters that have a digital readout (Figure 4–19). This makes the meter easier to read.

(See page 121 for step-by-step instructions on how to read a kilowatt-hour meter.)

Figure 4–19 A digital kilowatt-hour meter. This meter is programmable and uses a digital display to show the total number of kilowatt-hours used by a residential customer. *Courtesy of Landis+Gyr Energy Management Inc.*

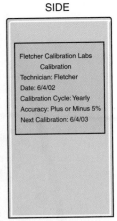

Figure 4–20 An example of a meter showing a calibration sticker. The calibration sticker will tell an electrician when the meter was last calibrated and when the next calibration should take place.

Safety and Meters

Safety is very important for a residential electrician when using test and measurement instruments. Many times a meter is used to make a test or measurement while the electrical circuit is energized. Be sure to follow regular safety procedures when working on energized circuits or equipment. A review of Chapter 1 may be appropriate at this time. Any meter that is not in good working order should not even be taken to the job site, where it could be used and somebody could get seriously injured or killed. Equipment could also become badly damaged or destroyed if an electrician does not follow proper safety procedures when using test and measurement instruments.

Meters that are used on electrical construction sites tend to lose their accuracy over time. This is usually because of the rough handling that sometimes occurs with these meters. Meters that are exposed to hot or cold temperature extremes are also likely to become inaccurate over time. Recalibration is necessary from time to time to bring a meter back to its intended level of accuracy. Stickers are placed on a meter so that an electrician can see when the meter was last calibrated (Figure 4–20). It is up to the electrician to make sure that the meters being used are recalibrated on a regular basis.

Although the electrical dangers encountered in residential wiring may not be as severe as those found in commercial and industrial wiring, they are still present and present a serious safety hazard. Residential electricians should always follow safe meter practices when using meters of any kind. Here are a few safety reminders:

- Always wear safety glasses when using test and measurement instruments.
- Wear rubber gloves when testing or measuring "live" electrical circuits or equipment.
- Never work on energized circuits unless absolutely necessary.
- If you must take measurements on energized circuits, make sure you have been properly trained to work with "live" circuits.
- Do not work alone, especially on "live" circuits. Have a buddy work with you.
- Keep your clothing, hands, and feet as dry as possible when taking measurements.
- Make sure that the meter you are using has a rating that is equal to or exceeds the highest value of electrical quantity you are measuring.

 FROM EXPERIENCE

A residential electrician who got a job in a local paper mill was seriously injured when he used a voltage tester with a 600-volt maximum voltage rating to do a test on a 4,160-volt motor circuit. The meter exploded in his hands, and he suffered serious burns on his face, neck, hands and arms that took many months to heal. He was lucky that he was not killed.

Meter Care and Maintenance

The final subject covered in this chapter has to do with the care and maintenance of test and measurement instruments. The meters previously discussed in this chapter run in price from around $20 for a basic noncontact voltage tester to over $400 for a good-quality true RMS DMM. To make sure that a meter lasts in the harsh environment of residential electrical construction, a few rules should be followed:

- Keep the meters clean and dry.
- Do not store analog meters next to strong magnets; the magnets can cause the meters to become inaccurate.
- All meters are very fragile and should be handled with care. Do not just throw them in your toolbox or on the dashboard of the company truck.
- Do not expose meters to large temperature changes. Too much heat or too much cold can damage a meter.
- Make sure you know the type of circuit you are testing (AC or DC).
- Never let the value being measured exceed the range of the meter.
- Multimeters and ohmmeters will need to have their batteries changed from time to time.

- Many meters have fuses to protect them from exposure to excessive voltage or current values. Replacement of these fuses may have to be done. Check the meter's owner's manual for replacement fuse sizes and the location of the fuses.
- Have measuring instruments recalibrated once a year by a qualified person.

Summary

This chapter presented information on the many measuring instrument types used in residential wiring. The proper use of these meters was also presented. Safe meter practices, as well as common meter care and maintenance techniques, were discussed in detail. Residential electricians must be familiar with a variety of test and measurement instruments. These instruments can provide a wealth of information to a residential electrician. All residential electricians need to know how to use test and measurement instruments in a safe and proper manner for a variety of residential wiring applications.

Procedures

Using a Continuity Tester

- In this example, we will determine if there is a break in a length of cable using a continuity tester.

- Put on safety glasses and observe regular safety procedures.

- Make sure the electrical power is off.

A Strip about 1 inch of insulation from the conductors at each cable end.

- On one end, wire-nut the conductors together.

- At the other end, connect the continuity tester across the conductors.

- If the continuity tester indicates continuity (this could be a light coming on or an audible tone), there is no break in the cable.

- If there is no indication of continuity, there is a break in the cable.

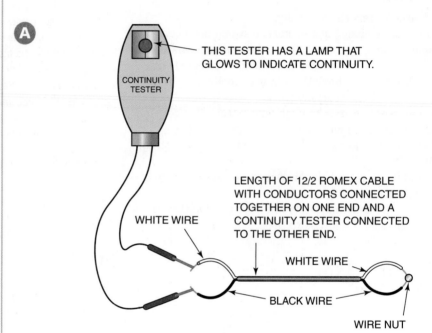

A

CONTINUITY TESTER

THIS TESTER HAS A LAMP THAT GLOWS TO INDICATE CONTINUITY.

LENGTH OF 12/2 ROMEX CABLE WITH CONDUCTORS CONNECTED TOGETHER ON ONE END AND A CONTINUITY TESTER CONNECTED TO THE OTHER END.

WHITE WIRE

WHITE WIRE

BLACK WIRE

WIRE NUT

Procedures   Using a Voltage Tester

- In this example, we will determine which conductor of a circuit is grounded using a voltage tester.

- Put on safety glasses and observe regular safety procedures.

A Connect the tester between one circuit conductor and a well-established ground.

- If the tester indicates a voltage, the conductor being tested is not grounded.

B Continue this procedure with each conductor until zero voltage is indicated between the tested conductor and the known ground. Zero voltage indicates that you have found a grounded circuit conductor.

A

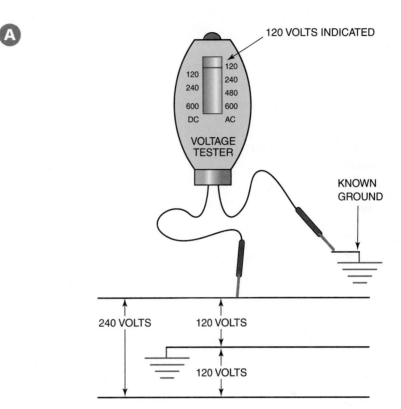

120 VOLTS INDICATED

120 120
240 240
600 480
600
DC AC

VOLTAGE TESTER

KNOWN GROUND

240 VOLTS   120 VOLTS

120 VOLTS

B

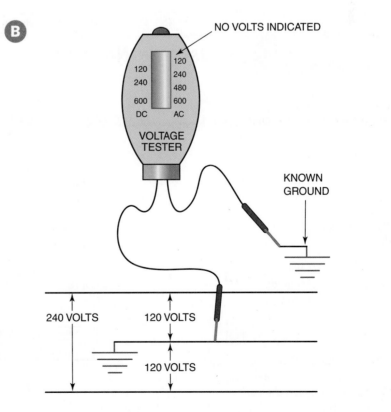

NO VOLTS INDICATED

120 120
240 240
600 480
600
DC AC

VOLTAGE TESTER

KNOWN GROUND

240 VOLTS   120 VOLTS

120 VOLTS

Procedures

Using a Voltage Tester (continued)

- In this example, we will determine the approximate voltage between two conductors using a voltage tester.

- Put on safety glasses and observe regular safety procedures.

C Connect the tester between the two conductors.

- Read the indicated voltage value on the meter. Note: With a solenoid type tester you should also feel a vibration that is another indication of voltage being present.

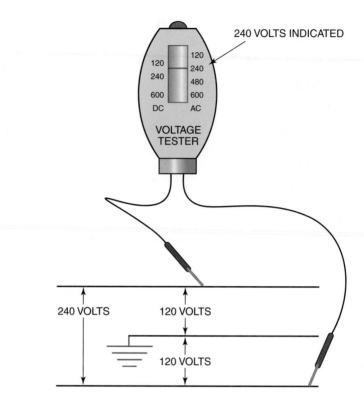

Procedures

Using a Noncontact Voltage Tester

- In this example, we will determine if an electrical conductor is energized using a non-contact voltage tester.

- Put on safety glasses and observe regular safety procedures.

- Identify the conductor to be tested.

A Bring the noncontact voltage tester close to the conductor. Note that some noncontact voltage testers may have to be turned on before using.

- Listen for the audible alarm, observe a light coming on, or feel a vibration to indicate that the conductor is energized.

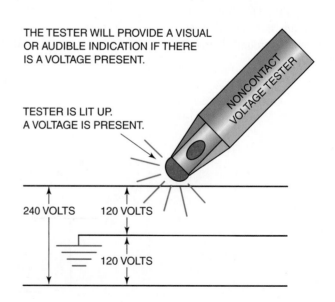

Procedures

Using an Analog Voltmeter

- In this example, we will measure the actual voltage between two conductors.

- Put on safety glasses and observe regular safety procedures.

- Plug in the leads to the meter.

- Set the range selector switch to the highest value available. If there is an AC/DC switch on the meter, put it on AC.

A Attach the leads across the conductors.

- Read the needle setting on the scale. If the indicated value is lower than the next-lower scale setting, change the range selector switch to that value. Work down until you are on the lowest possible scale. This will provide you with the most accurate reading.

- If necessary, multiply the final scale reading by the scale factor to get the actual voltage amount.

A THE METER WILL INDICATE A PRECISE MEASUREMENT OF THE VOLTAGE, FOR EXAMPLE, 123.5 VOLTS.

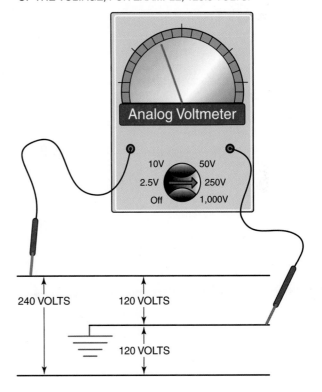

Procedures

Using a Digital Voltmeter

- In this example, we will measure the actual voltage between two conductors.

- Put on safety glasses and observe regular safety procedures.

- Plug in the leads to the meter.

A Attach the leads across the conductors.

- Read the value measured on the digital display.

A

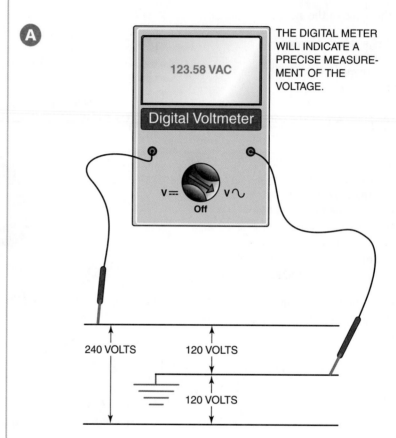

THE DIGITAL METER WILL INDICATE A PRECISE MEASUREMENT OF THE VOLTAGE.

123.58 VAC

Digital Voltmeter

V ⎓   V ∿

Off

240 VOLTS   120 VOLTS

120 VOLTS

Procedures

Using a Clamp-On Ammeter

- In this example, we will be measuring current flow through a conductor with a clamp-on ammeter. Note that you can take a current reading with a clamp-on ammeter clamped around only one conductor. For example, a clamp-on meter will not give a reading when clamped around a two-wire Romex cable.

- Put on safety glasses and observe regular safety procedures.

- If the meter is analog and has a scale selector switch, set it to the highest scale. Skip this step if the meter is digital and has an auto-ranging feature.

 Open the clamping mechanism and clamp it around the conductor.

- Read the displayed value.

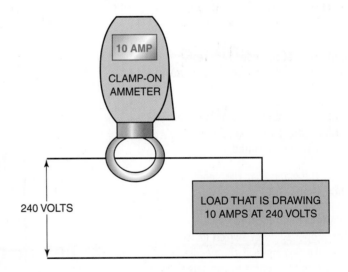

Procedures

Using an Analog Ohmmeter

- When measuring resistance, it is very important to make sure that the circuit is deenergized.

- Put on safety glasses and observe regular safety procedures.

- Plug in the meter leads.

- Set the range selector switch on the desired range.

- Connect the test leads together. The indicating needle should point to zero.

- If the needle is not aligned with the zero line, turn the zero adjustment knob until the needle and the zero line are aligned. Do not leave the test leads connected together because under zero resistance the batteries will deteriorate rapidly.

(A) Connect the test leads across (in parallel with) the component or circuit to be measured.

- Read the value that the needle is indicating.

- Multiply the indicated value by the proper selector switch multiplier. For example, if the needle indicates 20 ohms and the selector switch is on R × 1,000, the measured amount is 20,000 ohms.

(A)

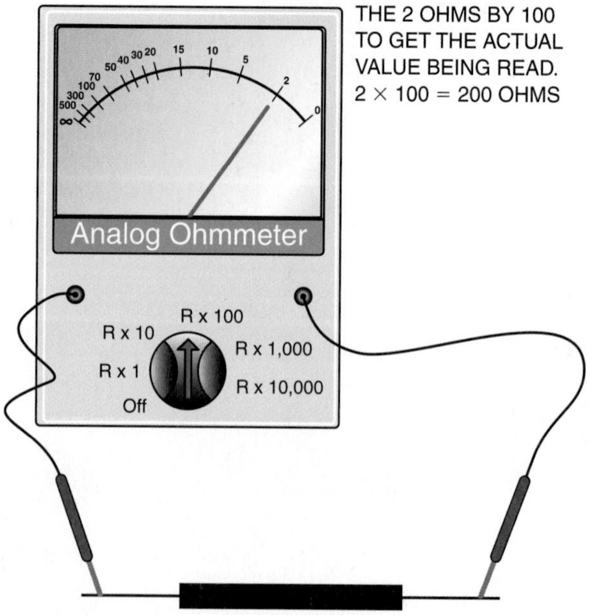

OHMMETER SELECTOR SWITCH IS ON R x 100 AND THE NEEDLE IS POINTING TO 2 OHMS. MULTIPLY THE 2 OHMS BY 100 TO GET THE ACTUAL VALUE BEING READ.
2 × 100 = 200 OHMS

Analog Ohmmeter

R x 10
R x 100
R x 1
R x 1,000
R x 10,000
Off

RESISTOR WITH A RESISTANCE VALUE OF 200 OHMS

Procedures

Using a Megohmmeter

- In this example, we will use a megohmmeter to test the insulation of a 230-volt well pump motor. Note that for motors, generators, transformers, and similar equipment, the minimum resistance for those designed to operate at 1,000 volts or less should be about 1 megohm (1,000,000 ohms).

- Put on safety glasses and observe regular safety procedures.

A Plug in the meter leads.

- Connect one lead to the frame of the motor.

- Connect the other lead to one of the motor conductors.

- Select the 1,000-volt setting and record the reading.

- If the value is 1 megohm or above, the insulation is fine. If the value is less than 1 megohm, the insulation may be deteriorating. Extremely low ohmic values indicate a short between the conductor and the motor frame.

- Continue to test the other motor leads and record the values.

A

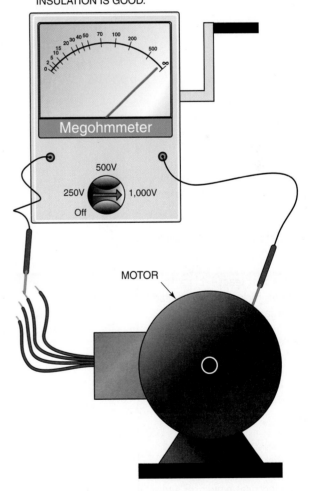

A HIGH VALUE INDICATES THAT THE MOTOR WINDING INSULATION IS GOOD.

Megohmmeter

MOTOR

Procedures

Using a Digital Multimeter

- Here we will measure alternating current voltage between two conductors.

- Put on safety glasses and observe regular safety procedures.

 Plug in the test leads—black to the common jack and the red lead to the volt/ohm jack.

- Set the selector switch to the AC voltage icon.

- Connect the test leads in parallel with the conductors.

- Read the displayed voltage.

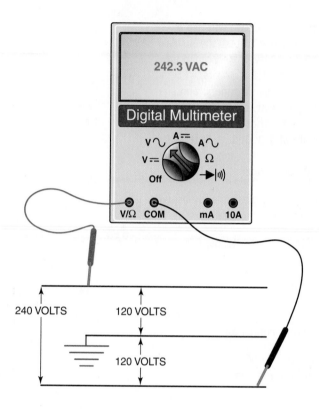

- Here we will measure the current draw of a small (200-milliamp or less) AC load.

- Put on safety glasses and observe regular safety procedures.

 Plug in the test leads—black to the common jack and the red lead to the mA jack.

- Set the selector switch to the AC current icon.

- Turn the electrical power off and connect the meter in series with the load.

- Turn the electrical power on and read the displayed current.

- Turn the electrical power off and disconnect the meter.

- Reconnect the circuit and reenergize.

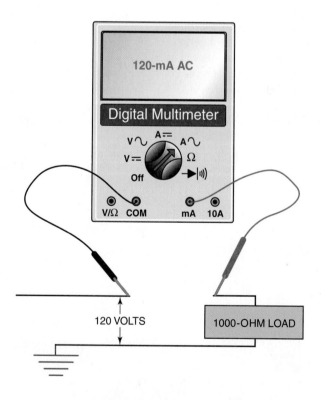

- Here we will measure the current draw of a 10-amp or less AC load using the in-line capabilities of the meter.

- Put on safety glasses and observe regular safety procedures.

 Plug in the test leads—black to the common jack and the red lead to the 10A jack.

- Set the selector switch to the AC current icon.

- Turn the electrical power off and connect the meter in series with the load.

- Turn the electrical power on and read the displayed current.

- Turn the electrical power off and disconnect the meter.

- Reconnect the circuit and reenergize.

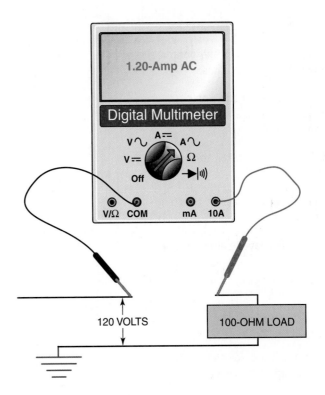

- Here we will measure the current draw of an AC load using a clamp-on ammeter accessory for the meter.

- Put on safety glasses and observe regular safety procedures.

 Plug in the clamp-on ammeter accessory according to the manufacturer's instructions.

- Set the selector switch to the AC current icon.

- Open the clamping mechanism and clamp it around the conductor.

- Read the displayed value.

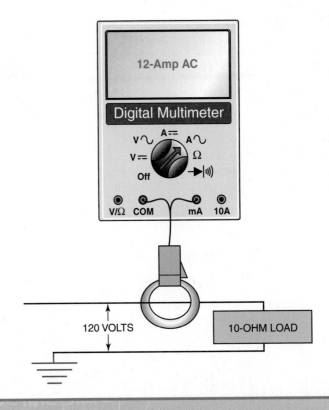

Procedures

Using a Digital Multimeter (continued)

- Here we will measure resistance.

- Put on safety glasses and observe regular safety procedures.

 Plug in the test leads—black to the common jack and the red lead to the volt/ohm jack.

- Set the selector switch to the "ohms" icon.

- Connect the leads across the item to be tested.

- Read the displayed value.

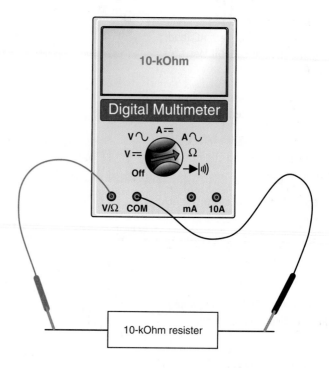

- Here we will test for continuity using a digital multimeter.

- Put on safety glasses and observe regular safety procedures.

 Plug in the test leads—black to the common jack and the red lead to the volt/ohm jack.

- Set the selector switch to the "continuity" icon.

- Connect the leads across the item to be tested.

- Listen for the audible tone. If there is a tone, there is continuity. If there is no tone, there is no continuity.

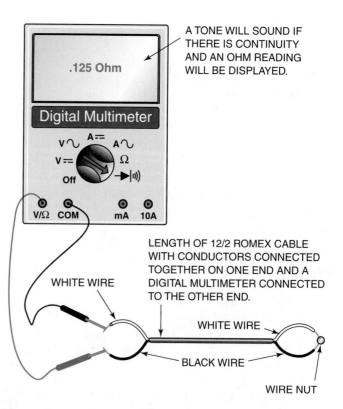

Procedures

Reading a Kilowatt-Hour Meter

- To calculate the kilowatt-hours used since the last meter reading, simply subtract the previous reading from the new reading. This is how local electric utility companies determine your kilowatt-hour usage per month. Your electric bill is based on this value.

- Put on safety glasses and observe regular safety procedures.

A Begin with the right-hand dial and, working to the left, identify the number indicated. The dials indicate units, tens, hundreds, thousands, and ten thousands. Remember, if the indicating needle is between two numbers, the smaller of the two numbers is always read.

- Once all of the numbers have been identified, read the number from left to right. This is the number of kilowatt-hours that the meter has registered.

A

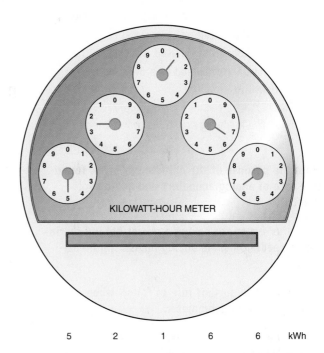

KILOWATT-HOUR METER

5   2   1   6   6   kWh

Review Questions

Directions: Answer the following items by circling the "T" for a true statement or the "F" for a false statement.

T F **1** An analog ohmmeter scale is read from left to right.

T F **2** A voltmeter is used to measure the amount of current flowing in a circuit.

T F **3** Multimeters can measure a variety of DC and AC electrical values.

T F **4** A voltage tester is a very accurate meter when compared to a voltmeter.

T F **5** A kilowatt-hour meter is used to measure the value of voltage delivered to a dwelling by an electric utility.

Directions: Answer the following questions with clear and complete responses.

6 Explain the purpose of the adjustment screw found on the face of analog ammeters and analog voltmeters.

7 Explain the reason for having a mirror behind the indicating needle on some analog measuring instruments.

8 State the most important rule to follow when using an ohmmeter or megohmmeter.

9 Describe the difference between an "In-line" ammeter and a "Clamp-on" ammeter.

10 List three important rules to follow in the care and maintenance of meters.

1.

2.

3.

Directions: Circle the letter of the word or phrase that best completes the statement.

11 **An ammeter is used to measure _____.**

a. resistance

b. current

c. voltage

d. ohms

12 **Voltage is measured using a(n) _____.**

a. ammeter

b. megohmmeter

c. continuity tester

d. voltmeter

13 **A megohmmeter measures high values of resistance in _____.**

a. megohms

b. milliohms

c. megawatts

d. kilovolts

14 **A Wiggy is an electrical trade name for a(n) _____.**

a. voltmeter

b. voltage tester

c. ammeter

d. ohmmeter

15 **Kilowatt-hour meters are used to measure _____.**

a. electrical power

b. electrical energy

c. large amounts of voltage

d. large amounts of amperage

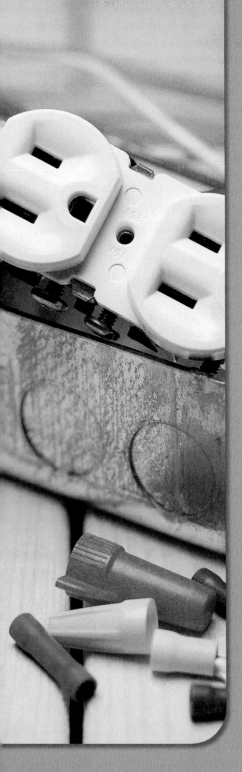

Chapter 5

Understanding Residential Building Plans

Every residential electrician must be familiar with residential building plans. Most new construction and remodel electrical wiring jobs require an electrician to follow a building plan when installing the electrical system. This chapter introduces you to residential building plans and the common architectural symbols found on them. Building plan specifications and basic residential framing are also covered.

OBJECTIVES

Upon completion of this chapter, the student should be able to:

- ✪ demonstrate an understanding of residential building plans.
- ✪ identify common architectural symbols found on residential building plans.
- ✪ demonstrate an understanding of residential building plan specifications.
- ✪ demonstrate an understanding of basic residential framing methods and components.

Glossary of Terms

architect a qualified person who creates and designs drawings for a residential construction project

balloon frame a type of frame in which studs are continuous from the foundation sill to the roof; this type of framing is found mostly in older homes

band joist the framing member used to stiffen the ends of floor joists where they rest on the sill

blueprint architectural drawings used to represent a residential building; it is a copy of the original drawings of the building

bottom plate the lowest horizontal part of a wall frame which rests on the subfloor

break lines lines used to show that part of the actual object is longer than what the drawing is depicting

ceiling joists the horizontal framing members that rest on top of the wall framework and form the ceiling structure; in a two-story house, the first-floor ceiling joists are the second floor's floor joists

centerline a series of short and long dashes used to designate the center of items, such as windows and doors

computer-aided drafting (CAD) the making of building drawings using a computer

detail drawing a part of the building plan that shows an enlarged view of a specific area

dimension a measurement of length, width, or height shown on a building plan

dimension line a line on a building plan with a measurement that indicates the dimension of a particular object

draft-stops also called "fire-stops"; the material used to reduce the size of framing cavities in order to slow the spread of fire; in wood frame construction, it consists of full-width dimension lumber placed between studs or joists

electrical drawings a part of the building plan that shows the electrical supply and distribution for the building electrical system

elevation drawing a drawing that shows the side of the house that faces in a particular direction; for example, the north elevation drawing shows the side of the house that is facing north

extension lines lines used to extend but not actually touch object lines and have the dimension lines drawn between them

floor joists horizontal framing members that attach to the sill plate and form the structural support for the floor and walls

floor plan a part of the building plan that shows a bird's-eye view of the layout of each room

footing the concrete base on which a dwelling foundation is constructed; it is located below grade

foundation the base of the structure, usually poured concrete or concrete block, on which the framework of the house is built; it sits on the footing

framing the building "skeleton" that provides the structural framework of the house

girders heavy beams that support the inner ends of floor joists

hidden line a line on a building plan that shows an object hidden by another object on the plan; hidden lines are drawn using a dashed line

leader a solid line that may or may not be drawn at an angle and has an arrow on the end of it; it is used to connect a note or dimension to a part of the building

legend a part of a building plan that describes the various symbols and abbreviations used on the plan

object line a solid dark line that is used to show the main outline of the building

platform frame a method of wood frame construction in which the walls are erected on a previously constructed floor deck or platform

rafters part of the roof structure that is supported by the top plate of the

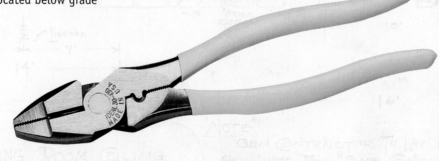

wall sections; the roof sheathing is secured to the rafters and then covered with shingles or other roofing material to form the roof

ribbon a narrow board placed flush in wooden studs of a balloon frame to support floor joists

scale the ratio of the size of a drawn object and the object's actual size

schedule a table used on building plans to provide information about specific equipment or materials used in the construction of the house

sectional drawing a part of the building plan that shows a cross-sectional view of a specific part of the dwelling

sheathing boards sheet material like plywood that is fastened to studs and rafters; the wall or roofing finish material will be attached to the sheathing

sill a length of wood that sets on top of the foundation and provides a place to attach the floor joists

specifications a part of the building plan that provides more specific details about the construction of the building

subfloor the first layer of floor material that covers the floor joists; usually 4-by-8-plywood or particleboard

symbol a standardized drawing on the building plan that shows the location and type of a particular material or component

top plate the top horizontal part of a wall framework

wall studs the parts that form the vertical framework of a wall section

Overview of Residential Building Plans

A residential building plan is prepared by an architect and consists of a set of drawings that craft people use as a guide to build the house. The drawings are usually drawn by a draftsman or a **computer-aided drafting (CAD)** operator. The building plan is often called various names, such as prints, **blueprints**, drawings, construction drawings, or working drawings. It is not that important what name is used to refer to the building plans, but it is very important for a residential electrician to know what the various parts of a typical building plan are and how to read and interpret the information found on them. The main parts of a building plan that a residential electrician should be familiar with are the floor plan, elevation drawings, sectional drawings, detail drawings, electrical drawings, schedules, and specifications. Understanding the types of drawing lines used on building plans and understanding the scale of a drawing is also very important.

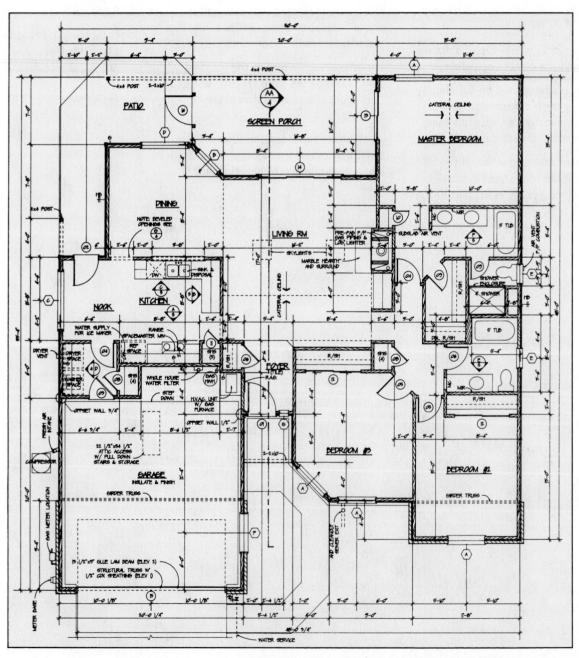

Figure 5–1 A floor plan shows the location of walls, partitions, windows, doors, and other building features. Building dimensions are found on the floor plan.

Floor Plans

The **floor plan** is a drawing that shows building details from a view directly above the house. A "bird's-eye" view is how it is often described. It is drawn to show the house as if a horizontal cut was made through the building at window height and then the top taken off. You are left with a view of the bottom half. Usually, each floor and the basement (if included) will have a floor plan drawing. The floor plans for each floor are called "first-floor plan," "second-floor plan," and so on. The floor plan for the basement is called the "basement plan." Figure 5–1 shows a typical floor plan.

Floor plans show the length and the width of the floor it is depicting. **Dimensions** are drawn on the floor plans and can be used by electricians to determine the exact size and location of various parts of the building structure, like doors, windows, and walls. This information is necessary when determining where to install wiring and electrical equipment.

Elevation Drawings

An **elevation drawing** shows the side of the house that is facing a certain direction. It may show the height, length, and width of the house. Electricians can use elevation drawings to determine the heights of windows, doors, porches, and other parts of the structure. This type of information is not available on floor plans but is needed when installing electrical items, such as outside lighting fixtures and outside receptacles. Figure 5–2 shows an example of an elevation drawing.

Sectional Drawings

A **sectional drawing** is a view that allows you to see the inside of a building. The view shown by a sectional drawing can be described as follows. Imagine that a Sawzall® has been used to cut off the side of a house. When the side of the house is moved away, you are left with a cutaway view of the rooms and the structural members of that part of the house. Figure 5–3 shows an example of a sectional drawing.

The point on the floor plan or the elevation drawing that is depicted by the sectional drawing is shown with a dashed line with arrows on the ends and is called a "section line." Because a building plan may have several sectional drawings, the section lines are distinguished with letters or numbers located at the end of the arrows on the section lines. A typical sectional drawing may be labeled as "Section A–A" or "Section B–B."

The sectional drawings contain information that is important to an electrician. For example, a wall section drawing can allow the electrician to determine how he may run cable, or a sectional drawing of a floor may show an electrician how thick the wood will be for her to drill.

Detail Drawings

A **detail drawing** shows very specific details of a particular part of the building structure. It is an enlarged view that makes details much easier to see than in a sectional drawing. They are usually located on the same plan sheet where the building feature appears. If they are shown on a separate sheet, they are numbered to refer back to a particular

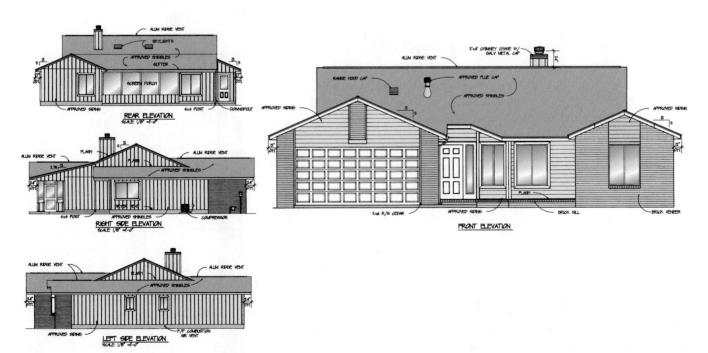

Figure 5–2 An elevation drawing shows a side view of the structure.

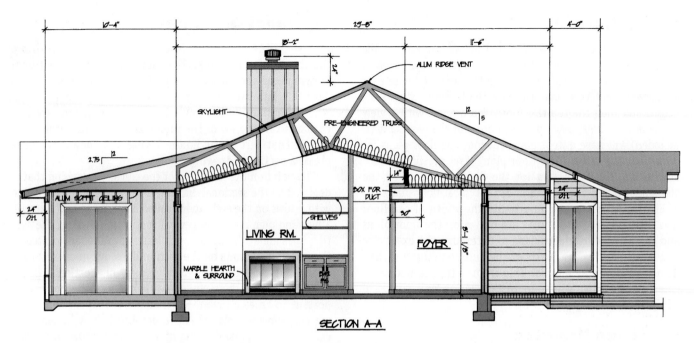

Figure 5–3 A sectional drawing shows a cutaway view of a certain part of the structure. *Courtesy of Barton Homes, Inc.*

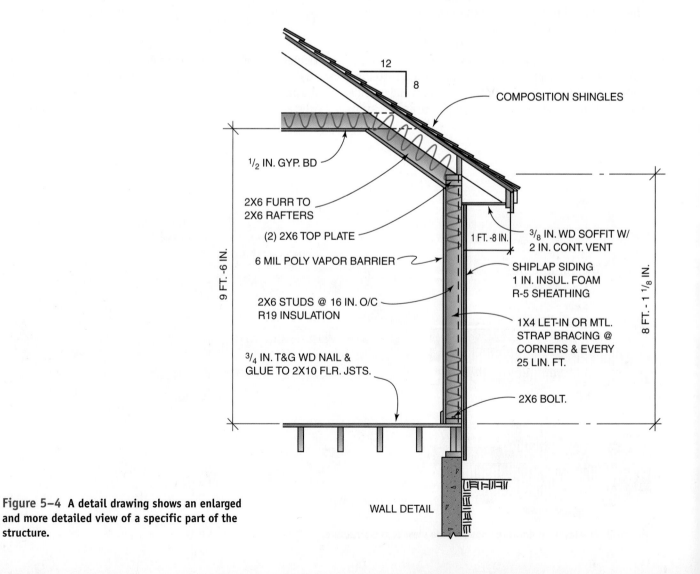

Figure 5–4 A detail drawing shows an enlarged and more detailed view of a specific part of the structure.

location on the building plan. Figure 5–4 shows an example of a detail drawing.

Electricians can use detail drawings to determine exact locations for the placement of electrical equipment. For example, a detail drawing of the kitchen cabinets can help an electrician locate where receptacles and switches should be located. Remember, the wiring and boxes will have to be installed before the cabinets are in, so it is very important to locate the electrical equipment properly from the plans so that when the cabinets are installed, they do not cover a switch or receptacle.

Electrical Drawings

The most important part of the building plan for electricians is the **electrical drawings**. They show exactly what is required of the electrician for the complete installation of the electrical system. Electrical symbols are used on the plans to depict electrical equipment and devices. They are used as a type of shorthand to show the electrician which electrical items are required and where they are located. Using the symbols makes the plan less cluttered and easier to read. Specific electrical symbols are covered later in this chapter. Figure 5–5 shows an example of an electrical plan.

Electrical contractors use the electrical plans to estimate the amount of material and labor needed to install the electrical system. This amount is used to project a total cost for the electrical system installation and is used in the bidding process. The electrical plans also provide a good map of the electrical system and can be consulted in the future if and when problems arise.

Schedules

Schedules are used to list and describe various items used in the construction of the building. The schedules are usually set up in table form. For example, door and window schedules list the sizes and other pertinent information about the various types of doors and windows used in the building.

Electricians rely on schedules to provide specific information about electrical equipment that needs to be installed in the building. For example, a lighting fixture schedule lists and describes the various types of lighting fixtures used in the house. This schedule also tells the electrician what type and how many lamps are used with each lighting fixture. Figure 5–6 shows an example of a lighting fixture schedule.

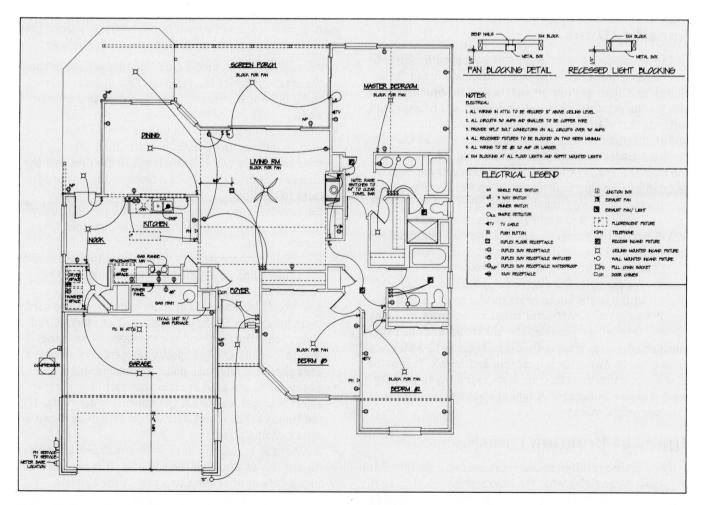

Figure 5–5 An electrical floor plan shows an electrician what is required for the electrical system being installed in the house.

Symbol	Number	Manufacturer and Catalog Number	Mounting	Lamps
A	2	Lightolier 10234	Wall	2 40-watt T-12 CWX
B	4	Lightolier 1234	Surface	4 40-watt T-12 WWX
C	1	Progress 32-486	Surface	1 100-watt medium base incandescent
D	1	Progress 63-8992	Surface	5 60-watt medium base incandescent
E	3	Lithonia 12002-10	Recessed	3 75-watt medium base reflector
F	3	Hunter Paddle Fan 1-3625-77	Surface	3 60-watt medium base incandescent
G	2	Nutone Fan/Light Model 162	Recessed	1 60-watt medium base incandescent

Figure 5–6 A lighting fixture schedule. Schedules provide more detailed information on certain pieces of equipment and materials being installed in a house. In this example, detailed information is given on the lighting fixtures to be installed as part of the electrical system.

Specifications

The building plan **specifications** help provide clarity to the building plans. Only so much information can be included in a floor plan or an electrical plan. Specifications provide the extra details about equipment and construction methods that are not in the regular building plans.

Specifications provide detailed information to all the construction trades involved with the building. Of course, it is the electrical specifications that we electricians are most interested in. The electrical "specs" often include the specific manufacturer's catalog numbers and other information so that the electrical items will be the right size and type as well as having the proper electrical rating. The specifications are also used by the electrical contractor to help with the estimate of what it will cost to install the proposed electrical system. For example, the electrical specifications may state that all wiring in the house be no smaller than 12-AWG copper. If an electrical contractor based his estimate on using 14-AWG wire where possible, the bid would end up being much smaller than what it should be because 12-AWG costs more than 14-AWG wire to purchase and install. It pays to read the specifications through from beginning to end. Figure 5–7 shows an example of typical electrical specifications for a residential wiring job.

Types of Drawing Lines

There are many different line types used on a set of building plans. Recognizing what the lines represent will make it easier for a residential electrician to understand the building

plan. Figure 5–8 shows several of the common drawing lines used on building plans. These lines are used as follows:

- An **object line** is a solid dark line that is used to show the main outline of the building. This includes exterior walls, interior partitions, porches, patios, and interior walls.
- **Dimension lines** are thin unbroken lines that are used to indicate the length or width of an object. Arrows are usually placed at each end of the line, and the dimension value is placed in a break of the line or just close to the line.
- **Extension lines** are used to extend but not actually touch object lines and have the dimension lines drawn between them.
- **Hidden lines** are straight dashed lines that are used to show lines of an object that are not visible from the view shown in the plan.
- A **centerline** is a series of short and long dashes used to designate the center of items, such as windows and doors. Sometimes you will see the dashed line going though the letter "C" to specify the center of an object. They provide a reference point for dimensioning.
- **Break lines** are used to show that part of the actual object is longer than what the drawing is depicting. The full length of the object may not be able to be drawn in some building plans.
- A **leader** is a solid line that is usually drawn at an angle and has an arrow on the end of it. It is used to connect a note or dimension to a part of the building shown in the drawing.

1. **GENERAL:** The "General Clause and Conditions" shall be and are hereby made a part of this division.

2. **SCOPE:** The electrical contractor shall furnish and install a complete electrical system as shown on the drawings and/or in the specifications. Where there is no mention of the responsible party to furnish, install, or wire for a specific item on the electrical drawings, the electrical contractor will be responsible completely for all purchases and labor for a complete operating system for this item.

3. **WORKMANSHIP:** All work shall be executed in a neat and workmanlike manner. All exposed conduits shall be routed parallel or perpendicular to walls and structural members. Junction boxes shall be securely fastened, set true and plumb, and flush with finished surface when wiring method is concealed.

4. **LOCATION OF OUTLETS:** The electrical contractor shall verify location, heights, outlet and switch arrangements, and equipment prior to rough-in. No additions to the contract sum will be permitted for outlets in wrong locations, in conflict with other work, and so on. The owner reserves the right to relocate any device up to 10 feet (3.0 m) prior to rough-in, without any charge by the electrical contractor.

5. **CODES:** The electrical installation is to be made in accordance with the latest edition of the National Electrical Code (NEC), all local electrical codes, and the utility company's requirements.

6. **MATERIALS:** All materials shall be new and shall be listed and bear the appropriate label of Underwriters Laboratories, Inc., or another nationally recognized testing laboratory for the specific purpose. The material shall be of the size and type specified on the drawings and/or in the specifications.

7. **WIRING METHOD:** Wiring, unless otherwise specified, shall be nometallic-sheathed cable, armored cable, or electrical metallic tubing (EMT), adequately sized and installed according to the latest edition of the NEC and local ordinances.

8. **PERMITS AND INSPECTION FEES:** The electrical contractor shall pay for all permit fees, plan review fees, license fees, inspection fees, and taxes applicable to the electrical installation and shall be included in the base bid as part of this contract.

9. **TEMPORARY WIRING:** The electrical contractor shall furnish and install all temporary wiring for handheld tools and construction lighting per latest OSHA standards and Article 527, NEC, and include all costs in base bid.

10. **NUMBER OF OUTLETS PER CIRCUIT:** In general, not more than 10 lighting and/or receptacle outlets shall be connected to any one lighting branch circuit. Exceptions may be made in the case of low-current-consuming outlets.

11. **CONDUCTOR SIZE:** General lighting branch circuits shall be 14-AWG copper-protected by 15-ampere overcurrent devices. Small appliance circuits shall be 12-AWG copper-protected by 20-ampere overcurrent devices. All other circuits: conductors and overcurrent devices as required by the NEC.

Figure 5–7 An example of a plan's electrical specifications. Specifications provide additional written information that help explain the building drawings to the craft people constructing the building. Electricians are interested mainly in the electrical specifications, but other areas of the plan specifications may be helpful.

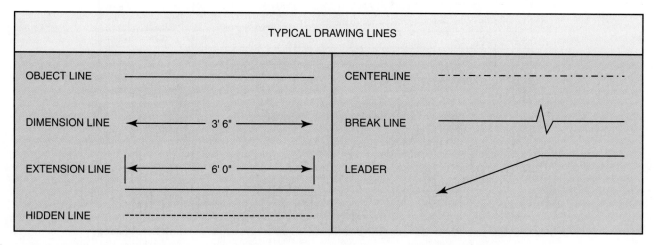

Figure 5–8 These lines represent some of the common drawing lines used on a set of building plans.

Scale

If a building were drawn to its true size, it obviously would not fit on a piece of paper. If a building were drawn just so it would fit on a piece of paper, the drawing would be very much out of proportion. To enable a drawing of a building to be put on a piece of paper and still keep everything in proportion, the building is drawn to some reduced **scale**. All dimensions of the building will be drawn smaller than the actual size and will be reduced in the same proportion. Most residential plans are drawn to a scale of 1/4-inch = 1 feet, 0 inches. This means that each 1/4-inch on the drawing would equal 1 foot on the actual building. If a part of the building plan used a 1/8-inch = 1 foot, 0 inch scale, 1/8-inch on the drawing would equal 1 foot of the actual building.

Electricians need to know what the scale of the drawing is so that they can get an accurate measurement for where electrical equipment will be located. The scale to which the drawing has been done can be usually found in the title block (Figure 5–9). The title block is usually located in the lower-right-hand corner of the drawing. In addition to the drawing scale, a title block contains other information, such as the name of the building project, the address of the project, the name of the architectural firm, the date of completion, the drawing sheet number, and a general description of the drawing.

Common Architectural Symbols

There are many architectural **symbols** used on building plans to depict everything from a kitchen sink to a window. Notes are used to provide additional information about a particular symbol. The symbols and notes help keep the plan

FROM EXPERIENCE

Because most residential construction plans are drawn to a 1/4-inch = 1 foot, 0 inch scale, electricians are able to use a folding rule or their tape measure to get accurate dimensions from the building plans. Each 1/4-inch on the tape or rule represents 1 foot, each 1/8-inch (half of 1/4 in.) equals 6-inch and each 1/16-inch (half of 1/8 in.) equals 3-inches. Going the other way, 1 inch (4 × 1/4 in.) on the tape or rule will equal 4 feet; 2 inches (8 × 1/4 in.) on the tape or rule will equal 8 feet.

from becoming so cluttered with information that it would be impossible to read. Each trade has its own set of symbols that are used to identify items associated with that trade. You should learn to recognize the common symbols used by other trades on residential building plans so that you can make sure that these items do not interfere with your installation of the electrical system. Figure 5–10 shows some common architectural symbols that residential electricians should be familiar with.

Electrical Symbols

As we discussed earlier in this chapter, the electrical drawing is the most important part of the building plan for an electrician. The electrical drawing contains many electrical symbols that show the location and type of electrical equipment required to be installed as part of the electrical system. Recognizing what each symbol represents is a very important skill for an electrician to have.

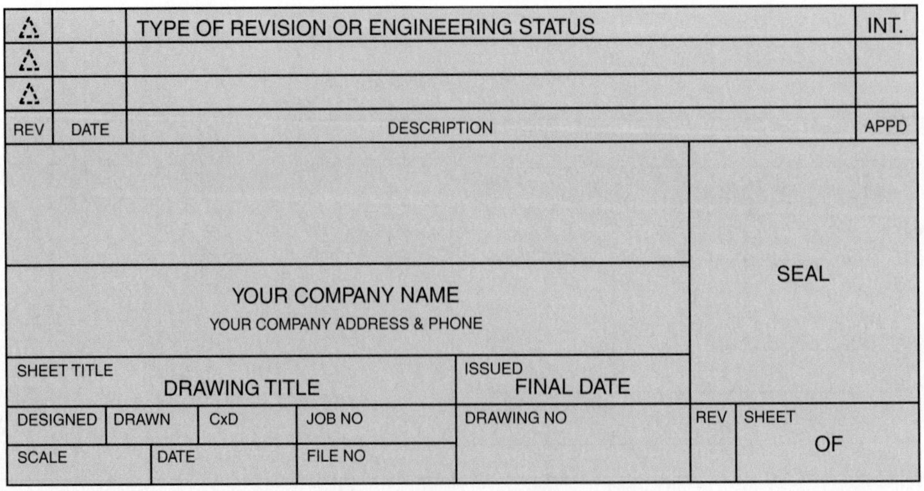

⚠	TYPE OF REVISION OR ENGINEERING STATUS					INT.
⚠						
⚠						
REV	DATE	DESCRIPTION				APPD
		YOUR COMPANY NAME YOUR COMPANY ADDRESS & PHONE			SEAL	
SHEET TITLE		DRAWING TITLE	ISSUED	FINAL DATE		
DESIGNED	DRAWN	CxD	JOB NO	DRAWING NO	REV	SHEET OF
SCALE		DATE	FILE NO			

Figure 5–9 A title block is usually located in the lower-right-hand corner of a building drawing. It contains some very important information for an electrician, such as the scale of the drawing.

COMMON ARCHITECTURAL BUILDING PLAN SYMBOLS

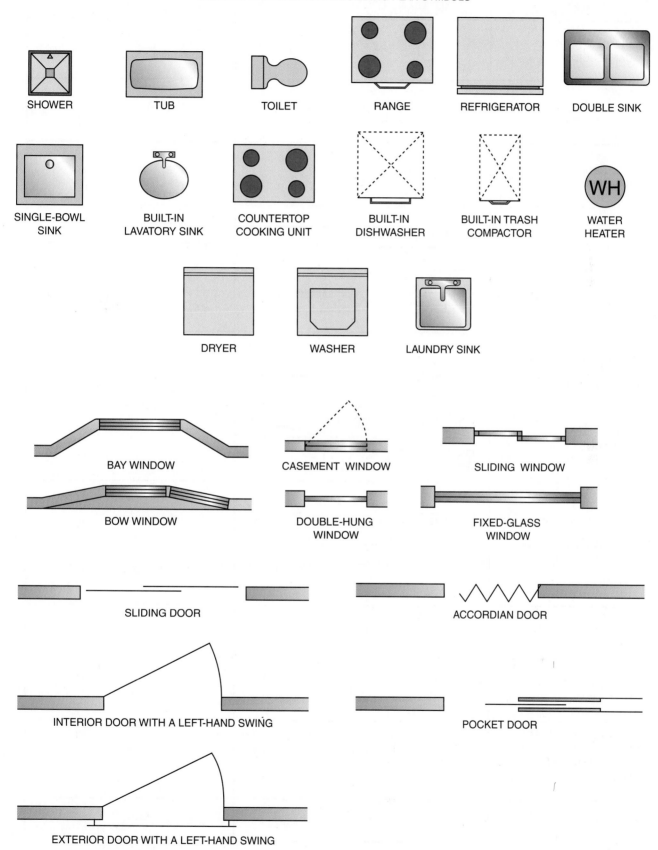

Figure 5–10 Common residential architectural symbols.

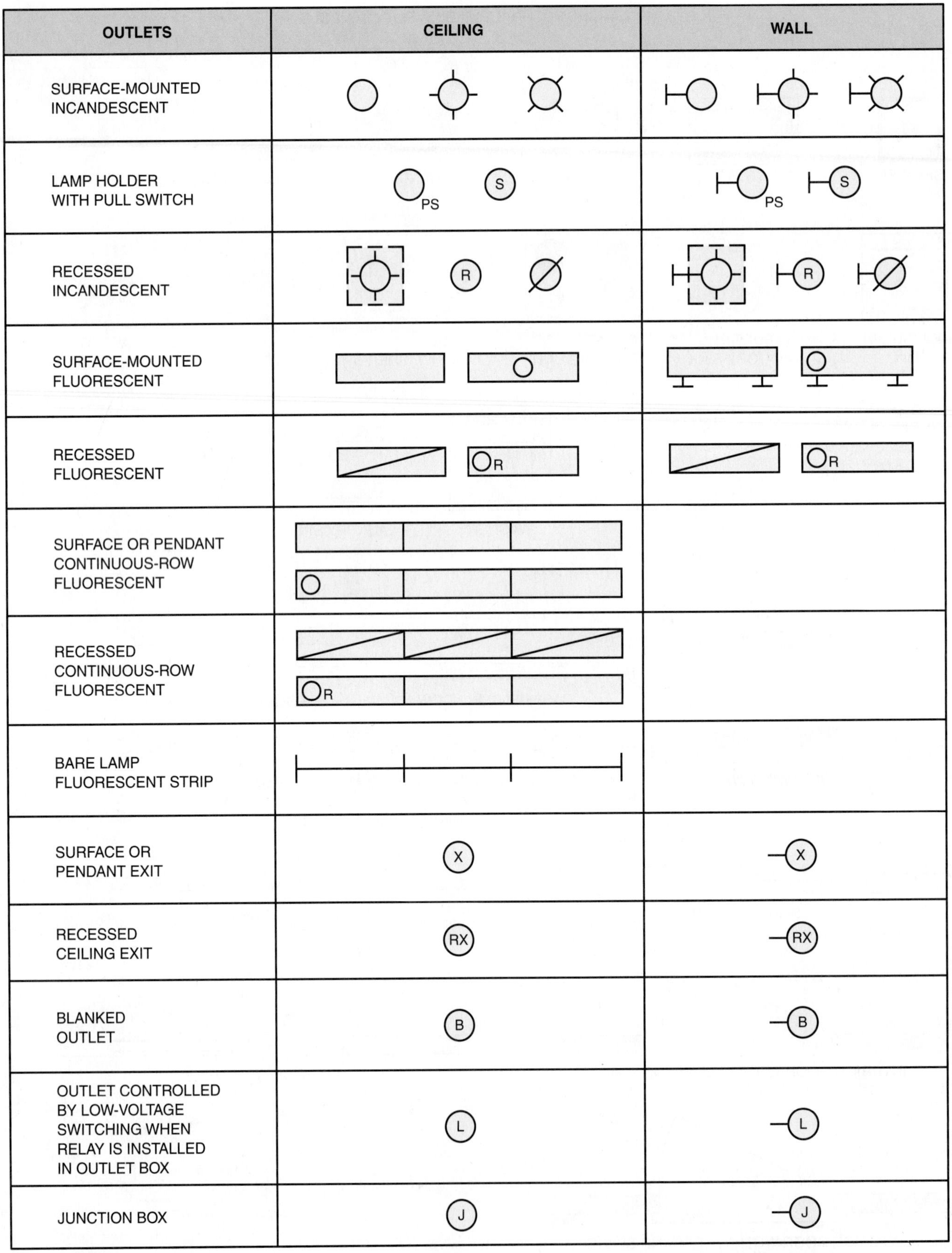
OUTLETS
CEILING
WALL
SURFACE-MOUNTED INCANDESCENT
LAMP HOLDER WITH PULL SWITCH
PS
S
PS
S
RECESSED INCANDESCENT
R
R
SURFACE-MOUNTED FLUORESCENT
RECESSED FLUORESCENT
R
R
SURFACE OR PENDANT CONTINUOUS-ROW FLUORESCENT
RECESSED CONTINUOUS-ROW FLUORESCENT
R
BARE LAMP FLUORESCENT STRIP
SURFACE OR PENDANT EXIT
X
X
RECESSED CEILING EXIT
RX
RX
BLANKED OUTLET
B
B
OUTLET CONTROLLED BY LOW-VOLTAGE SWITCHING WHEN RELAY IS INSTALLED IN OUTLET BOX
L
L
JUNCTION BOX
J
J

Figure 5–11 Lighting outlet symbols.

The American National Standards Institute (ANSI) has published a standard titled *Symbols for Electrical Construction Drawings*. This document shows standard electrical symbols for use on electrical drawings. Figure 5–11 shows the common electrical symbols for lighting outlets, Figure 5–12 shows the common electrical symbols for receptacle outlets, and Figure 5–13 shows common symbols used to show switch types. A number of other symbols used to represent common pieces of electrical equipment are shown in Figure 5–14. Most draftsmen or CAD operators who draw the electrical plans will use the ANSI electrical symbols. However, plans may have symbols that are not standard. If this is the case, a **legend** is usually included in the plans, listing the symbols used on the building plans and what they all mean.

There are a few symbols used on electrical plans to indicate electrical wiring. A curved dashed line on a plan that goes from a switch symbol to a lighting outlet symbol indicates that the outlet is controlled by that switch (Figure 5–15). This line is always curved to eliminate confusion about whether it is a hidden line or a switch leg. You may remember that a hidden line is also dashed, but it is always drawn straight. A curved solid line is used in a cabling diagram to represent the wiring method used. If the curved solid line has an arrow on the end of it, it is a "home run" and indicates that wiring from that point goes all the way to the load center. A number next to the "home run" symbol will indicate its circuit number. Slashes on the curved solid line are sometimes used in a cable diagram to indicate how many conductors are in the cable. A good cabling diagram can make the installation of the wiring very easy. It also is helpful in troubleshooting a circuit in the event of a problem anytime in the future. Figure 5–16 shows the difference between a cabling diagram and a wiring diagram.

Residential Framing Basics

Residential electricians need to become familiar with the basic structural framework of a house. During the rough-in stage of the residential wiring system installation, electrical boxes will need to be mounted on building **framing** members, and wiring methods will have to be installed on or through building framing members. Knowing how a house is constructed will allow an electrician to install the wiring system in a safe and efficient manner.

Wood construction is still the most often used framing type, but metal framing is being used more and more in residential construction. Whether the framing type is wood or metal, the structural parts that electricians should know

are pretty much the same. There are two construction framing methods that electricians will encounter most often. The **platform frame**, sometimes called the "western frame," is the most common method used in today's new home construction (Figure 5–17). In this type of construction, the floor is built first, and then the walls are erected on top of it. If there is a second floor, it is simply built the same way as the first floor and placed on top of the completed first floor. The other common framing method is called the **balloon frame** (Figure 5–18). This construction method is not often used today but was used extensively in the construction of many existing homes. In balloon framing, the wall studs and the first-floor joists both rest on the sill. If there is a second floor, the second-floor joists rest on a 1-by-4-inch **ribbon** that is cut flush with the inside edges of the studs.

Knowing the names of the structural parts and the role they play will help you better understand where and how to install the electrical system. The following major structural parts are used in houses that are built with either the platform framing or the balloon framing method:

- **Rafters** are used to form the roof structure of the building and are supported by the top plate.
- **Ceiling joists** are horizontal framing members that sit on top of the wall framing. They form the structural framework for the ceiling. If there is a floor above, these ceiling joists become **floor joists** and also form the structural framework for the floor above.
- **Draft-stops**, commonly called "fire-stops," are used to curtail the spread of fire in a house. They are usually pieces of lumber that are placed between studs or joists to block the path of fire through the cavities formed between studs or joists.
- The **top plate** is located at the top of the wall framework. It is usually a 2-by-4-inch or a 2-by-6-inch piece of lumber. Some construction methods call for two pieces of lumber to form the top plate. When two pieces of lumber are used, the common name for this structural member is the "double plate."
- **Wall studs** are used to form the vertical section of the wall framework. Interior walls are usually 2-by-4-inch lumber but may be 2-by-6-inch.
- The **bottom plate** is the bottom of the wall framework and rests on the top of the subfloor. It is usually a 2-by-4-inch or a 2-by-6-inch piece of lumber.
- The **subfloor** is the first layer of flooring that covers the floor joists and is usually made of 4-by-8-feet sheet of plywood or particleboard. Some areas of the country may still be using 1-by-6-inch boards for the subfloor.
- **Floor joists** are horizontal framing members that attach to the sill and form the structural support for the floor and walls. Floor joists are usually 2-by-8-, 2-by-10-, or 2-by-12-inch pieces of lumber.

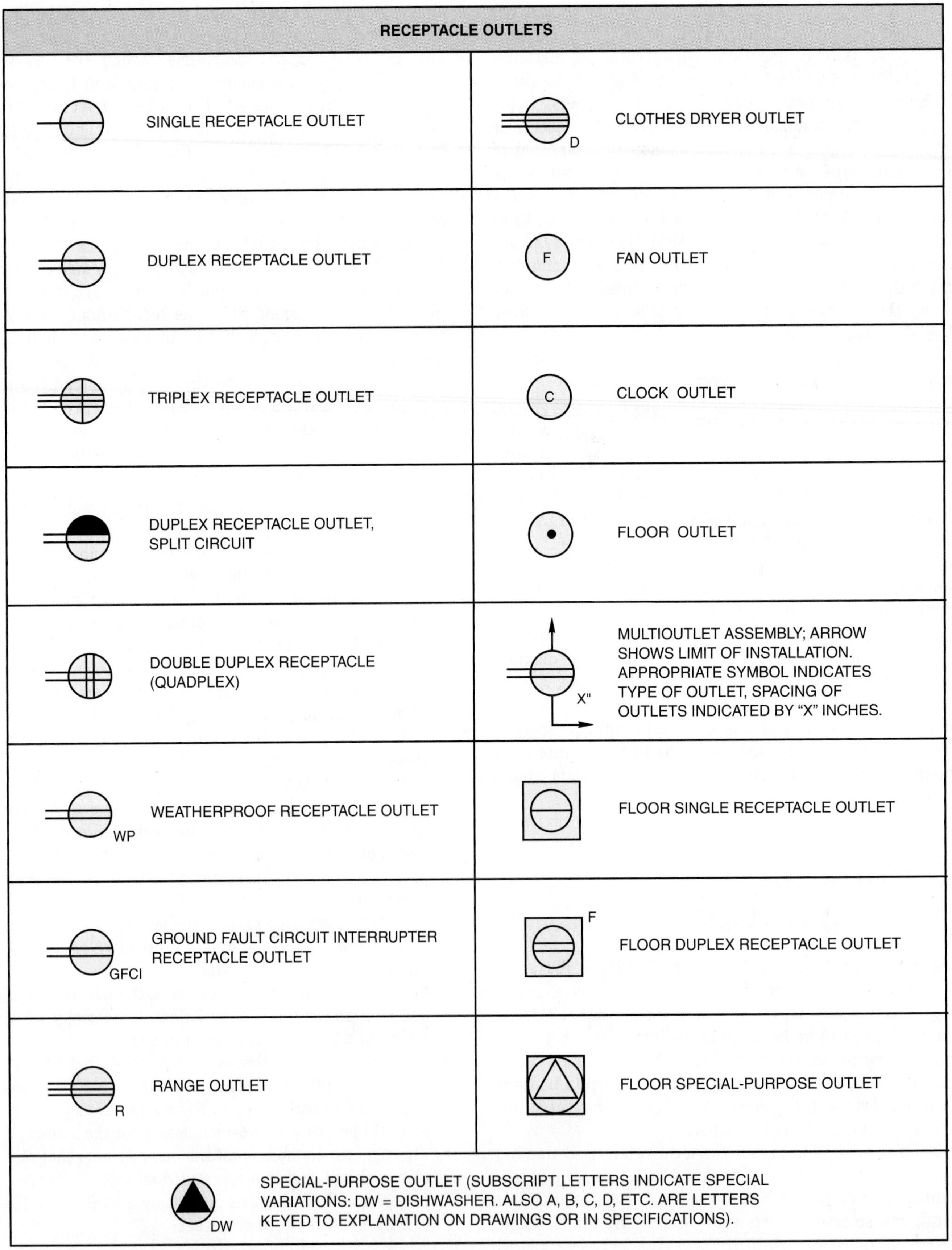

RECEPTACLE OUTLETS	
SINGLE RECEPTACLE OUTLET	CLOTHES DRYER OUTLET
DUPLEX RECEPTACLE OUTLET	FAN OUTLET
TRIPLEX RECEPTACLE OUTLET	CLOCK OUTLET
DUPLEX RECEPTACLE OUTLET, SPLIT CIRCUIT	FLOOR OUTLET
DOUBLE DUPLEX RECEPTACLE (QUADPLEX)	MULTIOUTLET ASSEMBLY; ARROW SHOWS LIMIT OF INSTALLATION. APPROPRIATE SYMBOL INDICATES TYPE OF OUTLET, SPACING OF OUTLETS INDICATED BY "X" INCHES.
WEATHERPROOF RECEPTACLE OUTLET	FLOOR SINGLE RECEPTACLE OUTLET
GROUND FAULT CIRCUIT INTERRUPTER RECEPTACLE OUTLET	FLOOR DUPLEX RECEPTACLE OUTLET
RANGE OUTLET	FLOOR SPECIAL-PURPOSE OUTLET
SPECIAL-PURPOSE OUTLET (SUBSCRIPT LETTERS INDICATE SPECIAL VARIATIONS: DW = DISHWASHER. ALSO A, B, C, D, ETC. ARE LETTERS KEYED TO EXPLANATION ON DRAWINGS OR IN SPECIFICATIONS).	

Figure 5-12 Receptacle outlet symbols.

SWITCH SYMBOLS	
S	SINGLE-POLE SWITCH
S_2	DOUBLE-POLE SWITCH
S_3	THREE-WAY SWITCH
S_4	FOUR-WAY SWITCH
S_D	DOOR SWITCH
S_{DS}	DIMMER SWITCH
S_G	GLOW SWITCH TOGGLE— GLOWS IN OFF POSITION
S_K	KEY-OPERATED SWITCH
S_{KP}	KEY SWITCH WITH PILOT LIGHT
S_{LV}	LOW-VOLTAGE SWITCH
S_{LM}	LOW-VOLTAGE MASTER SWITCH
S_{MC}	MOMENTARY-CONTACT SWITCH
 ◇M	OCCUPANCY SENSOR—WALL MOUNTED WITH OFF-AUTO OVERRIDE SWITCH
 ◇M P	OCCUPANCY SENSOR—CEILING MOUNTED "P" INDICATES MULTIPLE SWITCHES WIRE-IN PARALLEL
S_P	SWITCH WITH PILOT LIGHT ON WHEN SWITCH IS ON
S_T	TIMER SWITCH
S_R	VARIABLE-SPEED SWITCH
S_{WP}	WEATHERPROOF SWITCH

Figure 5–13 Electrical switch symbols.

- The **girders** in a house are the heavy beams that support the inner ends of the floor joists. A girder can be metal or wood. A wooden girder is usually made from several 2-by-10-inch or 2-by-12-inch lengths of lumber fastened together. There are two ways that a girder can support the floor joists: The joists can rest on top of the girder, or joist hangers can be nailed to the sides of the girder, and the hangers support the joists.
- The **band joist** is the framing member in platform framing that is used to stiffen the ends of the floor joists. It is normally the same size as the floor joists.
- The **sill** is a piece of wood that lies on the top of the foundation and provides a place to attach the floor joists.
- **Sheathing boards** are sheet material (like plywood) that are attached to the outside of studs or rafters and add rigidity to the framed structure. The roof and wall outside finish material is attached to the sheathing.
- A **foundation**, usually made of poured concrete or concrete block, is the part of the house that supports the framework of the building.
- The **footing** is located at the bottom of the foundation and provides a good base for the foundation to be constructed.

Summary

This chapter introduced you to the parts of a residential building plan that are most important to the electrician who will be installing the electrical system. Common residential architectural symbols were covered with special emphasis on electrical symbols. An overview of building specifications was given, and basic residential construction framing was covered. Reading and interpreting the information on a building plan are important skills for a residential electrician to have. As with most acquired skills, the more practice you have doing it, the more proficient you get at it. Your skill in reading and interpreting building plans will increase the more you practice. Remember the basics as covered in this chapter, and you should be able to figure out what most residential plans are showing. When in doubt, always ask your job site supervisor.

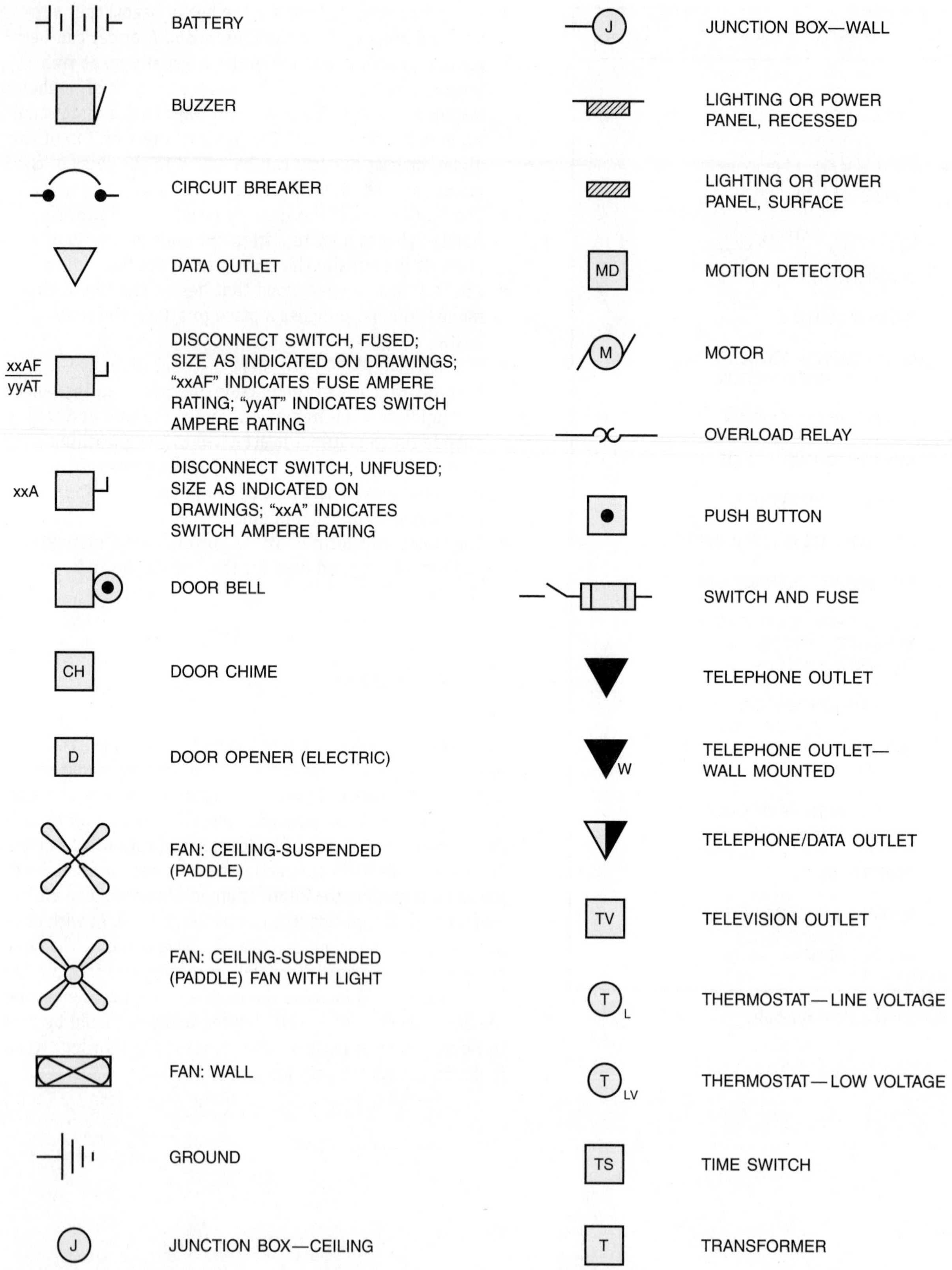
BATTERY
BUZZER
CIRCUIT BREAKER
DATA OUTLET
xxAF
yyAT
DISCONNECT SWITCH, FUSED; SIZE AS INDICATED ON DRAWINGS; "xxAF" INDICATES FUSE AMPERE RATING; "yyAT" INDICATES SWITCH AMPERE RATING
xxA
DISCONNECT SWITCH, UNFUSED; SIZE AS INDICATED ON DRAWINGS; "xxA" INDICATES SWITCH AMPERE RATING
DOOR BELL
CH
DOOR CHIME
D
DOOR OPENER (ELECTRIC)
FAN: CEILING-SUSPENDED (PADDLE)
FAN: CEILING-SUSPENDED (PADDLE) FAN WITH LIGHT
FAN: WALL
GROUND
J
JUNCTION BOX—CEILING
J
JUNCTION BOX—WALL
LIGHTING OR POWER PANEL, RECESSED
LIGHTING OR POWER PANEL, SURFACE
MD
MOTION DETECTOR
M
MOTOR
OVERLOAD RELAY
PUSH BUTTON
SWITCH AND FUSE
TELEPHONE OUTLET
W
TELEPHONE OUTLET—WALL MOUNTED
TELEPHONE/DATA OUTLET
TV
TELEVISION OUTLET
T
L
THERMOSTAT—LINE VOLTAGE
T
LV
THERMOSTAT—LOW VOLTAGE
TS
TIME SWITCH
T
TRANSFORMER

Figure 5–14 Miscellaneous electrical symbols.

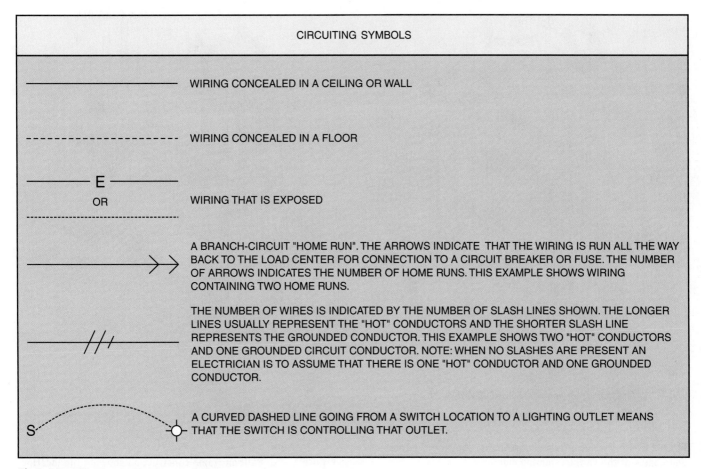

CIRCUITING SYMBOLS

WIRING CONCEALED IN A CEILING OR WALL

WIRING CONCEALED IN A FLOOR

E OR

WIRING THAT IS EXPOSED

A BRANCH-CIRCUIT "HOME RUN". THE ARROWS INDICATE THAT THE WIRING IS RUN ALL THE WAY BACK TO THE LOAD CENTER FOR CONNECTION TO A CIRCUIT BREAKER OR FUSE. THE NUMBER OF ARROWS INDICATES THE NUMBER OF HOME RUNS. THIS EXAMPLE SHOWS WIRING CONTAINING TWO HOME RUNS.

THE NUMBER OF WIRES IS INDICATED BY THE NUMBER OF SLASH LINES SHOWN. THE LONGER LINES USUALLY REPRESENT THE "HOT" CONDUCTORS AND THE SHORTER SLASH LINE REPRESENTS THE GROUNDED CONDUCTOR. THIS EXAMPLE SHOWS TWO "HOT" CONDUCTORS AND ONE GROUNDED CIRCUIT CONDUCTOR. NOTE: WHEN NO SLASHES ARE PRESENT AN ELECTRICIAN IS TO ASSUME THAT THERE IS ONE "HOT" CONDUCTOR AND ONE GROUNDED CONDUCTOR.

S A CURVED DASHED LINE GOING FROM A SWITCH LOCATION TO A LIGHTING OUTLET MEANS THAT THE SWITCH IS CONTROLLING THAT OUTLET.

Figure 5–15 Electrical wiring symbols.

CABLING DIAGRAM VERSUS A WIRING DIAGRAM

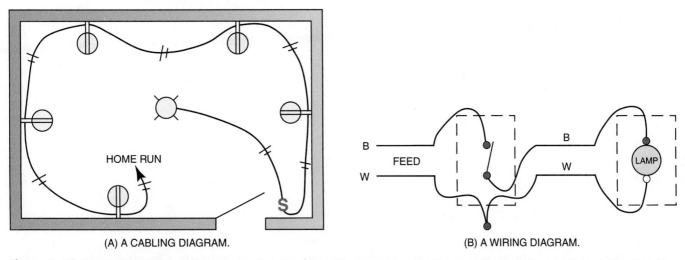

(A) A CABLING DIAGRAM.

(B) A WIRING DIAGRAM.

Figure 5–16 The cabling diagram (A) indicates that two-wire cable is used to wire the room. It should be very clear to the electrician that a two-conductor cable is to be run between the electrical boxes. The "home run" is also clearly indicated, and it will be up to the electrician to run the cable from the receptacle box indicated to the load center. Notice that the cabling diagram does not show the actual connections to be made at the electrical boxes. The wiring diagram (B) will show an electrician exactly where the conductors terminate. New electricians find wiring diagrams helpful when wiring circuits and equipment. As you become more experienced, you will not need to rely on wiring diagrams as much.

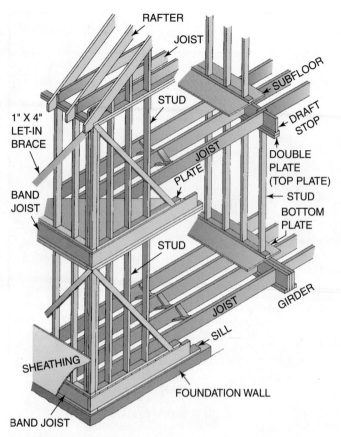

Figure 5–17 Platform frame construction showing the location and names of the various framing members.

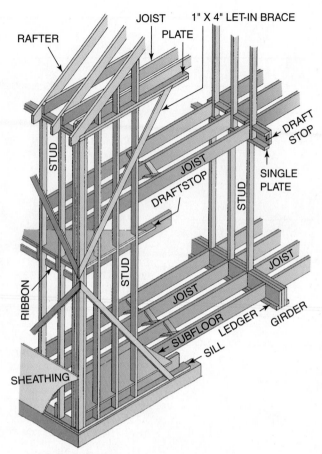

Figure 5–18 Balloon frame construction showing the location and names of the various framing members.

Review Questions

Directions: Answer the following items with clear and complete responses.

1. Describe an elevation drawing.

2. Describe a detail drawing.

3. Describe a floor plan.

4. Describe an electrical floor plan.

5. What is the purpose of a legend on a set of building plans?

6. The scale on a building plan is ¼-inch = 1 foot, 0 inch. An electrician uses a tape measure to measure the length of a wall on the plan. If the length on the tape measure is 4½ inches, determine the actual length of the wall.

7. Describe the purpose of a dimension line.

8. A hidden line shows part of the building that is not visible on the drawing. How are hidden lines drawn?

9. On an electrical floor plan, dashed lines are sometimes drawn from switches to various outlets. What do the dashed lines represent?

10. Describe the purpose of the plan specifications.

Directions: In the spaces provided, draw the electrical symbol for each of the items listed.

11. _____ Duplex receptacle outlet

12. _____ Single-pole switch

13. _____ Clock outlet

14. _____ Three-way switch

15. _____ Four-way switch

16. _____ Ceiling lighting outlet

17. _____ Ceiling junction box

18. _____ Floor receptacle outlet

19. _____ Single-receptacle outlet

20. _____ Special-purpose outlet (dishwasher)

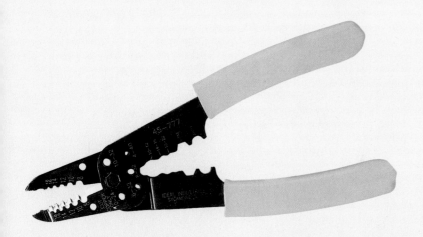

Residential Service Entrances and Equipment

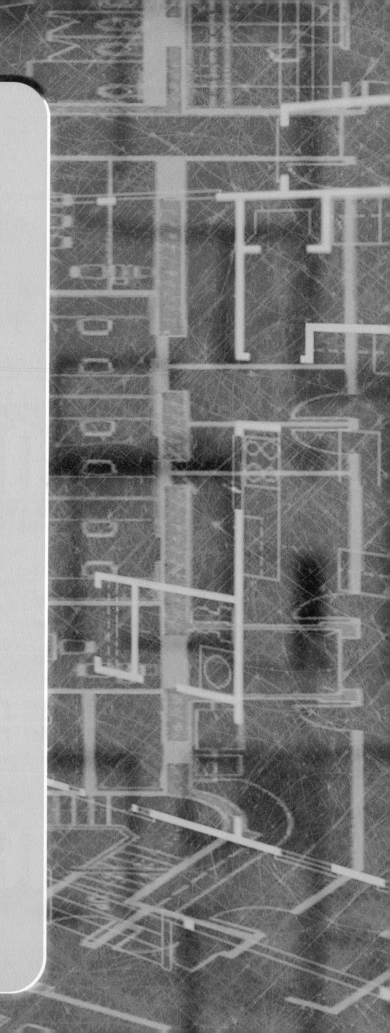

Chapter 6

Introduction to Residential Service Entrances

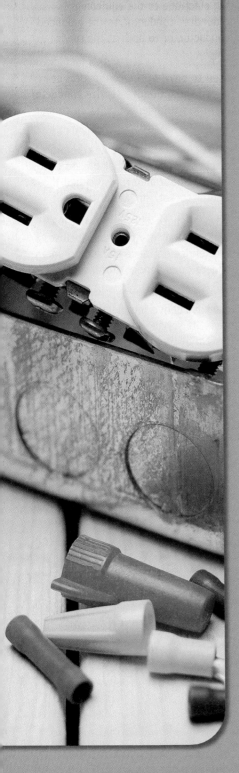

One of the most important parts of a residential electrical system is the service entrance. The service entrance provides a way for the home electrical system to get electrical power from the electric utility company. This chapter discusses the most common items involved in preparing and planning for the installation of a residential service entrance, service entrance terminology, and several important National Electrical Code® rules that electricians must apply when installing a residential service entrance.

OBJECTIVES

Upon completion of this chapter, the student should be able to:

- ⊗ demonstrate an understanding of an overhead and an underground residential service entrance.
- ⊗ demonstrate an understanding of how to establish temporary and permanent power with an electric utility company.
- ⊗ define common residential service entrance terms.
- ⊗ demonstrate an understanding of *National Electrical Code*® (*NEC*®) requirements for residential service entrances.
- ⊗ demonstrate an understanding of grounding and bonding requirements for residential service entrances.
- ⊗ list several *NEC*® requirements that pertain to residential service entrances.
- ⊗ demonstrate an understanding of common electric utility company requirements.

Glossary of Terms

accessible, readily (readily accessible) capable of being reached quickly for operation, renewal, or inspections without requiring a person to climb over or remove obstacles or to use portable ladders

bonding the permanent joining of metal parts to form an electrically conductive path that ensures electrical continuity and the capacity to conduct safely any current likely to be imposed on the metal object; the purpose of bonding is to establish an effective path for fault current that facilitates the operation of the overcurrent protective device

bonding jumper a conductor used to ensure electrical conductivity between metal parts that are required to be electrically connected

concentric knockout a series of removable metal rings that allow the knockout size to vary according to how many of the metal rings are removed; the center of the knockout hole stays the same as more rings are removed; some standard residential wiring sizes are ½, ¾, 1, 1¼, 1½, 2, and 2½ inches

drip loop an intentional loop put in service entrance conductors at the point where they extend from a weatherhead; the drip loop conducts rainwater to a lower point than the weatherhead, helping to ensure that no water will drip down the service entrance conductors and into the meter enclosure

eccentric knockout a series of removable metal rings that allow a knockout size to vary according to how many of the metal rings are removed; the center of the knockout hole changes as more metal rings are

removed; common sizes are the same as for concentric knockouts

equipment a general term including material, fittings, devices, appliances, luminaires (lighting fixtures), apparatus, and other parts used in connection with an electrical installation

equipment-grounding conductor the conductor used to connect the non-current-carrying metal parts of equipment, raceways, and other enclosures to the system-grounded conductor, the grounding electrode conductor, or both at the service equipment

ground a conducting connection, whether intentional or accidental, between an electrical circuit or equipment and the earth or to some conducting body that serves in place of the earth

grounded connected to earth or to some conducting body that serves in place of the earth

grounded conductor a system or circuit conductor that is intentionally grounded

grounding conductor a conductor used to connect equipment or the grounded conductor of a wiring system to a grounding electrode or electrodes

grounding electrode a part of the building service entrance that connects the grounded service (neutral) conductor to the earth

grounding electrode conductor the conductor used to connect the grounding electrode to the equipment-grounding conductor, to the grounded conductor, or to both at the service

main bonding jumper a jumper used to provide the connection between the grounded service conductor and the equipment-grounding conductor at the service

meter enclosure the weatherproof electrical enclosure that houses the kilowatt-hour meter; also called the "meter socket" or "meter trim"

riser a length of raceway that extends up a utility pole and encloses the service entrance conductors in an underground service entrance

service the conductors and equipment for delivering electric energy from the serving utility to the wiring system of the premises served

service conductors the conductors from the service point to the service disconnecting means

service drop the overhead service conductors from the last pole to the point connecting them to the service entrance conductors at the building

service entrance cable service conductors made up in the form of a cable

Glossary of Terms

service entrance conductors, overhead system the service conductors between the terminals of the service equipment and a point usually outside the building where they are joined by tap or splice to the service drop

service entrance conductors, underground system the service conductors between the terminals of the service equipment and the point of connection to the service lateral

service equipment the necessary equipment connected to the load end of the service conductors supplying a building and intended to be the main control and cutoff of the supply

service head the fitting that is placed on the service drop end of service entrance cable or service entrance raceway and is designed to minimize the amount of moisture that

can enter the cable or raceway; the service head is commonly referred to as a "weatherhead"

service lateral the underground service conductors between the electric utility transformer, including any risers at a pole or other structure, and the first point of connection to the service entrance conductors in a meter enclosure

service mast a piece of rigid metal conduit or intermediate metal conduit, usually 2 or 2½ inches in diameter, that provides service conductor protection and the proper height requirements for service drops

service point the point of connection between the wiring of the electric utility and the premises wiring

service raceway the rigid metal conduit, intermediate metal conduit,

electrical metallic tubing, rigid non-metallic conduit, or any other approved raceway that encloses the service entrance conductors

supplemental grounding electrode a grounding electrode that is used to "back up" a metal water pipe grounding electrode

transformer a piece of electrical equipment used by the electric utility to step down the high voltage of the utility system to the 120/240 volts required for a residential electrical system

utility pole a wooden circular column used to support electrical, video, and telecommunications utility wiring; it may also support the transformer used to transform the high utility company voltage down to the lower voltage used in a residential electrical system

Service Entrance Types

The *National Electrical Code®* (*NEC®*) defines a service as the conductors and equipment for delivering electric energy from the serving electric utility to the wiring system of the premises served. There are two types of service entrances used to deliver electrical energy to a residential wiring system: an overhead service and an underground service. The advantages and disadvantages of each service type need to be considered in the preparation and planning stage of a residential electrical system. Residential electricians must recognize the differences between the two service types as well as the specific installation techniques required for each of them.

Overhead Service

The overhead service is the service type most often installed in residential wiring. It is less expensive and takes less time to install than an underground service. Electric utility companies encourage the use of overhead services because of the ready access to the service conductors if there is ever a problem. An overhead service entrance includes the service conductors between the terminals of the service equipment main disconnect and a point outside the home where they are connected to overhead wiring, that is connected to the electric utility's electrical system. The overhead wiring is placed high enough to protect it from physical damage and to keep it away from contact with people. Figure 6–1 shows an example of a typical overhead service entrance.

Underground Service

An underground service entrance is often installed as an alternative to an overhead service. The service conductors between the terminals of the service main disconnect and the point of connection to the utility wiring are buried in the ground at a depth that protects the conductors from physical damage and also prevents accidental contact with the conductors by people. Since underground services have no exposed overhead wiring, some people consider that this service type makes a residence more attractive and worth the extra cost and time for the installation. If a problem arises with the underground wiring, the repair procedure usually means digging it up and fixing the problem. This is a more costly and time-consuming procedure for repair than for an overhead service. Figure 6–2 shows a typical underground service entrance.

Service Entrance Terms and Definitions

There is a lot of special terminology that applies to residential service entrances. Most of the terms commonly used are defined in Article 100 of the *NEC®*. However, there are many residential service entrance terms that are often used that the *NEC®* does not define. The following terms are those that residential electricians will find most important to be familiar with when working with residential services. Figure 6–3 illustrates the different service terms defined in the following list:

- The conductors and equipment for delivering electric energy from the electric utility to the house electrical system are called the service.
- The fitting that is placed on the service drop end of service entrance cable or service entrance raceway and is designed to minimize the amount of moisture that can enter the cable or raceway is called the service head. This part of a service entrance is usually called a 'weatherhead.'

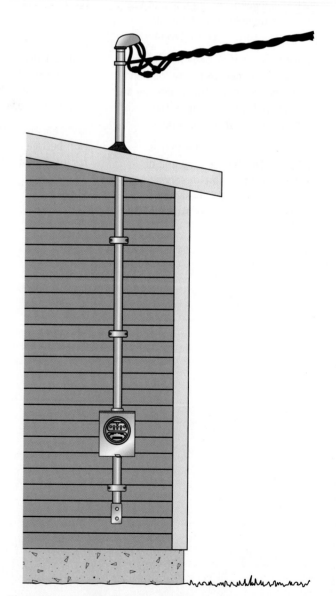

Figure 6–1 A typical overhead service entrance.

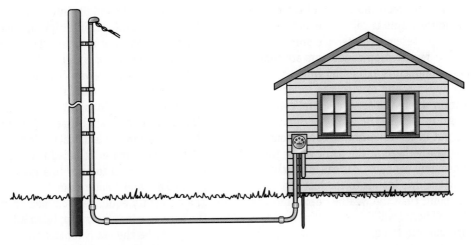

Figure 6–2 **A typical underground service entrance.**

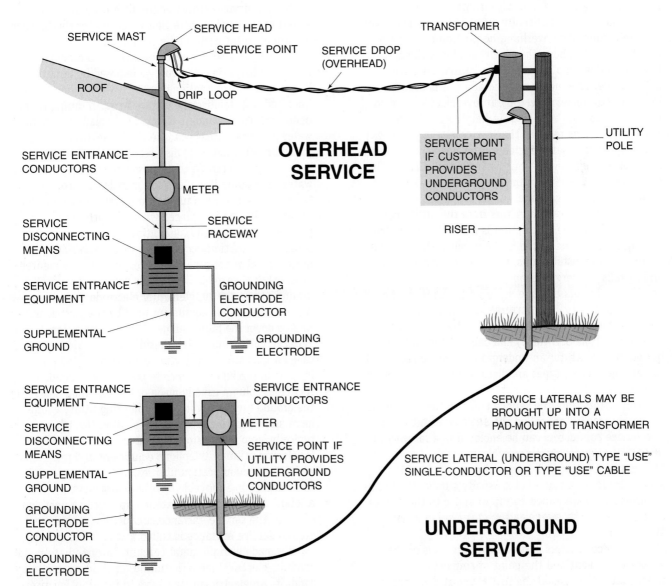

Figure 6–3 **Overhead and underground service entrance terms.**

- The point of connection between the wiring of the electric utility and the house wiring is called the **service point**. With an overhead service, the service point is at the location where the service drop conductors are connected to the service conductors extending from the weatherhead. In an underground service, the service point is at the location where the service lateral conductors are connected to the electric utility wiring.

- An intentional loop put in service entrance conductors that extends from a weatherhead at the point where they connect to the service drop conductors is called a **drip loop**. The purpose of the drip loop is to conduct any rainwater to a lower point than the weatherhead. This will help ensure that no water will drip down the service entrance conductors and into the meter enclosure.

- A piece of rigid metal conduit or intermediate metal conduit, usually 2 or 2½ inches in diameter, that provides protection for service conductors and the proper height requirements for service drops is called a **service mast**. The mast usually extends from a meter enclosure through the overhanging portion of a house roof to a height above the roof that allows the attached service drop to have the required distance above grade. This method of installation is used on lower-roofed homes served by an overhead service so that the minimum service drop heights can be met. Some electricians refer to the service raceway that extends up from a meter enclosure but does not go above the roofline as a "mast," but most electricians use the term "service mast" only to refer to the raceway that extends above the roofline.

- The overhead service conductors from the utility pole to the point where the connection is made to the service entrance conductors at the house is called the **service drop**. In most installations, the utility company owns and installs the service drop.

- Service conductors that are made up in the form of a cable are called **service entrance cable**. There is service entrance cable designed to be used outdoors on the side of a house (Type SEU cable) and cable designed to be buried in a trench for an underground service (Type USE cable). These cable types are discussed in more detail in Chapter 8.

- The conductors from the service point to the service disconnecting means are called the **service conductors**. The service conductors can be enclosed in a raceway or be part of a service entrance cable assembly.

- The service conductors between the terminals of the service equipment and a point usually outside the building where they are joined by tap or splice to the service drop are called the **overhead system service entrance conductors**.

- The service conductors between the terminals of the service equipment and the point of connection to the service lateral are called the **underground system service entrance conductors**.

- The rigid metal conduit, intermediate metal conduit, electrical metallic tubing, rigid nonmetallic conduit, or any other approved raceway that encloses the service entrance conductors is called the **service raceway**. Installation of the raceway must meet the *NEC®* requirements that are outlined in the raceway's specific article. For example, if rigid metal conduit is being used, the electrician must install it according to *NEC®* Article 344.

- The weatherproof electrical enclosure that houses the kilowatt-hour meter is called the **meter enclosure**. Other names used to refer to a meter enclosure are "meter socket" and "meter trim."

- The necessary equipment connected to the load end of the service conductors supplying a building and intended to be the main control and cutoff of the supply is called the **service equipment**. This equipment can consist of a fusible disconnect switch or a main breaker panel that also accommodates branch-circuit overcurrent protection devices (fuses or circuit breakers).

- A part of the service entrance that allows for the transfer of current to the earth under certain fault conditions is called the **grounding electrode**. The grounding electrode also helps limit the voltage imposed by lightning, line surges, or unintentional contact with higher-voltage lines and will stabilize the voltage to earth during normal operation. The grounding electrode is usually the metal water pipe that brings water to the home. Other types of grounding electrodes are discussed later in this chapter.

- A grounding electrode that is used to "back up" a metal water pipe grounding electrode is called a **supplemental grounding electrode**. If a metal water pipe electrode breaks and is replaced with a length of plastic plumbing pipe, grounding continuity will be lost. Because this situation occurs fairly often, the *NEC®* requires metal water pipe electrodes to be supplemented by another electrode. Electricians usually drive an 8-foot ground rod as the supplemental electrode.

- A piece of electrical equipment used by the electric utility to step down the high voltage of the utility system to the 120/240 volts required for a residential electrical system is called a **transformer**. It can be either mounted on a utility pole or placed on a concrete pad on the ground.

- A circular column usually made of treated wood and set in the ground for the purpose of supporting utility equipment and wiring is called a **utility pole**. Utility poles typically support transformers and electrical system wiring for electric utilities, telephone equipment and telephone wiring for communication utilities, and fiber-optic cable and coaxial cable for cable television providers.

- A length of raceway that extends up a utility pole and encloses the service entrance conductors in an underground service entrance is called a **riser**. The riser is usually made of rigid metal conduit, intermediate metal conduit, electrical metallic tubing, or rigid nonmetallic conduit. An electrician may have to install other risers on the utility pole for such things as telephone lines

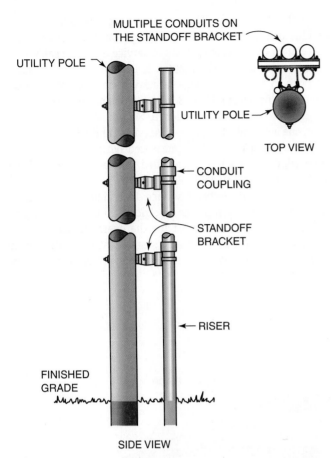

MULTIPLE CONDUITS ON THE STANDOFF BRACKET

UTILITY POLE

UTILITY POLE

TOP VIEW

CONDUIT COUPLING

STANDOFF BRACKET

RISER

FINISHED GRADE

SIDE VIEW

Figure 6–4 A standoff bracket is used to secure riser conduits on a utility pole for an underground service installation. Many electric utility companies allow electricians to secure one riser conduit directly to a pole but require the use of standoff brackets when securing multiple riser conduits to one pole.

and cable television lines. A piece of electrical equipment called a "standoff" allows ease of supporting multiple risers on one utility pole. Some utility companies require the use of standoffs whenever more than one riser is to be installed on the pole. Figure 6–4 shows an example of how a standoff on a utility pole is used.

- The underground service conductors between the electric utility transformer, including any risers at a pole or other structure, and the first point of connection to the service entrance conductors in a meter enclosure is called the **service lateral**. Electricians usually refer to the underground service conductors as simply the "lateral."

Residential Service Requirements (Article 230)

Article 230 of the *NEC*® covers many of the requirements for the installation of service entrances. The following discussion covers those *NEC*® rules that electricians installing residential services need to know.

Section 230.7 states that wiring other than service conductors must not be installed in the same service raceway or service cable. All other residential wiring system conductors must be separated from the service conductors. The reason is that service conductors are not provided with overcurrent protection where they receive their supply from the electric utility. They are protected against short circuits, ground faults, and overload conditions at their load end by the main service disconnect fuses or circuit breaker. The amount of current that could be imposed on the residential wiring system conductors, should they be in the same raceway or cable and should a fault occur, would be much higher than the ampacity of the wiring, and extreme damage to the wiring and unsafe conditions would result.

Section 230.8 requires an electrician to install a raceway seal where a service raceway enters a residential building from an underground distribution system. The sealant, such as duct seal or a bushing incorporating the physical characteristics of a seal, must be used to seal the end of service raceways (Figure 6–5). Sealing can take place at either end of the raceway or both ends if the electrician chooses. The intent of this requirement is to prevent water, usually the result of condensation due to temperature differences, from

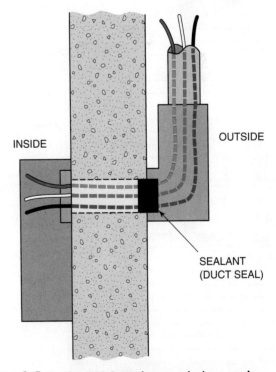

INSIDE

OUTSIDE

SEALANT (DUCT SEAL)

Figure 6–5 Section 230.8 requires a seal where service raceways enter a building from an underground distribution system. Sections 300.5(G) and 300.7 also require seals anytime a raceway containing electrical conductors is installed from a cold area (outside the house) to a warmer area (inside the house). The reason for the seal is to keep moisture away from the electrical equipment. Duct seal or a bushing incorporating the physical characteristics of a seal can be used to seal the ends of service raceways.

entering the service equipment through the raceway. The sealant material needs to be compatible with the conductor insulation and should not cause deterioration of the insulation over time. Any spare or unused raceways that are installed also must be sealed.

Service conductor clearance from building openings is covered in Section 230.9. Service conductors must have a minimum clearance of not less than 3 feet (900 mm) from windows that can be opened, doors, porches, balconies, ladders, stairs, fire escapes, or similar locations in a residential building (Figure 6–6). The intent is to protect the conductors from physical damage and protect people from accidental contact with the conductors. An exception states that conductors run above the top level of a window are permitted to be closer than 3 feet (900 mm). This exception permits service conductors, including drip loops and service drop conductors, to be located just above window openings because they are considered out of reach. It is important to

understand that the 3-foot clearance applies only to open conductors, such as those extending from the service head, and not to the service raceway or service entrance cable that encloses the service conductors. It is a common installation practice to have the service raceway or service entrance cable assembly closer than 3 feet (900 mm) to windows or doors on the side of a residence. Section 230.9(B) requires that the vertical clearance of service conductors above or within 3 feet (900 mm) of (measured horizontally) surfaces from which they might be reached (like a deck) be at least 10 feet above the surface (Figure 6–7).

Installing an overhead service to a home in a heavily wooded area may tempt the electrician to use a tree or several trees to provide support for the overhead service conductors. Section 230.10 clearly states that vegetation such as trees shall not be used to support overhead service conductors (Figure 6–8).

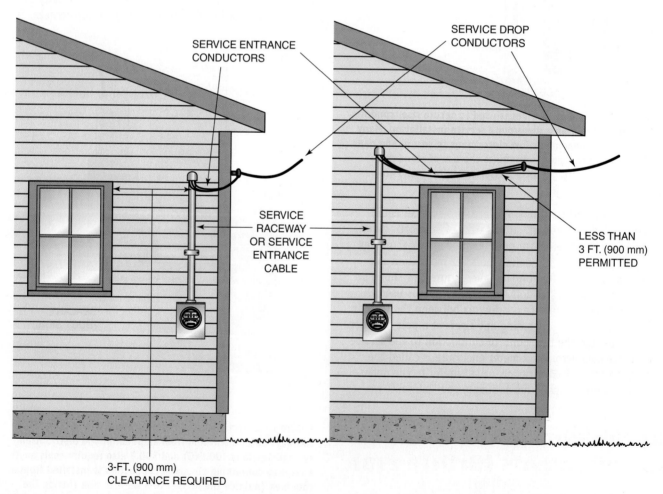

Figure 6–6 The required dimensions for service conductors located alongside a window (left) and service conductors above the top level of a window that is designed to be opened (right).

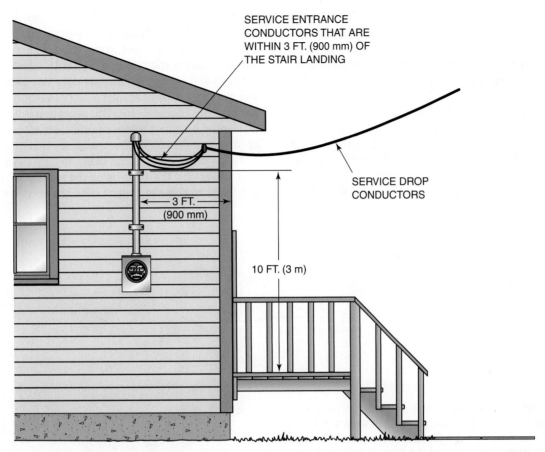

SERVICE ENTRANCE CONDUCTORS THAT ARE WITHIN 3 FT. (900 mm) OF THE STAIR LANDING

SERVICE DROP CONDUCTORS

3 FT. (900 mm)

10 FT. (3 m)

Figure 6–7 If service conductors are within 3 feet measured horizontally from a balcony, stair landing, or other platform, clearance to the platform of at least 10 feet must be maintained.

Figure 6–8 Trees are not allowed to support overhead service entrance conductors.

Section 230.22 requires individual service entrance conductors to be insulated. However, an exception allows the grounded (neutral) conductor of a multiconductor service entrance cable to be bare. This is the case with a three-wire service entrance cable like Type SEU. The intent of this section is to prevent problems created by weather, abrasion, and other effects that could reduce the insulating quality of the conductor insulation. The grounded conductor (neutral) of triplex service drop cable is usually bare and is used to mechanically support the other ungrounded conductors. Type SEU service entrance cable and triplex service drop cable are covered in more detail in Chapter 8.

Section 230.23 guides us in determining the requirements for the minimum size of the service entrance conductors. The service conductors must have sufficient ampacity to carry the current for the computed residential electrical load and must have adequate mechanical strength. Chapter 7 covers how to use Article 220 to calculate the electrical load for a residential wiring system.

Section 230.24 provides the dimensions for the minimum amount of vertical clearance for service drop conductors. When installing service drop conductors above a residential roof, conductors must have a vertical clearance of at least 8 feet (2.5 m) above the roof. This includes service drop conductors going over another building on the property, like a garage or storage shed. The 8-foot (2.5 m) vertical clearance above the roof must be maintained for a distance at least 3 feet (900 mm) in all directions from the edge of the roof (Figure 6–9). Exception No. 4 to 230.24(A) allows the requirement for maintaining the vertical clearance of 3 feet (900 mm) from the edge of the roof to not apply to

the final conductor length where the service drop is attached to the side of a building. Exception No. 2 to Section 230.24(A) permits a reduction in service drop conductor clearance above the roof from 8 feet to 3 feet if the voltage between the service conductors does not exceed 300 volts and the roof is sloped at least 4 inches vertically in 12 inches horizontally (Figure 6–10). This exception is applied often to residential buildings where peaked roofs with slopes of 4/12 or greater are common. The reason for this exception is that a steeply sloped roof is less likely to be walked on, and if a person were on the roof, he or she would be bent over and not likely to come in contact with the service drop conduc-

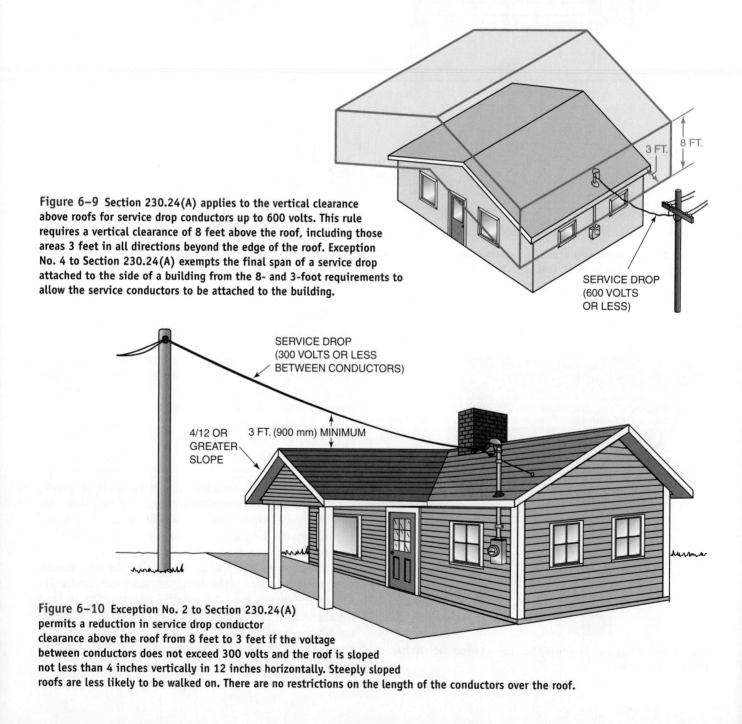

Figure 6–9 Section 230.24(A) applies to the vertical clearance above roofs for service drop conductors up to 600 volts. This rule requires a vertical clearance of 8 feet above the roof, including those areas 3 feet in all directions beyond the edge of the roof. Exception No. 4 to Section 230.24(A) exempts the final span of a service drop attached to the side of a building from the 8- and 3-foot requirements to allow the service conductors to be attached to the building.

Figure 6–10 Exception No. 2 to Section 230.24(A) permits a reduction in service drop conductor clearance above the roof from 8 feet to 3 feet if the voltage between conductors does not exceed 300 volts and the roof is sloped not less than 4 inches vertically in 12 inches horizontally. Steeply sloped roofs are less likely to be walked on. There are no restrictions on the length of the conductors over the roof.

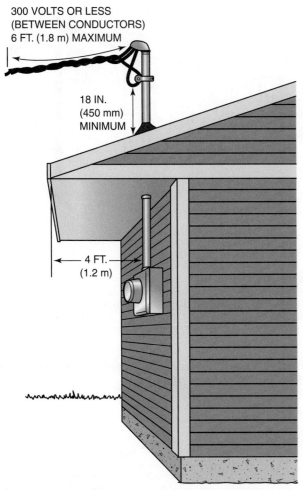

300 VOLTS OR LESS
(BETWEEN CONDUCTORS)
6 FT. (1.8 m) MAXIMUM

18 IN.
(450 mm)
MINIMUM

4 FT.
(1.2 m)

Figure 6–11 Exception No. 3 to Section 230.24(A) permits a reduction of service drop conductor clearances to 18 inches above the roof. This reduction is for service mast installations where the voltage between conductors does not exceed 300 volts and the mast is located within 4 feet of the edge of the roof, measured horizontally. Exception No. 3 applies to either a sloped or a flat roof. Not more than 6 feet of conductors is permitted to pass over the roof.

tors. There is no limit given for the length of the conductors over the roof. Exception No. 3 to Section 230.24(A) allows a reduction of service drop conductor clearance to 18 inches (450 mm) above the roof (Figure 6–11). This reduction is designed for residential services using a service mast (through-the-roof) installation where the voltage between conductors does not exceed 300 volts and the mast is located within 4 feet (1.2 m) of the edge of the roof. This exception applies to either sloped or flat roofs that are easily walked on. To use this exception, not more than 6 feet of service drop conductor is permitted to pass over the roof and attach to the service mast.

Section 230.24(B) sets the requirements for the vertical clearance from the ground for service drop conductors (Figure 6–12). Keep in mind that the vertical clearance is

dependent on how high the electrician has located the point of attachment for the service drop and that in most cases it has been preapproved by a utility company representative. The following minimum clearances must be maintained for residential service drop conductors:

- 10 feet (3.0 m): At the electric service entrance to buildings, also at the lowest point of the drip loop of the building electric entrance, and above areas or sidewalks accessible only to pedestrians, measured from final grade or other accessible surface for service drop cables supported on and cabled together with a grounded bare messenger where the voltage does not exceed 150 volts to ground
- 12 feet (3.7 m): Over residential property and driveways where the voltage does not exceed 300 volts to ground
- 18 feet (5.5 m): Over public streets, alleys, roads, parking areas subject to truck traffic, and driveways on other than residential property

▸▸▸▸ **CAUTION** ▸▸▸

CAUTION: Many electric utility companies require higher vertical clearances for service drop conductors than what the *NEC*® requires. Remember that the *NEC*® sets the minimum requirements and that if a utility company requires a vertical clearance over a residential driveway of at least 15 feet instead of the *NEC*® requirement of 12 feet, you must set up your service installation to meet the utility company's 15-foot requirement.

▸▸▸▸ **CAUTION** ▸▸▸

CAUTION: When taking a local or state electrical licensing exam, make sure to answer the questions based on the *NEC*® and not with information based on the local electric utility requirements. Licensing exams usually base their questions on the *NEC*® and not local electric utility requirements. If your answers are based on the utility requirements, you will probably get the answer wrong. Make sure you establish at the beginning of the exam whether the answers are to be based on the *NEC*® or the local electric utility requirements.

When a service mast is used for the support of the service drop conductors, Section 230.28 requires it to be of adequate strength or be supported by braces or guy wires to withstand the strain imposed by the service drop. Only power service

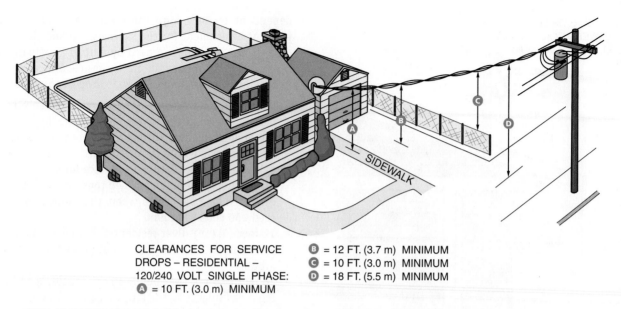

CLEARANCES FOR SERVICE
DROPS – RESIDENTIAL –
120/240 VOLT SINGLE PHASE:
Ⓐ = 10 FT. (3.0 m) MINIMUM

Ⓑ = 12 FT. (3.7 m) MINIMUM
Ⓒ = 10 FT. (3.0 m) MINIMUM
Ⓓ = 18 FT. (5.5 m) MINIMUM

NOTE: ELECTRIC UTILITIES FOLLOW THE NATIONAL ELECTRICAL SAFETY CODE (NESC). THE CLEARANCE REQUIREMENTS IN THE NESC ARE DIFFERENT THAN THOSE IN THE *NATIONAL ELECTRICAL CODE* (*NEC®*). THE DECIDING FACTOR MIGHT BE WHETHER THE INSTALLATION CUSTOMER IS INSTALLED, OWNED, AND MAINTAINED OR WHETHER THE INSTALLATION UTILITY IS INSTALLED, OWNED, AND MAINTAINED

Figure 6–12 Section 230.24(B) provides the minimum service drop vertical clearances over residential property. Be aware that some local electric utility companies may require clearances that are greater than those given in the *NEC®*.

drop conductors are permitted to be attached to an electrical service mast. Cable television or telephone service wires are not permitted to be attached to the service mast. Another mast for only the communications conductors must be installed (Figure 6–13).

Table 300.5 gives the minimum cover requirements for underground service conductors. The minimum depth for burying service entrance conductors is 24 inches (600 mm). Service conductors buried under residential driveways must be at least 18 inches (450 mm) below grade (Figure 6–14). Remember that the local electric utility may require greater burial depths than what Table 300.5 requires.

Section 300.5 also requires underground service conductors to be protected according to the following (Figure 6–15):

- Direct-buried conductors and enclosures emerging from the ground must be protected by enclosures or raceways extending from a minimum depth of 18 inches (450 mm) below grade to a point at least 8 feet (2.5 m) above finished grade.
- Conductors entering a building shall be protected to the point of entrance. For example, where the conductors emerge from the ground at the side of a house, the conductors must be protected up to the first enclosure, which is usually a meter socket. The meter socket is mounted at approximately 5 feet to the top of the meter enclosure from finished grade.

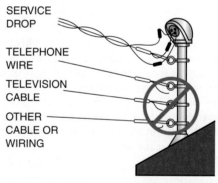

SERVICE DROP

TELEPHONE WIRE

TELEVISION CABLE

OTHER CABLE OR WIRING

Figure 6–13 Only power conductors can be attached to the service mast. Section 230.28 allows no cable television wiring or telephone wiring to be attached to the service mast.

- Underground service conductors that are buried 18 inches (450 mm) or more below grade shall have their location identified by a warning ribbon that is placed in the trench at least 12 inches (300 mm) above the underground installation. Providing a warning ribbon reduces the risk of an accident or electrocution during digging near underground service conductors that are not encased in concrete.
- Where the enclosure or raceway is subject to physical damage, the conductors shall be installed in rigid metal conduit, intermediate metal conduit, Schedule 80 rigid

Table 300.5 Minimum Cover Requirements, 0 to 600 Volts, Nominal, Burial in Millimeters (Inches)

Location of Wiring Method or Circuit	Type of Wiring Method or Circuit									
	Column 1 Direct Burial Cables or Conductors		Column 2 Rigid Metal Conduit or Intermediate Metal Conduit		Column 3 Nonmetallic Raceways Listed for Direct Burial Without Concrete Encasement or Other Approved Raceways		Column 4 Residential Branch Circuits Rated 120 Volts or Less with GFCI Protection and Maximum Overcurrent Protection of 20 Amperes		Column 5 Circuits for Control of Irrigation and Landscape Lighting Limited to Not More Than 30 Volts and Installed with Type UF or in Other Identified Cable or Raceway	
	mm	in.	mm	in.	mm	in.	mm	in.	mm	in.
All locations not specified below	600	24	150	6	450	18	300	12	150	6
One- and two-family dwelling driveways and outdoor parking areas, and used only for dwelling-related purposes	450	18	450	18	450	18	300	12	450	18

Notes:
1. Cover is defined as the shortest distance in millimeters (inches) measured between a point on the top surface of any direct-buried conductor, cable, conduit, or other raceway and the top surface of finished grade, concrete, or similar cover.
2. Raceways approved for burial only where concrete encased shall require concrete envelope not less than 50 mm (2 in.) thick.
3. Lesser depths shall be permitted where cables and conductors rise for terminations or splices or where access is otherwise required.

4. Where one of the wiring method types listed in Columns 1–3 is used for one of the circuit types in Columns 4 and 5, the shallower depth of burial shall be permitted.
5. Where solid rock prevents compliance with the cover depths specified in this table, the wiring shall be installed in metal or nonmetallic raceway permitted for direct burial. The raceways shall be covered by a minimum of 50 mm (2 in.) of concrete extending down to rock.

Figure 6–14 Table 300.5 in the *NEC®* shows the minimum burial depths for underground service conductors in a residential application. Be aware that some local electric utility companies may require burial depths that are greater than those given in this table. *Reprinted with permission from NFPA 70-2002, the National Electrical Code®, Copyright © 2001, National Fire Protection Association, Quincy, MA 02269. This reprinted material is not the referenced subject, which is represented only by the standard in its entirety. National Electrical Code ® and NEC® are registered trademarks of the National Fire Protection Association, Inc., Quincy, MA 02269.*

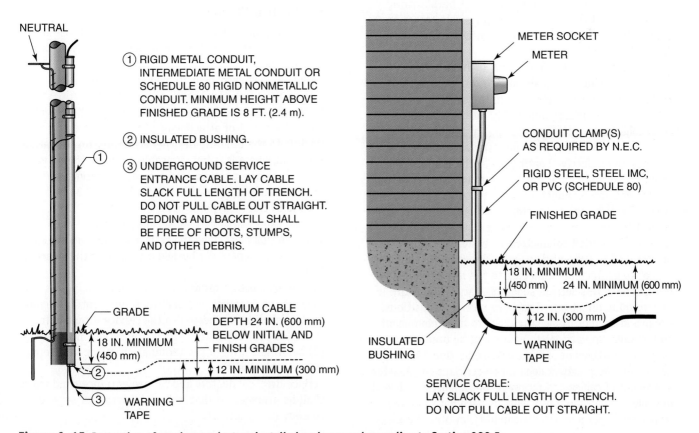

Figure 6–15 Protection of service conductors installed underground according to Section 300.5.

nonmetallic conduit, or a method that provides equivalent protection. Since any installation outdoors is subject to physical damage, underground service conductors will always be installed in one of the raceways listed in this section.

- Any cables or insulated service conductors installed in raceways in underground installations must be listed for use in wet locations. Table 310.13 is used to determine which general wiring conductor types are permitted to be installed in wet locations. Conductor insulations with a "W" in the insulation type name are permitted in wet locations. For example, an XHHW type of insulated conductor can be installed in a raceway buried in a trench as part of an underground service entrance.

- Backfill that contains large rocks, paving materials, cinders, large or sharply angular substances, or corrosive material shall not be placed in a trench on top of the service conductors, where they may damage raceways or cables. Where necessary to prevent physical damage to the raceway or cable, protection must be provided in the form of sand or other suitable material, suitable running boards, suitable sleeves, or other approved means. Electric utility companies usually will have their own requirements for the type of backfill used and the methods for backfilling an underground service trench.

- A bushing must be used at the end of a conduit or other raceway that terminates underground where the conductors or cables emerge as a direct burial wiring method. A seal that provides the physical protection characteristics of a bushing is permitted to be used instead of a bushing.

- All conductors of the same circuit, including the grounded conductor and all **equipment-grounding conductors**, must be installed in the same raceway or cable. If buried underground, they must be installed close together in the same trench. This requirement is designed to make sure that the installation does not produce any inductive heating under the ground, which could produce temperatures high enough to damage the service entrance conductors.

- Where direct-buried conductors, raceways, or cables are subject to movement by settlement of the ground or frost, direct-buried conductors, raceways, or cables must be arranged to prevent damage to the enclosed conductors or to equipment connected to the raceways. A fine-print note to this section suggests that "S" loops in underground direct-burial cable to raceway transitions, expansion fittings in raceway risers to fixed equipment, and flexible connections to equipment be used when subject to settlement or frost heaves. Section 300.5(J) points out the practical need for electricians to allow for movement of cables and raceways. Slack must be allowed in cables, expansion joints must be used with raceways (especially rigid nonmetallic conduit), or other measures must be taken if ground movement due to frost or settlement is anticipated.

Section 230.43 lists the wiring methods that could be used to install a residential service entrance. The various wiring methods must be installed in accordance with the applicable article covering that wiring method. For example, if using rigid metal conduit, you would install the conduit according to the requirements in Article 344. The following methods are the most common used for residential service installation:

- Rigid metal conduit
- Intermediate metal conduit
- Electrical metallic tubing
- Rigid nonmetallic conduit
- Service entrance cables. When using service entrance cables where they are subject to physical damage, they must be protected by any of the following:
  - Rigid metal conduit
  - Intermediate metal conduit
  - Schedule 80 rigid nonmetallic conduit
  - Electrical metallic tubing
  - Other approved means (remember that the authority having jurisdiction approves methods and materials)

Section 230.51 gives the requirements for supporting service entrance cable. Service entrance cables must be supported by straps, staples, or other approved means within 12 inches (300 mm) of the service head or any electrical enclosure and at intervals not exceeding 30 inches (750 mm) (Figure 6–16).

Section 230.54 lists some rules that apply to overhead service locations (Figure 6–17):

- Service raceways must be equipped with a rain-tight service head at the point of connection to service drop conductors.
- Service cables must be equipped with a rain-tight service head (weatherhead).
- Service heads in service entrance cables shall be located above the point of attachment of the service drop conductors to the building or other structure. An exception says that where it is impracticable to locate the service head above the point of attachment, the service head location shall be permitted not farther than 24 inches (600 mm) from the point of attachment.
- Service cables shall be held securely in place.
- Weaherheads must have the service conductors brought out through separately bushed openings in the weatherhead.
- Drip loops must be formed on individual conductors. To prevent the entrance of moisture, service entrance conductors shall be connected to the service drop conductors either (1) below the level of the service head or (2) below the level of the termination of the service entrance cable sheath.
- Service drop conductors and service entrance conductors shall be arranged so that water will not enter service raceway or equipment.

Section 230.70 applies to both an overhead and an underground service entrance type. A means must be provided

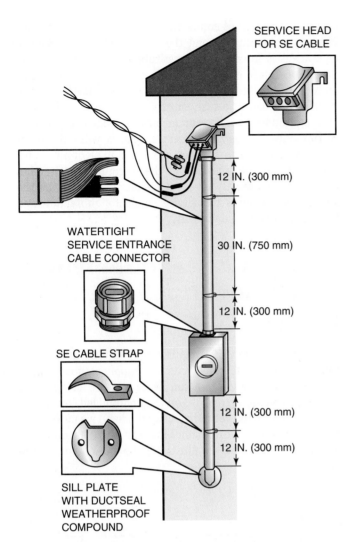

Figure 6–16 Service entrance cable support requirements are given in Section 230.51.

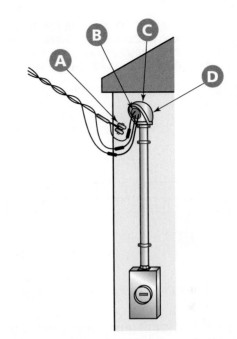

Figure 6–17 Overhead service entrances have several general rules that must be followed during the installation process. Section 230.54 contains many of them. (A) Drip loops must be formed on individual conductors. To prevent the entrance of moisture, service entrance conductors must be connected to the service drop conductors either (1) below the level of service head or (2) below the level of the termination of the service entrance cable sheath. (B) Service heads shall have conductors of different potential brought out through separately bushed openings. (C) Service heads must be located above the point of attachment of the service drop conductors to the building or other structure. However, where it is impracticable to locate the service head above the point of attachment, the service head location is permitted not farther than 24 inches (600 mm) from the point of attachment. (D) Service raceways and service cables must be equipped with a rain-tight service head at the point of connection to service drop conductors.

to disconnect all conductors in a building or other structure from the service entrance conductors (Figure 6–18). It must be installed at a **readily accessible** location either outside of a building or structure, or, if inside, as near as possible the point of entrance of the service conductors. No maximum distance is specified from the point of entrance of service conductors to a readily accessible location for the installation of a service disconnecting means. The authority enforcing this code has the responsibility for, and is charged with, making the decision as to how far inside the building the service entrance conductors are allowed to travel to the main disconnecting means. The length of service entrance conductors should be kept to a minimum inside buildings because power utilities provide limited overcurrent protection, and, in the event of a fault, the service conductors could ignite nearby combustible materials (Figure 6–19). In addition, the service disconnecting means cannot be installed in bathrooms, and each service disconnect must be permanently marked to identify it as a service disconnect.

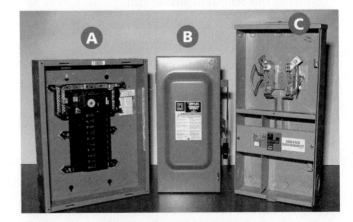

Figure 6–18 Section 230.70 requires a means to disconnect all conductors in a building from the service entrance conductors. This is accomplished with a main service disconnect switch. The main disconnect can consist of (A) the main circuit breaker in a load center, (B) a fusible disconnect switch, or (C) a combination meter socket/disconnect switch.

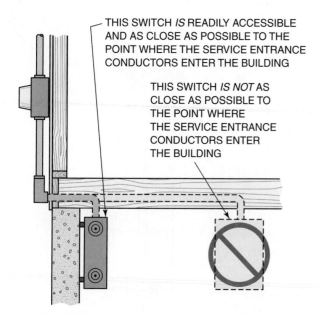

THIS SWITCH *IS* READILY ACCESSIBLE AND AS CLOSE AS POSSIBLE TO THE POINT WHERE THE SERVICE ENTRANCE CONDUCTORS ENTER THE BUILDING

THIS SWITCH *IS NOT* AS CLOSE AS POSSIBLE TO THE POINT WHERE THE SERVICE ENTRANCE CONDUCTORS ENTER THE BUILDING

Figure 6–19 **Section 230.70(A)(1) requires the service disconnect switch to be installed in a readily accessible location as soon as the service conductors enter the building. The length of service entrance conductors should be kept to a minimum inside buildings because power utilities provide limited overcurrent protection and, in the event of a fault, the service conductors could ignite nearby combustible materials.**

CAUTION: Some local electric utilities have rules that allow service entrance conductors to run within the building up to a specified length to terminate at the disconnecting means. They call this "limiting the length of inside run."

Section 230.71(A) covers the maximum number of disconnects permitted as the disconnecting means for the service conductors that supply a building. One set of service entrance conductors, either overhead or underground, is permitted to supply two to six service disconnecting means in lieu of a single main disconnect. A single-occupancy building can have up to six disconnects for each set of service entrance conductors.

Section 230.79 states that the service disconnecting means shall have a rating not less than the load to be carried as determined in accordance with Article 220. Three-wire services that supply one-family dwellings are required to be installed using wire with the capacity to supply a 100-ampere service for all single-family dwellings.

Section 230.90 requires each ungrounded service conductor to have overload protection. Service entrance conductors, overhead or underground, are the supply conductors

between the point of connection to the service drop or service lateral conductors and the service equipment. Service equipment is intended to constitute the main control and means of cutoff of the electrical supply to the premises wiring system. At this point, an overcurrent device, usually a circuit breaker or a fuse, must be installed in series with each ungrounded service conductor to provide overload protection only. The service overcurrent device will not protect the service conductors under short-circuit or ground-fault conditions on the line side of the disconnect. Protection against ground faults and short circuits is provided by the special requirements for service conductor protection and the location of the conductors.

Grounding Requirements for Residential Services (Article 250)

Article 250 of the *NEC*® requires that alternating current (AC) systems of 50 to 1,000 volts that supply premises wiring systems where the system can be **grounded** so that the maximum voltage to **ground** on the ungrounded conductors does not exceed 150 volts shall be grounded. This means that all residential electrical systems rated at 120/240 volts must be grounded. Figure 6–20 illustrates the grounding require-

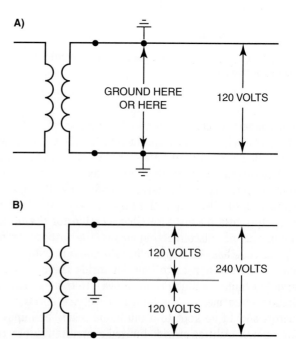

A)

GROUND HERE OR HERE   120 VOLTS

B)

120 VOLTS
                             240 VOLTS
120 VOLTS

Figure 6–20 **Article 250 in the *NEC*® requires certain electrical system conductors to be grounded. (A) A 120-volt, single-phase, two-wire electrical system must have one conductor grounded. (B) This shows which conductor must be grounded in a 120/240-volt, single-phase, three-wire electrical system.**

ments for a 120-volt, single-phase, two-wire system and for a 120/240-volt, single-phase, three-wire system. Since grounding a residential electrical system starts at the service entrance, the following grounding requirements must be followed when installing a residential service.

Section 250.24 states that a residential wiring system supplied by a grounded AC service must have a **grounding electrode conductor** connected to the grounded service conductor. The connection can be made at any accessible point from the load end of the service drop or service lateral to the terminal strip to which the grounded service conductor is connected at the service disconnecting means. Figure 6–21 illustrates the two most common connection points for connecting the **grounded conductor** of the service to the grounding electrode conductor and grounding electrode for a residential installation. Section 250.4 gives information that helps explain why the grounded service conductor must be connected to a grounding electrode:

- The grounded conductor of an AC service is connected to a grounding electrode system to limit the voltage to ground imposed on the system by lightning, line surges, and (unintentional) high-voltage crossovers.

- Another reason for requiring this connection is to stabilize the voltage to ground during normal operation, including short circuits.

Section 250.24(B) requires the grounded conductor of a residential service to be run to the service disconnecting means and be bonded (attached) to the disconnecting means enclosure. The grounded conductor must be routed with the phase conductors and cannot be smaller in size than the required grounding electrode conductor specified in Table 250.66. However, it never has to be larger than the largest ungrounded service entrance phase conductor.

Section 250.28 covers the requirements for the **main bonding jumper**. For a grounded system, an unspliced main bonding jumper must be used to connect the **equipment-grounding conductor(s)** and the service disconnect enclosure to the grounded conductor of the system within the enclosure for each service disconnect. Where the main bonding jumper specified in Section 250.28 is a wire or bus bar and is installed from the neutral bar or bus to the equipment grounding terminal bar in the service equipment, the grounding electrode conductor is permitted to be connected to the equipment-grounding terminal bar to which the main bonding jumper is connected (Figure 6–22):

- Main bonding jumpers can be made of copper or other corrosion-resistant material. A main **bonding jumper** can be a wire, bus, screw, or similar suitable conductor.
- Where a main bonding jumper is a screw, such as those used in most residential service equipment, the screw

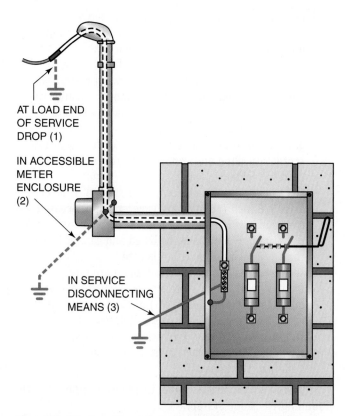

Figure 6–21 A residential service entrance supplied from an overhead distribution system illustrating three possible connection points where the grounded service conductor could be connected to the grounding electrode conductor according to 250.24(A)(1). Locations 2 and 3 are the most common locations for electricians to make the grounding electrode conductor connection.

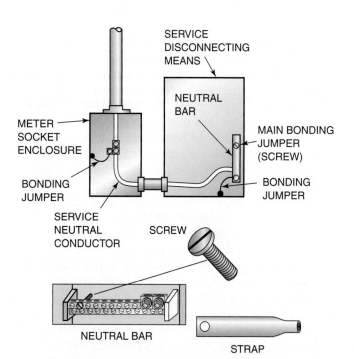

Figure 6–22 A main bonding jumper connects the grounded service conductor to the service entrance main disconnect enclosure and the equipment-grounding conductors. Section 250.28.

must have a green finish that shall be visible with the screw installed. The requirement of Section 250.28(B) for a green screw makes it possible to readily distinguish the main bonding jumper screw for inspection.

- The main bonding jumper cannot be smaller than the sizes shown in Table 250.66 for grounding electrode conductors. In residential applications, the size of the main bonding jumper has been sized at the factory and is included with every new loadcenter. Sizing it is not something that the electrician normally has to do.

Section 250.50 covers the grounding electrode system. If available on the premises at each residential building, each item in Section 250.52(A)(1) through (A)(6) must be bonded together to form the grounding electrode system. Where none of these electrodes are available, one or more of the electrodes specified in Section 250.52(A)(4) through (A)(7) must be installed and used. Section 250.50 introduces the important concept of a "grounding electrode system," in which all electrodes are bonded together. Rather than relying totally on a single electrode to perform its function over the life of the electrical installation, the *NEC®* encourages the formation of a system of electrodes "if available on the premises." There is no doubt that building a system of electrodes adds a level of reliability and helps ensure system performance over a long period of time. It is not the intent of Section 250.50 that reinforcing steel, if used in a residential building footing or foundation, must be made available for grounding.

Section 250.52 lists the electrodes permitted for grounding:

- A metal underground water pipe in direct contact with the earth for 10 feet (3 m) or more—Interior metal water piping located more than 5 feet (1.52 m) from the point of entrance to the building cannot be used as a part of the grounding electrode system or as a conductor to interconnect electrodes that are part of the grounding electrode system. This is the electrode type that is most often used in residential applications. Figure 6–23 shows the complete wiring for a typical residential service entrance with a metal water pipe being used as the grounding electrode. If a residence is located in a rural area not served by a water utility but instead has a drilled well with plastic pipe bringing water to the house, a ground rod (discussed later in this section) is often used as the grounding electrode. The metal well casing is not recognized as a grounding electrode by the *NEC®* but is required to be grounded when electrical conductors are used to supply power to a submersible pump located in the well casing.
- The metal frame of the building or structure, where effectively grounded can be used as a grounding electrode— This electrode type is rarely used in residential applications since residential construction methods usually do not include metal framing. However, in some areas of the country metal studs are being used for wall framing. It is unlikely that the metal wall framing would

be considered to be "effectively grounded" by the authority having jurisdiction and therefore would not be suitable for a grounding electrode.

- A concrete-encased electrode—Such an electrode is an excellent choice (Figure 6–24). It is an electrode encased by at least 2 inches (50 mm) of concrete and is located within and near the bottom of a concrete foundation or footing that is in direct contact with the earth. This electrode type must consist of at least 20 feet (6.0 m) of one or more electrically conductive steel reinforcing rods of not less than 1/2 inch (13 mm) in diameter. It can also be at least 20 feet (6.0 m) of bare copper conductor not smaller than 4 AWG buried in the footing. Electricians who use concrete-encased electrodes usually choose to install at least 20 feet of 4 AWG bare copper in the footing. Obviously, the electrician will have to be at the job site when the footing is being poured. A length of the wire is left coiled up at the location where the service disconnecting means will be located for connection in the future.
- A ground ring encircling the house, in direct contact with the earth, consisting of at least 20 feet (6.0 m) of bare copper conductor at least 2 AWG—This is a great grounding electrode, but it has been the author's experience that this electrode type is not used very often.
- Rod and pipe electrodes—Rod electrodes, commonly called "ground rods," are very popular with residential electricians. They must be at least 8 feet (2.5 m) in length and must be made of the following materials:
  - Electrodes of pipe or conduit cannot be smaller than 3/4-inch trade size (metric designator 21) and, if made of iron or steel, must have their outer surface galvanized or otherwise metal coated for corrosion protection. Pipe and conduit electrodes are rarely used in residential work.
  - Electrodes of rods made of iron or steel must be at least 5/8 inch (16 mm) in diameter. Copper-coated steel rods, which are very popular in residential work, or their equivalent must be listed and cannot be less than 1/2 inch (13 mm) in diameter.
- Plate electrodes can be used but again are rarely used in residential work. Each plate electrode must have at least 2 square feet (0.186 square m^2) of surface exposed to the soil. Electrodes of iron or steel plates must be at least 1/4 inch (6.4 mm) in thickness. Electrodes of nonferrous metal, like copper, must be at least 0.06 inch (1.5 mm) in thickness.

Section 250.53 covers some installation rules for the grounding electrode system:

- Where practicable, rod, pipe, and plate electrodes must be embedded below the permanent moisture level. They must also be free from nonconductive coatings, such as paint or enamel.
- Where more than one ground rod, pipe, or plate is used, each electrode type must be located at least 6 feet

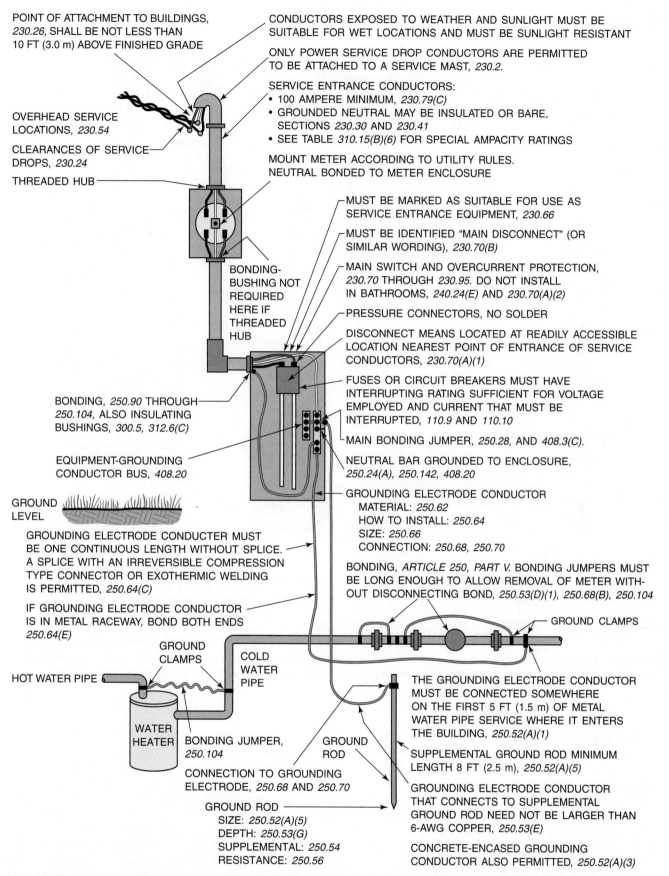

POINT OF ATTACHMENT TO BUILDINGS, *230.26*, SHALL BE NOT LESS THAN 10 FT (3.0 m) ABOVE FINISHED GRADE

CONDUCTORS EXPOSED TO WEATHER AND SUNLIGHT MUST BE SUITABLE FOR WET LOCATIONS AND MUST BE SUNLIGHT RESISTANT

ONLY POWER SERVICE DROP CONDUCTORS ARE PERMITTED TO BE ATTACHED TO A SERVICE MAST, *230.2*.

SERVICE ENTRANCE CONDUCTORS:
• 100 AMPERE MINIMUM, *230.79(C)*
• GROUNDED NEUTRAL MAY BE INSULATED OR BARE, SECTIONS *230.30* AND *230.41*
• SEE TABLE *310.15(B)(6)* FOR SPECIAL AMPACITY RATINGS

OVERHEAD SERVICE LOCATIONS, *230.54*

CLEARANCES OF SERVICE DROPS, *230.24*

THREADED HUB

MOUNT METER ACCORDING TO UTILITY RULES. NEUTRAL BONDED TO METER ENCLOSURE

MUST BE MARKED AS SUITABLE FOR USE AS SERVICE ENTRANCE EQUIPMENT, *230.66*

MUST BE IDENTIFIED "MAIN DISCONNECT" (OR SIMILAR WORDING), *230.70(B)*

MAIN SWITCH AND OVERCURRENT PROTECTION, *230.70* THROUGH *230.95*. DO NOT INSTALL IN BATHROOMS, *240.24(E)* AND *230.70(A)(2)*

PRESSURE CONNECTORS, NO SOLDER

BONDING-BUSHING NOT REQUIRED HERE IF THREADED HUB

DISCONNECT MEANS LOCATED AT READILY ACCESSIBLE LOCATION NEAREST POINT OF ENTRANCE OF SERVICE CONDUCTORS, *230.70(A)(1)*

FUSES OR CIRCUIT BREAKERS MUST HAVE INTERRUPTING RATING SUFFICIENT FOR VOLTAGE EMPLOYED AND CURRENT THAT MUST BE INTERRUPTED, *110.9* AND *110.10*

BONDING, *250.90* THROUGH *250.104*, ALSO INSULATING BUSHINGS, *300.5*, *312.6(C)*

MAIN BONDING JUMPER, *250.28*, AND *408.3(C)*.

NEUTRAL BAR GROUNDED TO ENCLOSURE, *250.24(A)*, *250.142*, *408.20*

EQUIPMENT-GROUNDING CONDUCTOR BUS, *408.20*

GROUNDING ELECTRODE CONDUCTOR
    MATERIAL: *250.62*
    HOW TO INSTALL: *250.64*
    SIZE: *250.66*
    CONNECTION: *250.68*, *250.70*

GROUND LEVEL

GROUNDING ELECTRODE CONDUCTOR MUST BE ONE CONTINUOUS LENGTH WITHOUT SPLICE. A SPLICE WITH AN IRREVERSIBLE COMPRESSION TYPE CONNECTOR OR EXOTHERMIC WELDING IS PERMITTED, *250.64(C)*

BONDING, *ARTICLE 250, PART V*. BONDING JUMPERS MUST BE LONG ENOUGH TO ALLOW REMOVAL OF METER WITH-OUT DISCONNECTING BOND, *250.53(D)(1)*, *250.68(B)*, *250.104*

IF GROUNDING ELECTRODE CONDUCTOR IS IN METAL RACEWAY, BOND BOTH ENDS *250.64(E)*

GROUND CLAMPS

GROUND CLAMPS

COLD WATER PIPE

HOT WATER PIPE

THE GROUNDING ELECTRODE CONDUCTOR MUST BE CONNECTED SOMEWHERE ON THE FIRST 5 FT (1.5 m) OF METAL WATER PIPE SERVICE WHERE IT ENTERS THE BUILDING, *250.52(A)(1)*

WATER HEATER

BONDING JUMPER, *250.104*

GROUND ROD

SUPPLEMENTAL GROUND ROD MINIMUM LENGTH 8 FT (2.5 m), *250.52(A)(5)*

CONNECTION TO GROUNDING ELECTRODE, *250.68* AND *250.70*

GROUNDING ELECTRODE CONDUCTOR THAT CONNECTS TO SUPPLEMENTAL GROUND ROD NEED NOT BE LARGER THAN 6-AWG COPPER, *250.53(E)*

GROUND ROD
    SIZE: *250.52(A)(5)*
    DEPTH: *250.53(G)*
    SUPPLEMENTAL: *250.54*
    RESISTANCE: *250.56*

CONCRETE-ENCASED GROUNDING CONDUCTOR ALSO PERMITTED, *250.52(A)(3)*

Figure 6–23 A typical residential service entrance installation with a water pipe being used as the grounding electrode. Notice that the water pipe is supplemented by a ground rod. The rod electrode will "back up" the water pipe electrode in the event that the water pipe breaks and is repaired with a length of plastic pipe.

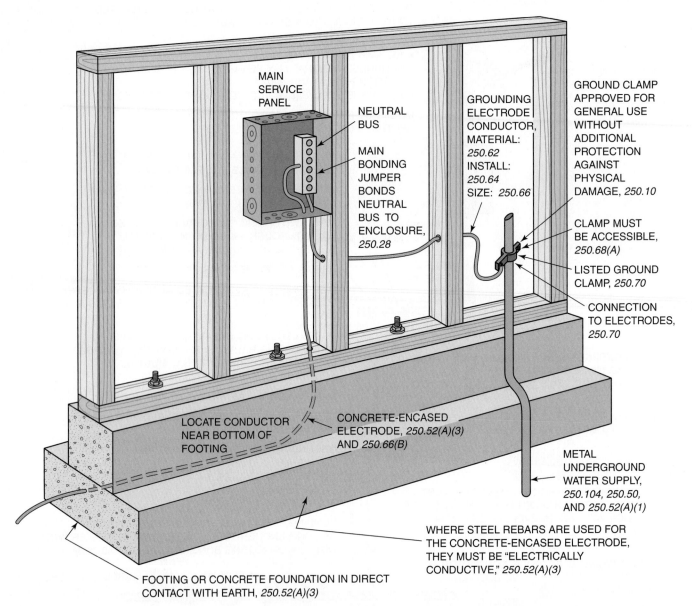

MAIN SERVICE PANEL

NEUTRAL BUS

MAIN BONDING JUMPER BONDS NEUTRAL BUS TO ENCLOSURE, 250.28

GROUNDING ELECTRODE CONDUCTOR, MATERIAL: 250.62 INSTALL: 250.64 SIZE: 250.66

GROUND CLAMP APPROVED FOR GENERAL USE WITHOUT ADDITIONAL PROTECTION AGAINST PHYSICAL DAMAGE, 250.10

CLAMP MUST BE ACCESSIBLE, 250.68(A)

LISTED GROUND CLAMP, 250.70

CONNECTION TO ELECTRODES, 250.70

LOCATE CONDUCTOR NEAR BOTTOM OF FOOTING

CONCRETE-ENCASED ELECTRODE, 250.52(A)(3) AND 250.66(B)

METAL UNDERGROUND WATER SUPPLY, 250.104, 250.50, AND 250.52(A)(1)

WHERE STEEL REBARS ARE USED FOR THE CONCRETE-ENCASED ELECTRODE, THEY MUST BE "ELECTRICALLY CONDUCTIVE," 250.52(A)(3)

FOOTING OR CONCRETE FOUNDATION IN DIRECT CONTACT WITH EARTH, 250.52(A)(3)

Figure 6–24 A concrete encased electrode is allowed by Section 250.52(A)(3). At least 20 feet of at least 4 AWG copper conductor can be located near the bottom of the foundation or footing as long as it is encased in at least 2 inches (51 mm) of concrete.

(1.8 m) from any other electrode. Two or more grounding electrodes, such as two ground rods, that are effectively bonded together are considered to be a single grounding electrode system.

- When used as a grounding electrode, the metal underground water pipe must meet the following requirements (Figure 6–25):
  - The continuity of the grounding path must not rely on water meters or filtering devices and similar equipment. **Bonding** around such equipment is required to ensure good grounding continuity.
  - A metal underground water pipe must be supplemented by an additional electrode. The supplemental electrode is permitted to be connected to the ground-

ing electrode conductor, the grounded service entrance conductor, the grounded service raceway, or any grounded service enclosure. Section 250.53(D)(2) specifically requires that rod, pipe, or plate electrodes used to supplement metal water piping be installed in accordance with Section 250.56. This requirement clarifies that the supplemental electrode system must be installed as if it were the sole grounding electrode for the system. One of the permitted methods of bonding a supplemental grounding electrode conductor to the primary electrode system is to connect it to the service enclosure. Another common method is to connect the supplemental grounding electrode to the meter enclosure. The requirement to supplement the metal water

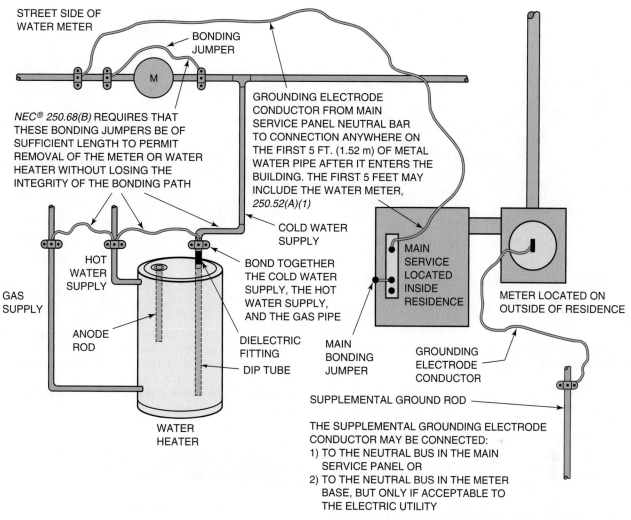

STREET SIDE OF
WATER METER

BONDING
JUMPER

NEC® 250.68(B) REQUIRES THAT
THESE BONDING JUMPERS BE OF
SUFFICIENT LENGTH TO PERMIT
REMOVAL OF THE METER OR WATER
HEATER WITHOUT LOSING THE
INTEGRITY OF THE BONDING PATH

M

GROUNDING ELECTRODE
CONDUCTOR FROM MAIN
SERVICE PANEL NEUTRAL BAR
TO CONNECTION ANYWHERE ON
THE FIRST 5 FT. (1.52 m) OF METAL
WATER PIPE AFTER IT ENTERS THE
BUILDING. THE FIRST 5 FEET MAY
INCLUDE THE WATER METER,
250.52(A)(1)

COLD WATER
SUPPLY

HOT
WATER
SUPPLY

GAS
SUPPLY

ANODE
ROD

BOND TOGETHER
THE COLD WATER
SUPPLY, THE HOT
WATER SUPPLY,
AND THE GAS PIPE

DIELECTRIC
FITTING

DIP TUBE

MAIN
SERVICE
LOCATED
INSIDE
RESIDENCE

METER LOCATED ON
OUTSIDE OF RESIDENCE

MAIN
BONDING
JUMPER

GROUNDING
ELECTRODE
CONDUCTOR

WATER
HEATER

SUPPLEMENTAL GROUND ROD

THE SUPPLEMENTAL GROUNDING ELECTRODE
CONDUCTOR MAY BE CONNECTED:
1) TO THE NEUTRAL BUS IN THE MAIN
   SERVICE PANEL OR
2) TO THE NEUTRAL BUS IN THE METER
   BASE, BUT ONLY IF ACCEPTABLE TO
   THE ELECTRIC UTILITY

Figure 6–25 Section 250.53 states that when using a metal water pipe as the grounding electrode, the continuity of the grounding path cannot rely on water meters, water filters, or similar equipment. For this reason, an electrician must bond around these items.

pipe is based on the practice of using a plastic pipe for replacement when the original metal water pipe fails. This type of replacement leaves the system without a grounding electrode unless a supplementary electrode is provided.

- Where the supplemental electrode is a rod, pipe, or plate electrode, that portion of the bonding jumper that is the sole connection to the supplemental grounding electrode shall not be required to be larger than 6 AWG copper wire or 4 AWG aluminum wire. For example, if a metal underground water pipe is used as the grounding electrode, Table 250.66 must be used for sizing the grounding electrode conductor, and the size may be required to be larger than a 6 AWG copper conductor. However, the size of the grounding electrode conductor for ground rod, pipe, or plate electrodes between the service equipment and the electrodes is not

required to be larger than 6 AWG copper or 4 AWG aluminum when the rod, pipe, or plate is being used either as a supplemental electrode or as the only grounding electrode.

- If a ground ring is installed as the grounding electrode, it must be buried at least 30 inches (750 mm).
- Rod and pipe electrodes must be installed so that at least 8 feet (2.44 m) of length is in contact with the soil. Where large rocks or ledge is encountered and a rod or pipe cannot be driven straight down into the ground, the electrode must be either driven at not more than a 45-degree angle or buried in a 2½-foot-deep trench (Figure 6–26). Ground clamps used on buried electrodes must be listed for direct earth burial. Ground clamps installed aboveground must be protected where subject to physical damage.
- Plate electrodes must be buried at least 30 inches (750 mm) below the surface of the earth.

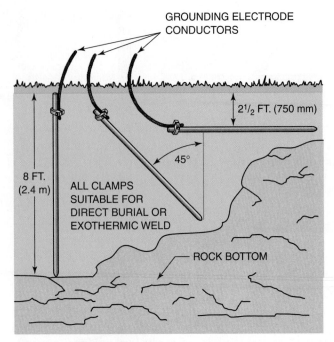

Figure 6–26 The installation requirements for rod and pipe electrodes as specified by Section 250.53(G).

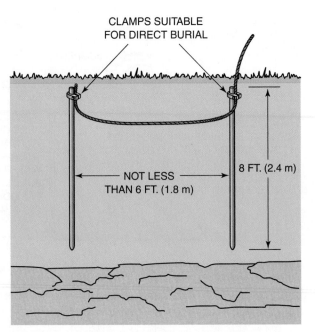

Figure 6–27 Section 250.56 states that a single electrode consisting of a rod, pipe, or plate that does not have a resistance to ground of 25 ohms or less must be augmented by one additional electrode. A supplementary rod, pipe, or plate electrode must be spaced at least 6 feet from any other rod, pipe, and plate electrode.

Section 250.56 requires a single electrode that is a rod, pipe, or plate and that does not have a resistance to ground of 25 ohms or less to be augmented by one additional electrode (Figure 6–27). Where multiple rod, pipe, or plate electrodes are installed to meet the requirements of this section, they must be installed so they are at least 6 feet (1.8 m) apart. It takes special instrumentation to determine if the resistance to ground of a rod, pipe, or plate electrode is 25 ohms or less. It is safe to say that most residential electricians do not have the necessary instruments.

Section 250.64 covers the installation of the grounding electrode conductor. The following rules must be followed when installing a grounding electrode conductor:

- Bare aluminum or copper-clad aluminum grounding electrode conductors must not be used in direct contact with masonry or the earth or where subject to corrosive conditions. Where used outside, aluminum or copper-clad aluminum grounding conductors must be terminated within 18 inches (450 mm) of the ground. For all practical purposes, aluminum or copper-clad aluminum grounding electrode conductors should not be used in residential applications because of this restriction.
- A grounding electrode conductor or its enclosure must be securely fastened to the surface on which it is carried. Staples, tie wraps, or other fastening means are used in residential construction to secure the grounding electrode conductor to the surface. A grounding electrode conductor that is 4 AWG or larger must be protected if it is exposed to severe physical damage. A 6 AWG grounding conductor that is free from exposure to physical damage is permitted

 FROM EXPERIENCE

Because of the fact that most residential electricians do not have the necessary measuring instrumentation to determine the resistance to ground of a rod, pipe, or plate electrode, many local electric utilities require the installation of an additional electrode automatically to help ensure that the resistance to ground is of a low enough value. Always check with the local electric utility and the local authority having jurisdiction to determine if additional grounding techniques for service entrances are required in your area.

to be run along the surface of the building construction without metal covering or protection where it is securely fastened to the construction. 8 AWG grounding electrode conductors must always be protected. Protection of grounding electrode conductors, when required, is accomplished by putting them in rigid metal conduit, intermediate metal conduit, rigid nonmetallic conduit, electrical metallic tubing, or cable armor.

- The grounding electrode conductor must be installed in one continuous length without a splice. Splicing is allowed only by irreversible compression-type connectors listed for the purpose or by the exothermic (CAD weld)

FROM EXPERIENCE

There is no *NEC®* requirement listed for how often you need to support the grounding electrode conductor. A good rule to follow is to support it within 12 inches from where it is terminated and no more than 24 inches between supports after that. If the grounding electrode conductor is placed in a raceway for protection, support of the raceway should be done according to the *NEC®* requirements located in the article that covers that raceway type.

welding process. These methods create a permanent connection that will not become loose after installation. It is very rare for the grounding electrode conductor to require splicing in a residential application, but it may be necessary to splice the grounding electrode conductor because of remodeling of the building or to add equipment.

• Metal raceways for grounding electrode conductors must be electrically continuous from the point of attachment to cabinets or equipment to the grounding

electrode. They also must be securely fastened to the ground clamp or fitting. Metal raceways that are not physically continuous from cabinet or equipment to the grounding electrode must be made electrically continuous by bonding each end to the grounding electrode conductor (Figure 6–28).

Section 250.66 specifies how to determine the size of the grounding electrode conductor. Table 250.66 (Figure 6–29) is used to size the grounding electrode conductor of a grounded AC system and is based on the largest-size service entrance conductor. For example, a 200-ampere residential service entrance can use a 2/0 copper or 4/0 aluminum service entrance conductor size. (Sizing a residential service entrance is covered in Chapter 7.) Table 250.66 tells us that the minimum-size copper grounding electrode conductor would be a 4 AWG copper conductor. Special sizing for connections to rods, pipes, plates, concrete-encased electrodes, and ground rings are listed as follows:

• Where the grounding electrode conductor is connected to rod, pipe, or plate electrodes, that portion of the conductor that is the only connection to the grounding electrode is not required to be larger than 6 AWG copper wire or 4 AWG aluminum wire.

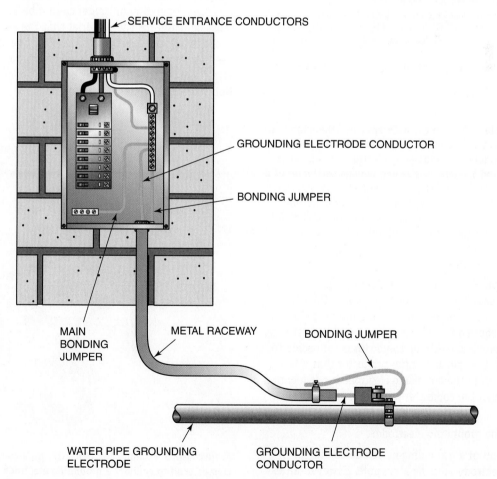

SERVICE ENTRANCE CONDUCTORS

GROUNDING ELECTRODE CONDUCTOR

BONDING JUMPER

MAIN BONDING JUMPER

METAL RACEWAY

BONDING JUMPER

WATER PIPE GROUNDING ELECTRODE

GROUNDING ELECTRODE CONDUCTOR

Figure 6–28 Bonding of a metal raceway to the grounding electrode conductor at both ends is required by Section 250.64(E).

Table 250.66 Grounding Electrode Conductor for Alternating-Current Systems

Size of Largest Ungrounded Service-Entrance Conductor or Equivalent Area for Parallel Conductors[a] (AWG/kcmil)		Size of Grounding Electrode Conductor (AWG/kcmil)	
Copper	Aluminum or Copper-Clad Aluminum	Copper	Aluminum or Copper-Clad Aluminum[b]
2 or smaller	1/0 or smaller	8	6
1 or 1/0	2/0 or 3/0	6	4
2/0 or 3/0	4/0 or 250	4	2
Over 3/0 through 350	Over 250 through 500	2	1/0
Over 350 through 600	Over 500 through 900	1/0	3/0
Over 600 through 1100	Over 900 through 1750	2/0	4/0
Over 1100	Over 1750	3/0	250

Notes:

1. Where multiple sets of service-entrance conductors are used as permitted in 230.40, Exception No. 2, the equivalent size of the largest service-entrance conductor shall be determined by the largest sum of the areas of the corresponding conductors of each set.

2. Where there are no service-entrance conductors, the grounding electrode conductor size shall be determined by the equivalent size of the largest service-entrance conductor required for the load to be served.

[a]This table also applies to the derived conductors of separately derived ac systems.

[b]See installation restrictions in 250.64(A).

Figure 6–29 **Table 250.66 in the *NEC*® specifies the minimum-size grounding electrode conductor for a certain size service entrance. The minimum-size grounding electrode conductor is determined by using the largest-size ungrounded conductor of a service entrance.** *Reprinted with permission from NFPA 70-2002, the* National Electrical Code ®, *Copyright © 2001, National Fire Protection Association, Quincy, MA 02269. This reprinted material is not the referenced subject, which is represented only by the standard in its entirety.*

- Where the grounding electrode conductor is connected to a concrete-encased electrode, that portion of the conductor that is the only connection to the grounding electrode is not required to be larger than 4 AWG copper wire.
- Where the grounding electrode conductor is connected to a ground ring, that portion of the conductor that is the only connection to the grounding electrode is not required to be larger than the conductor used for the ground ring.

Section 250.68 covers the grounding electrode conductor connection to the grounding electrode:

- The connection of a grounding electrode conductor to a grounding electrode must be accessible. However, an exception states that an encased or buried connection to a concrete-encased, driven, or buried grounding electrode shall not be required to be accessible. Ground clamps and other connectors suitable for use where buried in earth or embedded in concrete must be listed for such use, either by a marking on the connector or by a tag attached to the connector. Ground clamps that are suitable for direct burial will have the words "direct burial" written on the clamp.
- The connection of a grounding electrode conductor to a grounding electrode must be made in a manner that will ensure a permanent and effective grounding path. Where necessary to ensure the grounding path for a metal piping system used as a grounding electrode, effective bonding must be provided around insulated joints and around any equipment likely to be disconnected for repairs or replacement. Bonding conductors shall be of sufficient length to permit removal of the equipment while still maintaining the integrity of the bond. Examples of equipment likely to be disconnected for repairs or replacement are water meters and water filter systems.

Section 250.70 lists some methods of connecting the grounding conductor to an electrode. The grounding or bonding conductor must be connected to the grounding electrode by exothermic welding, listed lugs, listed pressure connectors, listed clamps, or other listed means (Figure 6–30). Connections depending on solder cannot be used. Ground clamps must be listed for the material that the grounding electrode is made of and the grounding electrode conductor. Where used on pipe, rod, or other buried electrodes, the clamp must also be listed for direct soil burial or concrete encasement. Not more than one conductor shall be connected to the

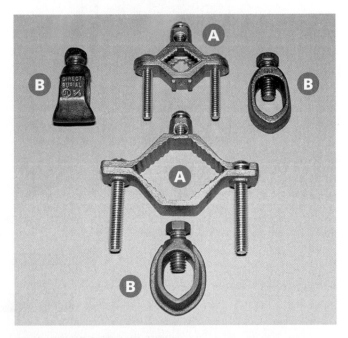

Figure 6–30 **Examples of (A) water pipe clamps and (B) rod clamps used to connect a grounding electrode conductor to a grounding electrode.**

grounding electrode by a single clamp or fitting unless the clamp or fitting is listed for multiple conductors.

Section 250.80 states that metal enclosures and raceways for service conductors and equipment must be grounded. An exception allows a metal elbow that is installed in an underground installation of rigid nonmetallic conduit and is isolated from possible contact by a minimum cover of 18 inches (450 mm) to any part of the elbow to not be required to be grounded. The exception to Article 250.80 recognizes that metal sweep elbows are often installed in underground installations of rigid nonmetallic conduit (PVC). The metal elbows are installed because nonmetallic elbows can be damaged by friction from the pulling ropes used during conductor installation. The elbows are isolated from physical contact by burial so that no part of the elbow is less than 18 inches below grade.

Section 250.92 requires the noncurrent-carrying metal parts of service equipment to be effectively bonded together. Figure 6–31 illustrates grounding and bonding at an individual service. The metal parts that require bonding together for a residential service entrance are the following:

- The service raceways.
- All service enclosures containing service conductors, including meter enclosures, boxes, or other metal electrical equipment connected to the service raceway.
- Any metallic raceway or armor enclosing a grounding electrode conductor as specified in Section 250.64(B). Bonding must apply at each end and to all intervening raceways, boxes, and enclosures between the service equipment and the grounding electrode. Section 250.92(A)(3) is intended to clarify that where metal race-

ways, boxes, or enclosures contain a grounding electrode conductor, both ends of the raceway, box, or enclosure must be bonded to the grounding electrode conductor.

Section 250.92(B) lists the allowed methods of bonding at the service. Electrical continuity at service equipment, service raceways, and service conductor enclosures must be ensured by one of the following methods:

- Bonding equipment to the grounded service conductor by exothermic welding, listed pressure connectors, listed clamps, or other listed means.
- Connections utilizing threaded couplings or threaded bosses on enclosures where made up wrenchtight.
- Threadless couplings and connectors where made up tight for metal raceways and metal-clad cables.
- Other approved devices, such as bonding-type locknuts and bushings. Standard locknuts or sealing locknuts are not acceptable as the "sole means" for bonding on the line side of service equipment. Grounding and bonding bushings for use with rigid or intermediate metal conduit are provided with means (usually one or more set screws that make positive contact with the conduit) for reliably bonding the bushing and the conduit on which it is threaded to the metal equipment enclosure or box. Grounding and bonding type bushings used with rigid or intermediate metal conduit, such as those shown in Figure 6–32, have provisions for connecting a bonding jumper or have means provided by the manufacturer for use in mounting a wire connector. This type of bushing may also have means (usually one or more set screws) to reliably bond the bushing to the conduit.
- Bonding jumpers must be used around **concentric knockouts** or **eccentric knockouts** that are punched or otherwise formed so as to impair the electrical connection to ground. Standard locknuts or bushings shall not be the sole means for the bonding required by this section. Both concentric- and eccentric-type knockouts can impair the electrical conductivity between the metal parts and may actually introduce unnecessary impedance

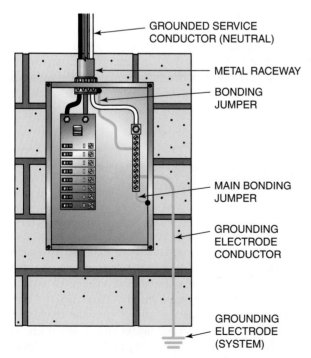

Figure 6–31 Grounding and bonding for a service with one disconnecting means.

GROUNDED SERVICE CONDUCTOR (NEUTRAL)

METAL RACEWAY

BONDING JUMPER

MAIN BONDING JUMPER

GROUNDING ELECTRODE CONDUCTOR

GROUNDING ELECTRODE (SYSTEM)

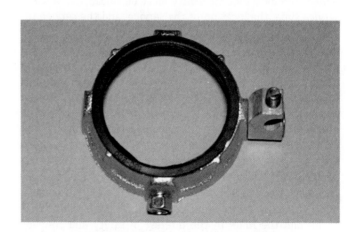

Figure 6–32 An example of a bonding-type bushing.

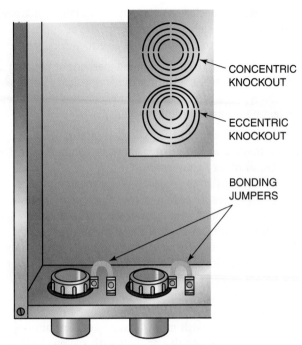

CONCENTRIC
KNOCKOUT

ECCENTRIC
KNOCKOUT

BONDING
JUMPERS

Figure 6–33 Bonding jumpers installed around concentric or eccentric knockouts.

into the grounding path. Installing bonding jumper(s) is one method often used between metal raceways and metal parts to ensure electrical conductivity. Figure 6–33 shows the difference between concentric- and eccentric-type knockouts and illustrates one method of applying bonding jumpers at these types of knockouts.

Working with the Local Utility Company

Once the type of service entrance being installed is determined, the local electric utility must be contacted, and the service type, the location of the service, and the service installation is coordinated with them. Electric utility companies have many rules governing the installation of service entrances that electricians have to follow. There is usually a publication from the utility company that is made available to electrical contractors that outlines the rules to follow when installing services. The intent of the publication is to provide information to electrical contractors, engineers, and architects in order that home electrical systems can be connected to the electric utility system in a safe and uniform manner. The utility publication usually states that the provisions of the publication are based on the *NEC®*. However, many times the utility requires additional requirements that go beyond the *NEC®* in the interest of safety and convenience. Some of the more common utility rules for service installation are covered in the following paragraphs.

CAUTION

CAUTION: The electric utility company rules discussed in this section are those that the author has encountered in the area of the country where he does electrical work. Be aware that the rules may be somewhat different in your area. Always be sure to check with the electric utility company in your area and the authority having jurisdiction before starting to install an overhead or underground residential service entrance.

An application for a new service connection must be initiated with the local electric utility as far in advance as possible. The location of the service entrance, the service entrance type, and the location of any transformers and poles must be reviewed and approved by the utility before any service wiring is installed. Most utilities will charge the customer whenever any wiring installed without prior company approval results in an additional expense to the utility. Electrical contractors should not start any service installation, purchase service equipment, or install wiring for additional electrical loads in existing installations until all negotiations have been completed with the utility company and it has been determined that the required service can be supplied.

A utility company representative, usually from the metering department, will work with the electrician to determine a suitable meter location, point of attachment for the service drop, and the location of the service entrance weatherhead. The meter enclosure will have to be located in a safe and readily accessible location.

The utility publication will usually tell what the electrical contractor is responsible for. For example, for an overhead service, it may say that the customer is responsible for purchasing and installing the service entrance, which includes all the wiring and parts from the weatherhead down to the main

 FROM EXPERIENCE

An electrical plan might call for a particular service type, but that type may not be able to be installed because of local electric utility restrictions. For example, an underground service may be called for in the building plan, but because of the existence of extensive amounts of ledge in the area where the service trench has to be dug, the minimum burial depth for the service entrance conductors cannot be met. In this case, unless an agreement can be worked out with the electric utility that allows a shallower burial depth, an overhead service will have to be installed.

service disconnecting equipment. The service drop, which brings the electrical energy from the utility's system overhead to the house, will be installed and owned by the utility.

In municipalities where electrical inspections are required, a certificate of approval must be given to the utility company before the service can be connected to the utility system. This certificate will be obtained from the authority having jurisdiction (electrical inspector) and forwarded to the utility by the electrical contractor. In those areas that do not require an inspection certificate from an electrical inspector, utility companies will accept a certification stating that the service installation and all wiring in the house is done according to the *NEC®*. A representative of the electrical contracting company signs this certification. Many utility companies also send a representative from their company to do an inspection of the service entrance to make sure that all company installation rules have been met. Once the service installation has been approved, the utility can be contacted to set up a time when the line crew (sometimes the metering department) can come to the house site and energize the service.

Sometimes it is necessary for an electrician to establish temporary service with a utility company during the initial construction phase of the house. Many times installing temporary service is the very first thing done by the electrical contractor. The temporary service will provide the electrical power for the various craft people, including electricians, to use power tools during the construction of the house. Electricity for temporary lighting can also be supplied by the temporary service. Most utility companies have certain requirements to be met when an electrician sets up a temporary service. The electrical contractor will need to contact the utility company to set up a temporary service and to have it inspected prior to connection. The process is very similar to what is described in the preceding paragraphs for establishing permanent service. An example of a temporary service and typical utility requirements are shown in Figure 6–34.

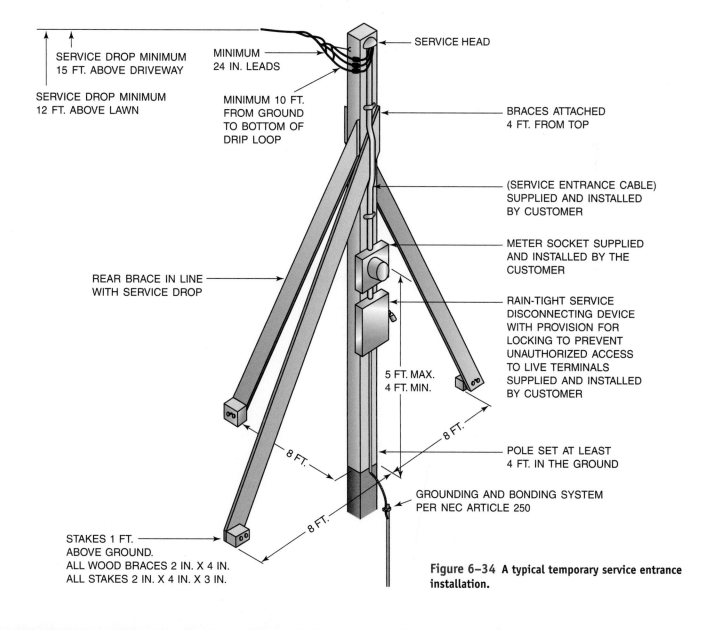

Figure 6–34 A typical temporary service entrance installation.

FROM EXPERIENCE

With the availability of powerful cordless power tools and reliable portable generators, many new residential construction sites do not require a temporary service. The author has installed many residential electrical systems where the building contractor has furnished electrical power through the use of a small electric generator.

Summary

In this chapter you were introduced to the different service entrance types used in residential wiring. Several *NEC*® rules that apply to residential service entrances were explained and illustrated. Working with the local electric utility to establish electrical service was also covered in detail. A lot of information was presented in this chapter, and all of it is important to an electrician getting ready to install the service entrance for a residential wiring system.

Review Questions

Directions: Answer the following items with clear and complete responses.

1. Define the term "service entrance."

2. Define the term "service drop."

3. Define the term "service lateral."

4. Describe a mast-type service entrance.

5. List the minimum vertical clearance for 120/240-volt service drop conductors over:

   a. a residential driveway _____
   b. a residential lawn _____
   c. a roof with a slope of 5/12 _____
   d. a garage roof with a slope of 3/12 _____

6. When the service conductors of an underground service exit the ground and terminate in a meter enclosure on the side of a house, there are three types of raceway that the *NEC*® recognizes as suitable for protecting the conductors. Name them.

7. Certain sizes of grounding electrode conductors that are subject to physical damage must be protected. Indicate whether the following size grounding electrode conductors need protection:

   a. 8 AWG copper grounding electrode conductor _____
   b. 6 AWG grounding electrode conductor _____
   c. 4 AWG grounding electrode conductor _____

8. List at least five items, other than a water pipe, that could be used as a grounding electrode for a residential service entrance.

9. Describe why it is required by the *NEC*® to install a supplemental grounding electrode when a metal water pipe is used as the main grounding electrode.

10. Name the table in the *NEC*® that you would use to size the minimum-size grounding electrode conductor.

11. Define the term "readily accessible."

12. List an advantage and a disadvantage of an underground service entrance versus an overhead service entrance.

13. Describe the purpose of a "drip loop."

14. Service entrance conductors, not the service raceway or service cable, must be kept a minimum of _____ feet from the sides and bottom of windows that can be opened.

15. The minimum burial depth, according to Table 300.5 of the *NEC*®, for underground service entrance cable serving a dwelling unit is _____ inches.

16. List four wiring methods that could be used by an electrician to install a residential service entrance.

17. Describe why the amount of "inside run" for the service entrance conductors must be as short as possible.

18. An electrician is driving an 8-foot ground rod and strikes ledge at about 3 feet. There is no way he or she can drive the rod all the way into the ground, as the *NEC*® requires. Describe what you might do if you were the electrician to meet *NEC*® requirements.

19. Explain why it is necessary to bond around eccentric and concentric knockouts when bonding service entrance equipment.

20. Describe the support requirements for service entrance cable.

Chapter 7
Residential Service Entrance Calculations

The National Electrical Code® (NEC®) outlines the procedures for calculating the minimum size of branch circuits, feeders, and service entrance conductors required for a residential electrical system. It can initially be a confusing task for new electricians to determine exactly which steps are required for the calculations. However, once the steps are identified, the calculations are not that difficult to perform. This chapter covers the calculation of branch-circuit and feeder loads as well as how to determine the minimum-size conductor and maximum-size overcurrent device for these circuits. It presents in detail the steps required to determine what size service entrance needs to be installed as part of a residential electrical system. Sizing service entrance main electrical panels and subpanels is also covered.

OBJECTIVES

Upon completion of this chapter, the student should be able to:

- ⊗ determine the minimum number and type of branch circuits required for a residential wiring system.
- ⊗ demonstate an understanding of the basic *NEC*® requirements for calculating branch-circuit sizing and loading.
- ⊗ calculate the minimum conductor size for a residential service entrance.
- ⊗ determine the proper size of the service entrance main disconnecting means.
- ⊗ determine the proper size for a panel board used to distribute the power in a residential wiring system.
- ⊗ calculate the minimum-size feeder conductors delivering power to a subpanel.
- ⊗ demonstrate an understanding of the steps required to calculate a residential service entrance using the standard or optional method as outlined in Article 220 of the *NEC*®.

Glossary of Terms

ambient temperature the temperature of the air that surrounds an object on all sides

ampacity the current, in amperes, that a conductor can carry continuously under the conditions of use without exceeding its temperature rating

bathroom branch circuit a branch circuit that supplies electrical power to receptacle outlets in a bathroom; lighting outlets may also be served by the circuit as long as other receptacle or lighting outlets outside the bathroom are not connected to the circuit; it is rated at 20 amperes

branch circuit the circuit conductors between the final overcurrent device (fuse or circuit breaker) and the outlets

dwelling unit one or more rooms for the use of one or more persons as a housekeeping unit with space for eating, living, and sleeping and permanent provisions for cooking and sanitation

feeder the circuit conductors between the service equipment and the final branch-circuit overcurrent protection device

general lighting circuit a branch-circuit type used in residential wiring that has both lighting and receptacle loads connected to it; a good example of this circuit type is a bedroom branch circuit that has both receptacles and lighting outlets connected to it

individual branch circuit a circuit that supplies only one piece of electrical equipment; examples are one range, one space heater, or one motor

laundry branch circuit a type of branch circuit found in residential wiring that supplies electrical power to laundry areas; no lighting outlets or other receptacles may be connected to this circuit

lug a device commonly used in electrical equipment used for terminating a conductor

nipple an electrical conduit of less than 2 feet in length used to connect two electrical enclosures

outlet a point on the wiring system at which current is taken to supply electrical equipment; an example is a lighting outlet or a receptacle outlet

service disconnect a piece of electrical equipment installed as part of the service entrance that is used to disconnect the house electrical system from the electric utility's system

small-appliance branch circuit a type of branch circuit found in residential wiring that supplies electrical power to receptacles located in kitchens and dining rooms; no lighting outlets are allowed to be connected to this circuit type

volt-ampere a unit of measure for alternating current electrical power; for branch-circuit, feeder, and service calculation purposes, a watt and a volt-ampere are considered the same

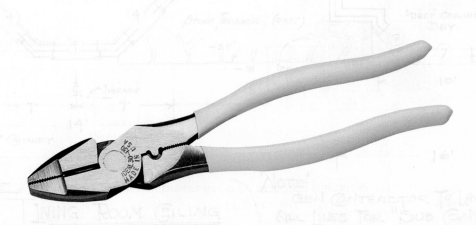

Determining the Number and Types of Branch Circuits

Before an electrician can determine the size of the service entrance required to supply the house electrical system with power, the number and types of circuits supplying power to the various electrical loads in the house must be determined. An understanding of the circuit types used in residential wiring is necessary for this. The *NEC®* defines a **branch circuit** as the circuit conductors between the final overcurrent device (fuse or circuit breaker) and the power and/or lighting outlets (Figure 7–1). Types of branch circuits used in residential wiring include the following:

- General lighting branch circuits
- Small-appliance branch circuits
- Laundry branch circuits
- Bathroom branch circuits
- Individual branch circuits

General Lighting Circuits

A **general lighting circuit** is a branch circuit that has both lighting and receptacle loads connected to it (Figure 7–2). A good example of this circuit type is a bedroom branch circuit that has several receptacles and a ceiling-mounted lighting fixture connected to it. This type of circuit makes up the majority of the branch circuits found in residential wiring.

To determine the minimum number of general lighting circuits required in a house, a calculation of the habitable floor area is required. The calculated floor area is then multiplied by the unit load per square foot for general lighting to get the total general lighting load in volt-amperes. Section 220.3(A) of the *NEC®* states that a unit load of not less than that specified in Table 220.3(A) (Figure 7–3) for occupancies listed in the table will be the minimum lighting load. The unit load for a **dwelling unit**, according to Table 220.3(A), is 3 **volt-amperes** per square foot. The floor area for each floor is to be computed from the *outside* dimensions of the dwelling unit. The computed floor area does not include open porches, garages, or unused or unfinished spaces not adaptable for future use. Examples of unused or unfinished spaces for dwelling units are attics, basements, or crawl spaces. A finished-off basement, such as for a family room, must be included in the floor area calculation. Let's look at an example. A house is determined to have 2,000 square feet of habitable living space. Once the habitable space has been determined, multiply it by the unit load per square foot: 2,000 square feet × 3 volt-amps per square foot = 6,000 volt-amperes of general lighting load.

To determine the minimum number of general lighting circuits required in a house, the total general lighting load

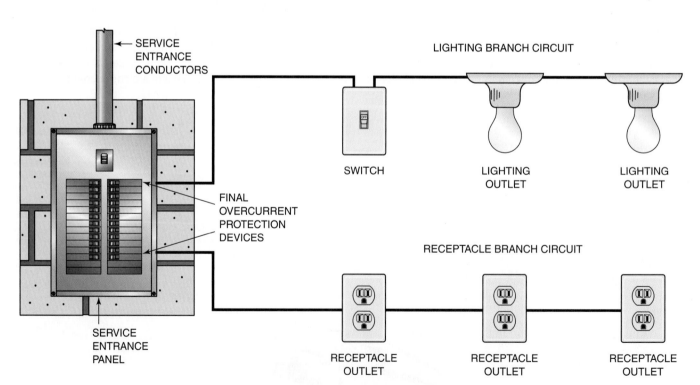

Figure 7–1 A branch circuit is defined as the circuit conductors between the final fuse or circuit breaker and the power or lighting outlets.

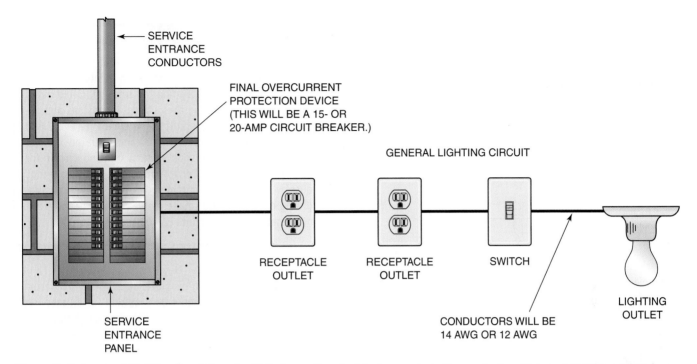

Figure 7–2 A general lighting circuit is a circuit that supplies electrical power to both receptacle outlets and lighting outlets in a residential wiring application. They will be installed with 14 AWG wire and rated at 15 amperes or 12 AWG wire rated at 20 amperes.

in volt-amperes is divided by 120 volts (the voltage of residential general lighting circuits). This will give you the total general lighting load in amperes. Now you only have to divide the total general lighting load in amperes by the size of the fuse or circuit breaker that is providing the overcurrent protection for the general lighting circuits. Let's continue to look at the example from the previous paragraph. We determined that a 2,000-square-foot house had a calculated general lighting load of 6,000 volt-amperes. Divide the 6,000 volt-amperes by 120 volts to get the general lighting load in amperes: 6,000 volt-amperes/120 volts = 50 amps. Now, if 15-ampere-rated general lighting circuits are being installed, divide the 50-ampere general lighting load by 15 amps: 50 amp/15 amp = 3.333. Therefore, a minimum of four 15-amp general lighting circuits must be installed. Always round up to the next-higher whole number when determining the minimum number of general lighting circuits. In our example, the calculated minimum number is 3⅓ circuits. Obviously, there is no such thing as 1/3 of a circuit—you either have a circuit or you do not—so to meet the *NEC*® requirement, we must round up to 4 as the minimum number of general lighting circuits. In our example, if we were installing 20-amp-rated general lighting circuits, a minimum of three 20-amp general lighting circuits would be required: 50 amp/20 amp = 2.5, then round up to 3.

 FROM EXPERIENCE

Most residential general lighting circuits are protected by 15-amp fuses or circuit breakers, although there are many applications where 20-amp overcurrent protection devices are used. Make sure to check the building plans, especially the electrical specifications, to determine whether the general lighting circuits are rated 15 amp or 20 amp.

The following formulas can be used to determine the minimum number of general lighting circuits in a dwelling:

$$\frac{\text{3 volt-amperes} \times \text{calculated square-foot area}}{\text{120 volts}}$$
= **Total general lighting load in amps**

$$\frac{\text{Total general lighting load in amps}}{\text{15 amp}}$$
= **Minimum number of 15-amp general lighting circuits**

$$\frac{\text{Total general lighting load in amps}}{\text{20 amp}}$$
= **Minimum number of 20-amp general lighting circuits**

Table 220.3(A) General Lighting Loads by Occupancy

Type of Occupancy	Unit Load	
	Volt-Amperes per Square Meter	Volt-Amperes per Square Foot
Armories and auditoriums	11	1
Banks	39[b]	3½[b]
Barber shops and beauty parlors	33	3
Churches	11	1
Clubs	22	2
Court rooms	22	2
Dwelling units[a]	33	3
Garages — commercial (storage)	6	½
Hospitals	22	2
Hotels and motels, including apartment houses without provision for cooking by tenants[a]	22	2
Industrial commercial (loft) buildings	22	2
Lodge rooms	17	1½
Office buildings	39	3½[b]
Restaurants	22	2
Schools	33	3
Stores	33	3
Warehouses (storage)	3	¼
In any of the preceding occupancies except one-family dwellings and individual dwelling units of two-family and multifamily dwellings:		
Assembly halls and auditoriums	11	1
Halls, corridors, closets, stairways	6	½
Storage spaces	3	¼

[a]See 220.3(B)(10).
[b]In addition, a unit load of 11 volt-amperes/m² or 1 volt-ampere/ft² shall be included for general-purpose receptacle outlets where the actual number of general-purpose receptacle outlets is unknown.

Figure 7–3 *NEC®* Table 220.3(A) provides the minimum general lighting load per square foot for a variety of building types, including 3 VA per square foot for dwelling units. *Reprinted with permission from NFPA 70-2002, the* National Electrical Code®, *Copyright © 2001, National Fire Protection Association, Quincy, MA 02269. This reprinted material is not the referenced subject, which is represented only by the standard in its entirety.*

Small-Appliance Branch Circuits

A small-appliance branch circuit is a type of branch circuit found in residential wiring that supplies electrical power to receptacles located in kitchens, dining rooms, pantries, and other similar areas (Figure 7–4). Section

FROM EXPERIENCE

Another way to determine the minimum number of general lighting circuits is to determine the total habitable living space for the house using the outside dimensions and assign one 15-ampere general lighting circuit for every 600 square feet. If you are using 20-ampere-rated general lighting circuits, assign one general lighting circuit for every 800 square feet. For example, a 2,400-square-foot house would require a minimum of four 15-amp general lighting circuits (2,400 square feet/600 = 4) or three 20-amp general lighting circuits (2,400 square feet/800 = 3).

210.11(C) states that two or more 20-ampere-rated small-appliance branch circuits must be provided for all receptacle outlets specified by Section 210.52(B). Section 210.52(B) requires two or more 20-ampere circuits for all receptacle outlets for the small-appliance loads, including refrigeration equipment, in the kitchen, dining room, pantry, and breakfast room of a dwelling unit. No fewer than two small-appliance branch circuits must supply the countertop receptacle outlets in kitchens. These circuits may also supply receptacle outlets in the pantry, dining room, and breakfast room as well as an electric clock receptacle and receptacles for gas-fired appliances, but these circuits are to have no other receptacle outlets. In addition, no lighting outlets are allowed to be connected to small-appliance branch circuits.

Section 220.16(A) states that in each dwelling unit, the load must be computed at 1,500 volt-amperes for each two-wire small-appliance branch circuit. This means that a minimum load for small appliance circuits in a house would be 3,000 volt-amperes (minimum of 2 small-appliance branch circuits × 1,500 volt-amperes = 3,000 volt-amperes). If a residential wiring system is to have more than two small-appliance branch circuits installed, each circuit over two must also be calculated at 1,500 volt-amperes. For example, a dwelling unit is to have four small-appliance branch circuits installed. The total load for the small-appliance branch circuits would be 6,000 volt-amperes (4 small-appliance branch circuits × 1,500 VA = 6,000 VA).

Laundry Branch Circuits

A laundry branch circuit is a type of branch circuit found in residential wiring that supplies electrical power to laundry areas (Figure 7–5). The area may be a separate laundry room, or it may be just an area in a basement or a garage. The purpose of this circuit type is to provide electrical power to a washing machine and other laundry-related items, like a clothes iron. If the house will have a gas dryer, the laundry circuit is used to also provide power (120 volt) for that appliance. An electric clothes dryer will require a separate electrical circuit (120/240 volt), and its load will be calcu-

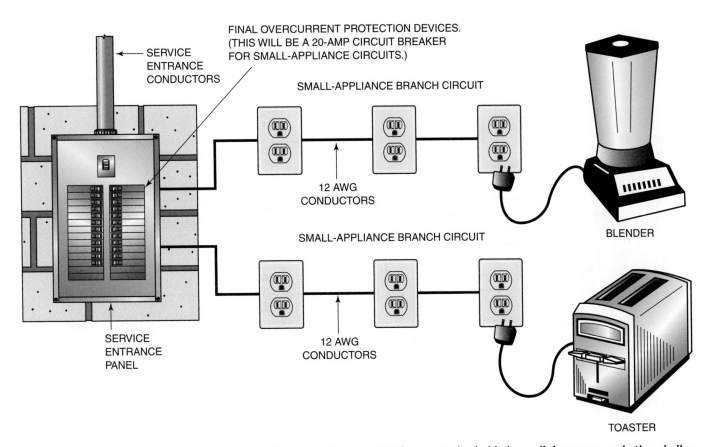

Figure 7–4 A small-appliance branch circuit provides electrical power to the receptacles in kitchens, dining rooms, and other similar areas. Section 210.11(C) requires the installation of at least two 20-ampere-rated small-appliance branch circuits in each dwelling unit.

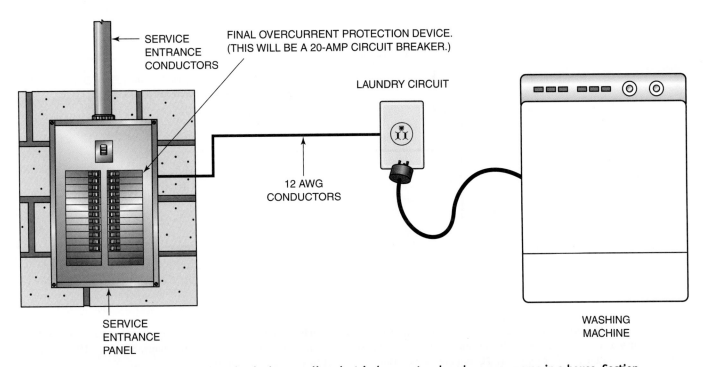

Figure 7–5 A laundry branch circuit is a circuit that supplies electrical power to a laundry room or area in a house. Section 210.11(C)(2) requires the installation of at least one 20-ampere-rated laundry circuit in each dwelling unit.

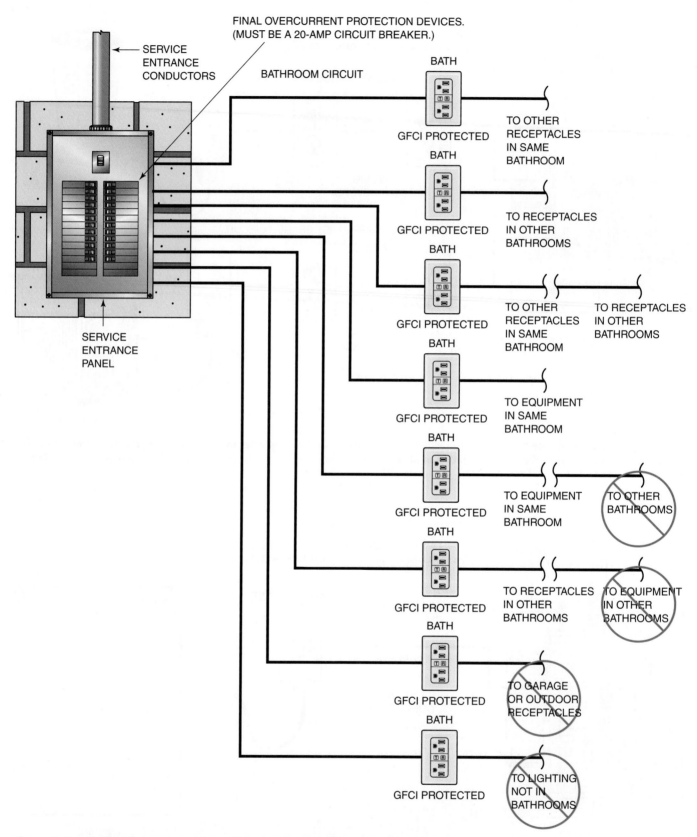

Figure 7–6 A bathroom branch circuit supplies electrical power for receptacles in a bathroom. Luminaires (lighting fixtures) in the bathroom may also be connected to the bathroom circuit as long as the circuit does not feed other bathrooms. Section 210.11(C)(3) requires the installation of at least one 20-ampere-rated bathroom circuit in each dwelling unit.

lated separately (we will look at electric clothes dryer loads later in this chapter). No lighting outlets or receptacles in other rooms may be connected to this circuit. Section 220.16(B) tells us that a load of not less than 1,500 volt-amperes is to be included for each two-wire laundry branch circuit installed. Usually, there is one two-wire laundry circuit installed in each dwelling unit, so the laundry circuit load is normally just 1,500 volt-amperes. Of course, if more two-wire laundry circuits are installed, 1,500 volt-amperes must be calculated for each circuit. Section 210.11(C)(2) states that the laundry circuit must have a 20-ampere rating.

Bathroom Branch Circuits

A bathroom branch circuit is a circuit that supplies electrical power to a bathroom in a residential application (Figure 7–6). A bathroom is defined as an area including a basin with one or more of the following: a toilet, a tub, or a shower. Section 210.11(C)(3) states that in addition to the number of general lighting, small-appliance, and laundry branch circuits, at least one 20-ampere branch circuit must be provided to supply the bathroom receptacle outlet(s). This circuit cannot have any other lighting or receptacle outlets con-

nected to it. An exception allows lighting outlets and other equipment in a bathroom to be connected to it as long as the 20-amp circuit feeds only items in that one bathroom. If the circuit is installed so that it feeds more than one bathroom, only receptacle outlets can be fed by the bathroom circuit.

There is no load allowance given by the *NEC*® to the required bathroom circuit. Be aware that at least one 20-ampere bathroom circuit must be provided for a residential electrical system, but there is no volt-ampere value for this circuit type included in your calculation for the total electrical load of a house.

Individual Branch Circuits

An individual branch circuit is a circuit that supplies only one piece of electrical equipment (Figure 7–7). An example is the circuit that supplies electrical power to one range or one clothes dryer. A circuit rating for a specific appliance is to be computed on the basis of the ampere rating of the appliance. This information is found on the appliance nameplate. For example, a garbage disposal being installed in a house may have a nameplate rating of 6 amps at 120 volts. By multiplying the amperage of the ap-

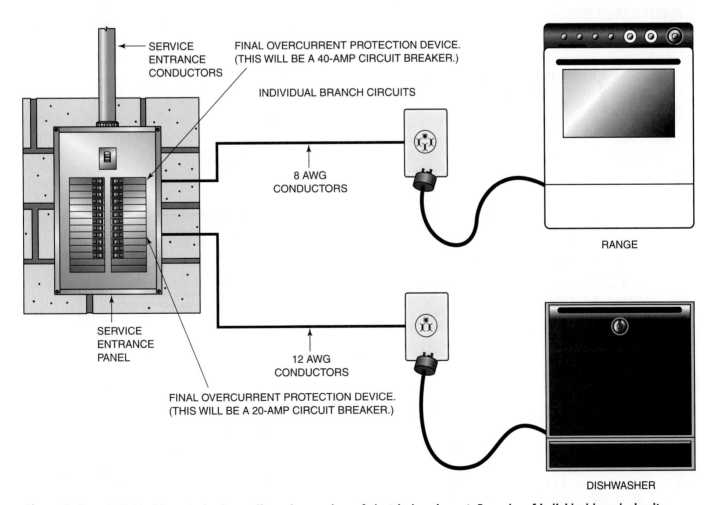

Figure 7–7 An individual branch circuit supplies only one piece of electrical equipment. Examples of individual branch circuits include a range circuit or a dishwasher circuit.

pliance by the voltage you will get the total load for that appliance in volt-amperes: 6 amps × 120 volts = 720 volt-amperes. The individual branch circuit for the garbage disposal in this example would need a capacity of at least 720 volt-amperes. This value will also be included in the service entrance calculation. Some appliances will have a volt-amp rating or wattage rating listed on their nameplate instead of an amperage rating, and this is the rating you would use.

Individual branch-circuit loads for electric clothes dryers and household electric cooking appliances are based on the information found in Section 220.18 for electric dryers and in Section 220.19 for electric ranges and other cooking appliances. Section 220.18 says that the load for household electric clothes dryers in a dwelling unit must be 5,000 watts (volt-amperes) or the nameplate rating, whichever is larger, for each dryer served. In other words, if the dryer nameplate says 4,500 watts, use 5,000 watts of load for that dryer. If the dryer nameplate says 6,000 watts, use 6,000 watts. Section 220.19 states that the demand load for household electric ranges, wall-mounted ovens, counter-mounted cooking units, and other household cooking appliances individually rated in excess of 1¾ kilowatts is permitted to be computed in accordance with Table 220.19. The alternative is to use the nameplate ratings for household cooking equipment, but electricians usually do not do this because Table 220.19 and the accompanying notes apply a demand factor that reduces the load from the nameplate rating. The reason for the reduction in load has to do with the idea that a cooking appliance will not usually have all parts cooking on the "high" setting at the same time. Kilovolt-amperes (kVA) is considered equivalent to kilowatts (kW) for loads computed under this section. The following examples show how to calculate common residential cooking equipment applications:

EXAMPLE 1 A house will have an electric range that has a nameplate rating of 11.5 kW. The first sentence in Note 4 to Table 220.19 says that it is permissible to use Table 220.19 to calculate the load for one range. Use column C of Table 220.19 to get the maximum demand in kilowatts of a range not over 12 kW in rating. The first row is used when there is one range in a house. Since the range being installed is rated at 11.5 kW, use row 1, column C, which says that 8 kW is the value to be used for one range that has a rated value of 12 kW or less (Figure 7–8).

EXAMPLE 2 A house will have an electrical range that has a nameplate rating of 14 kW. Since this value is over 12 kW, column C of Table 220.19 cannot be used without a slight modification. The modification comes from Note 1 to Table 220.19, which says that for every kilowatt over 12 kW, the maximum demand for one range (8 kW) must be increased by 5%. Here is how this is done. The nameplate rating of 14 kW is 2 kW over 12 kW. At an increase of 5% of each kilowatt over 12 kW, the total percentage increase will be 10% (2 × 5% = 10%). Increase the maximum demand for one range (8 kW) by 10% (8 kW × 1.10 = 8,800 W). So the actual value used for loading calculations for a 14-kW-rated range is 8.8 kW (Figure 7–9).

EXAMPLE 3 A house will have a countertop cook unit rated at 5 kW and a built-in oven rated at 7 kW. Both will be connected to the same branch circuit. In this case, Note 4 to Table 220.19 will be used and states that if a single branch circuit supplies a counter-mounted cooking unit and not more than two wall-mounted ovens, all of which are located

Table 220.19 Note 4 Branch-Circuit Load. It shall be permissible to compute the branch-circuit load for one range in accordance with Table 220.19. The branch-circuit load for one wall-mounted oven or one counter-mounted cooking unit shall be the nameplate rating of the appliance. The branch-circuit load for a counter-mounted cooking unit and not more than two wall-mounted ovens, all supplied from a single branch circuit and located in the same room, shall be computed by adding the nameplate rating of the individual appliances and treating this total as equivalent to one range.

MAXIMUM kW FOR ONE RANGE RATED 12 kW OR LESS ACCORDING TO COLUMN C IS 8 kW (8,000 W)

THE MINIMUM CONDUCTOR SIZE IS FOUND BY CALCULATING THE CURRENT AND MATCHING IT TO A WIRE SIZE THAT WILL HANDLE IT.
- 8,000 W ÷ 240 V = 33.33 AMPS
- 8 AWG COPPER IS THE MINIMUM WIRE SIZE ACCORDING TO TABLE 310.16

THE MAXIMUM OVERCURRENT PROTECTION DEVICE WILL BE RATED AT 40 AMPS ACCORDING TO SECTIONS 210.19(A)(3), 240.4, AND 240.6

120/240-VOLT 11.5-kW RATING

Figure 7–8 A range calculation for a single electric range rated not more than 12 kW.

Table 220.19 Note 1 Over 12kW through 27 kW ranges all of same rating. For ranges individually rated more than 12 kW but not more than 27 kW, the maximum demand in Column C shall be increased 5 percent for each additional kilowatt of rating or major fraction thereof by which the rating of individual ranges exceeds 12 kW.

MAXIMUM kW FOR A RANGE RATED MORE THAN 12 kW IS FOUND BY:

- INCREASING THE MAXIMUM kW FOR ONE RANGE NOT RATED MORE THAN 12 kW BY 5% FOR EACH kW OVER 12 kW:
  * 14 kW − 12 kW = 2 kW
  * 2 kW × 5% = 10%
  * 8 kW × 1.10 = 8.8 kW

THE MINIMUM CONDUCTOR SIZE IS FOUND BY CALCULATING THE CURRENT AND MATCHING IT TO A WIRE SIZE THAT WILL HANDLE IT.

- 8.8 kW ÷ 240 V = 36.67 AMPS
- 8 AWG CU IS THE MINIMUM WIRE SIZE ACCORDING TO TABLE 310.16

THE MAXIMUM OVERCURRENT PROTECTION DEVICE WILL BE RATED AT 40 AMPS ACCORDING TO SECTIONS 240.4 AND 240.6.

120/240-VOLT 14-kW RATING

Figure 7–9 A range calculation for a single electric range rated more than 12 kW.

in the same room, the nameplate ratings of these appliances can be added together and the total treated as the equivalent of one range. In this example, the 5 kW countertop unit and the 7 kW over are added together (5 kW + 7 kW = 12 kW). Now use column C of Table 220.19 to get the maximum demand in kilowatts of a range not over 12 kW in rating. Since the combined rating of the cooktop and oven being installed is 12 kW, use row 1, column C, which says that 8 kW is the value to be used for this combination (Figure 7–10).

For household electric ranges and other cooking appliances, the size of the conductors must be determined by the rating of the range. The rating is calculated as shown earlier. The minimum conductor size for the range in Example 1 would be an 8 AWG copper conductor with 60°C insulation. Note that Section 210.19(A)(3) does not permit the branch-circuit rating (fuse or circuit breaker) of a circuit supplying household ranges with a nameplate rating of 8¾ kW or more to be less than 40 amperes.

One other item to consider when getting ready to calculate the total electrical load for a residential service entrance is what the *NEC®* calls "noncoincident loads." Section 220.21 states that where it is unlikely that two or more loads will be in use simultaneously, it is permissible to use only the largest load that will be used at one time in computing the total load of a feeder or service. This can be applied to the individual branch-circuit loading for air-conditioning equipment and electric heating equipment. For example, if a house had 10 kW of electric heating load and 8 kW of air-conditioning load, the 8 kW of air-conditioning load can be eliminated from the total service entrance load calculation because the cooling load will not be used at the same time as the heating load. Therefore, if the

service entrance is sized using the larger of the two loads, there will be more than enough service capacity to supply the smaller cooling load when it is used.

Determining the Ampacity of a Conductor

Ampacity is defined as the current, in amperes, that a conductor can carry continuously under the conditions of use without exceeding its temperature rating. Section 210.19(A) requires branch-circuit conductors to have an ampacity at least equal to the maximum electrical load to be served. This rule makes a lot of sense when you think about it. For example, if you are wiring a branch circuit that will supply electrical power to a load that will draw 16 amperes, you need to use a wire size for the branch circuit that has an ampacity of at least 16 amps. Otherwise, the conductor will heat up and cause the insulation on the conductor to melt.

Electricians use Table 310.16 (Figure 7–11) to determine the ampacity of a conductor for use in residential wiring. You were introduced to this table in Chapter 2, but now we will look at it more closely and see how to use it. The table is set up so that the left half covers copper conductors and the right half covers aluminum and copper-clad aluminum conductors. Since the vast majority of the conductors used for branch-circuit wiring in residential applications are copper, we will concentrate on the copper half of the table. On the far-left-hand side of the table, you will see a column titled "Size AWG or kcmil." This column represents all the wire sizes you will encounter in residential wiring. The next three columns to the right are titled "60°C (140°F)," "75°C (167°F)," and "90°C

Table 220.19 **Note 4** Branch-Circuit Load. It shall be permissible to compute the branch-circuit load for one range in accordance with Table 220.19. The branch-circuit load for one wall-mounted oven or one counter-mounted cooking unit shall be the nameplate rating of the appliance. The branch-circuit load for a counter-mounted cooking unit and not more than two wall-mounted ovens, all supplied from a single branch circuit and located in the same room, shall be computed by adding the nameplate rating of the individual appliances and treating this total as equivalent to one range.

MAXIMUM kW FOR A COOKTOP AND NOT MORE THAN 2 OVENS IS FOUND BY:

    ADDING TOGETHER THE NAMEPLATE RATINGS AND
    TREATING THEM LIKE ONE RANGE.
    * 7 kW + 5 kW = 12 kW
    THE MAXIMUM kW FOR ONE RANGE RATED 12 kW OR
    LESS ACCORDING TO COLUMN C IN TABLE 220.19 IS 8 kW.

THE MINIMUM CONDUCTOR SIZE IS FOUND BY CALCULATING THE CURRENT AND MATCHING IT TO A WIRE SIZE THAT WILL HANDLE IT.

- 8,000 W ÷ 240 V = 33.33 AMPS
- 8 AWG COPPER IS THE MINIMUM WIRE SIZE ACCORDING TO TABLE 310.16

THE MAXIMUM OVERCURRENT PROTECTION DEVICE WILL BE RATED AT 40 AMPS ACCORDING TO SECTIONS 210.19(A)(3), 240.4, AND 240.6.

Figure 7–10 A calculation for a 5 kW cooktop and a 7 kW oven. Both appliances are connected to the same branch circuit.

120/240-V 7-kW OVEN

120/240-V 5 kW-COOKTOP

#8/3 NMSC

BOTH COOKING APPLIANCES ARE CONNECTED TO THE SAME BRANCH CIRCUIT.

(194°F)." These temperatures represent the maximum temperature ratings for the conductor insulation types listed in the columns under the temperature ratings. Temperature ratings for conductors, along with other insulation properties, are located in Table 310.13 of the *NEC®*. For example, a "TW" insulated conductor is rated at 60°C, a "THW" insulated conductor is rated at 75°C, and a "THHW" insulated conductor is rated at 90°C. The numbers in the table that correspond to a particular conductor size with a specific insulation type represent the ampacity of that conductor. For example, the ampacity of a 10 AWG copper conductor with a TW insulation is 30 amperes. The same conductor size with THW insulation has an ampacity of 35 amperes, and with THHW insulation it has an ampacity of 40 amperes. A conductor with a higher-temperature-rated insulation will have a higher-rated ampacity. Let's look at a few more examples:

- 12 AWG copper conductor with TW (60°C) insulation has an ampacity of 20 amps.
- 8 AWG copper conductor with TW (60°C) insulation has an ampacity of 40 amps.
- 6 AWG copper conductor with TW (60°C) insulation has an ampacity of 50 amps.
- 4 AWG copper conductor with THW (75°C) insulation has an ampacity of 85 amps.
- 4/0 copper conductor with THHN (90°C) insulation has an ampacity of 260 amps.

CAUTION

CAUTION: Next to the 14, 12, and 10 AWG wire sizes in Table 310.16, you will see an asterisk (*). At the bottom of Table 310.16, the asterisk tells us to refer to Section 240.4(D). This section states that the overcurrent protection must not exceed 15 amperes for 14 AWG, 20 amperes for 12 AWG, and 30 amperes for 10 AWG copper or 15 amperes for 12 AWG and 25 amperes for 10 AWG aluminum and copper-clad aluminum. So, no matter what the actual ampacity is for 14, 12, and 10 AWG conductors according to Table 310.16, the fuse or circuit breaker size can never be larger than 15 amp for 14 AWG, 20 amp for 12 AWG, and 30 amp for 10 AWG copper conductors for residential branch-circuit applications.

The ampacity of a conductor as determined in Table 310.16 and used in residential applications may have to be modified slightly. As the "conditions of use" change for a conductor, the ampacity is required to be adjusted so that at no time could the temperature of the conductor exceed the temperature of

Table 310.16 Allowable Ampacities of Insulated Conductors Rated 0 Through 2000 Volts, 60°C Through 90°C (140°F Through 194°F), Not More Than Three Current-Carrying Conductors in Raceway, Cable, or Earth (Directly Buried), Based on Ambient Temperature of 30°C (86°F)

	Temperature Rating of Conductor (See Table 310.13.)						
	60°C (140°F)	75°C (167°F)	90°C (194°F)	60°C (140°F)	75°C (167°F)	90°C (194°F)	
Size AWG or kcmil	Types TW, UF	Types RHW, THHW, THW, THWN, XHHW, USE, ZW	Types TBS, SA, SIS, FEP, FEPB, MI, RHH, RHW-2, THHN, THHW, THW-2, THWN-2, USE-2, XHH, XHHW, XHHW-2, ZW-2	Types TW, UF	Types RHW, THHW, THW, THWN, XHHW, USE	Types TBS, SA, SIS, THHN, THHW, THW-2, THWN-2, RHH, RHW-2, USE-2, XHH, XHHW, XHHW-2, ZW-2	Size AWG or kcmil
	COPPER			ALUMINUM OR COPPER-CLAD ALUMINUM			
18	—	—	14	—	—	—	—
16	—	—	18	—	—	—	—
14*	20	20	25	—	—	—	—
12*	25	25	30	20	20	25	12*
10*	30	35	40	25	30	35	10*
8	40	50	55	30	40	45	8
6	55	65	75	40	50	60	6
4	70	85	95	55	65	75	4
3	85	100	110	65	75	85	3
2	95	115	130	75	90	100	2
1	110	130	150	85	100	115	1
1/0	125	150	170	100	120	135	1/0
2/0	145	175	195	115	135	150	2/0
3/0	165	200	225	130	155	175	3/0
4/0	195	230	260	150	180	205	4/0
250	215	255	290	170	205	230	250
300	240	285	320	190	230	255	300
350	260	310	350	210	250	280	350
400	280	335	380	225	270	305	400
500	320	380	430	260	310	350	500
600	355	420	475	285	340	385	600
700	385	460	520	310	375	420	700
750	400	475	535	320	385	435	750
800	410	490	555	330	395	450	800
900	435	520	585	355	425	480	900
1000	455	545	615	375	445	500	1000
1250	495	590	665	405	485	545	1250
1500	520	625	705	435	520	585	1500
1750	545	650	735	455	545	615	1750
2000	560	665	750	470	560	630	2000

CORRECTION FACTORS

Ambient Temp. (°C)	For ambient temperatures other than 30°C (86°F), multiply the allowable ampacities shown above by the appropriate factor shown below.						Ambient Temp. (°F)
21–25	1.08	1.05	1.04	1.08	1.05	1.04	70–77
26–30	1.00	1.00	1.00	1.00	1.00	1.00	78–86
31–35	0.91	0.94	0.96	0.91	0.94	0.96	87–95
36–40	0.82	0.88	0.91	0.82	0.88	0.91	96–104
41–45	0.71	0.82	0.87	0.71	0.82	0.87	105–113
46–50	0.58	0.75	0.82	0.58	0.75	0.82	114–122
51–55	0.41	0.67	0.76	0.41	0.67	0.76	123–131
56–60	—	0.58	0.71	—	0.58	0.71	132–140
61–70	—	0.33	0.58	—	0.33	0.58	141–158
71–80	—	—	0.41	—	—	0.41	159–176

* See 240.4(D).

Figure 7–11 *NEC*® Table 310.16 is used by electricians to determine the minimum conductor size needed for a certain ampacity.

Reprinted with permission from NFPA 70-2002, the National Electrical Code®, Copyright © 2001, National Fire Protection Association, Quincy, MA 02269. This reprinted material is not the referenced subject, which is represented only by the standard in its entirety.

the insulation on the conductor. The following adjustments may have to be made for certain residential situations:

1. The ampacities given in Table 310.16 are based on an **ambient temperature** of no more than 30°C (86°F). When the ambient temperature that a conductor is exposed to exceeds 30°C (86°F), the ampacities from Table 310.16 must be multiplied by the appropriate correction factor shown at the bottom of Table 310.16. You will notice that the far-left-hand column for the correction factors is the ambient temperature in degrees Celsius. The far-right-hand column has temperatures listed in degrees Fahrenheit. Let's look at an example where the ambient temperature correction factors must be applied (Figure 7–12).

A 3 AWG copper conductor with THW insulation has an ampacity of 100 amperes according to Table 310.16. However, if an electrician installs the conductor in an area where the ambient temperature was 100°F, the actual ampacity of the wire can be no more than 88 amperes. Here is how you would determine this. First, find the 100°F ambient temperature in the far-right-side column. 100°F falls between 96 and 104, so we will use that row. Follow the row to the left until you find the number that is in the column where the 3 AWG copper conductor with THW insulation is located. You will see that the correction factor is 0.88. Now multiply the original ampacity by the correction factor to get the new allowable ampacity: 100 amps × .88 correction factor = 88 amps.

2. The ampacities listed in Table 310.16 are also based on not having any more than three current-carrying conductors in a raceway or cable. Section 310.15(B)(2)(a) states that where the number of current-carrying conductors in a raceway or cable exceeds three or where single conductors or multiconductor cables are stacked or bundled longer than 24 inches (600 mm) without maintaining spacing and are not installed in raceways, the allowable ampacity of each conductor must be reduced by the adjustment factor listed in Table 310.15(B)(2)(a) (Figure 7–13). This is called "derating" and it results in a lower conductor ampacity when there are more than three current-carrying conductors in a cable or raceway. The reason for this reduction in ampacity is the fact that when you have more current-carrying conductors close to each other, there is an overall increase in temperature, which could result in damage to the conductor insulation. Let's look at an example of conductor derating for more than three current-carrying conductors (Figure 7–14).

It is common wiring practice in residential situations to drill holes in building framing members and run several cables through the holes. If the length of these "bundled" cables exceeds 24 inches and the total number of current-carrying conductors exceeds three, the ampacity of the conductors in the cables must be derated. Let's say that an electrician bundles four 12/2 nonmetallic sheathed cables through the drilled holes of some wall studs in a house. This installation results in having eight current-carrying conductors. According to Table 310.16, the ampacity of a 12

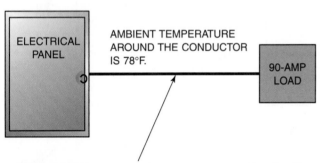

3 AWG COPPER WITH THW INSULATION INSTALLED IN AN AMBIENT TEMPERATURE NO GREATER THAN 30°C (86°F). THE AMPACITY OF THE CONDUCTOR IS 100 AMPS ACCORDING TO TABLE 310.16. THIS CONDUCTOR SIZE WILL HANDLE THE 90-AMP LOAD

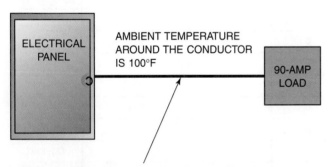

3 AWG COPPER WITH THW INSULATION INSTALLED IN AN AMBIENT TEMPERATURE THAT IS 100°F. THE AMPACITY OF THE CONDUCTOR IS NOW REQUIRED TO BE ADJUSTED. THE ADJUSTMENT FACTOR IS .88, SO 100 AMPS X .88 = 88 AMPS. THE NEW AMPACITY OF THE CONDUCTOR IS 88 AMPS, WHICH WILL NOT HANDLE THE 90-AMP LOAD. A LARGER WIRE WILL HAVE TO BE USED

Figure 7–12 An example showing the application of the ambient temperature correction factor.

Table 310.15(B)(2)(a) Adjustment Factors for More Than Three Current-Carrying Conductors in a Raceway or Cable

Number of Current-Carrying Conductors	Percent of Values in Tables 310.16 through 310.19 as Adjusted for Ambient Temperature if Necessary
4–6	80
7–9	70
10–20	50
21–30	45
31–40	40
41 and above	35

Figure 7–13 *NEC® Table 310.15(B)(2)(a) Reprinted with permission from NFPA 70-2002, the National Electrical Code®, Copyright © 2001, National Fire Protection Association, Quincy, MA 02269. This reprinted material is not the referenced subject, which is represented only by the standard in its entirety.*

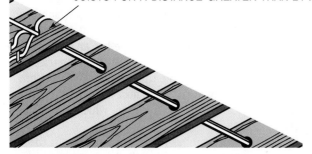

FOUR 12/2 NONMETALLIC SHEATHED CABLES INSTALLED THROUGH BORED HOLES IN JOISTS FOR A DISTANCE GREATER THAN 24 IN.

- TABLE 310.16 SHOWS AN AMPACITY OF 25 AMPS FOR A 12 AWG NMSC (60°C COLUMN).
- THERE ARE EIGHT CURRENT-CARRYING CONDUCTORS IN THE FOUR 12/2 CABLES IN THIS INSTALLATION. TABLE 310.15 (B)(2)(a) REQUIRES A DERATING FACTOR OF 70% BE APPLIED WHEN SEVEN TO NINE CONDUCTORS ARE RUN TOGETHER IN A RACEWAY, A CABLE, OR CABLES BUNDLED TOGETHER FOR LENGTHS OF MORE THAN 24 IN.
- 25 AMPS X .70 = 17.5 AMPS. THEREFORE, THE AMPACITY OF THE CONDUCTORS FOR THIS INSTALLATION IS 17.5 AMPERES.

Figure 7–14 An example of a residential wiring situation where there are more than three current-carrying conductors. This wiring technique is called "bundling."

AWG nonmetallic sheathed cable (using the "60°C" column) is 25 amperes. Table 310.15(B)(2)(a) tells us that when seven to nine current-carrying conductors are in the same cable or raceway, a reduction to 70% of the conductor ampacity is required. This will mean that the ampacity of the 12 AWG Romex cables is 17.5 amperes: 25 amps × .70 = 17.5 amps.

CAUTION

CAUTION: Sometimes a conductor is installed in an ambient temperature that is greater than 30°C (86°F) and there are more than three conductors in the same raceway or cable. If this is the case, a "double derating" will need to take place. Here is an example. Suppose an electrician installs five current-carrying conductors in a raceway. The raceway is located in an area where the ambient temperature is 100°F. If the conductor used is a 3 AWG copper with THW insulation, the normal ampacity from Table 310.16 is 100 amperes. However, you need to apply the ambient temperature correction factor of .88 *and* the derating factor of .80 for more than three current-carrying conductors: 100 amps × .88 × .80 = 70.4 amps.

FROM EXPERIENCE

Exception No. 3 to Section 310.15(B)(2)(a) states that derating factors shall not apply to conductors in nipples having a length not exceeding 24 inches (600 mm). This exception may apply to some residential wiring applications where a 24 inch or less length of electrical conduit, called a **nipple**, is placed between two electrical enclosures. In this case, more than three current-carrying conductors can be run in a nipple, and no derating will have to be done.

3. Section 110.14(C) requires an electrician to take a look at the temperature ratings of the equipment terminals that the electrical conductors will be connected to. The temperature rating (60°C, 75°C, or 90°C) associated with the ampacity of a conductor from Table 310.16 must be selected, so the temperature produced by that conductor cannot exceed the lowest temperature rating of any connected termination, conductor, or device. If the temperature produced by a current-carrying conductor at a termination does exceed the termination temperature rating, the termination will burn up. For example, the load on an 8 AWG THHN, 90°C copper wire is limited to 40 amperes (the 60°C ampacity) where connected to electrical equipment with terminals rated at 60°C. This same 8 AWG THHN, 90°C wire is limited to 50 amperes (the 75°C ampacity) where connected to electrical equipment with terminals rated at 75°C. Unless the electrical equipment is listed or marked otherwise, you must determine the Table 310.16 ampacity of a conductor for residential applications using the information outlined next.

CAUTION

CAUTION: Do not forget that any electrical circuit has a beginning and an end. You must always know what the termination temperature rating is at the panel and the temperature rating at the equipment (like switches and receptacles) at the end of the circuit. A typical residential circuit may have a 60°C rating at one end and a 75°C rating at the other. Conductor ampacity in this case would have to be based on the "60°C" column in Table 310.16.

Termination provisions of equipment for circuits rated 100 amperes or less, or marked for 14 AWG through 1 AWG conductors, must be used only for one of the following:

- Conductors rated 60°C (140°F): This means that for determining the ampacity of a conductor used in circuits rated 100 amps or less or wire sizes of 14 AWG through 1 AWG, always use the ampacity in the "60°C" column of Table 310.16 for the wire size you wish to use. The only conductor insulation shown in the "60°C" column is Type TW and Type UF. For example, it is determined that an electric dryer you have to install in a house requires a minimum 30-amp circuit. Because it is a circuit that is 100 amps or less, you go to the "60°C" column of Table 310.16 to find a wire size that has a minimum ampacity of 30 amps. In this case, you would use a 10 AWG TW insulated conductor (Figure 7–15).

- Conductors with higher temperature ratings, provided that the ampacity of such conductors is determined by the 60°C (140°F) ampacity of the conductor size used: This means that if you want to use conductors with in-

sulations whose temperature ratings are 75°C or 90°C, you may, but you must determine their ampacity from the "60°C" column in Table 310.16. For example; a 10 AWG copper conductor with a 75°C insulation has an ampacity of 35 amps, and the same 10 AWG conductor with a 90°C insulation type has an ampacity of 40 amps. Both wire insulation types could be used in the electric dryer example from the previous section, but the ampacity of the wire must be determined from the "60°C" column of Table 310.16, which is 30 amps (Figure 7–16).

- Conductors with higher temperature ratings if the equipment is listed and identified for use with such conductors: This means that if equipment terminations are rated for higher temperatures than 60°C,

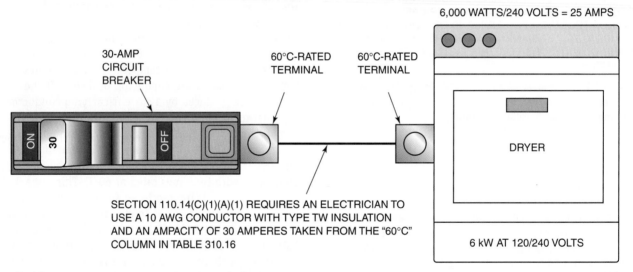

Figure 7–15 An example for determining the ampacity of a conductor when the circuit rating is 100 amps or less and the conductor size is no larger than 1 AWG. Section 110.14(C)(1)(a)(1).

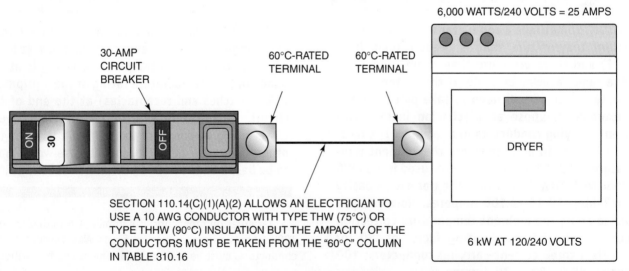

Figure 7–16 An example for determining the ampacity of a conductor when the circuit rating is 100 amps or less, the conductor size is no larger than 1 AWG, and the conductor insulation is 75°C or 90°C. Section 110.14(C)(1)(a)(2).

the ampacity for a conductor with a 75°C ampacity or 90°C ampacity can be used. For example, if the electric dryer from the previous section has terminals marked for 75°C, you can determine the ampacity of conductor using the "75°C" column of Table 310.16. The conductor would have to have either a 75°C or 90°C insulation type (Figure 7–17).

Termination provisions of equipment for circuits rated over 100 amperes or marked for conductors larger than 1 AWG must be used only for one of the following:

- Conductors rated 75°C (167°F): This means that for determining the ampacity of a conductor used in circuits rated more than 100 amps or when wire sizes larger than 1 AWG are used, always use the ampacity in the "75°C" column of Table 310.16 for the wire

size you wish to use. For example, it is determined that an electric furnace you have to install in a house requires a minimum 150-amp circuit. Because it is a circuit that is more than 100 amps and the terminals are marked 75°C, you go to the "75°C" column of Table 310.16 to find a wire size that has a minimum ampacity of 150 amps. In this case, you could use a 1/0 AWG THW insulated conductor (Figure 7–18).

- Conductors with higher temperature ratings, provided that the ampacity of such conductors does not exceed the 75°C (167°F) ampacity of the conductor size used or up to their ampacity if the equipment is listed and identified for use with such conductors: This means that if you want to use conductors with insulations whose temperature rating is 90°C, you may, but you

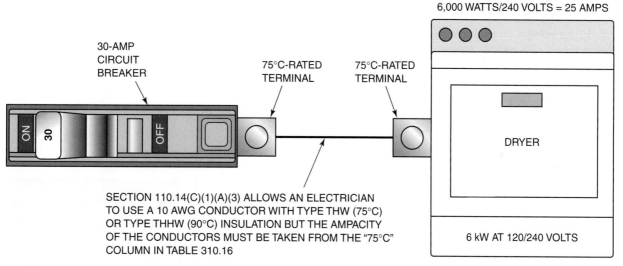

6,000 WATTS/240 VOLTS = 25 AMPS

30-AMP CIRCUIT BREAKER

75°C-RATED TERMINAL

75°C-RATED TERMINAL

ON 30 OFF

DRYER

6 kW AT 120/240 VOLTS

SECTION 110.14(C)(1)(A)(3) ALLOWS AN ELECTRICIAN TO USE A 10 AWG CONDUCTOR WITH TYPE THW (75°C) OR TYPE THHW (90°C) INSULATION BUT THE AMPACITY OF THE CONDUCTORS MUST BE TAKEN FROM THE "75°C" COLUMN IN TABLE 310.16

Figure 7–17 **An example for determining the ampacity of a conductor when the circuit rating is 100 amps or less, the conductor size is no larger than 1 AWG, and the termination ratings are higher than 60°C. Section 110.14(C)(1)(a)(3).**

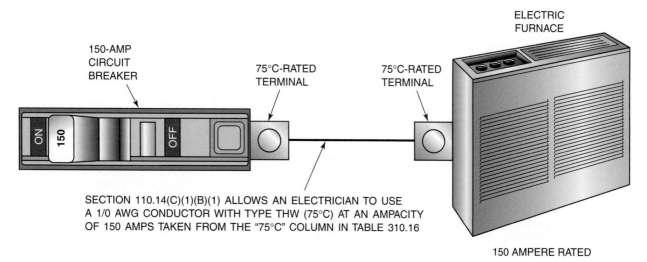

ELECTRIC FURNACE

150-AMP CIRCUIT BREAKER

75°C-RATED TERMINAL

75°C-RATED TERMINAL

ON 150 OFF

SECTION 110.14(C)(1)(B)(1) ALLOWS AN ELECTRICIAN TO USE A 1/0 AWG CONDUCTOR WITH TYPE THW (75°C) AT AN AMPACITY OF 150 AMPS TAKEN FROM THE "75°C" COLUMN IN TABLE 310.16

150 AMPERE RATED

Figure 7–18 **An example for determining the ampacity of a conductor when the circuit rating is more than 100 amps and the conductor size is larger than 1 AWG. Section 110.14(C)(1)(b)(1).**

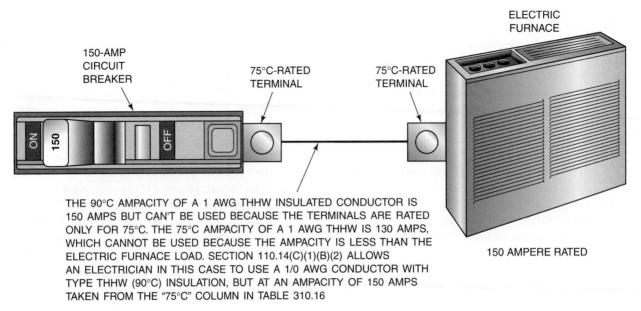

THE 90°C AMPACITY OF A 1 AWG THHW INSULATED CONDUCTOR IS 150 AMPS BUT CAN'T BE USED BECAUSE THE TERMINALS ARE RATED ONLY FOR 75°C. THE 75°C AMPACITY OF A 1 AWG THHW IS 130 AMPS, WHICH CANNOT BE USED BECAUSE THE AMPACITY IS LESS THAN THE ELECTRIC FURNACE LOAD. SECTION 110.14(C)(1)(B)(2) ALLOWS AN ELECTRICIAN IN THIS CASE TO USE A 1/0 AWG CONDUCTOR WITH TYPE THHW (90°C) INSULATION, BUT AT AN AMPACITY OF 150 AMPS TAKEN FROM THE "75°C" COLUMN IN TABLE 310.16

Figure 7–19 An example for determining the ampacity of a conductor when the circuit rating is more than 100 amps, the conductor size is larger than 1 AWG, and the conductor insulation is 90°C. Section 110.14(C)(1)(b)(2).

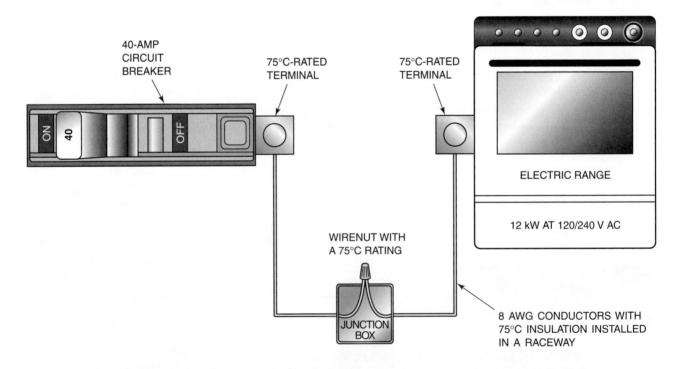

SECTION 110.14(C)(2) REQUIRES WIRE CONNECTORS, LIKE WIRENUTS, TO HAVE A TEMPERATURE RATING AT LEAST EQUAL TO THE TEMPERATURE RATING OF THE CONDUCTORS THEY ARE BEING USED ON. IN THIS CASE, AN ELECTRICIAN COULD *NOT* USE WIRE NUTS WITH A 60°C RATING BECAUSE THE CIRCUIT CONDUCTORS HAVE A 75°C RATING

Figure 7–20 Wirenuts and other wire connector types must have a temperature rating that is not less than the temperature rating of the conductors they are being used with. Section 110.14(C)(2).

must determine the ampacity from the "75°C" column in Table 310.16. For example, a 1 AWG copper conductor with a 90°C insulation has an ampacity of 150 amps. But the ampacity of the wire must be determined from the "75°C" column of Table 310.16, which is 130 amps. This ampacity is less than the required circuit ampacity. A 1/0 AWG copper conductor with a 90°C insulation could be used, but the ampacity is taken from the "75°C" column in Table 310.16 (Figure 7–19).

Separately installed pressure connectors must be used with conductors with an ampacity not exceeding the ampacity for the listed and identified temperature rating of the connector. This means that when using wire connectors such as wirenuts or split bolt connectors, the temperature rating of the connector must be equal to or greater than the temperature rating of the wire it is being used with. For example, say you are splicing wires in a junction box that has 75°C insulation on the wires and the ampacity of the wires has been determined using the "75°C" column in Table 310.16. The wirenuts you are using must have at least a 75°C temperature rating. If they had only a 60°C temperature rating, they could burn up (Figure 7–20).

Branch-Circuit Ratings and Sizing the Overcurrent Protection Device

The branch-circuit rating is based on the size of the fuse or circuit breaker protecting the circuit according to Section 210.3 in the *NEC®*. This section also states that standard branch-circuit ratings for receptacle and lighting circuits are 15, 20, 30, 40, and 50 amp. The circuit ratings for receptacles and lighting in a residential wiring system are usually 15 or 20 ampere. Individual branch-circuit ratings can be any am-

FROM EXPERIENCE

In residential wiring, it is highly unlikely that you will ever use equipment that has 90°C-rated terminations. The majority of the circuits in residential wiring are 100 ampere or less, and the circuit wire sizes are smaller than 1 AWG. This means that unless you know for sure that both ends (and the middle if spliced in a junction box) of your circuit conductors are to be terminated on equipment that is rated at 75°C, always determine your conductor ampacity based on the "60°C" column of Table 310.16. Also, nonmetallic sheathed cable, the most popular residential wiring method, always has its ampacity determined from the "60°C" column in Table 310.16.

pere rating, depending on the size of the load served. Standard sizes of fuses and circuit breakers are found in Section 240.6. There are many standard sizes available. A 200-ampere fuse or circuit breaker is usually the largest overcurrent protection device used in residential wiring. However, for very large homes, you may encounter service entrance main fuses or circuit breakers that could be larger than 200 ampere. The standard fuse or circuit breaker sizes through 200 ampere include ampere ratings of 15, 20, 25, 30, 35, 40, 45, 50, 60, 70, 80, 90, 100, 110, 125, 150, 175, and 200.

Table 210.24 (Figure 7–21) summarizes the requirements for the size of conductors and the size of the overcurrent protection for branch circuits where two or more outlets are required. The wire sizes listed in the table are for copper conductors. If the ampacity of a conductor in Table 310.16 does not match the rating of the standard overcurrent device,

Table 210.24 Summary of Branch-Circuit Requirements

Circuit Rating	15 A	20 A	30 A	40 A	50 A
Conductors (min. size):					
Circuit wires[1]	14	12	10	8	6
Taps	14	14	14	12	12
Fixture wires and cords — See 240.5					
Overcurrent Protection	**15 A**	**20 A**	**30 A**	**40 A**	**50 A**
Outlet devices:					
Lampholders permitted	Any type	Any type	Heavy duty	Heavy duty	Heavy duty
Receptacle rating[2]	15 max. A	15 or 20 A	30 A	40 or 50 A	50 A
Maximum Load	**15 A**	**20 A**	**30 A**	**40 A**	**50 A**
Permissible load	See 210.23(A)	See 210.23(A)	See 210.23(B)	See 210.23(C)	See 210.23(C)

[1]These gauges are for copper conductors.

[2]For receptacle rating of cord-connected electric-discharge luminaires (lighting fixtures), see 410.30(C).

Figure 7–21 *NEC®* **Table 210.24 Summary of Branch-Circuit Requirements.** *Reprinted with permission from NFPA 70-2002, the* National Electrical Code®, *Copyright © 2001, National Fire Protection Association, Quincy, MA 02269. This reprinted material is not the referenced subject, which is represented only by the standard in its entirety.*

Section 240.4 permits the use of the next-larger standard overcurrent device for those circuits found in residential wiring systems. However, if the ampacity of a conductor matches the standard rating of Section 240.6, that conductor must be protected at the standard size device. For example, in Table 310.16, a 3 AWG copper conductor with THW insulation has an ampacity of 100 amperes. This conductor would be protected by a 100-amp overcurrent device. On the other hand, a 6 AWG copper conductor with THW insulation has a Table 310.16 ampacity of 65 amperes. Since there is not a 65-amp standard size of fuse or circuit breaker, it is permissible to go up to the next-higher standard size, which is 70 amperes. Do not forget that based on the information in Section 240.4(D), residential branch circuits using 14, 12, or 10 AWG copper conductors must be protected by fuse or circuit breakers rated 15 amp for the 14 AWG, 20 amp for the 12 AWG, and 30 amp for the 10 AWG conductor sizes.

Sizing the Service Entrance Conductors

For dwelling units, wire sizes as listed in Table 310.15(B)(6) (Figure 7–22) are permitted as 120/240-volt, three-wire, single-phase service entrance conductors, service lateral conductors, and feeder conductors that serve as the main power feeder to a dwelling unit and are installed in raceway or cable. The wire sizes allowed in Table 310.15(B)(6) are smaller than the wire sizes given in Table 310.16 for a par-

ticular ampacity. This reduction in wire size reflects the fact that residential service entrances are not typically loaded very heavily and the heavier loads that are on the service are not operated for long time periods. Also, the total residential electrical load is not usually energized at the same time. To be able to use this table for feeder sizing, the feeder between the main disconnect and the lighting and appliance branch-circuit panelboard must carry the total residential load. An example of a feeder carrying the total load would be when a service entrance is installed on the side of an attached garage. The main service disconnect equipment would be located in the garage. A feeder can be taken from the garage and used to feed a loadcenter in a basement under the main part of the house. In this case, the feeder will carry the total residential load and could be sized according to Table 310.15(B)(6). The service entrance grounded conductor, or service neutral, is permitted to be smaller than the ungrounded conductors as long as it is never sized smaller than the grounding electrode conductor.

Now that you have an idea of the types of circuits found in a residential wiring system, the loading associated with those circuits, and how to size circuit conductors, the calculation of the dwelling service entrance can be covered. Two methods for calculating the minimum-size service entrance for dwellings are covered in Article 220. Article 220, Parts I and II, covers the "standard method." Article 220, Part III, covers the "optional method." It is easier to make the service entrance calculations when there is a series of steps to follow.

(See page 200 for the steps used for calculating a residential service entrance using the standard method.)

(See page 202 for the steps used for calculating a residential service entrance using the optional method.)

Table 310.15(B)(6) Conductor Types and Sizes for 120/240-Volt, 3-Wire, Single-Phase Dwelling Services and Feeders.

Conductor (AWG or kcmil)		Service or Feeder Rating (Amperes)
Copper	Aluminum or Copper-Clad Aluminum	
4	2	100
3	1	110
2	1/0	125
1	2/0	150
1/0	3/0	175
2/0	4/0	200
3/0	250	225
4/0	300	250
250	350	300
350	500	350
400	600	400

Figure 7–22 *NEC*® **Table 310.15(B)(6) Conductor Types and Sizes for 120/240-Volt, 3-Wire, Single-Phase Dwelling Services and Feeders.** *Reprinted with permission from NFPA 70-2002, the National Electrical Code®, Copyright © 2001, National Fire Protection Association, Quincy, MA 02269. This reprinted material is not the referenced subject, which is represented only by the standard in its entirety.*

Service Entrance Calculation Examples

The following examples show you how to apply the steps for calculating a residential service entrance using either the standard or the optional method. For this example, let's assume that a dwelling unit has a habitable floor area of 1,800 square feet and has the electrical equipment listed here:

1. Electric range		12 kW	120/240 volt
2. Electric water heater		4 kW	240 volt
3. Electric clothes dryer		6 kW	120/240 volt
4. Garbage disposal		6.2 amps	120 volt
5. Dishwasher			
	a. Motor	7.5 amps	120 volt
	b. Heater	1,100 watts	120 volt
1. Trash compactor		8 amps	120 volt
2. Freezer		6.8 amps	120 volt
3. Garage door opener		6.0 amps	120 volt
4. Attic exhaust fan		3.5 amps	120 volt
5. Central air conditioner		10 kW	240 volt
6. Electric furnace		20 kW	240 volt

By following the steps outlined next, you can easily calculate the minimum-size residential service entrance. Each step has a short explanation and a reference from the *NEC*®. There is also a math formula at the end of each step to help you do the calculation. A commentary written in *italics* is included with each step and will help you understand what is done in each step.

Calculation Steps: The Standard Method for a Single-Family Dwelling

❶ Calculate the square-foot area by using the outside dimensions of the dwelling. Do not include open porches, garages, and floor spaces unless they are adaptable for future living space. Include all floors of the dwelling. Refer to Section 220.3(A).

1,800 sq. ft.

Commentary: In this example, the habitable living area is given. Sometimes you will have to take a look at the building plans and determine the habitable living space yourself.

❷ Multiply the square-foot area by the unit load per square foot as found in Table 220.3(A). Use 3 VA per square foot for dwelling units. Refer to Section 220.3(A) and Table 220.3(A).

1,800 sq. ft. × 3 VA/sq. ft. = 5,400 VA

Commentary: The habitable living area is multiplied by 3 VA/sq. ft. to get the general lighting load in volt-amperes.

❸ Add the load allowance for the required two small-appliance branch circuits and the one laundry circuit. These circuits are computed at 1,500 VA each. Minimum total used is 4,500 VA. Be sure to add in 1500 VA each for any additional small-appliance or laundry circuits. Refer to Section 220.16(A) and (B).

5,400 VA + 4,500 VA + 0 VA (additional circuits) = 9,900 VA

Commentary: There will always be at least two small-appliance branch circuits and one laundry circuit rated at 1,500 VA each in every dwelling unit. Since the information for this example does not indicate that there are any more small-appliance branch circuits or laundry circuits, you will assume that only the minimum number required by the NEC® are to be installed.

❹ Apply the demand to the total of steps 1 through 3. According to Table 220.11, the first 3,000 VA is taken at 100%, 3,001 to 120,000 VA is taken at 35%, and anything over 120,000 VA is taken at 25%. Refer to Table 220.11.

If Step 3 is over 120,000 VA, use

$$[(\underline{\hspace{1.5cm}} - 120{,}000 \text{ VA}) \times 25\%] + (117{,}000 \text{ VA} \times 35\%) + 3{,}000 \text{ VA} = \underline{\hspace{0.8cm}} \text{ VA}$$

If step 3 is under 120,000 VA, use

$$[(\underline{9{,}900} \text{ VA} - 3{,}000 \text{ VA}) \times 35\%] + 3{,}000 \text{ VA} = \underline{2{,}415} \text{ VA}$$

Commentary: Applying the demand allows the service calculation to reflect the fact that not all the general lighting, small-appliance, or laundry circuits in a house will be fully loaded at any one time. By the way, it would be a tremendously large house that would have a 120,000-VA load at this point in the service calculation. You will use the "under 120,000" formula listed in this step for 99.9% of your residential service calculations.

❺ Add in the loading for ranges, counter-mounted cooking units, or wall-mounted ovens. Use Table 220.19 and the Notes for cooking equipment. Refer to Table 220.19 and the Notes.

2,415 VA + 8,000 VA (range) = 10,415 VA

Commentary: There is one 12,000-watt range in this example, and Note 4 to Table 220.19 tells us to use Table 220.19 to get the load for one range. Column C in Table 220.19 tells us to use 8,000 as the maximum demand load when the rating of the range does not exceed 12,000 watts and there is only one range.

❻ Add in the dryer load, using Table 220.18. Remember to use 5,000 VA or the nameplate rating of the dryer if it is greater than 5 kW. Refer to Section 220.18 and Table 220.18.

10,415 VA + 6,000 VA (dryer) = 16,415 VA

Commentary: In this example, the dryer has a nameplate rating of 6 kW. This is greater than the minimum of 5 kW, so the 6 kW is used in the calculation.

❼ Add the loading for electric heating (EH) or the air-conditioning (AC), whichever is greater. Remember that both loads will not be operating at the same time. Electric heating and air conditioning are taken at 100% of their nameplate ratings. Refer to Sections 220.15 and 220.21.

16,415 VA + 20,000 VA (EH or AC) = 36,415 VA

Commentary: In this example, the air-conditioning load is 10 kW, and the electric heating load is 20 kW. Since both loads will not be used at the same time, the larger of the two loads is included in the service entrance calculation. In this case, it is the electric heating load of 20 kW.

❽ Now add the loading for all other fixed appliances, such as dishwashers, garbage disposals, trash compactors, water heaters, and so on. Section 220.17 allows you to apply an additional demand of 75% to the fixed appliance load total if there are four or more of the appliances. If there are three or fewer, then take their total at 100%. Refer to Section 220.17.

APPLIANCES	VA OR W	
Electric water heater	4,000 W	
Garbage disposal	744 VA	6.2 amps x 120 volt
Dishwasher (heater)	1,100 W	(motor load is 7.5 amps x 120 volt = 900 VA; because the heater and the motor will not operate at the same time, use the larger of the two loads for the dishwasher)
Trash compactor	960 VA	8 amps x 120 volt
Freezer	816 VA	6.8 amps x 120 volt
Garage door opener	720 VA	6.0 amps x 120 volt
Attic exhaust fan	420 VA	3.5 amps x 120 volt

Total VA _____ × 1.0 = _____ VA (if three or fewer fixed appliances)

Total VA __8,760__ × 0.75 = __6,570__ VA (if four or more fixed appliances)

__36,415__ VA + __6,570__ Total VA (fixed appliances) = __42,985__ VA

Commentary: *In this step, all the fixed-appliance loads are considered. The number and types of appliance loads vary greatly from house to house. Many appliances will have a wattage rating on their nameplate, while others will have an amperage and voltage rating. Simply multiply the amperage rating by the voltage rating to get the volt-ampere load of the appliance. Remember, as far as service entrance calculations go, a watt and a volt-amp are considered to be the same thing. In this step, there is also a 75% demand that can be applied to the total fixed-appliance load if there are four or more of them. No demand factor is applied if there are three or fewer appliances.*

9 According to Section 220.14, which refers to Section 430.24, you must add an additional 25% of the largest motor load. Use the largest motor load from the fixed appliances listed in step 8. The largest motor load is based on the highest-rated full-load current for the motor listed. Refer to Sections 220.14 and 430.24.

__42,985__ VA + __240__ VA
(25% of the largest motor load) = __43,225__ VA

(trash compactor is the largest motor load:
960 VA × .25 = 240 VA)

Commentary: *Because service entrance conductors are feeding several load types, including motor loads, Section 430.24 requires us to increase the largest motor load by 25%. You choose the largest motor load based on the highest-rated amperage of a motor. In this example, the trash compactor has the largest motor based on its amperage and is multiplied by 25%. This amount is then added to the service calculation.*

10 The sum of steps 1 through 9 is called the "total computed load." Divide this figure by 240 volts to find the minimum ampacity required for the ungrounded service entrance conductors. Size them according to Table 310.15(B)(6). If there is no applicable listing in Table 310.15(B)(6), use Table 310.16. Refer to Tables 310.16 and 310.15(B)(6).

__43,225__ VA (total computed load)/240 volts = __180__ amps

Minimum service rating = __200__ ampere based on Table 310.15(B)(6)

Minimum ungrounded conductor size __2/0__ copper

__4/0__ aluminum

Commentary: *Remember that volt-amperes divided by voltage will give you amperage. Once the total computed load in volt-amperes is calculated, dividing by the service voltage of 240 volts will give you the minimum ampacity that the service conductors must be rated for. Table 310.15(B)(6) is used to find the minimum standard rating of the service and will also give you the minimum-size copper and aluminum conductors required for that service size.*

11 Calculate the minimum neutral ampacity by referring to Section 220.22. Add the volt-amp load for the 120-volt general lighting, small-appliance, and laundry circuits from step 4. Add the cooking and the drying loads and multiply the total by 70%. Now add in the loads of all appliances that operate on 120 volts. (Keep in mind that loads operating on two-wire 240 volts do not utilize a grounded neutral conductor.) Remember to apply the additional demand of 75% to the total neutral loading of the appliances if there are four or more of them. Refer to Sections 220.22, 220.14 and 430.24.

Step 4: 120-volt circuits load = __2,415__ VA

Cooking load from step 5 __8,000__ VA × .70 = __5,600__ VA

Drying load from step 6 __6,000__ VA × .70 = __4,200__ VA

120-volt appliance from step 8

APPLIANCES	VA
Garbage disposal	744 VA
Dishwasher (heater)	1,100 W
Trash compactor	960 VA
Freezer	816 VA
Garage door opener	720 VA
Attic exhaust fan	420 VA

Total VA _____ × 1.00 (if three or fewer) = _____ VA

Total VA __4,760__ × .75 (if four or more) = __3,570__ VA

Largest 120-volt motor load
__960__ VA × .25 = __240__ VA

Total (neutral) __16,025__ VA

FROM EXPERIENCE

Some electricians do not bother to calculate the minimum service neutral size. Instead, they will automatically use a neutral conductor size that is the same size as the ungrounded "hot" service conductors. While there is nothing wrong with this practice, you should be aware that a small amount of money, and possibly time, can be saved by using a smaller neutral service conductor. Another common practice is to automatically size the service neutral at two wire sizes smaller than the ungrounded service conductors. As a matter of fact, many service entrance cables (Type SEU) will come with the neutral wire already sized two wire sizes smaller than the ungrounded wires. Make sure the service entrance you are installing can use a smaller neutral conductor. Do not assume it can always be smaller.

Commentary: In most residential service entrance calculations the grounded neutral conductor will be smaller than the ungrounded "hot" service conductors. This is because the house 240-volt loads are connected across the two ungrounded service conductors. Since these load types (like an electric water heater) have no need for a connection to the service neutral conductor, their loads are not used in the calculation for finding the minimum neutral size. Electric cooking equipment and electric dryers use both 120- and 240-volt electricity. However, since they do use some 120-volt electricity (the dryer motor that turns the drum is connected to 120 volts), a portion (70%) of their load is used in the service neutral calculation.

⑫ Divide the total found in step 11 by 240 volts to find the minimum ampacity of the service neutral.

16,025 VA (step 11)/240 volts = **67** amperes (ampacity of neutral)

Commentary: Once the total volt-ampere load for the service entrance neutral conductor is determined, dividing it by 240 volts will give you the total calculated neutral load in amperes. From this number, you can then determine the minimum-size conductor that has an ampacity equal to or greater than the calculated number.

⑬ According to Section 310.15(B)(6), the neutral may be smaller than the ungrounded service entrance conductors, but in no case can it be smaller than the grounding electrode conductor as per Section 250.24(B)(1). With this in mind, size the neutral conductor, based on the ampacity found in step 12, by using Table 310.15(B)(6) or Table 310.16 (if the neutral load is less than 100 amps).

Minimum neutral size #4 copper #2 aluminum

Table 310.15(B)(6)

Minimum neutral size #4 Copper #3 aluminum

Table 310.16 (75°C)

Commentary: This step can be a little tricky when it comes to choosing the minimum neutral conductor size. In this example, the calculated neutral amperage is 67 amps. Since the lowest ampacity listed in Table 310.15(B)(6) is 100 amps, you can simply use the wire sizes that are listed in the table for 100 amperes (4 AWG copper, 2 AWG aluminum) You can also take a look at Table 310.16 to see if a smaller neutral conductor could be used. You can use the "75°C" column since service equipment terminations used for residential services usually have a 75°C temperature rating. In this example, it was determined that by using Table 310.16, 4 AWG copper can be used, which is the same size as permitted in Table 310.15(B)(6). However, if you want to, Table 310.16 would allow you to use a 3 AWG aluminum conductor, which is slightly smaller than the minimum aluminum conductor permitted in Table 310.15(B)(6).

⑭ Calculate the minimum-size copper grounding electrode conductor by using the largest-size ungrounded service entrance conductor and Table 250.66. Refer to Table 250.66.

Minimum grounding electrode conductor #4 AWG copper

Commentary: Based on the largest size service entrance conductor in this example, a 2/0 AWG copper or a 4/0 AWG aluminum, a 4-AWG copper grounding electrode conductor is the minimum size that must be used according to Table 250.66.

Shown next is the service entrance calculation for the same dwelling unit but using the optional method. Since sizing the service neutral conductor is done exactly the same way as in the standard method, only the steps used to get the minimum ungrounded service conductor size, the minimum service rating, and the grounding electrode conductor are shown.

Calculation Steps: The Optional Method for a Single-Family Dwelling

❶ Calculate the square-foot area by using the outside dimensions of the dwelling. Do not include open porches, garages, and floor spaces unless they are adaptable for future living space. Include all floors of the dwelling. Refer to Section 220.3(A).

1,800 sq. ft.

Commentary: Same first step as in the standard method.

❷ Multiply the square-foot area by the unit load per square foot as found in Table 220.3(A). Use 3 VA per square foot for dwelling units. Refer to Section 220.3(A) and Table 220.3(A).

1,800 sq. ft. × 3 VA/sq. ft. = 5,400 VA

Commentary: Same second step as in the standard method.

❸ Add the load allowance for the required two small-appliance branch circuits and the one laundry circuit. These circuits are computed at 1,500 VA each. Minimum total used is 4,500 VA. Be sure to add in 1,500 VA each for any additional small-appliance or laundry circuits. Refer to Section 220.16(A) and (B).

5,400 VA + 4,500 VA + 0 VA (additional circuits) = 9,900 VA

Commentary: Same third step as in the standard method.

❹ Add the nameplate ratings of all appliances that are fastened in place, permanently connected, or located to be on a specific circuit, such as ranges, wall-mounted ovens, counter-mounted cooktops, clothes dryers, trash compactors, dishwashers, and water heaters.

Example: 12-kW range = 12,000 W; a 3-kW water heater = 3,000 W. Refer to Section 220.30(B)(3).

9,900 VA + 26,760 VA (nameplate rating total of all appliances) = 36,660 VA

APPLIANCES	VA OR W	
Electric range	12,000 W	
Electric dryer	6,000 W	
Electric water heater	4,000 W	
Garbage disposal	744 VA	6.2 amps × 120 volts
Dishwasher (heater)	1,100 W	(motor load is 7.5 amps × 120 volts = 900 VA; because the heater and the motor will not operate at the same time, use the larger of the two loads for the dishwasher)
Trash compactor	960 VA	8 amps × 120 volts
Freezer	816 VA	6.8 amps × 120 volts
Garage door opener	720 VA	6.0 amps × 120 volts
Attic exhaust fan	420 VA	3.5 amps × 120 volts
Total VA	26,760	

Commentary: In the optional method, you add all the nameplate ratings of all the appliances together. Many electricians like this method because they do not have to bother with range or dryer calculations.

❺ Take 100% of the first 10,000 VA and 40% of the remaining load. Refer to Section 220.30(B).

[(36,660 VA − 10,000 VA) × .40] + 10,000 VA = 20,664 VA

Commentary: A demand still has to be applied in the optional method, but as you can see, it is slightly different than in the standard method. Remember, by applying the demand, your service entrance calculation is taking into account the fact that all the loads in the previous steps are not likely to be energized at exactly the same time.

❻ Add the value of the largest heating and air-conditioning load from the list included in Section 220.30(C). These loads include air conditioners, heat pumps, central electric heat (electric furnaces), and separately controlled electric space heaters, such as electric baseboard heaters. Refer to Section 220.30(C).

20,664 VA + 13,000 VA (largest from Section 220.30(C)) = 33,664 VA

Commentary: This step takes into account that both the electric heating load and the air-conditioning load will not be energized at exactly the same time. One of the tougher things to figure out in this step is what kind of electric heating system will be in the house. This example included an electric furnace. In most parts of the country, an electric furnace is not practical because of the high cost of electricity. However, there may be electric heat baseboard units or some other style of electric heating that must be considered in this step. This house has an air-conditioning load of 10 kW and an electric furnace rated at 20 kW. Section 220.30(C)(1) says to take the air-conditioning load at 100%. Section 220.3(C)(4) says to apply a factor of 65% to the central electric heating load. 10 kW of air conditioning × 100% = 10 kw; 20 kW of central electric heat × 65% = 13 kW. Therefore, the largest load is the electric heating load of 13 kW.

❼ The sum of steps 1 through 6 is called the "total computed load." Divide this Figure by 240 volts to find the minimum ampacity required for the ungrounded service entrance conductors. Size them according to Table 310.15(B)(6). If there is no applicable listing in Table 310.15(B)(6), use Table 310.16. Refer to Tables 310.16 and 310.15(B)(6).

33,664 VA (total computed load)/240 volts = 140 amps

Minimum service rating = 150 amperes based on Table 310.15(B)(6)

Minimum ungrounded conductor size #1 copper
2/0 aluminum

Commentary: Remember that volt-amperes divided by voltage will give you amperage. Once the total computed load in volt-amperes is calculated, dividing by the service voltage of 240 volts will give you the minimum ampacity that the service conductors must be rated for. Table 310.15(B)(6) is used to find the minimum standard rating of the service and will also give you the minimum-size copper and aluminum conductor sizes required for that service size.

FROM EXPERIENCE

Doing a service entrance calculation using the optional method will almost always result in a smaller service rating and smaller service conductor size than when using the standard method. It really does not matter which method is used. Both are considered adequate for service entrance sizing. You may want to check with the authority having jurisdiction and the local electric utility to see if they have a requirement on using either the standard or the optional method.

8 The neutral size is calculated exactly the same as for the standard method. There is no optional method for calculating the minimum neutral size. The calculation for the minimum neutral size for this service entrance was done in the previous standard method example.

Commentary: *Since calculating the service entrance neutral conductor is exactly the same as for the standard method, this example will jump to step 11 of the optional service entrance calculation for determining the minimum-size grounding electrode conductor.*

11 Calculate the minimum-size grounding electrode conductor by using the largest-size ungrounded service entrance conductor and Table 250.66.

Minimum grounding electrode conductor #6 AWG copper

Commentary: *Based on the largest-size service entrance conductor in this example, a 1 AWG copper or a 2/0 AWG aluminum, a 6 AWG copper grounding electrode conductor is the minimum size that must be used according to Table 250.66.*

Sizing the LoadCenter

Once the service entrance calculation has been done and the minimum-size service entrance has been calculated, a loadcenter size can be determined. Most residential loadcenters used today are designed to accommodate circuit breakers, and since circuit breaker loadcenters are used primarily in residential wiring, this section limits the discussion for the sizing of loadcenters to those that take circuit breakers. Loadcenters used in residential wiring are often called "panels" by electricians. Chapter 2 discussed loadcenters, or panels, and you discovered that a panelboard located in a cabinet makes up what many electricians call a "loadcenter." The *NEC*® does not use the term "loadcenter" but does use the term "panelboard" and states several rules that must be followed when installing them.

Panels used in residential wiring are available with a main circuit breaker already installed at the factory or in a style that has just wire lugs used to terminate the wires bringing electrical power to the panel. A panel with a main breaker is commonly called a "main breaker panel," and a panel with just wire lugs is called a "main lug only," or "MLO," panel (Figure 7–23). For example, a service calculation resulting in a minimum-size service entrance of 200 amperes would require a 200-amp main breaker panel. The 200-amp main breaker is the required main service disconnecting means used to disconnect all electrical power from the residential wiring system. Main-lug-only panels are used primarily as "subpanels," which are discussed in greater detail in the next part of this chapter.

(A)

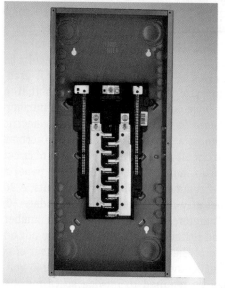

(B)

Figure 7–23 A typical 120/240-volt, single-phase, main breaker loadcenter (A) *Courtesy of Square D Company* **and a 120/240-volt, single-phase, main-lug-only (MLO) panel (B).** *Courtesy of Square D Company, Group Schneider.*

Section 408.13 of the *NEC®* says that all panelboards must have a rating not less than the minimum feeder capacity required for the load as computed in accordance with Article 220. The rating of a panelboard is based on the ampacity of the bus bar(s) in the panelboard. The bus bar is the part of the panelboard that is energized and distributes current to the individual circuit breakers located in the panel. In the most common style of panel used in residential wiring, the circuit breakers are installed by pushing them onto the bus bar. The circuit breakers are designed to clamp on to the bus bar once they have been pushed into place. Some panels require circuit breakers to be secured to the bus bar with screws, but this panel type is used mainly in commercial and industrial wiring systems. Panelboards must be durably marked by the manufacturer with the voltage, the current rating, and the number of phases (single phase or three phase) for which the panelboard is designed. The manufacturer's name or trademark must be visible after installation. Some panels are suitable for use as service equipment and are so marked. Only those panels that are marked can be used as service entrance equipment. Listed panelboards are used with copper conductors, unless marked to indicate which terminals are suitable for use with aluminum conductors. This marking must be independent of any marking on terminal connectors and must appear on a wiring diagram or other readily visible location. If all terminals are suitable for use with aluminum conductors as well as with copper conductors, the panelboard will be marked "Use Copper or Aluminum Wire." Unless the panelboard is marked to indicate otherwise, the termination provisions are based on the use of 60°C (140°F) ampacities for wire sizes 14 through 1 AWG and 75°C (167°F) ampacities for wire sizes 1/0 AWG and larger. As noted earlier in this chapter, most panels used in residential wiring will have terminals marked for 75°C.

In Chapter 2, you learned that there is a single-pole circuit breaker and a two-pole circuit breaker used in residential electrical work. The single-pole breaker is used to provide overcurrent protection to those circuits that operate on 120 volts. The two-pole circuit breaker is used to provide overcurrent protection to those circuits that operate on 240 volts. Once you have determined the number of circuits that will need to be installed for a house electrical system, you can choose a panel that is designed to accommodate the number of circuit breakers for the number of circuits you will install. For example, a 100-ampere-rated panel may be designed to accommodate 16, 20, or 24 circuits. A 200-ampere-rated panel may be designed to accommodate 30, 40, or 42 circuits. Therefore, based on the minimum number of circuits you plan on installing, you need to specify not only the ampere rating of the panel but also the number of circuits. For example, a 100-ampere main breaker, 120/240-volt, 24-circuit panel may be required to provide space for the minimum number of circuits needed in a house electrical system.

The number of circuits that each panel can handle is based on the 120-volt circuits. The circuit breaker for each 120-volt circuit you will be installing will take up one space

FROM EXPERIENCE

It is a good idea to oversize the panel when it comes to the number of circuits it is rated for. A good rule of thumb to follow is to never fill a panel to more than 80% of its capacity. For example, do not fill more than 16 spaces in a 20-circuit panel (20 × 80% = 16) or more than 32 spaces in a 40-circuit panel (40 × 80% = 32). This practice will help ensure that there is room in an existing panel for future expansion of the electrical system.

in the panel. On the other hand, each 240-volt (120 volt to ground from each "hot" conductor) circuit you install will require a two-pole circuit breaker, which will take up two spaces in the panel. For example, each small-appliance branch circuit required in a residential electric system is wired with 12 AWG wire and uses a 20-amp single-pole circuit breaker for overcurrent protection of the circuit. If you only install the *NEC®* minimum number of small-appliance branch circuits, two spaces in the panel will be taken up by the two 20-amp circuit breakers. If a 240-volt electric water heater is to be installed, a common 40-gallon size will call for it to be wired with 10 AWG wire with a two-pole 30-amp circuit breaker. The circuit breaker for this water heater circuit will take up two spaces in the panel.

Sizing Feeders and Subpanels

Some residential wiring situations may call for another loadcenter to be located in a house. This additional loadcenter is called a "subpanel." Subpanels may also be located in a detached garage. Garage wiring, including installing a subpanel in a detached garage, is covered in Chapter 15. The reason for installing a subpanel is usually to locate a load center closer to an area of the house where several circuits are required. This technique will eliminate having to run long distances back to the main panel with many branch circuits. For example, a kitchen in a house usually has more required circuits than other parts of the house. If the location of the main panel is a long distance away from the kitchen area, it is a wise practice to install a subpanel closer to the kitchen and feed the kitchen with circuits that originate in the subpanel.

The wiring from the main panel to the subpanel is called a "feeder." The feeder is made up of wiring in a cable or individual wires in an electrical conduit that are large enough to feed the required electrical load that the subpanel serves. Sizing the feeder to a subpanel is done exactly the same way as for sizing service entrance conductors feeding a main service panel. You can use either the standard or the optional method. Your calculation will include only those electrical loads fed from the subpanel.

Smaller loadcenters are typically used as subpanels. Although it is permissible to have a subpanel with a main circuit breaker, most residential subpanels will be of the MLO type. The overcurrent protection device for the subpanel feeder is located at the main service entrance panel. The ampere rating of a subpanel is determined by the rating of the bus bar in the subpanel. For example, a 125-ampere-rated MLO subpanel has a bus bar rating of 125 amps. This means that the subpanel could supply an electrical load of up to 125 amps. Many times, the feeder overcurrent protection device is sized less than the rating of the subpanel. For example, a 125-amp-rated MLO subpanel may be fed with an 8 AWG copper feeder with an ampacity of 40 amps. The feeder will be protected with a 40-amp circuit breaker in the main service panel.

Summary

This chapter has taken a look at the types of circuits found in a residential wiring system. Sizing the circuit conductors and calculating the conductor ampacity based on the conditions of use were discussed. Both the standard and the optional methods for calculating the minimum-size service for a house were presented. The last area covered was sizing a subpanel and sizing the feeder supplying a subpanel. Overall, there was a great amount of material for you to become familiar with. While many of these calculations are not done by beginning electricians, you should be aware of how to do them. Electrical licensing exams always have many questions over the material covered in this chapter. Knowing how to perform the calculations covered here will help you become a more professional electrician.

Procedures

Calculation Steps: The Standard Method for a Single-Family Dwelling

Steps:

1 Calculate the square foot area by using the outside dimensions of the dwelling. Do not include open porches, garages, and floor spaces unless they are adaptable for future living space. Include all floors of the dwelling. Refer to Section to 220.3(A).

_____ sq. ft.

2 Multiply the square-foot area by the unit load per square foot as found in Table 220.3(A). Use 3 VA per square foot for dwelling units. Refer to Section 220.3(A) and Table 220.3(A).

_____ sq. ft. × 3 VA/sq. ft. = _____ VA

3 Add the load allowance for the required two small-appliance branch circuits and the one laundry circuit. These circuits are computed at 1,500 VA each. Minimum total used is 4,500 VA. Be sure to add in 1,500 VA each for any additional small-appliance or laundry circuits. Refer to Section 220.16(A) and (B).

_____ VA + 4,500 VA + _____ VA (additional circuits)
= _____ VA

4 Apply the demand to the total of steps 1 through 3. According to Table 220.11, the first 3,000 VA is taken at 100%, 3,001 to 120,000 VA is taken at 35%, and anything over 120,000 VA is taken at 25%. Refer to Table 220.11.

If step 3 is over 120,000 VA, use:

[(_____ − 120,000 VA) × 25%] + (117,000 VA × 35%)
+ 3,000 VA 5 _____ VA

If step 3 is under 120,000 VA, use:

[(_____ VA − 3,000 VA) × 35%] + 3,000 VA = _____ VA

5 Add in the loading for ranges, counter-mounted cooking units, or wall-mounted ovens. Use Table 220.19 and the Notes for cooking equipment. Refer to Table 220.19 and the Notes.

_____ VA + _____ VA (range) = _____ VA

6 Add in the dryer load, using Table 220.18. Remember to use 5,000 VA or the nameplate rating of the dryer if it is greater than 5 kW. Refer to Section 220.18 and Table 220.18.

_____ VA + _____ VA (dryer) = _____ VA

7 Add the loading for electric heating (EH) or the air-conditioning (AC), whichever is greater. Remember that both loads will not be operating at the same time. Electric heating and air-conditioning are taken at 100% of their nameplate ratings. Refer to Sections 220.15 and 220.21.

_____ VA + _____ VA (EH or AC) = _____ VA

8 Now add the loading for all other fixed appliances, such as dishwashers, garbage disposals, trash compactors, water heaters, and so on. Section 220.17 allows you to apply an additional demand of 75% to the fixed appliance load total if there are four or more of the appliances. If there are three or fewer, then take their total at 100%. Refer to Section 220.17.

APPLIANCES	VA OR W

Total VA _____ × 1.0 = _____ VA (if three or fewer fixed appliances)

Total VA _____ × .75 = _____ VA (if four or more fixed appliances)

_____ VA + _____ Total VA (fixed appliances) = _____ VA

9 According to Section 220.14, which refers to Section 430.24, you must add an additional 25% of the largest motor load. Use the largest motor load from the fixed appliances listed in step 8. The largest motor load is based on the highest-rated full-load current for the motor listed. Refer to Sections 220.14 and 430.24.

_____ VA + _____ VA (25% of the largest motor load) = _____ VA

10 The sum of steps 1 through 9 is called the "total computed load." Divide this figure by 240 volts to find the minimum ampacity required for the ungrounded service entrance conductors. Size them according to Table 310.15(B)(6). If there is no applicable listing in Table 310.15(B)(6), use Table 310.16. Refer to Tables 310.16 and 310.15(B)(6).

_____ VA (total computed load)/240 volts = _____ amps

Minimum service rating = _____ ampere based on
                                         Table 310.15(B)(6)

Minimum ungrounded conductor size = _____ copper

                                                      _____ aluminum

⑪ Calculate the minimum neutral ampacity by referring to Section 220.22. Add the volt-amp load for the 120-volt general lighting, small-appliance, and laundry circuits from step 4. Add the cooking and the drying loads and multiply the total by 70%. Now add in the loads of all appliances that operate on 120 volts. (Keep in mind that loads operating on two-wire 240 volts do not utilize a grounded neutral conductor.) Remember to apply the additional demand of 75% to the total neutral loading of the appliances if there are four or more of them. Refer to Sections 220.22, 220.14, and 430.24.

Step 4: 120-volt circuits load   = _____ VA

Cooking load from step 5   _____ VA × .70 = _____ VA

Drying load from step 6   _____ VA × .70 = _____ VA

120-volt appliance from step 8

APPLIANCES	VA

Total VA _____   × .75 = _____ VA (if four or more)

Total VA _____   ×1.00 = _____ VA (if three or fewer)

Largest 120-volt
  motor load _____ VA × .25 = _____ VA

              Total _____ VA (neutral)

⑫ Divide the total found in step 11 by 240 volts to find the minimum ampacity of the service neutral.

_____ VA (step 11)/240 volts = _____ ampacity of neutral

⑬ According to Section 310.15(B)(6), the neutral may be smaller than the ungrounded service entrance conductors, but in no case can it be smaller than the grounding electrode conductor as per Section 250.24(B)(1). With this in mind, size the neutral conductor, based on the ampacity found in step 12, by using Table 310.15(B)(6) or Table 310.16 (if the neutral load is less than 100 amps).

(Table 310.15(B)(6))
Minimum neutral size = _____ copper
                                 _____ aluminum

(Table 310.16 (75°C))
Minimum neutral size = _____ copper
                                 _____ aluminum

⑭ Calculate the minimum-size copper grounding electrode conductor by using the largest-size ungrounded service entrance conductor and Table 250.66. Refer to Table 250.66.

Minimum grounding electrode conductor = _____ AWG copper

Procedures

Calculation Steps: The Optional Method for a Single-Family Dwelling

Steps:

① Calculate the square-foot area by using the outside dimensions of the dwelling. Do not include open porches, garages, and floor spaces unless they are adaptable for future living space. Include all floors of the dwelling. Refer to Section 220.3(A).

_____ sq. ft.

② Multiply the square-foot area by the unit load per square foot as found in Table 220.3(A). Use 3 VA per square foot for dwelling units. Refer to Section 220.3(A) and Table 220.3(A).

_____ sq. ft. × 3 VA/sq. ft. = _____ VA

③ Add the load allowance for the required two small-appliance branch circuits and the one laundry circuit. These circuits are computed at 1,500 VA each. Minimum total used is 4,500 VA. Be sure to add in 1,500 VA each for any additional small-appliance or laundry circuits. Refer to Sections 220.16(A) and (B).

_____ VA + 4,500 VA + _____ VA (additional circuits)
= _____ VA

④ Add the nameplate ratings of all appliances that are fastened in place, permanently connected, or located to be on a specific circuit, such as ranges, wall-mounted ovens, counter-mounted cooktops, clothes dryers, trash compactors, dishwashers, and water heaters.

Example: 12-kW range = 12,000 W; a 3-kW water heater = 3,000 W. Refer to Section 220.30(B)(3).

_____ VA + _____ VA (nameplate rating total of all appliances)
= _____ VA

APPLIANCES	VA
_____	____
_____	____
_____	____
_____	____
_____	____
_____	____
_____	____
_____	____

Total VA _____

⑤ Take 100% of the first 10,000 VA and 40% of the remaining load. Refer to Section 220.30(B).

[(_____ VA − 10,000 VA) × .40] + 10,000 VA = _____ VA

⑥ Add the value of the largest heating and air-conditioning load from the list included in Section 220.30(C). These loads include air conditioners, heat pumps, central electric heat (electric furnaces), and separately controlled electric space heaters, such as electric baseboard heaters. Refer to Section 220.30(C).

_____ VA + _____ VA (largest from Section 220.30(C)) =
_____ VA

⑦ The sum of steps 1 through 6 is called the "total computed load." Divide this figure by 240 volts to find the minimum ampacity required for the ungrounded service entrance conductors. Size them according to Table 310.15(B)(6). If there is no applicable listing in Table 310.15(B)(6), use Table 310.16. Refer to Tables 310.16 and 310.15(B)(6).

_____ VA (total computed load)/240 volts = _____ amps

Minimum service rating = _____ amperes based
on Table 310.15(B)(6)

Minimum ungrounded conductor size = _____ copper
_____ aluminum

⑧ The neutral size is calculated exactly the same as for the standard method. There is no optional method for calculating the minimum neutral size. Calculate the minimum neutral ampacity by referring to Section 220.22. Add the volt-amp load for the 120-volt general lighting, small-appliance, and laundry circuits from step 4. Add the cooking and the drying loads and multiply the total by 70%. Now add in the loads of all appliances that operate on 120 volts. (Keep in mind that loads operating on two-wire 240 volts do not utilize a grounded neutral conductor.) Remember to apply the additional demand of 75% to the total neutral loading of the appliances if there are four or more of them. Refer to Sections 220.22, 220.14, and 430.24.

*Step 4: 120-volt circuits load = _____ VA

*Cooking load from step 5 _____ VA × .70 = _____ VA

*Drying load from step 6 _____ VA × .70 = _____ VA

*From standard calculation steps

120-volt appliances:

APPLIANCES	VA
_____	
_____	
_____	
_____	
_____	
_____	
_____	
_____	

Total VA _____ × .75 = _____ VA (if four or more)

Total VA _____ × 1.00 = _____ VA (if three or fewer)

Largest 120-volt
motor load _____ VA × .25 = _____ VA

Total _____ VA (neutral)

9 Divide the total found in step 8 by 240 volts to find the minimum ampacity of the service neutral.

_____ VA (step 8)/240 volts = _____ amps of neutral

10 According to Section 310.15(B)(6), the neutral may be smaller than the ungrounded service entrance conductors, but in no case can it be smaller than the grounding electrode conductor as per Secton 250.24(B)(1). With this in mind, size the neutral conductor based on the ampacity found in step 9 by using Table 310.15(B)(6) or Table 310.16 (if the neutral load is less than 100 amps). Refer to Tables 310.16 and 310.15(B)(6).

Table 310.15(B)(6)
Minimum neutral size = _____ copper
_____ aluminum

Table 310.16 (75°C)
Minimum neutral size = _____ copper
_____ aluminum

11 Calculate the minimum-size grounding electrode conductor by using the largest-size ungrounded service entrance conductor and Table 250.66.

Minimum grounding electrode conductor = _____ AWG copper

Review Questions

Directions: Answer the following items with clear and complete responses.

1 The minimum size of service entrance for a one-family dwelling is 100 amperes. Name the section in the *NEC®* that states this rule.

2 The general lighting circuit unit load per square foot for a dwelling is _____ VA/sq. ft. This information is found in *NEC®* Table _____ .

3 The maximum demand for one electric range rated not over 12 kW is _____ . This information is found in *NEC®* Table _____ .

4 Calculate the computed load for an electric range that has a nameplate rating of 15 kW.

5 Name the standard sizes of circuit breakers up to and including 100 amperes. This information is found in *NEC®* Section _____ .

6 The *NEC®* states that there must be a minimum of _____ small-appliance branch circuits installed in each dwelling unit. This information is found in *NEC®* Section _____ .

7 Calculate the minimum number of 15-amp and 20-amp general lighting circuits required in an 1,800-square-foot house.

8 Define a branch circuit.

9 The load for each small-appliance and laundry circuit is calculated at _____ VA each. This information is found in *NEC®* Section _____ .

10 Explain why both the electric heating load and the air-conditioning load are not used in a service entrance calculation. This information is found in *NEC®* Section _____ .

11 Name the ampacity for the following copper conductors based on Table 310.16:

a. 10 AWG THW _____ amps

b. 8 AWG THWN _____ amps

c. 1 AWG THHN _____ amps

d. 4/0 TW _____ amps

12 Determine the ampacity of a 3 AWG THW copper conductor that is being installed in an ambient temperature of 110°F. Show all work.

13 Name the part that is used to determine the ampere rating of a panel board.

14 Explain why the 240-volt, two-wire circuits (like an electric water heater) are not used to calculate the service neutral conductor.

15 A 1,500-square-foot house will have a 12-kW electric range; a 5-kW electric clothes dryer; eight individually controlled 2-kW, 240-volt electric baseboard heaters; a 9-kW-rated air conditioner; a 3-kW electric water heater; a 6.8-amp, 120-volt garbage disposal; and a 1,200-W, 120-volt dishwasher. Use the standard method to calculate the following:

a. Minimum-size service entrance: _____ amp

b. Minimum-size service conductors: _____ copper; _____ aluminum

c. Minimum-size service neutral conductor: _____ copper; _____ aluminum

d. Minimum-size grounding electrode conductor: _____ copper

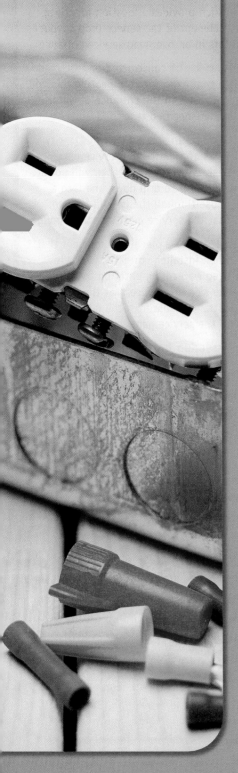

Chapter 8

Service Entrance Equipment and Installation

In previous chapters, you were introduced to many of the pieces of electrical equipment and materials commonly used in residential wiring, including some items that could be used as part of a service entrance. You have learned that a service entrance is the part of a residential wiring system that provides the means for electrical energy from the electric utility to be delivered to a residential electrical system. You have also learned how to do the calculations necessary to determine the minimum-size service entrance that will supply electricity to a house electrical system. This chapter brings together all the things you have learned about service entrances from prior chapters and covers service entrance equipment and materials in detail. Common installation techniques for overhead and underground service entrances are also covered.

OBJECTIVES

Upon completion of this chapter, the student should be able to:

- identify common overhead service entrance equipment and materials.
- identify common underground service entrance equipment and materials.
- demonstrate an understanding of common installation techniques for overhead services.
- demonstrate an understanding of common installation techniques for underground services.
- demonstrate an understanding of panelboard installation techniques.
- demonstrate an understanding of subpanel installation techniques.
- demonstrate an understanding of existing service entrance upgrade techniques.

Glossary of Terms

backboard the surface on which a service panel or subpanel is mounted; it is usually made of plywood and is painted a flat black color

backfeeding a wiring technique that allows electrical power from an existing electrical panel to be fed to a new electrical panel by a short length of cable; this technique is commonly used when an electrician is upgrading an existing service entrance

band joist the member used to stiffen the ends of the floor joists where they rest on the sill

bushing (insulated) a fiber or plastic fitting designed to screw onto the ends of conduit or a cable connector to provide protection to the conductors

cable hook also called an "eyebolt"; it is the part used to attach the service drop cable to the side of a house in an overhead service entrance installation

conduit "LB" a piece of electrical equipment that is connected in-line with electrical conduit to provide for a 90° change of direction

enclosure the case or housing of apparatus or the fence or walls surrounding an installation to prevent personnel from accidentally contacting energized parts or to protect the equipment from physical damage

line side the location in electrical equipment where the incoming electrical power is connected; an example is the line-side lugs in a meter socket where the incoming electrical power conductors are connected

load side the location in electrical equipment where the outgoing electrical power is connected; an example is the load-side lugs in a meter socket where the outgoing electrical power conductors to the service equipment are connected

mast kit a package of additional equipment that is required for the installation of a mast-type service entrance; it can be purchased from an electrical distributor

porcelain standoff a fitting that is attached to a service entrance mast, which provides a location for the attachment of the service drop conductors in an overhead service entrance

rain-tight constructed or protected so that exposure to a beating rain will not result in the entrance of water under specified test conditions

roof flashing/weather collar two parts of a mast-type service entrance that, when used together, will not allow water to drip down and into a house through the hole in the roof that the service mast extends through

sill plate a piece of equipment that, when installed correctly, will help keep water from entering the hole in the side of a house that the service entrance cable from the meter socket to the service panel goes through

threaded hub the piece of equipment that must be attached to the top of a meter socket so that a raceway or a cable connector can be attached to the meter socket

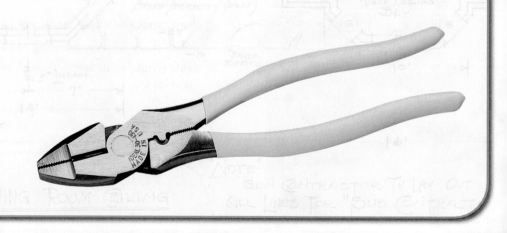

Overhead Service Equipment and Materials

ince an overhead service entrance is the most common service type installed in residential situations, we will cover the equipment and materials used to install them first. There are three different ways to install an overhead service entrance: (1) using service entrance cable installed on the side of a house, (2) using electrical conduit installed on the side of a house, and (3) using a mast-style service installation.

Overhead Service Entrance Using Service Entrance Cable

If the service will be installed with service entrance cable (Figure 8–1), the equipment will include the service drop conductors, a cable hook, a cable weatherhead, service entrance cable, clips for the service entrance cable, a meter socket with a threaded hub, a rain-tight service entrance cable connector, "regular" service entrance cable connectors, a sill plate, and the service equipment. The purpose for each part is explained in the following:

- Service drop conductors: As described in a previous chapter, the service drop conductors bring the electrical energy from the local electric utility system to the house. The overhead cable, typically called "triplex cable," is usually installed and owned by the electric utility company. So even though the service drop conductors are part of an overhead service, electricians typically do not have to purchase or install it. The area where you live may have different electric utility rules. Check with your local electric utility to see if they supply and install the service drop cable.
- Cable hook: The **cable hook,** or "eyebolt" as it is sometimes called, is used to attach the service drop cable to the side of the house. The electrician is usually required to install the cable hook even if the electric utility is installing the service drop. Many electric utilities will provide the cable hook to the electrician. The hook must be installed so that it is below the weatherhead.
- Weatherhead: You may remember that the weatherhead was introduced in a previous chapter. It is used to stop the entrance of water into the end of a service entrance cable. Each service entrance conductor must exit from a separate hole in the weatherhead.
- Service entrance cable: Service entrance cable has also been introduced in previous chapters. Overhead services will usually use an SEU-type service entrance cable to bring the electricity down the side of the house from the point of attachment to the meter socket and then on into the main service equipment.
- Clips for the service entrance cable: Clips are used to provide support for the cable. Section 230.51(A) of the

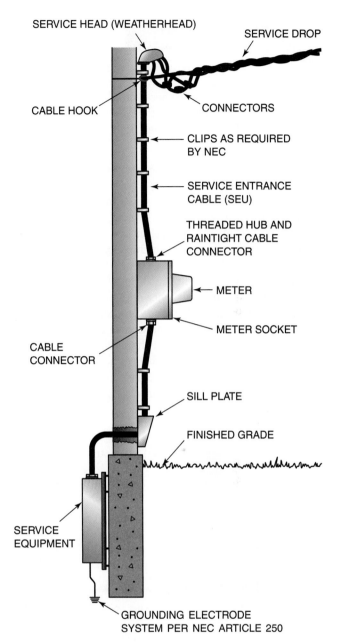

Figure 8–1 A typical overhead service entrance installed using Type SEU service entrance cable.

National Electrical Code® (*NEC®*) states that they must be located no more than 12 inches (300 mm) from the weatherhead and meter socket. Additional support is required at intervals no more than 30 inches (750 mm). Clip types that are often used include "fold-over" clips, one hole strap and two hole straps.

- Meter socket and threaded hub: The meter socket is used to provide a location in the service entrance for the local electric utility to install a kilowatt-hour meter that will measure the amount of electrical energy used by the house electrical system. When you buy the meter socket, a **threaded hub** for the top of the socket must

Some electric utilities will provide the meter socket to the electrician. Be sure to check with the electric utility in your area to determine whether you supply the meter socket or whether the local electric utility supplies the meter socket.

also be purchased. The hub is sized for the size of the service entrance cable connector coming into the top of the meter socket. For example, if you are using a 1¼-inch connector, use a 1¼-inch threaded hub on the top of the meter socket.

- Rain-tight service entrance cable connector: This connector is used to secure the service entrance cable to the top of the meter socket. It has a special design that allows it to provide a **rain-tight** connection so that no water can enter the meter socket around the service entrance cable.
- "Regular" service entrance cable connectors: This connector type is used to connect the service entrance cable to the bottom of the meter socket. It is also the

connector type used to connect the service cable to the main service equipment **enclosure** when the enclosure is located inside the house. A rain-tight connector is not required since the bottom of the meter socket and an inside-located enclosure are not subject to rain or snow.

- Sill plate: A **sill plate** is used to protect service entrance cable at the point where it enters a house through an outside wall. It is usually made of metal and comes with a couple of screws to attach it with and some duct seal to help make a good weatherproof seal at the point where the cable enters the outside wall of the house.
- Service equipment: The service equipment contains the main service disconnecting means, usually a main circuit breaker, that is used to disconnect the supply electricity from the house electrical system. There are basically two types of service equipment commonly used in residential wiring systems:

1. A main breaker panel that is installed inside the house: It also contains the circuit breakers (or fuses) that provide the overcurrent protection required for the branch circuits that supply electricity to the different electrical loads in a house. This is the most common style of service entrance equipment (Figure 8–2).

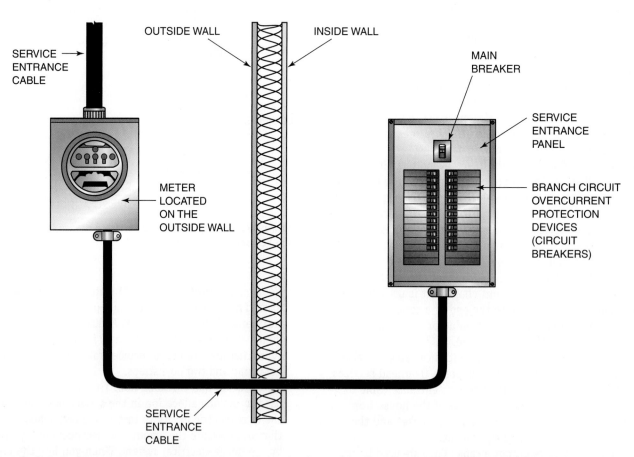

Figure 8–2 A service entrance panel with a main circuit breaker located inside a house. The service panel will contain circuit breakers that protect the branch-circuit wiring.

2. A combination meter socket/main breaker disconnect that is installed outside the house: This style of service equipment allows easy access to the main disconnecting means in the case of an emergency. A feeder is run from this type of service equipment to a subpanel located inside the house. The subpanel will contain the circuit breakers (or fuses) that provide the overcurrent protection required for the branch circuits supplying electricity to the different electrical loads in a house (Figure 8–3).

FROM EXPERIENCE

Some homes have an attached garage, and the service entrance is often installed on the side of the garage. In this case, the meter socket will be located on the outside wall of the garage. The main service disconnect may be either included with the meter socket (combination meter socket/main breaker disconnect) or located on the inside wall of the garage, usually directly behind the meter socket. In either case, it is common wiring practice to install a subpanel in the main part of the house and run the electrical system circuits from there.

FROM EXPERIENCE

It is common practice in some areas of the country, where house building styles do not provide for a good location inside a house for the service equipment, to use an enclosure located on the outside of a house that contains both a main circuit breaker and the branch-circuit overcurrent protection devices. In this case, branch-circuit wiring is run from the outside-located service equipment to the different electrical loads in a house. This installation will require a NEMA Type 3R enclosure. Homes with no attached garage or a basement or crawl space under the home are candidates for this type of installation.

Overhead Service Entrance Using Electrical Conduit

An overhead service entrance that utilizes electrical conduit as a service raceway on the side of a house will use slightly different equipment (Figure 8–4). This equipment will include the service drop conductors, a cable hook, the metal or nonmetallic electrical conduit used as the service

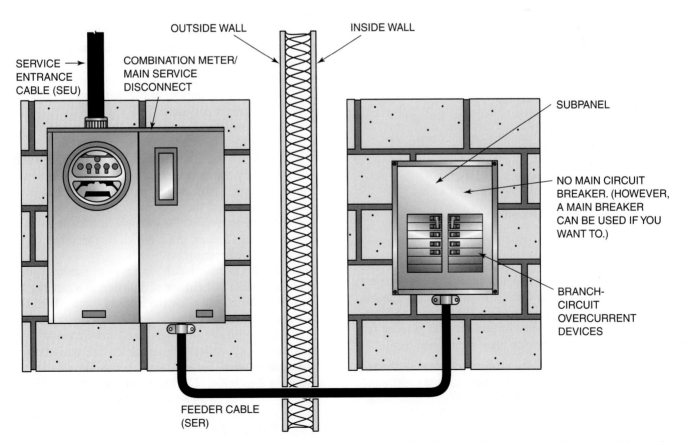

Figure 8–3 A combination meter socket/main service disconnect located on the outside wall of a house and a subpanel located inside the house. The subpanel will contain circuit breakers that protect the branch-circuit wiring.

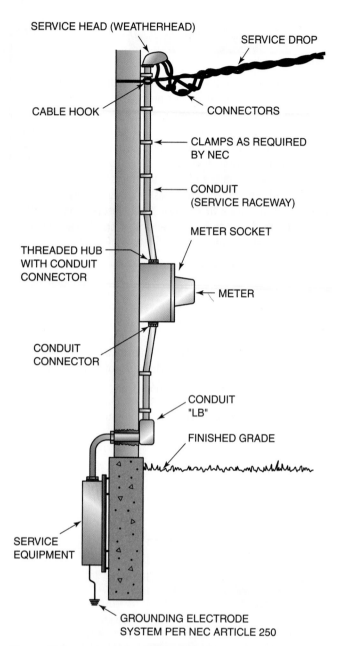

SERVICE HEAD (WEATHERHEAD)

SERVICE DROP

CABLE HOOK

CONNECTORS

CLAMPS AS REQUIRED BY NEC

CONDUIT (SERVICE RACEWAY)

METER SOCKET

THREADED HUB WITH CONDUIT CONNECTOR

METER

CONDUIT CONNECTOR

CONDUIT "LB"

FINISHED GRADE

SERVICE EQUIPMENT

GROUNDING ELECTRODE SYSTEM PER NEC ARTICLE 250

Figure 8–4 A typical overhead service entrance installed using a service raceway.

raceway, separate service entrance conductors of the proper size, a weatherhead for the raceway, conduit clamps for service entrance raceway, conduit connector fittings, a meter socket with a threaded hub, a conduit "LB," insulated bushings, and the service entrance equipment. The purpose for each part unique to this service type is explained here:

- Service raceway: The service raceway will provide protection for the service entrance conductors. It is usually installed with rigid nonmetallic conduit (RNC), rigid metal conduit (RMC), intermediate metal conduit (IMC), or electrical metallic tubing (EMT).
- Service entrance conductors: The service entrance conductors are installed in the service raceway as individual

conductors. They are used to bring the electricity down the side of the house from the point of attachment to the meter socket and then on into the main service panel.

- Service raceway weatherhead: The weatherhead must be designed to fit the size and type of electrical conduit being used for the service raceway. It is used to stop the entrance of water into the end of the service entrance raceway. Each service entrance conductor must exit from a separate hole in the weatherhead.
- Conduit clamps: The clamps are used to support the service raceway. They must be located no more than 3 feet (900 mm) from the weatherhead and meter socket. Additional support is required at intervals no more than what is required by the *NEC*® for that type of conduit. It is recommended that additional support be located so there is no more than 3 feet (900 mm) between additional supports. Conduit clamp types that are often used include one-hole straps, two-hole straps, and mineralic straps.
- Conduit connector fittings: The conduit connectors are used to connect the service raceway to the various enclosures and conduit fittings used in the service entrance installation. If RNC is used, connector fittings are attached to the conduit using a PVC cement, which allow for attachment to the meter socket and service equipment. If EMT is used, compression-type fittings are attached to the conduit ends, allowing the EMT to be attached to the meter socket threaded hub and the service entrance equipment. Compression-type fittings must be used for outside use of EMT. Inside the house, setscrew-type fittings could be used. RMC and IMC raceway is threaded on each end and will be directly tightened into the meter socket threaded hub. Locknuts will hold the RMC and IMC to the bottom of the meter socket and the service equipment enclosure.
- Meter socket with a threaded hub: The meter socket is used to provide a location in the service entrance for the local electric utility to install a kilowatt-hour meter. When you buy the meter socket, a threaded hub for the top of the socket must also be purchased. The hub is sized for the size of the service raceway coming into the top of the meter socket. For example, if you are using a 2-inch service raceway, use a 2-inch threaded hub on the top of the meter socket.
- Conduit "LB": **Conduit "LB"** is a type of conduit body that allows for a 90° change of direction in a raceway path. This fitting will allow the raceway coming from the bottom of the meter socket to go through the side of the house and continue to the service equipment enclosure. Each conduit type used will require an "LB" made from the same material. For example, a 2-inch RNC raceway will require a 2-inch RNC "LB" fitting, and a 1½-inch EMT raceway will require a 1½-inch EMT "LB" fitting.
- Insulated bushing: An insulated **bushing** is a fiber or plastic fitting designed to screw onto the ends of conduit or a cable connector to provide protection to the conductors. Section 300.4(F) in the *NEC*® states that where race-

ways containing ungrounded conductors 4 AWG or larger enter a cabinet, box enclosure, or raceway, the conductors must be protected by a substantial fitting providing a smoothly rounded insulating surface. The exception to this section says that where threaded hubs are used, such as at the top of the meter socket, an insulated bushing does not have to be used on the raceway end. For this type of service entrance, insulated bushings would have to be used where the raceway enters the bottom of the meter socket and where the raceway enters the service entrance panel. Conduit bushings constructed wholly of insulating material must not be used to secure the raceway to the enclosure. The insulating bushing must have a temperature rating not less than the insulation temperature rating of the installed conductors. The reason for this requirement for using insulated bushings on the ends of conduits

containing wires of 4 AWG or larger is that heavy conductors and cables tend to stress the conductor insulation at terminating points. Providing insulated bushings at raceway reduces the risk of insulation failure at conductor insulation "stress" points. The temperature rating of an insulating bushing must coordinate with the insulation of the conductor to ensure that the protection remains intact over the life cycle of the insulated conductor.

Overhead Service Entrance Using a Service Entrance Mast

A mast-type overhead service entrance uses a few different pieces of equipment in addition to the equipment listed in the preceding section (Figure 8–5). The additional pieces include a length of raceway used as the service entrance

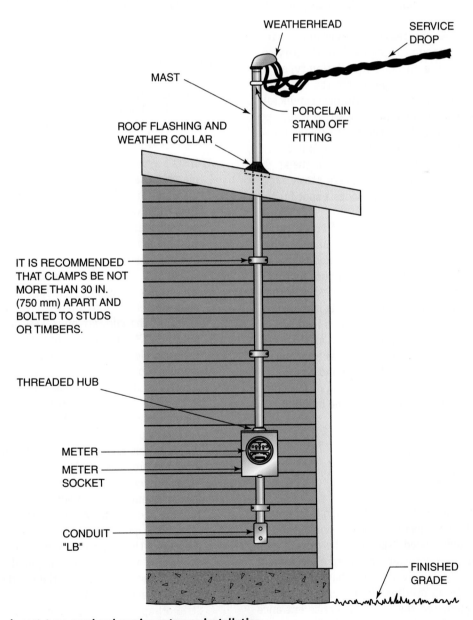

Figure 8–5 A typical mast-type overhead service entrance installation.

mast, a weatherhead to fit the mast, a porcelain standoff fitting, and a weather collar and roof flashing. A **mast kit** is often purchased through an electrical distributor and includes the required additional pieces of equipment for a mast-type service. The purpose for each part unique to this service type is explained next:

- Mast: As explained previously, a mast-type service extends above the roofline so that the minimum required service drop clearance can be achieved when bringing a service drop to a low building like a ranch-style house. The mast should be made of RMC or IMC to provide adequate strength. A minimum of a 2-inch or 2½-inch trade-size mast is usually used. Larger sizes may be needed if the service size requires conductors that are larger than what a 2½-inch conduit can handle. The size and type of raceway make the mast strong enough to resist the tendency to bend from the force placed on it from the service drop.
- Weatherhead: The weatherhead must be designed to fit the size of electrical conduit being used for the service mast. As in other service entrance installation types, it is used to stop the entrance of water into the end of the service entrance mast. Each service entrance conductor must exit from a separate hole in the weatherhead.
- Porcelain standoff fitting: Since the mast extends above the roof in this service entrance style, there is not a convenient place to install the cable hook for the attachment of the service drop. A **porcelain standoff** fitting is placed on the mast and provides a spot for the service drop to be attached.
- Weather collar and roof flashing: The **roof flashing** is placed around the hole in the roof that is made for the service mast to extend through. It is fastened in place with several screws or roofing nails. The house roofing material (like shingles) is eventually installed over the edges of the roof flashing. Once the mast is installed through the hole, a rubber weather collar is installed on the mast. The collar fits tightly around the mast and is placed at the point where the mast extends from the roof flashing. Together, the roof flashing and the **weather collar** provide a seal against rain or melting snow leaking around the mast and into the building.

Overhead Service Installation

Earlier we covered the preparation and planning for a service entrance installation. Remember that the location of the service entrance, commonly called "spotting the meter," is a joint effort between the electrician and a representative of the local electric utility company. Once you have determined that the service will be an overhead type, verified the service location, and purchased the required service entrance equipment, all that is left is the actual installation.

CAUTION

CAUTION: Electricians use ladders and scaffolding when they are installing service entrances. Always observe and follow the proper safety procedures when using ladders and scaffolds. Electricians will also need to wear the proper personal protection equipment when installing service entrances. This equipment should include at a minimum safety glasses, hard hats, rugged work shoes, and leather gloves to protect your hands from cuts and bruises.

Installing an Overhead Service Using Service Entrance Cable

The procedure for installing an overhead service entrance using service entrance cable will vary slightly depending on whether the house has a full basement or a crawl space or is built on a concrete slab. The installation steps outlined next assume that a full basement is available and that the service equipment will be located there. The steps used to install an overhead service entrance with service cable (Figure 8–6) installed on the side of the house are as follows:

1. Install a threaded hub and rain-tight connector to the meter socket (Figure 8–7). The threaded hub is attached to the meter socket using the four screws that come with the threaded hub. The rain-tight connector is screwed into the hub by hand and then tightened securely in place with an adjustable wrench or pump pliers.
2. Locate and install the meter socket. The meter socket is typically installed at a height that will provide a measurement of 4 to 5 feet above ground level to the center of the kilowatt-hour meter (Figure 8–8). This is not an *NEC*® requirement but rather a requirement by the local electric utility. The meter socket is surface mounted to the outside wall of the house using four screws inserted through the mounting holes provided in the back of the meter socket. Make sure to install the meter socket so it is level and plumb on the top, bottom, and sides. A torpedo level works well for this. Where the meter socket is located and how it is mounted provides a point of reference from which all other service entrance equipment installation measurements can be made.
3. Measure and cut the service entrance cable to the length required to go from the meter socket up to a height on the side of the house that will keep the service drop conductors at the minimum required clearance from grade. Allow 3 feet extra at the weatherhead end and 1 foot extra at the meter socket end.
4. Strip off approximately 3 feet of outside sheathing from one end of the SEU cable and 1 foot of outside

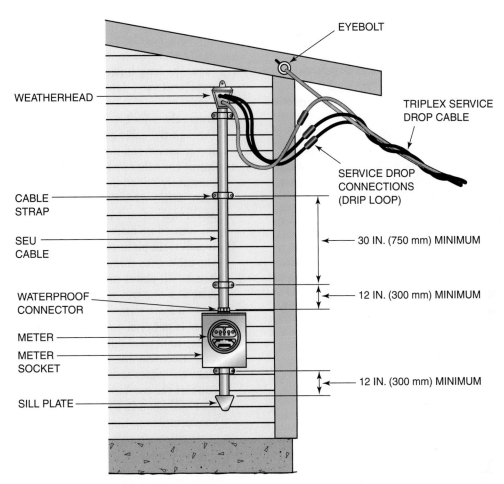

Figure 8–6 A service entrance installation using Type SEU cable attached to the side of a house.

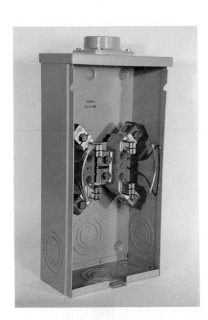

Figure 8–7 A meter socket enclosure with a threaded hub attached to the top. The rain-tight connector that secures the service entrance cable to the meter socket will be threaded into the hub. *Courtesy of Milbank Mfg. Co.*

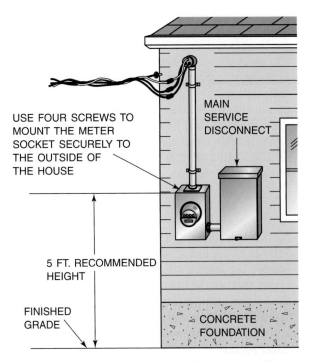

Figure 8–8 The meter socket is secured to the outside wall surface at approximately 5 feet to the top of the meter socket from grade.

sheathing from the other end. Twist together the individual conductors that make up the neutral conductor in Type SEU cable.

5. Install the cable weatherhead on the end that has 3 feet of free conductor. Insert the two black-colored ungrounded conductors and the bare neutral conductor through separate holes in the weatherhead so that there is approximately 3 feet of service conductor that can be formed into a drip loop and connected to the service drop conductors by the utility company. Secure the weatherhead to the cable by tightening the screws of the built-in clamping mechanism on the weatherhead.

6. Mount the assembled service entrance cable with attached weatherhead on the side of the house in a position that is directly above and vertically aligned with the rain-tight connector previously installed in the meter socket hub. A screw installed through a hole in the cable weatherhead will support the entire cable assembly at this point.

7. Insert the other stripped end into the meter socket through the rain-tight connector and tighten the connector around the cable.

8. Making sure the cable is vertically plumb, secure the cable to the side of the building with clips or straps. Locate a clip no more than 12 inches (300 mm) from the meter socket and no more than 12 inches (300 mm) from the weatherhead. Locate the other supports as needed so there is no more than 30 inches (750 mm) between additional supports.

9. Install the cable hook (eyebolt) at a proper location in relation to the weatherhead. Section 230.54(C) of the *NEC*® requires the hook to be located below the weatherhead. If this is not possible, an exception to 230.54(C) allows the hook to be above or to the side of the weatherhead but no more than 24 inches (600 mm) away.

10. Locate and drill a hole for the service entrance cable to enter the house and be attached to the service entrance main panel board enclosure. The hole is usually located directly below the center knockout in the bottom of the meter socket. The hole will have to be drilled at a spot that will provide access to the basement and the location of the panel board. This usually means that the hole will be drilled though the outside wall siding, the wall sheathing, and the **band joist** at a spot just above the sill.

11. Remove the knockout from the bottom of the meter socket that will be used to attach the service cable to the bottom of the meter socket. Install a cable connector in the knockout. The connector will secure the service entrance cable going to the panel, to the meter socket.

12. Measure and cut the service entrance cable to the length required to go from the bottom of the meter socket to the service panelboard enclosure. Allow

1 foot extra at the meter socket end and enough at the panelboard end to allow the conductors to be easily attached to the main circuit breaker and neutral lugs. The length necessary to make connections at the panel will vary, depending on the style of the panel used.

13. Insert the 1-foot stripped end of service entrance cable through the cable connector on the bottom of the meter socket and tighten the connector onto the cable.

14. Insert the other end of the cable through the hole you previously drilled in the wall. Make sure to secure the cable within 12 inches (300 mm) of the meter socket. If additional support is required, make sure the supports are located no more than 30 inches (750 mm) apart.

15. Install a sill plate over the cable at the point where the cable goes through the hole in the side of the house. Sill plates usually come with a small amount of duct seal and two screws. The duct seal is used to help make a more watertight seal around the sill plate. If there is not enough duct seal that comes with the sill plate, additional duct seal can be purchased at your local electrical distributor.

16. Make the meter socket connections (Figure 8–9). Connect the conductors from the weatherhead end to the **line-side** lugs in the meter socket. Connect the conductors from the service panel to the **load-side** lugs.

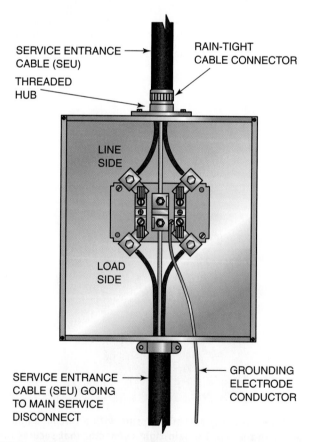

SERVICE ENTRANCE CABLE (SEU)

RAIN-TIGHT CABLE CONNECTOR

THREADED HUB

LINE SIDE

LOAD SIDE

SERVICE ENTRANCE CABLE (SEU) GOING TO MAIN SERVICE DISCONNECT

GROUNDING ELECTRODE CONDUCTOR

Figure 8–9 **Meter socket connections for a typical three-wire overhead service entrance installed with service entrance cable.**

17. Secure the service entrance cable to the service entrance panel and make the proper connections in the service entrance panel. (Panelboard installation is covered later in this chapter.)

Installing an Overhead Service Using Service Entrance Raceway

The procedure for installing an overhead service entrance using service entrance raceway will also vary slightly, depending on whether the house has a full basement or a crawl space or is built on a concrete slab. The installation steps outlined next assume that a full basement is available and that the service equipment will be located there. The steps used to install an overhead service entrance with service raceway (Figure 8–10) installed on the side of the house are as follows:

1. Install a threaded hub to the meter socket. The threaded hub is attached to the meter socket using the four screws that come with the threaded hub. It must be the same size as the size of the raceway being used for the service entrance conductors.

2. Locate and install the meter socket. The meter socket is typically installed at a height that will provide a measurement of 4 to 5 feet above ground level to the center of the kWH meter. This is not an *NEC®* requirement but rather a requirement by the local electric utility. It is surface mounted to the outside wall of the house using four screws inserted through the

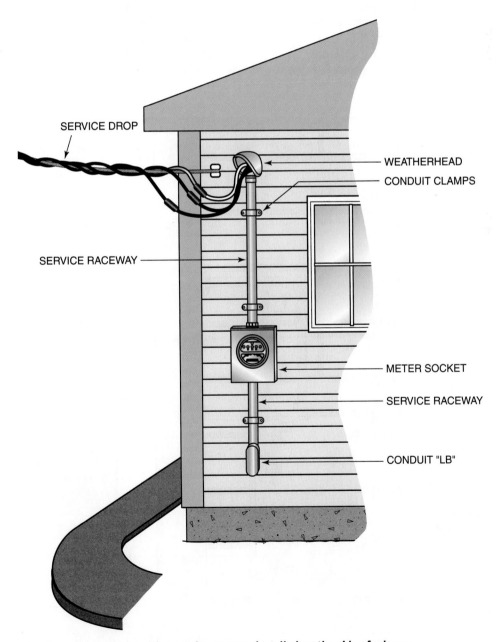

SERVICE DROP

SERVICE RACEWAY

WEATHERHEAD

CONDUIT CLAMPS

METER SOCKET

SERVICE RACEWAY

CONDUIT "LB"

Figure 8–10 An overhead service entrance with a service raceway installed on the side of a house.

mounting holes provided in the back of the meter socket. Make sure to install the meter socket so it is level and plumb on the top, bottom, and sides. A torpedo level works well for this. Where the meter socket is located and how it is mounted provides a point of reference from which all other service entrance equipment installation measurements can be made.

3. Measure and cut the electrical conduit to the length required to go from the meter socket up to a height on the side of the house that will keep the service drop conductors at the minimum required clearance from grade. Remember that electrical conduit will come in 10-foot lengths. In many conduit service entrance installations, a 10-foot length of conduit from the top of a meter socket located approximately 5 feet from the ground will provide all the height that is needed. If more height is needed, you will have to couple together the necessary raceway lengths.

4. Measure and cut the length of service entrance conductor wire needed. Cut enough length to leave

approximately 3 feet of wire on the weatherhead end and 1 foot of wire for the meter socket end. You will need to cut three equal lengths: two lengths for the ungrounded "hot" service conductors and one length for the grounded service neutral conductor. At this time, use white electrical tape to identify each end of the service neutral conductor.

5. Insert one end of the service raceway into the meter socket hub and tighten it. Mount the service entrance raceway on the side of the house in a position that is directly above and vertically aligned with the meter socket hub. Use your torpedo level for this. One-hole straps, two-hole straps, or mineralic straps can be used to provide the support. Make sure to support the raceway within 3 feet (900 mm) of the meter socket and the weatherhead. Additional support is required, depending on the *NEC®* requirements for the type of raceway you are using.

6. Install the raceway weatherhead on the top of the raceway and then remove the top cover of the weatherhead.

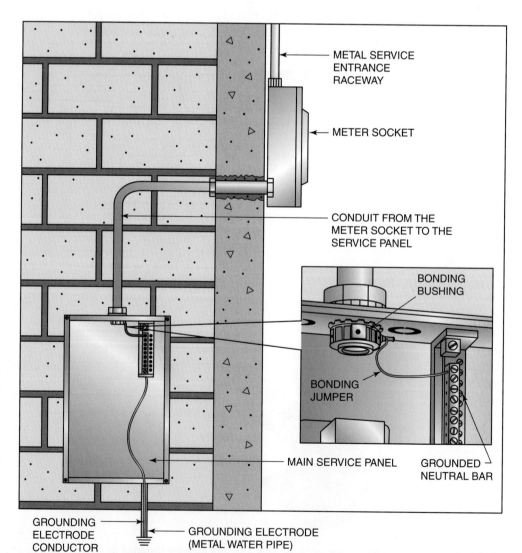

Figure 8-11 A bonding bushing must be installed on the end of a metal service raceway that runs from the meter socket to the service panel and encloses the service entrance conductors. The bonding bushing ensures that the metal conduit will provide a low-resistance path for current to flow in a fault condition. A bonding jumper is run from the bushing to the grounding bus in the panel.

7. At the meter socket end, insert all three service entrance conductors and push them up through the raceway until approximately 3 feet of the conductors come out of the top.

8. At the weatherhead end, insert the two black-colored ungrounded conductors and the white-tape-identified neutral conductor through separate holes in the weatherhead. There should be approximately 3 feet of service conductor coming out of the weatherhead for connection to the service drop conductors by the utility company. Reinstall the cover on the weatherhead.

9. Install the cable hook (eyebolt) at a proper location in relation to the weatherhead on the service raceway. Section 230.54(C) of the *NEC*® requires the hook to be located below the weatherhead. If this is not possible, an exception to 230.54(C) allows the hook to be above or to the side of the weatherhead but no more than 24 inches (600 mm) away.

10. Locate and drill a hole for the service entrance raceway to enter the house and be attached to the service entrance main panel board enclosure. The hole is usually located directly below the center knockout in the bottom of the meter socket. The hole will have to be drilled at a spot that will provide access to the basement and the location of the panel board. This usually means that the hole will be drilled though the outside wall siding, the wall sheathing, and the band joist at a spot just above the sill. A hole saw or some other tool for making larger holes can be used.

11. Remove the knockout from the bottom of the meter socket that will be used to attach the service raceway to the bottom of the meter socket.

12. Install the service raceway between the meter socket and the service panel. "LB" fittings will have to be used to make a 90-degree turn through the drilled hole and into the house. The location of the service panel inside the house may allow you to run the raceway directly from the "LB" to the back of the panel. If the location of the panel is in the basement, another "LB" fitting will be needed to change direction again to bring the raceway into the top of the panel. If you are using a metal raceway, Section 250.92 states that a bonding bushing is required at the connection point where the conduit is attached to the service entrance panel *or* at the connection point where the conduit is attached to the meter socket. The bonding bushing is normally located at the service panel end. A bonding jumper bonds this bushing to the neutral bus bar in the service panel (Figure 8–11). Make sure to install the insulated bushings required by Section 300.4(F) on the meter socket end of the raceway as well as the panel end. The bonding bushing required on one end of the metal service raceway is insulated, and a separate insulated bushing will not have to be used on the bonding bushing end.

13. Measure and cut the service entrance conductors to the length required to go from the bottom of the meter socket to the service panel board enclosure. Allow 1 foot extra at the meter socket end and enough at the panelboard end to allow the conductors to be easily attached to the main circuit breaker and neutral lugs. The length necessary to make connections at the panel will vary, depending on the style of the panel used. Remember to identify the service neutral conductor with white tape.

14. Install the conductors between the meter socket and the service panel. The cover(s) on the "LB" fitting(s) will have to be removed to install the wires through them. Make sure to reinstall the "LB" covers and weatherproof gaskets after you have installed the service conductors.

15. Make the meter socket connections (Figure 8–12). Connect the conductors from the weatherhead end to the line-side lugs in the meter socket. Connect the conductors from the service panel to the load-side lugs.

16. Make the proper connections in the service entrance panel. (Panelboard installation is covered later in this chapter.)

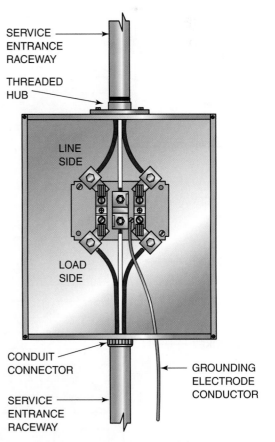

Figure 8–12 Meter socket connections for a typical three-wire overhead service entrance installed with a raceway.

Installing an Overhead Service Using a Mast

The installation of an overhead service entrance that utilizes a service mast (Figure 8–13) will require slightly different installation steps. The installation steps outlined next assume that a full basement is available and that the service equipment will be located there. The installation steps for a typical overhead service entrance using a mast are as follows:

1. Install a threaded hub to the meter socket. The threaded hub is attached to the meter socket using the four screws that come with the threaded hub. It must be the same size as the size of the raceway being used for the service mast.

2. Locate and install the meter socket. The meter socket is typically installed at a height that will provide a measurement of 4 to 5 feet above ground level to the center of the kilowatt-hour meter. This is not an *NEC*®

requirement but rather a requirement by the local electric utility. It is surface mounted to the outside wall of the house using four screws inserted through the mounting holes provided in the back of the meter socket. Make sure to install the meter socket so it is level and plumb on the top, bottom, and sides. A torpedo level works well for this. Where the meter socket is located and how it is mounted provides a point of reference from which all other service entrance equipment installation measurements can be made.

3. Locate and cut a hole in the roof that is aligned with the threaded hub of the meter socket. Make the hole as small as possible but big enough to accommodate the size of the mast pipe.

4. Measure and cut the electrical conduit to the length required to go from the meter socket up through the hole in the roof and to a height that will keep the service drop conductors at the minimum required clearance from grade. In most mast service entrance installations, a 10-foot length of conduit from the top of a meter socket will provide all the height that is needed.

5. Insert the mast pipe through the hole you cut in the roof and raise the mast into position. Attach it to the meter socket and secure the mast in position with the appropriate support straps or brackets.

6. Install the roof flashing and weather collar (rubber boot).

7. Measure and cut the length of service entrance conductor wire needed. Cut enough length to leave approximately 3 feet of wire on the weatherhead end and 1 foot of wire for the meter socket end. You will need to cut three equal lengths: two lengths for the ungrounded "hot" service conductors and one length for the grounded service neutral conductor. At this time, use white electrical tape to identify each end of the service neutral conductor.

8. At the meter socket end, insert all three service entrance conductors and push them up through the raceway (Figure 8–14) until approximately 3 feet of the conductors come out of the top of the mast.

9. Remove the top cover of the weatherhead and install the weatherhead body on the top of the mast. Insert the two black-colored ungrounded conductors and the white-tape-identified neutral conductor through separate holes in the weatherhead. There should be approximately 3 feet of service conductor coming out of the weatherhead for connection to the service drop conductors by the utility company. Reinstall the cover on the weatherhead (Figure 8–15).

10. Install the porcelain standoff onto the mast and align it in the direction that the service drop will be coming from. If necessary, install a guy wire to add additional support to the mast. The *NEC*® says that a guy wire *may* be needed. Some electric utilities will *require* a guy wire if, for example, the point of attachment of the service drop is 30 inches or higher on the mast (Figure 8–16).

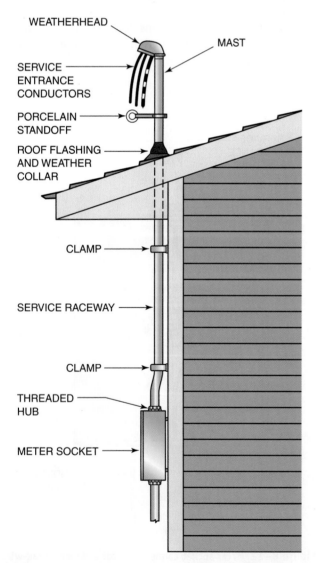

WEATHERHEAD

MAST

SERVICE ENTRANCE CONDUCTORS

PORCELAIN STANDOFF

ROOF FLASHING AND WEATHER COLLAR

CLAMP

SERVICE RACEWAY

CLAMP

THREADED HUB

METER SOCKET

Figure 8–13 The parts of a mast-type service entrance.

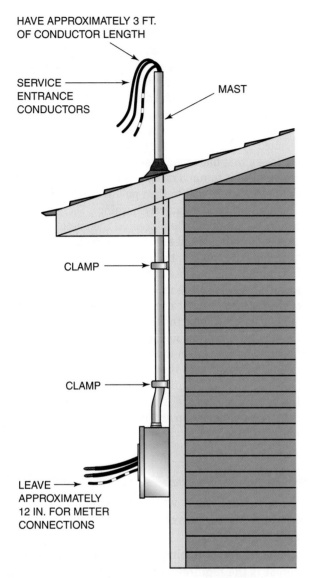

Figure 8–14 Approximately 3 feet of service entrance conductor must extend from the top of the service mast.

11. The installation of a mast-style service entrance from the meter socket to the service entrance panel is the same as the installation steps already outlined in the preceding sections of this chapter.

Underground Service Equipment, Materials, and Installation

An underground service entrance will consist of many of the same kinds of equipment and materials that an electrician uses in an overhead service installation. However, there are some pieces of equipment that are needed for the underground service that are not needed for an overhead service.

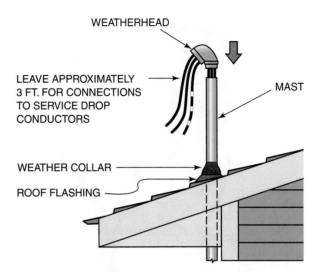

Figure 8–15 Once the service conductors are inserted in the proper holes of the weatherhead, the weatherhead must be secured to the top of the mast. Make sure to leave approximately 3 feet of service conductor for connection to the service drop conductors.

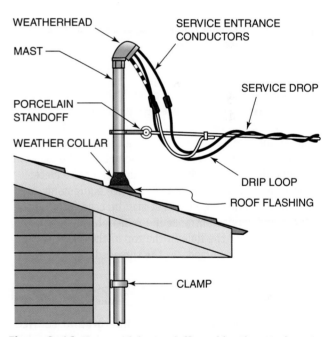

Figure 8–16 The porcelain standoff provides the attachment point for the service drop to the service mast.

These pieces include a meter socket designed for use in an underground service entrance, a service raceway type that conforms with Section 300.5(D)(4), bushings that conform with Section 300.5(H), underground service entrance cable, and a marking tape that conforms with Section 300.5(D)(3). The purpose for each of these parts unique to an underground service is explained here:

• Meter socket for an underground service: The meter socket for an underground service entrance installation

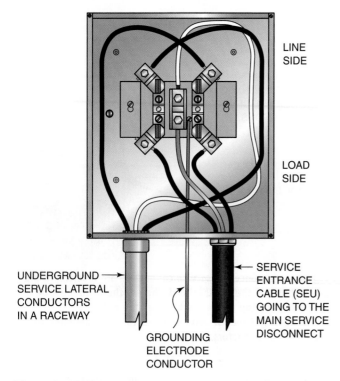

LINE SIDE

LOAD SIDE

UNDERGROUND SERVICE LATERAL CONDUCTORS IN A RACEWAY

SERVICE ENTRANCE CABLE (SEU) GOING TO THE MAIN SERVICE DISCONNECT

GROUNDING ELECTRODE CONDUCTOR

Figure 8–17 The meter connections for a typical three-wire underground service entrance lateral.

(Figure 8–17) must be large enough to allow a minimum of two service entrance raceways to enter the bottom of the socket: one raceway for the incoming electrical power and one raceway to come back out of the meter socket and to the service panel. Sometimes the load-side conductors exit directly out of the back of the meter socket and go directly into the service equipment. An overhead service meter socket may be used for an underground service entrance if a rain-tight plug is installed to cover the hole in the top of the meter socket that usually accommodates the threaded hub.

▶▶▶▶ **CAUTION** ◀◀◀◀

CAUTION: When installing an underground service entrance, remember that the "line" lugs are located at the top of the meter socket and the "load" lugs at the bottom of the meter socket. It is easy to mix them up in an underground service installation because the incoming service conductors enter the meter socket from the bottom.

• Service raceways used in an underground service: Where the underground service conductors exit the trench at the meter socket side or at the transformer side, Section 300.5(D)(4) requires the conductors to be protected from

physical damage with RMC, IMC, or Schedule 80 RNC. These conduit styles must extend upward to a height of 8 feet (2.4 m) above grade at the riser end of the service lateral or at the point where the conductors enter the meter socket on the house side of the service lateral.

• Underground service conductors: The underground service entrance conductors can be installed using Type USE service entrance cable, which can be direct buried at a depth required by the local electric utility and the *NEC®*. The other way that underground service entrance conductors are installed is to run a raceway for the entire length of the service lateral and to then pull in suitable conductors.

• Underground service raceway bushings: Section 300.5(H) requires a bushing to be placed on the end of a raceway that ends underground and provides protection for the underground service entrance conductors emerging from the ground. The purpose of the bushing is to provide a smooth opening to the raceway so that the sharp edges of the raceway will not cut the insulation of the underground service conductors.

• Marking tape: Section 300.5(D)(3) requires a warning ribbon to be used when installing an underground service. The warning tape is to be located at least 12 inches (300 mm) above the buried service entrance conductors. The purpose of the tape is to give someone a warning that they are digging in an area where live electrical wires are buried.

▶▶▶▶ **CAUTION** ◀◀◀◀

CAUTION: Electricians will have to work in trenches during the installation of an underground service entrance. OSHA Section 1926.651 sets out specific requirements for trenching and excavating. Of special interest is the requirement for sloping or shoring up the sides of trenches by means of bracing, underpinning, or piling. The work here has to be done by specialists (in engineering), but you, the electrician, have to be particularly concerned with the safety implications. Without shoring, the threat of an excavated hole caving in while you are in the hole is very real—and also very sobering. There have been many workers who have been buried alive. Do not put yourself in a position where this could happen to you.

Electrical power from the local electric utility can be delivered to the meter socket from a pole-mounted transformer or a pad-mounted transformer (Figure 8–18). Both delivery methods will require a trench to be dug to the house at the point where the meter socket is located. The trench must be dug to a depth that will provide the minimum burial requirements as stated in Table 300.5 of the *NEC®* or at the depth required by the local electric utility company. Figure 8–19 shows

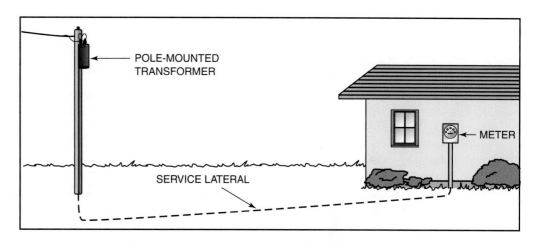

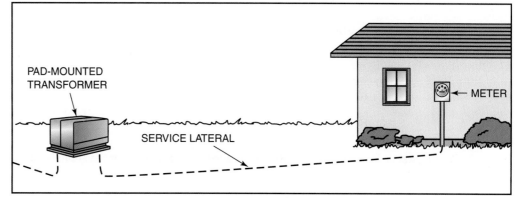

Figure 8–18 Electric power from the local electric utility can be delivered underground to a house from a pole-mounted transformer or a pad-mounted transformer.

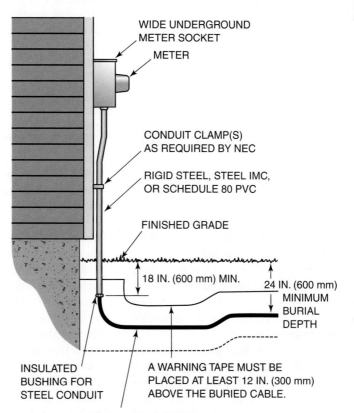

Figure 8–19 A typical underground service entrance installed using Type USE underground service entrance cable.

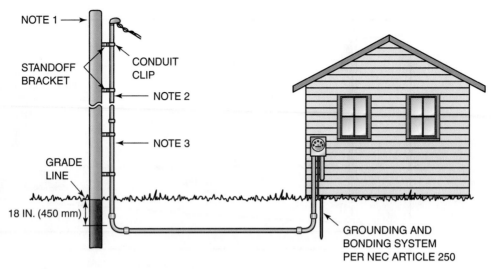

TYPICAL UTILITY COMPANY REQUIREMENTS
NOTES:
   1) LOCATION AND HEIGHT OF RISER POLE TO BE SPECIFIED BY THE UTILITY COMPANY.

   2) RISER SHOULD BE RIGID METAL, STEEL IMC, OR SCHEDULE 40 PVC (MINIMUM 2 IN.) RATED FOR OUTDOOR USE.

   3) RIGID METAL, STEEL IMC, OR SCHEDULE 80 PVC (MINIMUM 2 IN.). MINIMUM HEIGHT ABOVE FINISHED GRADE IS 8 FT (2.4 m).

Figure 8–20 A typical underground service entrance installed using a raceway for the entire length of the underground service lateral.

a common installation for an underground service entrance using underground service entrance (USE) cable. Figure 8–20 shows a common installation for an underground service entrance using a continuous conduit run. If you are installing an underground service entrance, always check with the local electric utility to see if there are any specific requirements that must be followed.

Service Panel Installation

The service entrance panel provides the means of disconnecting the house electrical system from the local electric utility electrical system. It will also usually contain the branch-circuit overcurrent protection devices that provide protection for the various circuits that make up the residential electrical system. As previously discussed, the service entrance panel can be located inside the house or, if in a NEMA Type 3R enclosure, outside the house. Since the most common location for the service entrance panel is inside a house, the discussion here on the installation of the service entrance panel is limited to an inside location.

In Chapter 1, you learned that the *NEC®* provides working clearance requirements around electrical equipment in Article 110. These requirements will have to be observed when installing an electrical panel in a house. Section 110.26 states that sufficient access and working space must be provided and maintained about all electric equipment to permit ready and safe operation and maintenance of such equipment. Storage of

materials that block access to electrical equipment or prevent safe work practices when an electrician is working on a piece of electrical equipment must be avoided at all times. Section 110.26(A)(1) tells us that the depth of the working space in front of residential electrical equipment must not be less than that specified in Table 110.26(A)(1), which specifies the space requirements for working space in front of electrical equipment. Distances are measured from the face of the panel board enclosure. Since residential electrical systems never exceed 150 volts to ground, the 0–150-volt row is used in the table. The dimensions given in the table are based on Conditions 1, 2, or 3. For all three conditions in the 0–150-volt row, the minimum free space that must be maintained in front of residential electrical equipment is 3 feet (900 mm). Notice that Conditions 1, 2, and 3 are explained under the table, but for residential wiring it really does not make any difference which condition we are under since they all have the same clearance for the 0–150-volt locations. Section 110.26(A)(2) states that the width of the working space in front of the electric equipment must be the width of the equipment or 30 inches (750 mm), whichever is greater. Residential electrical panels are usually no more than 14½ inches wide so that they can fit between wall studs that are installed at 16 inches center to center. In all cases, the work space must permit at least a 90-degree opening of equipment doors on the panels. So regardless of the width of the electrical equipment, the working space cannot be less than 30 inches (750 mm) wide. This allows an individual to have at least shoulder-width space in front of the equipment. This 30-inch (750 mm) measurement can be based on the center of the panel or made from either the left or the right edge of the equipment. Section 110.26(E) addresses the required headroom. The minimum headroom of working spaces about electrical service equipment is 6½ feet (2.0 m). An exception to this section states that in existing dwelling units, service equipment or panelboards that do not exceed 200 amperes are permitted in spaces where the headroom is less than 6½ feet (2.0 m). This exception was put in the *NEC®* to allow service upgrades that include a new panel in older basements that have less than a 6½-foot ceiling height. So, when you install an electrical panel for a residential service entrance, there must be at least 3 feet (900 mm) of free space in front of it, the space must be at least 30 inches (750 mm) wide, and the space must extend from the floor to a height of 6½ feet (2.0 m) (Figure 8–21). Nothing else can be allowed to exist in this space.

The preceding sections in this chapter outlined the steps needed to install the service entrance up to the service panel. Now let's take a look at the actual installation of the panel. The service panel (Figure 8–22) should be installed before any branch-circuit installations are made since all the branch-circuit wiring originates at the service panel. The following steps should be followed when installing the service panel:

1. Fasten the service panel in place. If it is a flush-mount installation, the panel is designed to fit between two studs that are 16 inches on center. Make sure to mount the panel enclosure so the front edge of the enclosure will be flush with the finished wall surface. Secure the panel enclosure to the studs with at least two screws per side. If it will be a surface installation, install a mounting backboard for the service entrance panel on the wall where you wish to locate the panel. Mounting **backboards** are often made of 3/4-inch plywood and

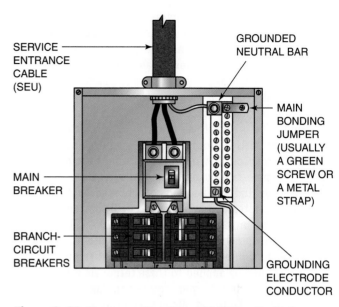

Figure 8–22 The connections for a 120/240-volt, single-phase, main circuit breaker service entrance panel.

are painted black. If the backboard is to be mounted on a concrete wall in a basement, an air gap of at least 1/4 inch behind the backboard should be maintained to allow for moisture to dry out. The backboard must be large enough for the panel and any other items installed on it, such as a surface-mount box with a duplex receptacle.

CAUTION

CAUTION: Section 230.70(A)(1) requires the service conductors to terminate at the main disconnecting means as soon as the conductors enter the house from the outside meter socket. Make sure that you install the service panel at a location that gives you the least amount of "inside run" of the service entrance conductors. Failure to do this will usually result in the electrical inspector turning down your installation.

CAUTION

CAUTION: Make sure to install the service entrance panel so that the requirements of Section 110.26 are met. Remember that a working space must be provided in front of the panel that is at least 30 inches (750 mm) wide, extends at least 3 feet (900 mm) in front of the panel, and extends from the floor up to a height of 6½ feet (2.0 m) or more.

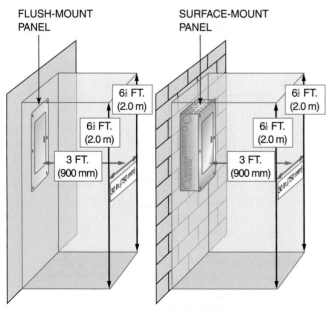

Figure 8–21 Minimum working space clearances in front of service entrance equipment (Section 110.26).

2. Attach the service conductors from the meter socket to the service panel. Make sure there is enough "free" conductor to make the necessary connections. Remember, service entrance cable or a service raceway will be coming to the panel from the meter socket. The cable will require a cable connector to secure it to the enclosure, and the raceway will require some type of conduit connector to secure it to the enclosure.

3. Connect both ungrounded ("hot") service entrance conductors to the main circuit breaker terminals: 100-ampere-rated main circuit breakers have slotted setscrews that will need to be tightened, and 150- and 200-amp-rated and larger main circuit breakers will have hexagonal countersunk screws that will require an Allen wrench or a ratchet wrench with a hexagonal socket to tighten them. Remember that the proper torque must be applied to these terminals with a torque wrench or torque screwdriver. The proper torque amount can be found, along with other information about the panel, inside the panel enclosure on a sticker installed by the panel manufacturer.

CAUTION

CAUTION: If aluminum wire is being used as service conductors, make sure to use an antioxidant and prepare the aluminum wires according to the instructions that come with the antioxidant.

4. Connect the grounded neutral service entrance conductor to the proper terminal on the neutral bus bar.

5. Bond the enclosure to the equipment grounding bus bar and the neutral bus bar using a main bonding jumper. The main bonding jumper can be a green screw, a short length of wire, or a strap. Most newer residential service panels use a green screw as the main bonding jumper.

6. Connect the grounding electrode conductor to the proper terminal on the neutral bus. Route this conductor to the grounding electrode. If a metal water pipe bringing water to the house is available, it must be used as the grounding electrode. Attach the grounding electrode conductor to the water pipe at a point that is 5 feet (1.5 m) or less from where the pipe enters the house. Use a water pipe grounding clamp for the attachment (Figure 8–23).

CAUTION

CAUTION: Remember that if you use a metal water pipe as the grounding electrode, you must supplement it with another electrode. Usually, a ground rod type of electrode is used as the supplement and is connected at the meter socket as shown in Figure 8–24.

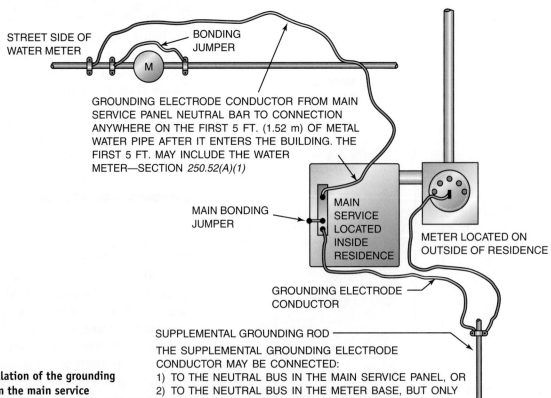

STREET SIDE OF WATER METER

BONDING JUMPER

GROUNDING ELECTRODE CONDUCTOR FROM MAIN SERVICE PANEL NEUTRAL BAR TO CONNECTION ANYWHERE ON THE FIRST 5 FT. (1.52 m) OF METAL WATER PIPE AFTER IT ENTERS THE BUILDING. THE FIRST 5 FT. MAY INCLUDE THE WATER METER—SECTION *250.52(A)(1)*

MAIN BONDING JUMPER

MAIN SERVICE LOCATED INSIDE RESIDENCE

METER LOCATED ON OUTSIDE OF RESIDENCE

GROUNDING ELECTRODE CONDUCTOR

SUPPLEMENTAL GROUNDING ROD

THE SUPPLEMENTAL GROUNDING ELECTRODE CONDUCTOR MAY BE CONNECTED:
1) TO THE NEUTRAL BUS IN THE MAIN SERVICE PANEL, OR
2) TO THE NEUTRAL BUS IN THE METER BASE, BUT ONLY IF ACCEPTABLE TO THE ELECTRIC UTILITY

Figure 8–23 The installation of the grounding electrode conductor from the main service panel to a water pipe electrode.

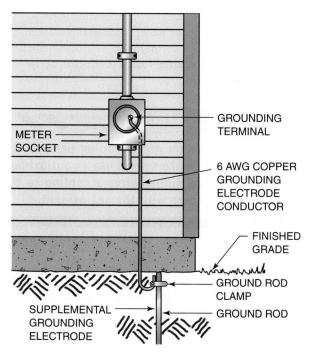

METER SOCKET

GROUNDING TERMINAL

6 AWG COPPER GROUNDING ELECTRODE CONDUCTOR

FINISHED GRADE

GROUND ROD CLAMP

GROUND ROD

SUPPLEMENTAL GROUNDING ELECTRODE

Figure 8–24 A water pipe electrode must be supplemented by another electrode. The most common wiring practice is to use a driven ground rod that is then connected to the meter socket with 6 AWG copper grounding electrode conductor.

Subpanel Installation

Subpanels are often used to eliminate the need for long branch-circuit runs from the main service entrance panel to the various loads of a residential electrical system. They also will provide an opportunity for the installation of more circuits than what the main service panel is designed to supply. The main service entrance panel is usually located inside the house, but as we learned previously, it can be located outside in a proper weatherproof enclosure if there is not a suitable location to install the panel inside. Subpanels will be installed inside the house at a location that is near the area it is intended to serve. If you are installing a subpanel solely for the extra circuits that it provides, you can install it alongside the main service entrance panel.

Electrical power is supplied to a subpanel using a feeder cable run from the main service panel. An electrical raceway, such as EMT, can also be used to run feeder conductors from the service panel to a subpanel. Most residential wiring systems use a feeder cable to bring power to a subpanel, so the discussion on the installation of a subpanel in this section assumes the use of a feeder cable. The feeder cable most often used is a Type SER, which includes two ungrounded black insulated wires, a white insulated grounded neutral conduc-

tor, and one uninsulated grounding conductor. Type SER cable is available with either copper or aluminum conductors. A two-pole circuit breaker at the main service panel provides the necessary overcurrent protection for the feeder cable and subpanel. An earlier chapter covered the sizing of subpanels and feeder conductors.

The subpanel is usually an MLO (main-lug-only) panel and does not have a main circuit breaker. This is because the disconnecting means for the subpanel is the two-pole circuit breaker located in the main service panel. However, if an electrician wants to, a panel with a main circuit breaker can be used as a subpanel.

Subpanels have some specific design differences as compared to a panel used as the main service panel. These differences include the following:

- A separate grounding bar and a separate grounded (neutral) bar are used. The grounded (neutral) bar is actually isolated from the metal subpanel enclosure by insulated standoffs. The grounding conductor bar is actually attached with screws directly to the metal subpanel enclosure. Main service entrance panels typically use a combined grounding/grounded conductor bus bar.
- A four-conductor feeder cable is used to provide electrical power to a subpanel. A three-wire service entrance cable or three wires in a raceway supply the electrical power to a main service entrance panel.
- Subpanels do not normally have a main circuit breaker. The ungrounded feeder conductors are terminated at the panel's main lugs.

The installation procedures for a subpanel are similar to the procedures for installing a main service panel; however, there are a few differences that a residential electrician must be aware of. The following steps should be followed when installing a subpanel:

1. Fasten the subpanel in place. If it is a flush-mount installation, the panel is designed to fit between two studs that are 16 inches on center. Make sure to mount the panel enclosure so the front edge of the enclosure will be flush with the finished wall surface. Secure the panel enclosure to the studs with at least two screws per side. If it will be a surface installation, install a mounting backboard for the subpanel on the wall where you wish to locate it. Mounting backboards are often made of 3/4-inch plywood and are painted black. If the backboard is to be mounted on a concrete wall in a basement, an air gap of at least 1/4 inch behind the backboard should be maintained to allow for moisture to dry out. The backboard must be large enough for the panel and any other items installed that need to be installed on it.

> ◤◤◤◤◤ **CAUTION** ◢◢◢◢◢
>
> **CAUTION: Make sure to install the subpanel so that the requirements of Section 110.26 are met. Remember that a working space must be provided in front of the panel that is at least 30 inches (750 mm) wide, extends at least 3 feet (900 mm) in front of the panel, and extends from the floor up to a height of 6½ feet (2.0 m) or more.**

2. Run the proper-size feeder cable from the main service panel to the subpanel location. Using a cable connector, attach the feeder cable to each panel. Make sure there is enough "free" conductor to make the necessary connections. Support the feeder cable within 12 inches (300 mm) of where it attaches to the main panel and the subpanel. Then make sure the cable is supported at intervals no more than 4½ feet (1.4 m) apart.
3. In the main service panel, first connect both ungrounded ("hot") feeder conductors to the two-pole circuit breaker terminals. Remember that the proper torque must be applied to these terminals with a torque wrench or a torque screwdriver. The proper torque amount can be found, along with other information

about the panel, inside the panel enclosure on a sticker installed by the panel manufacturer. Then connect *both* the feeder grounded neutral conductor and the feeder grounding (bare) conductor to the grounded neutral bus bar.

> ◤◤◤◤◤ **CAUTION** ◢◢◢◢◢
>
> **CAUTION: If aluminum wire is being used as the feeder conductors, make sure to use an antioxidant and prepare the aluminum wires according to the instructions that come with the antioxidant.**

4. In the subpanel (Figure 8–25), connect both ungrounded ("hot") feeder conductors to the main lug terminals. Remember that the proper torque must be applied to these terminals with a torque wrench or a torque screwdriver. The proper torque amount can be found, along with other information about the panel, inside the panel enclosure on a sticker installed by the panel manufacturer. Next, connect the grounded neutral feeder conductor to the neutral bus bar. Finally, connect the grounding (bare) conductor of the feeder to the equipment-grounding bar.

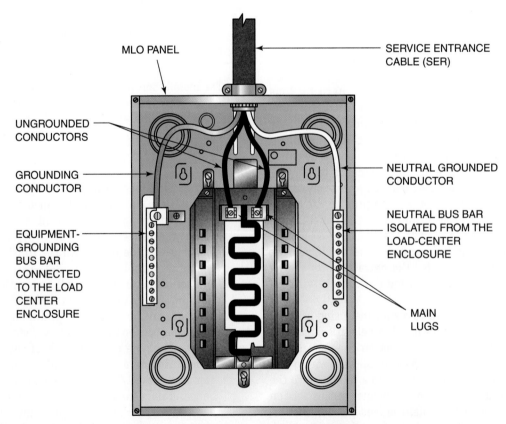

Figure 8–25 **The connections for a 120/240-volt, single-phase, main-lug-only (MLO) subpanel.**

Most panels used as a subpanel do not come from the factory with an equipment-grounding conductor bar already installed in the panel. The electrician must buy a separate grounding bar kit that is designed for the subpanel and install it. The installation of the grounding bar kit can actually take place at anytime during the installation process.

CAUTION

CAUTION: Do not bond the subpanel enclosure to the neutral bus bar using a main bonding jumper like you do with a main service panel. If a main bonding jumper (a green screw or a strap) is included with the subpanel, *throw it away!* Section 250.142(B) clearly states that the grounded circuit conductor cannot be used to ground non-current-carrying metal parts on the *load* side of the service entrance main disconnect. This is exactly what the situation is when using a subpanel and is the reason why the grounded neutral bar and the equipment-grounding bar must be separated in a subpanel. Failure to follow this *NEC*® requirement will result in an electrical safety hazard to the occupants of the house.

Service Entrance Upgrading

The preceding sections of this chapter discussed the installation of service entrances and equipment for new residential applications. However, many service entrance installations involve upgrading an existing service. The reason for the upgrade is usually to increase the service size because of an increase in the load being served by the service. Another reason for an upgrade may be because the condition of the service entrance equipment has deteriorated over time to a point where it is not safe to continue to use the service equipment in its present condition. Whatever the reason, there are some special installation procedures to consider that are not part of the installation of a service entrance in a new residential building.

There are three different service upgrade scenarios:

- A house where nobody lives, electricity has been disconnected, and the house is being remodeled for new occu-

pants: In this case, the old service entrance and equipment can be taken out and a new service entrance installed by following the same procedures as used for installing a service entrance in a new house.
- A house that is occupied, and the service entrance needs to be upgraded at a new location on the house: Some modification to the procedures for a new house service installation is needed.
- A house that is occupied, and the service entrance needs to be upgraded in the same location: Again, some modification to the procedures for a new house service installation is needed.

Upgrading an existing service entrance will require an electrician to work very closely with the local electric utility. If the upgrade requires a new location for the service, contact must be made with the local electric utility to establish the new location for the meter socket and other service equipment. Once the new location for the service has been established, the electrician will install the new service equipment. If the house is occupied, the electrician will typically need to continue electrical service to the house until the switch-over from the old service to the new service is accomplished. This is done by running a jumper cable from a load-side circuit breaker in the old service entrance main panel to a load-side circuit breaker in the new service entrance main panel. The new service entrance main disconnect must remain in the OFF or open position. This procedure is commonly called **backfeeding** and should be done only by an experienced electrician. Once the upgraded service is installed, an inspection will be required. Following approval of the installation, the electric utility will install the new service drop (assuming an overhead installation) and meter in the new meter socket. The utility workers will then deenergize the old service and energize the new service. The electrician will disconnect the jumper cable and then close the main disconnect for the new service panel. The electrician will then switch the circuit breakers in the new panel ON and energize the house branch circuits. The old service entrance and equipment can now be taken out.

If the service entrance upgrade requires an installation of the new service in the same location as the existing service, contact must still be made with the local electric utility. The local utility will "float" the meter socket so that the electrician can install a new meter socket in the same location on the building. The electrician will then install a new meter socket, new service panel, and new service conductors from the meter socket up the side of the house to the new point of attachment. If the house is occupied, the electrician will typically need to continue electrical service to the house until the switch-over from the old service to the new service is accomplished. This is done by running a jumper cable from a load-side circuit breaker in the old service entrance main panel to a load-side circuit breaker in the new service entrance main panel. The new service entrance main disconnect must remain in the OFF or open position. This procedure is commonly called "backfeeding" and should be done only by an experienced electrician. The electrician will then make an appointment with the electric utility company so that the utility can disconnect the old service drop while the electrician is removing the old service and installing new service conductors and associated equipment from the new meter socket to the new main panel through the same hole that the old service conductors used. The utility workers will then install a new meter in the meter socket and energize the new service. The electrician will disconnect the jumper cable and then turn on the main disconnect for the new service panel. The electrician will then switch the circuit breakers in the new panel ON and energize the house branch circuits.

Summary

This chapter covered residential service entrance equipment in detail and the procedures that are commonly used to install a service entrance. You learned that overhead and underground service entrances have many similarities and use many of the same parts. However, the few differences between the two service types greatly affect how each is installed. Installation of a service panel and a subpanel was presented in detail. Since many electricians find themselves doing electrical upgrade work in an existing house, upgrading an existing service was also covered.

Review Questions

Directions: Answer the following items with clear and complete responses.

1 Describe the purpose of a weatherhead.

2 Describe the purpose of the cable hook used in an overhead service entrance.

3 Name the part that is installed on a meter socket whose purpose is to provide a connection point to the meter socket for the service entrance conductors in an overhead service entrance.

4 Explain the difference between Type SEU, SER, and USE service entrance cable. Be as specific as possible.

5 Describe the purpose of a sill plate.

6 Name four common raceway types used in the installation of an overhead residential service entrance where the service entrance conductors are installed in a raceway installed on the side of a house.

7 Describe the purpose of an "LB" conduit body used in a residential service installation that uses electrical conduit.

8 Name two types of electrical conduit that are suitable as a service mast.

9 A mast-type service entrance will need a porcelain standoff fitting and a weather collar/roof flashing. Describe the purpose of each item.

10 The *NEC®* requires the cable hook to be located below the weatherhead in an overhead service installation. If this is not possible, the *NEC®* allows the hook to be above or to the side of the weatherhead, but no more than _____ away. Name the *NEC®* section that states this requirement.

11 Describe the difference between a surface-mount and a flush-mount installation of a service entrance panel.

12 Explain why backfeeding a panel is sometimes done during a service entrance upgrade.

13 Explain the difference between the load-side lugs and the line-side lugs in a meter socket.

14 If Type SER service entrance cable is used by an electrician to feed a subpanel from a main panel, the cable must be secured within _____ of the panel enclosures and at intervals not exceeding _____. Name the *NEC®* section that states this requirement.

15 An MLO panel is usually used as a subpanel in residential wiring systems. Explain what an MLO panel is.

Residential Electrical System Rough-In

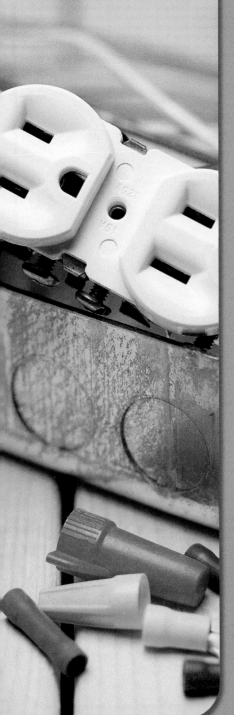

Chapter 9

General Requirements for Rough-In Wiring

Wiring is a term that electricians use to describe the process of installing an electrical system. The rough-in stage of wiring a residential electrical system involves mounting boxes and installing the circuit conductors using an appropriate wiring method. There are many National Electrical Code® (NEC®) requirements that need to be followed during the rough-in stage. This chapter looks at several general requirements that an electrician must consider when installing the rough-in wiring.

OBJECTIVES

Upon completion of this chapter, the student should be able to:

- discuss the selection of appropriate wiring methods, conductor types, and electrical boxes for a residential electrical system rough-in.

- demonstrate an understanding of general requirements for wiring as they apply to residential rough-in wiring.

- demonstrate an understanding of general requirements for conductors as they apply to residential rough-in wiring.

- demonstrate an understanding of general requirements for electrical box installation as they apply to residential rough-in wiring.

- list several general requirements that pertain to wiring methods, conductors, and electrical boxes installed during the rough-in stage of a residential wiring system.

Glossary of Terms

deteriorating agents a gas, fume, vapor, liquid, or any other item that can cause damage to electrical equipment

inductive heating the heating of a conducting material in an expanding and collapsing magnetic field; inductive heating will occur when current-carrying conductors of a circuit are brought through separate holes in a metal electrical box or enclosure

knockout plug a piece of electrical equipment used to fill unused openings in boxes, cabinets, or other electrical equipment

rough-in the stage in an electrical installation when the raceways, cable, boxes, and other electrical equipment are installed; this electrical work must be completed before any construction work can be done that covers wall and ceiling surfaces

sheetrock a popular building material used to finish off walls and ceilings in residential and commercial construction; it is available in standard sizes, such as 4 by 8 feet, and is constructed of gypsum sandwiched between a paper front and back

wiring a term used by electricians to describe the process of installing a residential electrical system

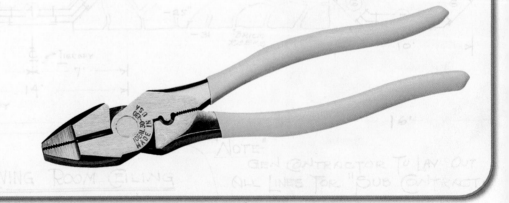

General Wiring Requirements

hapter 2 covered many of the types of equipment and material used to install a residential electrical system. It included an introduction to the different cable and raceway types used in residential wiring. Conductor types and sizes were also presented. Based on the *NEC®* and local electrical code requirements, an electrician must determine an appropriate wiring method and conductor type to be used for installing the residential electrical system. Residential wiring systems are usually installed using a cable-type wiring method. In most areas of the country, electricians use nonmetallic sheathed cable (Type NM), commonly called "Romex," to install residential electrical systems. In certain areas of the country, the authority having jurisdiction does not allow Type NM cable to be used in residential construction. If this is the case, armored-clad cable (Type AC) or metal-clad cable (Type MC) are great alternative wiring methods to Romex. Rarely are residential electrical systems installed using a raceway wiring method. However, some local electrical codes do require dwelling units to be wired using a raceway like electrical metallic tubing (EMT). We discuss raceway installation in a later chapter. In an effort to keep this book easier to understand for the new electrician, and since the vast majority of residential electrical system wiring is done using nonmetallic sheathed cable (NMSC), this chapter and subsequent chapters assume the use of NMSC as further discussion of the rough-in stage is presented.

Requirements for Electrical Installations

Article 110 of the *NEC®* includes several requirements that apply to rough-in wiring. These requirements will need to be followed by electricians during the rough-in stage of a residential wiring job. The following sections discuss in detail the Article 110 sections that must be considered during the rough-in stage.

Section 110.3(B) states that listed or labeled equipment must be installed and used in accordance with any instructions included in the listing or labeling. Manufacturers usually supply installation instructions with equipment for use by electrical contractors, electrical inspectors, and others concerned with an installation. It is important to follow the listing or labeling installation instructions when installing electrical equipment in the rough-in stage.

CAUTION

CAUTION: Never throw away the instructions that come with any piece of electrical equipment. According to Section 110.3(B), all electrical equipment must be installed according to the instructions that are included by the manufacturer of the equipment.

Section 110.7 addresses insulation integrity. It states that completed wiring installations must be free from short circuits and ground faults. Insulation is the material that prevents the short circuits and faults to ground. Failure of the insulation system is one of the most common causes of problems in residential electrical systems. The installing electrician must take care not to damage the conductor insulation in any way during the rough-in stage.

Section 110.11 covers **deteriorating agents** and states that, unless identified for use in the operating environment, no conductors or equipment can be located in damp or wet locations; be exposed to gases, fumes, vapors, liquids, or other agents that have a deteriorating effect on the conductors or equipment; or be exposed to excessive temperatures. When choosing a cable or conductor type, an electrical box style, or any other piece of electrical equipment used in a residential electrical system, an electrician must make sure that it is suitable for the location where you wish to install it. Otherwise, damage to the electrical equipment can occur that will cause problems—if not immediately, then down the road. Fine-Print Note 2 of Section 110.11 tells us that some cleaning and lubricating compounds can cause severe deterioration of plastic materials used for insulating and structural applications in electrical equipment. Equipment identified only as "dry locations," "NEMA Type 1," or "indoor use only" must be protected against permanent damage from the weather during building construction. This last sentence requires electricians to cover and protect any electrical equipment that is being used during the rough-in stage and that might be subject to rain or snow damage. This is especially true when electricians start to install the rough-in wiring and the doors and windows (and sometimes even the roof!) have not been installed yet.

Section 110.12 states that all electrical equipment must be installed in a neat and workmanlike manner. The "neat and workmanlike" installation requirement has appeared in the *NEC®* for more than 50 years. It stands as a basis for pride in one's work and helps make electrical work a profession and not just a "job." Electrical inspectors have cited many *NEC®* violations based on their interpretation of "neat and workmanlike manner." Many electrical inspectors use their own experience or common wiring practice in their local areas as the basis for their judgments. Examples of installations that do not qualify as "neat and workmanlike" include exposed runs of cables or raceways that are not properly supported and result in sagging between supports; field-bent and kinked, flattened, or poorly measured electrical conduit; or electrical boxes and enclosures that are not level or not properly secured.

Section 110.12(A) covers unused openings and states that any unused cable or raceway openings in boxes, cabinets, or other electrical equipment must be effectively closed. The material being used to cover the opening must be at least as strong as the material the electrical box is made of. A common piece of equipment used to meet this requirement is called a **knockout plug** and is inserted into any knockout opening that is open and not used (Figure 9–1).

Section 110.12(C) requires that any internal parts of electrical equipment not be damaged or contaminated by foreign materials such as paint, plaster, cleaners, abrasives, or corrosive residues. There must be no damaged parts that may adversely affect safe operation or mechanical strength of the equipment, such as parts that are broken, bent, cut, or deteriorated by corrosion, chemical action, or overheating. This rule will mean that after the rough-in wiring has been done, an electrician must look at the installed electrical equipment and determine if any of the equipment may have to be covered so that when the house finish work begins, no paint, plaster, or anything else could contaminate the insides of the electrical equipment. For example, a flush-mounted service panel located in an area of the house that will be finished off may require covering the enclosure so that contamination of the inside of the panel cannot take place by paint that is being applied to the walls with a spray gun.

Wiring Methods

Article 300 of the *NEC®* contains several requirements for the wiring methods used by an electrician when installing a residential electrical system. Several of the more important requirements that electricians need to be aware of are covered in the following sections.

Section 300.3(A) states that single conductors with an insulation type that is listed in *NEC®* Table 310.13 can be installed only as part of an *NEC®*-recognized wiring method. In other words, individual insulated conductors with, for example, a THHN insulation cannot be used in a wiring method that is not listed in the *NEC®* or simply installed by itself. Here is an example where an electrician probably wishes this rule were not in the *NEC®*. An electrician installs a switching circuit in a house and, after everything is installed, discovers that a two-wire cable was installed instead of the required three-wire cable. Some electricians believe that they are allowed to simply run one more individual insulated conductor in this situation to fix the problem. They are wrong! Section 300.3(A) does not allow one individual conductor to be run unless it is part of a recognized wiring method, like a Romex cable. In this case, a new three-wire cable would have to be installed.

Section 300.3(B) states that all conductors of the same circuit and, where used, the grounded conductor and all equipment-grounding conductors and bonding conductors must be contained within the same raceway, trench, cable, or cord. This is designed to eliminate **inductive heating.** By keeping all circuit conductors of an individual circuit grouped together, the magnetic fields around the conductors cancel each other out. This means that an expanding and collapsing magnetic field will not be present and will not cause the molecules in the metal enclosure to move around and produce heat, which could damage the conductor insulation (Figure 9–2).

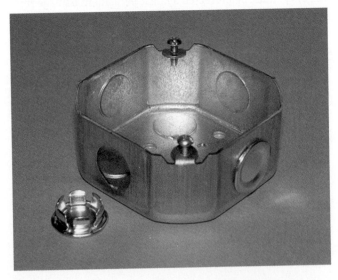

Figure 9–1 Section 110.12(A) requires unused cable or raceway openings in electrical boxes to be effectively closed.

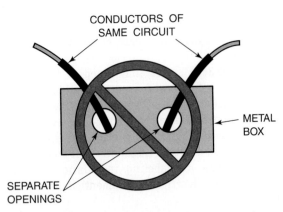

Figure 9–2 Section 300.3(B) requires all circuit conductors of an individual circuit to be grouped together and run through the same box opening to reduce inductive heating and to avoid increases in overall circuit impedance.

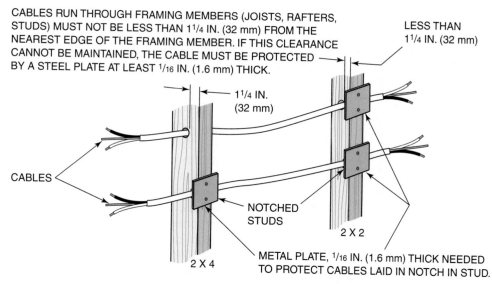

CABLES RUN THROUGH FRAMING MEMBERS (JOISTS, RAFTERS, STUDS) MUST NOT BE LESS THAN 1¼ IN. (32 mm) FROM THE NEAREST EDGE OF THE FRAMING MEMBER. IF THIS CLEARANCE CANNOT BE MAINTAINED, THE CABLE MUST BE PROTECTED BY A STEEL PLATE AT LEAST ¹/₁₆ IN. (1.6 mm) THICK.

LESS THAN 1¼ IN. (32 mm)

1¼ IN. (32 mm)

CABLES

NOTCHED STUDS

2 X 2

2 X 4

METAL PLATE, ¹/₁₆ IN. (1.6 mm) THICK NEEDED TO PROTECT CABLES LAID IN NOTCH IN STUD.

Figure 9–3 Cables or raceways installed through a wood framing member must be protected according to Section 300.4(A)(1) and (2). The intent of this section is to prevent nails and screws from being driven into cables and raceways. An exception to this section permits RMC, IMC, RNC, or EMT to be installed in wood framing members without additional protection.

Section 300.4 states that where a wiring method is subject to physical damage, conductors must be adequately protected. If a cable or raceway is going to be installed through a wood framing member, the following rules must be followed (Figure 9–3):

- Bored holes: In both exposed (like exposed studs in a garage) and concealed (wall studs inside with sheetrock on them) locations, where a cable or raceway type wiring method is installed through bored holes in joists, rafters, or wood members, holes must be bored so that the edge of the hole is not less than 1¼ inches (32 mm) from the nearest edge of the wood member. Where this distance cannot be maintained, the cable or raceway must be protected from penetration by screws or nails by a steel plate or bushing, at least 1/16 inch (1.6 mm) thick, and of appropriate length and width installed to cover the area of the wiring. An exception states that steel plates are not required to protect rigid metal conduit (RMC), intermediate metal conduit (IMC), rigid nonmetallic conduit (RNC), or EMT, that is installed through a wood framing member.
- Notches in wood: Where there is no objection because of weakening the building structure, in both exposed and concealed locations, cables or raceways are permitted to be laid in notches in wood studs, joists, rafters, or other wood members where the cable or raceway at those points is protected against nails or screws by a steel plate at least 1/16 inch (1.6 mm) thick installed before the building finish is applied. Again, an exception states that steel plates are not required to protect RMC, IMC, RNC, or EMT when these raceways are installed in a notch.

The intent of Section 300.4(A)(1) is to prevent nails and screws from being driven into cables and raceways. Keeping the edge of a drilled hole 1¼ inches (32 mm) from the nearest edge of a stud should prevent nails from penetrating the wooden framing member far enough to injure a cable. Building codes limit the maximum size of bored or notched holes in studs, and Section 300.4(A)(2) indicates that consideration should be given to the size of notches in studs so they do not affect the strength of the structure. Most electricians will bore a hole in the framing member rather than notch it.

Sometimes, metal framing members are encountered in residential construction. Section 300.4(B) covers the use of NMSC and electrical nonmetallic tubing (ENT) through metal

 FROM EXPERIENCE

Many building contractors in charge of the actual framing of a house will not want an electrician to "notch" framing members. They are concerned that the studs, joists, or rafters may not have the required strength after a notch has been cut into it. Always check with the building contractor to see if "notching" of building framing members is allowed in the house you are wiring. It is a good idea to also check local building codes to find out what the maximum hole size is that you can bore in a building framing member. Be aware that some premanufactured roof truss systems are designed and engineered so that drilling them for cable runs is not allowed.

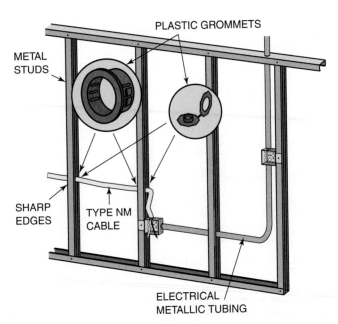

Figure 9–4 **Where metal framing members are encountered,** ***NEC*® Sections 300.4(B) and 334.17 require protection for NMSC when it is run through holes in the metal framing members. This protection is provided by using listed bushings, or grommets, in the holes that are field punched or drilled by an electrician or holes that are provided by the manufacturer. The grommet must cover all metal edges of the hole.**

framing members. If a Type NM cable or ENT raceway is going to be installed through a metal framing member, the following rules must be followed (Figure 9–4):

- NMSC: In both exposed and concealed locations where NMSC pass through either factory- or field-punched, -cut, or -drilled slots or holes in metal members, the cable must be protected by listed bushings or listed grommets covering all metal edges that are securely fastened in the opening prior to installation of the cable. The listed grommets or listed bushings must completely encircle Type NM cables as they pass through holes in metal studs. This requirement affords physical protection for NMSC as the cables are "pulled" through the openings in metal studs. Fastening the listed grommet or listed bushing in place prior to installing cable is mandatory.
- NMSC and ENT: Where nails or screws are likely to penetrate NMSC or ENT, a steel sleeve, steel plate, or steel clip not less than 1/16 inch (1.6 mm) in thickness shall be used to protect the cable or tubing.

Section 300.4(C) addresses the installation of cables or raceways in spaces behind panels. It states that cables or raceway-type wiring methods, installed behind panels designed to allow access, must be supported according to their applicable articles. Sometimes you will find yourself installing wiring above a suspended ceiling. This ceiling type is sometimes called a "dropped ceiling." Any cable- or

raceway-type wiring methods installed above suspended ceilings with lift-up panels must not be allowed to lay on the suspended ceiling panels or grid system. They are required to be supported according to Sections 300.11(A) and 300.23 and the requirements of the article applicable to the wiring method involved. This also applies to the installation of low-voltage cable for chime or thermostat wiring, telephone wiring, cable television wiring, or home computer network wiring. They are not permitted to block access to equipment above the suspended ceiling.

Section 300.4(D) covers the requirements for installing cables and raceways parallel to framing members. In both exposed and concealed locations, where a cable- or raceway-type wiring method is installed along framing members (such as joists, rafters, or studs), the cable or raceway must be installed and supported so that the nearest outside edge of the cable or raceway is not less than 1¼ inches (32 mm) from the nearest edge of the framing member where nails or screws are likely to penetrate. Where this distance cannot be maintained, the cable or raceway shall be protected from penetration by nails or screws by a steel plate, sleeve, or equivalent at least 1/16 inch (1.6 mm) thick (Figure 9–5). Exception No. 1 states that steel plates or sleeves are not re-

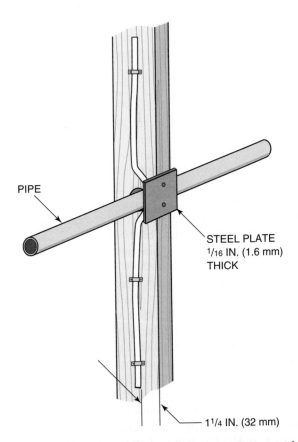

Figure 9–5 **Cables run parallel to framing members must have a clearance of at least 1¼ inches (32 mm) from the cable to the edge of the framing member. If it is not possible to maintain this clearance, a steel plate at least 1/16 inch (1.6 mm) must be installed.**

Figure 9–6 The intent of Section 300.4(D) is to prevent mechanical damage to cables and raceways from nails and screws. Cable is fastened to framing members so that it is at least 1¼ inches from the edge of the framing member. If the cable is installed closer than 1¼ inches from the edge of the framing member, physical protection, such as a steel plate or a sleeve, must be provided. An exception to this section says that this requirement does not apply to RMC, RNC, IMC, or EMT wiring methods because these methods provide physical protection for the conductors.

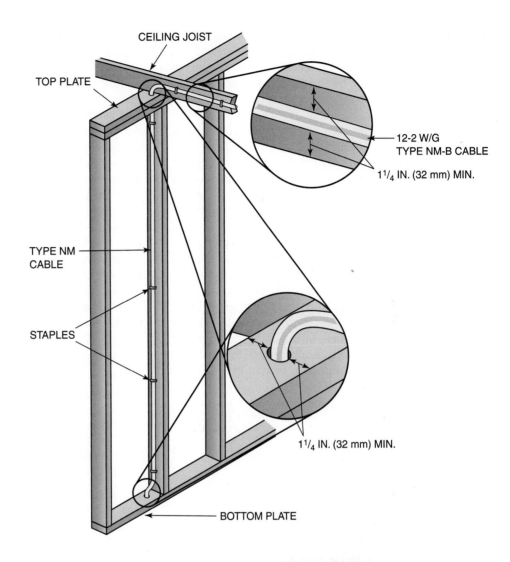

quired to protect RMC, IMC, RNC, or EMT where they have an edge that is less than 1¼ inches (32 mm) to the edge of the framing member. The intent of Section 300.4(D) is to prevent mechanical damage to cables and raceways from nails and screws. The *NEC*® offers two means of protection. The first method is to fasten the cable or raceway so that it is 1¼ inches (32 mm) from the edge of the framing member, as illustrated in Figure 9–6. This requirement generally applies to exposed and concealed work. The second method permits the cable or raceway to be installed closer than 1¼ inches (32 mm) from the edge of the framing member if physical protection, such as a steel plate or a sleeve, is provided (a steel plate is illustrated in Figure 9–7). Exception No. 2 states that for concealed work in a finished existing building, it is permissible to fish the cables between access points and not have the cable or raceway be at least 1¼ inches (32 mm) from the edge of a framing member.

Section 300.4(E) covers cables and raceways installed in shallow grooves. Cable- or raceway-type wiring methods installed in a groove and to be covered by wallboard, siding, paneling, carpeting, or similar finish must be protected by a steel plate or sleeve 1/16 inch (1.6 mm) thick or by not

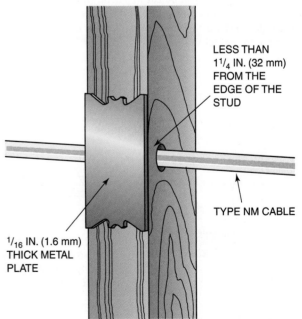

Figure 9–7 A steel plate used to protect an NMSC within 1¼ inches (32 mm) of the edge of a wood stud.

less than 1¼ inches (32 mm) free space for the full length of the groove in which the cable or raceway is installed. For example, an installation may require an electrician to groove out a solid wooden beam so that a Romex cable can be laid in it and run to a certain location (Figure 9–8). Before a wall covering can be placed over the beam with the cable in the groove, the groove must be covered with a metal plate at least 1/16 inch thick (1.6 mm). If the electrician makes the groove so that the cable sits at least 1¼ inches (32 mm) down into it, no steel plate would have to be used. Again, an exception says that steel plates or sleeves are not required to protect RMC, IMC, RNC, or EMT used in a groove.

Section 300.10 covers the electrical continuity of metal raceways and enclosures. It states that metal raceways, cable armor, and other metal enclosures for conductors must be metallically joined together into a continuous electric conductor and must be connected to all boxes, fittings, and cabinets to provide effective electrical continuity. Unless specifically permitted elsewhere in the *NEC®*, raceways and cable assemblies must be mechanically secured to boxes, fittings, cabinets, and other enclosures. Section 250.4(A) and

(B) state what must be accomplished by grounding and bonding the metal parts of the electrical system. These metal parts must form an effective low-resistance path to ground in order to safely conduct any fault current and facilitate the operation of overcurrent devices protecting the enclosed circuit conductors. If an electrician is installing metal boxes during the rough-in stage, a wiring method that can provide a way to ground all the metal boxes must be used. Type NM cable with a bare grounding wire connected to each box with a green grounding screw is very common and works well to meet the intent of this section.

Section 300.11(A) states that raceways, cables, boxes, cabinets, and fittings must be securely fastened in place (Figure 9–9). The specific *NEC®* article that covers a raceway or cable type will state the support requirements for that type of wiring method.

Section 300.11(B) states that raceways may be used as a means of support for other raceways, cables, or nonelectric equipment only under the following conditions:

- Where the raceway or means of support is identified for the purpose
- Where the raceway contains power supply conductors for electrically controlled equipment and is used to support Class 2 circuit conductors or cables that are solely for the purpose of connection to the equipment control circuits
- Where the raceway is used to support boxes or conduit bodies in accordance with Section 314.23 or to support luminaires (fixtures) in accordance with Section 410.16(F)

The purpose of Section 300.11(B)(3) is to prevent cables from being attached to the exterior of a raceway. Electrical, telephone, and computer cables wrapped around a raceway can prevent dissipation of heat from the raceway and affect the temperature of the conductors therein. This section also prohibits the use of a raceway as a means of support for nonelectric equipment, such as suspended ceilings, water pipes, nonelectric signs, and the like, which could cause a me-

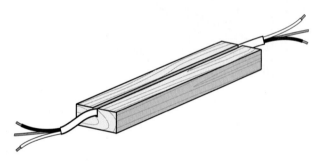

Figure 9–8 A groove can be cut in a solid wood framing member. After the cable has been laid in the groove, a metal plate not less than 1/16 inch (1.6 mm) thick must be installed over the groove to protect the cable.

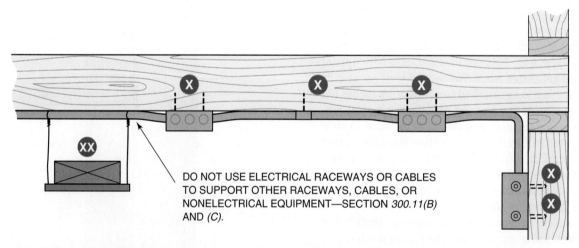

DO NOT USE ELECTRICAL RACEWAYS OR CABLES TO SUPPORT OTHER RACEWAYS, CABLES, OR NONELECTRICAL EQUIPMENT—SECTION *300.11(B)* AND *(C)*.

Figure 9–9 Section 300.11(A) requires all electrical boxes, raceways, and cable assemblies to be securely fastened in place (points X). Section 300.11(B) and (C) does not allow raceways or cables to support other raceways or cables, electrical boxes, or nonelectrical equipment (point XX).

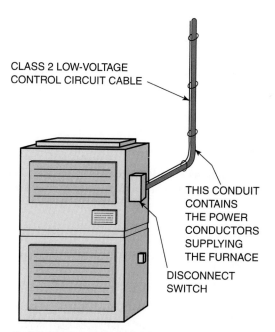

CLASS 2 LOW-VOLTAGE
CONTROL CIRCUIT CABLE

THIS CONDUIT
CONTAINS
THE POWER
CONDUCTORS
SUPPLYING
THE FURNACE

DISCONNECT
SWITCH

Figure 9–10 Sections 300.11(B)(2) and 725.54(D) do allow Class 2 control wiring to be supported by the raceway that supplies the power to the piece of equipment that the control wiring is associated with. A common residential wiring practice is to use the piece of EMT conduit that supplies a furnace to also provide support to the Class 2 thermostat control cable. The thermostat cable is usually secured to the raceway with electrical tape.

chanical failure of the raceway (Figure 9–9). However, Section 300.11(B)(2) does allow the installation of Class 2 thermostat conductors for a boiler or air-conditioning unit to be supported by the conduit supplying power to the unit, as shown in Figure 9–10.

Section 300.11(C) states that cable wiring methods must not be used as a means of support for other cables, raceways, or nonelectrical equipment. This section prohibits cables from being used as a means of support for other cables, raceways, or nonelectric equipment. Taking the requirements of both Section 300.11(B) and Section 300.11(C) together, the common practice (by some electricians) of using one supported cable or raceway to support other raceways and cables is clearly not allowed.

Section 300.12 requires that all metal or nonmetallic raceways, cable armors, and cable sheaths must be continuous between cabinets, boxes, fittings, or other enclosures or outlets. An electrician must install a complete length of cable or raceway from one box or enclosure to another. An exception allows an electrician to use short sections of raceways to provide support or protection of cable assemblies from physical damage. Used this way, a raceway is not required to be mechanically continuous.

Section 300.14 covers a very important requirement: the length of free conductors at outlets, junctions, and switch points. At least 6 inches (150 mm) of free conductor, measured from the point in the box where it emerges from its raceway or cable sheath, must be left at each outlet, junction, and switch point for splices or the connection of luminaires (fixtures) or devices (Figure 9–11). Where the opening to an outlet, junction, or switch point is less than 8 inches (200 mm)

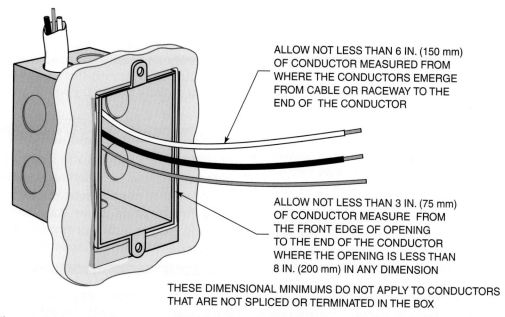

ALLOW NOT LESS THAN 6 IN. (150 mm)
OF CONDUCTOR MEASURED FROM
WHERE THE CONDUCTORS EMERGE
FROM CABLE OR RACEWAY TO THE
END OF THE CONDUCTOR

ALLOW NOT LESS THAN 3 IN. (75 mm)
OF CONDUCTOR MEASURE FROM
THE FRONT EDGE OF OPENING
TO THE END OF THE CONDUCTOR
WHERE THE OPENING IS LESS THAN
8 IN. (200 mm) IN ANY DIMENSION

THESE DIMENSIONAL MINIMUMS DO NOT APPLY TO CONDUCTORS
THAT ARE NOT SPLICED OR TERMINATED IN THE BOX

Figure 9–11 When roughing in the wiring to electrical boxes mounted in walls and ceilings, Section 300.14 requires a minimum of 6 inches (150 mm) of conductor length at each location. No maximum length is stated, but good wiring practice dictates that electricians leave no more than 8 inches (200 mm). Too much conductor left in a box makes it more difficult to place connected receptacles and switches into the box.

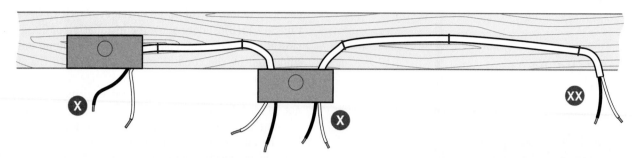

Figure 9–12 Section 300.15 requires that where the wiring method is conduit or cable, an electrical box must be installed at each conductor splice point, outlet point, switch point, junction point, or termination point as shown at points X. Point XX would be a code violation if no box were used.

in any dimension, each conductor must be long enough to extend at least 3 inches (75 mm) outside the opening. An exception states that conductors that are not spliced or terminated at the outlet, junction, or switch point are not required to have the minimum 6 inches (150 mm) of conductor. This is the case when conductors installed in a raceway may go straight through one box to get to another box. This section is very specific about the amount of free conductor length required at each splice point or device outlet.

Section 300.15 states that where the wiring method is conduit, tubing, Type AC cable, Type MC cable, or NMSC, a box or conduit body must be installed at each conductor splice point, outlet point, switch point, junction point, termination point, or pull point. There are some wiring methods used where a box is not required because the wiring method provides interior access to the wires by design or a built-in box is provided. For all practical purposes, all receptacle outlets, switch locations, or junction locations in residential wiring will require an electrical box be installed (Figure 9–12).

Section 300.21 covers the spread of fire or products of combustion. It states that electrical installations in hollow spaces, vertical shafts, and ventilation or air-handling ducts must be made so that the possible spread of fire or products of combustion (i.e., smoke) will not be substantially increased. Openings around electrical penetrations through fire-resistant-rated walls, partitions, floors, or ceilings must be fire-stopped using approved methods to maintain the fire-resistance rating (Figure 9–13). The intent of Section 300.21 is that cables and raceways must be installed through fire-rated walls, floors, or ceilings in such a manner that they do not contribute to the spread of fire or smoke. NFPA 221, Standard for Fire Walls and Fire Barrier Walls, defines fire-resistance rating as "the time, in minutes or hours, that materials or assemblies have withstood a fire exposure as established in accordance with the test procedures of NFPA 251, Standard Methods of Tests of Fire Endurance of Building Construction and Materials." The Fine-Print Note to Section 300.21 points out that directories of electrical construction materials published by qualified testing laboratories contain many listing installation restrictions necessary to maintain the fire-resistive rating of assemblies where penetrations or openings are made. Building codes also contain restrictions

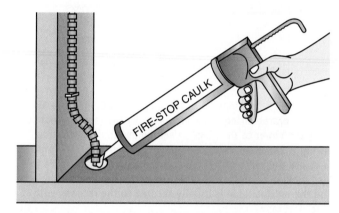

Figure 9–13 Section 300.21 requires all openings made to route cables or raceways through fire-rated walls and ceilings to be fire-stopped. This is to maintain the proper fire-resistance rating. Some state and local jurisdictions require that fire-rated as well as non-fire-rated penetrations be fire-stopped.

on penetrations on opposite sides of a fire-resistant—rated wall assembly. An example is the 24 inches (600 mm) of minimum horizontal separation that applies between boxes installed on opposite sides of the wall (Figure 9–14).

CAUTION

CAUTION: It is a good idea to always check with the authority having jurisdiction to determine which wiring method penetrations you have made will require fire-stopping. Some state and local inspectors will require both non-fire-rated and fire-rated penetrations to be fire-stopped.

Section 300.22 covers wiring in ducts, plenums, and other air-handling spaces. Usually, the information contained in this section is something that electricians doing commercial electrical work are more concerned with. However, Section 330.22(C) addresses wiring in spaces used for environmental air handling, such as those found in dwelling unit forced-

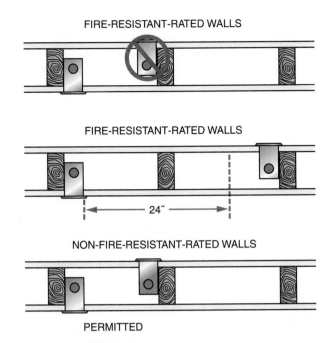

FIRE-RESISTANT-RATED WALLS

FIRE-RESISTANT-RATED WALLS

24″

NON-FIRE-RESISTANT-RATED WALLS

PERMITTED

Figure 9–14 Back-to-back electrical boxes in a fire-resistant wall. Boxes cannot be placed back-to-back in the same stud cavity unless the walls are non-fire rated. In fire-rated walls, the boxes must be placed a minimum of 24″ apart, even when they are in different stud cavities.

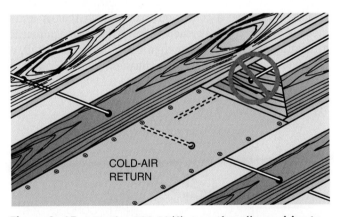

COLD-AIR RETURN

Figure 9–15 A Section 300.22(C) exception allows wiring to pass through in a direction that is perpendicular to the long dimension of air-handling spaces. This situation could present itself if the space between two joists is used as a cold-air return.

hot-air furnace systems. A common practice of heating system installers in a house is to install sheet metal across the bottom of two joists in a basement and use this newly created space as a cold-air return to the furnace. The exception to Section 300.22(C) states that this section does not apply to the joist or stud spaces of dwelling units where the wiring passes through such spaces perpendicular to the long dimension of such spaces. This exception permits cable to pass through joist or stud spaces of a dwelling unit, as illustrated in Figure 9–15. Equipment such as a junction box or device box is not permitted in this location.

General Requirements for Conductors

Article 310 of the *NEC*® contains several sections that cover general requirements for the conductors that an electrician will be installing during the rough-in stage. It is essential that a residential electrician understand these requirements. The most important requirements are presented in the following paragraphs.

Section 310.5 addresses the minimum size of conductors. The minimum size of conductors is shown in *NEC*® Table 310.5 (Figure 9–16). This table tells us that in the voltage range of 0 to 2,000 volts, the minimum-size conductor allowed is 14 AWG copper and 12 AWG aluminum. Since dwelling unit voltages (120/240 volts) fall into this range, we can say that the smallest wire size allowed in residential wiring is 14 AWG copper or 12 AWG aluminum. Exceptions to this section allow for the use of 16 or 18 AWG in residential wiring situations for such things as chime circuits, furnace thermostat circuits, lighting fixture wires, and flexible cords.

Section 310.7 states that conductors used for direct burial applications must be of a type identified for such use. If any of your rough-in wiring will involve installing conductors underground, the wiring method must be listed as suitable for use in an underground location. For example, if you are installing wiring for an outside-located pole light, NMSC could not be installed underground to feed the fixture. A wiring method listed for such use would have to be used. Type UF cable would be a good choice.

Section 310.8(D) addresses those locations exposed to direct sunlight. It states that insulated conductors and cables used where exposed to the direct rays of the sun must be of a type listed for sunlight resistance or listed and marked "sunlight resistant." Sometimes an electrician is required to install conductors outside a house where the wiring method is exposed to the direct rays of the sun; NMSC could not be

Table 310.5 Minimum Size of Conductors

Conductor Voltage Rating (Volts)	Minimum Conductor Size (AWG)	
	Copper	Aluminum or Copper-Clad Aluminum
0–2000	14	12

Figure 9–16 The minimum size of conductors shall be as shown in Table 310.5. For residential wiring this means that 14 AWG copper or 12 AWG aluminum are the minimum sizes allowed. The exceptions to Table 310.5 allow smaller wire sizes for such things as thermostat wiring, doorbell wiring, and lighting fixture wiring. *Reprinted with permission from NFPA 70-2002, the* National Electrical Code®, *Copyright © 2001, National Fire Protection Association, Quincy, MA 02269. This reprinted material is not the referenced subject, which is represented only by the standard in its entirety.*

used in this situation since it is not marked as "sunlight resistant." Too much exposure to the direct rays of the sun will cause the insulation on Romex cable to deteriorate quickly. Again, a Type UF cable, marked "sunlight resistant," would be a good choice in this application.

Section 310.10 covers the temperature limitation of conductors. No conductor can be used in such a manner that its operating temperature exceeds that designated for the type of insulated conductor involved. In no case shall conductors be associated together in such a way with respect to type of circuit, the wiring method employed, or the number of conductors that the limiting temperature of any conductor is exceeded. Residential terminations are normally designed for 60°C or 75°C maximum temperatures. Therefore, the higher-rated ampacities for conductors of 90°C cannot be used unless the terminals at which the conductors terminate have 90°C ratings. Table 310.16 has ampacity correction factors for ambient temperatures greater or less than the ambient temperature identified in the table heading. To assign the proper ampacity to a conductor in an ambient above 30°C (86°F), the appropriate temperature correction factor must be used. This correction factor is applied in addition to any adjustment factor, such as in Section 310.15(B)(2)(a). The information presented in this section was covered in a previous chapter.

Section 310.11(A) states that all conductors and cables must be marked to indicate the following information:

- The maximum rated voltage
- The proper type letter or letters for the type of wire or cable
- The manufacturer's name, trademark, or other distinctive marking by which the organization responsible for the product can be readily identified
- The AWG size or circular mil area
- Cable assemblies where the neutral conductor is smaller than the ungrounded conductors

This information is marked on the surface of the following conductors and cables: single-conductor (solid or stranded) insulated wire, NMSC, service entrance cable, and underground feeder and branch-circuit cable. The information is on a printed tag attached to the coil, reel, or carton for the following conductors and cables: Type AC cable and Type MC cable.

FROM EXPERIENCE

An electrician will often use NMSC from a partially used roll. Typically, the package that the cable originally came in and that has information about the cable written on it has been discarded. The conductor size, maximum rated voltage, and letter type for the cable will be clearly written on the sheathing.

Section 310.12 states that insulated grounded conductors must be identified in accordance with Section 200.6. Section 200.6 permits a grounded conductor of 6 AWG or smaller to be identified by a white or gray color along its entire length. An alternative method of identification is described as "three continuous white stripes on other than green insulation along the conductor's entire length." An insulated grounded conductor larger than 6 AWG must be identified either by a continuous white or gray outer finish, by three continuous white stripes on other than green insulation along its entire length, or, at the time of installation, by a distinctive white marking at its terminations. This marking must encircle the conductor or insulation. The general rule of Section 200.6(B) requires the insulated conductors to be white or gray for their entire length or to be identified by three continuous white stripes along the entire length of the insulated conductor. Another permitted method for these larger conductors is applying a distinctive white marking in the field, such as white electrical tape. The tape is applied at the time of installation at all the grounded conductor termination points. If field applied, the white marking must completely encircle the conductor in order to be clearly visible. This method of identification is shown in Figure 9–17.

Equipment-grounding conductor identification must be in accordance with Section 250.119. Equipment-grounding conductors are permitted to be bare or insulated. Individually insulated equipment-grounding conductors of 6 AWG or smaller must have a continuous outer finish that is either green or green with one or more yellow stripes. For equipment-grounding conductors larger than 6 AWG, it is permitted, at the time of installation, to be permanently identified as an

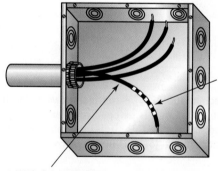

GROUNDED CONDUCTORS ARE USUALLY IDENTIFIED WITH WHITE TAPE

4 AWG OR LARGER CONDUCTORS

Figure 9–17 The general rule of Section 200.6(B) requires insulated grounded conductors to be white or gray for their entire length or to be identified by three continuous white stripes along the entire length of the insulated conductor. The most often used method to identify grounded conductors larger than 6 AWG having an insulation that is not white or gray in color is to field apply white marking tape at the time of installation at all the conductor termination points. If field applied, the white marking tape must completely encircle the conductor in order to be clearly visible.

Electricians typically use green marking tape to identify a grounding conductor. Make sure that the entire exposed length of conductor is reidentified with the tape. This requirement is slightly different than the reidentification of a grounded conductor in that as little as one wrap of white tape satisfactorily reidentifies a grounded conductor, while a grounding conductor must have its entire exposed length covered with green tape.

equipment-grounding conductor at each end and at every point where the conductor is accessible. Identification must encircle the conductor and must be accomplished by one of the following:

- Stripping the insulation or covering from the entire exposed length
- Coloring the exposed insulation or covering green
- Marking the exposed insulation or covering with green tape

Conductors that are intended for use as ungrounded or "hot" conductors, whether used as a single insulated in a cable conductor or in a multiconductor cable like Romex, must be colored to be clearly distinguishable from grounded and grounding conductors. In other words, the ungrounded conductors can be identified with any color other than white, gray, or green. An exception to Section 310.12 (C) says that ungrounded conductors with white or gray insulation in a cable are permitted if the conductors are permanently reidentified at termination points and if the conductor is visible and accessible. The normal method of reidentification is to use black-colored tape (Figure 9–18). Other applications where white conductors are permitted include flexible cords and circuits less than 50 volts. A white conductor used in single-pole, three-way and four-way switch loops also requires reidentification (a color other than white, gray, or green) if it is used as an ungrounded conductor. Switching circuits are covered in Chapter 13.

WHITE WIRE IS REIDENTIFIED AS A 'HOT' CONDUCTOR

Figure 9–18 A white insulated wire in a cable assembly can be reidentified for use as an ungrounded "hot" conductor by marking it with a piece of black electrical tape. Section 310.12(C) allows this.

Table 310.13, Conductor Applications and Insulations, lists and describes the insulation types recognized by the *NEC®*. These conductors are permitted for use in any of the wiring methods recognized in Chapter 3 of the *NEC®*. Table 310.13 also includes conductor applications and maximum operating temperatures. Some conductors have dual ratings. For example, Type XHHW is rated 90°C for dry and damp locations and 75°C for wet locations; Type THW is rated 75°C for dry and wet locations and 90°C for special applications within electric-discharge (fluorescent) lighting equipment. Types RHW-2, XHHW-2, and other types identified by the suffix "-2" are rated 90°C for wet locations as well as dry and damp locations. Conductors permitted to be identified by the suffix "-2," other than Types RHW-2 and XHHW-2, are identified in the table by footnote 4.

Section 310.15 covers the ampacities for conductors rated 0 to 2,000 volts. Table 310.16 is referenced as the place to look for determining the ampacity of a conductor used in residential wiring. A detailed description of ampacity and how to determine a conductor's ampacity using Table 310.16 was presented in Chapter 7. At this time, you should review the material in Chapter 7 that covers determining the ampacity of a conductor.

General Requirements for Electrical Box Installation

As we discussed earlier, installing electrical boxes is part of the rough-in stage for the wiring of a residential electrical system. Article 314 of the *NEC®* covers several requirements that an electrician must comply with when installing electrical boxes. Article 210 has several requirements for the actual location of receptacle outlets, lighting outlets, and switching locations. It is important for an electrician to understand the following sections that pertain to electrical boxes.

Installation and Use of Boxes Used as Outlet, Device, or Junction Boxes

Section 314.16 of the *NEC®* provides the guidelines for the calculation of the maximum number of conductors in outlet, device, and junction boxes. It states that boxes must be of sufficient size to provide free space for all enclosed conductors. In no case can the volume of the box, as calculated in Section 314.16(A), be less than the fill calculation, as calculated in Section 314.16(B). Calculating the maximum number of conductors in an electrical box is covered in detail in Chapter 10.

Section 314.17 states that any conductors entering electrical boxes must be protected from abrasion. This is true whether the box is metal or made of a nonmetallic material. Protection from abrasion can be accomplished by using bushings on sharp raceway ends, using connectors that are designed and built with a smooth opening for the conductors to go through, or simply by making sure that a short section of cable sheathing extends past the clamping mechanism of a cable clamp.

Section 314.17(B) applies to metal boxes. It states that where a raceway or cable is installed with metal boxes, the raceway or cable must be secured to such boxes. This is accomplished by using the proper cable or raceway connector. The connector may be an internal clamp or an external type of connector. Figure 9–19 shows internal cable clamps in both a metal device box and a metal octagon box.

CAUTION

CAUTION: Never install a Romex cable to a metal box by simply taking out the knockout and pushing the cable through the hole. Remember, there must always be a connector to secure the cable to the box.

Section 314.17(C) applies to nonmetallic boxes. It says that nonmetallic boxes must be suitable for the lowest-temperature-rated conductor entering the box. Where NMSC or Type UF cable is used, the sheath must extend not less than 1/4 inch (6 mm) inside the box and beyond any cable clamp. In all instances, all permitted wiring methods must be secured to the boxes. The exception to Section 314.17(C) states that where NMSC or Type UF cable is used with single-gang boxes not larger than a nominal size of 2¼ by 4 inches (57 by 100 mm) mounted in walls or ceilings and where the

cable is fastened within 8 inches (200 mm) of the box measured along the sheath and where the sheath extends through a cable knockout not less than 1/4 inch (6 mm), securing the cable directly to the box is not required. Multiple cable entries are permitted in a single-cable knockout opening (Figure 9–20). For nonmetallic boxes that are larger than

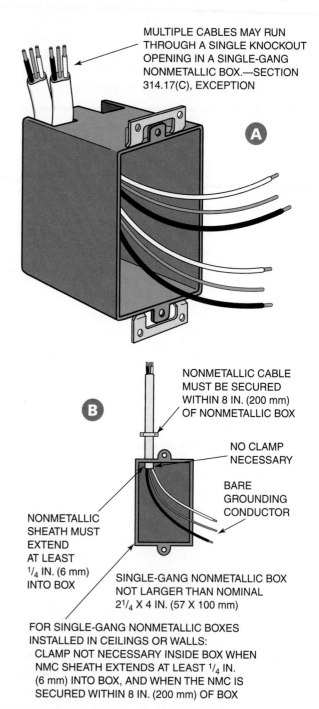

MULTIPLE CABLES MAY RUN THROUGH A SINGLE KNOCKOUT OPENING IN A SINGLE-GANG NONMETALLIC BOX.—SECTION 314.17(C), EXCEPTION

A

B

NONMETALLIC CABLE MUST BE SECURED WITHIN 8 IN. (200 mm) OF NONMETALLIC BOX

NO CLAMP NECESSARY

BARE GROUNDING CONDUCTOR

NONMETALLIC SHEATH MUST EXTEND AT LEAST 1/4 IN. (6 mm) INTO BOX

SINGLE-GANG NONMETALLIC BOX NOT LARGER THAN NOMINAL 2¼ X 4 IN. (57 X 100 mm)

FOR SINGLE-GANG NONMETALLIC BOXES INSTALLED IN CEILINGS OR WALLS: CLAMP NOT NECESSARY INSIDE BOX WHEN NMC SHEATH EXTENDS AT LEAST 1/4 IN. (6 mm) INTO BOX, AND WHEN THE NMC IS SECURED WITHIN 8 IN. (200 mm) OF BOX

Figure 9–20 (A) Section 314.17(C) allows single-gang nonmetallic electrical boxes to have more than one cable installed in one knockout opening. (B) Nonmetallic sheathed cable must be secured within 8 inches (200 mm) of a single-gang nonmetallic box, and the sheathing must extend into the box at least 1/4 inch (6 mm).

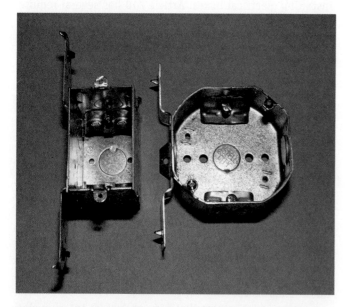

Figure 9–19 A metal device box and octagon box with internal clamps. Section 314.17(B) requires cables and raceways to be attached to all metal electrical boxes. External or internal clamps can be used.

2¼ by 4 inches, some type of cable-securing means is required (Figure 9–21). The requirement is based on the width of the box and the likelihood that the cable will be pushed back out of the box when the conductors and device, if any, are folded back into the box during installation of receptacles and switches.

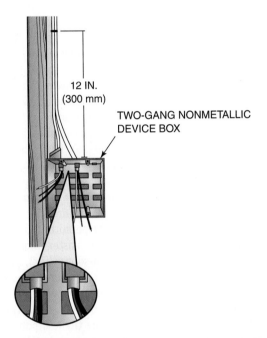

Figure 9–21 Nonmetallic electrical boxes, other than single gang, must have a way to secure a cable to the box. These boxes come from the manufacturer with internal clamps. Nonmetallic sheathed cable must be secured within 12 inches (300 mm) of this box type. Section 314.17(C).

Section 314.20 requires boxes that are installed in walls or ceilings with a surface of concrete, tile, gypsum (sheetrock), plaster, or other noncombustible material to be installed so that the front edge of the box will not be set back of the finished surface more than 1/4 inch (6 mm). In walls and ceilings constructed of wood or other combustible surface material, boxes must be installed flush with the finished surface (Figure 9–22). For example, a wall constructed of gypsum board fastened to the face of wood studs is permitted to contain boxes set back or recessed not more than 1/4 inch. Another example is a wall constructed of wood paneling fastened to the face of wood (or metal) studs; This requires that installed electrical boxes be mounted flush with the combustible finish.

CAUTION: Electrical inspectors may consider certain sheetrock, such as 1/4 inch and 1/2 inch, to be combustible. Check with the local inspector to make sure that it is okay for you to install electrical boxes that set back 1/4 inch from the finished surface. When in doubt as to whether the wall or ceiling surface is combustible or noncombustible, always mount your boxes flush with the finished surface.

Section 314.23 addresses support of electrical boxes. 314.23(A) states that an electrical box mounted on a building or other surface must be rigidly and securely fastened in place. If the surface does not provide rigid and secure support, additional support must be provided. Although there is no *NEC®* rule to address it, it is a common wiring practice to

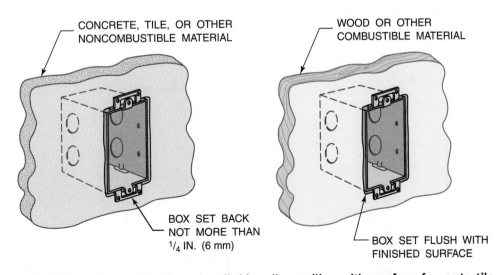

Figure 9–22 Section 314.20 requires electrical boxes installed in walls or ceilings with a surface of concrete, tile, gypsum, plaster, or other noncombustible material, to be installed so that the front edge of the box will not be set back of the finished surface more than 1/4 inch (6 mm). In walls and ceilings constructed of wood or other combustible surface material, boxes must be installed flush with the finished surface.

use at least two screws, nails, or other fastening means to properly secure a box to any surface.

Section 314.23(B) says that a box supported from a structural member of a building must be rigidly supported either directly or by using a metal, polymeric, or wood brace (Figure 9–23). If nails or screws are used as a fastening means, side-mounting brackets on the outside of the electrical box should be used. Some electricians still use nails that are driven through holes inside an electrical box. This practice is allowed as long as the nails are within 1/4 inch (6 mm) of the back, top, or bottom of the box (Figure 9–24). This requirement prevents the nails from interfering with the installation of switches and receptacles.

Section 314.23(C) allows mounting an electrical box in a finished surface as long as it is rigidly secured by clamps, anchors, or fittings identified for the application. This wiring practice is used in remodel work (old work) where boxes are cut into existing walls. Figure 9–25 shows one example of an acceptable mounting method. More information on old-work wiring is included in later chapters.

Section 314.27 has some requirements that must be observed when installing lighting outlet boxes. Section 314.27(A) states that boxes used at luminaire (lighting fixture) outlets must be designed for the purpose. At every outlet used exclusively for lighting, the box must be designed or installed so that a luminaire (lighting fixture) may be attached. Metal octagon boxes and nonmetallic round nail-on ceiling boxes are used most often in residential wiring at light fixture locations. These boxes are designed so that the 8-32 size screws used with these box types will allow attachment of a light fixture to the box (Figure 9–26). Device boxes, such as a 3- by 2- by 3½-inch metal box or a single-gang plastic nail-on box, are designed to have only devices like switches or receptacles attached to them. However, the exception to Section 314.27(A) allows a wall-mounted luminaire (fixture) weighing not more than 6 pounds (3 kg) to be permitted to be supported on a device box (Figure 9–27) or on plaster rings that are secured to other boxes (like a 4-inch-square box), provided the luminaire (fixture) or its supporting yoke is secured to the box with no fewer than two 6-32 or larger screws.

Section 314.27(B) allows outlet boxes to support luminaires (lighting fixtures) weighing no more than 50 pounds (23 kg). In other words, the 8-32 screws can support a lighting fixture to a box as long as the fixture weighs no more than 50 pounds (23 kg). A luminaire (lighting fixture) that weighs more than 50 pounds (23 kg) must be supported independently of the

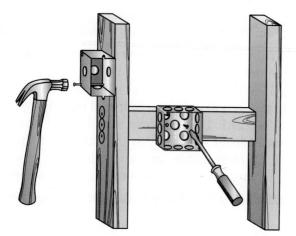

Figure 9–23 A building structural framing member must support electrical boxes. Attaching the box directly to the framing member or using a brace can accomplish this. Section 314.23(B)

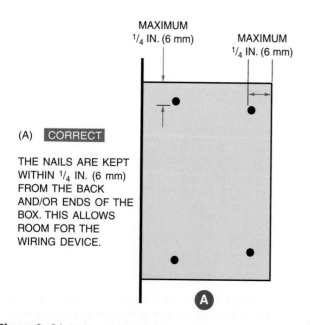

(A) CORRECT

THE NAILS ARE KEPT WITHIN 1/4 IN. (6 mm) FROM THE BACK AND/OR ENDS OF THE BOX. THIS ALLOWS ROOM FOR THE WIRING DEVICE.

(B) VIOLATION

THE NAILS HAVE BEEN INSTALLED MORE THAN 1/4 IN. (6 mm) FROM THE BACK AND/OR ENDS OF THE BOX. THESE NAILS WILL INTERFERE WITH THE WIRING DEVICE. IT MAY ALSO BE IMPOSSIBLE TO INSTALL A WIRING DEVICE DUE TO THE INTERFERENCE OF THESE NAILS.

Figure 9–24 Nails can be used to secure an electrical box to a building framing member, However, care must be taken to follow Section 314.23(B)(1), which requires the nails to be located at least 1/4 inch (6 mm) from the back and ends of the box.

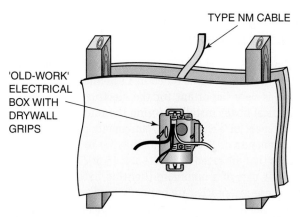

TYPE NM CABLE

'OLD-WORK' ELECTRICAL BOX WITH DRYWALL GRIPS

Figure 9–25 Section 314.23(C) allows an electrical box to be mounted in an existing wall or ceiling by using clamps, anchors, or other fittings identified for the application.

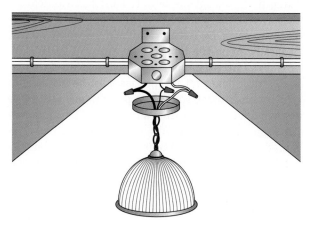

Figure 9–26 Section 314.27(A) states that electrical boxes used to support a luminaire (lighting fixture) must be designed specifically for that purpose. Section 314.27(B) states that any lighting fixture that weighs more than 50 pounds (23 kg) cannot rely on just the electrical box for support and must be independently supported.

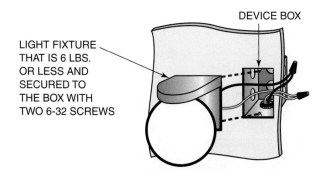

DEVICE BOX

LIGHT FIXTURE THAT IS 6 LBS. OR LESS AND SECURED TO THE BOX WITH TWO 6-32 SCREWS

Figure 9–27 Device boxes are not specifically designed to support luminaires (lighting fixtures). However, a Section 314.27(A) exception allows a luminaire (lighting fixture) of no more than 6 pounds (3 kg) to be supported by a device box when at least two 6-32 screws are used.

outlet box unless the outlet box is listed for the weight to be supported. Larger lighting fixtures often weigh over 50 pounds. Make sure to follow the installation instructions that come with the lighting fixture and support the fixture in such a way that the weight of the fixture is not supported by just the 8-32 screws.

Section 314.27(C) states that boxes listed specifically for floor installation must be used for receptacles located in the floor. No other box type can be used in a floor installation. Make sure to install only boxes that are specifically designed for floor installation (Figure 9–28).

```
CAUTION
```

CAUTION: Regular electrical device boxes are not suitable for installation in a floor. Only boxes that are listed for floor installation can be used. Floor boxes tend to be expensive and harder to install than a "regular" box. Try to avoid installing floor receptacles that require a special box whenever possible. Always try to install the box in a wall.

Section 314.27(D) tells us that where a box is used as the sole support for a ceiling-suspended (paddle) fan, the box must be listed for the application and for the weight of the fan to be supported. Outlet boxes specifically listed to adequately support ceiling-mounted paddle fans are available, as are several alternative and retrofit methods that can provide suitable support for a paddle fan (Figure 9–29). Section 422.18(A) permits boxes listed for the application (paddle fan boxes) as the sole support for fans that do not exceed 35 pounds (16 kg). Section 422.18(B) requires listed fans that exceed 35 pounds (16 kg), with or without accessories like a lighting fixture, to be supported independently from the outlet box. Ceiling-suspended paddle fan installation is covered in greater detail later in the text.

The last section covered here is Section 314.29, which states that boxes must be installed so that the wiring contained

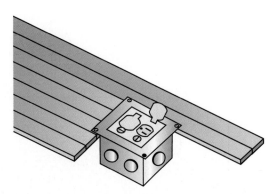

Figure 9–28 Floor boxes must be designed for the purpose. Section 314.27(C). Regular device boxes, square boxes, or octagon boxes cannot be installed in a floor of a dwelling unit.

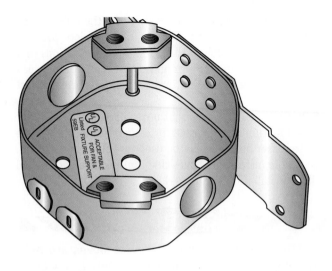

Figure 9–29 A box that is designed for support of a ceiling-suspended paddle fan. Special "beefed-up" mounting brackets to attach the box to a building framing member and larger tapped holes for attaching a ceiling-suspended paddle fan to the box are included.

in them can be rendered accessible without removing any part of the building. A box is permitted to be used at any point for the connection of conduit, tubing, or cable, provided it is not rendered inaccessible. See Article 100 for the definition of "accessible" (as applied to wiring methods).

▰▰▰◣ **CAUTION** ◢▰▰▰

CAUTION: Never install an electrical box in a location that is accessible during the rough-in stage but is rendered inaccessible once wall and ceiling materials have been installed. Remember, all electrical boxes that contain conductors must be accessible.

Dwelling Unit Required Receptacle Outlets

Section 210.52 of the *NEC*® tells an electrician where receptacle outlets must be installed in a dwelling unit. This information is very important for the electrician to know so that electrical boxes installed during the rough-in stage are located to meet or exceed the requirements of this section. The requirements of Section 210.52 apply to dwelling unit receptacles that are rated 125 volts and 15 or 20 amperes and that are not part of a luminaire (lighting fixture) or an appliance. These receptacles are normally used to supply lighting and general-purpose electrical equipment and are in addition to the ones that are 5½ feet (1.7 m) above the floor or located within cupboards and cabinets.

Section 210.52(A) states that in every kitchen, family room, dining room, living room, parlor, library, den, sunroom, bedroom, recreation room, or similar room or area of dwelling units, receptacle outlets must be installed in accordance with the general provisions specified as follows: Receptacles must be installed so that no point measured horizontally along the floor line in any wall space is more than 6 feet (1.8 m) from a receptacle outlet. This means that electrical boxes for receptacles must be installed during the rough-in stage so that no point in any wall space is more than 6 feet (1.8 m) from a receptacle. This rule means that an appliance or lamp with a flexible cord attached may be placed anywhere in the room near a wall and be within 6 feet (1.8 m) of a receptacle (Figure 9–30). This required placement of receptacles will eliminate the need for long extension cords running all over the place.

As used in this section, a wall space includes the following:

- Any space 2 feet (600 mm) or more in width (including space measured around corners) and unbroken along the floor line by doorways, fireplaces, and similar openings (Figure 9–31): Isolated, individual wall spaces 2 feet (600 mm) or more in width are considered usable for the location of a lamp or appliance, and a receptacle outlet

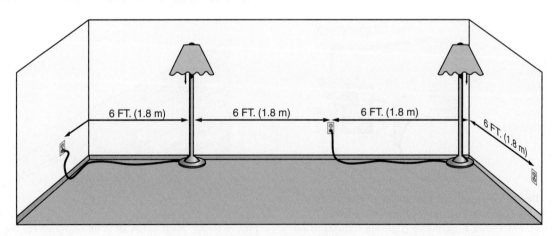

Figure 9–30 A good way to understand the placement of dwelling unit receptacles along a wall is to make sure that any piece of electrical equipment with a power cord that is 6 feet (1.8 m) long can be placed anywhere in a room and still be able to reach a receptacle.

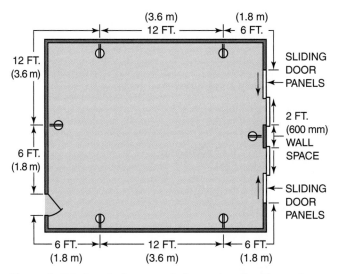

Figure 9–31 A typical receptacle layout in a dwelling unit room that meets the requirements of Section 210.52(A).

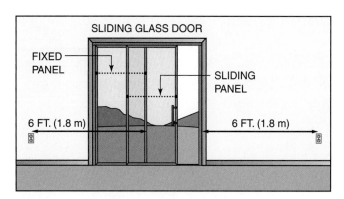

Figure 9–32 Fixed panels, like the part of sliding glass door that does not move, are considered wall space.

is required to be provided.

- The space occupied by fixed panels in exterior walls, excluding sliding panels: Fixed panels in exterior walls, such as the fixed glass section of a sliding glass door, are counted as regular wall space. A floor-type receptacle installed no more than 18 inches (450 mm) from the wall can be used if the spacing requirements might require a receptacle at the location of a glass fixed panel (Figure 9–32).
- The space afforded by fixed room dividers, such as freestanding bar-type counters or railings: Fixed room dividers, such as bar-type counters and railings, are to be included in the 6- foot (1.8-m) measurement (Figure 9–33).

Section 210.52(C) covers the required location of receptacles at countertop locations in a dwelling room kitchen or dining room. This information is extremely important for the electrician since electrical boxes for the receptacles will have to be installed and wire run to them before any of the kitchen cabinets and countertop have been installed. The correct placement of the electrical boxes during the rough-in stage is imperative so that electrical boxes do not end up hidden behind cabinets or other kitchen equipment. The following requirements must be met when installing boxes for receptacle outlets to serve countertops in a kitchen or dining room (Figure 9–34):

- A receptacle outlet must be installed at each wall counter space that is 12 inches (300 mm) or wider. Receptacle outlets must be installed so that no point along the wall line is more than 24 inches (600 mm) measured horizontally from a receptacle outlet in that space.
- At least one receptacle outlet must be installed at each island counter space with a long dimension of 24 inches (600 mm) or greater and a short dimension of 12 inches (300 mm) or greater.

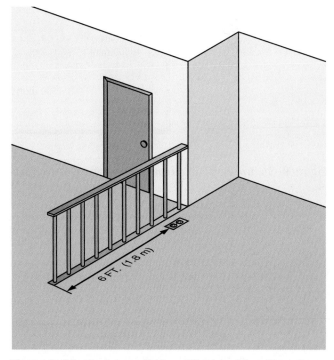

Figure 9–33 Fixed room dividers, like the railing shown here, are considered to be wall space and must meet the requirements of Section 210.52(A).

- At least one receptacle outlet must be installed at each peninsular counter space with a long dimension of 24 inches (600 mm) or greater and a short dimension of 12 inches (300 mm) or greater. A peninsular countertop is measured from the connecting edge.
- Countertop spaces separated by range tops, refrigerators, or sinks are considered as separate countertop spaces.
- Receptacle outlets must be located above, but not more than 20 inches (500 mm) above, the countertop. On island and peninsular countertops where the countertop is flat across its entire surface (no backsplashes, dividers, and so on) and there is no way to mount a receptacle within 20 inches (500 mm) above the countertop,

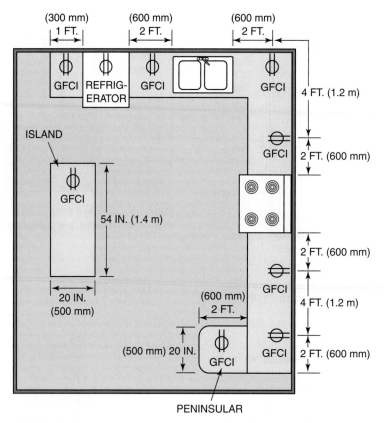

Figure 9–34 Dwelling unit receptacles serving countertop spaces in a kitchen and installed in accordance with Section 210.52(C).

receptacle outlets are permitted to be mounted not more than 12 inches (300 mm) below the countertop. However, receptacles mounted below a countertop cannot be located where the countertop extends more than 6 inches (150 mm) beyond its support base, such as at a bar-type eating area in a kitchen (Figure 9–35).

CAUTION

CAUTION: According to Section 406.4(E), receptacles cannot be installed in a face-up position in a countertop. Receptacles installed in a face-up position could collect crumbs, liquids, and other debris, resulting in a potential fire or shock hazard.

Section 210.52(D) requires one wall receptacle in each bathroom of a dwelling unit to be installed adjacent and within 36 inches (900 mm) of the sink (Figure 9–36). This receptacle is required in addition to any receptacle that may be part of any luminaire (lighting fixture) or medicine cabinet. If there is more than one sink, a receptacle outlet is required adjacent to each sink location. If the sinks are in close proximity, one duplex receptacle outlet installed between the two basins will satisfy this requirement (Figure 9–37).

Figure 9–35 Receptacles cannot be installed face-up in a countertop. However, the receptacles required by Section 210.52(C) can be mounted below countertops as long as they are located no lower than 12 inches (300 mm). If the countertop has an overhanging portion that is more than 6 inches (300 mm), receptacles are not allowed under that area.

Section 210.52(E) states that for a one-family dwelling and each unit of a two-family dwelling that is at grade level, at least one receptacle outlet accessible at grade level and not more than 6.5 feet (2.0 m) above grade must be installed at the front and back of the dwelling (Figure 9–38).

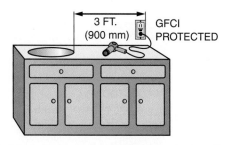

Figure 9–36 **At least one receptacle must be installed within 36 inches (900 mm) of a bathroom sink.**

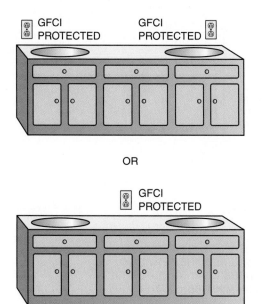

OR

Figure 9–37 **If a bathroom has more than one sink, receptacles can be placed within 36 inches (900 mm) of each sink or one receptacle may be placed so that it is within 36 inches (900 mm) of either sink.**

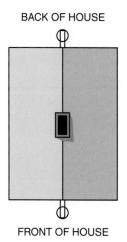

Figure 9–38 **Receptacles are required on the front and back of a one-family dwelling unit as well as each unit of a two-family dwelling. They must be located no more than 6½ feet (2 m) above grade.**

Dwelling Unit Required Lighting Outlets

During the rough-in stage, electrical boxes will have to be installed for the lighting outlets required in a house. Section 210.70 contains the minimum requirements for providing lighting in a dwelling unit. This information helps the electrician determine where a lighting outlet will have to be located. Some lighting fixtures are attached directly to electrical outlet boxes and will require a box to be mounted at the proper location. Other lighting fixtures are simply mounted to the surface, and electrical wiring is brought into the fixture wiring compartment where connections are made. Either way means that an electrician must install wiring with or without an electrical box at lighting outlet locations. Figure 9–39 shows the location of the required lighting outlets in a typical dwelling unit.

Section 210.70(A)(1) requires at least one wall switch-controlled lighting outlet to be installed in every habitable room and bathroom. Exception No. 1 to the general rule allows one or more receptacles controlled by a wall switch to be permitted in lieu of lighting outlets, but only in areas other than kitchens and bathrooms. A wall switch-controlled lighting outlet is required in the kitchen and bathroom. A receptacle outlet controlled by a wall switch is not permitted to serve as a lighting outlet in these rooms. Exception No. 2 allows lighting outlets to be controlled by occupancy sensors that are (1) in addition to wall switches or (2) located at a customary wall switch location and equipped with a manual override that will allow the sensor to function as a wall switch.

Section 210.70(A)(2) lists three additional locations where lighting outlets need to be installed:

- At least one wall switch–controlled lighting outlet must be installed in hallways, stairways, attached garages, and detached garages with electric power.

Section 210.52(F) requires at least one receptacle outlet to be installed for the laundry. Remember that a 20-ampere branch circuit, which can have no other outlets on the circuit, supplies the laundry receptacle outlet(s).

Section 210.52(G) states that for a one-family dwelling, at least one receptacle outlet, in addition to any provided for laundry equipment, must be installed in each basement and in each attached garage and in each detached garage with electric power. Where a portion of the basement is finished into one or more habitable rooms, each separate unfinished portion must have a receptacle outlet installed in accordance with this section.

Section 210.52(H) requires that in dwelling units, hallways of 10 feet (3.0 m) or more in length must have at least one receptacle outlet. In determining the hallway length, use the measured length along the centerline of the hall without passing through a doorway.

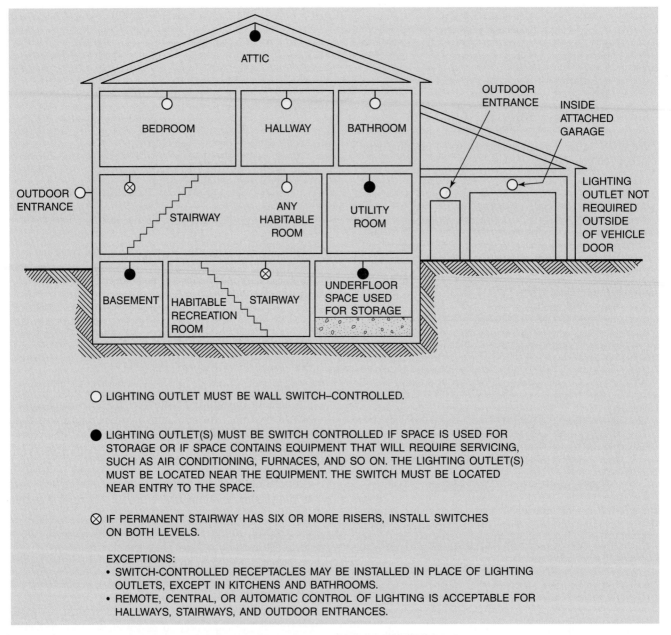

Figure 9–39 The required lighting outlets in a dwelling unit according to Section 210.70.

- In attached garages and detached garages with electric power, at least one wall switch–controlled lighting outlet must be installed to provide illumination on the exterior side of outdoor entrances or exits with grade-level access. A vehicle door in a garage is not considered as an outdoor entrance or exit.
- Where one or more lighting outlet(s) are installed for interior stairways, there must be a wall switch at each floor level at and each landing level that includes an entryway to control the lighting outlet(s) where the stairway between floor levels has six risers or more.

An exception states that in hallways, stairways, and at outdoor entrances, remote, central, or automatic control of lighting is permitted.

Section 210.70(A)(3) addresses storage or equipment spaces. It says that for attics, crawl spaces, utility rooms, and basements, at least one lighting outlet containing a switch or controlled by a wall switch must be installed where these spaces are used for storage or contains equipment requiring servicing. At least one point of control must be at the door to these spaces. The lighting outlet must be provided at or near equipment, like furnaces or other heating, ventilating, and air-conditioning equipment, that requires servicing. Installation of lighting outlets in attics, crawl spaces, utility rooms, and basements is required when these spaces are used for storage of items such as holiday decorations or luggage.

Summary

Before an electrician starts the rough-in stage of a residential wiring job, certain *NEC®* requirements must be considered. During the rough-in stage, these requirements are then applied. After completing the rough-in stage, an electrical in-spector will determine if you correctly met the *NEC®* require-ments with your installation. At this point in the installation of the wiring system, no ceiling or wall coverings have been installed, and the inspector will have easy access to the boxes and wire installed by the electrician. This chapter covered sev-eral common *NEC®* general requirements that must be consid-ered when wiring a house. Also discussed and presented were several requirements that pertain to the conductors and elec-trical boxes installed during the rough-in stage. Understand-ing and following the *NEC®* requirements presented in this chapter will result in the rough-in stage wiring being free of code violations.

Review Questions

Directions: Answer the following items with clear and complete answers.

1 Define the term "wiring" as it is used by electricians doing residential work.

2 Describe the "rough-in" stage of a residential wiring installation.

3 Name the most common wiring method used in residential wiring and which *NEC®* article covers this wiring method.

4 The *NEC®* requires electricians to install the electrical system in a "neat and workmanlike manner." Describe what this means and give an example of something not installed in a neat and workmanlike manner.

5 Name the table in the *NEC®* that lists the insulation types and their properties for those insulations recognized by the code.

6 When boring holes in a wooden framing stud for the installation of Romex cable, electricians must make sure that the distance from the edge of the hole to the face of the stud is no less than _____. If this distance cannot be maintained, a metal plate at least _____ thick must be used to protect the cable. Name the *NEC®* section that covers this wiring situation.

7 The minimum amount of free conductor required to be available at an electrical box used in residential wiring for termination purposes is _____. As long as the box has no dimension larger than 8 inches (200 mm), there must be at least _____ of free conductor extending out from the box opening. Name the *NEC®* section that covers this wiring practice.

8 Openings around electrical penetrations through fire-resistant-rated walls, partitions, floors, or ceilings must be fire-stopped using approved methods to maintain the fire-resistance rating. Explain why fire-stopping is required.

9 The minimum size of wire used in residential wiring is _____ AWG. (general rule)

10 The *NEC®* recognizes two wire sizes that are smaller than the answer to question 9. List the two sizes and give an example of where they might be used in a residential wiring system.

11 Name a cable wiring method that would be a good choice for an installation that would result in the cable being located outdoor where the direct rays of the sun could shine on it.

12 Name four items that must be written on the sheathing of a nonmetallic sheathed cable, according to the *NEC®*.

13 Describe how a 6 AWG or smaller insulated grounded conductor must be identified.

14 Describe how an insulated grounded conductor larger than 6 AWG is identified.

15 Describe how a white insulated conductor in a cable assembly is reidentified as a "hot" conductor.

16 Describe how a 6 AWG or smaller grounding conductor is identified.

17 Describe how a conductor that is larger than 6 AWG can be identified as a grounding conductor.

18 Conductors that are intended for use as ungrounded or "hot" conductors, whether used as a single insulated conductor or in a multiconductor cable like Romex, must be colored to be clearly distinguishable from grounded and grounding conductors. Describe how the identification is accomplished.

19 Explain what is required by Section 110.3(B) of the *NEC®*.

20 Name the common piece of electrical equipment that electricians use to insert into any knockout opening that is open and not used. Name the *NEC®* section that covers this wiring practice.

21 In single-gang nonmetallic boxes, the sheathing of a Romex cable must extend at least _____ into the box and be visible. Romex must also be secured no more than _____ from this same single-gang non-metallic box.

22 A wall constructed of sheetrock fastened to the face of wood studs is permitted to contain boxes set back or recessed not more than _____. Name the *NEC®* section that covers this requirement.

23 Some electricians use nails that are driven through holes inside an electrical box to attach the box to a wall stud. This practice is allowed as long as the nails are within _____ of the back, top, or bottom of the box. Name the *NEC®* section that covers this requirement.

24 A wall-mounted luminaire (fixture) weighing not more than _____ is permitted to be supported on a device box provided the luminaire (fixture) or its supporting yoke is secured to the box with no fewer than two 6-32 or larger screws. Name the *NEC®* section that covers this requirement.

25 The 8-32 screws used with a lighting outlet box can support a lighting fixture to the box as long as the fixture weighs no more than _____. Name the *NEC®* section that covers this requirement.

26 Switch-controlled receptacles, instead of lighting outlets, can be used in all areas of a dwelling unit except _____ and _____.

27 Receptacles located more than _____ above the floor are not counted in the required number of receptacles along a wall.

28 The maximum distance between wall receptacles in a dwelling unit is _____.

29 The maximum distance between countertop receptacles in a dwelling unit kitchen is _____.

30 Interior stairway lighting must be controlled by a wall switch placed at the top and bottom of the stairway when the stairway has _____ or more steps.

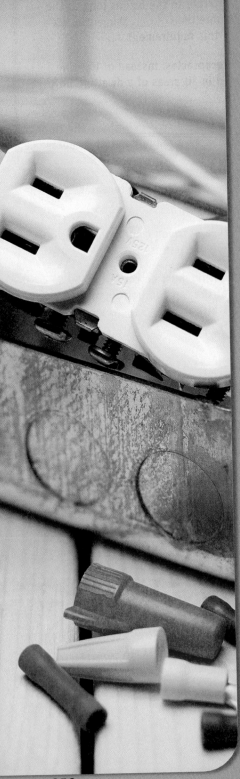

Chapter 10 | Electrical Box Installation

Electrical boxes are required at each location on an electrical circuit where a device or light fixture is connected to the circuit. An electrical box, called a junction box, is also required at all locations where circuit conductors are simply spliced together. During the rough-in stage, an electrician will install electrical boxes that are designed for the specific wiring application indicated on the building electrical plans. Switches and receptacles will require a device box to be installed, lighting fixtures will require an outlet box designed to allow the fixture to be secured to it, and junction box locations will require a box designed to accommodate a certain number of spliced conductors. This chapter covers how to select the appropriate electrical box type and how to properly size the electrical box according to the National Electrical Code® (NEC®). Installation methods for various electrical box types in both new construction and remodeling situations are covered in detail.

OBJECTIVES

Upon completion of this chapter, the student should be able to:

- ⊗ select an appropriate electrical box type for a residential application.
- ⊗ size electrical boxes according to the *NEC*®.
- ⊗ demonstrate an understanding of the installation of metal and non-metallic electrical device boxes in residential wiring situations.
- ⊗ demonstrate an understanding of the installation of lighting outlet and junction boxes in residential wiring situations.
- ⊗ demonstrate an understanding of the installation of electrical boxes in existing walls and ceilings.

Glossary of Terms

accessible (as applied to wiring methods) capable of being removed or exposed without damaging the building structure or finish or not permanently closed in by the structure or finish of the building

box fill the total space taken up in an electrical box by devices, conductors, and fittings; box fill is measured in cubic inches

lighting outlet an outlet intended for the direct connection of a lamp holder, a luminaire (lighting fixture), or a pendant cord terminating in a lamp holder

Madison hold-its thin metal straps that are used to hold "old-work" electrical boxes securely to an existing wall or ceiling

pigtail a short length of wire used in an electrical box to make connections to device terminals

setout the distance that the face of an electrical box protrudes out from the face of a building framing member; this distance is dependent on the thickness of the wall or ceiling finish surface material

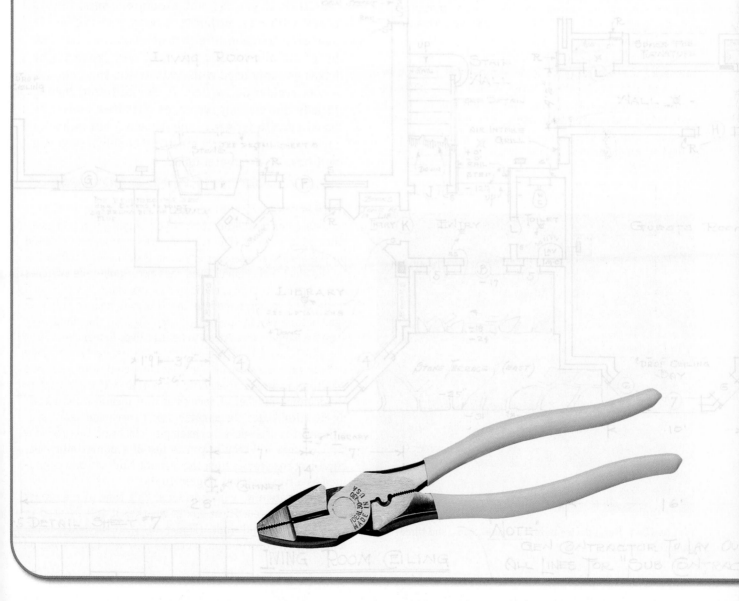

Selecting the Appropriate Electrical Box Type

Earlier you were introduced to several different electrical box types. You learned that there are metal and nonmetallic electrical boxes available for residential use. You also learned that electrical boxes used in residential wiring are often categorized as device boxes, outlet boxes, and junction boxes. Device boxes are designed to allow a device, like a switch or receptacle, to be attached to them. Device boxes come in specific sizes, such as a 3- by 2- by 3½-inch metal device box or a 20-cubic-inch plastic nail-on device box. Outlet boxes are designed to have lighting fixtures or ceiling-suspended paddle fans attached to them. This box style is a square or octagon metal box or a round nonmetallic box. Junction boxes are designed to safely contain conductors that are spliced together and are usually just outlet boxes with blank covers on them. Sometimes conductors are spliced together in a box that also has a device in it. In this case, the box serves as both a device box and a junction box.

Selecting the proper electrical box requires an electrician to check the electrical plans to determine what is being installed at a certain location. If the symbol for a duplex receptacle is shown on the plans, a device box will need to be installed. If the symbol for a ceiling-mounted incandescent lighting fixture is shown on the plans, an outlet box designed to accommodate a light fixture will have to be installed. If the symbol for a junction box is shown on the plans, a box type that is designed to accommodate the proper number of conductors coming into that box will have to be installed. I think you probably get the picture. The bottom line is that the electrician will have to install a box type that allows the item shown in the plans to be installed.

The determination of whether the boxes will be nonmetallic or metal will normally be left for the electrical contractor to decide. Sometimes a set of residential building plans will have specifications that call for the use of only metal boxes or only nonmetallic boxes. In today's residential construction market, most new houses are wired using nonmetallic boxes at those locations in the house where they are appropriate. A house may be wired using all nonmetallic electrical boxes or all metal electrical boxes, but usually a house is wired using a combination of these.

![CAUTION]

CAUTION: If you are not absolutely sure, always check with the authority having jurisdiction in your area to make sure that nonmetallic electrical boxes are allowed. For the most part, nonmetallic boxes are only used with nonmetallic sheathed cable in residential wiring. If the authority having jurisdiction in your area does not allow houses to be wired with Type NM cable (Romex) but rather an armored cable or metal electrical conduit, then you will have to use metal boxes.

An important item to consider when selecting an electrical device box is whether it must be a single gang, two gang, three gang, or more (Figure 10–1). This information is found on the electrical plan. For example, if you see an indication by the electrical symbols used that two single-pole switches are to be located at a specific spot, a two-gang box will need to be used to accommodate the two switches. If only one symbol for a switch or receptacle is shown on the plans, only a single-gang box will need to be installed. Remember, if you are using metal device boxes that are gangable, you may have to take the sides off the boxes and configure them in a way that gives you an electrical box that will accommodate the required number of devices at that location. Nonmetallic electrical boxes are manufactured to accommodate a specific number of devices. For example, a location that calls for two switches will require you to install a nonmetallic electrical box that comes from the manufacturer already configured as a two-gang box (Figure 10–2).

The last item we will consider at this time in the proper selection of an electrical box is whether it will be used in "new work" or "old work" (Figure 10–3). If the house being wired is new construction, so called new-work boxes will be

Figure 10–1 Metal device boxes in a one-, two-, and three-gang configuration.

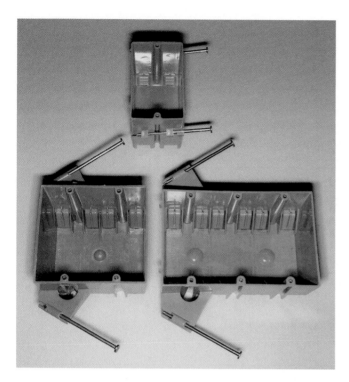

Figure 10–2 Plastic nonmetallic device boxes in a one-, two-, and three-gang configuration.

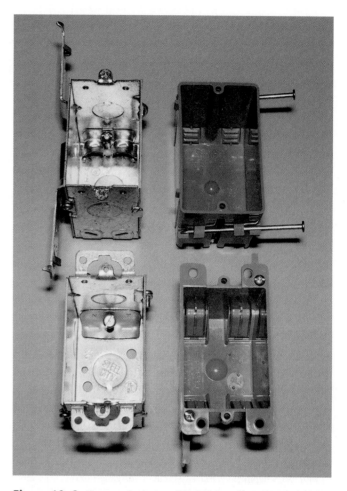

Figure 10–3 New-work versus old-work metal and nonmetallic electrical device boxes. Notice the plaster ears on the old-work boxes.

used. These boxes have no "ears" on them and are designed to be attached directly to building framing members like studs and joists. If the electrical wiring is to be done in an existing house, it is called "old work" or "remodel work" and will require old-work electrical boxes that have ears on them. In old work, electrical boxes are often secured directly in a wall or ceiling material such as sheetrock and require special mounting accessories to hold them in place. We discuss installing boxes in existing walls and ceilings at the end of this chapter. For now, we concentrate on the installation of electrical boxes in new construction.

Sizing Electrical Boxes

Once an electrician has selected the proper box type for a specific wiring application, he or she must make sure that the box can accommodate the required number of conductors and other electrical equipment that will be placed in the box. Knowing and understanding how to properly size electrical boxes is very important. An electrical box that is sized too small makes it harder to install electrical devices and other items, such as **pigtails** and wirenuts, in the box. It also presents a greater safety hazard. Article 314 of the *NEC®* shows us the proper procedures to follow to determine a minimum box size for the number of electrical conductors and

other items going in the box. Tables 314.16(A) and 314.16(B) should become very familiar to an electrician (Figure 10–4).

Section 314.16 provides the requirements and identifies the allowances for the number of conductors permitted in a box. This section requires that the total box volume be equal to or greater than the total **box fill**. The total box volume is determined by adding the individual volumes of the box components. The components include the box itself plus any attachments to it, such as a plaster ring, an extension ring, or a lighting fixture dome cover. Figure 10–5 shows an example of a 4- by 1½-inch square box and a 3/4-inch raised plaster ring being used together. The total volume results from adding the volume of the box and the raised plaster ring together. The volume of each box component is determined either from the volume marking on the box itself for nonmetallic boxes or from the standard volumes listed in Table 314.16(A) for metal boxes. If a box is marked with a larger volume than listed in

Table 314.16(A) Metal Boxes

Box Trade Size			Minimum Volume		Maximum Number of Conductors*						
mm	in.		cm³	in.³	18	16	14	12	10	8	6
100 × 32	(4 × 1¼)	round/octagonal	205	12.5	8	7	6	5	5	5	2
100 × 38	(4 × 1½)	round/octagonal	254	15.5	10	8	7	6	6	5	3
100 × 54	(4 × 2⅛)	round/octagonal	353	21.5	14	12	10	9	8	7	4
100 × 32	(4 × 1¼)	square	295	18.0	12	10	9	8	7	6	3
100 × 38	(4 × 1½)	square	344	21.0	14	12	10	9	8	7	4
100 × 54	(4 × 2⅛)	square	497	30.3	20	17	15	13	12	10	6
120 × 32	(4¹¹⁄₁₆ × 1¼)	square	418	25.5	17	14	12	11	10	8	5
120 × 38	(4¹¹⁄₁₆ × 1½)	square	484	29.5	19	16	14	13	11	9	5
120 × 54	(4¹¹⁄₁₆ × 2⅛)	square	689	42.0	28	24	21	18	16	14	8
75 × 50 × 38	(3 × 2 × 1½)	device	123	7.5	5	4	3	3	3	2	1
75 × 50 × 50	(3 × 2 × 2)	device	164	10.0	6	5	5	4	4	3	2
75 × 50 × 57	(3 × 2 × 2¼)	device	172	10.5	7	6	5	4	4	3	2
75 × 50 × 65	(3 × 2 × 2½)	device	205	12.5	8	7	6	5	5	4	2
75 × 50 × 70	(3 × 2 × 2¾)	device	230	14.0	9	8	7	6	5	4	2
75 × 50 × 90	(3 × 2 × 3½)	device	295	18.0	12	10	9	8	7	6	3
100 × 54 × 38	(4 × 2⅛ × 1½)	device	169	10.3	6	5	5	4	4	3	2
100 × 54 × 48	(4 × 2⅛ × 1⅞)	device	213	13.0	8	7	6	5	5	4	2
100 × 54 × 54	(4 × 2⅛ × 2⅛)	device	238	14.5	9	8	7	6	5	4	2
95 × 50 × 65	(3¾ × 2 × 2½)	masonry box/gang	230	14.0	9	8	7	6	5	4	2
95 × 50 × 90	(3¾ × 2 × 3½)	masonry box/gang	344	21.0	14	12	10	9	8	7	2
min. 44.5 depth	FS — single cover/gang (1¾)		221	13.5	9	7	6	6	5	4	2
min. 60.3 depth	FD — single cover/gang (2⅜)		295	18.0	12	10	9	8	7	6	3
min. 44.5 depth	FS — multiple cover/gang (1¾)		295	18.0	12	10	9	8	7	6	3
min. 60.3 depth	FD — multiple cover/gang (2⅜)		395	24.0	16	13	12	10	9	8	4

*Where no volume allowances are required by 314.16(B)(2) through 314.16(B)(5).

Table 314.16(B) Volume Allowance Required per Conductor

Size of Conductor (AWG)	Free Space Within Box for Each Conductor	
	cm³	in.³
18	24.6	1.50
16	28.7	1.75
14	32.8	2.00
12	36.9	2.25
10	41.0	2.50
8	49.2	3.00
6	81.9	5.00

Reprinted with permission from NFPA 70-2002.

Figure 10–4 *NEC*® **Tables 314.16 (A) and 314.16 (B).** *Reprinted with permission from NFPA 70-2002, the* National Electrical Code®, *Copyright © 2001, National Fire Protection Association, Quincy, MA 02269. This reprinted material is not the referenced subject, which is represented only by the standard in its entirety.*

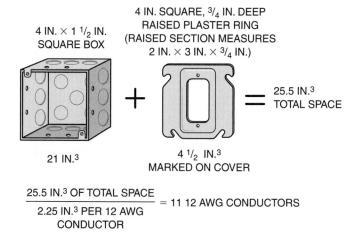

4 IN. × 1 ½ IN.
SQUARE BOX

21 IN.³

4 IN. SQUARE, ¾ IN. DEEP
RAISED PLASTER RING
(RAISED SECTION MEASURES
2 IN. × 3 IN. × ¾ IN.)

4 ½ IN.³
MARKED ON COVER

25.5 IN.³
TOTAL SPACE

$$\frac{25.5 \text{ IN.}^3 \text{ OF TOTAL SPACE}}{2.25 \text{ IN.}^3 \text{ PER 12 AWG CONDUCTOR}} = 11 \text{ 12 AWG CONDUCTORS}$$

Figure 10–5 Total volume in an electrical box includes the box itself and any raised cover that is attached to it. However, in order for the volume of the cover to be included, the cubic inch capacity of the cover must be marked on it.

Table 314.16(A), the larger volume can be used instead of the table value. Adding all the volume allowances for all items contributing to box fill determines the total box fill. The volume allowance for each item is based on the volume listed in Table 314.16(B) for the conductor size indicated. Table 10–1 summarizes the items that contribute to box fill and is based on the requirements of Section 314.16. Figure 10–6 is a Quick-Chek selection guide to determine the maximum number of conductors allowed for the more common box styles and sizes used in residential wiring.

The following examples illustrate the applicable requirements of Section 314.16 and Tables 314.16(A) and 314.16(B):

EXAMPLE ❶ An electrician has a case of 3- by 2- by 3½-inch metal device boxes with side-mounting brackets, but no internal clamps. He would like to know the maximum number of 12 AWG conductors that can be installed in this box. This information is very easy to find. The electrician simply needs to look at Table 310.16(A) and find the row for a 3- by 2- by 3½-inch device box. Once the proper row is found, find the

Table 10–1 This table summarizes the box fill rules from Sectioon 314.16 in the *NEC®*. Use the table to calculate the actual box fill for a specific electrical box situation and then choose a box with a volume in cubic inches that meets or exceeds the actual box fill.

(1) Items in the Electrical Box	(2) Volume Allowance	(3) Volume Based on Table 314.16(B)	(4) Actual Volume* (Cubic Inches)
Conductors that originate outside box	One for each conductor	Actual conductor size	
Conductors that pass through box without splice or connection	One for each conductor	Actual conductor size	
Conductors that originate within the box and do not leave the box (pigtails and jumpers)	None (These conductors are not counted.)		
Light fixture conductors (four or fewer that are smaller than #14 awg) [per 314.16(B)(1), Exception]	None (These conductors are not counted.)		
Internal cable clamps (one or more)	One only	Largest size conductor present	
Support fittings (such as fixture studs, hickeys)	One for each type of support fitting	Largest size conductor present	
Devices (such as receptacles, switches)	Two for each yoke or mounting strap	Largest size conductor Connected to device or equipment	
Equipment grounding Conductor (one or more)	One only	Largest equipment grounding conductor present	
Isolated equipment grounding Conductor (one or more) (see 250.74, Exception No. 4)	One only	Largest isolated and insulated equipment grounding conductor present	
		Total Volume	

*Actual volume is found by multiplying the volume allowance in column #2 by the volume based on Table 314.16(B) in column #3.

QUIK-CHEK BOX SELECTION GUIDE
FOR METAL BOXES GENERALLY USED FOR RESIDENTIAL WIRING

DEVICE BOXES

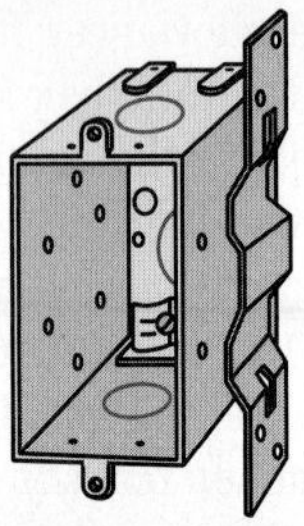

WIRE SIZE	3x2x2$\frac{1}{2}$ (12.5 IN.^{3})	3x2x2$\frac{3}{4}$ (14 IN.^{3})	3x2x3$\frac{1}{2}$ (18 IN.^{3})
14 AWG	6	7	9
12 AWG	5	6	8

SQUARE BOXES

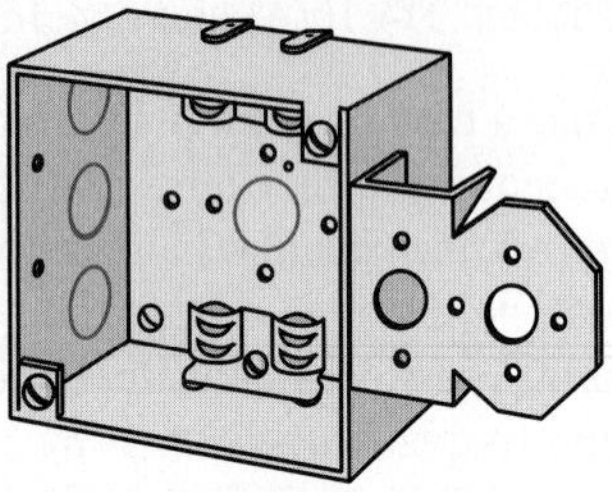

WIRE SIZE	4x4x1$\frac{1}{2}$ (21 IN.^{3})	4x4x2$\frac{1}{8}$ (30.3 IN.^{3})
14 AWG	10	15
12 AWG	9	13

OCTAGON BOXES

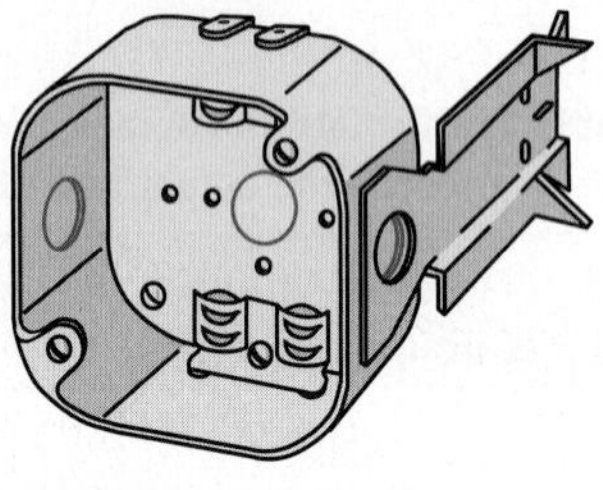

WIRE SIZE	4x1$\frac{1}{2}$ (15.5 IN.^{3})	4x2$\frac{1}{8}$ (21.5 IN.^{3})
14 AWG	7	10
12 AWG	6	9

HANDY BOXES

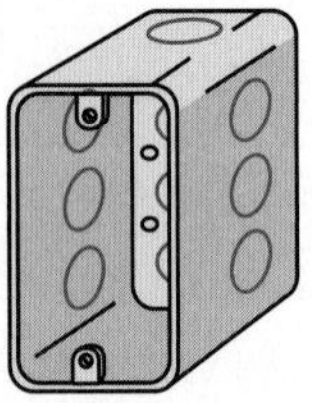

WIRE SIZE	4x2$\frac{1}{8}$x1$\frac{1}{2}$ (10.3 IN.^{3})	4x2$\frac{1}{8}$x1$\frac{7}{8}$ (13 IN.^{3})	4x2$\frac{1}{8}$x2$\frac{1}{8}$ (14.5 IN.^{3})
14 AWG	5	6	7
12 AWG	4	5	6

RAISED COVERS

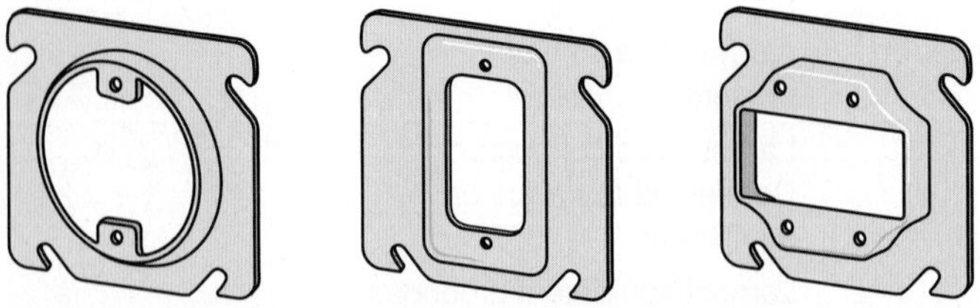

WHERE RAISED COVERS ARE MARKED WITH THEIR VOLUME IN CUBIC INCHES, THAT VOLUME MAY BE ADDED TO THE BOX VOLUME TO DETERMINE MAXIMUM NUMBER OF CONDUCTORS IN THE COMBINED BOX AND RAISED COVER.

NOTE: BE SURE TO MAKE DEDUCTIONS FROM THE ABOVE MAXIMUM NUMBER OF CONDUCTORS PERMITTED FOR WIRING DEVICES, CABLE CLAMPS, FIXTURE STUDS, AND GROUNDING CONDUCTORS. THE CUBIC INCH (IN.^{3}) VOLUME IS TAKEN DIRECTLY FROM *TABLE 314.16(A)* OF THE *NEC.*® NONMETALLIC BOXES ARE MARKED WITH THEIR CUBIC INCH CAPACITY.

Figure 10–6 Quick-Chek electrical box selection guide.

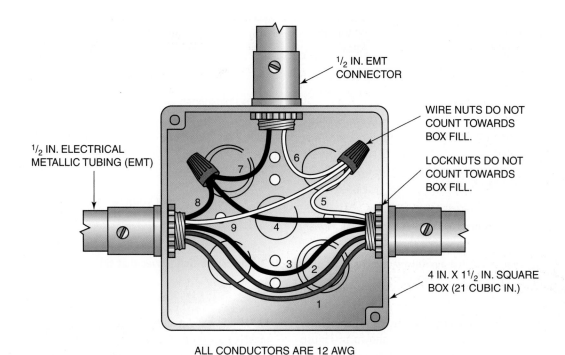

Figure 10–7 A standard-size 4- by 1½-inch square box (21.0 cubic inches) containing no fittings or devices, such as fixture studs, cable clamps, switches, receptacles, or equipment-grounding conductors.

maximum number of 12 AWG conductors by locating the number that is in the same row but under the column for 12 AWG conductors. In this case, eight 12 AWG conductors is the maximum allowed.

EXAMPLE ➋ An electrician needs to select a standard-size 4-inch square box for use as a junction box where all the conductors are the same size and the box does not contain any cable clamps, support fittings, devices, or equipment-grounding conductors (Figure 10–7). To determine the number of conductors permitted in the standard 4-inch square box, count the conductors in the box and compare the total to the maximum number of conductors permitted by Table 314.16(A). Each unspliced conductor running through the box is counted as one conductor, and every other conductor that originates outside the box is counted as one conductor. Therefore, the total conductor count for this box is nine. Now look at Table 314.16(A) to find a 4-inch square box that will accommodate nine 12 AWG conductors. A 4- by 1½-inch (21.0-cubic-inch) square box is the minimum size box that will meet the requirements for this application.

EXAMPLE ➌ A common method for determining proper box size is to find the total box volume in Table 314.16(A) and then subtract the actual total box fill from that amount. If the result is zero or some number greater than zero, the box is adequate for the application. Using this method, refer to Figure 10–8 and determine whether the box is adequately sized. For a standard 3- by 2- by 3½-inch device box (18 cubic inches), Table 314.16(A) allows up to a maximum of nine

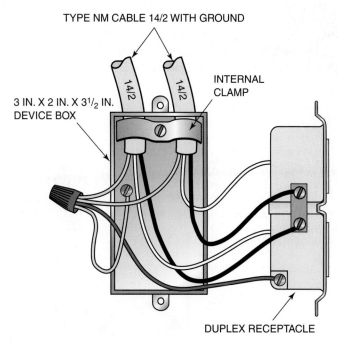

Figure 10–8 A 3- by 2- by 3½-inch device box with a device and 14 AWG conductors.

14 AWG conductors. Using Table 10–1 to calculate the total cubic inch volume of the box fill, we come up with 16 cubic inches (Figure 10–9). Therefore, because the total box fill of 16 cubic inches is less than the 18 cubic inch total box volume permitted, the box is adequately sized.

(1) Items in the Electrical Box	(2) Volume Allowance	(3) Volume Based on Table 314.16(B)	(4) Actual Volume* (Cubic Inches)
Conductors that originate outside box	4	2.0 cubic inches	8.0 cubic inches
Internal cable clamps (one or more)	1	2.0 cubic inches	2.0 cubic inches
Devices (such as receptacles and switches)	2	2.0 cubic inches	4.0 cubic inches
Equipment grounding conductor (one or more)	1	2.0 cubic inches	2.0 cubic inches
		Total Volume	16.0 cubic inches

*Actual volume is found by multiplying the volume allowance in column 2 by the volume based on Table 314.16(B) in column 3.

Figure 10–9 The total calculated box fill for the box used in example 3. The calculation proves that this box is adequately sized.

EXAMPLE ❹ Using the same method we used in example 3, determine if the electrical box application shown in Figure 10–10 complies with Section 314.16. This application has two 3- by 2- by 3½-inch device boxes ganged together to make a two-gang electrical box. The total box volume is 36 cubic inches (2 x 18 cubic inches) according to Table 314.16(A). Using Table 10–1, calculate the total cubic inch volume of the box fill. The total box fill for this application is shown in Figure 10–11. This electrical box application meets the requirements of the *NEC*® since only 26 cubic inches of the total allowed volume of 36 cubic inches is used.

EXAMPLE ❺ An electrician is installing some single-gang plastic nail-on boxes and is not sure how many 14 AWG conductors can be installed in the box. Nonmetallic boxes that are used in residential wiring will have their cubic inch volume marked on the box. A typical volume for a single-gang plastic box is 22 cubic inches. Table 314.16(B) allows 2.0 cubic inches for each 14 AWG conductor in a box, so all the electrician has to do in this situation is divide the 22 cubic inches by 2.0 cubic inches to get the maximum number of 14 AWG conductors allowed in this size box: 22.0 cubic inches/2.0 cubic inches = 11 14 AWG conductors.

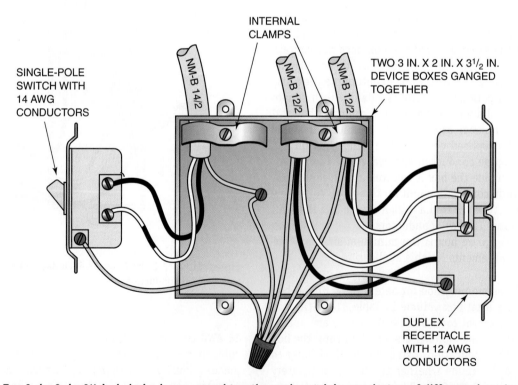

Figure 10–10 Two 3- by 2- by 3½-inch device boxes ganged together and containing conductors of different sizes. A switch and receptacle are to be installed in the box.

(1) Items in the Electrical Box	(2) Volume Allowance	(3) Volume Based on Table 314.16(B)	(4) Actual Volume* (Cubic Inches)
Conductors that originate outside box	Two 14 AWG Four 12 AWG	2.0 cubic inches 2.25 cubic inches	4.0 cubic inches 9.0 cubic inches
Internal cable clamps (one or more)	One 12 AWG	2.25 cubic inches	2.25 cubic inches
Devices (such as receptacles and switches)	Two 14 AWG Two 12 AWG	2.0 cubic inches 2.25 cubic inches	4.0 cubic inches 4.5 cubic inches
Equipment grounding conductor (one or more)	One 12 AWG	2.25 cubic inches	2.25 cubic-inches
		Total Volume	26.0 cubic inches

*Actual volume is found by multiplying the volume allowance in column 2 by the volume based on Table 314.16(B) in column 3.

Figure 10–11 The total calculated box fill for the box used in example 4. The calculation proves that this box is adequately sized.

FROM EXPERIENCE

Nonmetallic electrical boxes used in residential wiring usually have both their cubic inch volume marked on the box and the maximum number of 14, 12, and 10 AWG conductors that can be contained in the box. The box manufacturer has done the calculations for the electrician, which can save valuable time during the installation process.

Installing Nonmetallic Device Boxes

Since most device boxes used in residential rough-in wiring are nonmetallic, we cover their installation first. The most common method for installing nonmetallic boxes is to simply nail them directly to a wood framing member (Figure 10–12). Nonmetallic device boxes usually come equipped with nails and are ready for installation as soon as an electrician takes it out of the cardboard box they come in. Some styles of nonmetallic boxes come with a side-mounting bracket that is nailed or screwed to a framing member (Figure 10–13). Many

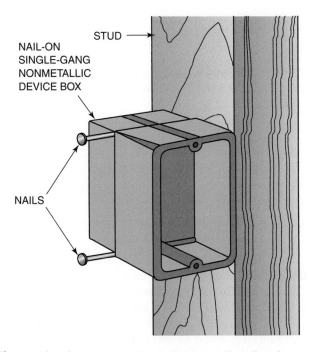

Figure 10–12 A nonmetallic device box nailed directly to a stud.

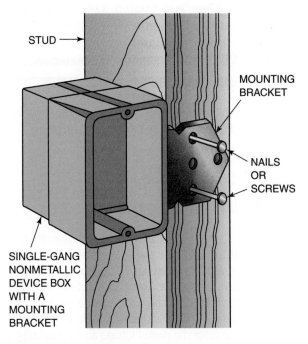

Figure 10–13 A nonmetallic device box with a side-mounting bracket attached to a stud.

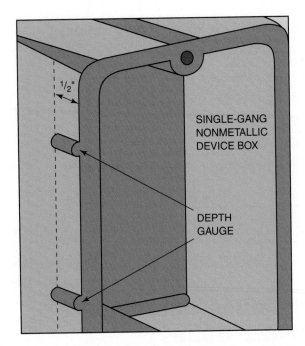

Figure 10–14 Many nonmetallic device boxes have a depth gauge of some type that helps the electrician mount the box with the proper setout.

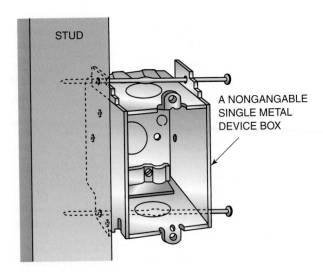

Figure 10–15 A common metal device box style with manufacturer-equipped nails.

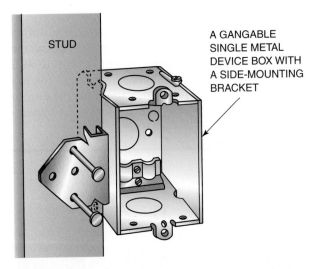

Figure 10–16 A metal device box with a side-mounting bracket.

nonmetallic box manufacturers provide indicators on the boxes (Figure 10–14) that an electrician can use to set the box out from the framing member at the proper depth, depending on what is being used as the wall covering.

CAUTION

CAUTION: Remember that Section 314.20 requires electrical boxes to be flush with the finished surface in combustible materials. Boxes may be set back no more than 1/4 inch in noncombustible materials.

(See pages 274–276 for step-by-step instructions on installing nonmetallic device boxes in new construction.)

Installing Metal Device Boxes

Like nonmetallic electrical device boxes, metal device boxes are usually nailed to building framing members. Sometimes screws, such as sheetrock screws, are used to attach them to studs, joists, or rafters. The best way to attach a metal device box with nails is to keep the nails outside the box. The box shown in Figure 10–15 is a common style of metal device box that comes from the manufacturer already equipped with nails. Many electricians use metal device boxes that are attached to framing members using a side-mounting bracket

as shown in Figure 10–16. Also like nonmetallic boxes, some metal device boxes are available with depth gauge markings (Figure 10–17) on the side of the box to help electricians determine the proper **setout** for the box.

(See page 277 for step-by-step instructions on installing metal device boxes in new construction.)

There are times when installing a box directly to a stud, joist, or rafter does not put it in a location that is called for in the building electrical plan. Once in a while, a box has to be mounted so that it is actually positioned between two studs, joists, or rafters. When this mounting situation is encountered, wood or metal strips can be placed between the framing members, and the electrical box can be secured to them (Figure 10–18). Wood straps can usually

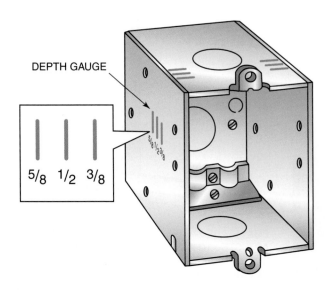

Figure 10–17 Some metal device boxes have depth gauge markings to help the electrician determine the proper box setout.

be made from some scrap wood pieces found at the building site. The carpenters at the job site will be more than willing to help you out and cut the wood pieces to fit. If you decide to cut the pieces of wood yourself, use proper safety equipment and follow proper safety procedures when using power saws. Metal supports, such as adjustable bar hangers, are readily available at the local electrical supplier (Figure 10–19).

The last item that we cover in this section is the mounting of handy boxes, or Utility boxes, as they are often called. You may remember that this box type is classified by the *NEC®* as a device box and is normally mounted directly on the surface of a framing member or existing wall. They are often used in places where a finished wall surface is not going to be used, like an unfinished basement, unfinished garage, or some agricultural buildings.

(See page 278 for step-by-step instructions on installing a handy box on a wood surface.)

(See page 278 for step-by-step instructions on installing a handy box on a masonry surface.)

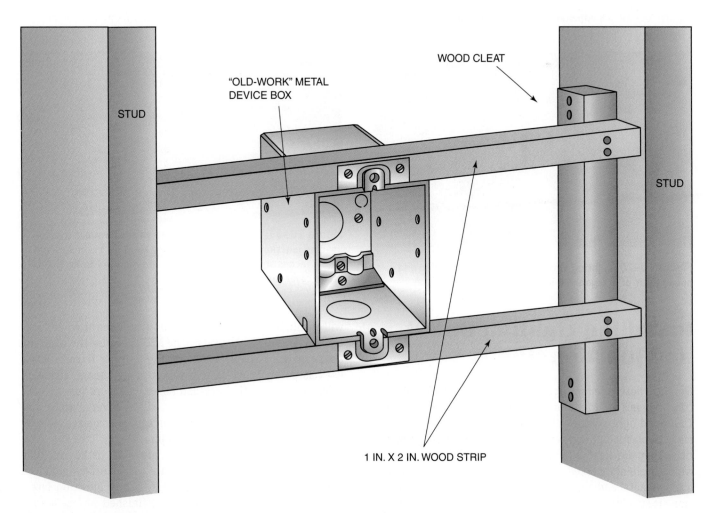

Figure 10–18 Wood cleats and strips can be used to mount electrical boxes between studs.

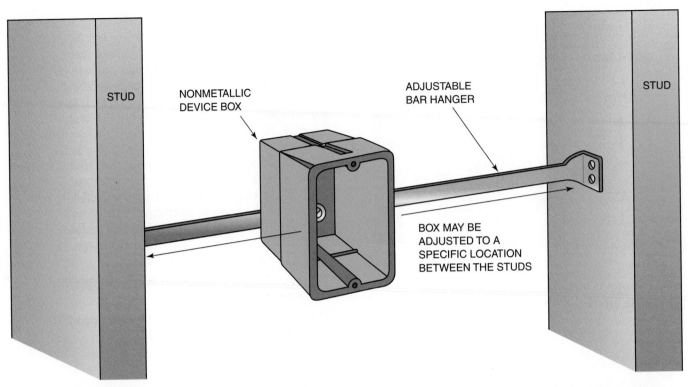

Figure 10–19 An adjustable metal bar hanger is a common method for mounting electrical boxes between two framing members.

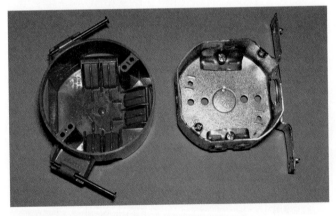

Figure 10–20 Lighting outlet boxes are usually round and nonmetallic or octagon and made of metal.

Figure 10–21 Square boxes like the ones shown are usually used as junction boxes. Common sizes are 4 by 1½ inches and 4 by 2⅛ inches square. Junction boxes are required to have a flat blank cover installed.

Installing Lighting Outlet and Junction Boxes

Electrical boxes used for **lighting outlets** in residential electrical systems are round or octagon in shape (Figure 10–20). The round lighting outlet boxes are usually nonmetallic and made of plastic but can be made of other nonmetallic materials. They are sized according to their cubic inch capacity, which is plainly marked on each box. The octagon boxes are made of metal and are used in residential wiring in two common sizes: 4 by 1½ inches and 4 by 2⅛ inches. Both of these boxes are listed in Table 314.16(A). Some wiring schemes will result in lighting outlet boxes also being used as junction boxes. This is allowed by the *NEC®* because the wires are **accessible** by simply removing the lighting fixture. Boxes used as purely junction boxes are usually square, although some electricians find octagon boxes may work just as well. Boxes used only as junction boxes will be covered with a flat blank cover. The two most common square box sizes are 4 by 1½ inches and 4 by 2⅛ inches (Figure 10–21). Like the octagon boxes, these square boxes are also listed in Table 314.16(A). Junction boxes that are to be surface mounted are installed the same way as handy boxes. Junction boxes that are to be mounted flush with a finished surface are installed the same way as lighting outlet boxes.

Lighting outlet boxes are mounted to the building framing members with side-mounting brackets or bar hangers that are adjustable (Figure 10–22). Installing outlet boxes with side-mounting brackets requires the box to be mounted directly to the stud, joist, or rafter. Using an adjustable bar hanger allows an outlet box to be properly positioned between two studs or joists. Both nonmetallic and metal outlet boxes are available with adjustable bar hangers.

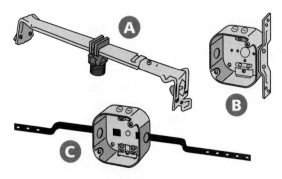

(See page 279 for step-by-step instructions on installing outlet boxes with a side-mounting bracket.)

(See page 280 for step-by-step instructions on installing outlet boxes with an adjustable bar hanger.)

Figure 10–22 **(A) Adjustable bar hanger with a fixture stud; (B) an octagon box with a side-mounting bracket; and (C) an octagon box with an offset-style bar hanger.**

If an electrician needs to position an outlet box between two framing members and there is no adjustable bar hanger available, an alternative method is to use some scrap wood from the construction site and fabricate your own bar hanger. Short lengths of 2- by 4-inch or 2- by 6-inch pieces of lumber will work well. Simply cut the piece of lumber so it will fit tightly between the two studs, joists, or rafters where you want to locate the outlet box. Place the wood box hanger between the framing members and nail it in place at a spot that, when the outlet box is attached to the hanger, results in an installation where the box face is flush with the finished surface (Figure 10–23).

Lighting fixtures that weigh more than 50 pounds (23 kilograms) or ceiling-suspended paddle fans weighing more than 35 pounds (16 kilograms) cannot be supported by just the electrical outlet box according to Sections 314.27 and 422.18(A) of the *NEC®*. For these special heavy-load installations, boxes that are rated for ceiling-suspended paddle fan and heavy-load installations are available in both metal and nonmetallic boxes. Figure 10–24 shows how a listed ceiling fan hanger could be installed in an existing ceiling or a new construction ceiling.

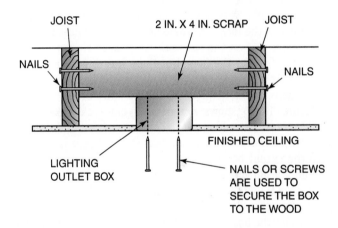

Figure 10–23 **Scrap wood can be used to fabricate an electrical box hanger when factory-made hangers are not available.**

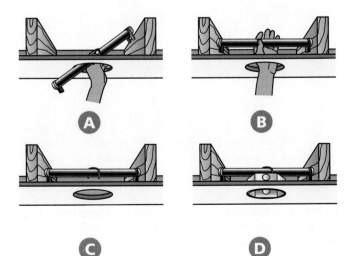

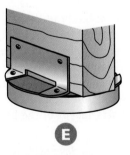

Figure 10–24 **How a listed ceiling fan hanger and box assembly is installed through a properly sized hole in a ceiling (A–D). This is shown for an existing installation, but the same equipment can be used in new construction as well. A popular box type that can be used to hang a heavy light fixture or ceiling paddle fan from a joist (E).** *Courtesy of Reiker Enterprises, Inc.*

Installing Boxes in Existing Walls and Ceilings

Up to now, we have discussed box installation in new construction. However, most remodel electrical jobs will require mounting boxes in existing walls and ceilings. Newer homes have existing walls and ceilings made of wood or metal framing with a drywall surface. This wall or ceiling type is relatively easy to cut and install an electrical box. Older homes typically have wood lath and plaster walls and ceilings, which are much harder to cut for the installation of an electrical box. Old-work boxes will have to be used when installing electrical boxes in an existing wall or ceiling. The old-work boxes are available in metal or nonmetal, but both have one common characteristic: They have plaster ears that help hold the box in the wall or ceiling.

To install old-work boxes, a hole must be cut into the wall or ceiling at the location where you wish to mount the box. The plaster ears on the old-work boxes keep the boxes from going too far into the hole, but there has to be some other device that will keep the box from coming back out of the hole. There are many ways this is accomplished.

Figure 10–25 shows nonmetallic old-work boxes with metal straps that, along with the plaster ears, hold the box in the wall or ceiling opening. Once this box type is inserted into the cutout box opening, the metal straps spread open and do not allow the box to come back out of the hole. The screws are tightened, and the box is held into the box opening by the plaster ears on the outside and the metal strap on the inside. The box shown in Figure 10–26 is a common style of old-work nonmetallic device box. It has ears like other old-work boxes and has brackets that swing up and into place once the box has been inserted into a wall. A Phillips screwdriver is used to swing the brackets into the correct holding position.

The box shown in Figure 10–27 shows another common old-work box with sheetrock grips. Once a hole has been cut, the box is pushed into the hole. The plaster ears keep the box from going into the hole too far, and the sheetrock grips are tightened to hold the box firmly in the wall or ceiling.

Another very common method for holding old-work electrical boxes in a wall or ceiling is by using **Madison hold-its** (Figure 10–28). Electricians commonly refer to this type of device as a "Madison strap" or "Madison hanger." These thin metal straps are installed on each side of an old-work box and provide a secure attachment of the box to the wall or ceiling material.

(See page 281 for step-by-step instructions on installing old-work electrical boxes in a wood lath and plaster wall or ceiling.)

(See page 282 for step-by-step instructions on installing old-work electrical boxes in a sheetrock wall or ceiling.)

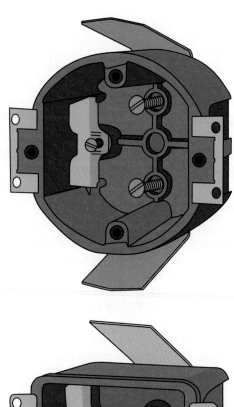
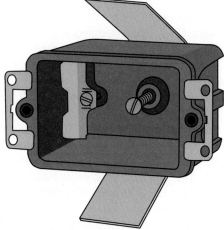

Figure 10–25 Two types of nonmetallic old-work boxes.

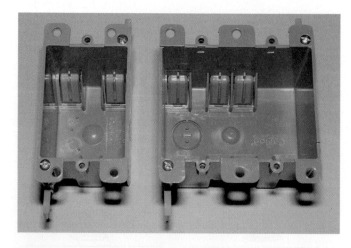

Figure 10–26 A common style of old-work plastic device box. This box has brackets that swing up behind the wall to help hold the box in place.

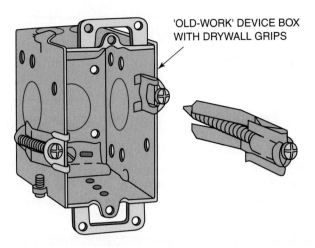

'OLD-WORK' DEVICE BOX WITH DRYWALL GRIPS

Figure 10–27 This box style is commonly used in remodel work. The box has drywall grips that, when properly tightened, will hold the box in a wall.

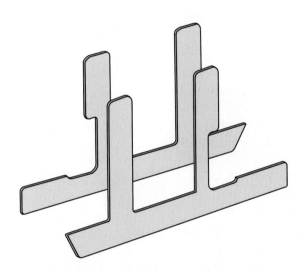

Figure 10–28 These thin metal old-work box supports are called Madison hold-its or Madison straps. They are used to help hold an old-work electrical box in a wall.

Summary

There was a lot of material covered in this chapter that has to do with the installation of electrical boxes during the rough-in stage of installing a residential electrical wiring system. The type of electrical equipment to be installed at a particular location in a house will determine what kind of electrical box will have to be installed at that location. Device boxes are used where switches or receptacles will be located and can be metal or nonmetal. Lighting outlet locations usually require a round nonmetallic box or an octagon metal box to be installed. Square metal boxes are normally installed in those areas where a junction box is required. All electrical boxes installed during the rough-in stage will have to be sized according to Section 314.16 so that adequate room will be available in the boxes for devices, conductors, and other items, such as cable clamps. This chapter also presented many installation techniques for mounting electrical boxes in both new construction and existing homes where remodeling work is being done. A good rough-in job starts with the proper installation of the electrical boxes. By following the information presented in this chapter, electricians should have a good understanding of electrical box selection, sizing, and mounting techniques.

Procedures

Installing Nonmetallic Device Boxes in New Construction

- Put on safety glasses and follow all applicable safety rules.

- Determine the depth at which the box is to be mounted. Remember the depth will depend on the material thickness that is used for the finished wall surface.

A

Determine the height from the finished floor that the box is to be mounted and make a mark using a permanent marker or pencil on the framing member at the proper mounting height.

CAUTION

CAUTION: Most boxes in a residential situation are mounted with a setout from the framing member of 1/2 inch, assuming the use of 1/2-inch sheetrock. However, always check the building plans and ask the construction supervisor to find out for sure what the required setout will need to be for the electrical boxes.

A

There is no *NEC*® rule for the height of electrical boxes in residential wiring. A common height for switches is 48 inches to the center of the box. A common height for receptacles is 16 inches to the center of the box. The table at right shows some common mounting heights for various box locations in a house. Check the building plans and the electrical supervisor to determine the required box height for the house you are working on. Also, some electricians like to make their box height measurement to the bottom or top of an electrical box instead of to the center. It really does not make any difference as long as you are consistent throughout the house so that some boxes are not higher or lower than others.

Common box mounting heights for receptacles, switches, and wall-mounted lighting fixtures in residential wiring.

Switches	
Regular	48 inches (1.12 m)
Between counter and kitchen cabinets—depends on backsplash	45 inches (1.125 m)

Receptacle Outlets	
Regular	16 inches (400 mm)
Between counter and kitchen cabinets—depends on backsplash	45 inches (1.125 m)
In garages	48 inches (1.2 m), minimum 18 inches (450 mm)
In unfinished basements	48 inches (1.2 m)
In finished basements	16 inches (400 mm)
Outdoors (above grade or deck)	18 inches (450 mm)

Wall Luminaire (Fixture) Outlets	
Outside entrances—depends on luminaire (fixture)	66 inches (1.7 m). If luminaire (fixture) is "upward" from box on luminaire (fixture), mount wall box lower. If luminaire (fixture) is "downward" from box on luminaire (fixture), mount wall box higher.
Inside wall brackets	5 feet (1.5 m)
Side of medicine cabinet or mirror Above medicine cabinet or mirror	You need to know the measurement of the medicine cabinet. Check the rough-in opening of medicine cabinet measurement of mirror. Mount electrical wall box approximately 6 inches (150 mm) to center above rough-in opening or mirror. Medicine cabinets that come complete with luminaires (fixtures) have a wiring compartment with a conduit knockout(s) in which the supply cable or conduit is secured using the appropriate fitting.
	Many strip luminaires (fixtures) have a backplate with a conduit knockout in which the supply cable or conduit is secured using the appropriate fitting. Where to bring in the cable or conduit takes careful planning.

Note: All dimensions are from finished floor to center of the electrical box. If possible, try to mount wall boxes for luminaires (lighting fixtures) based on the type of luminaire (fixture) to be installed. Verify all dimensions before "roughing in." If wiring for physically handicapped, the above heights may need to be lowered in the case of switches and raised in the case of receptacles.

Procedures

Installing Nonmetallic Device Boxes in New Construction (continued)

 B Nail the box to the framing member at the required height and the proper box setout.

Procedures

Installing Metal Device Boxes in New Construction

- Put on safety glasses and follow all applicable safety rules.

- Determine the depth at which the box is to be mounted. Remember the depth will depend on the material thickness that is used for the finished wall surface.

- Determine the height from the finished floor that the box is to be mounted and make a mark using a permanent marker or pencil on the framing member at the proper mounting height.

Ⓐ Nail the box to the framing member at the required height and the proper box setout.

Ⓐ

Procedures

Installing a Handy Box on a Wood Surface

- Put on safety glasses and follow all applicable safety rules.

- Determine the height from the finished floor that the box is to be mounted and make a mark using a permanent marker or pencil on the framing member at the proper mounting height.

 Nail or screw the box to the framing member at the required height.

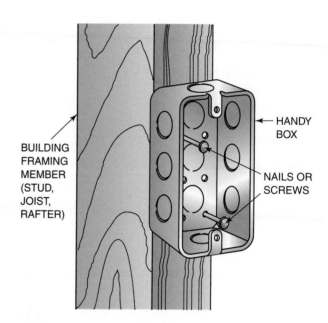

BUILDING FRAMING MEMBER (STUD, JOIST, RAFTER)

HANDY BOX

NAILS OR SCREWS

Procedures

Installing a Handy Box on a Masonry Surface

- Put on safety glasses and follow all applicable safety rules.

- Determine the height from the finished floor that the box is to be mounted and make a mark on the masonry at the proper mounting height.

- Hold the handy box at the location you want to mount it and, using a pencil, mark the masonry through the two mounting holes in the bottom of the box that you wish to use for mounting.

- Using a masonry bit, drill into the masonry at the locations marked in the previous step and install masonry anchors.

 Position the box at the proper location and secure the box to the masonry with the proper screws.

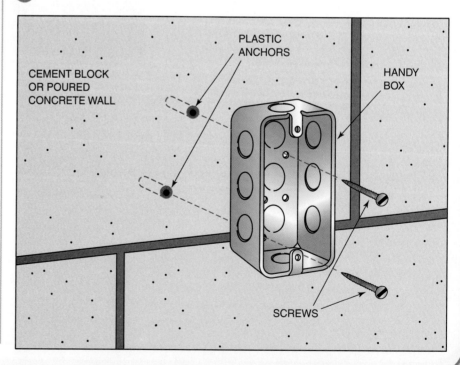

PLASTIC ANCHORS

CEMENT BLOCK OR POURED CONCRETE WALL

HANDY BOX

SCREWS

Procedures

Installing Outlet Boxes with a Side-Mounting Bracket

- Put on safety glasses and follow all applicable safety rules.

- Determine the depth at which the box is to be mounted. Remember the depth will depend on the material thickness that is used for the finished wall or ceiling surface.

- Determine which building framing member will provide the correct location for the outlet box as specified on the building electrical plan.

- Make a mark using a permanent marker or pencil on the framing member at the desired mounting location.

(A) Nail or screw the box to the framing member at the marked location using the proper box setout.

(A)

Procedures

Installing Outlet Boxes with an Adjustable Bar Hanger

- Put on safety glasses and follow all applicable safety rules.

- Determine the depth at which the box is to be mounted. Remember the depth will depend on the material thickness that is used for the finished wall or ceiling surface.

- Determine which building framing members will provide the correct location for the outlet box as specified on the building electrical plan.

- Make a mark using a permanent marker or pencil on the framing members at the desired mounting location. Since the bar hanger is secured between two different framing members, be sure to mark the location on both studs, joists, or rafters.

A Attach the outlet box to the bar hanger with the screw(s) and fitting supplied with the bar hanger. Do not completely tighten the screw(s) that hold the box to the bar at this time. This will allow the bar to be adjusted back and forth so it fits between the two framing members.

- Adjust the bar so it fits tightly between the two framing members and align the bar hanger with the marks at the mounting location.

B While making sure that the box depth will result in a flush box position with the finished surface, nail or screw the bar hanger in place.

C Move the box into the desired position between the framing members and tighten the screw(s) that hold the box on to the hanger.

A

B

C

Procedures

Installing Old-Work Electrical Boxes in a Wood Lath and Plaster Wall or Ceiling

- Put on safety glasses and follow all applicable safety rules.

- Determine the location where you want to mount the box and make a mark. Make sure there are no studs or joists directly behind where you want to install the box.

- Make a small cut through the plaster at the location you marked and carefully cut the plaster away until you have exposed one full lath.

A Turn the old-work box you are installing backward and place it at the mounting location so the center of the box is centered on the full lath. Trace around the box with a pencil. Do not trace around the plaster ears.

- Now, using a sharp tool such as a wood chisel, carefully remove all the plaster from inside the lines of the box tracing.

- Using a keyhole saw, carefully cut out the middle full lath.

- Still using a keyhole saw, carefully notch and remove only enough of the lath above and below so the box can be inserted through the hole.

- Assuming that a cable has been run to the box opening, secure the cable to the box and insert the box into the hole. Secure the box to the remaining lath with small screws placed through the plaster ears. Additional support can be gained by also using Madison hold-its.

A

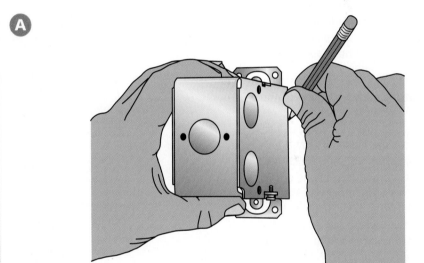

Procedures

Installing Old-Work Electrical Boxes in a Sheetrock Wall or Ceiling

- Put on safety glasses and follow all applicable safety rules.

- Determine the location where you want to mount the box and make a mark. Make sure there are no studs or joists directly behind where you want to install the box.

- Turn the old-work box you are installing backward and place it at the mounting location so the center of the box is centered on the box location mark. Trace around the box with a pencil. Do not trace around the plaster ears.

- Using a keyhole saw, carefully cut out the outline of the box. There are two ways to get the cut started. One way is to use a drill and a flat blade bit (say a 1/2-inch size), drill out the corners, and start cutting in one of the corners. The other way that is used often is to simply put the tip of the keyhole saw at a good starting location and, with the heel of your hand, hit the keyhole saw handle with enough force to cause the blade to go through the sheetrock. It usually does not require much force to start the cut this way.

 Assuming that a cable has been run to the box opening, secure the cable to the box and insert the box into the hole. Secure the box to the wall or ceiling surface with Madison hold-its or use a metal or nonmetallic box with built-in drywall grips.

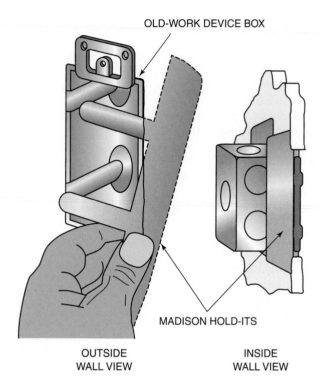

OLD-WORK DEVICE BOX

MADISON HOLD-ITS

OUTSIDE WALL VIEW

INSIDE WALL VIEW

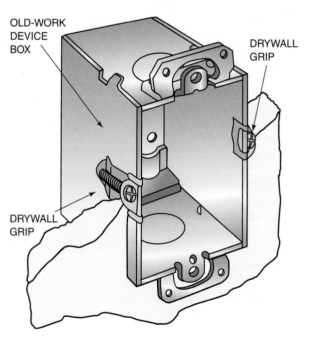

OLD-WORK DEVICE BOX

DRYWALL GRIP

DRYWALL GRIP

Review Questions

Directions: Answer the following items with clear and complete responses.

1 Who is it that determines whether metal or non-metallic electrical boxes will be used in a house?

2 Explain the difference between "new-work" and "old-work" electrical boxes.

3 What does "ganging" electrical boxes mean? Also, why would electricians need a "ganged" electrical box?

4 Describe why an electrician would use an adjustable bar hanger to mount a ceiling electrical box.

5 What is the size opening of a standard metal device box used to mount a switch or receptacle in a wall?

For the following problems, use Table 10–1 in this book and Table 314.16(A) or Table 314.16(B) of the NEC® to find the total conductor volume and correct box size for the situations listed in the following. List the box size that will be the smallest based on the volume of the box. Show all work in making the box size calculations.

6 A metal device box has four #12 conductors, two #12 grounding conductors, two internal clamps, and a duplex receptacle.
Total volume _____ Device box size _____

7 A metal device box has two #14 conductors, one #14 grounding conductor, two internal clamps, and a single-pole switch.
Total volume _____ Device box size _____

8 A metal octagon box has seven #14 conductors, three #14 grounding conductors, and two internal clamps.
Total volume _____ Octagon box size _____

9 A metal octagon box has eight #14 conductors, two #14 grounding conductors, and two external clamps.
Total volume _____ Octagon box size _____

10 A metal 4-inch square box has ten #12 conductors, four #12 grounding conductors, and two internal clamps.
Total volume _____ Square box size _____

11 A metal 4-inch square box has six #12 conductors, two #12 grounding conductors, and two internal clamps.
Total volume _____ Square box size _____

12 A plastic single-gang device box has a capacity of 25 cubic inches. Calculate the minimum number of 12 AWG conductors allowed.
_____ 12 AWG conductors

13 A plastic ceiling box has a capacity of 32 cubic inches. Calculate the minimum number of 10 AWG conductors allowed.
_____ 10 AWG conductors

14 A 4- by 2⅛-inch square box has a 31-cubic-inch capacity. A 4- by 3/4-inch raised plaster ring is used in conjunction with the square box and has a capacity of 4.5 cubic inches. Calculate the maximum number of 12 AWG conductors allowed in this box-and-cover combination.
_____ 12 AWG conductors

15 Determine the number of 8 AWG conductors allowed in a 4¹¹⁄₁₆- by 2⅛-inch square box.
_____ 8 AWG conductors

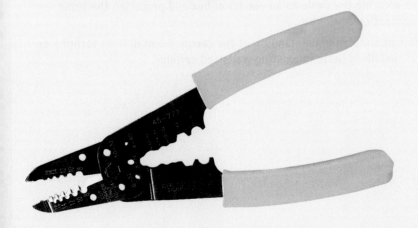

Chapter 11 | Cable Installation

In previous chapters, you were introduced to the different cable types used in residential wiring. Now it is time to take a look at how to properly install cables during the rough-in stage. An electrician must have a planned route for the cable before he or she starts to install it. A cable run starts at the service entrance panel (or subpanel if one is installed) and is then run from box to box to power lighting and receptacle outlets on the circuit. Switching locations on a circuit will require cabling to be installed between the switches and the outlets they control. Cables are installed along or through the framing members of a house. In this chapter, we take a look at common methods of installation for electrical cables in new residential construction as well as in existing homes.

OBJECTIVES

Upon completion of this chapter, the student should be able to:

- ❽ select an appropriate cable type for a residential application.
- ❽ state several *NEC*® requirements for the installation of the common cable types used in residential wiring.
- ❽ demonstrate an understanding of the proper techniques for preparing, starting, and supporting a cable run in a residential wiring application.
- ❽ demonstrate an understanding of the proper installation techniques for securing the cable to an electrical box and preparing the cable for termination in the box.
- ❽ demonstrate an understanding of the common installation techniques for installing cable in existing walls and ceilings.

Glossary of Terms

fish the process of installing cables in an existing wall or ceiling

home run the part of the branch-circuit wiring that originates in the loadcenter and provides electrical power to the first electrical box in the circuit

jug handles a term used to describe the type of bend that must be made with certain cable types; bending cable too tightly will result in damage to the cable and conductor insulation; bending them in the shape similar to a "handle" on a "jug" will help satisfy *NEC*® bending requirements

pulling in the process of installing cables through the framework of a house

redhead sometimes called a "red devil"; an insulating fitting required to be installed in the ends of a Type AC cable to protect the wires from abrasion

reel a drum having flanges on each end; reels are used for wire or cable storage

running boards pieces of board lumber nailed or screwed to the joists in an attic or basement; the purpose of using running boards is to have a place to secure cables during the rough-in stage of installing a residential electrical system

secured (as applied to electrical cables) fastened in place so the cable cannot move; common securing methods include staples, tie wraps, and straps

supported (as applied to electrical cables) held in place so the cable is not easily moved; common supporting methods include running cables horizontally through holes or notches in framing members or using staples, tie wraps, or straps after a cable has been properly secured close to a box according to the *NEC*®

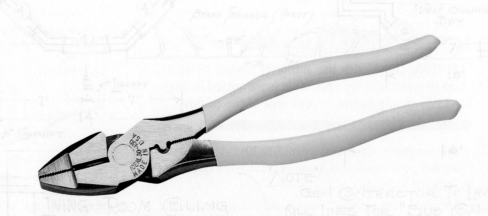

Selecting the Appropriate Cable Type

The common cable types used in residential wiring were introduced earlier in this text. While nonmetallic sheathed cable (Type NM) is the most often used cable, other types, such as underground feeder cable (Type UF), armored-clad cable (Type AC), metal-clad cable (Type MC), service entrance cable (Type SEU, SER, or USE), and even low-voltage cable for door chime and thermostat wiring, are used in residential wiring. Selecting the appropriate cable type depends on (1) the cable type allowed by the authority having jurisdiction in your area and (2) the specific wiring application. Many residential building plans will state either on the electrical plan or in the electrical specifications what the required cable type is. In those situations that do not have a specific type listed in the building plans, the electrical contractor will decide the cable type to be used on the basis of the two previously mentioned criteria.

Requirements for Cable Installation

No matter which cable type is used, there are certain installation requirements for that cable type that will have to be followed during the installation of the cable in the rough-in stage. These requirements are outlined in the specific article for that cable type in the *National Electrical Code®* (*NEC®*). Let's take a look at the most important installation requirements that residential electricians need to know.

Nonmetallic Sheathed Cable (Type NM)

Article 334 of the *NEC®* covers the installation requirements for nonmetallic sheathed cable. Section 334.10(1) states that Type NM cable is allowed to be used in one- and two-family homes. The following paragraphs summarize the most important installation requirements:

> **CAUTION**
>
> **CAUTION: Check with the authority having jurisdiction to make sure Type NM cable is permitted in one- and two-family homes in your area of the country.**

Section 334.15 states that when Type NM cable is run exposed and not concealed in a wall or ceiling, the cable must be installed as follows:

- The cable must closely follow the surface of the building finish or be secured to  running boards.

- The cable must be protected from physical damage where necessary by rigid metal conduit, electrical metallic tubing, Schedule 80 PVC rigid nonmetallic conduit, guard strips, or other means. Where passing through a floor, the cable must be enclosed in rigid metal conduit, intermediate metal conduit, electrical metallic tubing, or Schedule 80 PVC rigid nonmetallic conduit extending at least 6 inches (150 mm) above the floor (Figure 11–1).

- Where the cable is run across the bottom of joists in unfinished basements, it is permissible to secure cables not smaller than two 6 AWG conductors (such as a 6/2 Type NM cable) or three 8 AWG conductors (such as 8/3 Type NM cable) directly to the bottom edge of the joists. Smaller cables must be run either through bored holes in joists or on running boards. Figure 11–2 shows how nonmetallic sheathed cables are installed in an unfinished basement through holes drilled in joists, attached to the side or face of joists or beams, and on running boards. Section 300.4(D) requires cables that are run along the side of framing members to be installed at least 1¼ inches (32 mm) from the nearest edge of studs, joists, or rafters.

Section 334.17 covers running Romex through or parallel to framing members. It states that Type NM cable must be protected in accordance with Section 300.4 where installed

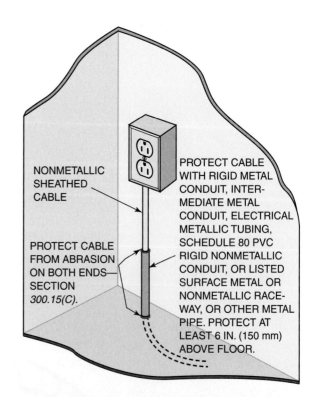

Figure 11–1 When Type NM cable is installed exposed where it passes through a floor, the cable must be protected for the entire exposed length or for at least 6 inches (150 mm) from where it comes through the floor.

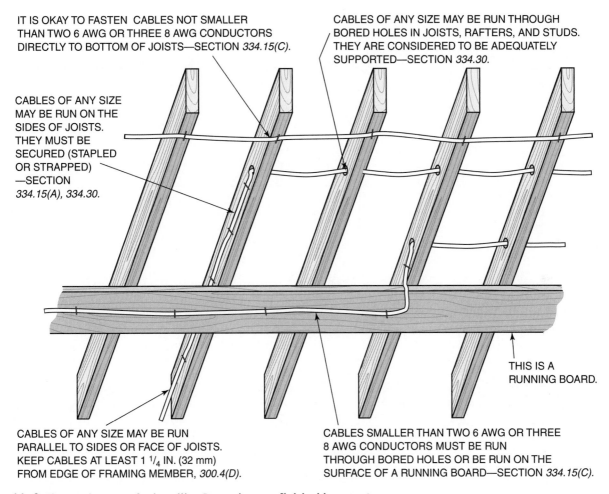

IT IS OKAY TO FASTEN CABLES NOT SMALLER THAN TWO 6 AWG OR THREE 8 AWG CONDUCTORS DIRECTLY TO BOTTOM OF JOISTS—SECTION *334.15(C).*

CABLES OF ANY SIZE MAY BE RUN THROUGH BORED HOLES IN JOISTS, RAFTERS, AND STUDS. THEY ARE CONSIDERED TO BE ADEQUATELY SUPPORTED—SECTION *334.30.*

CABLES OF ANY SIZE MAY BE RUN ON THE SIDES OF JOISTS. THEY MUST BE SECURED (STAPLED OR STRAPPED) —SECTION *334.15(A), 334.30.*

THIS IS A RUNNING BOARD.

CABLES OF ANY SIZE MAY BE RUN PARALLEL TO SIDES OR FACE OF JOISTS. KEEP CABLES AT LEAST 1 ¼ IN. (32 mm) FROM EDGE OF FRAMING MEMBER, *300.4(D).*

CABLES SMALLER THAN TWO 6 AWG OR THREE 8 AWG CONDUCTORS MUST BE RUN THROUGH BORED HOLES OR BE RUN ON THE SURFACE OF A RUNNING BOARD—SECTION *334.15(C).*

Figure 11–2 **The requirements for installing Romex in an unfinished basement.**

through or parallel to framing members. Section 300.4 requirements were covered in Chapter 8, and you should review that section of the text at this time for a more detailed explanation. Only a short overview of Section 300.4 is given in the following:

- If Type NM cable is run through a hole that has an edge closer than 1¼ inches (32 mm) to the edge of the framing member, a metal plate at least 1/16 inch (1.6 mm) thick must be used to protect that spot from nails or screws that could penetrate the cable sheathing.
- If the cable is placed in a notch in wood studs, joists, rafters, or other wood members, a steel plate must protect the cable and be at least 1/16 inch (1.6 mm) thick.
- If the cable is run along a framing member, where the cable is closer than 1¼ inches (32 mm) to the edge, metal protection plates at least 1/16 inch (1.6 mm) thick must be used to protect the cable.

Installing Romex in accessible attics is covered in Section 334.23. The installation of Type NM cable in accessible attics or roof spaces must comply with Section 320.23,

which also applies to armored-clad cable. This section tells us that if Romex is run across the top of floor joists, or within 7 feet (2.1 m) of an attic floor or floor joists, the cable must be protected by guard strips that are at least as high as the cable (Figure 11–3). The 7-foot (2.1 m) rule is also applied when Romex is run across the face of rafters or studding in an attic. Where the attic space is not accessible by permanent stairs or ladders, protection is only required within 6 feet (1.8 m) of the nearest edge of a scuttle hole or attic entrance (Figure 11–4). Where the cable is installed along the sides of rafters, studs, or floor joists in an attic, neither guard strips nor running boards are required, but the cable must be kept at least 1¼ inches (32 m) from the edge of the framing member.

Section 334.24 states that bends in nonmetallic sheathed cable must be made so that the cable will not be damaged. The radius of the curve of the inner edge of any bend during or after installation must not be less than five times the diameter of the cable. This means that when a Romex cable is brought through a hole in a building framing member, the cable must be formed in a **jug handle** to ensure that the

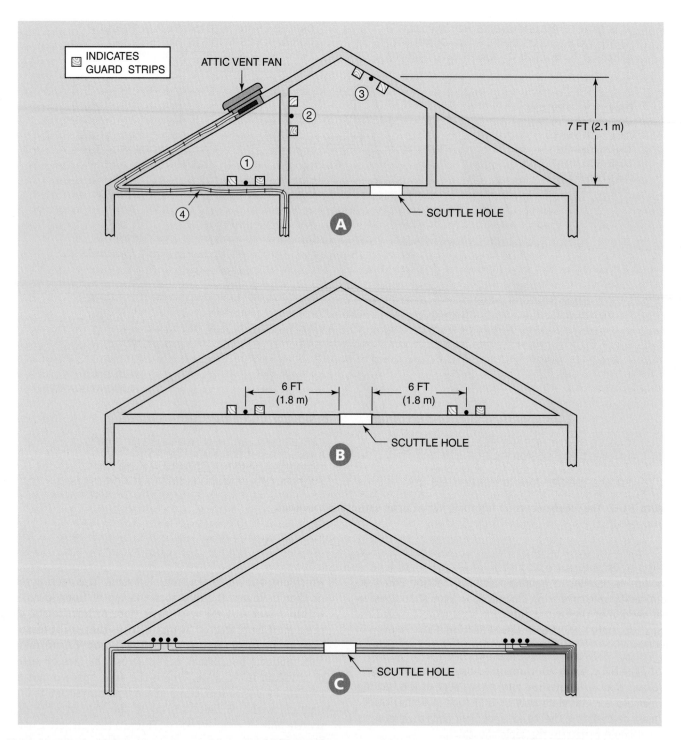

Figure 11–3 (A) In accessible attics, cables must be protected by guard strips when (1) they are run across the top of joists, (2) they are run across the face of studs and are within 7 feet (2.1 m) of the floor or floor joists, or (3) they are run across the face of rafters and are within 7 feet (2.1 m) of the floor or floor joists. Also, (4) guard strips are not required when the cable is run along the sides of joists or rafters in the attic. (B) In attics that are accessible only through a scuttle hole, protection of the cable is required only within 6 feet (1.8 m) of the scuttle hole. (C) An installation of cable in an attic where the cables are located close to the "eaves." This is considered a safe installation because it is unlikely that items can ever be stored on top of the cables located in this postion.

GUARD STRIPS REQUIRED WHEN CABLES
ARE RUN ACROSS THE TOP OF JOISTS.

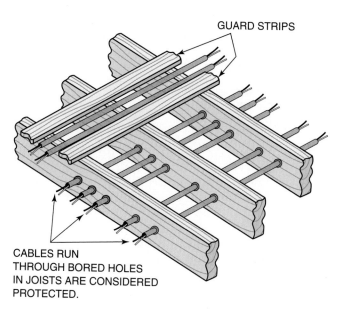

GUARD STRIPS

CABLES RUN
THROUGH BORED HOLES
IN JOISTS ARE CONSIDERED
PROTECTED.

Figure 11–4 Two methods for protecting cables in an accessible attic are to (1) use suitable guard strips and (2) install the cables through holes that are bored into the joists. Remember that there must be at least 1¼ inches (32 mm) from the edge of the bored holes to the edge of the joists.

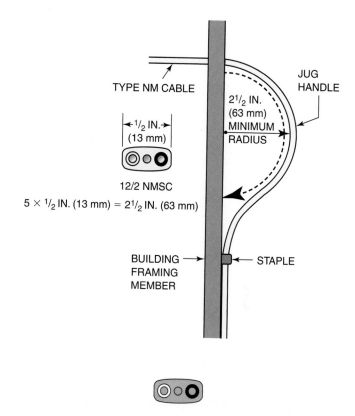

TYPE NM CABLE

½ IN.
(13 mm)

12/2 NMSC

$5 \times \frac{1}{2}$ IN. (13 mm) = 2½ IN. (63 mm)

2½ IN.
(63 mm)
MINIMUM
RADIUS

JUG
HANDLE

BUILDING
FRAMING
MEMBER

STAPLE

Figure 11–5 Type NM cable must be bent so that the radius of the bend is not less than five times the cable's diameter. As shown in this illustration, the cable will have to be installed using jug handles when it is bent to avoid damaging the cable.

cable is not bent too sharply. Bending the cable too much will weaken the outer cable sheath as well as the insulation on the individual conductors (Figure 11–5).

Securing and supporting Type NM cable is addressed in Section 334.30. It requires nonmetallic sheathed cable to be secured by staples, cable ties, straps, hangers, or similar fittings designed and installed so as not to damage the cable at intervals not exceeding 4½ feet (1.4 m) and within 12 inches (300 mm) of every cabinet, box, or fitting (Figure 11–6). The general requirement of Section 334.30 requires that the cable be secured. Simply draping the cable over air ducts, timbers, joists, pipes, and ceiling grid members is not permitted (Figure 11–7).

Section 334.30 also says that flat cables, like 14/2 or 12/2 Romex, must not be stapled on edge. The intent of this section is to prohibit the cable from being installed with its short dimension against a wood framing member. When stapled in this manner, two cables are placed side by side under the staple. If the staple is driven too far into the stud, damage to the insulation and conductors can occur. Flat cables can be installed on top of each other as long as the "flat side" of the cable is against the framing member and the next cable. There is no *NEC®* rule against installing round cables (like a 14/3 or 12/3 Romex) on top of each other, although this is generally considered to be not the best way to install these cables (Figure 11–8).

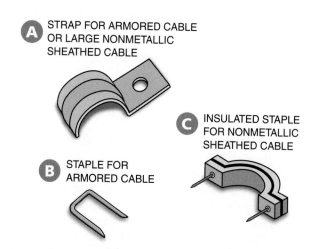

A STRAP FOR ARMORED CABLE OR LARGE NONMETALLIC SHEATHED CABLE

C INSULATED STAPLE FOR NONMETALLIC SHEATHED CABLE

B STAPLE FOR ARMORED CABLE

Figure 11–6 (A) A one-hole strap that is used to secure larger sizes of Type NM and Type AC cable. (B) A staple that can be used for securing cables in a residential electrical installation. (C) An insulated style of staple. Some local electrical codes require the use of only insulated staples to secure Type NM cable.

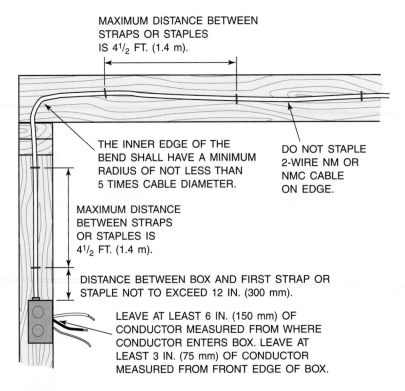

MAXIMUM DISTANCE BETWEEN
STRAPS OR STAPLES
IS 4$\frac{1}{2}$ FT. (1.4 m).

THE INNER EDGE OF THE
BEND SHALL HAVE A MINIMUM
RADIUS OF NOT LESS THAN
5 TIMES CABLE DIAMETER.

DO NOT STAPLE
2-WIRE NM OR
NMC CABLE
ON EDGE.

MAXIMUM DISTANCE
BETWEEN STRAPS
OR STAPLES IS
4$\frac{1}{2}$ FT. (1.4 m).

DISTANCE BETWEEN BOX AND FIRST STRAP OR
STAPLE NOT TO EXCEED 12 IN. (300 mm).

LEAVE AT LEAST 6 IN. (150 mm) OF
CONDUCTOR MEASURED FROM WHERE
CONDUCTOR ENTERS BOX. LEAVE AT
LEAST 3 IN. (75 mm) OF CONDUCTOR
MEASURED FROM FRONT EDGE OF BOX.

Figure 11–7 Securing and supporting requirements for nonmetallic sheathed cable.

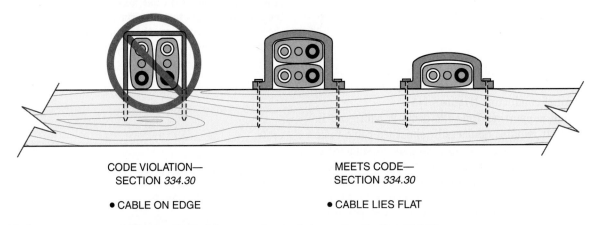

CODE VIOLATION—
SECTION *334.30*

● CABLE ON EDGE

MEETS CODE—
SECTION *334.30*

● CABLE LIES FLAT

Figure 11–8 Flat-type nonmetallic sheathed cable cannot be stapled on edge. Section 334.30.

Section 334.30(A) states that Type NM cables that run horizontally through framing members that are spaced less than 4½ feet (1.4 m) apart and that pass through bored or punched holes in framing members without additional securing are considered *supported* by the framing members. Cable ties or staples are not required as the cable passes through these members. However, the Romex cable must be *secured* within 12 inches (300 mm) of any outlet box. Where the cable terminates at a single-gang nonmetallic outlet box that does not contain cable clamps, the cable must be secured within 8 inches (200 mm) of the outlet box, according to an exception to Section 314.17(C) (Figure 11–9).

Section 334.30(B) allows Type NM cables to not be supported in these situations:

● Where the cable is fished between access points or concealed in finished buildings and supporting is impractical. This is the situation with remodel electrical work.

● Where the cable is not more than 4½ feet (1.4 m) from the last point of support for connections within an accessible ceiling to luminaires (lighting fixture). This allows short, unsupported lengths of Type NM cable for luminaire connections.

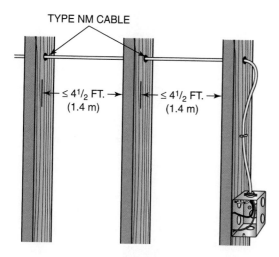

Figure 11–9 Horizontal (or diagonal) runs of Type NM cable through framing members are considered supported and secured as long as the distance between the framing members is not more than the minimum interval for support of 4½ feet (1.4 m).

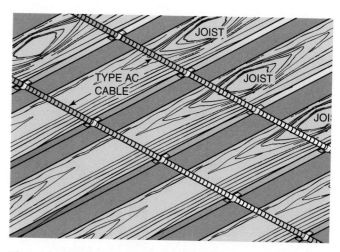

Figure 11–10 Exposed runs of Type AC cable on the underside of joists are allowed as long as the cable is secured at each joist.

Underground Feeder Cable (Type UF)

Article 340 of the *NEC®* allows Type UF cable to be used in residential wiring for direct burial applications or for installation in the same wiring situations where you would use nonmetallic sheathed cable. For underground installation requirements, Section 300.5 must be followed. Underground residential branch-circuit installation is discussed in Chapter 15. When installed as nonmetallic sheathed cable, the installation and conductor requirements must comply with the provisions of Article 334.

Section 340.24 states that bends in Type UF cable must be made so that the cable is not damaged. The radius of the curve of the inner edge of any bend must not be less than five times the diameter of the cable. This is the same rule as for nonmetallic sheathed cable and is intended to prevent damage to the cable and individual conductor insulation.

Armored-Clad Cable (Type AC)

Armored-clad cable, sometimes referred to as "BX" cable by electricians, can be installed exposed or concealed in a house. Section 320.15 requires exposed runs of Type AC cable to closely follow the surface of the building finish or be secured to running boards. Exposed runs are permitted to be installed on the underside of joists, like in a basement, where supported at each joist and located so as not to be subject to physical damage (Figure 11–10).

Section 320.17 allows Type AC cable to be run through or along framing members in a house. When Type AC cable is run this way, it must be protected in accordance with Section 300.4. In other words, if Type AC cable is run

through a hole that has an edge closer than 1¼ inches (32 mm) to the edge of the framing member, a metal plate at least 1/16 inch (1.6 mm) thick must be used to protect that spot from nails or screws that could penetrate the cable sheathing. If the cable is placed in a notch in wood studs, joists, rafters, or other wood members, a steel plate must protect the cable and be at least 1/16 inch (1.6 mm) thick. Also, if the cable is run along a framing member, where the cable is closer than 1¼ inches (32 mm) to the edge, metal protection plates at least 1/16 inch (1.6 mm) thick must be used to protect the cable. These requirements are the same as for Type NM cable.

> ### ◄◄◄ CAUTION ►►►
>
> **CAUTION: Armored-clad cable must be protected in the same manner as nonmetallic sheathed cable. Many electricians believe that since Type AC cable has an armored sheathing, no additional protection for the cable is required. They are wrong.**

If installed in accessible attics, Section 320.23 requires Type AC cable to be installed exactly the same way as nonmetallic sheathed cable in that situation. The installation requirements for Type NM cable installed in an accessible attic were presented earlier in this chapter.

Also, like Type NM cable, Type AC cable must not be bent too much. Section 320.24 states that bends in Type AC cable must be made so that the cable will not be damaged. The radius of the curve of the inner edge of any bend must not be less than five times the diameter of the Type AC cable. Again, the intent of this requirement is to prevent damage to the cable sheathing and the individual conductor insulation (Figure 11–11).

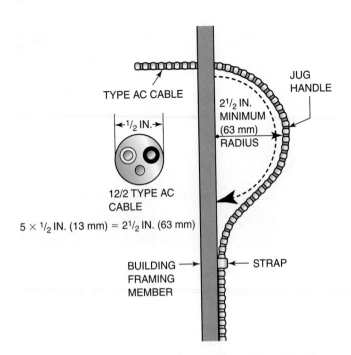

Figure 11–11 Type AC cable must be bent so that the radius of the bend is not less than five times the cable's diameter. As shown in this illustration, the cable will have to be installed using jug handles when it is bent to avoid damaging the cable.

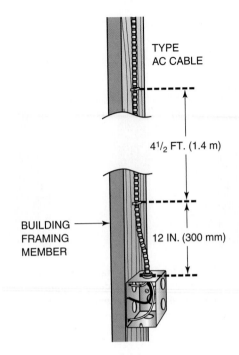

Figure 11–12 Type AC cable securing and supporting requirements.

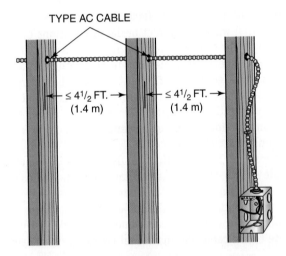

Figure 11–13 Horizontal (or diagonal) runs of Type AC cable through framing members are considered supported and secured as long as the distance between the framing members is not more than the minimum interval for support of 4½ feet (1.4 m).

Type AC cable must be secured by staples, cable ties, straps, hangers, or similar fittings designed and installed so as not to damage the cable at intervals not exceeding 4½ feet (1.4 m) and within 12 inches (300 mm) of every outlet box, junction box, cabinet, or fitting according to Section 320.30. These are the same securing and supporting requirements as for Type NM cable (Figure 11–12).

Section 320.30(A) addresses Type AC being run horizontally through holes and notches in framing members. In other than vertical runs, cables installed in accordance with Section 300.4 are to be considered supported and secured where the support does not exceed 4½-foot (1.4-m) intervals and the armored cable is securely fastened in place by an approved means within 12 inches (300 mm) of each box, cabinet, conduit body, or other armored cable termination (Figure 11–13).

According to 320.30(B), Type AC cable is permitted to be unsupported in these situations:

- Where the cable is fished between access points or concealed in finished buildings or structures and supporting is impractical such as in remodel work in existing houses
- Where the cable is not more than 2 feet (600 mm) in length at terminals where flexibility is necessary, like a connection to an electric motor
- Where the cable is not more than 6 feet (1.8 m) from the last point of support for connections within an accessible ceiling to a luminaire (lighting fixture)

The last installation requirement to be looked at for Type AC cable is in Section 320.40. It states that at all points where the armor of Type AC cable terminates, a fitting must be provided to protect wires from abrasion, unless the design of the outlet boxes or fittings provides equivalent protection. The required insulating fitting is called many names by electricians in the field, including **redhead** or "red devil." It gets its name from the red color

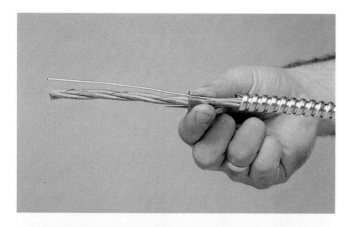

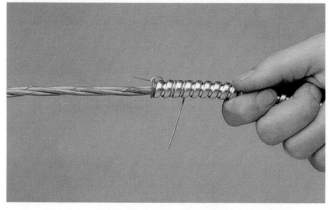

Figure 11–14 Antishort bushings used with Type AC cable, commonly called "redheads," prevent the cable conductors from being cut by the sharp edges of the metal armor sheathing. *Courtesy of AFC Cable Systems, Inc.*

that the fitting has. The connector or clamp by which the Type AC cable is fastened to boxes or cabinets must be of such design that the insulating bushing (red head) will be visible for inspection (Figure 11–14).

Metal-Clad Cable (Type MC)

Metal-clad cable is another cable type that some electricians use to wire a house. This cable type is very popular in some parts of the country. It has gotten to a point that many local electrical distributors are stocking only Type MC cable and not Type AC cable any longer. Article 330 of the *NEC®* provides some installation requirements for Type MC cable.

CAUTION: While there is no *NEC®* requirement that red heads be used with Type MC cable, it is a good idea to do so. Rolls of Type MC cable come from the manufacturer with small bags that contain a few insulated fittings.

Section 330.17 requires Type MC cable to be protected in accordance with Section 300.4 where installed through or parallel to framing members. If Type MC cable is run through a hole that has an edge closer than 1¼ inches (32 mm) to the edge of the framing member, a metal plate at least 1/16 inch (1.6 mm) thick must be used to protect that spot from nails or screws that could penetrate the cable sheathing. If the cable is placed in a notch in wood studs, joists, rafters, or other wood members, a steel plate must protect the cable and be at least 1/16 inch (1.6 mm) thick. Also, if the cable is run along a framing member where the cable is closer than 1¼ inches (32 mm) to the edge, metal protection plates at least 1/16 inch (1.6 mm) thick must be used to protect the cable. These requirements are the same as for Type NM cable and Type AC cable.

Section 330.23 requires the installation of Type MC cable in accessible attics or roof spaces to comply with Section 320.23. In accessible attics, Type MC cable installed across the top of floor joists or within 7 feet (2.1 m) of the floor or floor joists must be protected by guard strips. Where the attic is not accessible by a permanent ladder or stairs, guard strips are required only within 6 feet (1.8 m) of the scuttle hole or opening. This requirement is the same as for Type NM and Type AC cable.

Bends in Type MC cable must be made so that the cable will not be damaged. Section 330.24(B) states that the radius of the curve of the inner edge of any bend must not be less than seven times the external diameter of the metallic sheath for the style of Type MC used by electricians in residential wiring. This style has an interlocking-type armor or corrugated sheathing and looks very similar on the outside to Type AC cable. This requirement simply means that when installing Type MC cable, bigger turns and jug handles will have to be used.

Securing and supporting Type MC cable is addressed in Section 330.30. The general rule states that Type MC cable must be supported and secured at intervals not exceeding 6 feet (1.8 m). Section 330.30(A) covers horizontal installation through framing members and allows Type MC cables installed in accordance with Section 300.4 to be considered supported and secured where the framing members are no more than 6 feet (1.8 m) apart. Cable ties or other securing methods are not required as the cable passes through these members. According to Section 330.30(C), Type MC cable containing four or fewer conductors of 10 AWG or less is required to be secured within 12 inches (300 mm) from every box, cabinet, or fitting. This means that in residential work, Type MC for branch circuits of 14, 12, and 10 AWG must be secured within 12 inches (300 mm) of each box (Figure 11–15). Section 330.30(C) allows Type MC cable to be installed unsupported in these situations:

- when it is fished between access points or concealed in finished buildings or structures and supporting is impracticable (Figure 11–16)

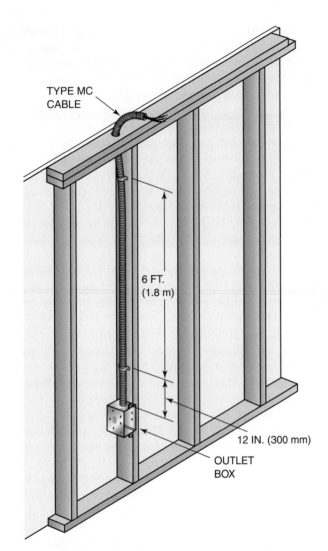

Figure 11–15 Type MC cable must be supported and secured at intervals not exceeding 6 feet (1.8 meters) and within 12 inches (300 mm) of an electrical box.

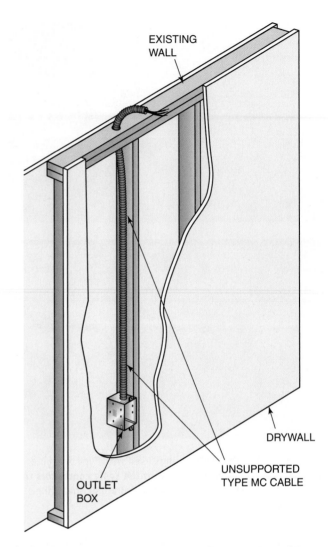

Figure 11–16 The *NEC®* permits Type NM, Type AC, and Type MC cable to be fished in walls, floors, or ceilings of existing buildings without support. This illustration shows a Type MC cable fished into the wall of an existing building. It is not required to be secured and supported.

- when it is not more than 6 feet (1.8 m) from the last point of support for connections within an accessible ceiling to a luminaire (lighting fixture)

Fittings used for connecting Type MC cable to boxes, cabinets, or other equipment must be listed and identified for such use according to Section 330.40. Connectors should be selected in accordance with the size and type of cable for which they are designated. Some Type AC cable connectors are also acceptable for use with Type MC cable when specifically indicated on the device or the shipping carton.

Service Entrance Cable (Type SEU and SER)

Service entrance cable installation, when used as a service entrance wiring method, was covered in Chapter 8. However, there are certain situations in residential wiring when service

entrance cable is used as the branch-circuit wiring method for individual appliances, such as electric furnaces, electric ranges, and electric clothes dryers, or as the wiring method for a feeder going to a subpanel. Section 338.10(B)(1) of the *NEC®* allows us to use service entrance cable as branch-circuit or feeder conductors. Branch circuits using service entrance cable as a wiring method are permitted only if all circuit conductors within the cable are fully insulated with an insulation recognized by the *NEC®* and listed in Table 310.13. The equipment-grounding conductor is the only conductor permitted to be bare within service entrance cable used for branch circuits. Section 338.1(B)(2) states that service entrance cable containing a bare grounded (neutral) conductor (Type SEU) is not permitted for new installations where it is used as a branch circuit to supply appliances such as ranges, wall-mounted ovens, counter-mounted cooking units, or clothes dryers. Type SER cable would have to be installed for these situations. The exception to this section does permit a bare neutral service entrance cable for existing installations only and is coordinated with Sections 250.140 and 250.142 of the 2002 *NEC®*. Prior to the 1996 *NEC®*, it was permissible to use Type SEU with a bare grounded (neutral) conductor as the wiring method used to supply electric ranges and electric clothes dryers.

Section 338.10(B)(4) says that in addition to the provisions of Article 338, Type SE cable used for interior wiring must comply with the installation requirements of Article 334 for nonmetallic sheathed cable. This means that when using Type SE for branch-circuit or feeder runs inside a dwelling unit, the electrician must install the Type SEU or Type SER service entrance cable the same way as for nonmetallic sheathed cable.

Preparing the Cable for Installation

The first thing that needs to be done in preparation for installing branch-circuit and feeder cables during the rough-in stage is to simply get them ready for installation. It is important that a residential electrician know the correct procedure for setting up and then unrolling cables from a box, roll, or **reel**. When electrical contractors purchase non-metallic sheathed cable (Type NM) or underground feeder cable (Type UF) from the local electrical supplier, they usually get it in one of two ways: (1) as a 250-foot (75 m) roll packaged in plastic or in a cardboard box or (2) on a 1,000-foot (300 m) reel. Type AC and Type MC cables are purchased in much the same way. Each is available in rolls of 250 feet (75 m) but do not come in a plastic wrapping or in a cardboard box. Rolls of Type AC or Type MC cable are usually held together with wire ties that must be cut away before the cables may be unrolled. Type AC and Type MC cable is also available on reels of 1,000 feet (300 meters), and many electrical contractors choose to purchase these cable types on reels. Service entrance cable (Type SE) is usually purchased from the electrical supplier in the lengths needed for the service installation of the house being wired. However, there are many electrical contractors who purchase larger reels of service entrance cable, especially if they have many homes to wire that all require service entrances installed with service entrance cable.

When cables are used from a roll, the best way to begin unrolling them is to start from the inside of the coil. If the roll was purchased in a box, many manufacturers provide a round cutout on the box that is used to indicate the starting point for unrolling the cable. If the cable is not unrolled properly, twists and kinks of the cable will develop that could damage the cable insulation (Figure 11–17).

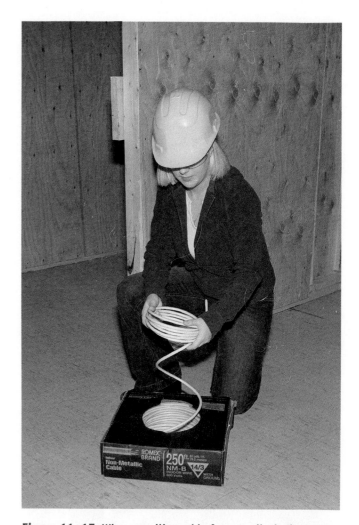

Figure 11–17 When unrolling cable from a coil, the best way is to pull the cable out of the coil from the inside. In this picture, an electrician is pulling Type NM cable from the inside of a 250-foot roll.

Rolls of cable can also be unrolled easily with a device commonly called a "spinner" (Figure 11–18). The spinner is usually attached to a wall stud in the building being wired. A roll of cable is placed on the spinner, and the electrician simply pulls the cable off the roll from the outside of the coil. As the cable is pulled, the spinner spins, allowing the cable to be unrolled in a manner that does not produce twists or kinks.

For cable that comes on reels, a commercial reel holder helps simplify unrolling cable from the reel. These reel holders are readily available from the local electrical supplier. Some electricians build their own by simply using a length of pipe placed between two framing members or between two step ladders. The reel is placed on the pipe, and the cable is unrolled from the reel. Either using a commercial reel holder or a "home made" reel holder will result in cable that is unrolled without twists or kinks (Figure 11–19).

Figure 11–18 Unrolling cable from a coil is very easy when using a device commonly called a "spinner." This picture shows an electrician pulling a length of Type NM cable from a spinner.

Figure 11–19 Large cable sizes and long lengths of smaller cable are usually purchased on reels. This picture shows an electrician pulling a length of cable from a reel.

Installing the Cable Runs

Installing the cable in the framework of a house is often referred to as **pulling in** the cable. The cables must be routed through or along studs, joists, and rafters to all the receptacle, switch, or lighting outlet locations that make up a residential circuit. There are a few common installation techniques and procedures that electricians need to become familiar with so that when the cables are pulled in, the job will be done in a safe, neat, and workmanlike manner. The installation must also meet or exceed the installa-

 FROM EXPERIENCE

Many electricians have found that the best way for them to unroll cable from a roll is to bend over and pick up the roll and, with their hands and wrists placed though the opening in the roll, rotate the roll and play out the cable as they walk across a room. This technique will allow two-wire "flat" cable, like 14/2, 12/2, and 10/2 Romex, to be unrolled with no twists or kinks. The cable is then ready for installation. This technique also works well with three wire Romex and armored cables (Figure 11–20).

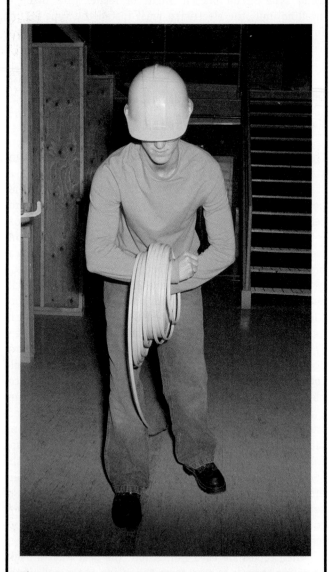

Figure 11–20 Many electricians prefer to unroll cables from the outside of the coil. This picture is showing an electrician unrolling Type NM from the outside of the coil. This technique will result in very few kinks or twists in the cable.

tion requirements of the *NEC®*. The following items should always be considered by an electrician when installing a cable run:

- Plan the cable route ahead of time and visually line it all up in your mind.

- If holes have to be drilled in the building framing members, drill them all, making sure to maintain a minimum clearance of 1¼ inches (32 mm) from the edge of the hole to the edge of the framing member.

- When drilling holes through consecutive framing members for longer cable runs, align the holes as close as possible. This will result in making it easier to pull cables through the holes and will make the installation neat and organized.

- Once the cable has been pulled to all the electrical boxes on the circuit run, double check your work for the following:
  - All locations on the circuit have the correct cable installed
  - The cable is correctly secured and supported according to the *NEC®*
  - There are no twists or kinks in the cable
  - Cable turns have not been bent too tightly
  - There is enough cable at each electrical box location for the required connections

Installing Cables through Studs and Joists

There are several ways for electricians to install the cable runs through the wall studs of a house; however, there are four methods commonly used:

- Drilling holes that are all a certain distance from the floor and in line. This method will allow the easiest pulling of the cables.

FROM EXPERIENCE

Drawing a cable diagram for the circuits you are running in a house helps keep the cable installation organized. Use the common electrical symbols discussed in Chapter 5 when doing a cable diagram. The cable diagram will quickly tell a new electrician where to run a cable as well as how many conductors must be in each cable run. As each cable run is installed, an electrician can cross off the completed runs on the cable diagram. This will help ensure that all the wiring has been roughed in that is supposed to be roughed in. It is a very bad feeling when the walls and ceilings are installed and an electrician discovers that some cables were not run.

- When there is a need to make changes in elevation along the cable run, there are two ways to accomplish it:
  - Drill holes parallel to the floor at a certain height, keeping the holes in line, and then repeat the drilling process at a new elevation from the floor. The cable will be run with small turns to get from one elevation to the next (Figure 11–21).
  - Drill holes that are gradually higher from one stud to the next (Figure 11–22).
- Drilling holes close to the bottom of the studs on outside walls. This is a technique that keeps the cable run close to the bottom plate of a wall and allows for a better and easier fit for insulation that will be placed in the wall between the studs (Figure 11–23).

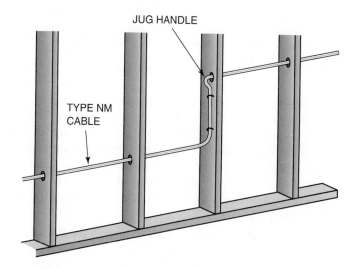

Figure 11–21 A cable run through studs with a change in elevation.

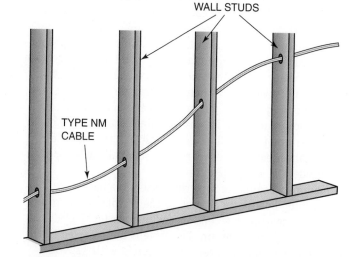

Figure 11–22 A cable run through studs with a gradual elevation change.

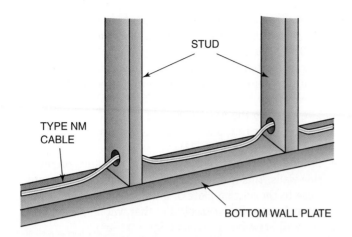

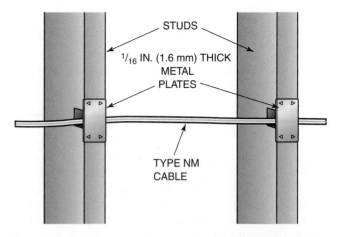

Figure 11–23 Drilling the holes for installing cables at the bottom of wall studs is a good practice. It allows the cables to be placed along the bottom of the wall and results in an easier installation of thermal insulation placed in the wall spaces between the studs.

Figure 11–24 The *NEC*® allows cutting notches in wall studs for cable installation, but not all local codes allow it.

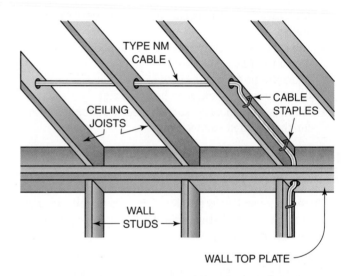

Figure 11–25 Drilling holes in joists and routing cable through the holes.

- Some electricians notch the studs instead of drilling them. The cable is simply laid into the notches, and then a 1/16-inch (1.6 mm) or thicker metal plate is placed over the notch to protect the cables from screws and nails and also to keep the cable from coming out of the notch. Not all areas allow notching of studs. Check with the local electrical inspector to find out for sure if notching studs is allowed in your area (Figure 11–24).

Installing cables through ceiling or floor joists is similar to installing cables through wall studs, and some of the methods previously described will work just fine. However, some cable installations in joists will require the use of one of the installation methods discussed here:

- Drill holes through the joists. This is similar to drilling holes through wall studs. Try to line up the holes as you drill them so the cables will pull through easier (Figure 11–25).
- Running the cables over the joists. This is accomplished by drilling a hole through the top plate of a wall section and running the cable above the ceiling joists. This technique will work when the joists are in an attic and no flooring is to be put on top of them. No additional support is required in this case since the distance between the joists will be much less than the maximum distance required between supports of all cables used in residential wiring (Figure 11–26).
- Running boards are sometimes used in a basement. The running boards are secured to the bottom of the joists in a basement in an end-to-end manner for the full length of the cable run. The cable is then stapled to the running boards at intervals consistent with *NEC*® requirements. This technique eliminates the need to drill through joists (Figure 11–27).

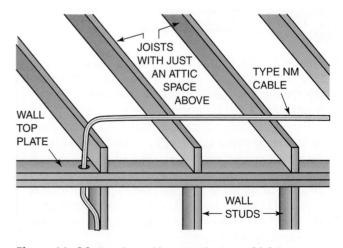

Figure 11–26 Running cables over the tops of joists.

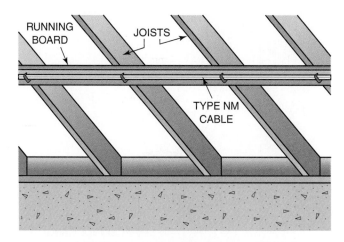

Figure 11–27 Running boards can be used to route cables along the undersides of joists and rafters.

Metal Framing Members

As mentioned earlier, electricians are encountering metal framing more and more in residential construction. Several of the cable installation methods we described previously in this section will work with metal framing members. However, there are a few special considerations. The following methods and materials for installing cable and electrical boxes with metal framing are recommended:

- Drilled or punched holes in metal studs should be in line for easier cable pulling.
- Snap-in grommets or bushings made of an insulating material must be installed at each hole in the metal framing member. They are required to protect the cable sheathing from the sharp edges of metal.
- Electrical device boxes should be mounted with side-mounting brackets and be attached to the stud with a no. 6 × 3/4-inch sheet metal screw or self-drilling screws. Boxes do not have to be metal but usually are when mounted on metal studs.
- Since staples cannot be used to secure a cable to a metal stud within the required distance stated in the *NEC®*, plastic tie wraps are often used. They are installed by drilling two small holes into the stud at the proper location, passing the tie wrap through the holes, and then securing the cable in place by tightening the tie wrap (Figure 11–28).

Starting the Cable Run

Although it is not a specific requirement, the best place to start a cable run for a branch circuit is at the service panel or subpanel. This is considered to be the starting point of a circuit because it is the location on the circuit where electrical power is provided. It is also where the circuit overcur-

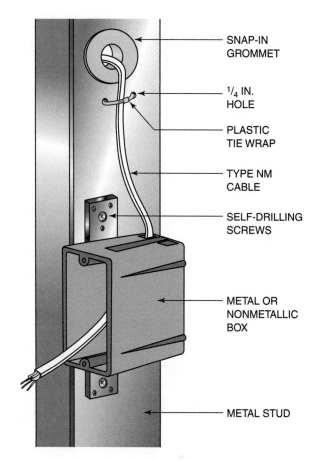

Figure 11–28 A cable installed in a metal stud application.

rent protection devices are located. The branch-circuit cable run that goes from the service panel or subpanel to the first electrical box of the circuit is called a **home run.** Once an electrician has more experience and understands the wiring process better, starting the circuit run at other locations is possible. However, for now we will assume the starting point to be at the load center. The procedures discussed in this section assume the use of nonmetallic sheathed cable. Other wiring methods may require some modification of the procedures shown.

(See pages 305–307 for step-by-step instructions on starting a cable run from a load center.)

Securing and Supporting the Cable Run

Once the cable has been run to each electrical outlet box on the circuit, an electrician must make sure that the cable is properly **secured** and **supported.** Some electricians secure and support the cable as they pull the cable into place. Other electricians prefer to make the full cable run for the circuit and then go back and properly secure and support the run. There is no right or wrong way. It is up to the electrician to decide which technique to use.

At the Device and Outlet Box

The *NEC®* requirements for securing and supporting cables attached to nonmetallic or metal electrical boxes were discussed earlier in this book. This section provides a quick review of common securing techniques.

If you are using single-gang nonmetallic electrical device boxes, at least 1/4 inch (6 mm) of the cable sheathing must extend into the box. This technique provides extra protection for the cable conductors from cuts to the insulation that could occur from the sharp edges of the box (Figure 11–29). Also, make sure to secure the cable to the building framing member no more than 8 inches (200 mm) from the box with a staple or some other approved method (Figure 11–30).

If you are using nonmetallic round ceiling boxes, nonmetallic device boxes of two gang or more, or any metal boxes, you must use an approved connector to secure the cable to the box. Built-in cable clamps or a separate cable connector can be used. Although it is not an *NEC®* rule, approximately 1/4 inch (6 mm) of cable sheathing should be exposed inside the box (Figure 11–31). Again, the technique of having the cable sheath extend into the electrical box a short distance will help keep the conductor insulation from being cut from sharp edges on the box. The cable must be secured with staples or other approved method within 12 inches (300 mm) of any of these box types (Figure 11–32).

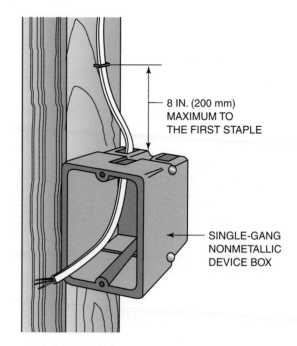

Figure 11–30 The cable must be secured within 8 inches (200 mm) of a single-gang nonmetallic device box.

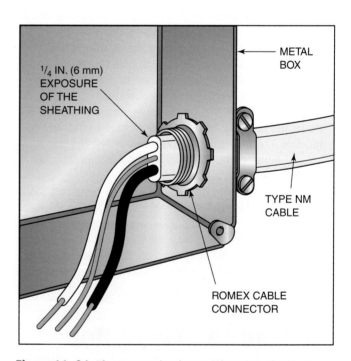

Figure 11–31 If you are using boxes other than single-gang nonmetallic device boxes, it is recommended that you extend at least 1/4 inch (6 mm) of Type NM cable into the box. This illustration shows a metal box with an external cable clamp securing the cable to the box. Notice that 1/4 inch (6 mm) of cable is exposed in the box.

Figure 11–29 The Type NM cable must extend at least 1/4 inch into a single-gang nonmetallic device box.

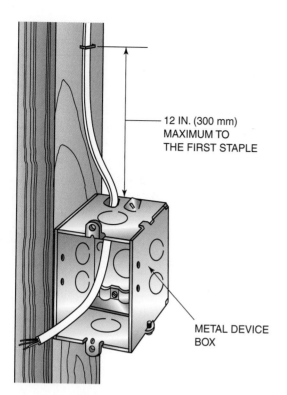

12 IN. (300 mm)
MAXIMUM TO
THE FIRST STAPLE

METAL DEVICE
BOX

Figure 11–32 The cable must be secured within 12 inches (300 mm) of all metal boxes types and all nonmetallic boxes other than single-gang nonmetallic device boxes.

Supporting the Cable Run

Once the cable has been properly secured to the box according to the *NEC®* requirements, additional support of the cable must be done. Horizontal runs of cable through holes in building framing members is adequately supported as long as the distance between framing members does not exceed the *NEC®* support requirements for a particular wiring method. For example, a Romex cable installed through holes drilled in studs that are placed 16 inches O.C. (on center) (400 mm) means that the cable is adequately supported because the distance between supports is less than the 4½-foot (1.4-meters) requirement for supporting Type NM cable. When the cable is run along the sides of a building framing member, it must be supported according to the *NEC®* requirements. For example, Romex cable running along the side of a stud must be supported with staples or other approved means not more than every 4½ feet (1.4 meters) (Figure 11–33).

Sometimes a suspended ceiling (often called a "dropped ceiling") is encountered by an electrician doing the rough-in wiring in a house. Suspended ceilings are often used in the basement of a home where a finished family room or recreation room is to be located. The suspended ceiling panels can be removed to provide access to electrical, plumbing, heating, and other system equipment installed in the ceiling. Section 300.11(A) states that suspended ceiling support

wires that do not provide secure support must not be permitted as the only support for boxes or wiring methods. However, support wires and associated fittings that provide secure support and that are installed *in addition to* the ceiling grid support wires are permitted as the support. Where independent support wires are used, they must be secured at both ends. Suspended ceiling grids cannot support cables. Wiring methods of any type and all luminaires (lighting fixtures) are not allowed to be supported or secured to the support wires or "T-bars" of a suspended ceiling assembly unless the assembly has been tested and listed for that use. If support wires are selected as the supporting means for cables within the ceiling cavity, they must be distinguishable from the ceiling support wires and secured at both ends (Figure 11–34). A common method used by electricians to distinguish a cable support wire from a ceiling support wire is to color code the cable support wire with tape or paint.

If an electrical box is to be mounted to the structural frame or supporting wires of a suspended ceiling, Section 314.23(D) requires it to be not more than 100 cubic inches (1,650 cn³) in size and be securely fastened in place by either of the following methods:

- Fastening to the framing members by mechanical means, such as bolts, screws, or rivets, or by the use of clips or other securing means identified for use with the type of ceiling framing member(s) and enclosure(s) employed. The framing members must be adequately supported and securely fastened to each other and to the building structure (Figure 11–35).
- Using methods identified for the purpose, to ceiling support wire(s), including any additional support wire(s) installed for that purpose. Support wire(s) used for enclosure support must be fastened at each end so as to be taut within the ceiling cavity. Some manufacturers make specially designed fittings that attach to a support wire and provide a way to secure an electrical box or cable to the support wire.

Installing Cable in Existing Walls and Ceilings

In remodel work, an electrician often needs to install cables in existing walls and ceilings. It is not an easy task and usually will require two people to help **fish** the cable from spot to spot. Sometimes an electrician can remove the baseboard molding around a room, notch or drill holes in the exposed studs, and run cable through. Molding around doors and windows may be removed and expose a space where a cable could be run. An electrician running cables in old work must be ingenious in discovering ways to get the cable to each circuit electrical box location.

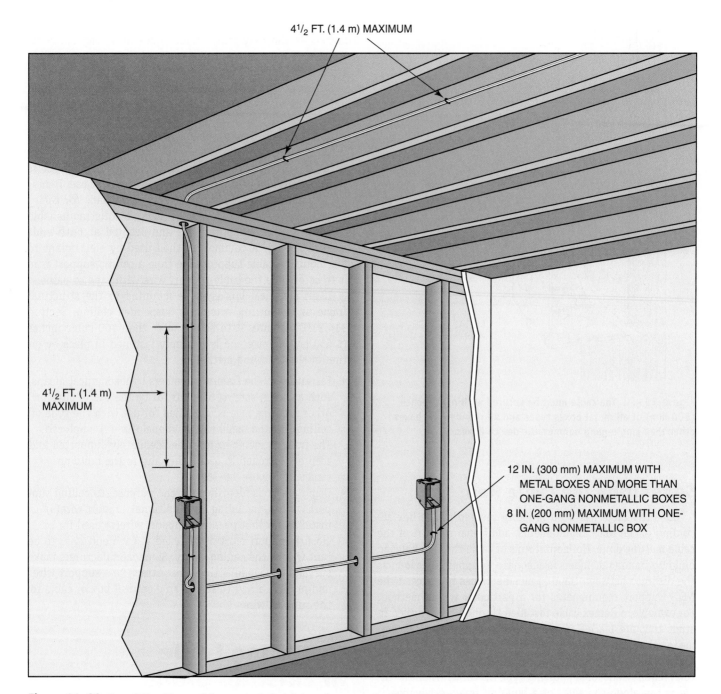

4½ FT. (1.4 m) MAXIMUM

4½ FT. (1.4 m) MAXIMUM

12 IN. (300 mm) MAXIMUM WITH METAL BOXES AND MORE THAN ONE-GANG NONMETALLIC BOXES
8 IN. (200 mm) MAXIMUM WITH ONE-GANG NONMETALLIC BOX

Figure 11–33 Type NM cable must be supported at intervals no greater than 4½ feet (1.8 m) when it is run along studs, joists, and rafters.

If the service entrance panel is located in a basement, new cable runs can start at the panel and be run along the exposed basement framing members until the cable is under the wall section in which you want to install an outlet. Drill a hole from the basement up and into the wall cavity as shown in Figure 11–36. A long auger bit works well for this operation. The drilling must be done at the proper angle to get from the basement into the wall. If you are drilling from the basement up through the floor and into an inside wall partition, there

is no need to drill at an angle. Be prepared to drill through several items, including flooring, bottom wall plates, studs, joists, and so on. This same drilling technique can be used when running cable from an unfinished attic space into the wall cavities of the rooms below. The only difference is that you will be drilling *down* into the wall cavities from the attic instead of drilling *up* from the basement. Be sure to measure accurately to avoid drilling into the wrong wall cavity or missing the wall cavity altogether and drilling out into the room.

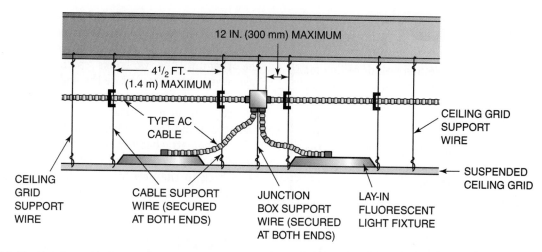

Figure 11–34 The framing grid of a suspended ceiling or the ceiling grid support wires cannot provide support for cables and electrical boxes installed above this type of ceiling. Support wires specifically installed for support of boxes or cables can be used, provided they are in addition to the regular ceiling grid support wires and are attached at both ends.

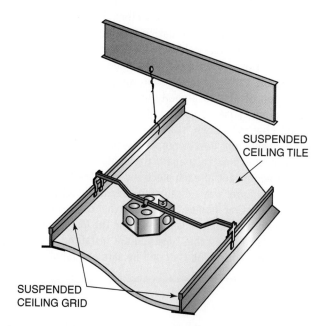

Figure 11–35 Attaching an electrical box to the grid of a suspended ceiling must be done using mechanical means identified for the purpose. This illustration shows an electrical box attached to the suspended ceiling grid with a special box hanger.

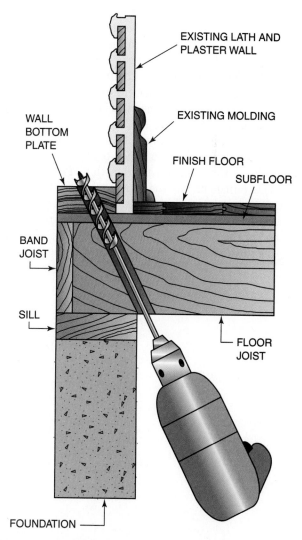

Figure 11–36 Drilling into a wall from a basement in an existing dwelling often involves angling the drill bit so you can get into the proper wall cavity.

Once the holes have been drilled for the cable, use a fish tape to determine if the wall cavity in which you wish to install the cable is free from obstructions. Firestops and other pieces of lumber used as bracing may block the route that you wish to run the cable. Usually an electrician can make a proper decision as to whether the cable can get to the desired box location by listening carefully to the sound that the fish tape produces when in the wall or ceiling space.

FROM EXPERIENCE

Before actually cutting the hole in a wall or ceiling for the installation of an old-work electrical box, be sure you can get the wiring method you are using to the box location. Many electricians have cut holes in existing walls and ceilings and then found out that there is no available route for the circuit wiring to get to the box. The result of this poor planning is a hole in a wall or ceiling that will now have to be patched.

Once the electrical outlet box hole has been cut, the cable can be installed through the walls or ceiling to the box location. This is usually a two-person job. One electrician will push a fish tape through the bored holes toward the electrical box location. Another electrician will look for the fish tape at the electrical box location and catch the end of the fish tape when it gets to the box hole. The electrician at the box hole will attach a cable to the end of the fish tape and feed it carefully through the box hole as the electrician at the other end carefully pulls the fish tape with the attached cable back into the basement or attic space. The installed cable can then be attached to a junction box or to the service entrance panel.

Sometimes a cable will have to be installed in a location that does not allow one end of a fish tape to be easily seen and caught at the electrical box location. Two fish tapes with hooks formed on the ends are required for this installation. One electrician will feed fish tape 1 into the wall or ceiling space from one end of the desired cable run. At the other end of the run, another electrician will feed fish tape 2 into the same area. By twisting and moving the fish tapes around, an electrician can feel and hear when contact is made between the two fish tapes. Once contact is established, the electricians will slowly pull back on one end or the other until the two hooks "catch" each other. This may take several attempts before the ends are hooked together. While keeping steady pressure on the fish tapes so the hooked ends do not come apart, one electrician will pull fish tape 1 and the "hooked" fish tape 2 back to them. A cable is then attached to the fish tape 2 end, and the other electrician simply pulls fish tape 2 and the newly attached cable back to the new location, completing the cable installation between the two points.

Summary

Once an electrician has determined the routing for the various branch circuits in a house, the first step is to locate and install the appropriate electrical boxes. The next step is to install the appropriate circuit conductors. In this chapter, we took a look at common installation techniques for cables. Selecting the appropriate cable for the job and installing it according to the proper *NEC*® requirements was discussed. Preparing the cable for installation was covered and included a discussion of how to properly unroll cables from reels and other packaging methods. Installation techniques discussed included how to actually start a cable run, how to properly support a cable run, how to properly secure a cable to an electrical box, and how to prepare the cable for connection to electrical equipment at each electrical box. The last item covered in the chapter was a discussion of cable installation techniques in existing homes. Following the cable installation techniques presented in this chapter and the *NEC*® installation requirements for the cable type you are installing in a house will result in a wiring job that will be safe and reliable.

Procedures

Starting a Cable Run from a Loadcenter

- Put on safety glasses and observe all applicable safety rules.

- Remove a knockout in the loadcenter cabinet and install a cable connector that is appropriate for the size and type of cable you are using.

- Place one end of the cable for the circuit you are installing through the connector and into the loadcenter. Leave enough cable in the loadcenter so the wires in the cable will have plenty of length to get to the circuit breaker, grounded neutral bar, and grounding bar connection points. It is up to each individual electrician whether the cable sheathing is stripped away at this time or during the trim-out stage of the loadcenter. The trimming out of a loadcenter is when the circuit breakers are installed and the circuit wires are properly terminated. We discuss trimming out the loadcenter in Chapter 19.

- Secure the cable to the cabinet by tightening the cable connector onto the cable.

- Identify each circuit cable in the loadcenter cabinet with the circuit number, type, and location as it is shown on the wiring plan. This will help eliminate any confusion about which cable goes to which circuit breaker during the trim-out stage of the loadcenter.

FROM EXPERIENCE

There is a good rule to follow for determining how much cable to leave in the loadcenter so you will have enough conductor length to make all the necessary connections. Measure the height and width of the cabinet, add them together, and use that dimension for the minimum amount of cable to be left in the cabinet. For example, if a loadcenter is 24 inches (600 mm) high and 14.5 inches (363 mm) wide, the minimum length of cable that should be left in the cabinet would be 24 inches (600 mm) + 14.5 inches (363 mm) = 38.5 inches (963 mm).

FROM EXPERIENCE

A permanent marker may be used to write directly on the cable sheathing. Other methods of marking include using tags or marker tape.

FROM EXPERIENCE

The cables can enter the loadcenter from the top, bottom, or sides. Most cables are brought into the loadcenter from the top or bottom. If the loadcenter is located in a basement, the cables are almost always brought into the loadcenter through the top so the cables can easily be routed to the floors above. If a loadcenter is flush mounted in a wall of a house that has no basement, cables are commonly brought into the loadcenter from both the top and the bottom. Manufacturers of loadcenters provide knockouts in the back of the panel cabinet as well. Sometimes it is necessary to bring your home-run cable through the back of the loadcenter. An example of this situation would be when the loadcenter is a weatherproof type located on the outside of a house and electricians route the circuit cables from the back of the loadcenter into the house and to the various outlets.

Procedures

Starting a Cable Run from a Loadcenter (continued)

A Once the cable run is started from the loadcenter, it is continued along or through building framing members until you reach the first electrical box on the circuit.

A

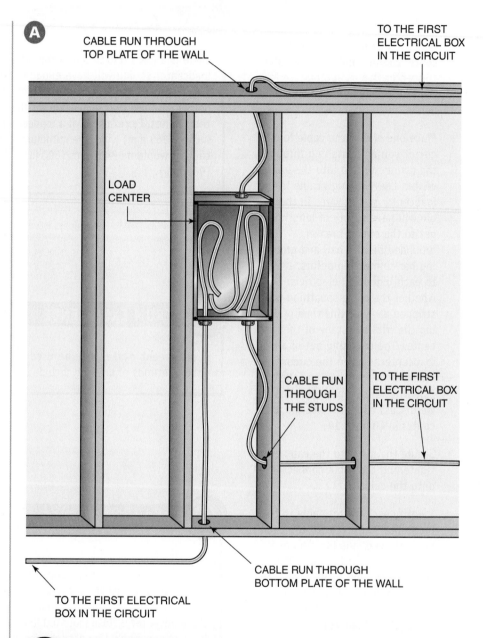

CABLE RUN THROUGH
TOP PLATE OF THE WALL

TO THE FIRST
ELECTRICAL BOX
IN THE CIRCUIT

LOAD
CENTER

CABLE RUN
THROUGH
THE STUDS

TO THE FIRST
ELECTRICAL BOX
IN THE CIRCUIT

CABLE RUN THROUGH
BOTTOM PLATE OF THE WALL

TO THE FIRST ELECTRICAL
BOX IN THE CIRCUIT

FROM EXPERIENCE

Some electricians install the cable into the electrical outlet box without stripping away the outside sheathing first. This technique will require the sheathing to be stripped later. Stripping the outside sheathing from a cable while it is in an electrical box can be tricky, especially with deeper boxes, and many electricians (including the author) prefer to strip the outside sheathing before the cable is secured to the electrical box. There is no absolutely right or wrong way to do this. The best way to do it is the way your instructor or supervisor suggests.

B At each outlet box, leave enough cable so that approximately 8 inches (200 mm) of conductor will be available in each box for connection purposes. Remember that the *NEC®* requires at least 6 inches (150 mm) of free conductor in electrical boxes for connection to devices. At least 3 inches (75 mm) of conductor must extend from the front of each box.

C Install another cable into the electrical box and continue the cable run to the next box in the circuit. Again, be sure to leave approximately 8 inches (200 mm) of cable (or conductor) in the box.

D Fold the cable (or conductors if the cable is stripped) back into the box and continue to pull the cable to the rest of the boxes in the circuit.

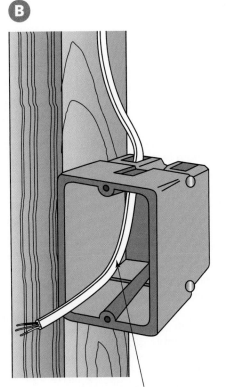

B

AT LEAST 8 IN. (200 mm) OF CABLE LEFT IN EACH BOX IS RECOMMENDED

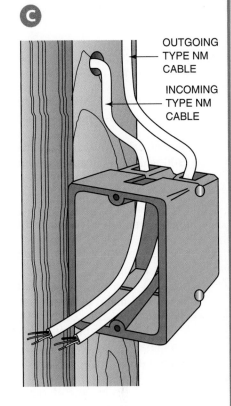

C

OUTGOING TYPE NM CABLE

INCOMING TYPE NM CABLE

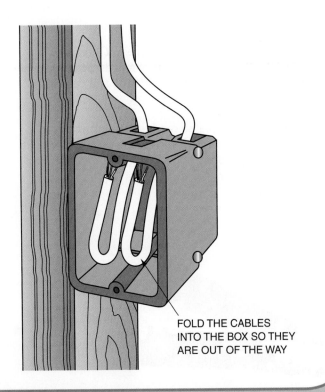

D

FOLD THE CABLES INTO THE BOX SO THEY ARE OUT OF THE WAY

Review Questions

Directions: Answer the following items with clear and complete responses.

1. Name five common cable types used in residential wiring and the *NEC®* article that covers each type.

2. Nonmetallic sheathed cable must be secured within _____ of a two-gang plastic nail-on device box.

3. Type AC cable must be supported within 12 inches (300 mm) of each electrical box and then no less than every _____.

4. The minimum bending radius for both Type NM and Type AC cable is _____ times the cable diameter.

5. When a Type NM cable is installed through a bored hole in a wood stud that is less than 1¼ inches (32 mm) from the edge of the stud, it must be protected by a metal plate at least _____ thick.

6. A 12/2 Type MC cable must be secured within _____ of a metal device box and then supported at intervals not exceeding _____.

7. When Type NM or Type AC cable is installed in an accessible attic space used for storage, the cable must be protected within _____ of the attic scuttle hole.

8. The maximum distance permitted between supports for a run of Type NM cable is _____.

9. When Type AC cable is used, the *NEC®* requires a special fitting to be used in the ends. State the common trade name for the fitting and describe its purpose.

10. Service entrance cable can be installed as an interior wiring method, such as for the branch-circuit wiring to an electric range. Name the article in the *NEC®* that covers the installation requirements for this situation.

11. Horizontal runs of Type NM cable through holes in building framing members are adequately supported as long as the distance between framing members does not exceed the *NEC®* support requirements. True or False (Circle the correct response.)

10. If you are using single-gang nonmetallic electrical device boxes, at least _____ of the Romex cable sheathing must extend into the box.

13. Although it is not a specific requirement, most electricians feel that the best place to start a cable run for a branch circuit is at the service panel or sub-panel. Explain why this is so.

14. When electricians use Romex cable that comes in a plastic wrapping or in a box, it usually is bought in _____ lengths. When Romex cable is bought on reels, it usually comes in _____ lengths.

15. Flat cable, like 14/2 or 12/2 Romex, cannot be stapled on edge. Explain why this is true and state the *NEC®* section that prohibits it.

16. If an electrical box is to be mounted to the structural frame or supporting wires of a suspended ceiling, the box may not be more than _____ cubic inches in size and be securely fastened in place. Name the *NEC®* section that requires this.

17. Electricians sometimes drill holes and run cables at the bottom of wall studs. Discuss the benefit of drilling holes and routing cables in this manner.

18. When drilling holes through consecutive framing members for longer cable runs, electricians align the holes as close as possible. Explain why this is done.

19. Discuss the advantages of having a cable diagram of the circuits that you have to install.

20. Name the two main criteria for selecting an appropriate cable type to be used in house wiring.

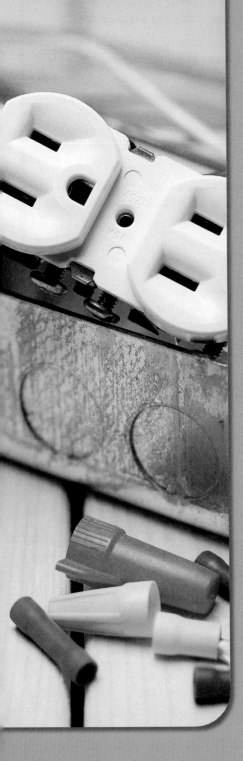

Chapter 12 Raceway Installation

Raceway is not used very often as the primary wiring method in residential wiring. However, certain service entrance installations use raceway, and there are some local electrical codes that require all wiring in houses to be installed in a raceway. When a raceway is used in residential wiring, it has a circular cross section and is called "conduit." Conduit may be made of metal or nonmetallic materials, and it may be rigid or flexible. It is up to the electrician to select the proper type of conduit for the wiring application that is encountered. This chapter discusses the common raceway types used in residential wiring. The selection of the proper raceway type and size is covered, as are the applicable National Electrical Code® (NEC®) installation requirements for each raceway type. An introduction to proper cutting, threading, and bending techniques is presented. The last area covered is the actual installation of electrical conductors in a completely installed raceway system.

OBJECTIVES

Upon completion of this chapter, the student should be able to:

- ⊗ select an appropriate raceway size and type for a residential application.
- ⊗ demonstrate an understanding of the proper techniques for cutting, threading, and bending electrical conduit for residential applications.
- ⊗ demonstrate an understanding of the proper installation techniques for common raceway types used in residential wiring.
- ⊗ demonstrate an understanding of the common installation techniques for installing conductors in an installed raceway system.

Glossary of Terms

back-to-back bend a type of conduit bend that is formed by two 90-degree bends with a straight length of conduit between the two bends

box offset bend a type of conduit bend that uses two equal bends to cause a slight change of direction for a conduit at the point where it is attached to an electrical box

conduit a raceway with a circular cross section, such as electrical metallic tubing, rigid metal conduit, or intermediate metal conduit;

field bend any bend or offset made by installers, using proper tools and equipment, during the installation of conduit systems

offset bend a type of conduit bend that is made with two equal-degree bends in such a way that the conduit changes elevation and avoids an obstruction

raceway an enclosed channel of metal or nonmetallic materials designed expressly for holding wires or cables; raceways used in residential wiring include rigid metal conduit, rigid nonmetallic conduit, intermediate metal conduit, liquid-tight flexible conduit, flexible metal conduit, electrical nonmetallic tubing, and electrical metallic tubing

saddle bend a type of conduit bend that results in a conduit run going over an object that is blocking the path of the run; there are two styles: a three-point saddle and a four-point saddle

stub-up a type of conduit bend that results in a 90-degree change of direction

take-up the amount that must be subtracted from a desired stub-up height so the bend will come out right; the take-up is different for each conduit size

thinwall a trade name often used for electrical metallic tubing

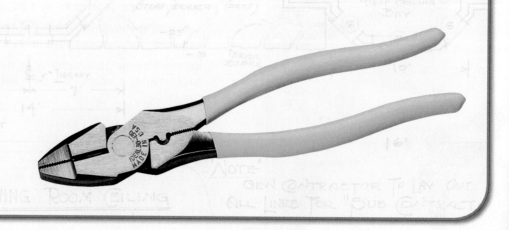

Selecting the Appropriate Raceway Type and Size

The raceway types used in residential wiring installations were introduced in Chapter 2. While service entrance installations using conduit of some type are very common, branch-circuit installation using a **raceway** wiring method is seldom used. There is no question that cable wiring methods are easier to install than raceway wiring methods, and this is the main reason why most houses are wired using as little **conduit** as possible. However, as we have mentioned before, some areas of the country require that all wiring in a house be installed in a raceway wiring method.

When installing a circuit in a raceway wiring method, individual conductors can be used. If the raceway type is made of metal, the raceway itself may serve as the equipment-grounding conductor so that there is no need to run a grounding wire in the raceway. However, it is common wiring practice for many electricians to install a green insulated equipment-grounding conductor in every raceway. Those circuits that operate on 120 volts will require a black insulated ungrounded conductor and a white insulated grounded conductor to be installed in the raceway. Those circuits that operate on 240 volts will require two black insulated ungrounded conductors or possibly one black and one red insulated ungrounded conductor. Circuits that operate on 120/240 volts will need one white insulated grounded conductor and two insulated ungrounded conductors of any color other than white, gray, or green. As mentioned previously, a green insulated equipment-grounding conductor can be installed with the regular circuit conductors, or an electrician can use the raceway itself as the grounding conductor, provided that the raceway is a listed grounding conductor in Section 250.118 of the *NEC*®. Although it is usually done only when added physical protection is desired, nonmetallic sheathed cable is allowed to be run in a raceway if an electrician so chooses.

Selecting the type of raceway to be used in a residential wiring installation will depend on the actual wiring application, an electrician's personal preference of raceway types, and what any local electrical codes will or will not allow. As mentioned in Chapter 8, rigid metal conduit (RMC) or intermediate metal conduit (IMC) are the raceway types used for a mast-style service entrance installation. Service entrance installations that use a raceway for the installation of service conductors that do not require the raceway to extend above a roof (a mast-type service entrance) will be able to use electrical metallic tubing (EMT) or rigid nonmetallic conduit (RNC) in addition to RMC and IMC. The most common raceway type used for branch-circuit installation is EMT, mainly because of its relatively easy bending and connection techniques. It is also much less expensive than other metal raceways. Installations where flexibility is desired, such as connections to air-conditioning systems, often require a flexible raceway like flexible metal conduit (FMC) for use indoors or liquid-tight flexible metal conduit (LFMC) for use outdoors.

Installation requirements for the various raceways are located in the appropriate articles of the *NEC*®. The following paragraphs cover the most important installation requirements for the raceway types used most often in residential wiring.

RMC

Article 344 covers the installation requirements for RMC. Section 344.2 defines RMC as a threadable raceway of circular cross section designed for the physical protection and routing of conductors and cables and for use as an equipment-grounding conductor when installed with appropriate fittings (Figure 12–1). RMC is generally made of steel with a protective galvanized coating and can be used in all atmospheric conditions and occupancies.

Section 344.22 states that the number of conductors or cables allowed in RMC must not exceed that permitted by the percentage fill specified in Table 1 in Chapter 9 of the *NEC*® (Figure 12–2). Table 1 in Chapter 9 specifies the maximum fill percentage of a conduit or tubing. It says that for an application where only two conductors are installed, the conduit cannot be filled

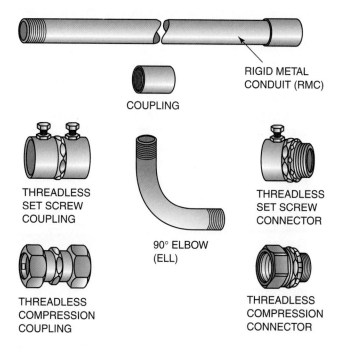

Figure 12–1 RMC and associated fittings.

Number of Conductors	All Conductor Types
1	53
2	31
Over 2	40

Figure 12–2 Table 1 in Chapter 9 of the *NEC*® specifies the maximum fill percentage for conduit and tubing. *Reprinted with permission from NFPA 70-2002, National Electrical Code®, Copyright © 2001, National Fire Protection Association, Quincy, MA 02269. This reprinted material is not the complete and official Position of the NFPA on the referenced subject, which is represented only by the standard in its entirety.*

Annex C is used to determine the maximum number of conductors that are all of the same size and have the same insulation type for *all* conduit types that are typically used in residential wiring.

to more than 31% of the conduit's cross-sectional area. If three or more conductors are to be installed (a common practice), the conduit cannot be filled to more than 40% of its cross-sectional area. If the conductors being installed are all of the same wire size and have the same insulation type, Annex C is used. Annex C is located in the back of the *NEC®* and, through 12 sets of tables, very accurately indicates the maximum number of conductors permitted in a conduit or tubing. The maximum number of conductors allowed in a conduit according to Annex C takes into account the fill percentage requirements of Table 1 in Chapter 9. No additional calculations are necessary.

For example, to select the proper trade size of RMC for an application calling for the installation of sixteen 12 AWG conductors with THHN insulation, the following procedure should be followed:

1. Go to Table C8 in Annex C (Figure 12–3).
2. Look in the far-left-hand column and locate the THHN insulation type.
3. Look in the next column to the right to find the conductor size, which in this example is 12 AWG.
4. Continue to look to the right until you find the number of conductors you will be installing *or* the next higher number to the number of conductors you are installing. In this case, the number of installed conductors is 16.
5. Look at the top of the table for the conduit size. It will be at the top of the column where you found the number of installed conductors in the previous step. In this example, you will find that a 3/4-inch (21 metric designator) minimum trade size is required for this installation.

When determining the maximum number of conductors of different sizes and of different insulations in a conduit, Table 4 in Chapter 9 will provide the usable area within the selected conduit or tubing, and Table 5 in Chapter 9 will provide the required area for each conductor. The calculation procedure is a little complicated but, once explained, is easily applied by an electrician. Let's look at an example. An electrician needs to

There is an index at the beginning of Annex C that will tell you which table applies to the specific type of conduit you are using.

Table C8   *Continued*

| Type | Conductor Size (AWG/kcmil) | Metric Designator (Trade Size) | | | | | | | | | | | |
		16 (½)	21 (¾)	27 (1)	35 (1¼)	41 (1½)	53 (2)	63 (2½)	78 (3)	91 (3½)	103 (4)	129 (5)	155 (6)
THHN, THWN, THWN-2	14	13	22	36	63	85	140	200	309	412	531	833	1202
	12	9	(16)	26	46	62	102	146	225	301	387	608	877
	10	6	10	17	29	39	64	92	142	189	244	383	552
	8	3	6	9	16	22	37	53	82	109	140	221	318
	6	2	4	7	12	16	27	38	59	79	101	159	230
	4	1	2	4	7	10	16	23	36	48	62	98	141
	3	1	1	3	6	8	14	20	31	41	53	83	120
	2	1	1	3	5	7	11	17	26	34	44	70	100
	1	1	1	1	4	5	8	12	19	25	33	51	74
	1/0	1	1	1	3	4	7	10	16	21	27	43	63
	2/0	0	1	1	2	3	6	8	13	18	23	36	52
	3/0	0	1	1	1	3	5	7	11	15	19	30	43
	4/0	0	1	1	1	2	4	6	9	12	16	25	36
	250	0	0	1	1	1	3	5	7	10	13	20	29
	300	0	0	1	1	1	3	4	6	8	11	17	25
	350	0	0	1	1	1	2	3	5	7	10	15	22
	400	0	0	1	1	1	2	3	5	7	8	13	20
	500	0	0	0	1	1	1	2	4	5	7	11	16
	600	0	0	0	1	1	1	1	3	4	6	9	13
	700	0	0	0	1	1	1	1	3	4	5	8	11
	750	0	0	0	0	1	1	1	3	4	5	7	11
	800	0	0	0	0	1	1	1	2	3	4	7	10
	900	0	0	0	0	1	1	1	2	3	4	6	9
	1000	0	0	0	0	1	1	1	1	3	4	6	8
FEP, FEPB, PFA, PFAH, TFE	14	12	22	35	61	83	136	194	300	400	515	808	1166
	12	9	16	26	44	60	99	142	219	292	376	590	851
	10	6	11	18	32	43	71	102	157	209	269	423	610
	8	3	6	10	18	25	41	58	90	120	154	242	350
	6	2	4	7	13	17	29	41	64	85	110	172	249
	4	1	3	5	9	12	20	29	44	59	77	120	174
	3	1	2	4	7	10	17	24	37	50	64	100	145
	2	1	1	3	6	8	14	20	31	41	53	83	120
PFA, PFAH, TFE	1	1	1	2	4	6	9	14	21	28	37	57	83
PFA, PFAH, TFE, Z	1/0	1	1	1	3	5	8	11	18	24	30	48	69
	2/0	1	1	1	3	4	6	9	14	19	25	40	57
	3/0	0	1	1	2	3	5	8	12	16	21	33	47
	4/0	0	1	1	1	2	4	6	10	13	17	27	39
Z	14	15	26	42	73	100	164	234	361	482	621	974	1405
	12	10	18	30	52	71	116	166	256	342	440	691	997
	10	6	11	18	32	43	71	102	157	209	269	423	610
	8	4	7	11	20	27	45	64	99	132	170	267	386
	6	3	5	8	14	19	31	45	69	93	120	188	271
	4	1	3	5	9	13	22	31	48	64	82	129	186
	3	1	2	4	7	9	16	22	35	47	60	94	136
	2	1	1	3	6	8	13	19	29	39	50	78	113
	1	1	1	2	5	6	10	15	23	31	40	63	92
XHH, XHHW, XHHW-2, ZW	14	9	15	25	44	59	98	140	216	288	370	581	839
	12	7	12	19	33	45	75	107	165	221	284	446	644
	10	5	9	14	25	34	56	80	123	164	212	332	480
	8	3	5	8	14	19	31	44	68	91	118	185	267
	6	1	3	6	10	14	23	33	51	68	87	137	197
	4	1	2	4	7	10	16	24	37	49	63	99	143
	3	1	1	3	6	8	14	20	31	41	53	84	121
	2	1	1	3	5	7	12	17	26	35	45	70	101

Figure 12–3 Table C8 in Annex C of the *NEC®* lists the maximum number of conductors allowed in RMC that are the same size and have the same insulation. *Reprinted with permission from NFPA 70-2002, National Electrical Code®, Copyright © 2001, National Fire Protection Association, Quincy, MA 02269. This reprinted material is not the complete and official Position of the NFPA on the referenced subject, which is represented only by the standard in its entirety.*

determine the minimum-size RMC that can be used for the installation of eight 14 AWG THHN conductors, five 12 AWG THW conductors, and two 8 AWG THWN conductors. (This is not a common situation in residential wiring, but it will help you understand conduit sizing when different wire sizes with differ-

ent insulation types are used.) The procedure for calculating the minimum size of RMC for this application is as follows:

1. From Table 5 in Chapter 9, determine the cross-sectional area (in.²) of each conductor size with its insulation type (Figure 12–4):

Table 5 Dimensions of Insulated Conductors and Fixture Wires

Type	Size (AWG or kcmil)	Approximate Diameter mm	Approximate Diameter in.	Approximate Area mm²	Approximate Area in.²
Type: FFH-2, RFH-1, RFH-2, RHH*, RHW*, RHW-2*, RHH, RHW, RHW-2, SF-1, SF-2, SFF-1, SFF-2, TF, TFF, THHW, THW, THW-2, TW, XF, XFF					
RFH-2,	18	3.454	0.136	9.355	0.0145
FFH-2	16	3.759	0.148	11.10	0.0172
RHW-2, RHH,	14	4.902	0.193	18.90	0.0293
RHW	12	5.385	0.212	22.77	0.0353
	10	5.994	0.236	28.19	0.0437
	8	8.280	0.326	53.87	0.0835
	6	9.246	0.364	67.16	0.1041
	4	10.46	0.412	86.00	0.1333
	3	11.18	0.440	98.13	0.1521
	2	11.99	0.472	112.9	0.1750
	1	14.78	0.582	171.6	0.2660
	1/0	15.80	0.622	196.1	0.3039
	2/0	16.97	0.668	226.1	0.3505
	3/0	18.29	0.720	262.7	0.4072
	4/0	19.76	0.778	306.7	0.4754
	250	22.73	0.895	405.9	0.6291
	300	24.13	0.950	457.3	0.7088
	350	25.43	1.001	507.7	0.7870
	400	26.62	1.048	556.5	0.8626
	500	28.78	1.133	650.5	1.0082
	600	31.57	1.243	782.9	1.2135
	700	33.38	1.314	874.9	1.3561
	750	34.24	1.348	920.8	1.4272
	800	35.05	1.380	965.0	1.4957
	900	36.68	1.444	1057	1.6377
	1000	38.15	1.502	1143	1.7719
	1250	43.92	1.729	1515	2.3479
	1500	47.04	1.852	1738	2.6938
	1750	49.94	1.966	1959	3.0357
	2000	52.63	2.072	2175	3.3719
SF-2, SFF-2	18	3.073	0.121	7.419	0.0115
	16	3.378	0.133	8.968	0.0139
	14	3.759	0.148	11.10	0.0172
SF-1, SFF-1	18	2.311	0.091	4.194	0.0065
RFH-1, XF, XFF	18	2.692	0.106	5.161	0.0080
TF, TFF, XF, XFF	16	2.997	0.118	7.032	0.0109
TW, XF, XFF, THHW, THW, THW-2	14	3.378	0.133	8.968	0.0139
TW, THHW, THW, THW-2	12	3.861	0.152	11.68	0.0181
	10	4.470	0.176	15.68	0.0243
	8	5.994	0.236	28.19	0.0437
RHH*, RHW*, RHW-2*	14	4.140	0.163	13.48	0.0209
	12	4.623	0.182	16.77	0.0260

Table 5 *Continued*

Type	Size (AWG or kcmil)	Approximate Diameter mm	Approximate Diameter in.	Approximate Area mm²	Approximate Area in.²
Type: RHH*, RHW*, RHW-2*, THHN, THHW, THW, THW-2, TFN, TFFN, THWN, THWN-2, XF, XFF					
THHW, THW, AF, XF, XFF	10	5.232	0.206	21.48	0.0333
RHH*, RHW*, RHW-2*	8	6.756	0.266	35.87	0.0556
TW, THW, THHW, THW-2, RHH*, RHW*, RHW-2*	6	7.722	0.304	46.84	0.0726
	4	8.941	0.352	62.77	0.0973
	3	9.652	0.380	73.16	0.1134
	2	10.46	0.412	86.00	0.1333
	1	12.50	0.492	122.6	0.1901
	1/0	13.51	0.532	143.4	0.2223
	2/0	14.68	0.578	169.3	0.2624
	3/0	16.00	0.630	201.1	0.3117
	4/0	17.48	0.688	239.9	0.3718
	250	19.43	0.765	296.5	0.4596
	300	20.83	0.820	340.7	0.5281
	350	22.12	0.871	384.4	0.5958
	400	23.32	0.918	427.0	0.6619
	500	25.48	1.003	509.7	0.7901
	600	28.27	1.113	627.7	0.9729
	700	30.07	1.184	710.3	1.1010
	750	30.94	1.218	751.7	1.1652
	800	31.75	1.250	791.7	1.2272
	900	33.38	1.314	874.9	1.3561
	1000	34.85	1.372	953.8	1.4784
	1250	39.09	1.539	1200	1.8602
	1500	42.21	1.662	1400	2.1695
	1750	45.11	1.776	1598	2.4773
	2000	47.80	1.882	1795	2.7818
TFN, TFFN	18	2.134	0.084	3.548	0.0055
	16	2.438	0.096	4.645	0.0072
THHN, THWN, THWN-2	14	2.819	0.111	6.258	0.0097
	12	3.302	0.130	8.581	0.0133
	10	4.166	0.164	13.61	0.0211
	8	5.486	0.216	23.61	0.0366
	6	6.452	0.254	32.71	0.0507
	4	8.230	0.324	53.16	0.0824
	3	8.941	0.352	62.77	0.0973
	2	9.754	0.384	74.71	0.1158
	1	11.33	0.446	100.8	0.1562
	1/0	12.34	0.486	119.7	0.1855
	2/0	13.51	0.532	143.4	0.2223
	3/0	14.83	0.584	172.8	0.2679
	4/0	16.31	0.642	208.8	0.3237
	250	18.06	0.711	256.1	0.3970
	300	19.46	0.766	297.3	0.4608

Figure 12–4 Table 5 in Chapter 9 of the *NEC*® lists the dimensions of insulated conductors. *Reprinted with permission from NFPA 70-2002, National Electrical Code®, Copyright © 2001, National Fire Protection Association, Quincy, MA 02269. This reprinted material is not the complete and official Position of the NFPA on the referenced subject, which is represented only by the standard in its entirety.*

Table 4 *Continued*

		Nominal Internal Diameter		Total Area 100%		2 Wires 31%		Over 2 Wires 40%		1 Wire 53%		60%	
Metric Designator	Trade Size	mm	in.	mm²	in.²	mm²	in.²	mm²	in.²	mm²	in.²	mm²	in.²
12	⅜	—	—	—	—	—	—	—	—	—	—	—	—
16	½	16.1	0.632	204	0.314	63	0.097	81	0.125	108	0.166	122	0.188
21	¾	21.2	0.836	353	0.549	109	0.170	141	0.220	187	0.291	212	0.329
27	1	27.0	1.063	573	0.887	177	0.275	229	0.355	303	0.470	344	0.532
35	1¼	35.4	1.394	984	1.526	305	0.473	394	0.610	522	0.809	591	0.916
41	1½	41.2	1.624	1333	2.071	413	0.642	533	0.829	707	1.098	800	1.243
53	2	52.9	2.083	2198	3.408	681	1.056	879	1.363	1165	1.806	1319	2.045
63	2½	63.2	2.489	3137	4.866	972	1.508	1255	1.946	1663	2.579	1882	2.919
78	3	78.5	3.090	4840	7.499	1500	2.325	1936	3.000	2565	3.974	2904	4.499
91	3½	90.7	3.570	6461	10.010	2003	3.103	2584	4.004	3424	5.305	3877	6.006
103	4	102.9	4.050	8316	12.882	2578	3.994	3326	5.153	4408	6.828	4990	7.729
129	5	128.9	5.073	13050	20.212	4045	6.266	5220	8.085	6916	10.713	7830	12.127
155	6	154.8	6.093	18821	29.158	5834	9.039	7528	11.663	9975	15.454	11292	17.495

Article 344 — Rigid Metal Conduit (RMC)

Figure 12–5 Table 4 in Chapter 9 of the *NEC®* lists the dimensions and percent area of conduit and tubing. Each conduit or tubing type has its own section in Table 4. *Reprinted with permission from NFPA 70-2002,* National Electrical Code®, *Copyright © 2001, National Fire Protection Association, Quincy, MA 02269. This reprinted material is not the complete and official Position of the NFPA on the referenced subject, which is represented only by the standard in its entirety.*

- 14 AWG THHN = 0.0097 in.²
- 12 AWG THW = 0.0181 in.²
- 8 AWG THWN = 0.0366 in.²

2. Multiply each cross-sectional area by the number of conductors to be installed in the conduit:
- 14 AWG THHN = 0.0097 in.² × 8 = 0.0776 in.²
- 12 AWG THW = 0.0181 in.² × 5 = 0.0905 in.²
- 8 AWG THWN = 0.0366 in.² × 2 = 0.0732 in.²

3. Add all the conductor cross-sectional areas together to get a total cross-sectional area for all the conductors to be installed:
- 0.0776 in.² + 0.0905 in.² + 0.0732 in.² = 0.2413 in.²

4. Using Table 4 in Chapter 9, go to the section that covers RMC and find the minimum conduit size that will work (Figure 12–5). Use the "Over 2 Wires 40%" column and look for 0.2413 in.² *or* the number that is the next-higher-listed number to this number. In this example, you will not find 0.02413, but you will find the next-higher-listed number, which is 0.355. Now look to the left to find the conduit size that corresponds with this number. It is 1 inch, and this is the minimum-size RMC that could be used for this application.

Section 344.24 requires that when bending RMC, the bends must be made so that the conduit is not damaged and the internal diameter of the conduit is not effectively reduced. Section 344.26 limits the number of bends in one conduit run from one box to another to no more than 360 degrees total (Figure 12–6). Limiting the number of bends in a conduit run will reduce the pulling tension on conductors and help ensure easy insertion and removal of conductors.

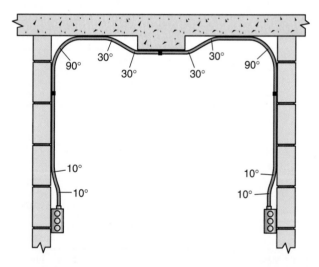

Figure 12–6 No more than 360 degrees of bends are allowed in a conduit run (Section 344.26). Because the total bends for this application is only 340 degrees, it meets the requirements of the *NEC®*.

When cutting RMC, Section 344.28 requires all cut ends to be reamed or otherwise finished to remove rough edges. Conduit is cut using a saw or a pipe cutter. Care should be taken to ensure a straight cut. After the cut is made, the conduit must be reamed. Proper reaming removes any burrs from the interior of the cut conduit so that as wires and cables are pulled through the conduit, no chafing of the insulation or cable jacket can occur. Finally, the conduit is threaded. Where conduit is threaded in the field, a standard cutting die with a 1 in 16 taper (3/4-inch taper per foot) must be used.

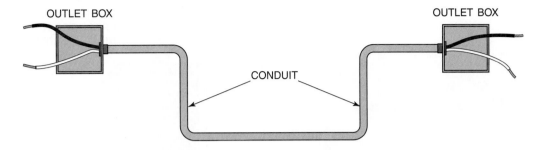

INSTALL CONDUIT COMPLETELY BETWEEN BOXES BEFORE PULLING
IN CONDUCTORS—SECTION *300.18(A).*

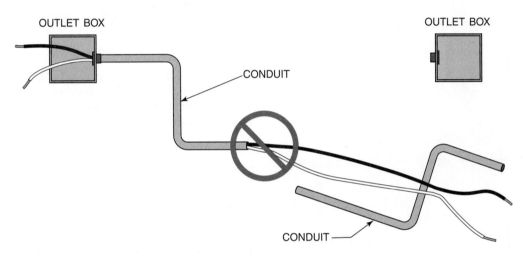

DO NOT PULL CONDUCTORS INTO A PARTIALLY INSTALLED
CONDUIT SYSTEM, AND THEN ATTEMPT TO SLIDE THE
REMAINING SECTION OF RACEWAY OVER THE CONDUCTORS.
THIS IS A VIOLATION OF SECTION *300.18(A).*

Figure 12–7 Conduit systems must be completely installed before the electrical conductors are installed.

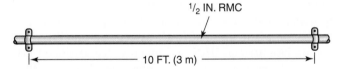

Figure 12–8 The maximum support interval for RMC is 10 feet (3 m). However, Section 344.30 (B)(1), Section 344.30 (B)(2), and Table 344.30 (B)(2) allow greater distances between supports for straight runs of conduit made up with threaded couplings.

RMC must be installed as a complete system as required in Section 300.18(A) (Figure 12–7) and must be securely fastened in place and supported in accordance with Section 344.30(A) and (B). Straight runs of RMC in 1/2- and 3/4-inch trade sizes are required to be securely fastened at intervals of no more than 10 feet (3 m) (Figure 12–8). As the trade size of RMC gets larger, Table 344.30(B)(2) allows support distances of more than 10 feet (3 m). Secure fastening is also required

within 3 feet (900 mm) of outlet boxes, junction boxes, cabinets, and conduit bodies (Figure 12–9). However, where structural support members do not permit fastening within 3 feet (900 mm), secure fastening may be located up to 5 feet (1.5 m) away (Figure 12–10). Section 344.30(B)(4) permits lengths of RMC to be supported by framing members at 10-foot (3-m) intervals, provided the RMC is secured and supported at least 3 feet (900 mm) from the box or enclosure. This would apply when RMC is run through holes drilled in framing members of a residential structure.

Where a rigid metal conduit enters a box, fitting, or other enclosure, a bushing must be provided to protect the wire from abrasion unless the design of the box, fitting, or enclosure is made so it already gives protection, according to Section 344.46. Section 300.4(F) requires that an insulated bushing be used for the protection of conductor sizes 4 AWG and larger that is installed in conduit. Usually a plastic bushing is installed (Figure 12–11).

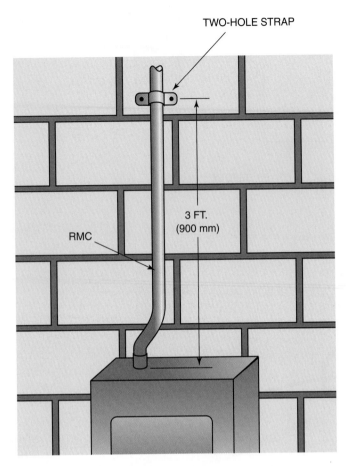

Figure 12–9 RMC must be securely fastened within 3 feet (900 mm) of each conduit termination point.

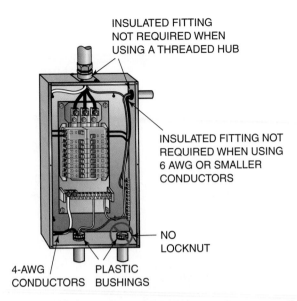

Figure 12–11 Bushings are required with RMC to protect the wires from abrasion. Insulated bushings are required when the conductors are 4 AWG or larger. At no time can an insulated bushing be used to secure a conduit to a box.

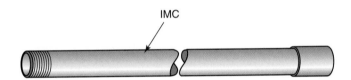

Figure 12–12 IMC is lighter in weight than RMC and has slightly thinner walls. It can be used with the same fittings as RMC.

Figure 12–10 If structural members are not available for securing RMC within 3 feet (900 mm), a distance of 5 feet (1.5 m) is acceptable.

The last installation requirement mentioned here for RMC is that according to Section 344.60, RMC is permitted as an equipment-grounding conductor. This means that if the RMC is attached correctly at each electrical enclosure and any couplings used to connect lengths of RMC together are properly tightened, the conduit itself can be the equipment-grounding conductor, and there is no need to run an additional grounding conductor in the conduit along with the regular circuit conductors.

IMC

IMC is covered in Article 342. Section 342.2 defines IMC as a steel threadable raceway of circular cross section designed for the physical protection and routing of conductors and cables and for use as an equipment-grounding conductor when installed with associated couplings and appropriate fittings. IMC is a thinner-walled version of RMC that can be used in all locations where RMC is permitted to be used (Figure 12–12). Also, threaded fittings, couplings, connectors, and other items used with IMC can be used with RMC.

Section 342.22 covers the requirements for determining the maximum number of conductors allowed in a specific size of IMC. The procedures used are exactly the same as for RMC with the exception of using Table C4 in Annex C when determining the maximum number of conductors that are all the same size and with the same insulation type. Also, you must use the IMC section of Table 4 in Chapter 9 when determining the minimum-size IMC for conductors of different sizes that have different insulation types. All other *NEC*® installation requirements for IMC, including support, are the same as for RMC and were presented earlier in this chapter.

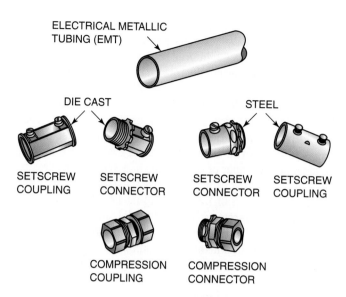

Figure 12–13 **EMT and associated fittings.**

FROM EXPERIENCE

Reaming tools are commercially available. One style that is very popular with electricians fits on a larger screwdriver shank and can ream 1/2-, 3/4-, and 1-inch trade-size EMT.

- In addition, each EMT run between termination points must be securely fastened within 3 feet (900 mm) of each outlet box, junction box, device box, cabinet, conduit body, or other tubing termination.
- Exception No. 2 to Section 358.30(A) states that for concealed work in finished buildings or prefinished wall panels where such securing is impractical, unbroken lengths (without a coupling) of EMT are permitted to be fished.
- Section 358.30(B) allows horizontal runs of EMT supported by openings through framing members at intervals not greater than 10 feet (3 m) and securely fastened within 3 feet (900 mm) of termination points to be permitted.

Section 358.42 requires couplings and connectors used with EMT to be made up tight. EMT is also permitted as an equipment-grounding conductor according to Section 358.60.

EMT

Article 358 covers the installation requirements for EMT. Section 358.2 defines EMT as an unthreaded **thinwall** raceway of circular cross section designed for the physical protection and routing of conductors and cables and for use as an equipment-grounding conductor when installed utilizing appropriate fittings (Figure 12–13). EMT is generally made of steel.

Section 358.22 covers the requirements for determining the maximum number of conductors allowed in a specific size of EMT. The procedures used are exactly the same as for RMC and IMC with the exception of using Table C1 in Annex C when determining the maximum number of conductors that are all the same size and with the same insulation type. Also, you must use the EMT section of Table 4 in Chapter 9 when determining the minimum-size EMT for conductors of different sizes that have different insulation types.

Section 358.24 requires bends in EMT to be made so that the tubing is not damaged and the internal diameter of the tubing is not effectively reduced. Like all the other circular raceways discussed in this chapter, Section 358.26 does not allow more than the equivalent of 360 degrees total bending between termination points of EMT.

Like RMC and IMC, Section 358.28 requires all cut ends of EMT to be reamed or otherwise finished to remove rough edges. A half-round file has proved practical for removing rough edges. However, many electricians find that the nose of lineman pliers, the nose of diagonal cutting pliers, or an electrician's knife can be an effective reaming tool on the smaller sizes of EMT.

Securing and supporting requirements of EMT is similar to both RMC and IMC. Section 358.30 requires EMT to be installed as a complete system and to be securely fastened in place and supported in accordance with the following:

- EMT must be securely fastened in place at least every 10 feet (3 m).

FMC

Article 348 covers the installation requirements for FMC (Figure 12–14). Many electricians often refer to this raceway type as "Greenfield." FMC is appropriate for use indoors where a need for flexibility at the connection points is re-

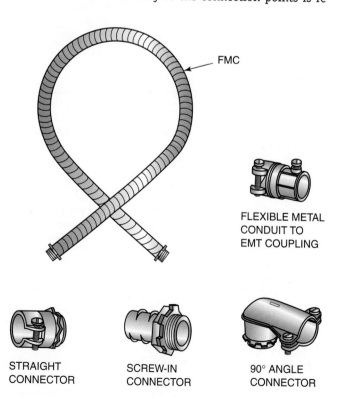

FMC

FLEXIBLE METAL CONDUIT TO EMT COUPLING

STRAIGHT CONNECTOR

SCREW-IN CONNECTOR

90° ANGLE CONNECTOR

Figure 12–14 **FMC and associated fittings.**

quired. Section 348.2 defines FMC as a raceway of circular cross section made of helically wound, formed, interlocked metal strip. On the outside, it looks a lot like Type AC cable and Type MC cable. However, unlike the armored cables, FMC requires an electrician to install electrical wires in the raceway. Although FMC can be used in longer lengths if desired, electricians usually use it in lengths of 6 feet (1.8 m) or less where a flexible connection is necessary (Figure 12–15).

Section 348.22 covers the requirements for determining the maximum number of conductors allowed in a specific

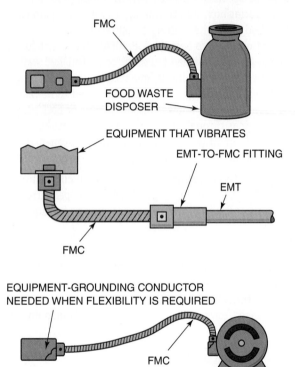

Figure 12–15 Some common FMC applications in residential wiring.

size of FMC. The procedures used are exactly the same as for the raceways we have previously discussed, with the exception of using Table C3 in Annex C when determining the maximum number of conductors that are all the same size and with the same insulation type. Also, you must use the FMC section of Table 4 in Chapter 9 when determining the minimum-size FMC for conductors of different sizes that have different insulation types. There is one other difference, and that is the use of Table 348.22 when determining the maximum number of conductors in a 3/8-inch trade-size FMC. This table must be used because the 3/8-inch size is not included in Table C3 of Annex C (Figure 12–16).

As with other raceways, a run of flexible metal conduit installed between boxes, conduit bodies, and other electrical equipment is not permitted to contain more than the equivalent of 360 degrees total. Proper shaping and support of this flexible wiring method will ensure that conductors can be easily installed or taken out at any time. These requirements are found in Sections 348.24 and 348.26.

Section 348.28 requires all cut ends to be trimmed or otherwise finished to remove rough edges, except where fittings that thread into the convolutions (so-called inside fittings) are used. Many electricians believe that an antishort bushing similar to the "redhead" style used with Type AC cable must be used with FMC. While this is not an *NEC®* requirement, it is a good idea to use a redhead in each end of FMC.

The securing and supporting requirements are given in Section 348.30. They are exactly the same as for Type AC cable and Type NM cable. FMC must be securely fastened in place and supported by an approved means within 12 inches (300 mm) of each box, cabinet, conduit body, or other conduit termination and be supported and secured at intervals not to exceed 4½ feet (1.4 m). The supporting and securing rules do not have to be followed when FMC is fished in a wall or ceiling or lengths do not exceed 3 feet (900 mm) at terminals where flexibility is required. Horizontal runs of FMC are considered supported by

Table 348.22 Maximum Number of Insulated Conductors in Metric Designator 12 (Trade Size ⅜) Flexible Metal Conduit*

| Size (AWG) | Types RFH-2, SF-2 | | Types TF, XHHW, TW | | Types TFN, THHN, THWN | | Types FEP, FEBP, PF, PGF | |
	Fittings Inside Conduit	Fittings Outside Conduit	Fittings Inside Conduit	Fittings Outside Conduit	Fittings Inside Conduit	Fittings Outside Conduit	Fittings Inside Conduit	Fittings Outside Conduit
18	2	3	3	5	5	8	5	8
16	1	2	3	4	4	6	4	6
14	1	2	2	3	3	4	3	4
12	—	—	1	2	2	3	2	3
10	—	—	1	1	1	1	1	2

*In addition, one covered or bare equipment-grounding conductor of the same size shall be permitted.

Figure 12–16 Table 348.22 in the *NEC®* lists the maximum number of insulated conductors allowed in 3/8-inch trade-size FMC and LFMC. *Reprinted with permission from NFPA 70-2002,* National Electrical Code®, *Copyright © 2001, National Fire Protection Association, Quincy, MA 02269. This reprinted material is not the complete and official Position of the NFPA on the referenced subject, which is represented only by the standard in its entirety.*

SECTION 250-118 (6) ALLOWS FMC THAT IS NOT LISTED FOR GROUNDING TO BE USED AS AN EQUIPMENT GROUNDING CONDUCTOR WHEN:

(1) CONNECTORS ARE LISTED FOR GROUNDING

(2) THE MAXIMUM USED LENGTH IS 6 FT. (1.8 m)

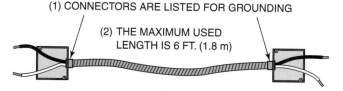

(3) OVERCURRENT PROTECTION DEVICES PROTECTING THE ENCLOSED CONDUCTORS ARE 20 AMPS OR LESS

Figure 12–17 FMC used as a grounding path.

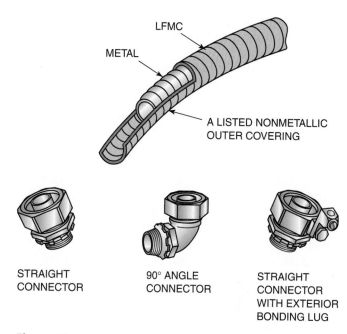

LFMC

METAL

A LISTED NONMETALLIC OUTER COVERING

STRAIGHT CONNECTOR

90° ANGLE CONNECTOR

STRAIGHT CONNECTOR WITH EXTERIOR BONDING LUG

Figure 12–18 LFMC and associated fittings.

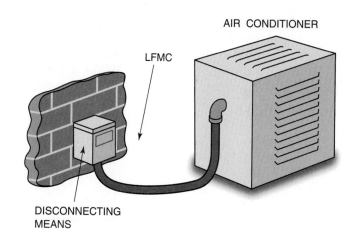

AIR CONDITIONER

LFMC

DISCONNECTING MEANS

Figure 12–19 A common LFMC application in residential wiring.

openings through framing members when the framing members are located at intervals not greater than 4½ feet (1.4 m).

FMC can be used as a grounding means, but there are a few requirements in Section 348.60 that have to be met first. According to Underwriters Laboratories, FMC longer than 6 feet (2 m) has not been judged to be suitable for grounding purposes. Therefore, any installation of FMC over 6 feet (2 m) in length will require an installed grounding conductor (Figure 12–17). The general rules for permitting or not permitting FMC for grounding purposes are found in Section 250.118(5) and (6). One specific exception is where FMC is used for flexibility. An additional equipment-grounding conductor is *always* required where FMC is used for flexibility. Examples of such installations include using FMC to minimize the transmission of vibration from equipment such as motors or to provide flexibility for floodlights, spotlights, or other equipment that requires adjustment. According to Section 250.118(6), where the length of the total ground-fault return path does not exceed 6 feet (2 m) and the circuit overcurrent protection does not exceed 20 amperes and connectors that are approved for grounding are used on each termination end, a separate equipment-grounding conductor does not have to be installed with the circuit conductors.

LFMC

Article 350 covers the installation requirements for LFMC. Section 350.2 defines LFMC as a raceway of circular cross section having an outer liquid-tight, nonmetallic, sunlight-resistant jacket over an inner flexible metal core with associated couplings, connectors, and fittings for the installation of electric conductors (Figure 12–18). LFMC is intended for use in wet locations for connections to equipment located outdoors, such as air-conditioning equipment (Figure 12–19). LFMC may be installed in unlimited lengths, provided it meets the other requirements of Article 350 and a separate equipment-grounding conductor is installed with the circuit conductors.

Section 350.22 covers the requirements for determining the maximum number of conductors allowed in a specific size of LFMC. The procedures used are exactly the same as for the raceways we have previously discussed, with the exception of using Table C7 in Annex C when determining the maximum number of conductors that are all the same size and with the same insulation type. Also, you must use the LFMC section of Table 4 in Chapter 9 when determining the minimum-size LFMC for conductors of different sizes that have different insulation types. There is one other difference, and that is the use of Table 348.22 when determining the maximum number of conductors in a 3/8-inch trade-size LFMC. This table must be used because the 3/8-inch size is not included in Table C7 of Annex C. For LFMC only, use the "Fittings Outside Conduit" columns in Table 348.22.

As with FMC, a run of LFMC installed between boxes, conduit bodies, and other electrical equipment is not permitted to contain more than the equivalent of 360 degrees total. Proper shaping and support of this flexible wiring method will ensure that conductors can be easily installed or taken out at any time. These requirements are found in Sections 350.24 and 350.26.

The securing and supporting requirements are given in Section 350.30. They are exactly the same as for FMC. LFMC must be securely fastened in place and supported by an approved means within 12 inches (300 mm) of each box, cabinet, conduit body, or other conduit termination and be supported and secured at intervals not to exceed 4½ feet (1.4 m). The supporting and securing rules do not have to be followed when LFMC is fished in a wall or ceiling or lengths do not exceed 3 feet (900 mm) at terminals where flexibility is required. Horizontal runs of LFMC are considered supported by openings through framing members when the framing members are located at intervals not greater than 4½ feet (1.4 m).

Section 350.60 allows LFMC to be used as a grounding means, but there are a few requirements that have to be met first (Figure 12–20). Like FMC, Underwriters Laboratories has determined that lengths of LFMC longer than 6 feet (2 m) have not been judged to be suitable for grounding. Therefore, any installation of LFMC over 6 feet (2 m) in length will require an installed grounding conductor. The general rules for permitting or not permitting FMC for grounding purposes are found in Section 250.118(7). One specific exception is where LFMC is used for flexibility. An additional equipment-grounding conductor is *always* required where LFMC is used for flexibility. According to Section 250.118(7)(b), where the LFMC is size 3/8 through 1/2 inch and the circuit overcurrent

protection does not exceed 20 amperes, the raceway itself is allowed to be the equipment-grounding conductor, provided the connectors used on each end are approved for grounding. According to Section 250.118(7)(c), where the LFMC is size 3/4 through 1¼ inches, and the circuit overcurrent protection does not exceed 60 amperes, the raceway itself is allowed to be the equipment-grounding conductor, provided the connectors used on each end are approved for grounding.

RNC

Article 352 covers the installation requirements for RNC. Section 352.2 defines RNC as a nonmetallic raceway of circular cross section, with integral or associated couplings, connectors, and fittings for the installation of electrical conductors (Figure 12–21). Two types are commonly used in residential wiring. Underwriters Laboratories recognizes the two types as: (1) Rigid Nonmetallic Schedule 40 and (2) Schedule 80 PVC conduit. Rigid nonmetallic Schedule 40 PVC conduit is suitable for underground use by direct burial or encasement in concrete. Unless marked "Underground Use Only" or equivalent wording, Schedule 40 conduit is also suitable for aboveground use indoors or outdoors exposed to sunlight and weather where not subject to physical damage. Schedule 80 conduit is suitable for use wherever Schedule 40 conduit may be used. The marking "Schedule 80" identifies the conduit as suitable for use where exposed to physical damage. Unless marked for a higher temperature, RNC is in-

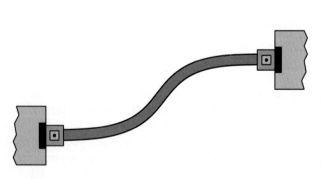

A PERMITTED AS A GROUNDING MEANS IF IT IS NOT OVER TRADE SIZE 1¼, IS NOT OVER 6 FT. (1.8 m) LONG, AND IS CONNECTED BY FITTINGS LISTED FOR GROUNDING. THE 6 FT. (1.8 m) LENGTH INCLUDES THE TOTAL LENGTH OF ANY AND ALL FLEXIBLE CONNECTIONS IN THE RUN.

B MINIMUM TRADE SIZE ½. TRADE SIZE 3/8 PERMITTED AS A LUMINAIRE (FIXTURE) WHIP NOT OVER 6 FT. (1.8 m) LONG.

C WHEN USED AS THE GROUNDING MEANS, THE MAXIMUM OVERCURRENT DEVICE IS 20 AMPERES FOR TRADE SIZES 3/8 AND ½, AND 60 AMPERES FOR TRADE SIZES 3/4, 1, AND 1¼.

B NOT SUITABLE AS A GROUNDING MEANS UNDER ANY OF THE FOLLOWING CONDITIONS:
• TRADE SIZES 1½ AND LARGER
• TRADE SIZES 3/8 AND ½ WHEN OVERCURRENT DEVICE IS GREATER THAN 20 AMPERES
• TRADE SIZES 3/8 AND ½ WHEN LONGER THAN 6 FT. (1.8 m)
• TRADE SIZES 3/4, 1 AND 1¼ WHEN OVERCURRENT DEVICE IS GREATER THAN 60 AMPERES
• TRADE SIZES 3/4, 1 AND 1¼ WHEN LONGER THAN 6 FT. (1.8 m)
FOR THESE CONDITIONS, INSTALL A SEPARATE EQUIPMENT-GROUNDING CONDUCTOR SIZED PER *TABLE 250.122.*

Figure 12–20 LFMC used as a grounding means.

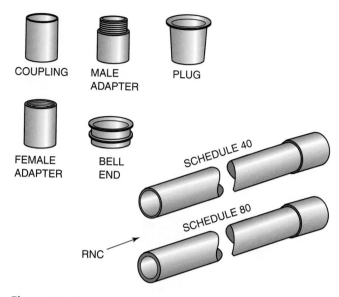

Figure 12–21 RNC and associated fittings.

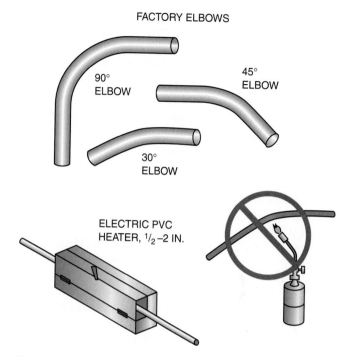

Figure 12–22 When electricians make field bends with RNC, they must use equipment that is designed for the purpose. Some electricians find it easier to simply buy factory elbows already bent.

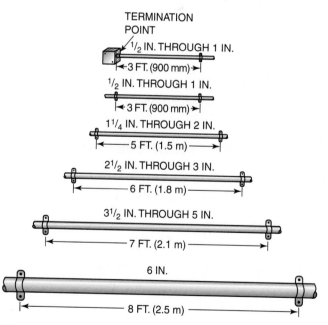

Figure 12–23 Support requirements for RNC.

tended for use with wires rated 75°C or less. PVC conduit is designed for connection to couplings, fittings, and boxes by the use of a suitable solvent-type cement. Instructions supplied by the solvent-type cement manufacturer describe the method of assembly and precautions to be followed.

Section 352.22 covers the requirements for determining the maximum number of conductors allowed in a specific size of RNC. The procedures used are exactly the same as for the other solid-length conduits discussed in this chapter with the exception of using Table C9 in Annex C when determining the maximum number of conductors that are all the same size and with the same insulation type. Also, you must use the RNC section of Table 4 in Chapter 9 when determining the minimum-size RNC for conductors of different sizes that have different insulation types.

Section 352.24 requires that when bending RNC, the bends must be made so that the conduit is not damaged and the internal diameter of the conduit is not effectively reduced. **Field bends** must be made only with bending equipment identified for the purpose (Figure 12–22).

Section 352.26 limits the number of bends in one conduit run from one box to another to no more than 360 degrees total. Limiting the number of bends in a conduit run will reduce the pulling tension on conductors and help ensure easy insertion or removal of conductors.

When cutting RNC, Section 352.28 requires all cut ends to be reamed or otherwise finished to remove rough edges. The rough edges should be removed on both the inside and outside of the RNC.

Section 352.30 requires RNC to be installed as a complete system and to be fastened so that movement from thermal expansion or contraction is permitted. Expansion and contraction caused by temperature changes can cause damage to the raceway, its supports, and the electrical boxes the RNC is attached to. Expansion fittings should be used, and the sup-

ports must be installed to allow expansion or contraction cycles without damage. RNC must be securely fastened and supported in accordance with the following (Figure 12–23):

- RNC must be securely fastened within 3 feet (900 mm) of each outlet box, junction box, device box, conduit body, or other conduit termination.

- RNC must be supported as required in Table 352.30(B).
- Horizontal runs of RNC supported by openings through framing members at intervals not exceeding those in Table 352.30(B) and securely fastened within 3 feet (900 mm) of termination points are permitted.

Expansion fittings for RNC are covered in Section 352.44 and are required to compensate for thermal expansion and contraction where the length change, in accordance with Table 352.44(A) or (B), is expected to be 1/4 inch (6 mm) or greater in a straight run between securely mounted items such as boxes, cabinets, elbows, or other conduit terminations (Figure 12–24). Expansion fittings are generally provided in exposed runs of rigid nonmetallic conduit where (1) the run is long, (2) the run is subjected to large temperature variations during or after installation, or (3) expansion and contraction measures are provided for the building or other structures.

RNC exhibits a considerably greater change in length per degree change in temperature than do metal raceway systems. In some parts of the United States and other countries, outdoor temperature variations of over 100°F are common. According to Table 352.44(A), a 100-foot run of RNC PVC will change 4.1 inches in length if the temperature change is 100°F. The normal expansion range of most larger sizes of rigid nonmetallic conduit expansion couplings is generally 6 inches. Information concerning installation and application of expansion couplings is found in the manufacturer's instructions.

Where RNC enters a box, fitting, or other enclosure, a bushing must be provided to protect the wire from abrasion unless the design of the box, fitting, or enclosure is such as to afford equivalent protection, according to Section 352.46. Section 300.4(F) requires that an insulated bushing (like plastic) be used for the protection of conductors sizes 4 AWG and larger that is installed in conduit.

Since RNC is made of a nonconductive material (PVC), Section 352.60 requires a separate equipment-grounding conductor to always be installed in the conduit.

Electrical Nonmetallic Tubing

Article 362 covers electrical nonmetallic tubing (ENT). Section 362.2 defines ENT as a nonmetallic pliable corrugated raceway of circular cross section with integral or associated couplings, connectors, and fittings for the installation

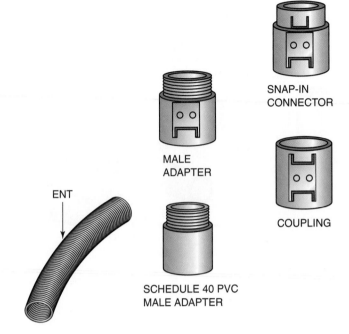

Figure 12–25 **ENT with associated fittings.**

of electric conductors (Figure 12–25). ENT is composed of a material that is resistant to moisture and chemical atmospheres and is flame retardant. A pliable raceway is a raceway that can be bent by hand with a reasonable force and does not require a special bending tool. Because of the corrugations, the raceway can be bent by hand and has some degree of flexibility. ENT is made of the same material (PVC) used for RNC. The outside diameter of ENT for 1/2- through 2-inch trade size is made so that standard couplings and other fittings for RNC can be used if an electrician so desires. ENT is suitable for the installation of conductors having a temperature rating as indicated on the ENT. However, the maximum allowable ambient temperature is 122°F.

Section 362.22 covers the requirements for determining the maximum number of conductors allowed in a specific size of ENT. The procedures used are exactly the same as for EMT with the exception of using Table C2 in Annex C when determining the maximum number of conductors that are all the same size and with the same insulation type. Also, you must use the ENT section of Table 4 in Chapter 9 when determining the minimum-size ENT for conductors of different sizes that have different insulation types.

Section 362.24 requires bends in ENT to be made so that the tubing is not damaged and the internal diameter of the tubing is not effectively reduced, and, like all the other circular raceways discussed in this chapter, Section 362.26 does not allow more than the equivalent of 360 degrees total bending between termination points of ENT.

Like the other raceway types discussed in this chapter, Section 362.28 requires all cut ends of ENT to be reamed or otherwise finished to remove rough edges.

EXPANSION FITTING

Figure 12–24 **Expansion fittings must be used when there will be thermal expansion or contraction of 1/4 inch or more in a run of RNC.**

Section 362.30 covers securing and supporting of ENT. It states that ENT must be installed as a complete system and be securely fastened in place and supported according to the following:

- ENT must be securely fastened at intervals not exceeding 3 feet (900 mm). In addition, ENT must be securely fastened in place within 3 feet (900 mm) of each outlet box, device box, junction box, cabinet, or fitting where it terminates. Where ENT is run on the surface of framing members, it is required to be fastened to the framing member every 3 feet (900 mm) and within 3 feet (900 mm) of every box (Figure 12–26).
- Horizontal runs of ENT supported by openings in framing members at intervals not exceeding 3 feet (900 mm) and securely fastened within 3 feet (900 mm) of termination points is permitted (Figure 12–27).

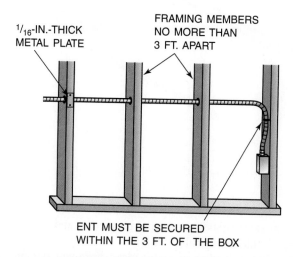

Figure 12–26 Securing and support requirements for ENT.

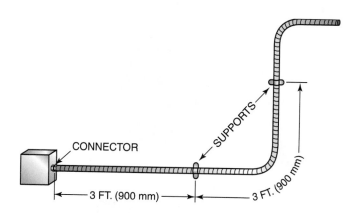

ENT MUST BE SECURED
WITHIN THE 3 FT. OF THE BOX

Figure 12–27 ENT run horizontally through framing members is considered supported as long as the framing members are no more than 3 feet (900 mm) apart. Section 300.4(A)(1) requires a 1/16-inch-thick metal plate if the ENT is run through a bored hole that is less that 1¼ inches (31 mm) from the edge of the framing member.

Where ENT enters a box, fitting, or other enclosure, a bushing must be provided to protect the wire from abrasion unless the design of the box, fitting, or enclosure is such as to afford equivalent protection, according to Section 362.46. Section 300.4(F) requires that an insulated bushing (like plastic) be used for the protection of conductors sizes 4 AWG and larger that is installed in conduit.

Since Electrical Nonmetallic Tubing (ENT) is made of a nonconductive material (PVC), Section 362.60 requires a separate equipment-grounding conductor to always be installed in the conduit.

Introduction to Cutting, Threading, and Bending Conduit

Cutting Conduit

Electricians will need to know how to cut to length the various conduit types used in residential wiring. Solid-length metal conduit, like RMC, IMC, and EMT, is cut to length using a pipe cutter, a tubing cutter, a portable band saw or simply a hacksaw. The ends of cut lengths of RMC and IMC will also have to be threaded. Figure 12–28 shows some of the tools used to cut and thread RMC and IMC. RNC can be cut using a hacksaw or with saws equipped with special blades for PVC (Figure 12–29). FMC and LFMC are usually cut with a hacksaw. ENT can also be cut with a hacksaw, but special nonmetallic tubing cutters that look like big scissors are available.

No matter what raceway type you encounter and have to cut, the following procedures should always be followed:

- Wear safety glasses and observe all applicable safety rules.
- Secure the raceway before you attempt to cut it.
- Mark the locations where you wish to make the cut clearly and accurately on the raceway.
- Choose a cutting tool that is appropriate for the raceway type you are using.
- Make the cut as straight as possible.
- File or ream the ends of the raceway after making the cut to remove any sharp or rough edges that could damage the conductor insulation.

 FROM EXPERIENCE

Some electricians cut FMC by intentionally bending it so tightly that it actually breaks open. A pair of diagonal cutting pliers is then used to complete cutting off the length needed. This method is not recommended because of the damage it can cause the raceway. Check with your supervisor to see if this is an acceptable method for cutting FMC in your area.

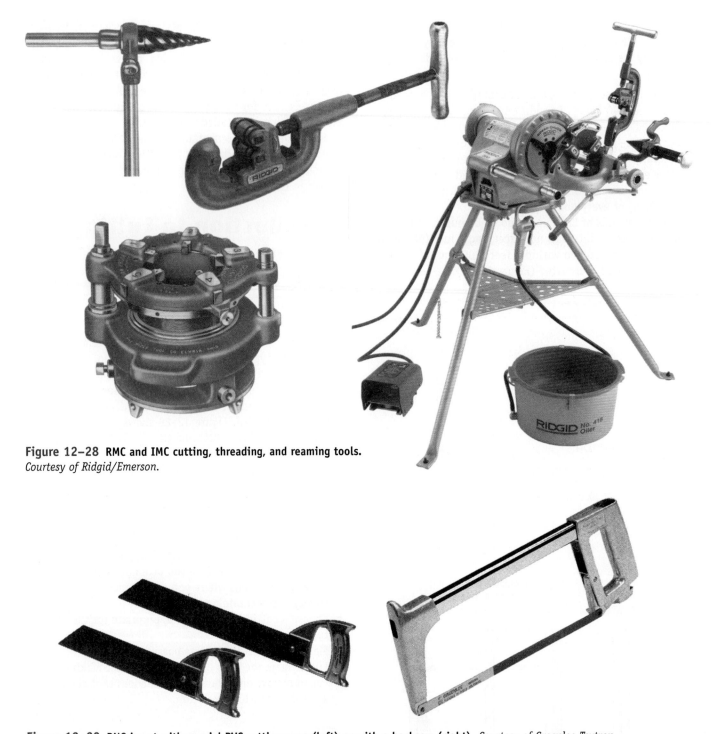

Figure 12–28 RMC and IMC cutting, threading, and reaming tools.
Courtesy of Ridgid/Emerson.

Figure 12–29 RNC is cut with special PVC cutting saws (left) or with a hacksaw (right). *Courtesy of Greenlee Textron.*

Threading Conduit

It does not happen very often in residential wiring, but once in a while an electrician will have to cut to length a piece of RMC or IMC and then thread the ends. Either hand threaders or power threading equipment like those shown in Figure 12–28 can be used. When cutting new threads on a conduit end, the following items should always be observed:

- Wear safety glasses and hand protection and observe all applicable safety rules.
- Secure the conduit in a pipe vise before you attempt to thread it.
- Choose the proper threading die for the size of conduit you are threading.
- Always use plenty of cutting oil during the threading process.

CAUTION: Always use plenty of cutting oil when threading a raceway. The cutting oil will allow the cutting die to cut the threads into the pipe easier. It extends the life of the cutting die and, when applied properly, will flush away the metal shavings produced during the threading process.

(See page 330 for the proper procedure to follow for cutting and then threading a piece of 1/2-inch RMC.)

Bending Conduit

Bending conduit is definitely a skill that improves with practice. The more you bend conduit, the better conduit bender you become. The most common electrical conduit installed in houses is EMT, and for this reason, the discussion that follows focuses on EMT. However, the bending techniques described also apply to the other types of circular metal raceway, such as RMC and IMC.

There are a few common bends that electricians will need to know. The most common bend is a 90-degree bend, commonly called a **stub-up** (Figure 12–30). When electricians refer to "stubbing up" a pipe, they mean that a 90-degree bend is to be made. A **back-to-back bend** is a type of bend where the distance is measured between the outside diameters of two sections of the pipe (Figure 12–31). This type of bend consists of putting a 90-degree bend on each end of one length of pipe, resulting in a straight section of pipe between the two 90-degree bends. This bending technique allows the conduit to fit precisely between two objects. An **offset bend** requires two equal bends in a conduit that results in the direction of the conduit being changed so that it can avoid an obstruction blocking the conduit run (Figure 12–32). A **saddle bend** is similar to an offset bend in that it results in a conduit run going around an object that is blocking the path of the run (Figure 12–33). The difference with a saddle bend is that it ac-

tually goes *over* the obstruction rather than *around* it. There are two styles of saddle bends: a three-point saddle and a four-point saddle. Later in this section, we discuss the actual bending of a three-point saddle since it is considered to be a little easier to do than the four-point saddle. The last type of bend we discuss is one that is used when conduits enter a box or other electrical enclosure that is surface mounted. It is called a **box offset bend** and is really just a smaller version of the regular offset bend (Figure 12–34). An installation where the

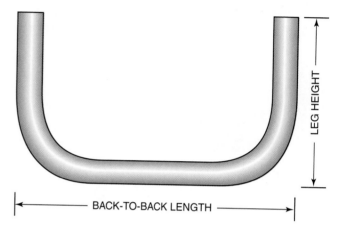

Figure 12–31 A back-to-back bend.

Figure 12–32 An offset bend.

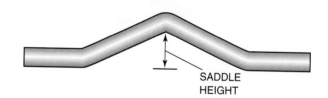

Figure 12–33 A three-point saddle bend.

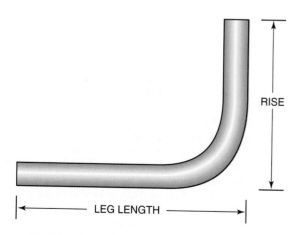

Figure 12–30 A stub-up bend.

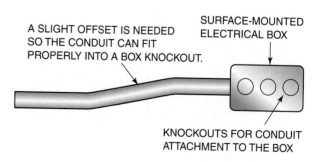

Figure 12–34 A box offset bend.

electrician has used box offsets at each box location is considered to be a neater and more professional installation.

EMT is bent in the field using either a hand bender or a hydraulic bender. Sizes from 1/2 inch through 1¼ inches can be bent with a hand bender (Figure 12–35). Larger sizes

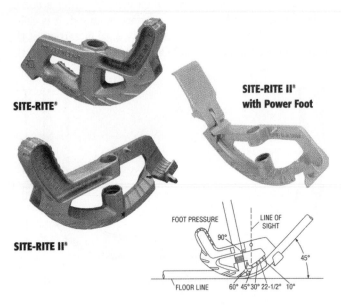

Figure 12–35 An example of a hand bender that can be used for EMT, RMC, and IMC. A handle is used with each bender shown. *Courtesy of Greenlee Textron.*

- STRONG PORTABLE PIN-ASSEMBLED ALUMINUM COMPONENTS.
- EASY-TO-USE RAM TRAVEL SCALE AND BENDING CHARTS.
- CONDUIT SUPPORTS INDEX TO SUIT ALL SIZES.
- MODEL AVAILABLE TO BEND PVC-COATED RIGID CONDUIT.

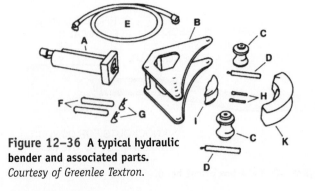

Figure 12–36 A typical hydraulic bender and associated parts. *Courtesy of Greenlee Textron.*

will require the use of a hydraulic or electric power bender (Figure 12–36). Since most EMT installed in houses will be 1/2-, 3/4-, or 1-inch trade sizes, we will keep our discussion focused on bending with a hand bender.

An EMT bender has many features (Figure 12–37). A hook at the front of the bender is used to help secure and correctly place the bender onto a piece of conduit. The opposite end of the bender will have a foot pedal that is designed to allow an electrician to exert foot pressure on the bender while making a bend. The foot pressure does two things: It helps keep the bender seated against the conduit, and it actually causes the bend to be made. A handle is attached to the bender and is used more for allowing the hands and arms to guide the bender during the bending operation than it is for bending the pipe. An EMT bender also has many markings on it, and an electrician will need to know what each of the markings indicate. The bender will usually have a degree scale marked on the bender in 22-, 30-, 45-, 60-, and 90-degree increments. The degree indicator will help an electrician determine the amount of bend put into a piece of EMT. There is also an arrow mark on the bender that is located toward the hook end. The arrow is aligned with bending marks placed on the conduit at the locations needed for a specific type of bend, like a 90-degree stub bend. There will be a mark that indicates the location of the back of a bend. On most EMT benders, this mark is a star with one of the star points being longer than the other points. The longer star

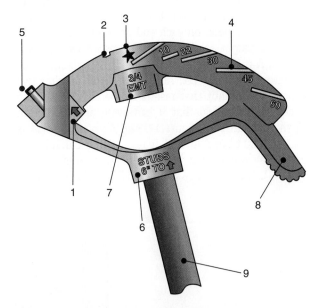

Figure 12–37 EMT hand bender features. (1) Arrow: used with stub-up, offset, and the outer marks of saddle bends; (2) rim notch: used to bend the center bend of a three-point saddle; (3) star point: indicates the back of a 90-degree bend; (4) degree scale: used to indicate the amount of angle in the bend; (5) hook: used to help secure the bender to the pipe being bent; (6) take-up: indicates the amount of take-up for the bender head; (7) pipe size; indicates the sizes of conduit the bender can bend; (8) foot pedal; (9) handle.

point is aligned with the marking on the conduit where the back of a bend will be. The last marking on an EMT bender that you need to be aware of is commonly called a "rim notch." This mark is aligned with the marking on a conduit that represents the center of a three-point saddle bend. The amount of **take-up** for the bender is also marked on it. Take-up is the amount that must be subtracted from a desired stub up height so the bend will come out right. Take-up must be taken into account because in a 90-degree bend, the actual conduit bend is in an arc and not a true right angle. The arc, or sweep, of the bend is due to the shape of the bender. The take-up for a 1/2-inch EMT bender is 5 inches and for a 3/4-inch bender is 6 inches.

When actually making a bend using a hand bender, the following procedures should always be followed:

- Wear safety glasses and observe all applicable safety rules.
- Bend on a flat surface that is not slippery.
- Mark the locations on the conduit where you wish to make the bends clearly and accurately.
- Apply heavy foot pressure on the foot pedal to keep the conduit tightly in the bender.
- When making multiple bends on the same pipe length, keep all bends in the same plane.

(See page 333 for the proper procedure to follow for bending a stub-up in a length of 1/2-inch EMT.)

(See page 335 for the proper procedure to follow for bending a back-to-back bend in a length of 1/2-inch EMT.)

(See page 337 for the proper procedure to follow for bending an offset bend in a length of 1/2-inch EMT.)

(See page 340 for the proper procedure to follow for bending a three-point saddle in a length of 1/2-inch EMT.)

(See page 343 for the proper procedure to follow for bending box offsets in a length of 1/2-inch EMT.)

Installation of Raceway in a Residential Wiring System

In this section, we discuss the installation of raceway in a residential wiring system. The raceway will contain the conductors for the various branch circuits used throughout the house. As we mentioned earlier, EMT is the most commonly used raceway type in residential work, and for this reason we limit our discussion in this section to the installation of EMT.

When EMT is used, only metal outlet and device boxes can be used. The EMT will be connected to the boxes with approved fittings, called "connectors" (Figure 12–38). The setscrew type of connector is used most often because of its ease of installation and lower cost. Some electricians like to use the compression type of connector, which is a little more

Figure 12–38 Setscrew and compression-type EMT connectors.

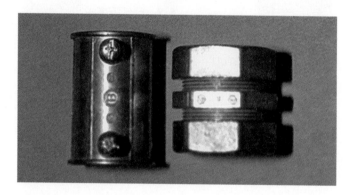

Figure 12–39 Setscrew and compression-type EMT couplings.

expensive to buy. Either connector type will provide a solid and secure connection of the conduit to an electrical box. When lengths of EMT need to be coupled together, only approved couplings may be used. Just like the connectors, couplings are available in a setscrew or a compression type (Figure 12–39). In locations where EMT is used outdoors, the compression connectors and couplings must be used for a watertight connection. Setscrew connectors and couplings are not allowed in wet locations.

As we learned earlier, EMT is an approved grounding method and, as such, does not always require an equipment-grounding conductor to be run in the raceway with the other circuit conductors. Most electricians will run a green insulated grounding conductor in the raceway. Always check with your local electrical inspector to determine if an insulated equipment-grounding conductor is always required or if it is left up to the electrician to decide. Electricians choose the conductor insulation color for the conductors they install in the raceway. Usually, the same color coding found in cables is used. For a 120-volt branch circuit, an electrician will use a white insulated wire and a black insulated wire. For a straight 240-volt circuit (like an electric water heater), two black conductors or a black and a red conductor will usually be used. If the circuit is a 120/240-volt circuit (like an electric clothes dryer), a white insulated wire, a black insulated wire, and a red insulated wire will be run. Remember, a green insulated equipment-grounding wire is not always

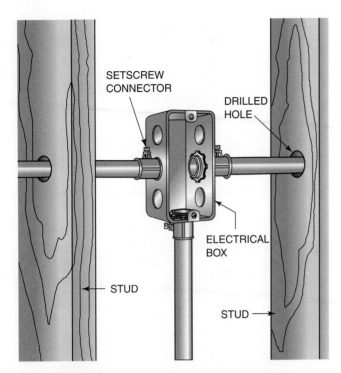

Figure 12-40 EMT installed through holes in building framing members.

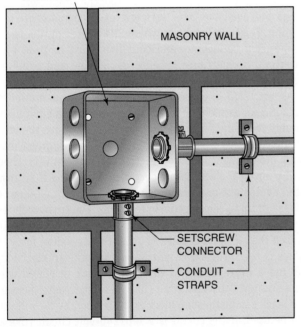

Figure 12-41 EMT run on the surface.

required when running conductors in an EMT conduit, but it is recommended.

EMT can be installed though drilled holes in studs, joists, and rafters like Type NM cable (Figure 12-40). It can also be installed or on a wall or on a ceiling's finished surface (Figure 12-41). When installing an EMT raceway system, it is common practice to not fully tighten boxes to the framing members until after the conduit has been attached to them. This allows for some fine-tuning of the boxes, conduit, and fittings. After any adjustments have been made, the boxes are then fully secured to the framing members.

Raceway Conductor Installation

Once the raceway system has been installed, the electrician must install the various branch-circuit conductors in the conduit. The conductors are usually pulled into the conduit, but in shorter runs between electrical boxes, the conductors may be pushed through the raceway.

The conductors will be taken off spools. The electrician will need to set up the pull by arranging the spools of the required circuit conductors in such a way that the conductors on the spools can be easily pulled off without becoming tangled with each other. One of the easiest ways to do this is to use a commercially available wire cart that allows several spools of wire to be put on them at one time (Figure 12-42). Once the electrician has determined how many wires must be installed in the conduit run, a few feet of each wire are pulled from the spools on the wire cart, and the ends are taped together. The taped end of the wires is then inserted into the raceway and pushed to the electrical boxes. If the length of conduit between boxes is fairly long, a fish tape must be used (the fish tape was introduced in Chapter 3). It

 FROM EXPERIENCE

Some residential electrical contractors do not encounter raceway wiring very often. As such, they probably do not have a commercially available wire cart. If this is the case, an electrician can cut a length of 1/2-inch EMT conduit, put the pipe through the holes in the wire spools, and then use a step ladder (or two step ladders) to support the pipe with the spools on it. The electrician can then pull the wires from the spools as needed. Another common practice is to support the pipe length with the spools on it by a couple of building framing members.

Figure 12–42 A commercially available wire cart that can hold several spools of wire at one time. The cart allows the wires to be easily pulled off the spools without tangling.

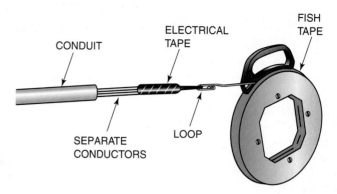

Figure 12–43 A fish tape is used to pull wires through a length of conduit.

CAUTION

CAUTION: Do not use a metal fish tape to pull or push conductors in a raceway that is connected to an energized load center. Nonconductive fish tapes are available for this wiring situation.

Summary

In this chapter, we have taken a look at the different raceway types that are used in residential wiring. The *NEC®* installation requirements for these raceways were covered in detail, and although there are many similarities in the raceway installation requirements, it is important for a residential electrician to recognize the specific rules for the type of conduit being installed. Since EMT is the most often used raceway system in a residential wiring installation, the proper bending and installation techniques for it were explained and examples given. The last thing covered in this chapter was some common installation techniques for installing the conductors in a completed raceway system.

Even though many electricians who do residential wiring seldom have the opportunity to install raceway, they must still be aware of the common raceway types and installation practices for those times that they do encounter it. State and local electrical licensing exams also have many questions about raceways and their installation requirements. For electricians who live and work in an area of the country where they have to install a residential electrical system using a raceway wiring method or for residential electricians who work with raceways only once in a while, following the information presented in this chapter will make for a professionally installed and *NEC®*-compliant raceway installation.

is made of a flexible metal or nonmetal tape that is enclosed in a metal or plastic enclosure. The tape can be pulled out of the enclosure and inserted into a raceway and pushed through it until it comes out at a box location. The fish tape will have a hook on the end of it, and the conductors are attached by the electrician to the fish tape end. While one electrician pulls the conductors slowly off the spools, another electrician will pull the fish tape with the attached conductors back through the raceway (Figure 12–43). This process of either pushing the conductors through the conduit or pulling them in with a fish tape is repeated until all the required conductors are installed in the raceway system.

Procedures

- Wear safety glasses and hand protection and observe all applicable safety rules.

A Secure a length of 1/2-inch RMC in a pipe vise.

B Measure and clearly mark the pipe at the location you wish to make the cut.

C Prepare to cut the conduit using a pipe cutter. Do this by loosening the cutter up so it will easily go over the pipe. Place the cutter at the cutting mark and tighten the cutter onto the conduit until the cutting blade is tight against the pipe.

Cutting and Threading

Follow these procedures for cutting and then threading a length of 1/2-inch RMC.

A

B

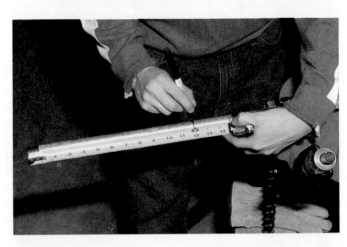

C

- Rotate the pipe cutter around the conduit. Tighten the pipe cutter onto the conduit a little more each time you make a complete rotation around the pipe.

 Continue to alternately rotate and tighten the cutter until you have completely cut off the end of the conduit.

 Using an appropriate reaming tool, ream the conduit and, if necessary, use a file to eliminate all burrs on the inside and outside of the conduit ends.

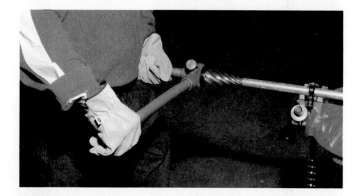

 Using the hand threader with the proper size die, place the cutting die on the end of the conduit. Most hand threader handles are of the ratchet type. Make sure it is set to turn in the right direction.

Procedures   Cutting and Threading (continued)

G To get it started, use one hand to exert some force on the end of the die and at the same time use your other hand to rotate the threader handle in a clockwise direction. Once the cutting die starts to cut the threads on the end of the conduit, there will be no need to continue pushing the die onto the conduit end.

• Continue to cut the new threads onto the conduit until the length of the cut thread is the same as the length of the cutting threads of the die. Do not forget to apply a good amount of cutting oil during the threading process.

H Now change the direction of the ratcheting handle and slowly remove the cutting die from the end of the conduit. Be careful, as the newly cut threads are sharp.

Procedures

Bending a 90-Degree Stub-Up

- Wear safety glasses and observe all applicable safety rules.

- Subtract the "take-up" from the required finished stub height. The take-up for a 1/2-inch bender is 5 inches, so 12 inches − 5 inches = 7 inches.

 Measure back from the end of the EMT a distance of 7 inches. Mark this dimension clearly on the conduit.

- Place the conduit on a flat surface, such as the floor.

B Line up the arrow on the bender head with the mark on the conduit.

Follow these procedures for bending a 12-inch-high 90-degree stub-up in a 1/2-inch EMT conduit using a hand bender.

A
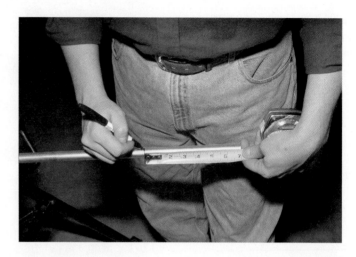

B
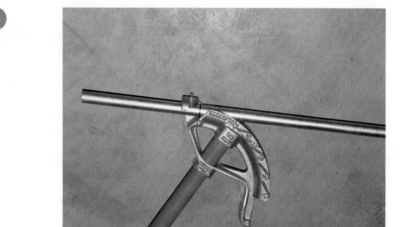

 FROM EXPERIENCE

It is a good idea is to use a pencil or a permanent marker for marking conduit. Be sure to wrap your mark all the way around the pipe in case a pipe is not bent correctly and the bender has to be placed back on the conduit for additional bending. This is called "fine-tuning" by some electricians. Adding or taking away some angle from a bend must be done at exactly the same spot as the original bend, or damage to the conduit will result.

Procedures

Bending a 90-Degree Stub-Up (continued)

 While applying constant foot pressure to the bender, bend the conduit to 90 degrees.

 Measure the stub height. You should have a stub that is 12 inches high.

C

D

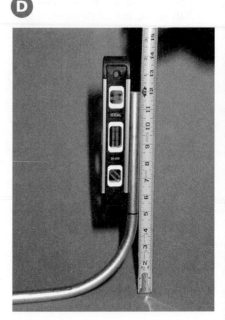

FROM EXPERIENCE

You can use the degree indicator on the bender to tell if you have a 90-degree bend, but a Torpedo level placed on the side of the pipe works well. Remember, a Torpedo level used in electrical work will have one side magnetized so that it will stick to the side of a metal conduit made of steel.

FROM EXPERIENCE

Do not be surprised if you come out a little over or a little under your target measurement when using a bender. They are fairly accurate but are not precise. Also, each bender will have its own characteristic that you will have to get used to. For example, if a bender keeps making bends that are consistently 1/4 inch *under* the target measurement, *add* 1/4 inch to the measurement that is marked on the conduit and then make the bend.

Procedures

- Wear safety glasses and observe all applicable safety rules.

- Subtract the "take-up" from the finished stub height for leg 1. Since the first leg is to be 25 inches high, 25 inches − 5 inches (1/2-inch EMT bender take-up) = 20 inches.

- Measure 20 inches back from one end of the conduit and mark this dimension clearly on the conduit.

- Place the conduit on a flat surface, such as the floor.

 Line up the arrow on the EMT bender with the mark on the conduit.

 Apply constant foot pressure to the bender and bend the conduit to 90 degrees. You should now have leg 1 bent to a 90-degree angle with a height of 25 inches.

Bending a Back-to-Back Bend

Follow these procedures for bending a back-to-back bend in a length of 1/2-inch EMT with a hand bender. For this example, leg 1 of the bend will be 25 inches high and leg 2 will be 30 inches high. The actual length of the bend from the outside of one leg to the outside of the other leg will be 48 inches.

| Procedures | Bending a Back-to-Back Bend (continued) |

C Now lay the conduit on a flat surface and measure 48 inches from the outside of leg 1 to the point where the back of the second bend is to be. Mark the conduit.

C
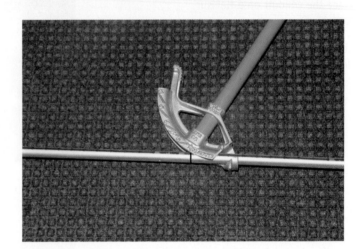

D Align this mark on the conduit with the starpoint (or the diamond) of the bender and bend to 90 degrees.

- Leg 2 must be 30 inches high. Simply cut off any extra conduit until the leg is the correct length.

D

E Measure the leg heights and the distance from the back of one leg to the back of the other leg. You should have a back-to-back bend with one leg that is 25 inches high, the other leg should be 30 inches high, and the back-to-back distance should be 48 inches. Adjust the pipe back-to-back bend as needed.

E

Procedures

Bending an Offset Bend

Follow these procedures for bending an offset bend in a 5-foot length of 1/2-inch EMT with a hand bender. The offset will be 4 inches high and will use 30-degree bends. It will be located in the center of the pipe length.

- Wear safety glasses and observe all applicable safety rules.

 Cut the conduit to the proper length.

 Make sure to ream the end of the conduit after cutting it to length.

Procedures

Bending an Offset Bend (continued)

C Determine the center of the pipe length and mark it.

C

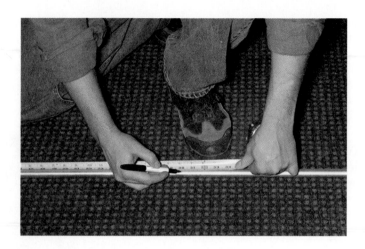

D Determine the distance between the two bends of the offset. Consult an offset table like the one shown. In this example, the distance is 8 inches.

D

Offset Bending Chart

Offset Depth in Inches	Degree of Offset Bends			
	22.5°	**30°**	**45°**	**60°**
2"	5¼"			
3"	7¾"	6"		
4"	10½"	**8"**		
5"	13"	10"	7"	
6"	15½"	12"	8½"	7¼"
7"	18¼"	14"	9¾"	8⅜"
8"	20¾"	16"	11¼"	9⅝"
9"	23½"	18"	12½"	10⅞"
10"	26"	20"	14"	12"

E Mark the pipe with the dimensions determined from the offset table on the conduit. Do this by making a mark an equal distance on each side of the center mark. The distance from the center of each mark is found by dividing the dimension from the offset table in half. For this example, 8 inches ÷ 2 = 4 inches.

● Place the conduit on a flat surface, such as the floor.

● Line up the arrow on the bender with the first mark on the conduit.

E

Apply heavy foot pressure to the bender and bend the conduit to 30 degrees. Note that when the bender handle is straight up, you have a 30-degree bend (this is true for most but not all hand benders).

 While keeping the conduit in the bender, invert the bender and place the handle end on the floor. Rotate the conduit 180 degrees, slide it ahead in the bender head, and align the arrow with the next mark. Then while standing up, bend the conduit to 30 degrees. Use the degree scale on the bender to determine when you have bent to 30 degrees.

- Make sure to keep the bends in the same plane. If they are not, fine-tune the offset as necessary.

 Check to be sure that the offset amount is correct. If it is not enough or is too much, add or subtract some angle from the bends. When adding or taking away angle from the bends, make sure to always bend on exactly the same spot on the conduit. Proper marking of the pipe at the beginning of the process is extremely important, especially if you need to make bending adjustments.

Procedures

Bending a Three-Point Saddle

- Wear safety glasses and observe all applicable safety rules.

- Cut the conduit to the proper length.

- Make sure to ream the end of the conduit after cutting it to length.

- Determine the mark spacing for the 4-inch three-point saddle. There will be three marks made on the conduit: a center mark and a mark placed on either side of the center.

A First determine the location on the conduit where you want the center of the three-point bend and mark it.

B Next determine the dimension of the two outer marks from the center mark. The multiplier is 2.5 for all pipe sizes when using a 45-degree center bend and two 22.5-degree outer bends or a 60-degree center bend and two 30-degree outer bends. Figure B shows actual mark spacing for saddle heights of 1 inch through 6 inches. In this example, 10 inches is the distance from the center mark for each outside mark. This is found by multiplying the desired saddle height by 2.5: 4 inches × 2.5 = 10 inches.

C Mark this dimension on the conduit by making a mark 10 inches on each side of the center mark.

- Place the conduit on a flat surface, such as the floor.

Follow these procedures for bending a three-point saddle bend in a 5-foot length of 1/2-inch EMT with a hand bender. The saddle must go over a 4-inch obstruction, like a plumbing pipe, and the saddle will be bent using a 45-degree center bend and two 22.5-degree side bends. This is considered the most common three-point saddle type.

A
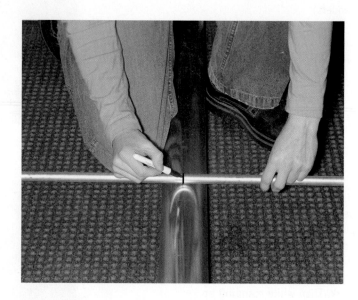

B

Saddle Bending Marks

If the obstruction to be saddled over is:	Make the outside marks on either side of the center mark:
1"	2½"
2"	5"
3"	7½"
4"	10"
5"	12½"
6"	15"

C
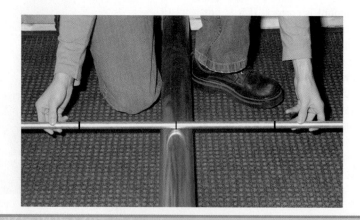

D Line up the *rim notch* or *saddle mark* on the bender with the center mark on the conduit.

- With the conduit placed on the floor, apply heavy foot pressure to the bender and bend the conduit to 45 degrees.

D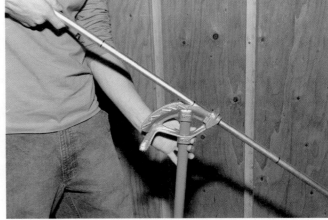

E While keeping the conduit in the bender head, invert the bender so the handle is now on the floor and rotate the conduit 180 degrees.

E

Procedures

Bending a Three-Point Saddle (continued)

F Slide the conduit ahead and align the *arrow* with the first outer mark.

G While standing up, bend the conduit to 22.5 degrees.

- Remove the conduit from the bender and reverse it end to end.

H Insert the other end of the conduit into the bender. Align the remaining outside mark at the arrow and make another 22.5-degree bend.

I Check the saddle bend and fine-tune as necessary. Be sure to have all bends in the same plane.

F

G

H

I

Procedures

- Wear safety glasses and observe all applicable safety rules.

- Cut the conduit to the proper length for attachment to an electrical box or enclosure.

- Make sure to ream the end of the conduit.

 Mark the pipe at 2 inches and 8 inches from the end of the pipe.

 Invert the bender by placing the handle on the floor and placing the 2-inch mark at the arrow; make a small bend (approximately 5 degrees).

Box Offsets

Follow these procedures for bending a box offset at the end of a 1/2-inch EMT with a hand bender.

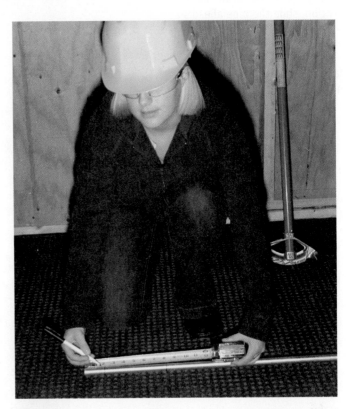

Procedures

Box Offsets (continued)

C Rotate the pipe 180 degrees and, keeping the bender in the same position, slide the pipe ahead so that the 8-inch mark now lines up with the arrow.

- Make a small bend that is equal to the first bend made.

C

 FROM EXPERIENCE

After electricians have installed box offsets enough times, they find that there is no need to make marks on the conduit for the location of the small bends. However, it is a good idea for electricians to always mark the pipe for the box offset bends so that if some adjustment has to be made, the pipe can be aligned in the bender at the same spots as the original bends. This will help eliminate kinking at the bends.

Check the box offset by placing the conduit at the knockout opening where you want to secure it and see if it will slide into the opening without hitting the sides of the box. If it does not, fine-tune the pipe by adding or taking away offset height until the pipe easily fits into the knockout opening.

 FROM EXPERIENCE

Some electricians use a commercially available mechanical box offset bending tool. It works on 1/2- and 3/4-inch trade-size EMT. The conduit is simply inserted in the bender, and the bender handle is depressed, which causes a perfect box offset to be formed in the pipe.

Courtesy of Greenlee Textron.

Review Questions

Directions: Answer the following items with clear and complete responses.

1 Name the article in the *NEC®* that covers RMC.

2 What does the *NEC®* have to say about reaming and threading of RMC?

3 Describe the support requirements of RMC as outlined in the *NEC®*.

4 Discuss the reasons for using cutting oil when threading RMC.

5 List the minimum trade size of EMT for each of the following applications.

Assume the use of conductors with THHN insulation.

a. Six 14 AWG _____

b. Six 12 AWG _____

c. Three 10 AWG _____

d. Four 8 AWG _____

6 Name the article of the *NEC®* that covers EMT.

7 List the minimum trade size EMT for each of the following applications.

a. Three 14 AWG THHN, four 12 AWG THHN _____

b. Three 12 AWG THHN, four 8 AWG TW _____

c. Five 12 AWG THW, three 10 AWG THWN, three 8 AWG THHN _____

8 Describe the support requirements for EMT as outlined in the *NEC®*.

9 Name the article in the *NEC®* that covers IMC.

10 IMC can be used in all the locations where RMC can be used. (Circle the correct answer.) True or False?

11 Describe the support requirements for IMC.

12 Name the article in the *NEC®* that covers RNC.

13 Describe the support requirements for RNC as outlined in the *NEC®*.

14 Name the article in the *NEC®* that covers ENT.

15 Describe the support requirements for ENT as outlined in the *NEC®*.

16 Name the article in the *NEC®* that covers FMC.

17 Describe the support requirements for FMC as outlined in the *NEC®*.

18 Name and describe the four basic conduit bend types used in residential wiring.

19 An electrician must install branch-circuit conductors in a metal raceway that will supply a 120/240-volt electric range. The raceway will not be used as the equipment-grounding conductor. How many conductors must the electrician install? What would be the colors for the required conductors?

20 Name the article in the *NEC®* that covers LFMC.

Chapter 13 | Switching Circuit Installation

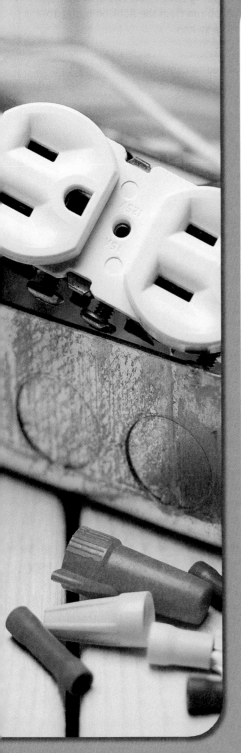

In residential wiring systems, switches are used to control various electrical loads. Understanding the common connection techniques for each switch type is very important for residential electricians because they must know how many conductors to include in the wiring method installed during the rough-in stage between switches and the loads they control. This chapter presents an overview of the different types of switches commonly used in residential wiring as well as the common connection techniques for each switch type. Lighting outlets are the most common loads controlled by switches in a residential wiring system. Since this is the case, the 120-volt switching circuits covered in this chapter are limited to common switching circuits with lighting outlets as the load. Double-pole switching of 240-volt loads is also covered.

OBJECTIVES

Upon completion of this chapter, the student should be able to:

- select an appropriate switch type for a specific residential switching situation.
- select a switch with the proper rating for a specific switching application.
- list several *National Electrical Code®* (*NEC®*) requirements that apply to switches.
- demonstrate an understanding of the proper installation techniques for single-pole, three-way, and four-way switches.
- demonstrate an understanding of the proper installation techniques for switched duplex receptacles, combination switches, and double-pole switches.
- demonstrate an understanding of the proper installation techniques for installing dimmer switches and ceiling-suspended paddle fan/light switches.

Glossary of Terms

combination switch a device with more than one switch type on the same strap or yoke

dimmer switch a switch type that raises or lowers the lamp brightness of a lighting fixture

double-pole switch a switch type used to control two separate 120-volt circuits or one 240-volt circuit from one location

four-way switch a switch type that, when used in conjunction with two three-way switches, will allow control of a 120-volt lighting load from more than two locations

single-pole switch a switch type used to control a 120-volt lighting load from one location

switch loop a switching arrangement where the feed is brought to the lighting outlet first and a two-wire loop run from the lighting outlet to the switch

three-way switch a switch type used to control a 120-volt lighting load from two locations

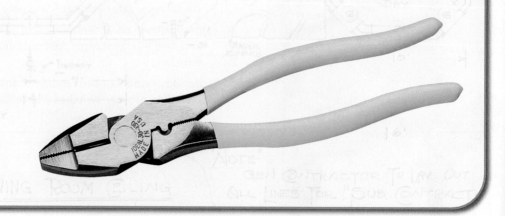

Selecting the Appropriate Switch Type

In Chapter 2, you were introduced to the different types of switches used in residential wiring. Single-pole, double-pole, three-way, and four-way switches were all discussed, and an explanation was given about how each of these common switch types work. Specialty switches, such as dimmer switches and combination switches, were also discussed. A quick review of the switch types introduced in Chapter 2 is appropriate at this time.

The most common switch type used in residential wiring is a **single-pole switch** (Figure 13–1). This switch type is used in 120-volt residential circuits to control a lighting outlet load from one specific location. A **double-pole switch** (Figure 13–2) is used to control a 240-volt load in a residential wiring system, such as an electrical water heater. Like the single-pole switch, the double-pole switch can control the load from only one specific location. **Three-way switches** (Figure 13–3) are used to control lighting outlet loads from two separate locations, say, at the top and bottom of a stairway. In combination with three-way switches, **four-way switches** (Figure 13–4) allow for control of lighting outlet loads from three or more locations, such as in a room with three doorways and the wiring plan calls for a switch controlling the room lighting to be located at each doorway. A **dimmer switch** (Figure 13–5) can be found in both a single-pole and a three-way configuration and is used to brighten or dim a lamp or lamps in a lighting fixture. **Combination switch** (Figure 13–6) devices consist of two switches on one strap or yoke. This allows the placement of multiple switches in a single-gang electrical box. For example, the wiring plans may call for two switches—a single-pole switch and a three-way switch—at a location next to a doorway. However, when the electrician attempts to install a two-gang electrical box for the two required switches at that location, it is found that there is not enough space between the building framing members to attach the electrical box. An alternative installation would be to use a single-gang box and install a combination switch that has a single-pole switch and a three-way switch on the same strap.

When selecting the proper switch type for a residential lighting application, there are many factors that contribute to the decision. One factor that electricians base their switch selection on is the voltage and current rating of the circuit the switch is being used in. According to Section 404.15(A) of the *NEC®*, each switch must be marked with the current and voltage rating for the switch. The switch rating must be matched to the voltage and current you encounter with the

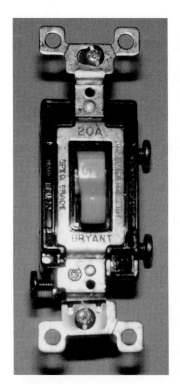

Figure 13–1 A single-pole switch is used to control a lighting load from one location.

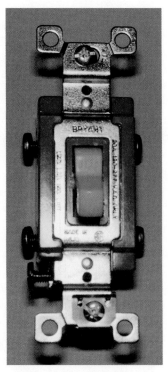

Figure 13–2 A double-pole switch is sometimes used in residential wiring to control a 240-volt load.

Figure 13–3 Three-way switches are used in pairs to control lighting loads from two separate locations.

Figure 13–4 Four-way switches, in combination with three-way switches, can control a lighting load from three or more locations.

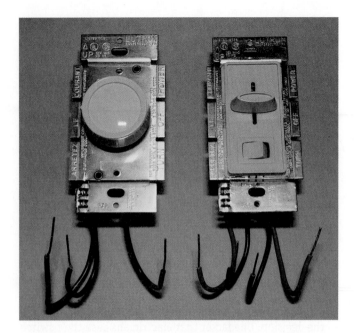

Figure 13–5 A dimmer switch is used to raise or lower the brightness of the light in a specific area of a house. Notice that dimmer switches have insulated pigtail wires rather than terminal screws.

circuit you are using the switch on. For example, a 20-amp-rated, 120-volt circuit wired with 12 AWG conductors having 16 amps of lighting load must have a switch with a voltage rating of at least 120 volts and a current rating of 20 amps (a standard switch rating). Many residential lighting circuits are wired with 14 AWG conductors protected with a 15-amp circuit breaker and will require switches with a 15-amp, 120-volt rating. This switch rating is the most common found in residential wiring.

Another factor to consider is the number of switching locations for the lighting load on the circuit. Knowing this information will allow the electrician to choose a single-pole switch when the load is controlled from one location, two three-way switches when the lighting load is controlled from two separate locations, or two three-way switches and the required number of four-way switches when the load is controlled from three or more locations.

Section 404.2(A) of the *NEC®* requires switches to be wired so that all switching is done in the ungrounded circuit conductor (Figure 13–7). No switching in the grounded (neutral) conductor is allowed. The switching techniques described later in this chapter will show common connections where only the "hot" ungrounded circuit conductor is switched. Section 404.2(B) actually says that switches must not disconnect the grounded conductor of a circuit. There is no need to ever connect a white insulated grounded conductor to any switch in a residential switching circuit.

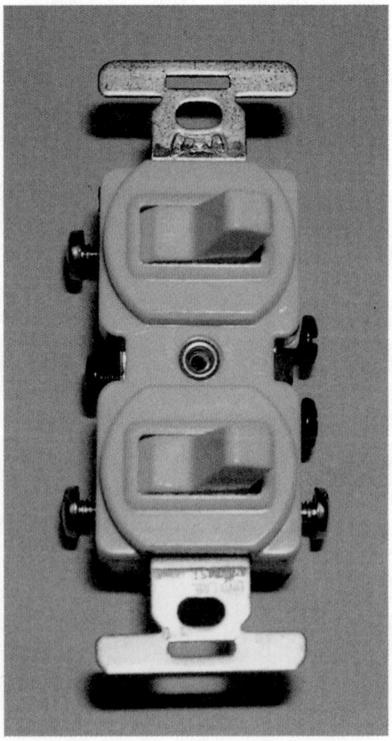

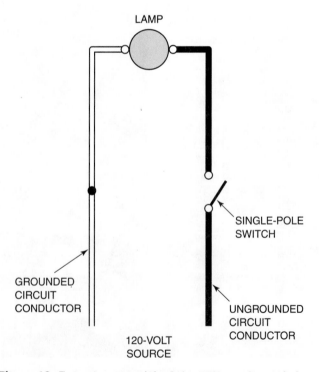

Figure 13–6 A combination switch consists of two switches on one strap or yoke.

Figure 13–7 Section 404.2(A) of the *NEC®* requires switches to be wired so that all switching is done in the ungrounded circuit conductor.

The *NEC®* and Underwriters Laboratories refer to the switches used to control lighting outlets in residential applications as "snap switches." However, most electricians refer to them as "toggle switches" or simply as "switches." Switches that are used to control lighting circuits are classified as "general-use snap switches," and the requirements for these switches are the same for both the *NEC®* and Underwriters Laboratories. The requirements are found in Article 404 of the *NEC®*. Section 404.14 (A) and (B) covers the requirements and recognizes two distinct switch categories. Section 404.14(A) categorizes one switch type as an "alternating current general-use snap switch" and states that it is suitable for use only on alternating current circuits for controlling the following:

- Resistive and inductive loads, including fluorescent lamps, not exceeding the ampere rating of the switch at the voltage involved.
- Tungsten-filament lamp loads (incandescent lamps) not exceeding the ampere rating of the switch at 120 volts
- Motor loads not exceeding 80% of the ampere rating of the switch at its rated voltage.

Section 404.14(B) categorizes the other switch type as an "alternating current or direct current general-use snap switch" and states that it can be used on either alternating current or direct current circuits for controlling the following:

- Resistive loads not exceeding the ampere rating of the switch at the voltage applied.
- Inductive loads not exceeding 50% of the ampere rating of the switch at the applied voltage. Switches rated in horsepower are suitable for controlling motor loads within their rating at the voltage applied.
- Tungsten-filament lamp loads not exceeding the ampere rating of the switch at the applied voltage if T-rated.

After an electrician has selected the appropriate switch type with the proper current and voltage rating for the switching circuit encountered, installing the wiring required for the actual switching connections is the next step. Installing a cable wiring system was presented in Chapter 11, and installing a raceway wiring system was presented in Chapter 12. Knowing and understanding common switch connection techniques will allow an electrician to install the appropriate number of conductors between switches and the loads they control during the rough-in stage. Once the wiring method has been installed, the switching connections are made.

To help electricians understand switching circuit connections, there are three things to always remember when making the switch and load connections:

1. First, at all electrical box locations in the switching circuit, connect the circuit grounding conductors to
   - metal electrical boxes with a green grounding screw,
   - the green screw on the switching devices, and
   - the lighting fixture ground screw or grounding conductor.
2. Next, at the lighting outlet locations, connect the white insulated grounded conductor to the silver screw terminal or wire identified as the grounded wire on the lighting fixture. Then connect the "hot" ungrounded conductor at the lighting fixture to the brass screw or the wire identified as the ungrounded conductor on the lighting fixture.
3. Finally, at the switch locations, determine the conductor connections to the switch, depending on the specific switching situation, and make those connections.

Installing Single-Pole Switches

Since single-pole switching is the most common switching found in residential wiring, we start our discussion of switching connections with single-pole switches. On a single-pole switch, two wires will be connected to the two terminal screws on the switch. Both of these wires are considered "hot" ungrounded conductors. The following three single-pole switching applications assume the use of nonmetallic sheathed cable with nonmetallic boxes.

Single-Pole Switching Circuit 1

The most common single-pole switching wiring arrangement is when the electrical power feed is brought to the switch and then the cable continues on to the lighting load. Follow these steps when installing a single-pole switch for a lighting load with the power source feeding the switch (Figure 13–8):

1. Always wear safety glasses and observe all applicable safety rules.
2. At the lighting fixture box, connect the bare grounding conductors to the grounding connection of the lighting fixture. If the fixture does not have any metal parts that must be grounded, simply fold the grounding conductor into the back of the box. Do not cut it off. Remember that a nonmetallic box does not require the circuit-grounding conductor to be connected to it.

FROM EXPERIENCE

Electricians doing residential wiring will only be using the alternating current general-use snap switch. When switches are purchased at the electrical equipment distributor, they will be for use only on alternating current circuits unless an electrician specifically asks for an AC/DC general-use snap switch.

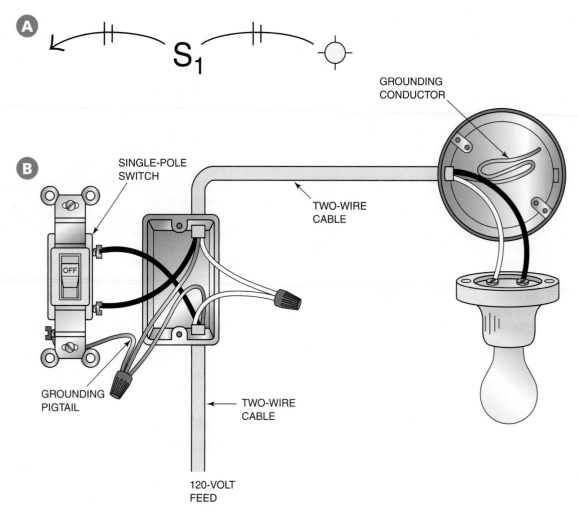

Figure 13–8 A cabling diagram (A) and a wiring diagram (B) for a switching circuit that has a single-pole switch controlling a lighting load. The power source is feeding the switch.

3. At the lighting fixture box, connect the white insulated grounded conductor to the silver screw terminal or wire identified as the grounded wire on the lighting fixture.
4. At the lighting fixture box, connect the black "hot" ungrounded conductor to the brass screw or a wire identified as the ungrounded conductor on the lighting fixture.
5. At the switch box, connect the bare grounding conductors together and, using a bare pigtail, connect all the grounding conductors to the green grounding screw on the switch. Remember that a nonmetallic box does not require the circuit-grounding conductor to be connected to it.
6. At the switch box, connect the white insulated grounded conductors together using a wirenut.
7. At the switch box, connect the black "hot" conductor in the incoming power source cable to one of the terminal screws on the single-pole switch.

8. At the switch box, connect the black "hot" conductor in the two-wire cable going to the lighting outlet to the other terminal screw on the single-pole switch.

Single-Pole Switching Circuit 2

It is very common for residential electricians to run the power source to the lighting outlet first and to then run a two-wire cable to the single-pole switching location. This is called a **switch loop.** When wiring a house with non-metallic sheathed cable, a two-wire cable is available with only a white insulated conductor and a black insulated conductor. The switch loop arrangement will require the white wire in the cable to be used as a "hot" ungrounded conductor. When this situation arises, the white wire must be reidentified as a "hot" conductor. This is usually accomplished by wrapping some black electrical tape on the conductor, although another color tape (like red) may also be used.

FROM EXPERIENCE

When wiring a house with a conduit wiring method, a white insulated wire cannot be used in a switch loop switching arrangement as a "hot" conductor. Section 200.7(C)(2) of the *NEC®* allows the reidentification of only a white insulated conductor as a "hot" conductor in a cable assembly, such as nonmetallic sheathed cable. Use two black insulated conductors or one black and one red insulated conductor as the switch loop wiring in a conduit system.

Follow these steps when installing a (switch loop) single-pole switch for a lighting load with the power source feeding the lighting outlet first (Figure 13–9):

1. Always wear safety glasses and observe all applicable safety rules.
2. At the lighting fixture box, connect the bare grounding conductors together using a wirenut. Remember that a nonmetallic box does not require the circuit-grounding conductor to be connected to it.

3. At the lighting fixture box, connect the white insulated grounded conductor from the incoming power source cable to the silver screw terminal or wire identified as the grounded wire on the lighting fixture.
4. At the lighting fixture box, use a wirenut to connect the black "hot" conductor from the incoming power source cable to the white insulated conductor of the two-wire cable going to the single-pole switch. Reidentify the white wire as a "hot" conductor with a piece of black electrical tape.
5. At the lighting fixture box, connect the black conductor from the two-wire switch loop cable to the brass screw or a wire identified as the ungrounded conductor on the lighting fixture.
6. At the switch box, connect the bare grounding conductor to the green grounding screw on the switch. Remember that a nonmetallic box does not require the circuit-grounding conductor to be connected to it.
7. At the switch box, reidentify the incoming white insulated conductor as a "hot" conductor with a piece of black tape and connect it to one of the terminal screws on the single-pole switch.
8. At the switch box, connect the black conductor in the switch loop cable to the other terminal screw on the single-pole switch.

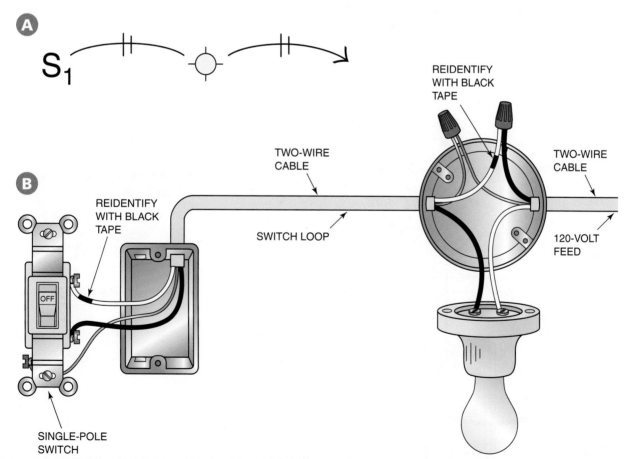

Figure 13–9 A cabling diagram (A) and a wiring diagram (B) for a switching circuit that has a single-pole switch controlling a lighting load with the power source feeding the lighting outlet. This switching arrangement is called a "switch loop."

Single-Pole Switching Circuit 3

Sometimes an electrician will decide to run the power source cable into a lighting outlet box, run a switch loop cable to a single-pole switch, and run a cable for control of other lighting outlets, all out of the same box. Follow these steps when installing a single-pole switch controlling two lighting outlets with the power source feeding one of the lighting outlets (Figure 13–10):

1. Always wear safety glasses and observe all applicable safety rules.
2. At lighting fixture box 1, connect the bare grounding conductors together using a wirenut. Remember that a nonmetallic box does not require the circuit-grounding conductor to be connected to it.
3. At lighting fixture box 1, use a wirenut to connect together a white pigtail jumper, the white insulated grounded conductor from the incoming power source cable, and the white insulated conductor of the cable feeding light fixture box 2. Then connect the white pigtail jumper to the silver screw or the wire identified as the grounded conductor on the lighting fixture.
4. At lighting fixture box 1, use a wirenut to connect the black "hot" conductor from the incoming power source cable to the white insulated conductor of the two-wire cable going to the single-pole switch. Reidentify the white wire as a "hot" conductor with a piece of black electrical tape.
5. At lighting fixture box 1, use a wirenut to connect together a black pigtail jumper, the black conductor from the two-wire switch loop cable, and the black conductor from the cable feeding light box 2. Then connect the black pigtail jumper to the brass screw or the wire identified as the ungrounded conductor on the lighting fixture.

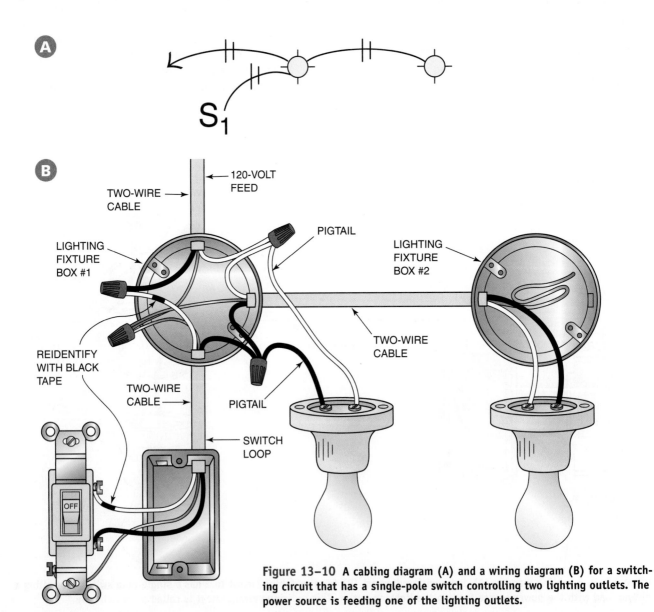

Figure 13–10 A cabling diagram (A) and a wiring diagram (B) for a switching circuit that has a single-pole switch controlling two lighting outlets. The power source is feeding one of the lighting outlets.

6. At lighting fixture box 2, fold the grounding conductor into the back of the box, connect the white insulated grounded conductor to the silver screw terminal or wire identified as the grounded wire on the lighting fixture, and connect the black "hot" ungrounded conductor to the brass screw or a wire identified as the ungrounded conductor on the lighting fixture.

7. At the switch box, connect the bare grounding conductor to the green grounding screw on the switch, reidentify the incoming white insulated conductor as a "hot" conductor with a piece of black tape and connect it to one of the terminal screws on the single-pole switch, and connect the black conductor in the switch loop cable to the other terminal screw on the single-pole switch.

Installing Three-Way Switches

Three-way switches are used to control a lighting load from two different locations. You may remember from Chapter 2 that three-way switches have three terminal screws on them. One of the screws is called the "common" terminal and is colored black. The other two terminal screws are the same color, usually brass or bronze, and are called the "traveler terminals." Beginning electricians often find the connections for three-way switches confusing. As a matter of fact, even some experienced electricians find themselves confused when wiring three-way switches after they have not had to wire them for a while. Learning some common rules will make the process much easier whether you wire three-way switches all the time or only once in a while. Common rules to keep in mind when wiring switching circuits with three-way switches include the following:

- Three-way switches must always be installed in pairs. There is no such thing as an installation with only one three-way switch.
- A three-wire cable must always be installed between the two three-way switches. If you are wiring with conduit, three separate wires must be pulled into the conduit between the two three-way switches.
- The black colored "common" terminal on a three-way switch should always have a black insulated wire attached to it. One three-way switch will have a black "hot" feed conductor attached to it, and the other three-way switch will have the black insulated conductor that will be going to the lighting load attached to it.
- Assuming the use of a nonmetallic sheathed cable, when the power source feed is brought to the first three-way switch, the traveler wires that interconnect the traveler terminals of both switches will be black and red in color.
- Assuming the use of a nonmetallic sheathed cable, when the power source feed is brought to the lighting outlet first, the traveler wires will be red and white in color.

The white traveler conductors will need to be reidentified with black tape at each switch location.

- There is no marking for the "On" or "Off" position of the toggle on a three-way switch, so it does not make any difference which way it is positioned in the electrical device box.

The following applications will assume the use of nonmetallic sheathed cable with nonmetallic boxes.

Three-Way Switching Circuit 1

Follow these steps when installing a three-way switching circuit with the power source feeding the first three-way switch location (Figure 13–11):

1. Always wear safety glasses and observe all applicable safety rules.
2. At the lighting fixture box, connect the bare grounding conductors to the grounding connection of the lighting fixture. If the fixture does not have any metal parts that must be grounded, simply fold the grounding conductor into the back of the box. Remember that a nonmetallic box does not require the circuit-grounding conductor to be connected to it.
3. At the lighting fixture box, connect the white insulated grounded conductor to the silver screw terminal or wire identified as the grounded wire on the lighting fixture.
4. At the lighting fixture box, connect the black ungrounded conductor to the brass screw or a wire identified as the ungrounded conductor on the lighting fixture.
5. At the first three-way switch location, connect the bare grounding conductors together and, using a bare pigtail, connect all the grounding conductors to the green grounding screw on the switch.
6. At the first three-way switch box, connect the white insulated grounded conductors together using a wirenut.
7. At the first three-way switch box, connect the black "hot" conductor in the incoming power source cable to the common terminal screw on three-way switch 1.
8. At the first three-way switch box, connect the black and red traveler wires in the three-wire cable going to three-way switch 2 to the traveler terminal screws on the three-way switch.
9. At the second three-way switch box, connect the bare grounding conductors together and, using a bare pigtail, connect all the grounding conductors to the green grounding screw on the switch.
10. At the second three-way switch box, connect the white insulated grounded conductors together using a wirenut.
11. At the second three-way switch box, connect the black conductor in the two-wire cable going to the lighting outlet to the common terminal screw on three-way switch 2.

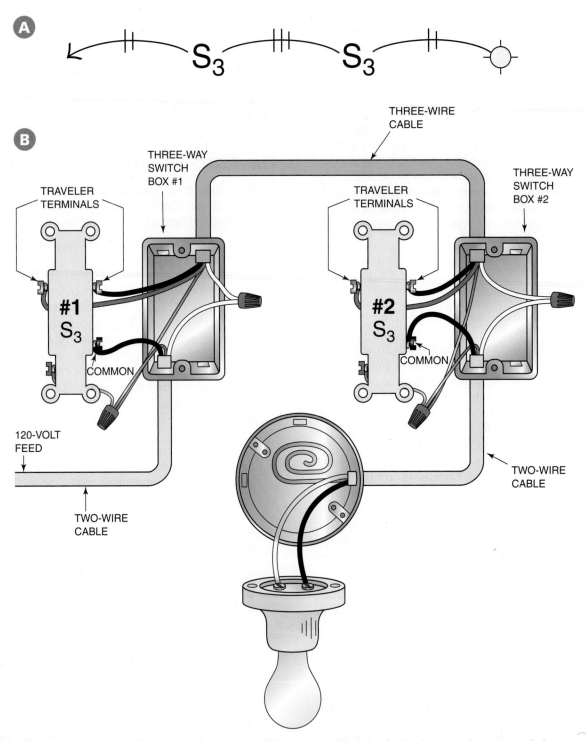

A

B

THREE-WIRE CABLE

THREE-WAY SWITCH BOX #1

TRAVELER TERMINALS

TRAVELER TERMINALS

THREE-WAY SWITCH BOX #2

#1 S₃

COMMON

#2 S₃

COMMON

120-VOLT FEED

TWO-WIRE CABLE

TWO-WIRE CABLE

Figure 13–11 A cabling diagram (A) and a wiring diagram (B) for a switching circuit that has two three-way switches controlling a lighting load. The power source is feeding the first three-way switch location.

12. At the second three-way switch box, connect the black and red traveler wires in the three-wire cable coming from three-way switch 1 to the traveler terminal screws on three-way switch 2.

Three-Way Switching Circuit 2

Follow these steps when installing a three-way switching circuit with the power source feeding the lighting outlet location (Figure 13–12):

1. Always wear safety glasses and observe all applicable safety rules.

2. At the lighting fixture box, connect the bare grounding conductors to the grounding connection of the lighting fixture. If the fixture does not have any metal parts that must be grounded, simply use a wirenut to connect the grounding conductors together and place them in the back of the box.

3. At the lighting fixture box, connect the white insulated grounded conductor of the incoming power feed cable to the silver screw terminal or wire identified as the grounded wire on the lighting fixture.

4. At the lighting fixture box, use a wirenut to connect the black "hot" conductor from the incoming power

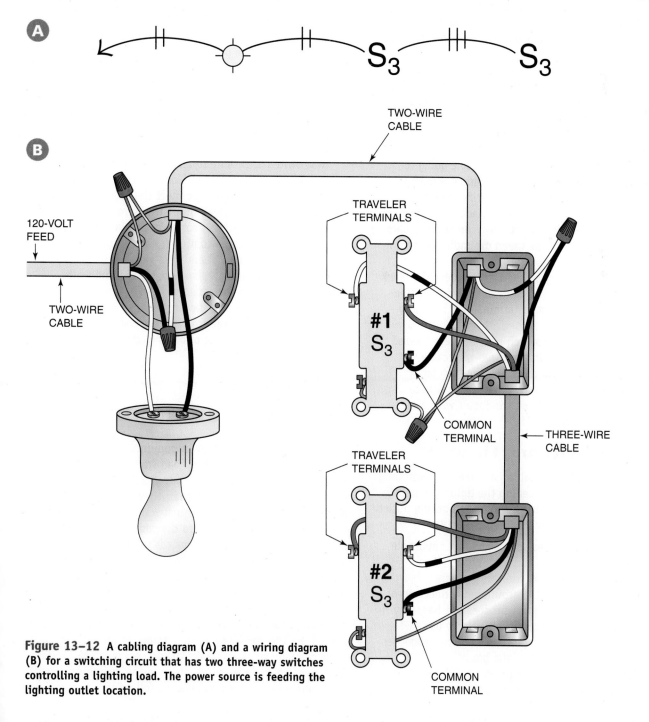

Figure 13–12 A cabling diagram (A) and a wiring diagram (B) for a switching circuit that has two three-way switches controlling a lighting load. The power source is feeding the lighting outlet location.

source cable to the white insulated conductor of the two-wire cable going to three-way switch 1. Reidentify the white wire as a "hot" conductor with a piece of black electrical tape.

5. At the lighting fixture box, connect the black conductor from the switch loop cable going to three-way switch 1 to the brass screw or a wire identified as the ungrounded conductor on the lighting fixture.

6. At the first three-way switch location, connect the bare grounding conductors together and, using a bare pigtail, connect all the grounding conductors to the green grounding screw on the switch.

7. At the first three-way switch box, use a wirenut to connect the white insulated conductor from the two-wire cable coming from the lighting outlet to the black insulated conductor in the three-wire cable going to three-way switch 2. Be sure to reidentify the white insulated conductor with a piece of black electrical tape.

8. At the first three-way switch box, connect the black conductor in the two-wire cable from the lighting outlet to the common terminal screw on three-way switch 1.

9. At the first three-way switch box, connect the red and white traveler wires in the three-wire cable going to three-way switch 2 to the traveler terminal screws on three-way switch 1. Be sure to reidentify the white insulated conductor with a piece of black electrical tape.

10. At the second three-way switch box, connect the bare grounding conductor to the green grounding screw on the switch.

11. At the second three-way switch box, connect the black conductor in the three-wire cable coming from three-way switch 1 to the common terminal on three-way switch 2.

12. At the second three-way switch box, connect the red and white traveler wires in the three-wire cable coming from three-way switch 1 to the traveler terminal screws on three-way switch 2. Be sure to reidentify the white insulated conductor with a piece of black electrical tape.

Three-Way Switching Circuit 3

Follow these steps when installing a three-way switching circuit with the power source feeding the lighting outlet location and a three-wire cable is run from the lighting outlet to each of the three-way switches (Figure 13–13):

1. Always wear safety glasses and observe all applicable safety rules.

2. At the lighting fixture box, connect the bare grounding conductors to the grounding connection of the lighting fixture. If the fixture does not have any metal parts that must be grounded, simply use a wirenut to connect the grounding conductors together and place them in the back of the box.

3. At the lighting fixture box, connect the white insulated grounded conductor from the incoming

power source cable to the silver screw terminal or wire identified as the grounded wire on the lighting fixture.

4. At the lighting fixture box, use a wirenut to connect the black "hot" conductor from the incoming power source cable to a black insulated conductor in the three-wire cable going to three-way switch 1.

5. At the lighting fixture box, connect the two red traveler wires together and the two white traveler wires together using wirenuts. Remember to reidentify the white traveler wires with black tape.

6. At the lighting fixture box, connect the black conductor from the three-wire cable going to three-way switch 2 to the brass screw or a wire identified as the ungrounded conductor on the lighting fixture.

7. At three-way switch 1, connect the bare grounding conductor to the green grounding screw on the switch.

8. At three-way switch 1, connect the black conductor in the three-wire cable coming from the lighting outlet to the common terminal on three-way switch 1.

9. At three-way switch 1, connect the red and white traveler wires in the three-wire cable coming from the lighting outlet to the traveler terminal screws on three-way switch 1. Be sure to reidentify the white insulated conductor with a piece of black electrical tape.

10. At three-way switch 2, connect the bare grounding conductor to the green grounding screw on the switch.

11. At three-way switch 2, connect the black conductor in the three-wire cable coming from the lighting outlet to the common terminal on three-way switch 2.

12. At three-way switch 2, connect the red and white traveler wires in the three-wire cable coming from the lighting outlet to the traveler terminal screws on three-way switch 2. Be sure to reidentify the white insulated conductor with a piece of black electrical tape.

Installing Four-Way Switches

Four-way switches are used in conjunction with three-way switches to control a lighting load from more than two different locations. Chapter 2 introduced you to four-way switches, and you learned that they have four terminal screws on them. All four terminal screws are called traveler terminals. The four screws are divided into two pairs. Each pair is of the same color. One pair is usually brass or bronze, and the other pair is some other color like black. Once you understand the common three-way switching connections, four-way connections should be relatively easy to understand. Four-way switches are simply inserted into the wiring between three-way switches. Only traveler wires are connected to the four-way switch screw terminals. Learning some common rules will make the process much easier. Common rules to keep in mind when wiring switching circuits with four-way switches include the following:

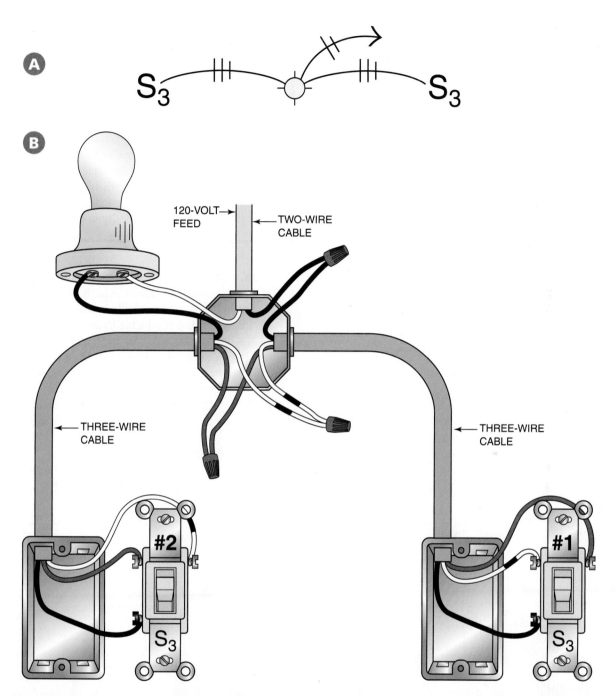

Figure 13–13 A cabling diagram (A) and a wiring diagram (B) for a switching circuit that has two three-way switches controlling a lighting load. The power source is feeding the lighting outlet location, and a three-wire cable is run from the lighting outlet to each of the three-way switches. No grounding wires are shown.

- Four-way switches must always be installed between two three-way switches. There is no such thing as an installation with only one four-way switch or a switching circuit that has only four-way switches. For example, if you wire a switching situation that requires four switching locations for the same lighting load, you would need two three-way switches and two four-way switches.
- A three-wire cable must always be installed between all four-way and three-way switches. If you are wiring with

conduit, three separate wires must be pulled into the conduit between all the four-ways and three-ways.
- Assuming the use of a nonmetallic sheathed cable, when the power source feed is brought to the first three-way switch in the circuit, the traveler wires that interconnect the traveler terminals of all four-way and three-way switches will be black and red in color.
- Assuming the use of a nonmetallic sheathed cable, when the power source feed is brought to the lighting outlet

first, the traveler wires will be red and white in color. The white traveler conductors will need to be reidentified with black tape at each switch location.

- Most four-way switch traveler terminals are vertically configured. This means that when the four-way switch is positioned in a vertical position, the top two screws have the same color and are a traveler terminal pair. The bottom two screws are the other traveler terminal pair and have the same color. Remember that the colors of each traveler pair is different.
- Like a three-way switch, there is no marking for the "On" or "Off" position of the toggle on the four-way switch, so it does not make any difference which way it is positioned in the electrical device box.

The following applications will assume the use of nonmetallic sheathed cable with nonmetallic boxes.

Four-Way Switching Circuit 1

Follow these steps when installing a four-way switching circuit with the power source feeding the first three-way switch location (Figure 13–14):

1. Always wear safety glasses and observe all applicable safety rules.
2. At the lighting fixture box, connect the bare grounding conductors to the grounding connection of the lighting fixture. If the fixture does not have any

FROM EXPERIENCE

Some four-way switches are made with the traveler terminals in a horizontal configuration. They are not very common, and because they are not encountered very often, they can confuse an electrician who is making connections at four-way switch locations. With this type of switch configuration, the traveler terminal pairs having the same color are located on the same side of the switch. Always look at the manufacturer's instructions when using four-way switches to determine whether the traveler terminals are configured vertically or horizontally.

metal parts that must be grounded, simply fold the grounding conductor into the back of the box. Remember that a nonmetallic box does not require the circuit-grounding conductor to be connected to it.
3. At the lighting fixture box, connect the white insulated grounded conductor to the silver screw terminal or wire identified as the grounded wire on the lighting fixture.
4. At the lighting fixture box, connect the black "hot" ungrounded conductor to the brass screw or a wire identified as the ungrounded conductor on the lighting fixture.

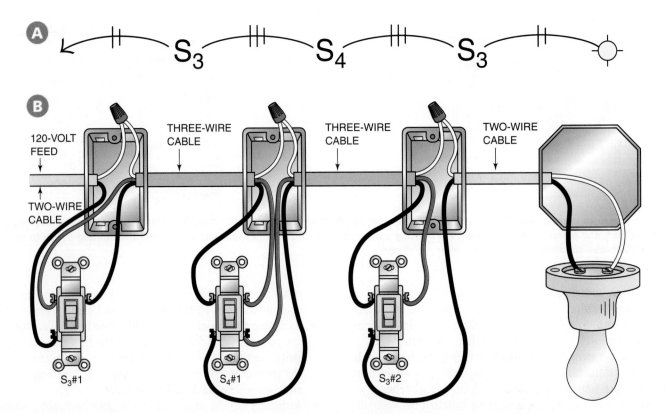

Figure 13–14 A cabling diagram (A) and a wiring diagram (B) for a switching circuit that has a four-way switch and two three-way switches controlling a lighting load. The power source is feeding the first three-way switch location. No grounding conductors are shown.

5. At three-way switch 1, connect the bare grounding conductors together and, using a bare pigtail, connect all the grounding conductors to the green grounding screw on the switch.

6. At three-way switch 1, connect the white insulated grounded conductors together using a wirenut.

7. At three-way switch 1, connect the black "hot" conductor in the incoming power source cable to the common (black) terminal screw on the three-way switch.

8. At three-way switch 1, connect the black and red traveler wires in the three-wire cable going to four-way switch 1 to the traveler terminal screws on three-way switch 1.

9. At four-way switch 1, connect the bare grounding conductors together and, using a bare pigtail, connect all the grounding conductors to the green grounding screw on the switch.

10. At four-way switch 1, connect the white insulated grounded conductors together using a wirenut.

11. At four-way switch 1, connect the black and red traveler wires in the three-wire cable coming from three-way switch 1 to a pair of traveler terminal screws on the four-way switch. It really does not make any difference if it is the top pair or the bottom pair.

12. At four-way switch 1, connect the black and red traveler wires from the three-wire cable coming from three-way switch 2 to the other pair of traveler terminal screws on four-way switch 1.

13. At three-way switch 2, connect the bare grounding conductors together and, using a bare pigtail, connect all the grounding conductors to the green grounding screw on the switch.

14. At three-way switch 2, connect the white insulated grounded conductors together using a wirenut.

15. At three-way switch 2, connect the black conductor in the two-wire cable going to the lighting outlet to the common (black) terminal screw on three-way switch 2.

16. At three-way switch 2, connect the black and red traveler wires in the cable coming from four-way switch 1 to the traveler terminal screws on three-way switch 2.

Four-Way Switching Circuit 2

Follow these steps when installing a four-way switching circuit with the power source feeding the lighting outlet (Figure 13–15):

1. Always wear safety glasses and observe all applicable safety rules.

2. At the lighting fixture box, connect the bare grounding conductors to the grounding connection of the lighting fixture. If the fixture does not have any metal parts that must be grounded, simply use a wirenut to connect the grounding conductors together and place them in the back of the box.

3. At the lighting fixture box, connect the white insulated grounded conductor from the incoming power source cable to the silver screw terminal or wire identified as the grounded wire on the lighting fixture.

4. At the lighting fixture box, use a wirenut to connect the black "hot" conductor from the incoming power source cable to the white insulated conductor in the three-wire cable going to three-way switch 1. Reidentify the white wire as a "hot" conductor with a piece of black electrical tape.

5. At the lighting fixture box, connect the black "hot" conductor from the three-wire cable going to three-way switch 1 to the brass screw or a wire identified as the ungrounded conductor on the lighting fixture.

6. At three-way switch 1, connect the bare grounding conductors together and, using a bare pigtail, connect all the grounding conductors to the green grounding screw on the switch.

7. At three-way switch 1, use a wirenut to connect the white insulated conductor from the two-wire cable coming from the lighting outlet to the black insulated conductor in the three-wire cable going to four-way switch 1. Be sure to reidentify the white insulated conductor with a piece of black electrical tape.

8. At three-way switch 1, connect the black conductor in the two-wire cable from the lighting outlet to the common (black) terminal screw on three-way switch 1.

9. At three-way switch 1, connect the red and white traveler wires in the three-wire cable going to four-way switch 1 to the traveler terminal screws on three-way switch 1. Be sure to reidentify the white insulated conductor with a piece of black electrical tape.

10. At four-way switch 1, connect the bare grounding conductors together and, using a bare pigtail, connect all the grounding conductors to the green grounding screw on the switch.

11. At four-way switch 1, connect the black insulated conductors together using a wirenut.

12. At four-way switch 1, connect the red and white traveler wires in the three-wire cable coming from three-way switch 1 to a pair of traveler terminal screws on the four-way switch. It really does not make any difference if it is the top pair or bottom pair. Reidentify the white traveler wire with black tape.

13. At four-way switch 1, connect the red and white traveler wires from the three-wire cable coming going to three-way switch 2 to the other pair of traveler terminal screws on the four-way switch. Reidentify the white traveler wire with black tape.

14. At three-way switch 2, connect the bare grounding conductor to the green grounding screw on the switch.

15. At three-way switch 2, connect the black conductor in the three-wire cable coming from four-way switch 1 to the common (black) terminal on three-way switch 2.

16. At three-way switch 2, connect the red and white traveler wires in the three-wire cable coming from four-way switch 1 to the traveler terminal screws on three-way switch 2. Be sure to reidentify the white insulated conductor with a piece of black electrical tape.

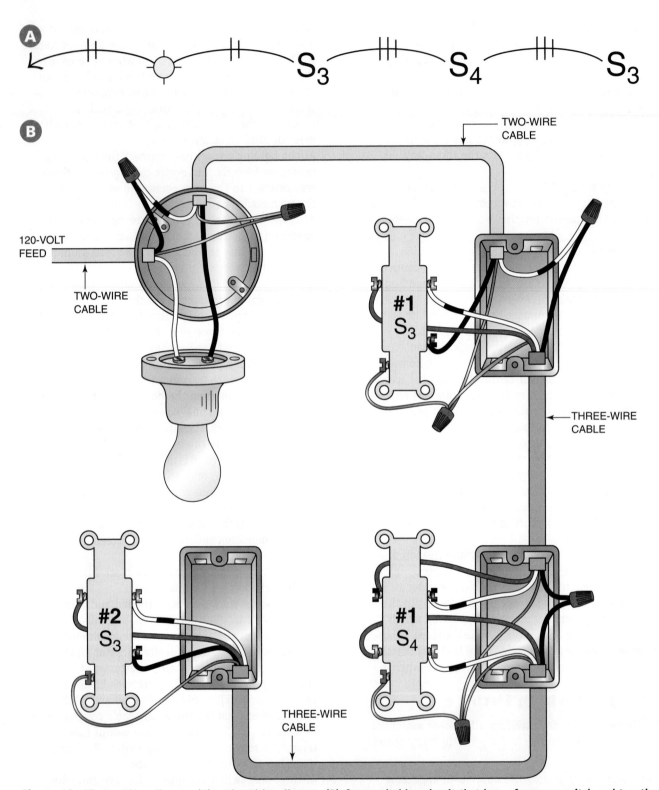

Figure 13–15 A cabling diagram (A) and a wiring diagram (B) for a switching circuit that has a four-way switch and two three-way switches controlling a lighting load. The power source is feeding the lighting outlet.

Installing Switched Duplex Receptacles

Switched receptacles are often found in areas such as bedrooms, living rooms, and family rooms of homes. In these areas, lamps are often the primary lighting source, and electricians place switches next to the doorways of these areas for control of the receptacles that the lamps are plugged into. The switch on the lamp is left in the "On" position, so when a switch is activated and energizes the receptacle that the lamp is plugged into, the lamp comes on. Switching of receptacles can be done so the whole receptacle is switched on or off, or the wiring can be installed so that half a duplex receptacle is energized with the switch while the other half remains "hot" at all times. This is accomplished by "splitting" a duplex receptacle. Splitting a duplex receptacle means removing the tab between the two brass screw terminals on the ungrounded side of a duplex receptacle. It is common wiring practice to switch the bottom half of the receptacle and leave the top half "hot" at all times because having a lamp cord plugged into the bottom half of the split-

duplex receptacle allows the top half to be clear for plugging in another piece of electrical equipment. On the other hand, if the top half were switched and a lamp cord plugged into it, the cord would hang down in front of the "hot" half of the duplex receptacle and be in the way when another piece of electrical equipment needed to be plugged into that half. A lamp can be plugged into the bottom half of the duplex receptacle, and an electrical load that needs to be energized at all times (like a television) can be plugged into the top half. Wiring connections for both of these wiring applications are presented in this section.

The following applications will assume the use of non-metallic sheathed cable with nonmetallic boxes.

Switched Duplex Receptacle Circuit 1

Follow these steps when installing a duplex receptacle that is controlled by a single-pole switch with the power source feeding the switch location (Figure 13–16):

1. Always wear safety glasses and observe all applicable safety rules.

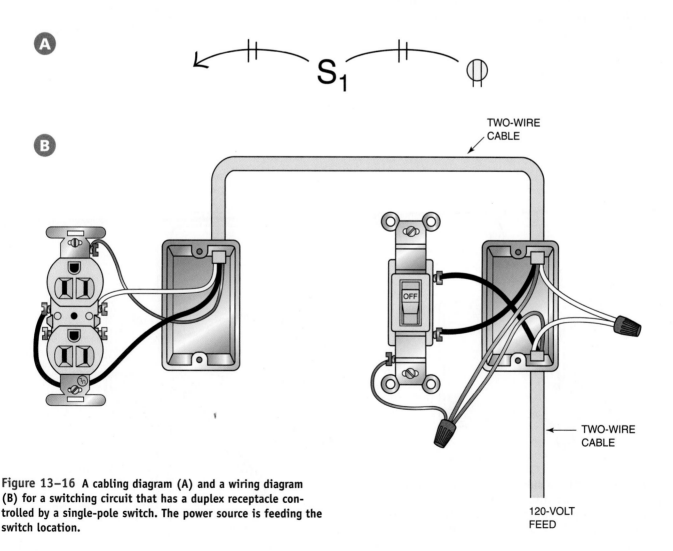

Figure 13–16 A cabling diagram (A) and a wiring diagram (B) for a switching circuit that has a duplex receptacle controlled by a single-pole switch. The power source is feeding the switch location.

2. At the single-pole switch location, connect the bare grounding conductors together and, using a bare pigtail, connect all the grounding conductors to the green grounding screw on the switch. Remember that a nonmetallic box does not require the circuit-grounding conductor to be connected to it.
3. At the single-pole switch location, connect the white insulated grounded conductors together using a wirenut.
4. At the single-pole switch location, connect the black "hot" conductor in the incoming power source cable to one of the terminal screws on the single-pole switch.
5. At the single-pole switch location, connect the black conductor in the two-wire cable going to the duplex receptacle outlet to the other terminal screw on the single-pole switch.
6. At the duplex receptacle location, connect the bare grounding conductor to the green grounding screw on the duplex receptacle.
7. At the duplex receptacle location, connect the white grounded conductor to one of the silver terminal screws on the duplex receptacle. It does not make a difference which of the two silver screws you use.
8. At the duplex receptacle location, connect the black conductor to one of the brass terminal screws on the duplex receptacle. Again, it does not make a difference which of the two brass screws you use.

Switched Duplex Receptacle Circuit 2

Follow these steps when installing a duplex receptacle that is controlled by a single-pole switch with the power source feeding the receptacle location (a switch loop situation) (Figure 13–17):

1. Always wear safety glasses and observe all applicable safety rules.
2. At the single-pole switch location, connect the bare grounding conductor to the green grounding screw on the single-pole switch.
3. At the single-pole switch location, connect the white conductor to one of the terminal screws on the single-pole switch. It does not make a difference which of the two screws you use. Make sure to reidentify the white wire with black electrical tape.
4. At the single-pole switch location, connect the black conductor to the other terminal screw on the single-pole switch.
5. At the duplex receptacle location, connect the bare grounding conductors together and, using a bare pigtail, connect all the grounding conductors to the green grounding screw on the receptacle.

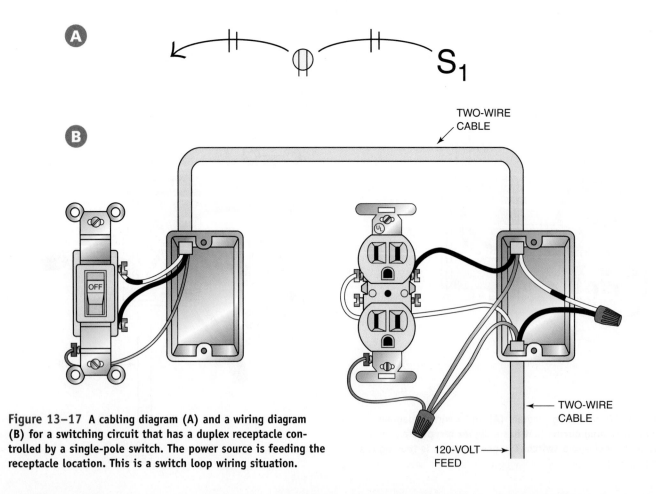

Figure 13–17 A cabling diagram (A) and a wiring diagram (B) for a switching circuit that has a duplex receptacle controlled by a single-pole switch. The power source is feeding the receptacle location. This is a switch loop wiring situation.

6. At the duplex receptacle location, connect the white grounded conductor of the incoming power source cable to one of the silver terminal screws on the duplex receptacle. It does not make a difference which of the two silver screws you use.

7. At the duplex receptacle location, connect the black "hot" conductor of the incoming power source cable to the white insulated conductor in the two-wire cable going to the single-pole switch with a wirenut. Reidentify the white conductor with a piece of black electrical tape.

8. At the duplex receptacle, connect the black insulated wire from the two-wire cable coming from the switch to one of the brass terminal screws.

Switched Duplex Receptacle Circuit 3

Follow these steps when installing a split-duplex receptacle so that the bottom half is controlled by a single-pole switch and the top half is "hot" at all times. The power source feeds the switch location (Figure 13–18):

1. Always wear safety glasses and observe all applicable safety rules.

2. At the single-pole switch location, connect the bare grounding conductors together and, using a bare pigtail, connect all the grounding conductors to the green grounding screw on the switch.

3. At the single-pole switch location, connect the white insulated grounded conductors together using a wirenut.

4. At the single-pole switch location, using a wirenut, connect together a black pigtail, the black conductor of the incoming power source cable, and the black conductor of the three-wire cable going to the split-duplex receptacle. Connect the black pigtail end to one of the terminal screws on the single-pole switch.

5. At the single-pole switch location, connect the red conductor in the three-wire cable going to the split-duplex receptacle outlet to the other terminal screw on the single-pole switch.

6. Split the duplex receptacle by using a pair of pliers to take off the tab between the two brass terminals on the ungrounded side of the receptacle.

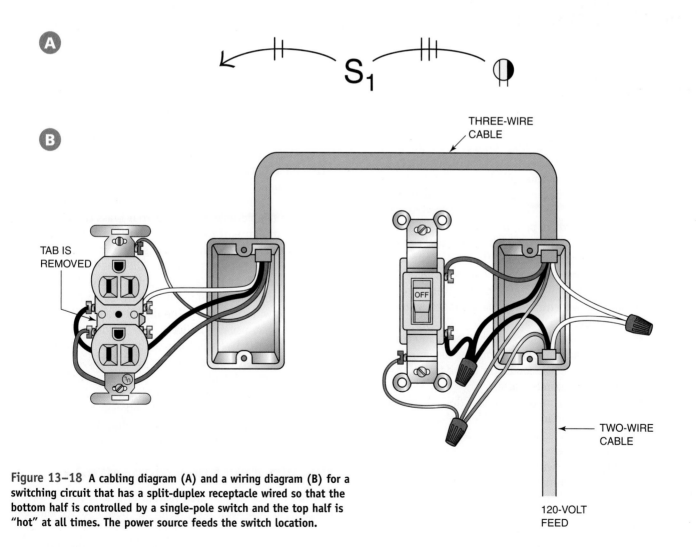

Figure 13–18 A cabling diagram (A) and a wiring diagram (B) for a switching circuit that has a split-duplex receptacle wired so that the bottom half is controlled by a single-pole switch and the top half is "hot" at all times. The power source feeds the switch location.

7. At the split-duplex receptacle location, connect the bare grounding conductor to the green grounding screw on the duplex receptacle.

8. At the split-duplex receptacle location, connect the white grounded conductor to one of the silver terminal screws on the duplex receptacle. It does not make a difference which of the two silver screws you use.

9. At the split-duplex receptacle location, connect the black conductor from the three-wire cable to the top brass terminal screw.

10. At the split-duplex receptacle location, connect the red conductor from the three-wire cable to the bottom brass terminal screw.

In Chapter 11, you learned that in a residential wiring system it is common practice to install wiring from outlet to outlet. Sometimes it is desired to continue an unswitched part of a circuit "downstream" of a switched lighting outlet. To accomplish this, a three-wire cable is used between the switch location and the lighting outlet. A two-wire cable is then continued from the lighting outlet to other non-switched loads, such as receptacle outlets.

Switched Duplex Receptacle Circuit 4

Follow these steps when installing a single-pole switch controlling a lighting load with a continuously "hot" receptacle located downstream of the lighting outlet. The power source will feed the switch (Figure 13–19):

1. Always wear safety glasses and observe all applicable safety rules.

2. At the lighting fixture box, connect the bare grounding conductors together using a wirenut. Remember that a nonmetallic box does not require the circuit-grounding conductor to be connected to it.

3. At the lighting fixture box, use a wirenut and connect together a white pigtail, the white conductor of the incoming three-wire cable, and the white conductor of the two-wire cable going to the duplex receptacle. Connect the white pigtail end to the silver screw terminal or wire identified as the grounded wire on the lighting fixture.

4. At the lighting fixture box, use a wirenut and connect together the black conductor from the incoming three-wire cable to the black conductor of the two-wire cable going to the duplex receptacle.

5. At the lighting fixture box, connect the red conductor from the three-wire incoming cable to the brass screw or a wire identified as the ungrounded conductor on the lighting fixture.

6. At the single-pole switch location, connect the bare grounding conductors together and, using a bare pigtail, connect all the grounding conductors to the green grounding screw on the switch.

7. At the single-pole switch location, connect the white insulated grounded conductors together using a wirenut.

8. At the single-pole switch location, using a wirenut, connect together a black pigtail, the black conductor of the incoming power source cable, and the black conductor of the three-wire cable going to the lighting outlet. Connect the black pigtail end to one of the terminal screws on the single-pole switch.

9. At the single-pole switch location, connect the red conductor in the three-wire cable going to the lighting outlet to the other terminal screw on the single-pole switch.

10. At the duplex receptacle location, connect the bare grounding conductor to the green grounding screw on the duplex receptacle.

11. At the duplex receptacle location, connect the white grounded conductor to one of the silver terminal screws on the grounded side of the duplex receptacle. It does not make a difference which of the two silver screws you use.

12. At the duplex receptacle location, connect the black "hot" conductor to one of the brass terminal screws on the ungrounded side of the duplex receptacle. Again, it does not make a difference which of the two brass screws you use.

Installing Double-Pole Switches

As you have learned, electrical equipment in a house can operate on 120 volts, 120/240 volts, or 240 volts. For those wiring situations when a switch is needed to control a 240-volt load, double-pole (also called two-pole) switches are used. The double-pole switch has four brass terminals on it, and at first glance, it looks like a four-way switch. However, unlike a four-way (or three-way) switch, the toggle on the double-pole switch has the words "On" and "Off" written on it. This means that like a single-pole switch, there is a correct mounting position for the switch, so when it is "On," the toggle will indicate it. The double-pole switch also has markings that usually indicate the "load" and the "line" sides of the switch. Many times a double-pole switch is used to control a 240-volt-rated receptacle. These receptacles types have special configurations (covered in Chapter 2) for the specific current and voltage rating of the receptacle. The 240-volt piece of electrical equipment will be plugged into the 240-volt-rated receptacle using a plug that has the same current and voltage rating as the receptacle.

Double-Pole Switching Circuit 1

Follow these steps when installing a double-pole switch controlling a 240-volt receptacle. The power source will feed

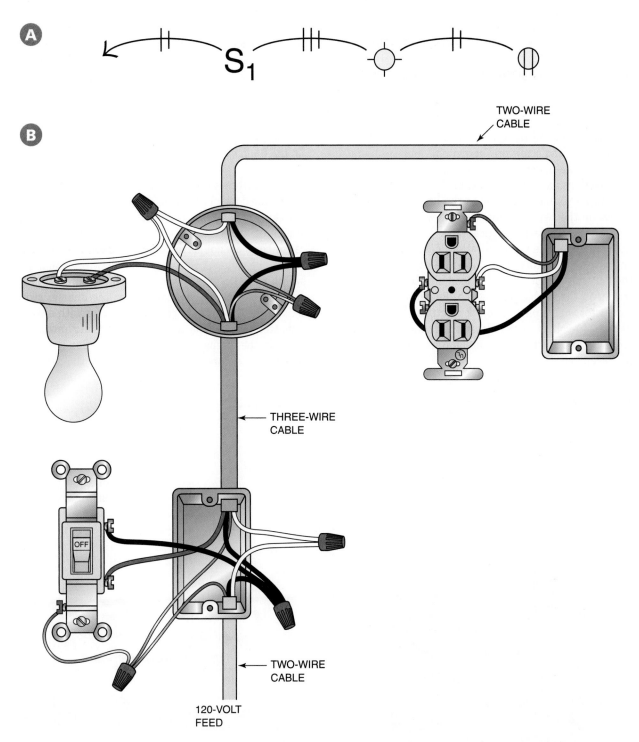

A

S₁

B

TWO-WIRE CABLE

THREE-WIRE CABLE

OFF

TWO-WIRE CABLE

120-VOLT FEED

Figure 13–19 A cabling diagram (A) and a wiring diagram (B) for a switching circuit that has a single-pole switch controlling a lighting load with a continuously "hot" receptacle located downstream of the lighting outlet. The power source is feeding the switch.

the switch. For this circuit, assume the use of nonmetallic boxes and nonmetallic sheathed cable (Figure 13–20):

1. Always wear safety glasses and observe all applicable safety rules.
2. At the double-pole switch location, connect the bare grounding conductors together and, using a bare pigtail, connect all the grounding conductors to the green grounding screw on the switch.
3. At the double-pole switch location, connect the white and black insulated conductors of the incoming two-wire power source feed cable to the line-side terminal screws on the switch. Reidentify the white conductor with a piece of black electrical tape.
4. At the double-pole switch location, connect the white and black conductors of the two-wire cable going to the 240-volt receptacle to the load-side terminal screws on the double-pole switch. Reidentify the white conductor with a piece of black electrical tape.
5. At the 240-volt receptacle location, connect the bare grounding conductor to the green grounding screw on the receptacle.

6. At the 240-volt receptacle location, connect the white and black conductors to the terminal screws on the receptacle. Reidentify the white conductor with a piece of black electrical tape.

Installing Dimmer Switches

Dimmer switches are used in residential wiring applications to provide control for the brightness of the lighting in a specific area of a house. They are available in both single-pole and three-way models. They differ from regular switches because they do not have terminal screws on them. Instead, they have colored insulated pigtail wires coming off the switch that are installed by the manufacturer. It is simply a matter of an electrician connecting the dimmer switch pigtails to the appropriate circuit conductor with a wirenut. As a matter of fact, the wirenuts are usually included in the package the dimmer switches come in, along with the 6-32 screws that secure the dimmer switch to the device box.

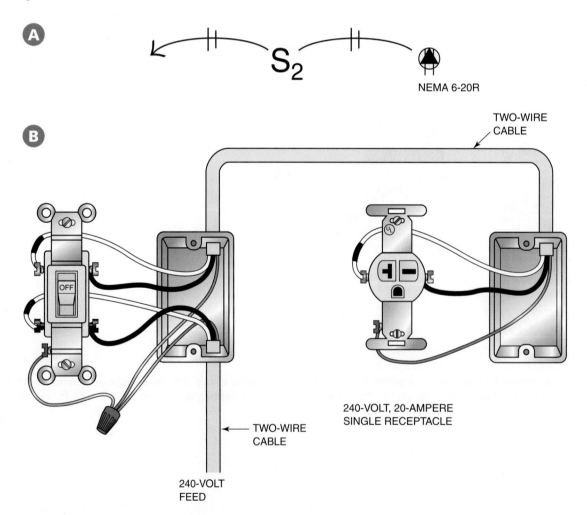

Figure 13–20 A cabling diagram (A) and a wiring diagram (B) for a switching circuit that has a double-pole switch controlling a 240-volt receptacle. The power source feeds the switch.

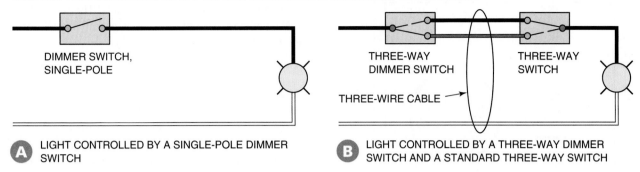

CAUTION: THE POWER MUST BE OFF BEFORE THE DIMMER IS CONNECTED TO THE CIRCUIT

DIMMER SWITCH, SINGLE-POLE

THREE-WAY DIMMER SWITCH

THREE-WAY SWITCH

THREE-WIRE CABLE

A LIGHT CONTROLLED BY A SINGLE-POLE DIMMER SWITCH

B LIGHT CONTROLLED BY A THREE-WAY DIMMER SWITCH AND A STANDARD THREE-WAY SWITCH

Figure 13–21 Single-pole and three-way dimmer switches are connected in a circuit the same way as regular single-pole and three-way switches.

Dimmer switches are larger in size than regular single-pole and three-way switches and take up more room in an electrical box. They also tend to generate more heat and should be installed in electrical device boxes that do not have many other circuit wires in them.

Single-pole or three-way dimmer switches are connected in a switching circuit exactly the same way as regular single-pole and three-way switches (Figure 13–21). Follow these rules when connecting dimmer switches:

- Single-pole dimmers have two black insulated conductors coming off them. Like the two brass screw terminals on a regular single-pole toggle switch, it does not make any difference which of the two wires is connected to the incoming "hot" feed wire or the outgoing switch leg.
- Three-way dimmer switches typically come from the manufacturer with two red insulated pigtails and one black insulated pigtail. The two red conductors are the "traveler" connections, and the black conductor is the "common" connection.
- When using three-way dimmers, it is common wiring practice to use one three-way dimmer and one three-way regular toggle switch when controlling and dimming a lighting load from two locations. The reason for this is that, if you use two three-way dimmers, the *highest* level the lighting load will be able to be brightened is whatever the *lowest* level is on the other three-way dimmer switch. Be aware that some dimmer switch manufacturers may require the use of two three-way dimmer switches when wiring a switching circuit that controls a lighting load from two locations. Always check the manufacturer's instructions to find out for sure whether two three-way dimmer switches must be used.
- Both single-pole and three-way dimmer switches will also have a green insulated grounding pigtail. Connect this pigtail to the circuit grounding conductor.
- Dimmer switches should never be connected to a "live" electrical circuit. The solid-state electronics that allow a dimmer switch to operate will be severely damaged. Always deenergize the electrical circuit first before installing a dimmer switch.

Installing Ceiling Fan Switches

In today's homes, ceiling-suspended paddle fans, with or without an attached lighting kit, are very popular. They are designed to provide movement of the air in a room so it will be more comfortable for the room's occupants. During the summer, the direction of the fan paddles should direct the air down. During the winter, the fan paddle direction should be reversed so the air is pulled upward. There is a switch on the paddle fan to switch the direction of rotation. The switching to provide electrical power to the paddle fan and any light fixture attached to it is normally placed in a wall-mounted switch box for easy access by the home owner. In this section, we look at the switching connections needed to control a ceiling-suspended paddle fan/light.

There are several styles of paddle fan control switches available for an electrician to install. A common switch style is shown in Figure 13–22. The rotary switch is used to control the paddle fan. It has three specific speed settings: low, medium, and high. The single-pole switch provides control of the attached lighting fixture. A two-gang electrical device box must be installed for this switch combination. Another common switching installation to control a ceiling-suspended paddle fan/light is to use two single-pole switches. One switch controls the paddle fan, and the other switch controls the lighting fixture. The speed of the paddle fan will need to be set by a switch located on the fan itself in this switch arrangement. Like the other switching circuits described in this chapter, the power source feed for a ceiling-suspended paddle fan switching circuit can originate at the paddle fan itself or at the switching location.

Ceiling-Suspended Paddle Fan/Light Circuit 1

Follow these steps when installing individual single-pole switch control for a ceiling-suspended paddle fan with an attached light fixture. The power source will feed the switch

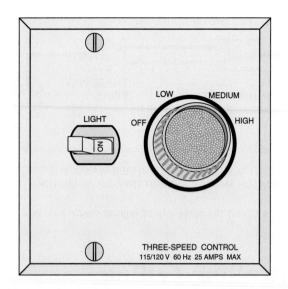

Figure 13–22 A common switch style for a ceiling-suspended paddle fan/light. The single-pole switch controls the lighting fixture attached to the paddle fan, and the rotary switch provides three-speed control of the paddle fan.

location. For this circuit, assume the use of nonmetallic boxes and nonmetallic sheathed cable. The outlet box used for the paddle fan/light must be listed and identified as suitable for the purpose (Figure 13–23):

1. Always wear safety glasses and observe all applicable safety rules.
2. At the switch location, using a wirenut, connect the bare grounding conductors together and, using two bare pigtails, connect all the grounding conductors to the green grounding screws on the two single-pole switches. Remember that a nonmetallic box does not require the circuit-grounding conductor to be connected to it.
3. At the switch location, connect the white insulated grounded conductors together with a wirenut.
4. At the switch location, using a wirenut, connect the black "hot" conductor of the incoming power source cable to one end of two black pigtails. Connect the other end of the black pigtails to a terminal screw on each single-pole switch. It does not matter which screw is used on each switch.

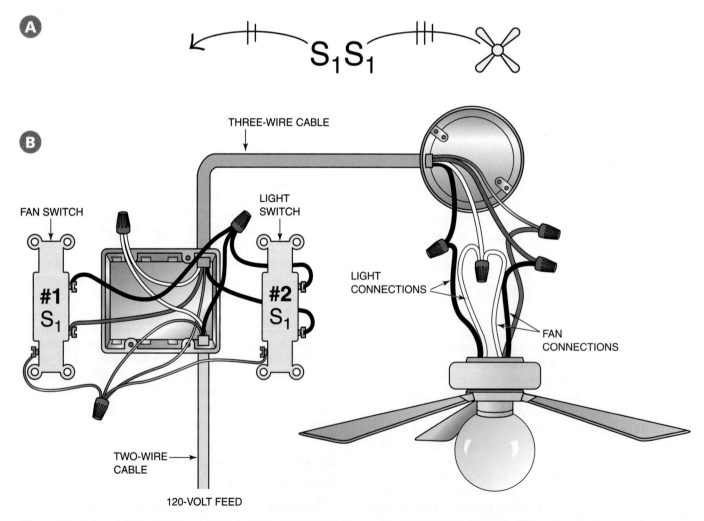

Figure 13–23 A cabling diagram (A) and a wiring diagram (B) for a switching circuit that has individual single-pole switch control for a ceiling-suspended paddle fan with an attached light fixture. The power source will feed the switch location.

5. At the switch location, connect the black conductor in the three-wire cable going to the paddle fan/light to the other terminal screw on single-pole switch 2.

6. At the switch location, connect the red conductor in the three-wire cable going to the paddle fan/light to the other terminal screw on single-pole switch 1.

7. At the paddle fan/light location, use a wirenut to connect the bare grounding conductor to the green insulated grounding conductor on the paddle fan/light. Typically, the paddle fan/light mounting bracket also has a green insulated grounding pigtail that must be connected to the incoming grounding conductor.

8. At the paddle fan/light location, using a wirenut, connect the white insulated grounded conductor to the white insulated grounded conductor of the paddle fan/light.

9. At the paddle fan/light location, using a wirenut, connect the black insulated conductor of the three-wire cable to the insulated conductor of the paddle fan/light that is labeled "Light Fixture."

10. At the paddle fan/light location, using a wirenut, connect the red insulated conductor of the three-wire cable to the insulated conductor of the paddle fan/light that is labeled "Paddle Fan."

Summary

This chapter presented the information that residential electricians need to know when they are roughing-in the wiring for the switching circuits found in residential wiring. An electrician must know the common connection techniques for single-pole, three-way, and four-way switching. Only by knowing how these switch types are connected into a circuit will the electrician know how many wires to run between switches and their loads. Much of the information presented will also help an electrician understand how to connect switches in electrical device boxes during the trim-out stage after the wall and ceiling finish material has been installed.

An overview of the common switch types was presented, and some rules to follow when selecting the appropriate switch type were covered. *NEC*® requirements from Article 404 that apply to switch selection were also presented. The procedures for making several common switch connections found in residential wiring were covered in detail. The switching situations covered in this chapter cover most of the switching situations that a residential electrician will encounter. A good understanding of the switching connection techniques described in this chapter will allow an electrician to correctly wire any switch situation that may arise.

Review Questions

Directions: Answer the following items with clear and complete responses.

1. Explain how the *NEC*® and Underwriters Laboratories categorize general-use snap switches.

2. Determine the switch type and rating for a single-pole switch controlling eight 100-watt incandescent lamps in a 120-volt residential switching circuit.

3. Describe the type of wiring situation that would require the use of a single-pole switch.

4. Describe the type of wiring situation that would require the use of three-way switches.

5. The *NEC*® requires all switching to be done in the _____ circuit conductor. The section of the *NEC*® that states this rule is _____.

6. Draw a line from the switch type to the correct number of terminal screws for the switch.

   single-pole switch   four terminal screws
   three-way switch   three terminal screws
   four-way switch   two terminal screws

7. The number of conductors required to be installed between three-way and four-way switches is _____.

8. Explain why it is common wiring practice to switch the bottom half of a split-duplex receptacle.

9. Describe a situation where you would need to use a combination single-pole/three-way switch.

10. A switching circuit used to control a lighting load from six locations would require _____ three-way switches and _____ four-way switches.

11. Complete the connections in the following diagram. Assume the use of nonmetallic sheathed cable and nonmetallic electrical boxes. A 120-volt feed is brought to the switch box, and the single-pole switch will control the lighting outlet.

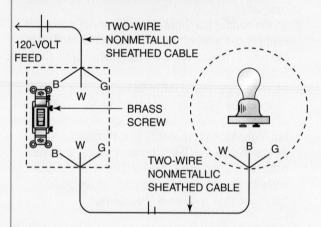

12. Complete the connections in the following diagram. Assume the use of nonmetallic sheathed cable and nonmetallic electrical boxes. The single-pole switch will control the lighting outlet, but notice that the power source feed is brought to the lighting outlet. This is a switch loop wiring situation.

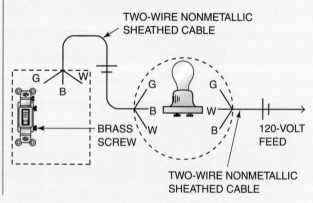

13 Complete the connections in the following diagram. Assume the use of nonmetallic sheathed cable and nonmetallic electrical boxes. Either three-way switch will control the lighting outlet. The 120-volt feed is brought to the first three-way switch.

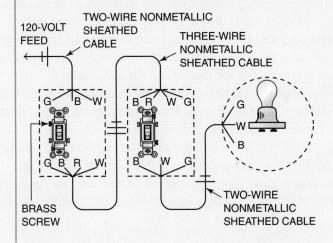

15 Complete the connections in the following diagram. Assume the use of nonmetallic sheathed cable and nonmetallic electrical boxes. The 120-volt feed is brought to the lighting outlet. All three switches will control the lighting outlet.

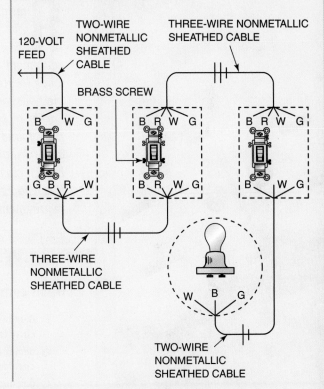

14 Complete the connections in the following diagram. Assume the use of nonmetallic sheathed cable and nonmetallic electrical boxes. The single-pole switch controls the lighting outlet but not the duplex receptacle. The duplex receptacle is "hot" at all times.

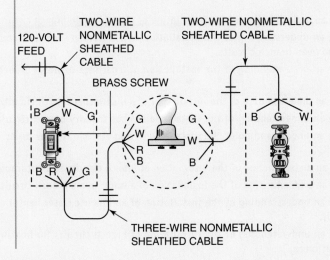

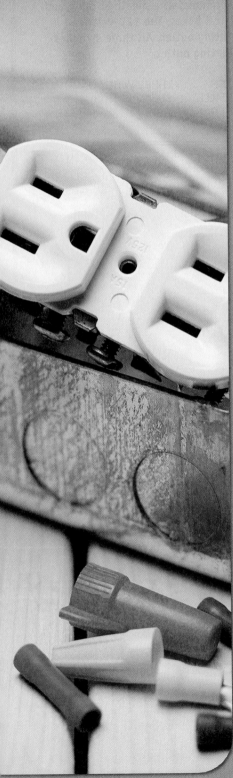

Chapter 14

Branch-Circuit Installation

A residential electrical system branch circuit is run from the main service panel or a subpanel to the various lighting and power outlets in the circuit. It consists of the conductors and other necessary electrical equipment needed to supply and control electrical power to the electrical loads located on the circuit. Branch circuits are installed using a variety of methods to achieve the goal of getting electrical power to the electrical loads. Earlier in this book, you learned that there are several types of branch circuits found in a residential electrical system. These branch circuits include general lighting branch circuits, small-appliance branch circuits, laundry branch circuits, bathroom branch circuits, and individual branch circuits. In this chapter, we take a close look at common installation methods and requirements for specific types of branch circuits found in residential wiring.

OBJECTIVES

Upon completion of this chapter, the student should be able to:

- demonstrate an understanding of the installation of general lighting branch circuits.
- demonstrate an understanding of the installation of small-appliance branch circuits.
- demonstrate an understanding of the installation of electric range branch circuits.
- demonstrate an understanding of the installation of countertop cook unit and wall-mounted oven branch circuits.
- demonstrate an understanding of the installation of a garbage disposal branch circuit.
- demonstrate an understanding of the installation of a dishwasher branch circuit.
- demonstrate an understanding of the installation of the laundry branch circuit.
- demonstrate an understanding of the installation of an electric clothes dryer branch circuit.
- demonstrate an understanding of the installation of branch circuits in a bathroom.
- demonstrate an understanding of the installation of a water pump branch circuit.
- demonstrate an understanding of the installation of an electric water heater branch circuit.
- demonstrate an understanding of the installation of branch circuits for heating and air conditioning.
- demonstrate an understanding of the installation of branch circuits for electric heating.
- demonstrate an understanding of the installation of a branch circuit for smoke detectors.
- demonstrate an understanding of the installation of a low-voltage chime circuit.

Glossary of Terms

cord-and-plug connection an installation technique where electrical appliances are connected to a branch circuit with a flexible cord with an attachment plug; the attachment plug end is plugged into a receptacle of the proper size and type

counter-mounted cooktop a cooking appliance that is installed in the top of a kitchen countertop; it contains surface cooking elements, usually two large elements and two small elements

dual-element time-delay fuses a fuse type that has a time-delay feature built into it; this fuse type is used most often as an overcurrent protection device for motor circuits

electric range a stand-alone electric cooking appliance that typically has four cooking elements on top and an oven in the bottom of the appliance

hardwired an installation technique where the circuit conductors are brought directly to an electrical appliance and terminated at the appliance

heat pump a reversible air-conditioning system that will heat a house in cool weather and cool a house in warm weather

hydronic system a term used when referring to a hot water heating system

interconnecting the process of connecting together smoke detectors so that if one is activated, they all will be activated

inverse time circuit breaker a type of circuit breaker that has a trip time that gets faster as the fault current flowing through it gets larger; this is the circuit breaker type used in house wiring

jet pump a type of water pump used in home water systems; the pump and electric motor are separate items that are located away from the well in a basement, garage, crawl space, or other similar area

nameplate the label located on an appliance that contains information such as amperage and voltage ratings, wattage ratings, frequency, and other information needed for the correct installation of the appliance

submersible pump a type of water pump used in home water systems; the pump and electric motor are enclosed in the same housing and are lowered down a well casing to a level that is below the water line

thermostat a device used with a heating or cooling system to establish a set temperature for the system to achieve; they are available as line voltage models or low-voltage models; some thermostats can also be programmed to keep a home at a specific temperature during the day and another temperature during the night

transformer the electrical device that steps down the 120-volt house electrical system voltage to the 16 volts a chime system needs to operate correctly

wall-mounted oven a cooking appliance that is installed in a cabinet or wall and is separated from the counter-mounted cooktop

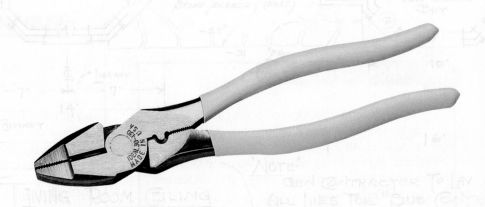

Installing General Lighting Branch Circuits

Since general lighting branch circuits make up the largest portion of the branch circuits installed in a residential wiring job, we start our discussion with them. A residential general lighting branch circuit will include the ceiling- and wall-mounted lighting outlets as well as most of the receptacles in a house. The only receptacles not found on a general lighting branch circuit will be those receptacles on small-appliance branch circuits, a bathroom circuit, or those located on an individual branch circuit for a specific appliance.

General lighting branch circuits will be wired with either a 14 AWG copper conductor and protected with a 15-ampere fuse or circuit breaker or a 12 AWG copper conductor protected with a 20-ampere fuse or circuit breaker. The general lighting branch circuit will originate in the service entrance panel or a subpanel and will be routed to the various lighting outlets, switching points for the lighting outlets, and receptacle outlets on the circuit.

In Chapter 7, you were taught how to determine the minimum number of general lighting branch circuits for a specific size of house. It is a good idea to review that information at this time. Since there is no maximum number, as long as you meet the *National Electrical Code® (NEC®)* requirements for the minimum number installed, you can install as many general lighting branch circuits as you want. It is common practice to install more general lighting branch circuits than the minimum number required by the *NEC®*. It is up to the electrician installing the residential electrical system just how many general lighting branch circuits are ultimately installed.

It is also up to the electrician installing a general lighting branch circuit to determine how many lighting outlets and how many receptacle outlets will be on each circuit. Electricians should try to divide the number of lighting and receptacle outlets on each circuit evenly among all the general lighting branch circuits. This is not always practical, but the effort should be made. It is common practice for some general lighting branch circuits to have more (or less) lighting

and receptacle outlets than on other general lighting branch circuits. As we mentioned before, most general lighting branch circuits will have both lighting outlets and receptacle outlets on the circuit; however, sometimes a general lighting branch circuit will have only lighting outlets *or* receptacle outlets. While there is no limit to the number of receptacles that can be connected to a general lighting branch circuit in a residential installation, a good rule of thumb is to assign each receptacle outlet a value of 1.5 amperes. Using this assumption, the maximum number of receptacles on a 15-amp-rated general lighting branch circuit would be 10 (15 amps/1.5 amps for each receptacle = 10). The maximum number of receptacles on a 20-amp-rated general lighting branch circuit would be 13 (20 amps/1.5 amps for each receptacle = 13). For those general lighting branch circuits that have both lighting outlets and receptacle outlets, it is recommended that not more than a total of 8 outlets (either lighting or receptacle) be put on one 15-amp-rated general lighting branch circuit and not more than 10 outlets (either lighting or receptacle) on a 20-amp-rated general lighting branch circuit. These numbers take into account the possibility of multiple lamp lighting fixtures being installed or the installation of high-wattage lamps in the lighting fixtures.

CAUTION

CAUTION: The *NEC®* permits loading an overcurrent protection device to 100% of its rating only if the device has a listing that allows a loading of 100%. Underwriters Laboratories (UL) has no 100% listing for molded-case circuit breakers of the type used in residential wiring applications. Normally, they are listed only for loading of up to 80% of their ratings. Fuses, on the other hand, can be loaded to 100% of their ratings if an electrician so chooses.

FROM EXPERIENCE

It is really up to the electrical contractor as to whether 15- or 20-amp-rated general lighting branch circuits are to be installed. Also, some building plan electrical specifications may require all wiring in a house to be a minimum of 12 AWG copper with 20-amp-rated overcurrent protection devices.

FROM EXPERIENCE

It is a good idea for electricians planning and installing general lighting branch circuits to include some lighting and receptacle outlets in a room on different circuits. This practice can help balance the load more evenly and may result in using less cable or conduit when wiring the installation. Another benefit to this practice is that should a circuit breaker trip on one circuit, the other outlets protected by a circuit breaker on another circuit will still provide some light and electrical power in the room.

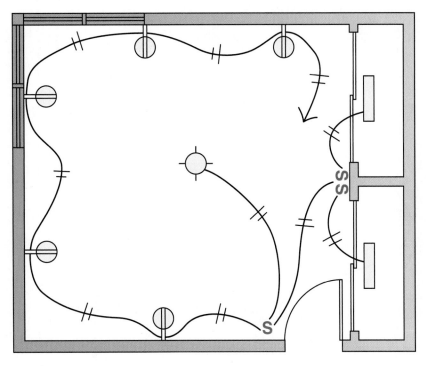

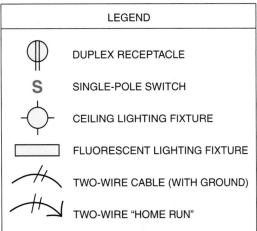

Figure 14–1 **A partial electrical plan showing the cabling diagram of a general lighting circuit serving this area of the house. Notice that both lighting outlets and receptacle outlets are included on this branch circuit.**

When roughing in general lighting branch circuits, there are a few steps to always follow:

1. Determine whether you will be installing 15- or 20-amp-rated general lighting branch circuits. Remember to use 14 AWG wire for 15-amp-rated branch circuits and 12 AWG wire for 20-amp-rated branch circuits.
2. Determine the number of lighting and receptacle outlets to be included on the general lighting branch circuit (Figure 14–1).
3. If they have not been installed previously, install the required lighting outlet boxes, switching location boxes, and receptacle outlet boxes at the appropriate locations.

4. Starting at the electrical panel where the circuit overcurrent protection device is located, install the wiring to the first electrical box on the circuit. Be sure to mark the cable or other wiring method with the circuit number and type and the area served at the panel. For example, general lighting branch circuit 1 serving the master bedroom can be marked "GL1MBR."
5. Continue roughing in the wiring for the rest of the circuit, following the installation practices described in previous chapters. Make sure to tuck the wiring in the back of all electrical boxes on the circuit.
6. Double-check to make sure all circuit wiring for the general lighting branch circuit you are installing has been roughed in.

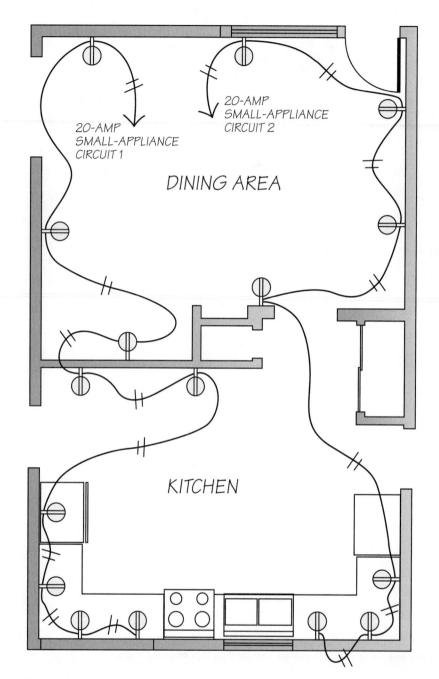

Figure 14–2 A partial electrical plan showing the cabling diagram of two small-appliance branch circuits serving a kitchen and dining room area. No lighting outlets are allowed on these circuits.

Installing Small-Appliance Branch Circuits

Small-appliance branch circuits are the circuits that supply electrical power to receptacles in kitchens, dining rooms, pantries, and other similar areas of a house. The *NEC®* requires that they must be protected with 20-ampere overcurrent protection devices. The *NEC®* also requires a minimum of two small-appliance branch circuits be installed in each house. However, in larger homes, it is normal wiring practice to install more than two small-appliance branch circuits to adequately feed all the small-appliance loads. There is no maximum number that may be installed. When small-appliance branch circuits are used to feed the countertop receptacles in a kitchen or dining area, at least two different small-appliance branch circuits must be used. It is common wiring practice to split the countertop receptacles as evenly as possible on two different small-appliance branch circuits. The *NEC*® allows no other outlets (lighting or receptacle) on a small-appliance branch circuit except a clock receptacle and/or the receptacle behind a gas-fired range. Therefore, ceiling lighting outlets, under-cabinet lighting, hood fan/lights, outside receptacles, garbage disposals, dishwashers, trash compactors, or any other electrical load in a kitchen may not be connected to a small-appliance branch circuit.

Electricians wire small-appliance branch circuits with 12 AWG copper conductors, although a larger-size wire, such as a 10 AWG, could be used to compensate for voltage drop problems when the distance back to the electrical panel is very long. Like general lighting branch circuits, the small-appliance branch circuits originate in the service entrance panel or a subpanel and are routed to the various receptacle outlet boxes on the circuit. Also, like general lighting branch circuits, there is no maximum number of receptacles that can exist on a small-appliance branch circuit, but again it is good wiring practice to limit the number of receptacles on them to 13 (20 amps/1.5 amps per receptacle = 13).

When roughing in small-appliance branch circuits, there are a few steps to always follow:

1. Use 12 AWG wire and 20-amp overcurrent protection devices for all small-appliance branch circuits.
2. Always install a minimum of two small-appliance branch circuits.
3. Determine the number of receptacle outlets to be included on each small-appliance branch circuit (Figure 14–2).
4. If they have not been installed previously, install the required receptacle outlet boxes at the appropriate locations.
5. Starting at the electrical panel where the overcurrent protection device is located, install the wiring to the first electrical box on the circuit. Be sure to mark the cable or other wiring method with the circuit number, type, and the area served at the panel. For example, small-appliance branch circuit 1 serving the kitchen countertop can be marked "SA1KCT."
6. Continue roughing in the wiring for the rest of the circuit, following the installation practices described in previous chapters. Make sure to tuck the wiring in the back of all electrical boxes on the circuit.
7. Double-check to make sure all circuit wiring for the small-appliance branch circuit you are installing has been roughed in.

Installing Electric Range Branch Circuits

In addition to the small-appliance branch circuits and the general lighting branch circuits serving the kitchen area in a house, there are several other branch-circuit types required.

Electric ranges, counter-mounted cooktops, built-in ovens, garbage disposals, dishwashers, and other appliances are commonly found in residential kitchens and must have branch-circuit wiring installed with the proper voltage and ampere ratings. In this section, a detailed look at the installation requirements for electric ranges is presented.

To install the branch circuit for a range, an electrician must first determine the minimum size of the wire and maximum overcurrent protection device required. In Chapter 7, you were shown how to determine this information (see Figures 7–8 and 7–9 for a review of the procedure). The branch circuit will be rated at 120/240 volts and will need a conductor size that will handle, at a minimum, the calculated load. Electric ranges are usually connected to the electrical system in a house through a **cord-and-plug connection.** An 8/3 copper cable with ground protected by a 40-ampere circuit breaker or a 6/3 copper cable with ground protected by a 50-ampere circuit breaker is often used. Nonmetallic sheathed cable, Type AC cable, Type MC cable, or Type SER service entrance cable may be used for the branch-circuit wiring method. Three conductors (with a fourth grounding conductor) of the proper size installed in a raceway may also be used. An electrician will install the wiring method between the electrical panel and the location of the receptacle for the range. The receptacle will be either surface mounted or flush mounted. Flush mounting the receptacle will require an electrical box. The receptacle location is usually close to the floor at the rear of the final position of the electric range. Access to the receptacle is by removing the bottom drawer of the range or simply by pulling the range out from the wall. The wiring method is attached to either the electrical box for flush mounting or the receptacle body itself for surface mounting. Assuming the use of a cable wiring method, the wires of the branch circuit will be connected to the receptacle terminals as follows (Figure 14–3):

1. Connect the white grounded neutral conductor to the terminal marked "W."
2. Connect the black ungrounded conductor to the terminal marked "Y."
3. Connect the red ungrounded conductor to the terminal marked "X."
4. Connect the green or bare equipment-grounding conductor to the terminal marked "G."

The cord that connects the range to the receptacle is usually not installed on the range by the appliance dealer, and an electrician often will have to install it. Make sure that the cord is a four-wire cord and has a plug that will properly match the receptacle that it will be plugged into. Follow the manufacturer's instructions, often found on the back of a range, for the correct wiring connections for the cord.

Sometimes the electrical installation calls for the connection to a range to be **hardwired.** This is done by bringing the branch-circuit wiring method directly to the back of a

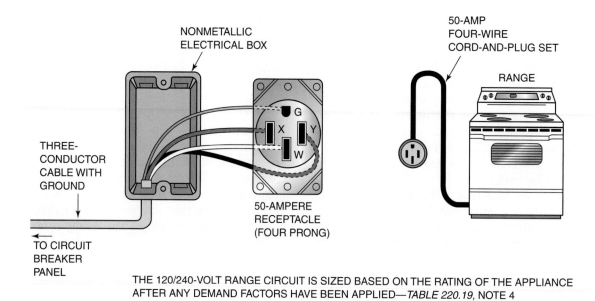

THE 120/240-VOLT RANGE CIRCUIT IS SIZED BASED ON THE RATING OF THE APPLIANCE
AFTER ANY DEMAND FACTORS HAVE BEEN APPLIED—*TABLE 220.19*, NOTE 4

Figure 14–3 A four-wire range installation using a cord-and-plug assembly.

range and making the terminal connections on the terminal block on the range. Assuming a cable wiring method, the connections are made as follows (Figure 14–4):

1. Connect the white grounded neutral conductor to the silver screw terminal on the range terminal block.
2. Connect the black ungrounded conductor to the brass screw on the range terminal block.
3. Connect the red ungrounded conductor to the other brass screw on the range termninal block.
4. Connect the green or bare equipment-grounding conductor to the green grounding screw on the range terminal block.

Although three-conductor range installations have not been allowed by the *NEC®* in new residential construction since 1996, electricians may run into this wiring situation when installing new ranges in homes built prior to 1996. If this is the case, the *NEC®* allows the existing three-wire branch circuit to stay in place, and a three-wire cord-and-plug connection will have to be made to the range. The connections for attaching the three-wire cord to the range (or dryer) are made as follows (Figure 14–5):

1. Connect the black (or one of the outside conductors of the cord assembly) to one of the brass screws on the terminal board.
2. Connect the red (or the other outside conductor of the cord assembly) to the other brass terminal screw on the terminal board.
3. Connect the white (or middle wire on the three-wire cord assembly) to the silver screw on the terminal board.
4. Connect a jumper, bare or green, from the silver screw grounded neutral terminal on the terminal board to

the green screw grounding terminal. Check to make sure that the green screw location is bonded (connected) to the frame of the range. Note 1: This jumper is usually installed at the factory, but if it is not, you will have to install one. Note 2: This jumper must be removed and discarded if this range is installed on a four-wire system in a new installation and the jumper was installed at the factory.

Installing the Branch Circuit for Counter-Mounted Cooktops and Wall-Mounted Ovens

Many home owners choose a separate **counter-mounted cooktop** and either one or two **wall-mounted ovens** instead of a single stand-alone range. Installing a branch circuit to these appliances is very similar to installing a branch circuit to a stand-alone range. Each countertop cook unit and wall-mounted oven normally comes from the factory equipped with a short length of flexible metal conduit enclosing the feed wires for the appliance. The feed wires are already connected to the terminal block in the appliance, and the only electrical connection that an electrician will need to make is the connection of the feed wires to the branch-circuit wiring method brought from the electrical panel to the appliance.

There are three common methods used to install the branch circuit and make the necessary connections to counter-mounted cooktops and wall-mounted ovens. The

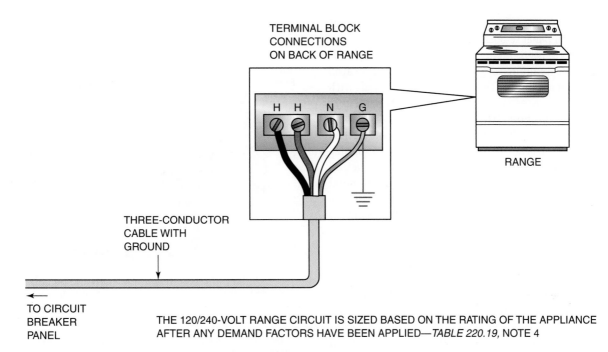

THE 120/240-VOLT RANGE CIRCUIT IS SIZED BASED ON THE RATING OF THE APPLIANCE AFTER ANY DEMAND FACTORS HAVE BEEN APPLIED—*TABLE 220.19*, NOTE 4

Figure 14–4 A four-wire range installation using the hardwired method.

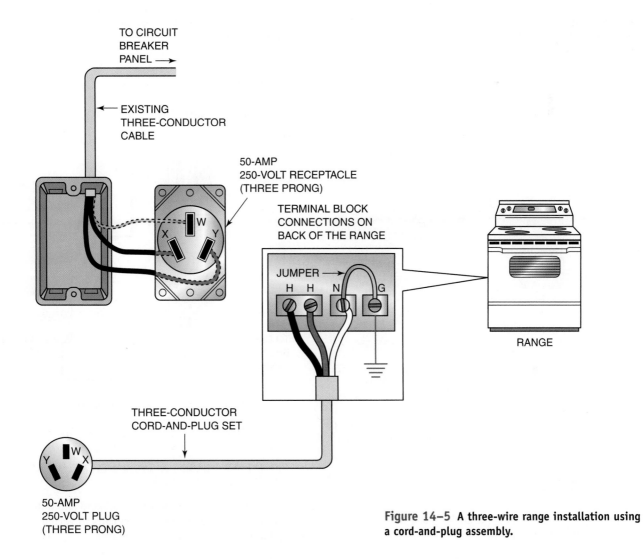

Figure 14–5 A three-wire range installation using a cord-and-plug assembly.

first method involves running a separate branch circuit from the electrical panel to both the countertop cook unit and the wall-mounted oven (Figure 14–6). The branch-circuit wiring method is brought to a metal junction box located near the cooking appliance. The flexible metal conduit with the appliance feed wires is installed into the junction box, and the proper connections are made using wirenuts. A blank cover is then put on the junction box. In the second method (Figure 14–7), an electrician runs one larger conductor-size branch circuit from the electrical panel to a metal junction box located close to both appliances. The flexible metal conduit–enclosed appliance wires from both appliance types are connected to the junction box, and the proper connections to the branch circuit wiring are made in the junction box using wirenuts. In the third method (Figure 14–8), individual branch-circuit wiring is installed to a receptacle outlet box located in a cabinet near the cooking appliance. The flexible metal conduit–enclosed wiring installed at the factory is disconnected, and a properly sized cord-and-plug assembly is attached to each unit. A receptacle with the proper voltage and amperage rating, which correctly matches the plug on the end of the newly installed appliance cord, will be installed in the electrical box. Each appliance will then be plugged into the receptacle to get electrical power. One advantage to this last method is the fact that if one of the appliances malfunctions and requires servicing, it is relatively easy to unplug it and remove the unit from its location and take it to a service center for repair.

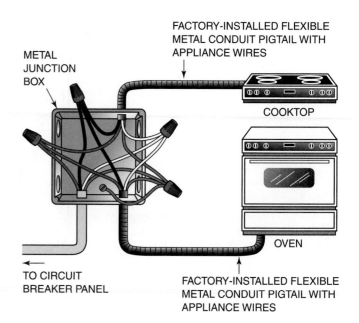

CIRCUIT IS SIZED LARGE ENOUGH FOR BOTH THE COOKTOP AND THE OVEN—*TABLE 220.19*, NOTE 4

Figure 14–7 A counter-mounted cooktop and wall-mounted oven installation with both appliances fed from the same branch circuit.

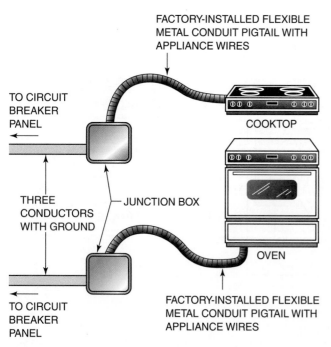

CIRCUIT IS SIZED LARGE ENOUGH FOR EACH APPLIANCE. USE THE NAMEPLATE RATING—*TABLE 220.19*, NOTE 4

Figure 14–6 A counter-mounted cooktop and wall-mounted oven installation using separate branch circuits to each unit.

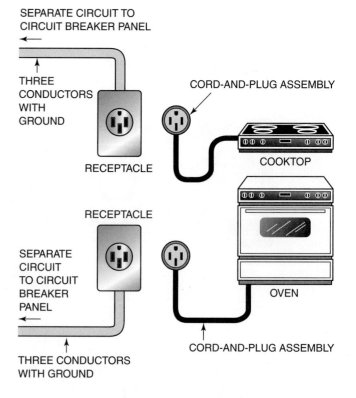

CIRCUIT SIZED BASED ON THE NAMEPLATE RATING OF EACH APPLIANCE—*TABLE 220.19*, NOTE 4

Figure 14–8 A counter-mounted cooktop and wall-mounted oven installation using separate circuits to each unit. The appliances are cord-and-plug connected.

With any of the three methods, three conductor cables or three conductors in a raceway will be required in the branch circuit since these appliances will need to be supplied with 120/240 volts to operate correctly. As usual, an equipment-grounding conductor will also be needed to properly ground the frames of the cooking appliances. Nonmetallic sheathed cable, Type AC cable, Type MC cable, and Type SER service entrance cable are common cable wiring methods used for the branch-circuit installation.

To install the branch circuit for a countertop cooking unit and/or a wall-mounted oven, an electrician must first determine the minimum size of the wire and maximum overcurrent protection device required for the installation method used. In Chapter 7, you were shown how to determine this information if the second method described previously is used and one larger branch circuit is sized to feed both a countertop cooking unit and one or two wall-mounted ovens (see Figure 7–10 for a review of the calculation). With both the first and the third installation methods, Table 220.19, Note 4, of the *NEC®* tells us that the **nameplate** wattage rating of each individual appliance is used. This rating is divided by 240 volts to get the current rating of the appliance. It is this amperage that is used to size the branch-circuit conductors and the branch-circuit overcurrent protection device. Here is an example: A countertop cook unit with a nameplate rating of 6.6 kilowatts and a wall-mounted oven with a nameplate rating of 7.5 kilowatts is to be installed in a house with separate branch circuits run to each unit.

- Cooktop load in amps: 6,600 watts/240 volts = 27.5 amps.
- Cooktop wire size: The minimum conductor size according to Table 310.16 would be a 10 AWG copper.
- Cooktop fuse or circuit breaker size: The overcurrent protection device would be a 30-amp-rated fuse or circuit breaker.
- Wall-mounted oven load in amps: 7,500 watts/240 volts = 31.25 amps.
- Wall-mounted oven wire size: The minimum conductor size according to Table 310.16 would be an 8 AWG copper.
- Wall-mounted oven fuse or circuit breaker size: The overcurrent protection device would be a 40-amp-rated fuse or circuit breaker.

Installing the Garbage Disposal Branch Circuit

Most new and existing homes have a garbage disposal installed under the kitchen sink. It is up to the electrical contractor to install a branch circuit to feed the disposal unit and to make the electrical connections at the unit itself. It is normally the plumbing contractor who actually installs the garbage disposal under the sink. The branch-circuit size and the overcurrent protection device are sized on the basis

of the nameplate rating of the disposal. Residential garbage disposals are rated at 120 volts and are normally installed on a separate 20-amp-rated branch circuit. The branch-circuit wire size will be 12 AWG copper.

CAUTION: Remember that appliances such as garbage disposals may not be connected to a small-appliance branch circuit supplying receptacles in a kitchen.

Two installation methods are used by electricians to connect electrical power to the garbage disposal. The first method involves running the wiring method to a single-pole switch and then hardwiring directly to the garbage disposal. The switch is usually located above the kitchen countertop near the kitchen sink (Figure 14–9). The second method involves installing a receptacle outlet under the kitchen sink during the rough-in stage that will be controlled by a single-pole switch located above the countertop and near the sink (Figure 14–10). A cord-and-plug assembly that according to Section 422.16 cannot be shorter than 18 inches (450 mm) and not longer than 3 feet (900 mm) is connected by the electrician to the disposal unit. The cord is plugged into the receptacle, and when the switch is on, electrical power is supplied to the disposal (Figure 14–11). This method has one major advantage over the hardwired method in that if a garbage disposal malfunctions and requires servicing, it is a simple job to unplug it and remove the appliance for servicing.

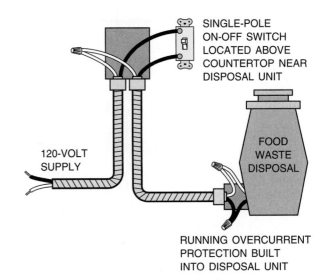

Figure 14–9 The branch-circuit wiring for a hardwired garbage disposal operated by a separate switch located above the countertop and near the sink. The wiring method shown is Type AC cable (BX).

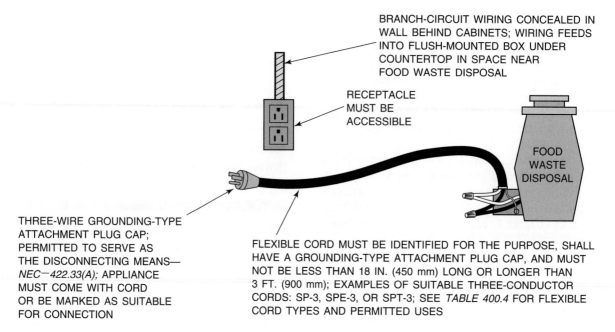

BRANCH-CIRCUIT WIRING CONCEALED IN
WALL BEHIND CABINETS; WIRING FEEDS
INTO FLUSH-MOUNTED BOX UNDER
COUNTERTOP IN SPACE NEAR
FOOD WASTE DISPOSAL

RECEPTACLE
MUST BE
ACCESSIBLE

FOOD
WASTE
DISPOSAL

THREE-WIRE GROUNDING-TYPE
ATTACHMENT PLUG CAP;
PERMITTED TO SERVE AS
THE DISCONNECTING MEANS—
NEC—422.33(A); APPLIANCE
MUST COME WITH CORD
OR BE MARKED AS SUITABLE
FOR CONNECTION

FLEXIBLE CORD MUST BE IDENTIFIED FOR THE PURPOSE, SHALL
HAVE A GROUNDING-TYPE ATTACHMENT PLUG CAP, AND MUST
NOT BE LESS THAN 18 IN. (450 mm) LONG OR LONGER THAN
3 FT. (900 mm); EXAMPLES OF SUITABLE THREE-CONDUCTOR
CORDS: SP-3, SPE-3, OR SPT-3; SEE *TABLE 400.4* FOR FLEXIBLE
CORD TYPES AND PERMITTED USES

Figure 14–10 A typical garbage disposal installation using a cord-and-plug assembly.

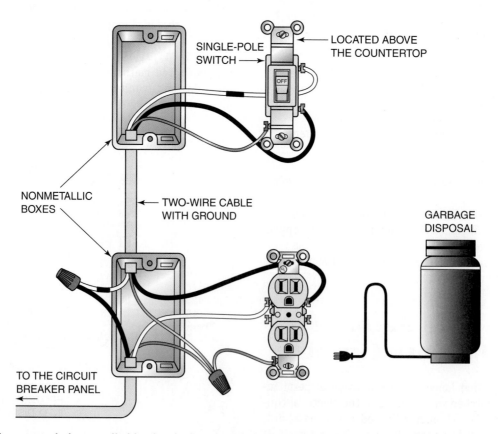

SINGLE-POLE
SWITCH

LOCATED ABOVE
THE COUNTERTOP

OFF

NONMETALLIC
BOXES

TWO-WIRE CABLE
WITH GROUND

GARBAGE
DISPOSAL

TO THE CIRCUIT
BREAKER PANEL

Figure 14–11 The receptacle is controlled by the single-pole switch. The garbage disposal is connected to the receptacle with a cord-and-plug assembly.

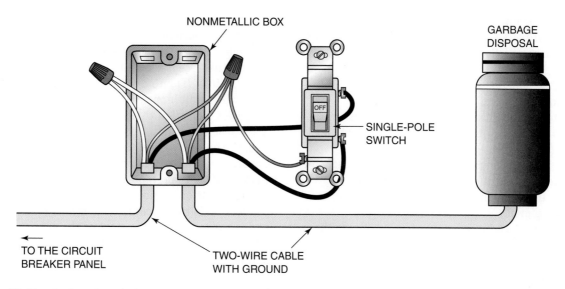

Figure 14–12 The single-pole switch controls the garbage disposal, which is hardwired into the branch circuit.

Figure 14–12 shows a typical branch-circuit installation and switch connection for a garbage disposal. The electrical connections at the garbage disposal are not very complicated. Assuming the use of a cable wiring method, the wiring connections in the disposal junction box are done as follows:

1. Loosen the screw that secures the blank cover to the junction box and take off the cover.
2. Strip the outside sheathing back about 6 inches and attach the cable to the disposal junction box though the 1/2-inch knockout located on the unit using an approved connector.
3. Connect the incoming equipment-grounding conductor to the green screw or green grounding jumper.
4. Wire-nut the white incoming grounded conductor to the white pigtail.
5. Wire-nut the black incoming ungrounded wire to the black pigtail.
6. Carefully push the wire connections into the junction box area and put the blank cover back on.

Installing the Dishwasher Branch Circuit

There are very few homes today that do not have automatic dishwashers installed. The plumbing contractor usually installs the dishwasher along with the necessary water and drain connections. It is up to the electrical contractor to supply and install a branch circuit to the dishwasher and make the necessary electrical connections at the dishwasher itself. Also, like a garbage disposal, there are two common ways to electrically connect a dishwasher: (1) hardwire it in the branch circuit or (2) use a cord-and-plug connection. It is this author's experience that hardwiring dishwashers is

the most often used method, but you may find in your area that cord-and-plug connections are normally done.

The branch-circuit size and the overcurrent protection device are sized on the basis of the nameplate rating of the dishwasher. Residential dishwashers are rated at 120 volts and are normally installed on a separate 20-amp-rated branch circuit. The branch-circuit wire size will be 12 AWG copper. Nonmetallic sheathed cable, Type AC cable, or Type MC cable are common cable wiring methods used for the branch-circuit wiring.

When wiring the dishwasher, an electrician will install the branch-circuit wiring from the electrical panel to the dishwasher location. The wiring method used will be brought to the dishwasher location from underneath (through the basement) or from behind (through a wall). Each dishwasher will have a junction box on the appliance where (for the hardwired method) the branch-circuit wiring is terminated. During the rough-in stage of the electrical system installation, enough cable needs to be left at the dishwasher location so that when the kitchen cabinets are finally installed and the dishwasher is moved into its final position, there will be enough wire to make the necessary connections. The electrical connections will be made in the dishwasher junction box. This junction box is normally located on the right side at the front of the appliance. The bottom "skirt" on the dishwasher

 FROM EXPERIENCE

Trash compactors are often installed in homes. The branch-circuit installation and connection procedure is the same for a trash compactor as it is for a dishwasher.

will have to be removed to gain access to the junction box. Assuming the use of a cable wiring method, the wiring connections are done as follows (Figure 14–13):

1. Take off the junction box cover and take out the 1/2-inch knockout on the junction box.
2. Attach the wiring method used to the dishwasher junction box though the 1/2-inch knockout with an approved connector.
3. Connect the incoming equipment-grounding conductor to the green screw or green grounding jumper.
4. Wire-nut the white incoming grounded conductor to the white pigtail.
5. Wire-nut the black incoming ungrounded wire to the black pigtail.
6. Carefully push the wire connections into the junction box area and put the blank cover back on.

If a dishwasher is to be cord-and-plug connected, the installation is similar to that of a cord-and-plug-connected garbage disposal. The difference is that there is no need to switch the receptacle that the dishwasher cord will be plugged into (Figure 14–14). In this method, an electrician will install branch-circuit wiring to a receptacle outlet box at a location that is adjacent to the location of the dishwasher. It is not a good idea to locate the receptacle behind the dishwasher location since it will be virtually impossible to access it once the dishwasher is installed. Also, locating the receptacle in a basement below the dishwasher location is not allowed since the cord-and-plug connection is the disconnecting means for the dishwasher and is not located within sight of the appliance. The cord-and-plug assembly must be at least 3 feet (900 mm) long and no more than 4 feet (1.2 m) long according to Section 422.16.

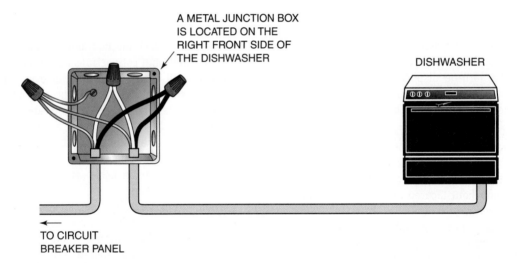

Figure 14–13 A hardwired dishwasher branch circuit.

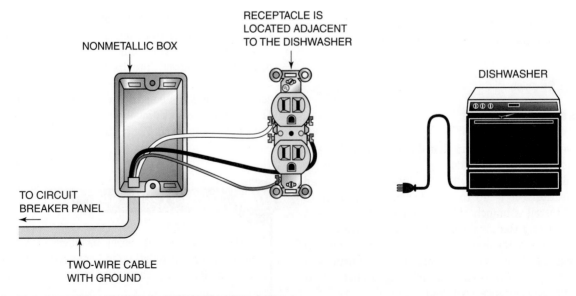

Figure 14–14 A cord-and-plug-connected dishwasher branch circuit.

Installing the Laundry Area Branch Circuits

A laundry branch circuit is used to supply electrical power to the laundry area of a house. The laundry area may be a location in a basement or garage where the clothes washer and clothes dryer may be located. Some houses have a specific laundry room that has the clothes washer, clothes dryer, and other appliances that assist the home owner when doing laundry.

Section 210.11(C)(2) of the *NEC*® requires the laundry branch circuit to be rated at 20 amps, which results in wiring it with 12 AWG wire. The laundry circuit may consist of only one receptacle outlet, such as when the clothes washer is located in the basement of a home and requires a receptacle to plug into for power. It can also consist of several receptacles located in a specific laundry room (Figure 14–15). In either case, no other outlets (lighting or receptacle) can be connected to the 20-amp laundry circuit.

An electrician will usually start the rough-in wiring of a laundry circuit from the panel or subpanel and then route the wiring to the laundry area. Section 210.50(C) of the *NEC*® requires that the receptacle for the clothes washer be located

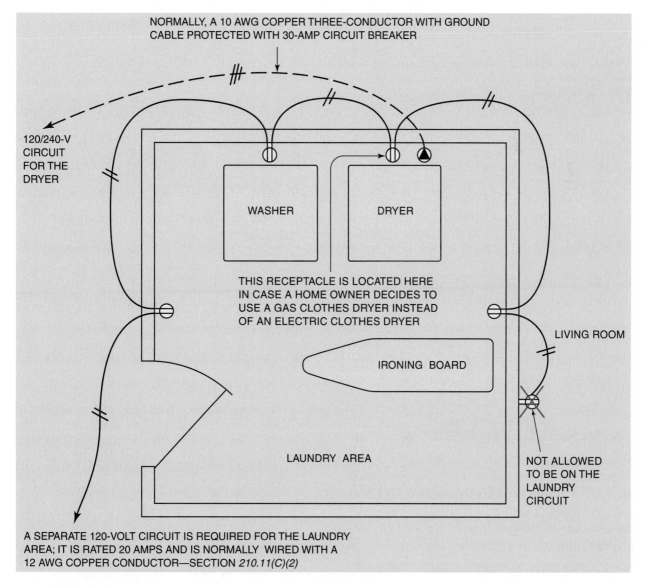

Figure 14–15 The laundry area in a house must be served by at least one 20-amp-rated laundry circuit. Only receptacles located in the laundry area can be connected to the circuit. An electric clothes dryer requires a separate 120/240-volt branch circuit.

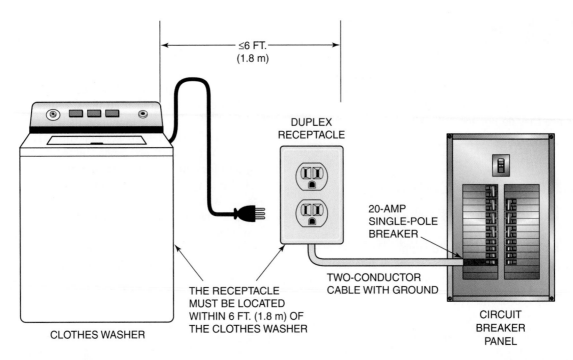

Figure 14–16 The receptacle on the laundry circuit that serves the clothes washer must be located no more than 6 feet (1.8 m) away from the washer.

within 6 feet (1.8 m) from the appliance location because most clothes washer cord-and-plug assemblies are 6 feet (1.8 m) long (Figure 14–16). If the clothes washer is located in a basement or garage area and is the only laundry appliance to be plugged into a receptacle, a surface-mounted box is often installed a little higher and just behind the location of the clothes washer, and the wiring method used is brought to that box. A single receptacle is recommended for this situation. If a laundry room is used, a determination of how many receptacle outlets required in the room is done and the wiring is roughed-in from box to box in that room. Remember that in a laundry room, regular receptacle spacing is not required. In other words, a laundry room can have receptacle outlets located so there is more than 12 feet between them if an electrician so chooses.

> ◤◤◤◤ **CAUTION** ◢◢◢◢
>
> **CAUTION: If the laundry area is located in an unfinished basement and a duplex receptacle is used to serve the washing machine, it must be GFCI protected. If a single receptacle is used and only the clothes washer is plugged into it, no GFCI protection is required.**

Installing the Electric Dryer Branch Circuit

Electric clothes dryers are also located in the laundry area of a house. However, they are not supplied by the 20-ampere laundry circuit discussed in the previous section. Instead, they are supplied by an individual branch circuit that is installed similar to the branch circuit for an electric range (See Figure 14–15).

To install the branch circuit for an electric clothes dryer, an electrician must first determine the minimum size of the wire and maximum overcurrent protection device required. The branch circuit will be rated at 120/240 volts and will need a conductor size that will handle, at a minimum, the calculated load. The nameplate of a dryer will give the load in watts. An electrician will simply need to divide the wattage rating by 240 volts to get the dryer load in amperes. For example, a dryer has a nameplate that says it is rated at 5.5 kilowatts. Therefore; 5,500 watts/240 volts = 23 amps. In this example, a 10/3 copper cable with ground and protected by a 30-ampere circuit breaker is required. This is the size of wire and circuit breaker that is most often used for an electric dryer branch circuit installation. Three conductors (with a fourth grounding conductor) of the proper size installed in a raceway may also be used.

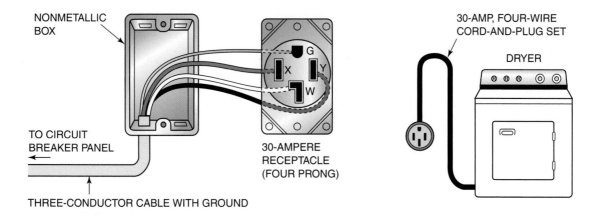

NONMETALLIC BOX

30-AMP, FOUR-WIRE CORD-AND-PLUG SET

DRYER

TO CIRCUIT BREAKER PANEL

30-AMPERE RECEPTACLE (FOUR PRONG)

THREE-CONDUCTOR CABLE WITH GROUND

THE 120/240-VOLT DRYER CIRCUIT IS SIZED BASED ON THE RATING OF THE APPLIANCE—*SECTION 220.18*

Figure 14–17 A cord-and-plug-connected clothes dryer branch-circuit installation. A four-wire connection is shown.

Electric clothes dryers are usually connected to the electrical system in a house through a cord-and-plug connection. An electrician will install the wiring method between the electrical panel and the location of the receptacle for the dryer. The receptacle location is usually behind the dryer and slightly higher than the top of the dryer for easy access. The receptacle will either be surface mounted or flush mounted. Flush mounting the receptacle will require an electrical box. The wiring method is attached to either the electrical box for flush mounting or the receptacle body itself for surface mounting. Assuming the use of a cable wiring method, the wires of the branch circuit will be connected to the receptacle terminals as follows (Figure 14–17):

1. Connect the white grounded neutral conductor to the terminal marked "W."
2. Connect the black ungrounded conductor to the terminal marked "Y."
3. Connect the red ungrounded conductor to the terminal marked "X."
4. Connect the green or bare equipment-grounding conductor to the terminal marked "G."

The cord-and-plug assembly that connects the dryer to the receptacle is often not installed by the appliance dealer. If this is the case, an electrician will have to install it. Make sure that the cord is a four-wire cord and has a plug that will properly match the receptacle that it is to be plugged into. Follow the manufacturer's instructions, often found on the back of the dryer, for the correct wiring connections for the cord to the terminal block at the back of the dryer.

Sometimes the electrical installation calls for the connection to a dryer to be hardwired. This is done by bringing the branch-circuit wiring method directly to the back of the clothes dryer and making the terminal connections on the terminal block on the dryer. Assuming the use of a 10/3 cable wiring method, the connections are made as follows (Figure 14–18):

1. Connect the white grounded neutral conductor to the silver screw terminal on the dryer terminal block.
2. Connect the black ungrounded conductor to the brass screw on the dryer terminal block.
3. Connect the red ungrounded conductor to the other brass screw on the dryer terminal block.
4. Connect the green or bare equipment-grounding conductor to the green grounding screw on the dryer terminal block.

Although three-conductor dryer installations have not been allowed by the *NEC®* in new residential construction since 1996, electricians may run into this wiring situation when installing a new dryer in homes built prior to 1996. If this is the case, the *NEC®* allows the existing three-wire branch circuit to stay in place, and a three-wire cord-and-plug connection will have to be made to the dryer. The connections for attaching the three-wire cord to the dryer are made as follows (Figure 14–19):

1. Connect the black (or one of the outside conductors of the cord assembly) to one of the brass screws on the terminal board.
2. Connect the red (or the other outside conductor of the cord assembly) to the other brass terminal screw on the terminal board.
3. Connect the white (or middle wire on the three-wire cord assembly) to the silver screw on the terminal board.
4. Connect a jumper, bare or green, from the silver screw grounded neutral terminal on the terminal board to the green screw grounding terminal. Check to make sure that the green screw location is bonded (connected) to the frame of the dryer. Note 1: This jumper is usually installed at the factory, but if it is not, you will have to install one. Note 2: This jumper must be removed and discarded if this dryer is installed on a four-wire system in a new installation and the jumper was installed at the factory.

Figure 14–18 A hardwired clothes dryer branch-circuit installation. A four-wire connection is shown.

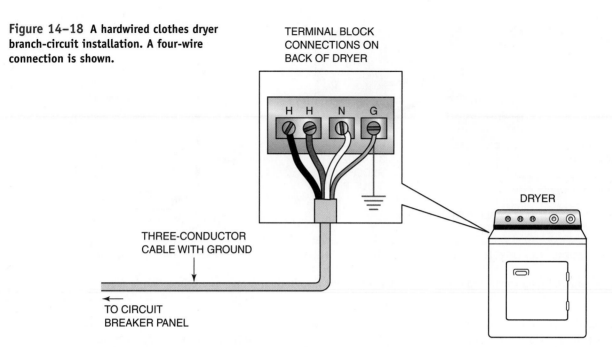

THE 120/240-VOLT DRYER CIRCUIT IS SIZED BASED ON THE RATING OF THE APPLIANCE—*SECTION 220.18*

Figure 14–19 A three-wire dryer branch-circuit installation.

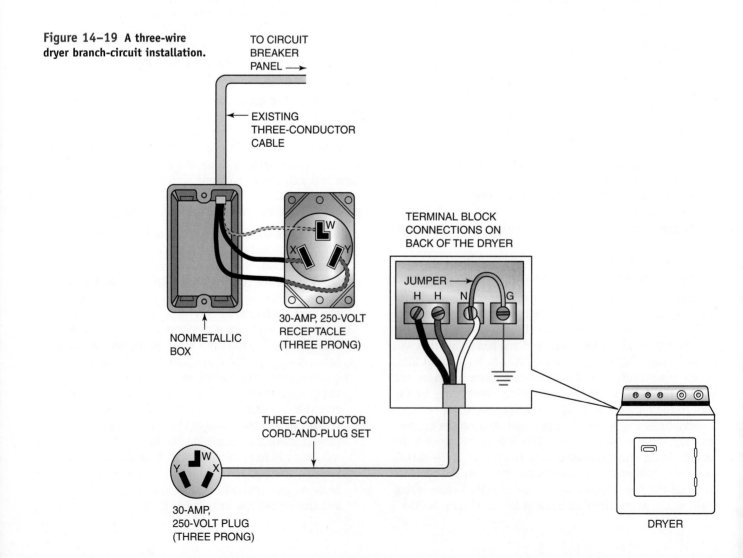

Installing the Bathroom Branch Circuit

Bathrooms are defined by the *NEC*® as an area that has a basin and one or more toilets, tubs, or showers. A bathroom circuit is a circuit that supplies electrical power to this area in a residential installation. The *NEC*® requires at least one 20-amp-rated branch circuit be installed to supply the bathroom receptacle or receptacles. You will remember that the *NEC*® requires only one receptacle in a bathroom and that it must be located no more than 3 feet (900 mm) from the edge of the basin. Additionally, each 15- and 20-amp 125-volt-rated receptacle in a bathroom must be GFCI protected (Figure 14–20).

The *NEC*® does allow one 20-amp-rated branch circuit to supply two (or more) bathrooms if certain conditions are met. These conditions were covered in Chapter 7 (see Figure 7–6 for a review). When roughing in bathroom branch circuits, there are a few steps to always follow:

1. Use 12 AWG wire and 20-amp overcurrent protection devices for all bathroom branch circuits.
2. Install at least one bathroom branch circuit for each bathroom.
3. Determine the number of receptacle outlets to be included on each bathroom branch circuit.
4. If they have not been installed previously, install the required lighting outlet and receptacle outlet boxes at the appropriate locations in the bathroom.
5. Starting at the electrical panel where the circuit overcurrent protection device is located, install the wiring to the first electrical box on the circuit. Be sure to mark the cable or other wiring method with the circuit number and type and the area served at the panel. For example, bathroom branch circuit 1 serving the first-floor bathroom can be marked "BATH1FF."
6. Continue roughing in the wiring for the rest of the circuit, making sure to follow the installation practices described in previous chapters. Make sure to tuck the wiring in the back of all electrical boxes on the circuit.

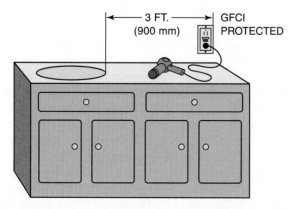

Figure 14–20 A bathroom receptacle must be located adjacent to the basin and no more than 3 feet (900 mm) away.

7. Double-check to make sure all circuit wiring for the bathroom branch circuit you are installing has been roughed in.

Installing a Water Pump Branch Circuit

Houses built in a town or city generally get their supply of water from a city water system. The water supply piping for the house is installed by the plumbing contractor, and the electricians have no circuitry to run for the water system. However, in rural areas where water system piping is not available, a well is used to supply water to the house. Getting the water from the well to the house will require a pump. It is usually up to the electricians wiring the house to supply a correctly sized water pump circuit to the control equipment for the pump. It is usually the plumbing contractor who, along with installing the necessary water piping throughout the house, will actually install the pump (with attached electric motor) and the control equipment used with the pump and water system. There are two types of water pumps commonly used: the deep-well jet pump and the submersible pump. In this section, we take a look at some common characteristics of each and what kind of electrical wiring must be done for these pumps.

Jet Pump

The **jet pump** and electric motor that turns it are usually located in a basement or crawl-space area, although other locations may be used. The electric motors are usually around 1 horsepower, single phase, capacitor start and operate on either 115 or 230 volts. (Note: Electric motors use the voltage designations of 115 and 230 volts instead of 120 and 240 volts.) Capacitor start electrical motors are designed to produce a high starting torque that will be enough to get the pump motor started. A 115/230-volt motor is called a dual-voltage electric motor and can operate at the same speed and produce the same horsepower at either voltage. When the motor is connected at the higher voltage (230 volts), the motor draws half as much current as when it is connected to operate on the lower voltage (115 volts). For this reason, it is common wiring practice to connect water pump motors to the higher voltage. You should remember that at the higher voltage of 230 volts (actually 240 volts from the circuit breaker panel), a two-pole circuit breaker is required.

The size of the circuit breaker and the circuit conductors is determined by the current draw of the motor. Article 430 of the *NEC*® tells us how to size the circuit conductors, the circuit breaker or fuses protecting the branch circuit, and the overload protection for the motor itself. Figure 14–21 shows a typical electrical circuit for a jet pump. The illustration shows the circuit wiring originating in a circuit breaker panel, going to the control equipment for the pump motor and then to the pump motor itself. In this example, the pump

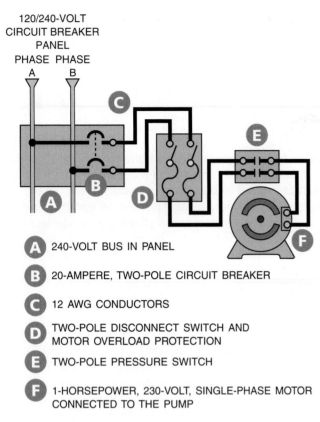

120/240-VOLT
CIRCUIT BREAKER
PANEL
PHASE   PHASE
A   B

A 240-VOLT BUS IN PANEL

B 20-AMPERE, TWO-POLE CIRCUIT BREAKER

C 12 AWG CONDUCTORS

D TWO-POLE DISCONNECT SWITCH AND
MOTOR OVERLOAD PROTECTION

E TWO-POLE PRESSURE SWITCH

F 1-HORSEPOWER, 230-VOLT, SINGLE-PHASE MOTOR
CONNECTED TO THE PUMP

Figure 14–21 A typical installation for a jet pump.

Table 430.148 Full-Load Currents in Amperes, Single-Phase Alternating-Current Motors
The following values of full-load currents are for motors running at usual speeds and motors with normal torque characteristics. Motors built for especially low speeds or high torques may have higher full-load currents, and multispeed motors will have full-load current varying with speed, in which case the nameplate current ratings shall be used.

The voltages listed are rated motor voltages. The currents listed shall be permitted for system voltage ranges of 110 to 120 and 220 to 240 volts.

Horsepower	115 Volts	200 Volts	208 Volts	230 Volts
⅙	4.4	2.5	2.4	2.2
¼	5.8	3.3	3.2	2.9
⅓	7.2	4.1	4.0	3.6
½	9.8	5.6	5.4	4.9
¾	13.8	7.9	7.6	6.9
1	16	9.2	8.8	8.0
1½	20	11.5	11.0	10
2	24	13.8	13.2	12
3	34	19.6	18.7	17
5	56	32.2	30.8	28
7½	80	46.0	44.0	40
10	100	57.5	55.0	50

Figure 14–22 Table 430.148 of the *NEC®. Reprinted with permission from NFPA 70-2002, National Electrical Code®, Copyright © 2001, National Fire Protection Association, Quincy, MA 02269. This reprinted material is not the complete and official Position of the NFPA on the referenced subject, which is represented only by the standard in its entirety.*

motor will be a 1-horsepower, single-phase motor connected at 230 volts. When calculating the size of branch-circuit conductors and the size of the fuse or circuit breaker for the circuit, Section 430.6 requires electricians to look at Table 430.148 (Figure 14–22). This table lets us determine the full-load current of a single-phase motor. For a 1-horsepower motor connected at 230 volts, Table 430.148 says that the full-load current is 8 amperes. Section 430.22 requires electricians to now multiply the full-load current of the motor by 125% and use that number to determine the minimum-size branch-circuit conductor. In this case, 8 amps × 125% = 10 amps. Table 310.16 is now used to size the minimum-size branch circuit. Using the "60°C" column, you will see that 14 AWG copper wire can be used. However, it is common wiring practice to install 12 AWG copper conductors to a water pump in case a larger pump may be needed in the future as the water supply requirements change. So, an electrician would install a 12/2 nonmetallic sheathed cable or a 12/2 armored cable or run 12 AWG conductors in a raceway system to the pump motor. The size of the circuit breaker or fuse is determined by following Section 430.52. This section tells us to take a look at Table 430.52 (Figure 14–23) and to multiply the full-load current that we got from Table 430.148 by the percentage found in Table 430.52 for a specific type of overcurrent protection device. We will limit our discussion to using either **dual-element time-delay fuses** or **inverse time circuit breakers** (the type of circuit breaker used in residential electrical panels) since these two overcurrent de-

vice types are commonly used in residential electrical system installation. The calculations are done as follows:

- Dual-element time-delay fuses: 175% × 8 amperes (from Table 430.148) = 14 amps.
- Since 14 amps is not a standard fuse size (see Section 240.6[A]), the maximum-size dual-element time-delay fuse we can use is 15 amps.

and

- Inverse time circuit breaker: 250% × 8 amps (from Table 430.148) = 20 amps.
- Since 20 amps is a standard size of circuit breaker (see Section 240.6[A]), the maximum-size inverse time circuit breaker we can use is 20 amps.

Overload protection for an electric motor is required by Article 430 so that the motor will not overheat from dangerous current overload levels and burn up. The overload protection can be thermal overloads (commonly called "heaters"), electronic sensing devices, thermal devices built in to the motor, or time-delay fuses. Section 430.32 requires the vast majority of motor applications to be protected by overload devices sized not more than 125% of the motor's nameplate full-load ampere rating, not the full-load current rating from Table 430.148. In this example, we assume that the full-load ampere rating listed on the motor's nameplate is the same as the full-load current rating found in Table 430.148. Therefore, 125% × 8 amps (full-load ampere rating from the nameplate) = 10 amps. This is the maximum size in amperes of the motor overload protection device. In the "real world," the overload protection devices nor-

Table 430.52 Maximum Rating or Setting of Motor Branch-Circuit Short-Circuit and Ground-Fault Protective Devices

Type of Motor	Percentage of Full-Load Current			
	Nontime Delay Fuse[1]	Dual Element (Time-Delay) Fuse[1]	Instantaneous Trip Breaker	Inverse Time Breaker[2]
Single-phase motors	300	175	800	250
AC polyphase motors other than wound-rotor				
Squirrel cage — other than Design E or Design B energy efficient	300	175	800	250
Design E or Design B energy efficient	300	175	1100	250
Synchronous[3]	300	175	800	250
Wound rotor	150	150	800	150
Direct current (constant voltage)	150	150	250	150

Note: For certain exceptions to the values specified, see 430.54.

[1]The values in the Nontime Delay Fuse column apply to Time-Delay Class CC fuses.

[2]The values given in the last column also cover the ratings of nonadjustable inverse time types of circuit breakers that may be modified as in 430.52(C), Exception No. 1 and No. 2.

[3]Synchronous motors of the low-torque, low-speed type (usually 450 rpm or lower), such as are used to drive reciprocating compressors, pumps, and so forth, that start unloaded, do not require a fuse rating or circuit-breaker setting in excess of 200 percent of full-load current.

Figure 14–23 **Table 430.52 of the** *NEC®.* *Reprinted with permission from NFPA 70-2002,* National Electrical Code®, *Copyright © 2001, National Fire Protection Association, Quincy, MA 02269. This reprinted material is not the complete and official Position of the NFPA on the referenced subject, which is represented only by the standard in its entirety.*

mally are sized at the factory and are included in the controller equipment that comes with the pump.

Provided that the pump is in sight of the enclosure that contains the fuse or circuit breaker protecting the water pump branch circuit, no additional disconnecting means is required next to the pump. An electrician will install a branch circuit from the electrical panel to the controller or, if the controller has already been connected to the pump circuit by the pump installer, to the two-pole pressure switch. If the pump motor is located in another room—more than 50 feet (15 m) away from the fuse or circuit breaker enclosure—or is just not in sight of the enclosure, a disconnecting means must be installed and located next to the pump.

Submersible Pump

A **submersible pump** is a pump type where both the electric motor and the pump are enclosed in the same housing. The housing is lowered into a well casing and placed below the water level in the well. The standard parts of a submersible pump system are shown in Figure 14–24. Many submersible pumps are of the "two-wire" type, and the housing contains all the parts of a motor starter and overload protection. This means there will be no need to have an aboveground controller box. Other submersible pump types, called "three-wire" pumps, will require a controller to be installed above the ground next to the precharged water tank. The controller will contain items such as a starting relay,

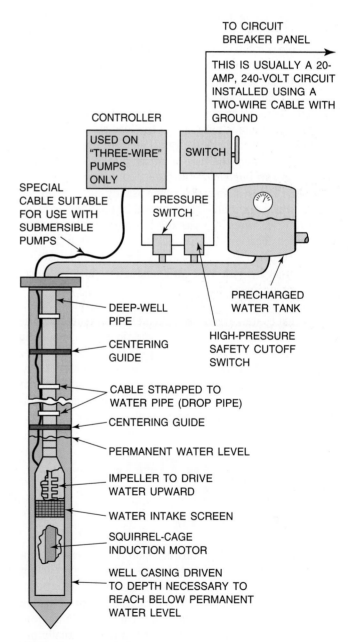

Figure 14–24 A typical submersible pump installation.

overload protection, starting and running capacitors, and the necessary connection terminals for connection of the pump branch-circuit wiring.

The calculations for sizing the branch-circuit conductor size, the fuse or circuit breaker size, and the overload protection size are the same as for the jet pump. One of the differences in wiring a submersible pump versus a jet pump is that a special submersible pump cable is used. It is buried in the ground, usually in the same trench with the incoming water pipe from the well. The cable is attached to the submersible pump, and electrical connections are made, then the pump is lowered down the well casing and into the water. It is the submersible pump installer who makes the connection of the submersible pump cable to the submersible pump and lowers it all into the well casing. The other end of the submersible pump cable is then left at the location of the pressure switch or the controller for the proper electrical connections to be made later. The size of the submersible pump cable is often larger than the calculated minimum size to overcome voltage drop problems that may be encountered because of the long distances from the house to the actual location of the submersible pump. It is not uncommon for the pump installer to do most if not all of the wiring to the controller (if there is one) and to the load side of the pressure switch. The electrician often is required only to bring a branch circuit from the fuse or circuit breaker enclosure to the pressure switch, where the proper connections are made on the line side. A submersible pump usually operates on 240 volts, and as a result an electrician will install a two-wire cable or run two wires in a raceway to the controller or pressure switch location. The conductor size is often 10 AWG with a two-pole, 30-amp circuit breaker, but some installations can get by with a 12 AWG conductor and a two-pole, 20-amp circuit breaker.

Similar to a jet pump, if the controller is in sight of the enclosure that contains the fuse or circuit breaker protecting the water pump branch circuit, no additional disconnecting means is required next to the pump. If the pump controller motor is located in another room—more than 50 feet (15 m) away from the fuse or circuit breaker enclosure—or is just not in sight of the enclosure, a disconnecting means must be installed and located next to the controller location.

Installing an Electric Water Heater Branch Circuit

Electric water heaters are quite common and are used to supply a house with domestic hot water. Residential electric hot water heaters come in several sizes, including 30, 32, 40, 42, 50, 52, 80, 82, 100, and 120 gallons. The 40- or 42-gallon size is the most common size used for most residential applications. Electric water heaters used in homes usually operate on 240 volts and normally will require a 10 AWG

conductor size with a 30-ampere overcurrent protection device. Some smaller electrical water heaters may require 120 volts and will be wired with a 12 AWG conductor with a 20-ampere overcurrent protection device.

Section 422.13 of the *NEC*® requires the branch-circuit rating to not be less than 125% of the nameplate rating of the water heater. You will remember that the rating of the branch circuit is based on the size of the overcurrent protection device. To calculate the size of the overcurrent protection device, refer to Section 422.11(E), which says to size the overcurrent device no larger than the protective device rating on the nameplate. If the water heater has no size listed on the nameplate, use the following:

- If the water heater current rating does not exceed 13.3 amps, use a 20-amp overcurrent protective device.
- If the water heater draws more than 13.3 amps, use 150% of the nameplate rating.

Let's look at an example: A water heater nameplate shows a rating of 4,800 watts at 240 volts. What minimum-size conductor and maximum-size overcurrent protection device are required?

1. Use the formula I = W/E, or 4,800 watts/240 volts = 20 amps.
2. Section 422.13 requires the branch-circuit rating to not be less than 125% of the water heater's nameplate rating. So, 20 amps × 125% = 25 amps.
3. Therefore, a 25-amp fuse or circuit breaker (Section 240.6[A]) would be the minimum size used and results in a 25-amp rating for the water heater branch circuit.
4. Since the water heater draws more than 13.3 amps, the maximum size overcurrent protective device would be 20 amps × 150% = 30 amps.
5. Using Table 310.16, we would use a 10 AWG conductor and a 30-amp fuse or circuit breaker for the installation of this electric water heater.

The individual branch circuit to an electric water heater is usually made directly from the overcurrent protection device in an electrical panel to the junction box or terminal block on the appliance. The overcurrent protection device in the panel will satisfy Section 422.30, which requires a disconnecting means for the water heater appliance. However, if the overcurrent protection device in the panel is not visible from the location of the water heater or is more than 50 feet (15 m) away, a separate disconnect must be installed within 50 feet (15 m) and within sight of the water heater. This could be a two-pole toggle switch with the proper voltage and ampere rating or a nonfusible safety switch type of disconnect. Assuming the use of a 10/2 nonmetallic sheathed cable wiring method, the wiring connections are done as follows (Figure 14-25):

1. Take off the junction box cover and take out the knockout on the junction box.
2. Attach the wiring method used to the water heater junction box though the knockout with an approved connector.

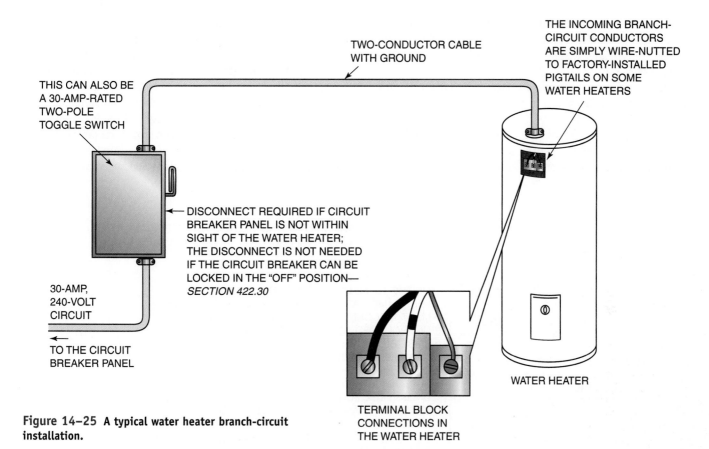

THE INCOMING BRANCH-
CIRCUIT CONDUCTORS
ARE SIMPLY WIRE-NUTTED
TO FACTORY-INSTALLED
PIGTAILS ON SOME
WATER HEATERS

TWO-CONDUCTOR CABLE
WITH GROUND

THIS CAN ALSO BE
A 30-AMP-RATED
TWO-POLE
TOGGLE SWITCH

DISCONNECT REQUIRED IF CIRCUIT
BREAKER PANEL IS NOT WITHIN
SIGHT OF THE WATER HEATER;
THE DISCONNECT IS NOT NEEDED
IF THE CIRCUIT BREAKER CAN BE
LOCKED IN THE "OFF" POSITION—
SECTION 422.30

30-AMP,
240-VOLT
CIRCUIT

TO THE CIRCUIT
BREAKER PANEL

WATER HEATER

TERMINAL BLOCK
CONNECTIONS IN
THE WATER HEATER

Figure 14–25 **A typical water heater branch-circuit installation.**

3. Connect the incoming equipment-grounding conductor to the green screw or green grounding jumper.
4. Wire-nut the white incoming conductor to one of the black pigtail wires in the junction box. Reidentify the white wire as a "hot" conductor with black tape.
5. Wire-nut the black incoming ungrounded wire to the other black pigtail.
6. Carefully push the wire connections into the junction box area and put the blank cover back on.

Note: Some electric water heaters may have a terminal block that is used to terminate the branch-circuit conductors.

Section 400.7(A)(8) and Section 422.16(C) of the *NEC®* describe the uses for flexible cords. Flexible cords can be used only where the appliance is designed for ready removal for maintenance or repair and the appliance is identified for use with flexible cord connections. Electric water heaters are not allowed to be cord-and-plug connected because they are not considered portable and cannot be easily removed for maintenance and repair. They will always be hardwired.

Installing Branch Circuits for Electric Heating

There are many different styles of electric heating available to a home owner, and electricians should be familiar with the most common types and installation techniques. Electric fur-

naces used to heat an entire house, individually controlled baseboard electric heaters used to heat specific rooms, and unit heaters used to heat a specific area like a basement or garage are the most common types installed. Other types of electric heating are beyond the scope of this book.

Electric heat has many advantages. It is very convenient to control since **thermostats** are easily installed in each room. It is considered safer than using fossil fuels like oil or gas because there is no storage of fuel on the premises and the explosion and fire hazard is certainly not as great. Electric heat is very quiet since there are few if any moving parts. It is relatively inexpensive and easy to install. Once it is installed, there is little maintenance required to keep it working. And one last advantage to consider is that no chimney is required to exhaust the combustion gasses into the outside air. This results in a more economical heating installation since the home owner does not have to pay for a chimney installation; it is also an environmentally cleaner heating system. You would think that with all the advantages of electric heat, the result would be electric heating installations in all homes. However, one big disadvantage of heating with electricity, which is the relatively high cost of electricity, severely limits the electric heating installations done by electricians. In some parts of the country that have lower electricity rates, you will see an increase in the number of installations of electric heat. In those areas of the country where electricity rates are high, you will not see as many electric heat installations. For example, in the northeastern area of the country, electricians almost never install electric heating. In the Pacific Northwest, electric heating is very popular, and electricians install electric heating systems on a regular basis.

When making an electric furnace installation (Figure 14–26), an electrician needs to do the following:

- Install the electric furnace on a separate circuit. Section 422.12 specifies that a central heating unit must be supplied by a separate branch circuit.

- Determine the size and type of the individual branch circuit required. Section 422.11 requires the overcurrent protective device to be sized according to the nameplate on the furnace. The nameplate will also list the minimum-size branch-circuit conductor required. Section 422.62(B)(1) states that the nameplate must specify the minimum size of the supply conductor's ampacity and the maximum size of the branch-circuit overcurrent protection device. In other words, always install the electric furnace according to the manufacturer's instructions and specifications.

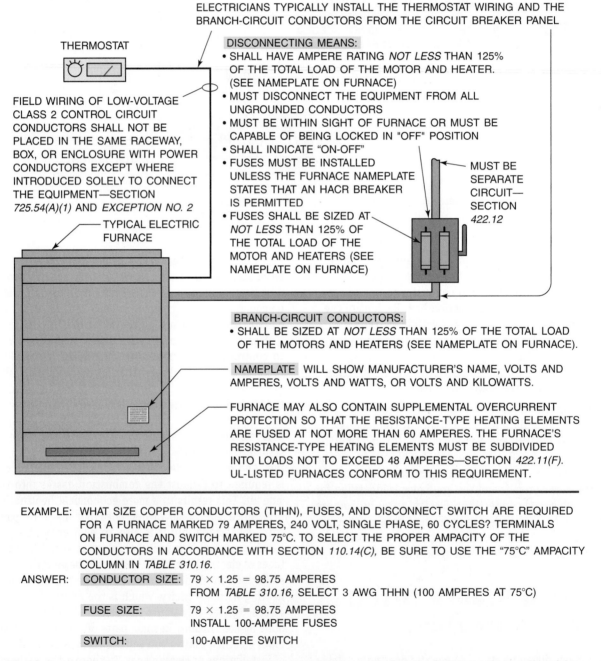

ELECTRICIANS TYPICALLY INSTALL THE THERMOSTAT WIRING AND THE BRANCH-CIRCUIT CONDUCTORS FROM THE CIRCUIT BREAKER PANEL

THERMOSTAT

FIELD WIRING OF LOW-VOLTAGE CLASS 2 CONTROL CIRCUIT CONDUCTORS SHALL NOT BE PLACED IN THE SAME RACEWAY, BOX, OR ENCLOSURE WITH POWER CONDUCTORS EXCEPT WHERE INTRODUCED SOLELY TO CONNECT THE EQUIPMENT—SECTION *725.54(A)(1)* AND *EXCEPTION NO. 2*

TYPICAL ELECTRIC FURNACE

DISCONNECTING MEANS:
- SHALL HAVE AMPERE RATING *NOT LESS* THAN 125% OF THE TOTAL LOAD OF THE MOTOR AND HEATER. (SEE NAMEPLATE ON FURNACE)
- MUST DISCONNECT THE EQUIPMENT FROM ALL UNGROUNDED CONDUCTORS
- MUST BE WITHIN SIGHT OF FURNACE OR MUST BE CAPABLE OF BEING LOCKED IN "OFF" POSITION
- SHALL INDICATE "ON-OFF"
- FUSES MUST BE INSTALLED UNLESS THE FURNACE NAMEPLATE STATES THAT AN HACR BREAKER IS PERMITTED
- FUSES SHALL BE SIZED AT *NOT LESS* THAN 125% OF THE TOTAL LOAD OF THE MOTOR AND HEATERS (SEE NAMEPLATE ON FURNACE)

MUST BE SEPARATE CIRCUIT—SECTION *422.12*

BRANCH-CIRCUIT CONDUCTORS:
- SHALL BE SIZED AT *NOT LESS* THAN 125% OF THE TOTAL LOAD OF THE MOTORS AND HEATERS (SEE NAMEPLATE ON FURNACE).

NAMEPLATE WILL SHOW MANUFACTURER'S NAME, VOLTS AND AMPERES, VOLTS AND WATTS, OR VOLTS AND KILOWATTS.

FURNACE MAY ALSO CONTAIN SUPPLEMENTAL OVERCURRENT PROTECTION SO THAT THE RESISTANCE-TYPE HEATING ELEMENTS ARE FUSED AT NOT MORE THAN 60 AMPERES. THE FURNACE'S RESISTANCE-TYPE HEATING ELEMENTS MUST BE SUBDIVIDED INTO LOADS NOT TO EXCEED 48 AMPERES—SECTION *422.11(F)*. UL-LISTED FURNACES CONFORM TO THIS REQUIREMENT.

EXAMPLE: WHAT SIZE COPPER CONDUCTORS (THHN), FUSES, AND DISCONNECT SWITCH ARE REQUIRED FOR A FURNACE MARKED 79 AMPERES, 240 VOLT, SINGLE PHASE, 60 CYCLES? TERMINALS ON FURNACE AND SWITCH MARKED 75°C. TO SELECT THE PROPER AMPACITY OF THE CONDUCTORS IN ACCORDANCE WITH SECTION *110.14(C)*, BE SURE TO USE THE "75°C" AMPACITY COLUMN IN *TABLE 310.16*.

ANSWER: CONDUCTOR SIZE: 79 × 1.25 = 98.75 AMPERES
FROM *TABLE 310.16*, SELECT 3 AWG THHN (100 AMPERES AT 75°C)

FUSE SIZE: 79 × 1.25 = 98.75 AMPERES
INSTALL 100-AMPERE FUSES

SWITCH: 100-AMPERE SWITCH

Figure 14–26 **An electric furnace installation. The furnace components are prewired at the factory so that an electrician usually has to just provide the branch-circuit wiring to the unit. The thermostat wiring may also be installed by an electrician.**

- Determine the size and location of the electric furnace disconnecting means. Section 422.30 requires a disconnecting means. Section 422.31(B) requires it to be within sight of the furnace. A common wiring practice is to mount the disconnecting means on the side of the electric furnace or on a wall space adjacent to the furnace.
- Determine the location of the thermostat and install the Class 2 wiring from the furnace controller box to the thermostat.

When making an electric baseboard heater installation (Figure 14–27), an electrician needs to do the following:

1. Determine whether a line-voltage thermostat or a low-voltage thermostat will be used to control the electric baseboard heating (Figure 14–28). The most common thermostat type used is the line-voltage type. This thermostat works at the full 240 volts required for the baseboard heating units and is connected directly in the electric baseboard heater branch circuit. Line-voltage thermostats are rated in watts. Some common sizes include 2,500, 3,000, and 5,000 watts. If a 5000-watt line-voltage thermostat is used, the total connected load of electric baseboard heat on it must not exceed 20.8 amps (5,000 watts/240 volts = 20.8 amperes). A low-voltage thermostat is used in conjunction with a relay when the load that a thermostat needs to control is more than what the line-voltage thermostat is rated for (Figure 14–29).

2. Determine the size and type of the individual branch circuit required. Most electric baseboard heating installations are done on 240-volt branch circuits with a 20-amp-rated, two-pole circuit breaker as the overcurrent protection device. This means that an electrician would use 12 AWG conductors to wire the

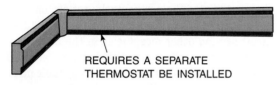
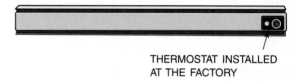

REQUIRES A SEPARATE THERMOSTAT BE INSTALLED

THERMOSTAT INSTALLED AT THE FACTORY

Figure 14–27 **Typical electric baseboard heating units.**

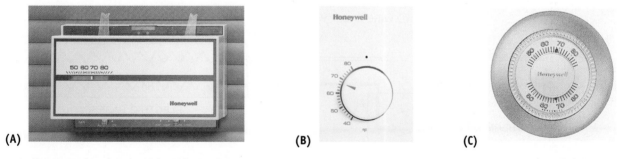

(A)   (B)   (C)

Figure 14–28 **Thermostats for electric heating systems. (A) and (C): Low-voltage thermostats; (B) Line-voltage thermostat.**
Courtesy of Honeywell, Inc.

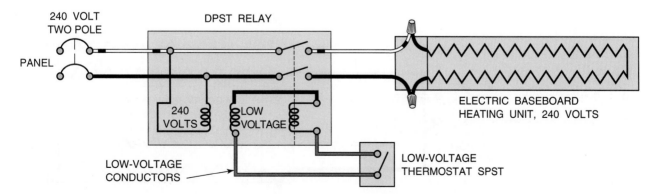

Figure 14–29 **Typical wiring for electric baseboard heating controlled by a low-voltage thermostat and relay. This wiring arrangement is used when the electrical load of the electric heating is greater than the rating of a high-voltage thermostat. The relay contacts are sized to switch "on" and "off" the heavy current drawn by the heating units.**

electric baseboard branch circuits. Larger groups of electric baseboard heaters may require 10 AWG conductors with 30-amp, two-pole circuit breakers. The wattage rating of electric baseboard heaters is on the nameplate of each unit. By comparing the wattage ratings of each baseboard unit with a line-voltage thermostat's wattage rating, a determination can be made as to the proper size of a line-voltage thermostat for a specific electric baseboard heating load. For example, three 1,500-watt electric baseboard units (a total load of 4,500 watts) can be connected to one circuit and controlled by a line-voltage thermostat with a rating of 4,500 watts or greater.

3. Starting at the location where the branch-circuit overcurrent protection device is located, install the wiring to the electrical box on the circuit that will contain the line-voltage thermostat. Be sure to mark the cable or other wiring method with the circuit number and type and the area served at the panel. For example, electric heating circuit 1 serving the master bedroom can be marked "EH1MBR." Continue roughing in the wiring to the electric heating baseboard unit (Figure 14–30) or to a group of baseboard units (Figure 14–31), making sure to follow the installation practices described in previous chapters.

CAUTION

CAUTION: When using a cable wiring method like nonmetallic sheathed cable, make sure you reidentify the white insulated conductor in the cable as a "hot" conductor by wrapping some black electrical tape on the conductor wherever it is visible, such as in the circuit breaker panel, line-voltage thermostat box, and at the electric baseboard heater itself.

4. Double-check to make sure all circuit wiring for the electric baseboard heating branch circuit you are installing has been roughed in.

When installing electric baseboard heating units along a wall space that requires receptacle outlets, Section 210.52 allows factory installed receptacles that are built in to the baseboard unit to count as the required receptacle outlets (Figure 14–32). These receptacle outlets must not be connected to the heater circuits but rather to a 120-volt branch circuit serving other receptacles in the room. The fine print note to Section 210.52 says that listed baseboard heaters include instructions that may not permit their installation below receptacle outlets (Figure 14–33). The correct positioning of electric baseboard heating units is shown in Figure 14–34.

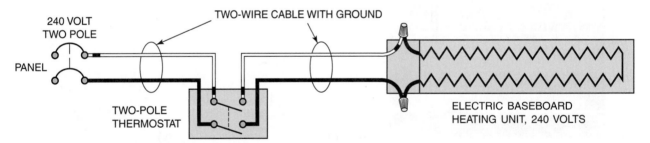

Figure 14–30 The electrical connection for a high-voltage thermostat controlling one section of electric baseboard heating. Remember to reidentify the white conductor in the cable as a "hot" conductor.

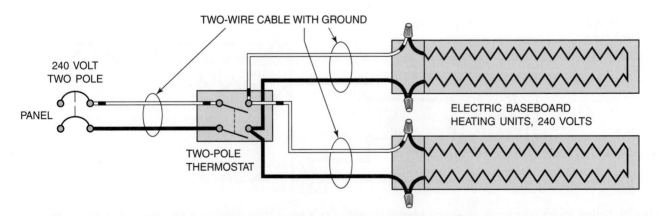

Figure 14–31 The electrical connection for a high-voltage thermostat controlling a group of electric baseboard heating units. Remember to reidentify the white conductor in the cable as a "hot" conductor.

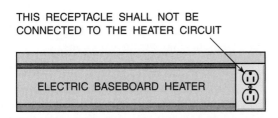

Figure 14–32 **A length of electric baseboard heater with a factory-installed receptacle outlet. The receptacle will meet Section 210.52 requirements for receptacle placement in a house. The receptacle(s) may not be connected to the electric baseboard heating branch circuit. A separate 120-volt circuit must be used.**

Figure 14–33 **Unless the manufacturer's installation instructions specifically permit the electric baseboard to be installed directly below receptacles, it is a violation of the *NEC*® to do so.**

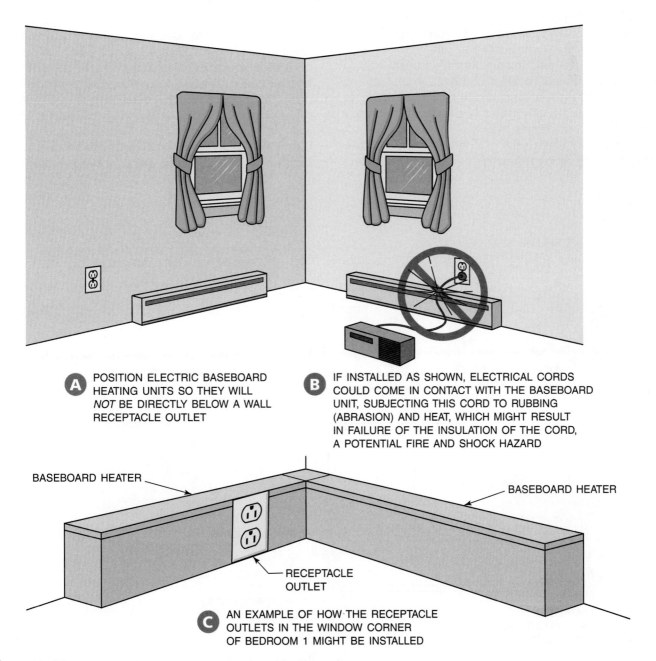

(A) POSITION ELECTRIC BASEBOARD HEATING UNITS SO THEY WILL *NOT* BE DIRECTLY BELOW A WALL RECEPTACLE OUTLET

(B) IF INSTALLED AS SHOWN, ELECTRICAL CORDS COULD COME IN CONTACT WITH THE BASEBOARD UNIT, SUBJECTING THIS CORD TO RUBBING (ABRASION) AND HEAT, WHICH MIGHT RESULT IN FAILURE OF THE INSULATION OF THE CORD, A POTENTIAL FIRE AND SHOCK HAZARD

BASEBOARD HEATER

BASEBOARD HEATER

RECEPTACLE OUTLET

(C) AN EXAMPLE OF HOW THE RECEPTACLE OUTLETS IN THE WINDOW CORNER OF BEDROOM 1 MIGHT BE INSTALLED

Figure 14–34 **Correct positioning of electric baseboard heating units.**

Installing Branch Circuits for Air Conditioning

Air conditioning a house is accomplished in one of two ways: (1) a central air conditioner system that blows cooled air from one centrally located unit into various areas of a house through a series of ducts or (2) individual room air conditioners located in different rooms throughout a house. Like a central heating system, when installing the wiring for a central air-conditioning system, an electrician usually just provides the rough-in wiring, and an air-conditioning technician who works for the heating and refrigeration contractor will make the final connections. If room air conditioners are going to be used in a house, the electrician may have to install specific branch circuits that will supply the proper amperage and voltage to the location where the air-conditioning unit will be located. Some homes are heated *and* cooled with a **heat pump.** During cooler weather, the heat pump supplies heat to a house. In warmer weather, the heat pump is simply run in reverse and, as a result, can supply cool air to the house as well. A heat pump is installed similar to a central air-conditioning system, and the wiring required to be installed by an elec-

trician is basically the same. In this section, we look at the branch-circuit installation for both central air-conditioner and room air-conditioner wiring situations.

Central Air Conditioners

If the system is to be a central air-conditioning system, the nameplate on the unit will give the branch-circuit ratings, the minimum conductor size, the overcurrent protection device size and type, and what the amperage and voltage ratings are for the unit. The manufacturer's instruction booklet will contain more information concerning the installation and should be consulted by an electrician installing the supply wiring to the system. Figure 14–35 shows a typical central air-conditioning system installation. An electrician will need to install branch-circuit wiring from the electrical panel to the main air-handling unit, which is located inside the home. Another branch circuit will need to be installed from the electrical panel to the disconnect switch of the compressor unit, which is located outside the house. The air-handling unit contains the fan that blows the cool air throughout a house and will need either a 120-volt or a 240-volt circuit. Be sure to consult the manufacturer's instructions to verify which voltage and amperage must be supplied to the air-handling unit. The circuit supplying the

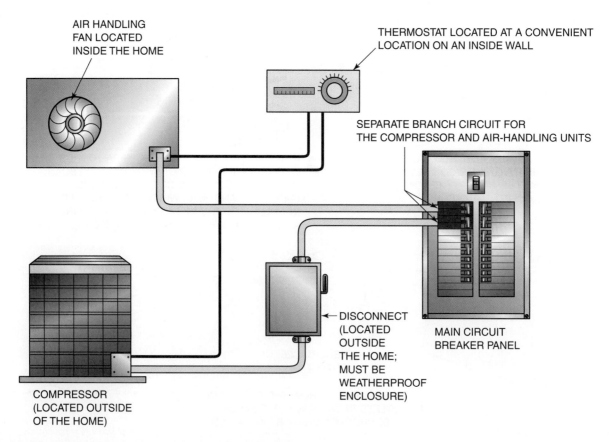

AIR HANDLING FAN LOCATED INSIDE THE HOME

THERMOSTAT LOCATED AT A CONVENIENT LOCATION ON AN INSIDE WALL

SEPARATE BRANCH CIRCUIT FOR THE COMPRESSOR AND AIR-HANDLING UNITS

DISCONNECT (LOCATED OUTSIDE THE HOME; MUST BE WEATHERPROOF ENCLOSURE)

MAIN CIRCUIT BREAKER PANEL

COMPRESSOR (LOCATED OUTSIDE OF THE HOME)

Figure 14–35 A typical installation of a central air conditioner.

compressor unit will require a 240-volt, two-wire circuit with a grounding conductor. If the disconnect located outside next to the compressor contains fuses or a circuit breaker, the circuit supplying it is a "feeder." If the disconnect has no overcurrent protection devices in it and is a nonfusible disconnect, the circuit supplying it is considered a "branch circuit." Article 440 of the *NEC®* contains most of the electrical installation requirements for air-conditioning systems.

Figure 14–36 shows the basic circuit requirements for the wiring for the compressor unit.

When selecting the type of overcurrent protection device for an air-conditioning circuit (or heating circuit), the nameplate on the unit must be consulted (Figure 14–37). If it says "maximum size fuse," the overcurrent protection device *must* be a fuse. If the nameplate says "maximum size fuse or HACR-type circuit breaker," a fuse *or* a HACR circuit

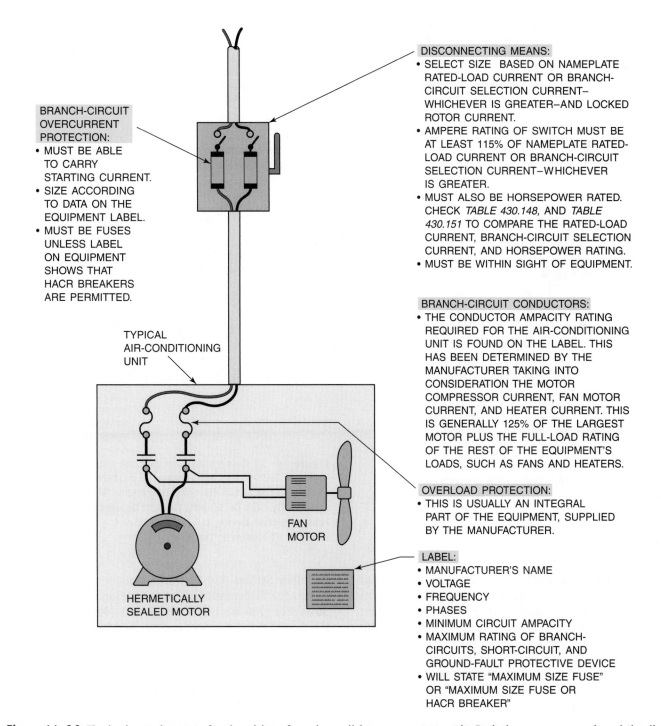

BRANCH-CIRCUIT OVERCURRENT PROTECTION:
• MUST BE ABLE TO CARRY STARTING CURRENT.
• SIZE ACCORDING TO DATA ON THE EQUIPMENT LABEL.
• MUST BE FUSES UNLESS LABEL ON EQUIPMENT SHOWS THAT HACR BREAKERS ARE PERMITTED.

TYPICAL AIR-CONDITIONING UNIT

HERMETICALLY SEALED MOTOR

FAN MOTOR

DISCONNECTING MEANS:
• SELECT SIZE BASED ON NAMEPLATE RATED-LOAD CURRENT OR BRANCH-CIRCUIT SELECTION CURRENT—WHICHEVER IS GREATER—AND LOCKED ROTOR CURRENT.
• AMPERE RATING OF SWITCH MUST BE AT LEAST 115% OF NAMEPLATE RATED-LOAD CURRENT OR BRANCH-CIRCUIT SELECTION CURRENT—WHICHEVER IS GREATER.
• MUST ALSO BE HORSEPOWER RATED. CHECK *TABLE 430.148*, AND *TABLE 430.151* TO COMPARE THE RATED-LOAD CURRENT, BRANCH-CIRCUIT SELECTION CURRENT, AND HORSEPOWER RATING.
• MUST BE WITHIN SIGHT OF EQUIPMENT.

BRANCH-CIRCUIT CONDUCTORS:
• THE CONDUCTOR AMPACITY RATING REQUIRED FOR THE AIR-CONDITIONING UNIT IS FOUND ON THE LABEL. THIS HAS BEEN DETERMINED BY THE MANUFACTURER TAKING INTO CONSIDERATION THE MOTOR COMPRESSOR CURRENT, FAN MOTOR CURRENT, AND HEATER CURRENT. THIS IS GENERALLY 125% OF THE LARGEST MOTOR PLUS THE FULL-LOAD RATING OF THE REST OF THE EQUIPMENT'S LOADS, SUCH AS FANS AND HEATERS.

OVERLOAD PROTECTION:
• THIS IS USUALLY AN INTEGRAL PART OF THE EQUIPMENT, SUPPLIED BY THE MANUFACTURER.

LABEL:
• MANUFACTURER'S NAME
• VOLTAGE
• FREQUENCY
• PHASES
• MINIMUM CIRCUIT AMPACITY
• MAXIMUM RATING OF BRANCH-CIRCUITS, SHORT-CIRCUIT, AND GROUND-FAULT PROTECTIVE DEVICE
• WILL STATE "MAXIMUM SIZE FUSE" OR "MAXIMUM SIZE FUSE OR HACR BREAKER"

Figure 14–36 The basic requirements for the wiring of an air-conditioner compressor unit. Both the compressor unit and the disconnect switch are typically located outside the house.

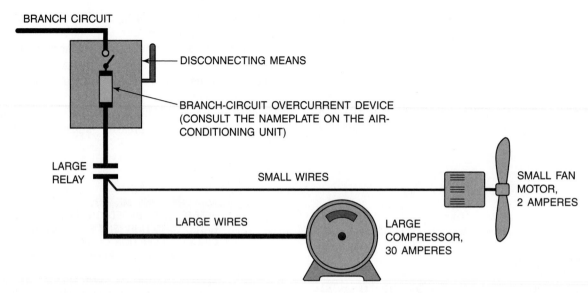

Figure 14–37 Consult the nameplate on an air-conditioning unit to determine the correct size of the overcurrent protection device. The nameplate will also specify whether the overcurrent device can be a fuse, circuit breaker, or either one.

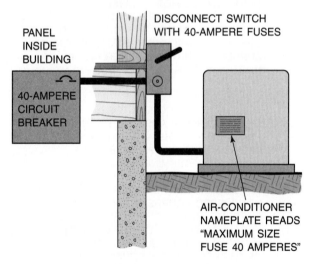

Figure 14–38 This installation meets the requirements of Section 440.14. The disconnect switch is within sight of the air-conditioning unit and contains the 40-amp fuses that the nameplate specified.

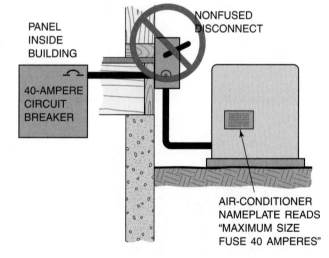

Figure 14–39 This installation does not meet the requirements of Section 110.3(B) because a nonfused disconnect switch is used as the disconnecting means. The nameplate specifically calls for 40-amp *fuses* to be used as the overcurrent protection device, but an electrician has installed a 40-amp circuit breaker in the electrical panel.

breaker could be used. (HACR is an acronym for "heating, air conditioning, and refrigeration.") Always look for the HACR label on a circuit breaker before using it to protect a heating or air-conditioning circuit. (Figures 14–38, 14–39, and 14–40). The disconnecting means must be within sight of the unit it disconnects (Section 440.14), and it must be readily accessible. The disconnecting means enclosure is normally mounted outside on the side of the house and in a spot that is on side of and not directly behind the air-conditioning unit. Remember that you have

to follow Section 110.26 working space requirements, so install the disconnect so that a working space of at least 30 inches (750 mm) in width, 3 feet (900 mm) out from the face of the disconnect enclose, and a height of 6.5 feet (2 m) is maintained.

A thermostat is required for the system control. It allows home owners to set the temperature to a value they want, and the air-conditioning system will automatically adjust the temperature in the area of the thermostat to the desired set temperature. Thermostat wire will need to be run from the

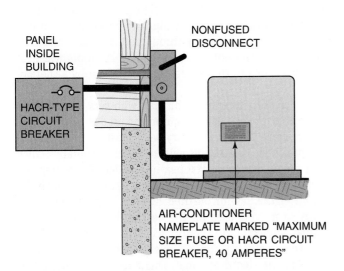

PANEL
INSIDE
BUILDING

NONFUSED
DISCONNECT

HACR-TYPE
CIRCUIT
BREAKER

AIR-CONDITIONER
NAMEPLATE MARKED "MAXIMUM
SIZE FUSE OR HACR CIRCUIT
BREAKER, 40 AMPERES"

Figure 14–40 This installation meets the requirements of the *NEC*® because even though a nonfused disconnect switch has been installed, a 40-amp HACR circuit breaker has also been installed in the electrical panel. This satisfies the nameplate requirement that fuses *or* an HACR circuit breaker must be used to provide the overcurrent protection for the air-conditioning unit.

compressor and the air handler to the thermostat. This low-voltage wiring is called a Class 2 circuit. Article 725 covers Class 2 wiring. Some information electricians need to be familiar with concerning Class 2 wiring includes the following:

- Class 2 circuits are power limited because of the **transformers** that supply the low voltage they operate on. As such, they are considered to be safe from electric shock and do not present much, if any, of a fire hazard. The transformers used to supply the low voltage for a Class 2 circuit in residential areas are normally marked "Class 2 transformer."
- The Class 2 conductors do not need separate fuses or circuit breakers to protect them because they are inherently current limited by the transformers they originate from.
- Class 2 wiring can be run exposed as a cable or, if an electrician so chooses, in a raceway. Section 725.8(A) and (B) allows this because any failure in the wiring would not result in a fire or shock hazard.
- Section 725.54(A)(1) states that Class 2 wiring can't be run in the same raceway, cable, compartment or electrical box with regular light and power conductors.

Thermostat wire comes in a cable form and is usually 18 or 20 AWG solid copper wire. Some installations may require the electrician to install the thermostat cable to and from the proper locations in the system; however, most central air-conditioning systems thermostats and thermostat wiring is done by the air-conditioning technician.

Room Air Conditioners

When homes have no central air conditioning but home owners still want cooler temperatures inside a house during days when the outside temperature is very hot, room air conditioners are used. They are available in 120-volt and 240-volt models and typically fit into a window opening or are installed in a more permanent way directly in an outside wall. These air conditioners are cord-and-plug connected. The circuits that feed them must be sized and installed according to the installation requirements as outlined in Article 440. Sections 440.60 through 440.64 cover room air conditioners. These code rules are summarized as follows:

- The air conditioners must be grounded and connected with a cord-and-plug set.
- The air conditioner may not have a rating that is greater than 40 amps at 250 volts.
- The rating of the branch circuit overcurrent protection device must not exceed the branch-circuit conductor rating or the receptacle rating, whichever is less.
- On an individual branch-circuit where no other items are served by the circuit, the air-conditioner load cannot exceed 80% of the branch-circuit ampacity.
- On a branch circuit where other loads are also served, the air-conditioner load cannot exceed 50% of the branch-circuit ampacity.
- The plug on the end of the cord can serve as the air conditioner's disconnecting means.
- The maximum length of the cord on a 120-volt room air conditioner is 10 feet (3 m).
- The maximum length of the cord on a 240-volt room air conditioner is 6 feet (1.8 m).

If a room air conditioner is rated at 120 volts and is purchased by a home owner after the electrical system has been installed, the room air conditioner is typically just plugged into a receptacle located next to the window or wall location where the air conditioner is placed. This receptacle will most likely be on a branch circuit that supplies electricity to other lighting and receptacle outlets in that particular room. If an electrician knows where room air conditioners will be located during the installation of the house electrical system, they can rough in the wiring to a receptacle location that is dedicated to a specific room air conditioner. The electrician will need to know the amperage and voltage rating of the room air conditioner so that the proper wire size and overcurrent protection device can be installed. Knowing the specific amperage and voltage rating of the unit will also enable the electrician to install the proper receptacle for the cord-and-plug connection of the air conditioner. In earlier chapters, we discussed different receptacle configurations based on specific voltage and amperage ratings. Care must be taken by the electrician to install a receptacle that has the proper configuration for the plug on the end of the room air-conditioner cord. Figure 14–41 shows a typical installation of a room air conditioner.

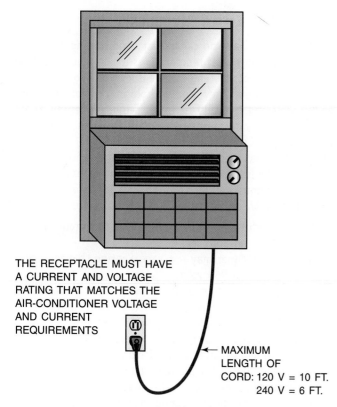

THE RECEPTACLE MUST HAVE
A CURRENT AND VOLTAGE
RATING THAT MATCHES THE
AIR-CONDITIONER VOLTAGE
AND CURRENT
REQUIREMENTS

MAXIMUM
LENGTH OF
CORD: 120 V = 10 FT.
240 V = 6 FT.

Figure 14–41 A typical installation of a room air conditioner.

Installing the Branch Circuit for Gas and Oil Central Heating Systems

Not all residential heating systems use electricity to produce the heat. In fact, the majority of central heating systems installed in houses use gas- or oil-fired furnaces. There are two methods for the heat to be transferred throughout a house. The first and most popular method uses what is called a forced-hot-air furnace. The combustion of gas or oil is used to heat air up to a specific temperature, and then a blower is used to "force" the warm air through air ducts to various parts of a house where the warm air forced into a room causes the colder air to vacate the area. The result is a room that has a rise in temperature. A cold-air return duct is used to bring the cool air from the rooms back to the furnace for reheating. A thermostat located in a specific area of the house will monitor the temperature in that area and, when the temperature gets below a certain point, will send a signal to the furnace that more hot air is needed in that area. Different zones installed throughout a house allow for more precise control of the air temperature in those parts of the house. The other method used to distribute the heat throughout a house is a hot water heating system, sometimes called a hydronic system. In this system, combustion of the gas or oil will heat up water. When the water

reaches a specific temperature, a circulating pump will force the water through a series of pipes to baseboard-mounted hot water radiators located throughout a house. The heat given off from the hot water baseboard radiators will heat the room or area to the temperature called for by a thermostat located in that area.

It is usually a heating technician who sets up and wires the controls for the heating equipment used in a home. New furnaces come from the factory with most of their components prewired. Electricians will install the separate branch circuit required for a central heating system. The wiring will normally be run from the electrical panel to a point over the location of the gas- or oil-fired furnace, leaving enough of the wiring method used for the heating technician to make the final connection to the furnace. The electrician will also make the necessary connections to the fuses or circuit breaker in the electrical panel. This circuit most often consists of a 12 AWG conductor size with a 20-amp-rated overcurrent protection device but could be a 14 AWG conductor with a 15-amp overcurrent protection device. Electricians may also install the Class 2 control wiring from a furnace controller box to the thermostat location (Figure 14–42). Electricians installing the Class 2 thermostat wiring must make sure to leave enough extra wire for the heating technician to make the necessary connections at the furnace controller box. Because electricians usually do not make the actual electrical connections for the wiring associated with a central heating system furnace, this book will not cover the specific wiring for a gas- or oil-fired furnace.

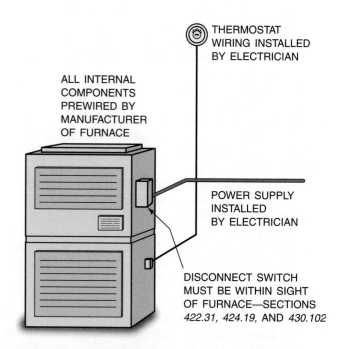

THERMOSTAT
WIRING INSTALLED
BY ELECTRICIAN

ALL INTERNAL
COMPONENTS
PREWIRED BY
MANUFACTURER
OF FURNACE

POWER SUPPLY
INSTALLED
BY ELECTRICIAN

DISCONNECT SWITCH
MUST BE WITHIN SIGHT
OF FURNACE—SECTIONS
422.31, 424.19, AND *430.102*

Figure 14–42 An electrician typically installs the branch-circuit power supply to the furnace, a disconnect switch, and the thermostat wiring. The other components of a furnace are usually prewired at the factory.

Local oil burner codes often require a safety switch to be installed at the head of the basement stairway or at the entrance to a room or crawl space that houses an oil-fired furnace. This allows a home owner to shut off the electrical power to the furnace should a fire break out at the furnace. This will stop the oil burner pump from continuing to pump oil and feed the fire with fuel. If this is the case, an electrician installing the branch-circuit wiring for the furnace will route the circuit to the switch location and then to the location above the furnace for future connections. Once the single-pole switch is installed, a red cover must be used on the switch that identifies it as the oil burner safety switch. Check with the authority having jurisdiction in your area to see if this requirement applies to you.

Figure 14–43 A typical smoke detector.

Installing the Smoke Detector Branch Circuit

Since fire in homes is the third leading cause of accidental death, some type of fire alarm system is required in all new home construction. Fire alarm systems can become quite complex. Article 760 of the *NEC®* covers the installation requirements for a fire alarm system. It defines a fire alarm circuit as the portion of the wiring system between the load side of the overcurrent device or the power-limited supply and the connected equipment of all circuits powered and controlled by the fire alarm system. Fire alarm circuits are classified as either non–power limited or power limited. Fire protective signaling systems installed in residential installations are power-limited. That is, the electrical power of the circuits is limited to a specific low level by the power supply for the system. The National Fire Protection Association (NFPA) publishes the National Fire Alarm Code, called NFPA 72. (You might remember that NFPA 70 is actually the *NEC®*.) NFPA 72 outlines the minimum requirement for the selection, installation, and maintenance of fire alarm equipment. Chapter 2 of NFPA 72 covers residential fire alarm systems. It defines a household fire alarm system as a system of devices that produces an alarm signal in the house for the purpose of notifying the occupants of the house of a fire so that they will evacuate the house. NFPA 72 is usually adopted by the building codes that must be followed when building a house and, as such, must be followed by electricians when installing a residential electrical system. The most common fire warning device used in a house is a smoke detector (Figure 14–43). In this section, we take a look at the installation requirements and methods for smoke detectors.

When wiring the smoke detector circuit for a new house, a detector unit should be placed in each bedroom and in the

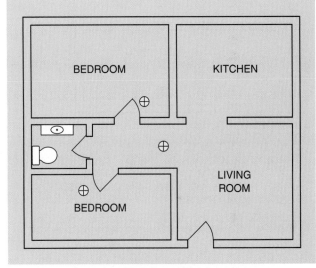

⊕: SMOKE DETECTOR

Figure 14–44 Recommended locations for smoke detectors in a house. Make sure a smoke detector is located in each bedroom of any new house.

area just outside the bedroom areas (Figure 14–44). They also should be installed on each level of a house (Figure 14–45). The smoke detectors must be installed in new residential construction so that when one detector is operated, all other detectors in the house will also operate. This is called **interconnecting.** The following installation requirements should always be followed when installing smoke detectors in a house:

- Install the smoke detectors on the ceiling at a location where there is no "dead" airspace (Figure 14–46).
- Install smoke detectors at the top of a stairway instead of the bottom since smoke will rise. An exception is in the basement, where it is better to install a smoke detector on the ceiling but close to the stairway to the first floor.
- Install smoke detectors in new houses that are hardwired directly to a 120-volt circuit. Dual powered

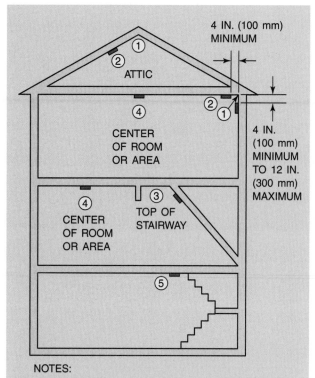

NOTES:

① DO *NOT* INSTALL DETECTORS IN *DEAD* AIRSPACES.

② MOUNT DETECTORS ON THE BOTTOM EDGE OF JOISTS OR BEAMS. THE SPACE BETWEEN THESE JOISTS AND BEAMS IS CONSIDERED TO BE *DEAD* AIRSPACE.

③ DO *NOT* MOUNT DETECTORS IN *DEAD* AIRSPACE AT THE TOP OF A STAIRWAY IF THERE IS A DOOR AT THE TOP OF THE STAIRWAY THAT CAN BE CLOSED. DETECTORS *SHOULD* BE MOUNTED AT THE TOP OF AN OPEN STAIRWAY BECAUSE HEAT AND SMOKE TRAVEL UPWARD.

④ MOUNT DETECTORS IN THE CENTER OF A ROOM OR AREA.

⑤ BASEMENT SMOKE DETECTORS MUST BE LOCATED IN CLOSE PROXIMITY TO THE STAIRWAY LEADING TO THE FLOOR ABOVE.

Figure 14–45 Smoke detectors are required on each level of a house. Locate them carefully so they will be sure to work correctly.

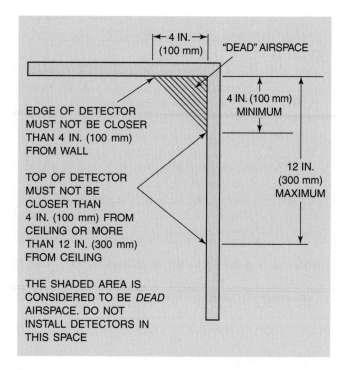

Figure 14–46 Never install smoke detectors in the "dead" airspace where a wall meets the ceiling.

smoke detectors that have a battery backup for when the electrical power is off are a good idea but are not normally required. Wire the smoke detectors so that they are interconnected so when one goes off, they all go off.

- Do not install smoke detectors to branch-circuit wiring that is controlled by a wall switch. The electrical power to the smoke detectors must be on at all times.
- Do not install smoke detectors on circuits that are GFCI protected. If the GFCI trips off, you have lost power to the smoke detectors.

When wiring smoke detectors, electricians will often run a two-wire cable (or two conductors in a raceway) to the first smoke detector location. A three-wire cable (or three wires in a raceway) is then run to each of the other smoke detector locations. The black and white conductors in the circuit are used to provide 120 volts to each smoke detector. The red conductor (or third wire in a raceway) is used as the interconnection between all the smoke detectors. Each smoke detector has a yellow wire that is used to connect to the red wire in a three-wire cable. The maximum number of interconnected smoke detectors varies by manufacturer, but a good rule of thumb is to not connect more than 10 smoke detectors on circuit. See Figure 14–47 for a typical smoke detector branch-circuit installation.

Installing the Low-Voltage Chime Circuit

A chime system is used in homes to signal when someone is at the front or rear door. A chime system will consist of the chime, the momentary contact switch buttons, a transformer, and the wire used to connect the system together. This system is often referred to as the "doorbell system" even though bells (and buzzers) have not been used in homes for many years. However, the term "doorbell" has stayed with us, and many electricians still refer to this part of the electrical system in-

SMOKE DETECTORS

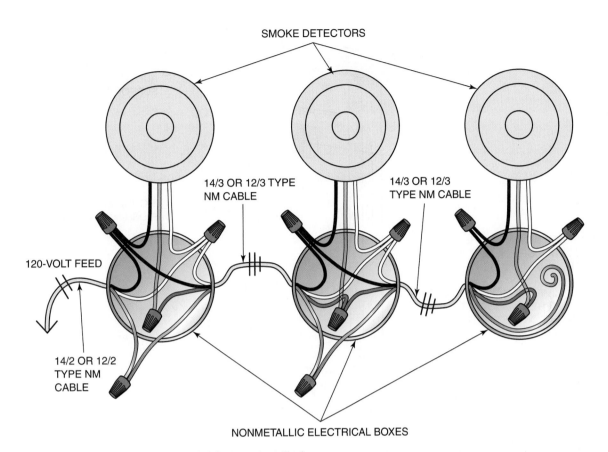

14/3 OR 12/3 TYPE NM CABLE

14/3 OR 12/3 TYPE NM CABLE

120-VOLT FEED

14/2 OR 12/2 TYPE NM CABLE

NONMETALLIC ELECTRICAL BOXES

Figure 14–47 A typical interconnected smoke detector installation.

stallation as "installing the doorbell." A chime system sounds a musical note or a series of notes when a button located next to an outside entrance or exit door is pushed. The signal from the door buttons is delivered to the chime itself through low-voltage Class 2 wiring. In this section, we look at common installation practices for a chime circuit.

The chime should be located in an area of the house where once it "chimes" and sounds a tone, the people in the house can hear it. Sometimes chimes are located on each floor of a larger home to make sure that everyone in the house will hear the chime when a tone is sounded. Chimes are available in a variety of styles (Figure 14–48). If a chime is to be surface mounted, it is common wiring practice to just bring the low-voltage wiring into the back of the chime enclosure and use hollow wall anchors to hold the chime onto the wall. Sometimes a flush-mounted chime installation is needed, and the installation requires adequate backing and an electrical box for attachment of the chime enclosure to the wall (Figure 14–49).

A chime button will need to be located next to the front door and rear door of a house. Chime buttons at any other exterior doorways may be included in the chime system if desired. Chime buttons also come in a variety of styles (Figure 14–50). Chime buttons are of the momentary contact type. This means that when someone's finger pushes the

Figure 14–48 Chimes are available in a variety of styles.
Courtesy of NuTone, Inc.

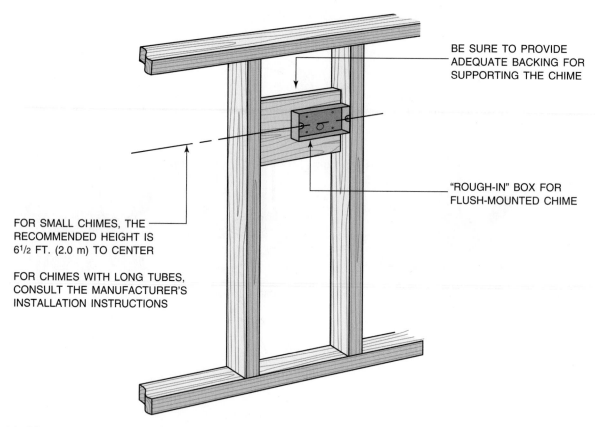

BE SURE TO PROVIDE
ADEQUATE BACKING FOR
SUPPORTING THE CHIME

"ROUGH-IN" BOX FOR
FLUSH-MOUNTED CHIME

FOR SMALL CHIMES, THE
RECOMMENDED HEIGHT IS
6½ FT. (2.0 m) TO CENTER

FOR CHIMES WITH LONG TUBES,
CONSULT THE MANUFACTURER'S
INSTALLATION INSTRUCTIONS

Figure 14–49 The rough-in for a flush-mounted chime installation. *Courtesy of NuTone, Inc.*

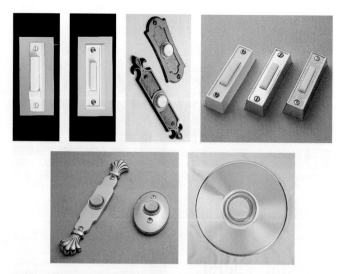

Figure 14–50 Various styles of chime buttons. *Courtesy of NuTone, Inc.*

button, contacts are closed and current can pass through the switch to the chime, causing a tone to be sounded. When finger pressure on the button is taken away, springs cause the contacts to come apart, and the switch will not pass current. This deenergizes the chime, and the tone stops. In other words, current can pass through the switch only as long as someone is holding the button in.

The chime transformer is used to transform the normal residential electrical system voltage of 120 volts down to the value that a chime system will work on, usually around 16 volts. These transformers have built-in thermal overload protection, and no additional overcurrent protection is required to be installed for them. The transformer is installed in a separate metal electrical box or right at the service entrance panel or subpanel. A 1/2-inch knockout is removed, and the high voltage side of the transformer (120 volts) fits through the knockout. A setscrew or locknut is used to hold the transformer in place. The high-voltage side has two black pigtails and a green grounding pigtail. The two black pigtail wires are connected across a 120-volt source, and the green pigtail is connected to an equipment-grounding conductor. The body of the transformer with the low-voltage side is located on the outside of the electrical box or panel. Two screws on the low-voltage side (16 volts) of the transformer are used to connect the chime wiring to the transformer (Figure 14–51).

The wire used to connect the buttons, chime, and transformer is often called "bell wire" or simply "thermostat wire" since it is the same type as that used to wire heating and cooling system thermostats. It is usually 18 or 20 AWG solid copper with an insulation type that limits it to use on 30-volt-or-less circuits. It comes in a cable assembly with two, three, or more single conductors covered with a protective outer sheathing. Each conductor is color coded in the cable

120-VOLT PRIMARY WIRES

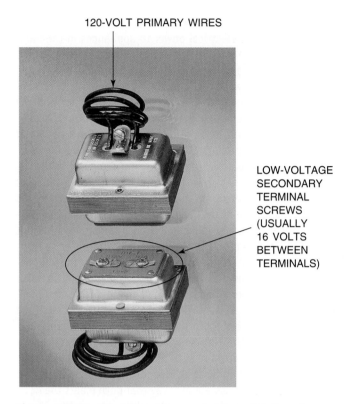

LOW-VOLTAGE SECONDARY TERMINAL SCREWS (USUALLY 16 VOLTS BETWEEN TERMINALS)

Figure 14–51 Chime transformers. *Courtesy of NuTone, Inc.*

assembly. The cable is run through bored holes in the building framing members or on the surface. It is supported with small insulated staples or cleats. It can also be installed in a raceway for added protection from physical damage. When running the cable, keep the following items in mind:

- Section 725.54(A)(1) tells us to not install Class 2 wiring in the same raceway or electrical box as the regular power and lighting conductors.
- Section 725.54(A)(3) tells us to keep the Class 2 bell wire at least 2 inches (50 mm) from light and power wiring. However, when the wiring method is nonmetallic sheathed cable, Type AC cable, Type MC cable, or a raceway wiring method, there is no problem with running the bell wire right next to these wiring methods.
- Try not to run the chime wiring through the same bored holes as regular light and power wiring methods. While this is not against the *NEC®*, it will ensure that there will not be any physical damage to the chime wiring.
- Do not strap or otherwise support the chime wiring to other electrical raceways or cables (Section 725[D]).
- The cable insulation on bell wire is very thin. Be very careful when installing this cable type so as not to damage it.

There are two common wiring schemes for a chime circuit. The first is to run a two-wire thermostat cable from each doorbell button location and from the transformer location to the chime. This means that at the chime location, there will be three two-wire cables for a total of six conductors.

The wiring connections for this arrangement are shown in Figure 14–52. The second scheme that electricians often use is to install a two-wire thermostat cable from each doorbell button location to the transformer location. Next, install a three-wire thermostat cable from the chime location to the transformer location. Make the connections as shown in Figure 14–53. The advantage of the second method is

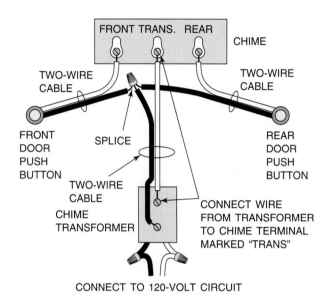

Figure 14–52 A typical chime installation showing the wiring connections when a two-wire cable is brought from the transformer and each chime button to the door chime.

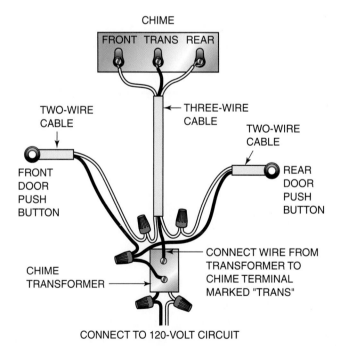

Figure 14–53 A typical chime installation showing the wiring connections when a two-wire cable is brought from each door chime button to the transformer and a three-wire cable is brought from the transformer to the door chime.

that there will be only three wires at the chime, making connections easier. The disadvantage is that there will be several wires at the transformer, making the connection more complicated. Another disadvantage is that an electrician will need to keep both a spool of two-wire and three-wire thermostat cable. Only a spool of two-wire thermostat cable is needed for the first method.

Summary

This chapter presented many of the branch-circuit installations commonly found as part of a residential electrical system. General lighting branch-circuit installation was discussed, and you learned that the majority of the branch circuits installed in a house are 15- and 20-amp-rated general lighting circuits. These circuits will supply the electrical power required for general illumination and convenience receptacles throughout a house and are wired with 14 or 12 AWG conductors. Small-appliance branch-circuit installation was presented. These circuits are 20-amp rated and are used to supply receptacle loads with electrical power in kitchens, dining rooms, and similar locations. There must be a minimum of two small-appliance branch circuits installed in each residential electrical system installation. In addition to the 20-amp-rated small-appliance circuit, at least one 20-amp-rated laundry cir-

cuit must also be installed. The main purpose of the laundry circuit is to supply electrical power to appliances in the laundry area, specifically a clothes washer. Bathroom circuits are required to be 20-amp rated and are installed much the same way as small-appliance and laundry branch circuits. The conductor size used for both the laundry branch circuit and the bathroom branch circuits is 12 AWG. This chapter also presented branch-circuit installation information for a variety of individual branch circuits to appliances commonly found in kitchens, such as electric ranges, counter-mounted cooktops, wall-mounted ovens, dishwashers, and garbage disposals. The wire sizes and overcurrent protection sizes are determined from the nameplate ratings and applying the proper *NEC®* section information. Other individual branch-circuit installations presented were electric clothes dryers, electric water heaters, electric heaters, air-conditioning systems, and gas or oil central heating systems. The wire sizes and fuse or circuit breaker sizes were also determined by the nameplate ratings and *NEC®* requirements. Installation of smoke detector circuits and chime systems were the last areas covered in this chapter.

By following the recommended installation practices for the branch circuits covered in this chapter, electricians will be able to determine and install the required branch circuits of a residential electrical system in a way that both meets the *NEC®* minimum requirements and adequately serves the home owner's needs for electrical power.

Review Questions

Directions: Answer the following items with clear and complete responses.

1 Explain the difference between a "two-wire" submersible pump and a "three-wire" submersible pump.

2 Determine the full-load current for a 2-horsepower, 230-volt, single-phase motor. Hint: Use Table 430.148

3 The *NEC*® requires electric water heaters that are 120 gallons (450 L) or less to be considered continuous duty and as such must have a circuit rating of not less than _____ % of the water heater rating.

4 Name and describe the three installation methods for a countertop cook unit and a wall-mounted oven.

5 A wall-mounted oven has a nameplate rating of 8 kilowatts at 240 volts. The rated load is _____ watts and _____ amps.

6 Name two common installation methods for garbage disposal units.

7 A receptacle located under a kitchen sink so that a cord-and-plug-connected garbage disposal can be plugged into it must be GFCI protected. True or False

8 Name the section of the *NEC*® that limits the length of the attachment cord to a dishwasher. Specify the minimum and maximum length of the cord.

9 List several advantages of electric heating.

10 Most electric heating equipment operates on 240 volts. When the equipment is fed using a 10/2 nonmetallic sheathed cable, what has to be done to the white conductor in the cable at the connection points?

11 The total load of a single room air conditioner on an individual branch circuit cannot exceed _____ % of the circuit rating.

12 The total load of a single-room air conditioner on a branch circuit that also supplies lighting in a room cannot exceed _____ % of the circuit rating.

13 The nameplate on an air-conditioner unit states "maximum size fuse or HACR circuit breaker." Describe what an HACR circuit breaker is.

14 The disconnecting means for an air conditioner or heat pump must be installed _____ of the unit.

15 Name the *NEC*® section that prohibits the use of low-voltage Class 2 thermostat wires from being in the same raceway or electrical enclosure with power and lighting circuit conductors.

16 Name the *NEC*® section that requires a separate branch circuit for a central heating system furnace.

17 The low-voltage thermostat wiring between a furnace and a thermostat is considered to be Class 1 wiring. True or False

18 Doorbell buttons used with a chime system are called "momentary contact switches." Describe the operation of this kind of switch.

19 Describe the purpose of the transformer as used in a residential chime system.

20 Assuming the use of a cable wiring method, describe the reason for running a three-wire cable between each smoke detector location in a house.

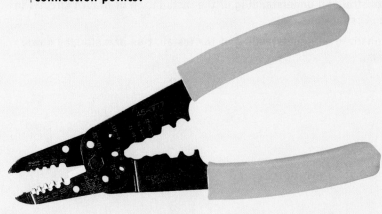

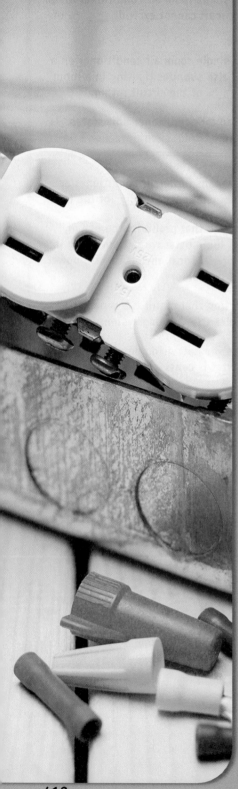

Chapter 15

Special Residential Wiring Situations

There are many special wiring situations that arise during the installation of the residential electrical system. Today most new homes are built with either an attached or a detached garage that will require electrical wiring. Many homes have swimming pools, hot tubs, and/or hydromassage bathtubs installed either at the time of construction or sometime after the house is built. All these items require an electrician to install electrical wiring for associated equipment such as pumps and pool lighting. Outside wiring may be required for items such as outdoor receptacles and outdoor lighting. Some homes are located in an area where the electrical power supply from the local electric utility may not be dependable, and a portable generator system may need to be installed. This chapter looks at several of these special wiring situations that electricians encounter on a regular basis.

OBJECTIVES

Upon completion of this chapter, the student should be able to:

- demonstrate an understanding of the installation of garage feeders and branch circuits.
- demonstrate an understanding of the installation of branch circuits for a swimming pool.
- demonstrate an understanding of the installation of branch circuits in outdoor situations.
- demonstrate an understanding of the installation of a standby power system.

Glossary of Terms

critical loads the electrical loads that are determined to require electrical power from a standby power generator when electrical power from the local electric utility company is interrupted

dry-niche luminaire a lighting fixture intended for installation in the wall of a pool that goes in a niche; it has a fixed lens that seals against water entering the niche and surrounding the lighting fixture

exothermic welding a process for making bonding connections on the bonding grid for a permanently installed swimming pool using specially designed connectors, a form, a metal disk, and explosive powder; this process is sometimes called "Cad-Weld"

forming shell the support structure designed and used with a wet-niche lighting fixture; it is installed in the wall of a pool

generator a rotating machine used to convert mechanical energy into electrical energy

hydromassage bathtub a permanently installed bathtub with recirculating piping, pump, and associated equipment; it is designed to accept, circulate, and discharge water each use

no-niche luminaire a lighting fixture intended for above or below the water-level installation; it does not have a forming shell that it fits into but rather sits on the surface of the pool wall; it can be located above or below the waterline

permanently installed swimming pool a swimming pool constructed totally or partially in the ground with a water depth capacity of greater than 42 inches (1 m); all pools, regardless of depth, installed in a building are considered permanent

pool cover, electrically operated a motor-driven piece of equipment designed to cover and uncover the water surface of a pool

self-contained spa or hot tub a factory-fabricated unit consisting of a spa or hot tub vessel having integrated water-circulating, heating, and control equipment

spa or hot tub a hydromassage pool or tub designed for immersion of users; usually has a filter, heater, and motor-driven pump; it can be installed indoors or outdoors, on the ground, or in a supporting structure; a spa or hot tub is not designed to be drained after each use

standby power system a backup electrical power system that consists of a generator, transfer switch, and associated electrical equipment; its purpose is to provide electrical power to critical branch circuits when the electrical power from the utility company is not available

storable swimming pool a swimming pool constructed on or above the ground with a maximum water depth capacity of

42 inches (1 m) or a pool with non-metallic, molded polymeric walls (or inflatable fabric walls) regardless of size or water depth capacity

transfer switch a switching device for transferring one or more load conductor connections from one power source to another

twistlock receptacle a type of receptacle that requires the attachment plug to be inserted and then turned slightly in a clockwise direction to lock the plug in place; the attachment plug must be turned slightly counter-clockwise to release the plug so it can be removed from the receptacle

wet location installations underground or in concrete slabs or masonry in direct contact with the earth in locations subject to saturation with water or other liquids, such as in unprotected areas exposed to the weather

wet-niche luminaire a type of lighting fixture intended for installation in a wall of a pool; it is accessible by removing the lens from the forming shell; this luminaire type is designed so that water completely surrounds the fixture inside the forming shell

Installing Garage Feeders and Branch Circuits

Attached Garages

Garages are either attached to the main house during the construction process or built detached at some distance away from the main house. If the garage is attached, Section 210.52(G) of the *National Electrical Code®* (*NEC®*) requires at least one 120-volt receptacle be located there, and Section 210.70(A)(2)(a) requires at least one wall switch–controlled lighting outlet in the garage (Figure 15–1). While these minimum requirements may be satisfactory for a small attached garage, most garages are large enough to require several lighting outlets and several receptacle outlets. There is no *NEC®* requirement concerning the maximum distance between receptacle outlets in a garage. Therefore, it is up to the electrician as to how many and how far apart the receptacles are located. The branch circuits supplying the receptacle(s) and lighting outlet(s) may be on the same branch circuit and can be rated either 15 or 20 amp. A good wiring practice is to install the lighting outlets on a separate 15- or 20-amp branch circuit and to install the receptacle(s) on a separate 20-amp-rated branch circuit. Section 210.8 (A)(2) requires all the receptacles installed in a garage to be GFCI protected. However, Exception No. 1 to this section allows receptacles that are not readily accessible, like a receptacle located in the ceiling for the connection of a garage door opener, to not require GFCI protection (Figure 15–2). Exception No. 2 to 210. (A)(2) allows duplex receptacles installed to serve cord-and-plug-connected appliances that are not easily moved, like a refrigerator or a freezer, to not require GFCI protection as long as the receptacle location is behind or at least blocked by the appliance. Exception No. 2 also states that if only a single cord-and-plug-connected appliance, such as a food freezer, occupies the dedicated space, then a single receptacle can be used and is not required to be GFCI protected (Figure 15–3).

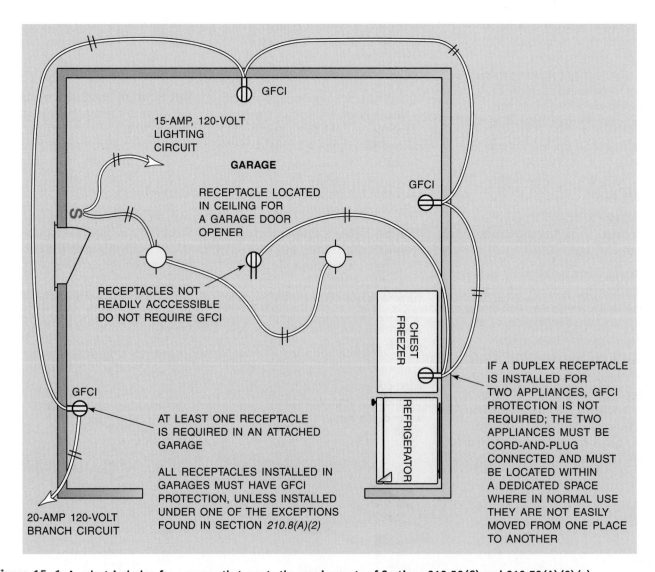

Figure 15–1 An electrical plan for a garage that meets the requirements of Sections 210.52(G) and 210.70(A)(2)(a).

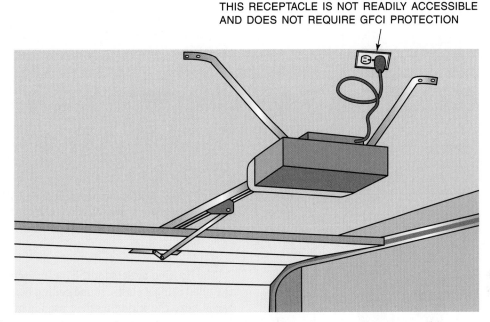

THIS RECEPTACLE IS NOT READILY ACCESSIBLE
AND DOES NOT REQUIRE GFCI PROTECTION

Figure 15–2 Garage door opener receptacle is not readily accessible and does not require GFCI protection.

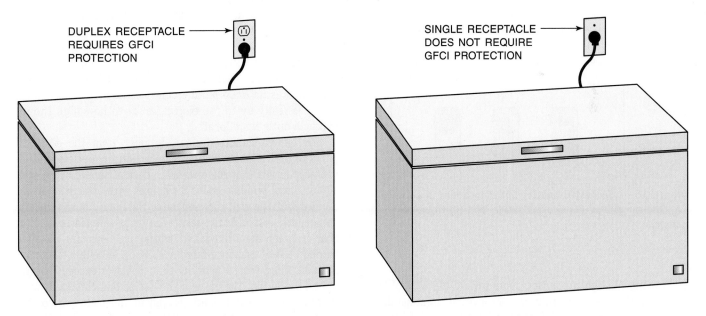

DUPLEX RECEPTACLE
REQUIRES GFCI
PROTECTION

SINGLE RECEPTACLE
DOES NOT REQUIRE
GFCI PROTECTION

Figure 15–3 A single receptacle in a garage that serves one appliance does not have to be GFCI protected.

It is important for an electrician to make sure that an adequate amount of light is installed in a garage. Many electricians install lighting outlets directly over a vehicle. This location will light up the roof of a car or truck but does not properly put the light where it is needed most—along the sides of the vehicles where the doors open and people enter and exit them. Good wiring practice requires electricians to install lighting outlets as follows (Figure 15–4):

- Two lighting outlets minimum in a one-car garage located on each side of the vehicle parking area

- Three lighting outlets minimum in a two-car garage located in the middle and to the outside of each vehicle parking location
- Four lighting outlets minimum in a three-car garage located so light shines down into the areas on each side of the vehicle parking areas

Locate the lighting outlets at a location that is a little bit off center and toward the front of the vehicle parking area so that if work is needed to be done on a car or truck and the hood is up, enough light will be available to see. Also,

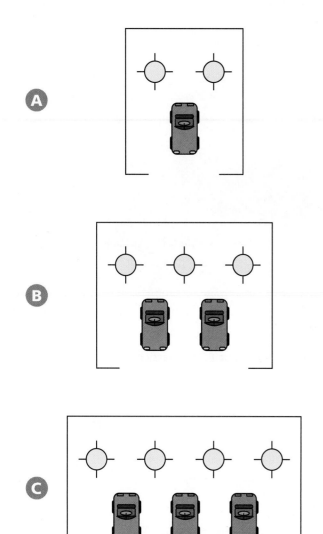

Figure 15–4 Suggested lighting outlet locations in a garage. (A) One-car garage, (B) two-car garage, and (C) three-car garage.

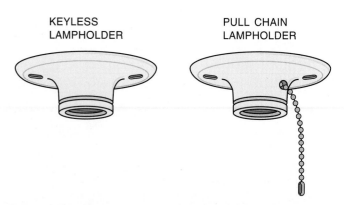

Figure 15–5 Typical incandescent lamp holders. They can be made of porcelain or plastic and are used in garages, attics unfinished basements, and crawl spaces.

The lighting outlets for the incandescent lamp holders are installed using a plastic round ceiling box or metal 4-inch octagon box. The wiring method used is brought to the outlet boxes, and the proper connections are made to the lighting fixtures. If fluorescent lighting is used, the wiring method is normally attached directly to surface-mounted fluorescent lighting fixtures, and all the wiring connections are done inside the fixtures.

When installing the electrical boxes for receptacle outlets in a garage, the usual mounting height is around 48 inches (1.2 m) to center from the finished garage floor. Remember that there is no *NEC®* requirement for the mounting height, and it is really up to the electrician as to how high the receptacle outlets are located.

The branch-circuit wiring is brought from the electrical panel to the various switching points, lighting outlets, and receptacle outlets that the electrician decides should be on each circuit. Local building codes for houses with attached garages require that the wall that separates the main house from the garage have a specific fire-resistance rating. We learned earlier that if holes are drilled and wiring methods are installed through a fire-resistance-rated wall, floor, or ceiling, the holes must be filled with a material that meets or exceeds the required fire rating. Therefore, if the garage branch circuits originate at an electrical panel located in the main part of the house and the circuit wiring penetrates the fire-resistant separating wall, be sure to plug the holes with an approved method.

Some interiors of a garage are finished off. That is, sheetrock or some other wall covering is installed on the wall studs and ceiling joists. When a garage is finished off in this manner, the lighting outlet and receptacle outlet boxes are installed in the same manner as inside the house. The proper box setout must be maintained so that the box front edge is flush with the finished wall surface. Most garages are unfinished, and proper box setout from the studs and joists is not as important. However, in case the garage may be finished off in the future, it is a good wiring practice to install the electrical boxes with a 1/2-inch (13-mm) setout.

by locating the lighting outlets more toward the back of the vehicle parking areas, a fully opened overhead garage door may block the light and defeat the whole purpose of having adequate lighting outlets to begin with.

In many areas of the country, garage lighting is usually done with 100-watt incandescent lamps used with porcelain or plastic lamp holders (Figure 15–5). (Note: Common lamp types used in residential wiring are discussed in Chapter 17.) In some areas, especially those where the outside temperature is not likely to fall below 50°F, if fluorescent lamps are used as the lighting source. Without special ballasts, a fluorescent lamp will usually not start when exposed to temperatures below 50°F. Of course, if the home has a heated garage, fluorescent lamps may be used even in the coldest of climates.

CAUTION: Garages that are unfinished result in exposed runs of the wiring method used to install the branch-circuit wiring. Some electrical inspectors may not allow wiring methods such as nonmetallic sheathed cable to be used exposed in a garage because it is considered to be exposed to physical damage. In these instances, a metal-clad cable or conduit wiring system may have to be installed in an unfinished garage. Always check with the authority having jurisdiction to make sure the wiring method you are using is allowed in an unfinished garage (or basement).

Detached Garage

If a garage is detached from the main house, the *NEC®* does not require that electrical power be brought to it. However, if electrical power is brought to a detached garage, the same rules apply as those for an attached garage (Figure 15–6). In other words, at least one 120-volt receptacle outlet and at least one wall switch–controlled lighting outlet must be installed. The same GFCI requirement for an attached garage must be applied to a detached garage with electrical power brought to it. The wiring practices described for an attached garage are also followed when installing the branch-circuit wiring in a detached garage.

If a detached garage is going to be supplied with electrical power, there are two ways to accomplish this: either install overhead conductors from the main house to the detached garage or install underground wiring from the main house to the detached garage. If there is not much of an

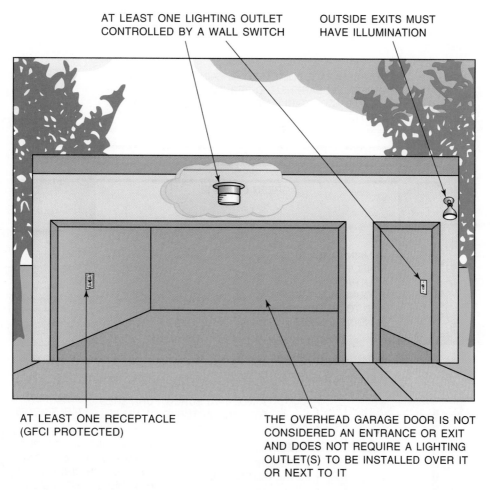

AT LEAST ONE LIGHTING OUTLET
CONTROLLED BY A WALL SWITCH

OUTSIDE EXITS MUST
HAVE ILLUMINATION

AT LEAST ONE RECEPTACLE
(GFCI PROTECTED)

THE OVERHEAD GARAGE DOOR IS NOT
CONSIDERED AN ENTRANCE OR EXIT
AND DOES NOT REQUIRE A LIGHTING
OUTLET(S) TO BE INSTALLED OVER IT
OR NEXT TO IT

Figure 15–6 A detached garage that meets the minimum *NEC®* requirements when electrical power is brought to it. Remember that a detached garage is not required to have electrical power brought to it, but if it is, the *NEC®* requirements are the same as those for an attached garage.

electrical load requirement in a detached garage, one circuit can be installed to serve the required receptacle outlet and switched lighting outlet. Electricians can install this single branch circuit with a Type UF cable containing an equipment-grounding conductor. It will need to be buried according to the requirements of Table 300.5 of the *NEC®*, which tells us that the minimum burial depth for this type of circuit installation is 24 inches (600 mm). Another way to install a single circuit would be to bury a rigid nonmetallic conduit from the main house to the detached garage and then pull in the required circuit conductors, including a green grounding conductor. One last method of installation worth mentioning would be to bury a metal raceway system, like rigid metal conduit, between the main house and the detached garage. The circuit conductors would then be pulled into the metal raceway, and if approved fittings are used to attach the raceway at each end, Section 250.118 would allow us to use the metal raceway as a grounding means and an additional green grounding conductor would not have to be installed. The Section 250.32(A) Exception says that if a detached garage is supplied by only one branch circuit with an equipment-grounding conductor, there is no requirement to establish a grounding electrode system or to connect to one if one exists.

CAUTION

CAUTION: Whenever burying a cable or raceway underground, consult Table 300.5 for the minimum burial depths.

Usually the electrical loads required to be served in a detached garage require more electrical power than a single branch circuit can supply. A common wiring practice is to install feeder wiring from the main service entrance panel to a subpanel located in the garage. Several different branch circuits can originate in the garage subpanel and supply the electrical load requirements of the garage. The feeder can be brought to the detached garage as either a cable, typically Type UF cable, or a raceway wiring method, typically rigid nonmetallic conduit.

If the feeder supplying the detached garage is installed with an equipment-grounding conductor, Section 250.32(A) requires that a grounding electrode system be established, unless one already exists (Figure 15–7). The most common method of establishing a grounding electrode system at the detached garage is to simply drive an 8-foot (2.5-m) ground

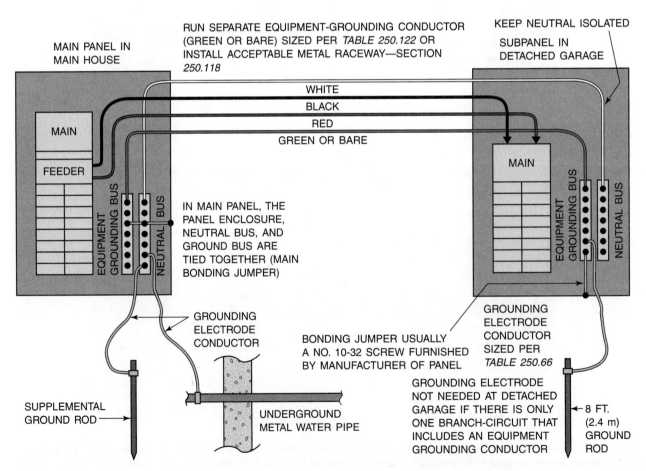

Figure 15–7 Proper grounding connections at a detached garage when a grounding conductor is included with the feeder from the main house to a detached garage.

rod. The equipment-grounding bus bar in the subpanel located in the detached garage must be bonded to the grounding electrode system. The disconnecting means must also be bonded to the grounding electrode system. Additionally, Section 250.32(B) requires that if a feeder supplying a detached garage from the service of the main house has an equipment-grounding conductor run with the feeder, the grounded conductor (neutral) is not permitted to be connected to the equipment-grounding conductor or to the grounding electrode system (Figure 15–8). The grounding electrode conductor in the detached garage is sized according to Table 250.66, the same way the grounding electrode conductor for the service entrance in the main house is sized.

When the feeder to the detached garage does not have an equipment-grounding conductor run with the feeder to the garage and there are no continuous metallic paths bonded to the grounding system in both the main house and the detached garage (like metal water pipes or metal raceways), the grounded neutral circuit conductor run with the supply to the detached garage must be connected to the subpanel en-

FROM EXPERIENCE

If you are required to install wiring for a storage shed, install the wiring the same as for a detached garage. There is no requirement that electrical power be brought to a storage shed, but if it is, at least one wall switch–controlled lighting outlet and at least one receptacle outlet must be installed. The GFCI requirements for receptacles as outlined in the *NEC®* for garages must be followed.

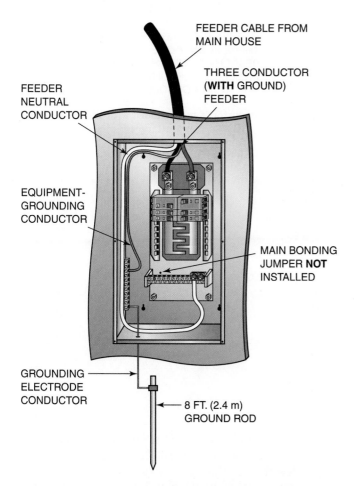

Figure 15–8 The neutral conductor of the feeder is not allowed to be connected to the equipment-grounding bar or to the grounding electrode system in the panel when a grounding conductor is brought to the detached garage with the feeder from the main house.

closure and to the grounding electrode (Figure 15–9). In this case, the connections in the garage subpanel are the same as for the main service panel in the main house. That is, use the main bonding jumper and connect the equipment-grounding bus bar and the grounded neutral bus bars together. The size of the grounded neutral conductor serving the detached garage must not be smaller than the sizes required by Table 250.122. Common wiring practice calls for the grounded neutral conductor to be the same size as the ungrounded "hot" circuit conductors feeding the garage subpanel.

Installing Branch-Circuit Wiring for a Swimming Pool

Swimming pools are often installed at the same time a new home is built. Many owners of existing homes have swimming pools installed after the house has been built for a few years. Whether it is during the initial installation of the electrical system or sometime after, electricians find themselves installing the branch-circuit wiring required for a swimming pool quite often. In this section, we look at the most important *NEC®* rules and common wiring practices for swimming pools and other similar equipment, such as hot tubs and hydromassage baths.

Article 680 covers swimming pools, hydromassage bathtubs, hot tubs, and spas. This article applies to both permanent and storable equipment and includes rules for auxiliary equipment such as pumps and filters. These are all items that an electrician may encounter during a residential electrical system installation.

Section 680.2 has many terms and definitions that relate specifically to the items covered in Article 680. The following are considered the most important to know for residential wiring:

• A **spa or hot tub** is a hydromassage pool or tub designed for immersion of users and usually has a filter, heater, and motor-driven pump. It can be installed indoors or outdoors on the ground or in a supporting

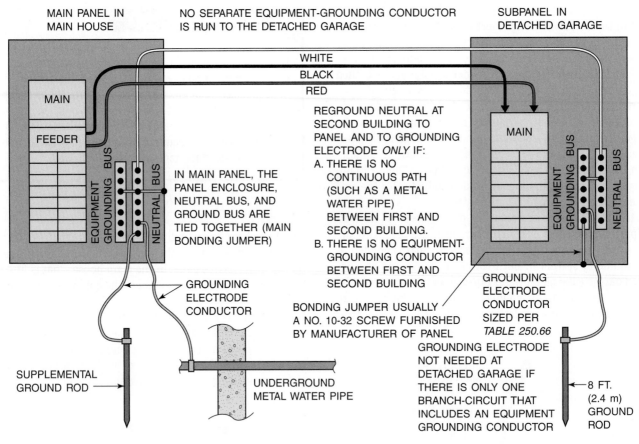

MAIN PANEL IN
MAIN HOUSE

NO SEPARATE EQUIPMENT-GROUNDING CONDUCTOR
IS RUN TO THE DETACHED GARAGE

SUBPANEL IN
DETACHED GARAGE

WHITE
BLACK
RED

MAIN

FEEDER

EQUIPMENT GROUNDING BUS

NEUTRAL BUS

IN MAIN PANEL, THE
PANEL ENCLOSURE,
NEUTRAL BUS, AND
GROUND BUS ARE
TIED TOGETHER (MAIN
BONDING JUMPER)

REGROUND NEUTRAL AT
SECOND BUILDING TO
PANEL AND TO GROUNDING
ELECTRODE *ONLY* IF:
A. THERE IS NO
   CONTINUOUS PATH
   (SUCH AS A METAL
   WATER PIPE)
   BETWEEN FIRST AND
   SECOND BUILDING.
B. THERE IS NO EQUIPMENT-
   GROUNDING CONDUCTOR
   BETWEEN FIRST AND
   SECOND BUILDING

MAIN

EQUIPMENT GROUNDING BUS

NEUTRAL BUS

GROUNDING
ELECTRODE
CONDUCTOR

SUPPLEMENTAL
GROUND ROD

BONDING JUMPER USUALLY
A NO. 10-32 SCREW FURNISHED
BY MANUFACTURER OF PANEL

UNDERGROUND
METAL WATER PIPE

GROUNDING
ELECTRODE
CONDUCTOR
SIZED PER
TABLE 250.66

GROUNDING ELECTRODE
NOT NEEDED AT
DETACHED GARAGE IF
THERE IS ONLY ONE
BRANCH-CIRCUIT THAT
INCLUDES AN EQUIPMENT
GROUNDING CONDUCTOR

8 FT.
(2.4 m)
GROUND
ROD

Figure 15–9 Proper grounding connections at a detached garage when a grounding conductor is not included with the feeder from the main house to a detached garage.

structure. Generally, a spa or hot tub is not designed to be drained after each use (Figure 15–10).

- A **self-contained spa or hot tub** is a factory-fabricated unit consisting of a spa or hot tub vessel having integrated water-circulating, heating, and control equipment (Figure 15–11).
- A **hydromassage bathtub** is a permanently installed bathtub with recirculating piping, pump, and associated equipment. It is designed to accept, circulate, and discharge water on each use (Figure 15–12).
- A **wet-niche luminaire** (lighting fixture) is a type of lighting fixture intended for installation in a wall of a pool. It is accessible by removing the lens from the forming shell. This luminaire type is designed so that water completely surrounds the fixture inside the forming shell (Figure 15–13).
- A **dry-niche luminaire** is a lighting fixture intended for installation in the wall of a pool that goes in a niche. However, unlike the wet-niche luminaire, the dry-niche luminaire has a fixed lens that seals against water entering the niche and surrounding the lighting fixture (Figure 15–14).
- A **no-niche luminaire** is a lighting fixture intended for above or below the water-level installation. It does

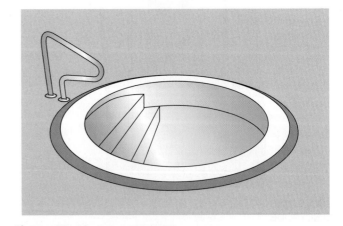

Figure 15–10 A spa or hot tub.

not have a forming shell that it fits into. It simply sits on the surface of the pool wall. It can be located above or below the waterline (Figure 15–15).

- A **permanently installed swimming pool** is one constructed totally or partially in the ground with a water depth capacity of greater than 42 inches (1 m). All pools, regardless of depth, installed in a building are considered permanent (Figure 15–16).

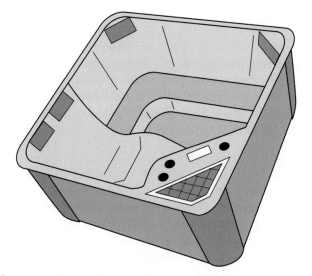

Figure 15–11 A self-contained spa or hot tub.

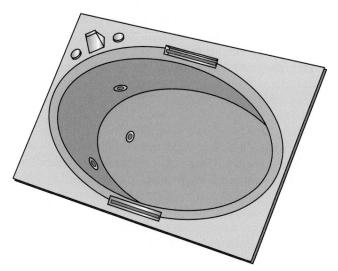

Figure 15–12 A hydromassage bathtub.

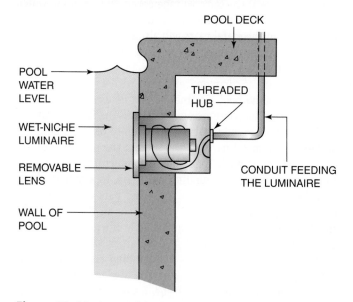

Figure 15–13 A wet-niche luminaire (lighting fixture).

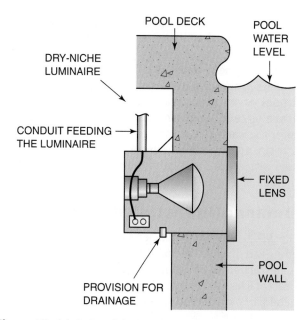

Figure 15–14 A dry-niche luminaire (lighting fixture).

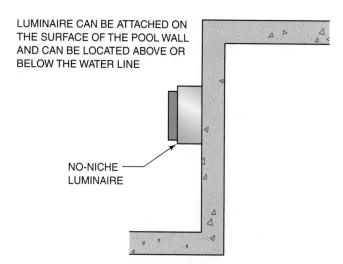

LUMINAIRE CAN BE ATTACHED ON THE SURFACE OF THE POOL WALL AND CAN BE LOCATED ABOVE OR BELOW THE WATER LINE

NO-NICHE LUMINAIRE

Figure 15–15 A no-niche luminaire (lighting fixture).

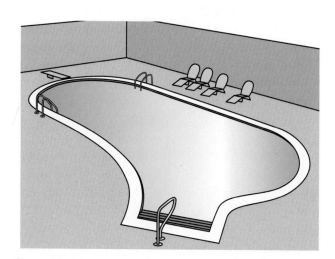

Figure 15–16 A permanently installed swimming pool.

- A **storable swimming pool** is one constructed on or above the ground and has a maximum water depth capacity of 42 inches (1 m) or a pool with nonmetallic, molded polymeric walls (or inflatable fabric walls) regardless of size or water depth capacity (Figure 15–17).
- A **pool cover, electrically operated,** is a motor-driven piece of equipment designed to cover and uncover the water surface of a pool (Figure 15–18).
- A **forming shell** is the support structure designed and used with a wet-niche lighting fixture. It is installed in the wall of a pool (Figure 15–19).

Permanently Installed Pools

The *NEC®* rules for permanently installed pools are located in Part II of Article 680. Part II provides installation rules for clearances of overhead lighting, underwater lighting, receptacles, switching, associated equipment, bonding of metal parts, grounding, and electric heaters.

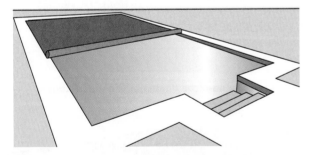

Figure 15–17 A storable swimming pool.

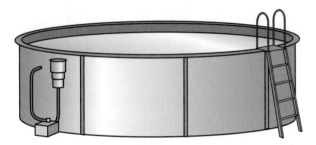

Figure 15–18 A pool cover, electrically operated.

Figure 15–19 A forming shell.

The electrical panel in a house or a subpanel in a garage or other building located on the premises can supply the electrical power to the pool. Branch-circuit wiring will be brought from the appropriate electrical panel to serve the various swimming pool loads. When running branch-circuit wiring from an electrical panel inside a house or any other associated building, like a garage, to pool-associated motors, the wiring method can be any of the *NEC®* Chapter 3 recognized methods. If a raceway is used, whether it is nonmetallic or metal, an insulated equipment-grounding conductor must be used. If a cable assembly is used, like a nonmetallic sheathed cable, an uninsulated equipment-grounding conductor can be used, but it must be enclosed within the cable's outer sheathing. Many permanently installed swimming pool pump motors are installed using a cord-and-plug connection. If so, the pump motor must have an approved double-insulation system and must provide a grounding means for the pump's internal and inaccessible non-current-carrying metal parts. Section 680.21(A)(1) requires the branch-circuit wiring for pool-associated motors to be installed in rigid metal conduit, intermediate metal conduit, rigid nonmetallic conduit, or metal-clad (Type MC) cable listed for the location conditions encountered. Any wiring method used must contain a copper equipment-grounding conductor that is sized according to Table 250.122, but in no case can it be smaller than 12 AWG. Section 680.21(A)(2) allows electrical metallic tubing to be used when the wiring will be installed on or within a building. Rigid nonmetallic conduit is often used by electricians to install the wiring for pool motors and other pool-associated equipment. A disconnecting means must be installed that will disconnect all ungrounded conductors. It must be located within sight of the equipment.

Probably the most important installation practice for a permanently installed swimming pool installation is the proper method of grounding and bonding together all metal parts in and around the pool. This will ensure that all the metal parts are at the same voltage potential to ground, which will help minimize the shock hazard. Proper grounding and bonding of all the metal parts will also facilitate the operation of the overcurrent protection devices protecting the circuit wiring that serves the pool. Figure 15–20 shows a typical grounding path of common items used with a permanently installed swimming pool. Pertinent *NEC®* requirements and the proper section location are also shown in the illustration.

Section 680.26 requires an electrician to properly bond together all the metal parts in and around the permanently installed swimming pool. The bonding conductor is not required to be connected to a grounding electrode, to any part of the house service entrance, or to any other part of the building electrical system. It is simply used to tie all the metal parts together. Figure 15–21 shows an example of what items should be connected together. An 8 AWG or larger *solid* copper conductor must be used to make the bonding connections. The bonding conductor can be bare or insulated, and the connection to the metal parts must be made with **exothermic welding** or by pressure connectors labeled for the purpose

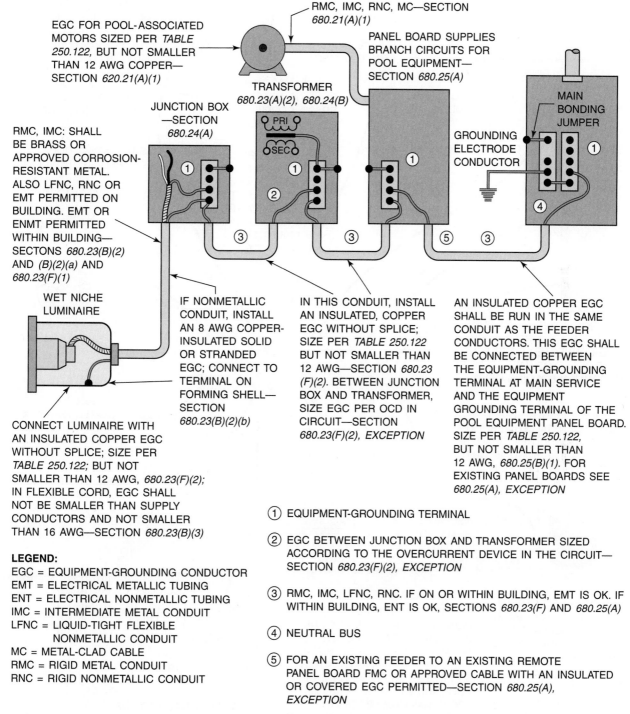

Figure 15–20 An overview of the grounding requirements for a typical permanently installed swimming pool installation.

① EQUIPMENT-GROUNDING TERMINAL

② EGC BETWEEN JUNCTION BOX AND TRANSFORMER SIZED ACCORDING TO THE OVERCURRENT DEVICE IN THE CIRCUIT—SECTION *680.23(F)(2), EXCEPTION*

③ RMC, IMC, LFNC, RNC. IF ON OR WITHIN BUILDING, EMT IS OK. IF WITHIN BUILDING, ENT IS OK, SECTIONS *680.23(F)* AND *680.25(A)*

④ NEUTRAL BUS

⑤ FOR AN EXISTING FEEDER TO AN EXISTING REMOTE PANEL BOARD FMC OR APPROVED CABLE WITH AN INSULATED OR COVERED EGC PERMITTED—SECTION *680.25(A), EXCEPTION*

and made of stainless steel, brass, copper, or a copper alloy. Exothermic welding is a process for making grounding and bonding connections using specially designed connectors, a form, a metal disk, and explosive powder. The copper bonding conductor is placed into the connector, and a form is placed around the connector and the part that the conductor is being connected to, like another piece of grounding conductor. The electrician puts the metal disk and explosive powder into

the form and then ignites it. The violent chemical reaction that takes place "welds" the copper conductor to the other conductor. This process is sometimes called "Cad-Weld." Other methods acceptable to the *NEC*® for bonding the swimming pool metal parts include the following:

- The structural reinforcing steel of a concrete pool where the rods are bonded together by steel tie wires or the equivalent

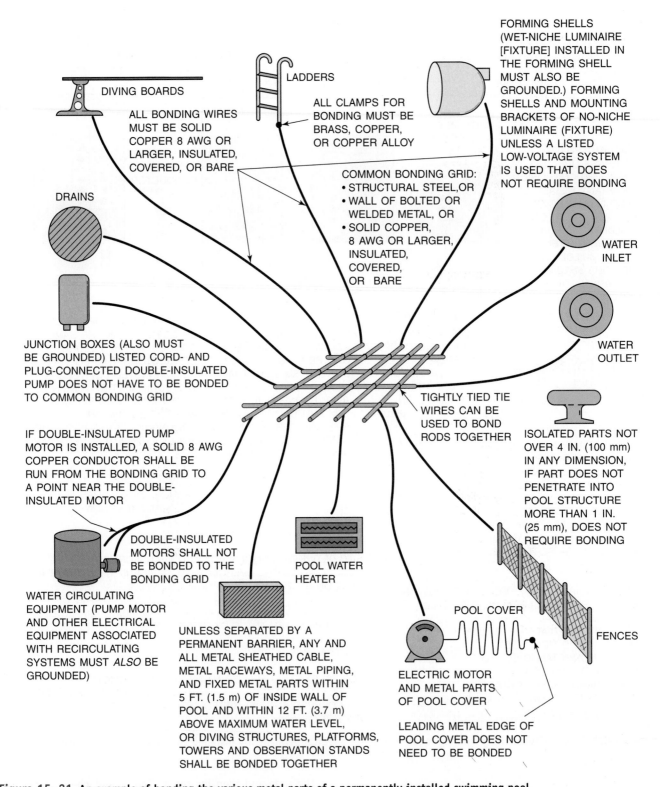

DIVING BOARDS

ALL BONDING WIRES MUST BE SOLID COPPER 8 AWG OR LARGER, INSULATED, COVERED, OR BARE

LADDERS

ALL CLAMPS FOR BONDING MUST BE BRASS, COPPER, OR COPPER ALLOY

FORMING SHELLS (WET-NICHE LUMINAIRE [FIXTURE] INSTALLED IN THE FORMING SHELL MUST ALSO BE GROUNDED.) FORMING SHELLS AND MOUNTING BRACKETS OF NO-NICHE LUMINAIRE (FIXTURE) UNLESS A LISTED LOW-VOLTAGE SYSTEM IS USED THAT DOES NOT REQUIRE BONDING

COMMON BONDING GRID:
• STRUCTURAL STEEL,OR
• WALL OF BOLTED OR WELDED METAL, OR
• SOLID COPPER, 8 AWG OR LARGER, INSULATED, COVERED, OR BARE

DRAINS

WATER INLET

WATER OUTLET

JUNCTION BOXES (ALSO MUST BE GROUNDED) LISTED CORD- AND PLUG-CONNECTED DOUBLE-INSULATED PUMP DOES NOT HAVE TO BE BONDED TO COMMON BONDING GRID

IF DOUBLE-INSULATED PUMP MOTOR IS INSTALLED, A SOLID 8 AWG COPPER CONDUCTOR SHALL BE RUN FROM THE BONDING GRID TO A POINT NEAR THE DOUBLE-INSULATED MOTOR

TIGHTLY TIED TIE WIRES CAN BE USED TO BOND RODS TOGETHER

ISOLATED PARTS NOT OVER 4 IN. (100 mm) IN ANY DIMENSION, IF PART DOES NOT PENETRATE INTO POOL STRUCTURE MORE THAN 1 IN. (25 mm), DOES NOT REQUIRE BONDING

DOUBLE-INSULATED MOTORS SHALL NOT BE BONDED TO THE BONDING GRID

POOL WATER HEATER

WATER CIRCULATING EQUIPMENT (PUMP MOTOR AND OTHER ELECTRICAL EQUIPMENT ASSOCIATED WITH RECIRCULATING SYSTEMS MUST *ALSO* BE GROUNDED)

UNLESS SEPARATED BY A PERMANENT BARRIER, ANY AND ALL METAL SHEATHED CABLE, METAL RACEWAYS, METAL PIPING, AND FIXED METAL PARTS WITHIN 5 FT. (1.5 m) OF INSIDE WALL OF POOL AND WITHIN 12 FT. (3.7 m) ABOVE MAXIMUM WATER LEVEL, OR DIVING STRUCTURES, PLATFORMS, TOWERS AND OBSERVATION STANDS SHALL BE BONDED TOGETHER

POOL COVER

FENCES

ELECTRIC MOTOR AND METAL PARTS OF POOL COVER

LEADING METAL EDGE OF POOL COVER DOES NOT NEED TO BE BONDED

Figure 15–21 An example of bonding the various metal parts of a permanently installed swimming pool.

- The walls of a bolted together or welded together metal pool
- Brass rigid metal conduit, intermediate metal conduit, or other approved corrosion-resistant metal conduit

Swimming pool electrical installation usually includes installing one or more receptacles. Section 680.22(A) includes the installation requirements for receptacles and must be followed closely by an electrician (Figure 15–22). Receptacles installed for a permanently installed swimming pool's water pump motor or other loads directly related to the circulation and sanitation system for the pool can be located between 5 feet (1.5 m) and 10 feet (3 m) from the inside walls of the pool. If this receptacle is used and located accordingly, it must be a single receptacle of the proper voltage and amperage rating, a locking and grounding type, and GFCI protected. A permanently installed swimming pool must have at least one 125-volt, 15- or 20-ampere-rated receptacle located a minimum of 10 feet (3 m), but no more than 20 feet (6 m), from the pool's inside wall installed by an electrician. This receptacle cannot be located more than 6 feet, 6 inches (2 m) above the floor, platform, or grade level serving the pool. This requirement is in Section 680.22(A)(3). Any 125-volt receptacles located within 20 feet (6 m) of the pool's inside wall must be GFCI protected. One last thing for an electrician to consider when installing receptacles to serve a permanently installed swimming pool is that, according to Section 406.8(B)(2)(a), any receptacle installed in a **wet location** (indoors or outdoors) and is left unattended when something is plugged into it must have a type of cover or enclosure that keeps it weatherproof at all times whether a plug is inserted in it or not (Figure 15–23).

Some permanently installed pool installations will require luminaires (lighting fixtures) or ceiling-suspended paddle fans to be installed. Section 680.22(B) lists the installation requirements for luminaires over a pool. In an outdoor pool area (Figure 15–24), Section 680.22(B)(1) requires luminaires, ceiling-suspended paddle fans, or lighting outlets to not be installed over the pool or over the area extending 5 feet (1.5 m) horizontally from the inside wall of the pool.

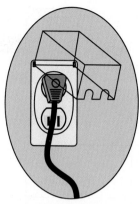
SECTION 406.8 (B)(2)(A)

Figure 15–23 Any receptacle installed in a wet location and is left unattended when something is plugged into it must have a type of cover or enclosure that keeps it weatherproof at all times.

This rule does not need to be followed if the lighting fixtures are located 12 feet (3.7 m) or more above the maximum water level in the pool. Section 680.22(B)(3) says that if there are existing lighting fixtures (such as when a pool is installed sometime after the house has been built and there are outside lighting fixtures on the house) and they are located less than 5 feet (1.5 m) measured horizontally from the inside wall of the pool, they must be at least 5 feet (1.5 m) above the maximum water level of the pool, rigidly attached to the existing structure they are on, and GFCI protected. In the area that is more than 5 feet (1.5 m) to 10 feet (3 m) measured horizontally from the inside wall of the pool, Section 680.22(B)(4) requires lighting fixtures and ceiling-suspended paddle fans to be GFCI protected. No GFCI protection is required if the luminaires or paddle fans are located more than 5 feet (1.5 m) above the maximum water level of the pool and are rigidly attached to an adjacent structure. Section 680.22(C) states that any switching device used for luminaires or ceiling-suspended paddle fans in or above a pool must be located at least 5 feet (1.5 m) horizontally from the inside wall of the pool, unless the switch location is separated from the pool by a solid fence, wall, or other permanent barrier. If the pool is to be installed indoors (Figure 15–25), Section 680.22(B)(2) does not require electricians to follow the rules described above if the following are true:

- The luminaires are of a totally enclosed type,
- Ceiling-suspended paddle fans are identified for use beneath ceiling structures, such as provided on porches or patios,
- The branch circuit supplying the equipment is GFCI protected,
- The bottom of the luminaire (lighting fixture) or ceiling-suspended paddle fan is at least 7 feet, 6 inches (2.3 m) above the maximum water level of the pool.

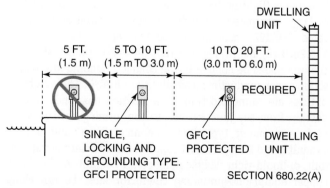

Figure 15–22 Required receptacles and their locations for a permanently installed swimming pool.

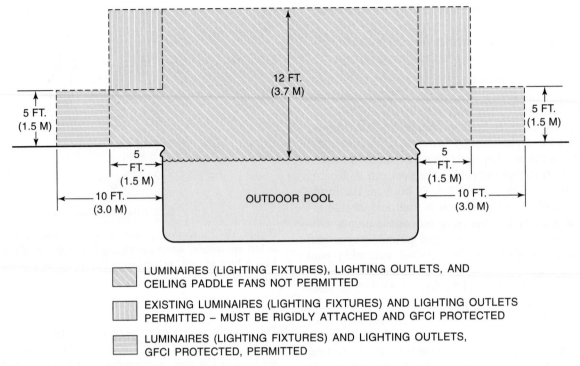

LUMINAIRES (LIGHTING FIXTURES), LIGHTING OUTLETS, AND CEILING PADDLE FANS NOT PERMITTED

EXISTING LUMINAIRES (LIGHTING FIXTURES) AND LIGHTING OUTLETS PERMITTED – MUST BE RIGIDLY ATTACHED AND GFCI PROTECTED

LUMINAIRES (LIGHTING FIXTURES) AND LIGHTING OUTLETS, GFCI PROTECTED, PERMITTED

Figure 15–24 Outdoor pool lighting fixture locations.

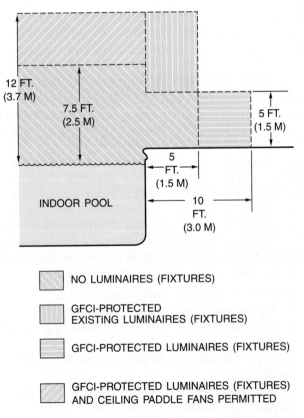

NO LUMINAIRES (FIXTURES)

GFCI-PROTECTED EXISTING LUMINAIRES (FIXTURES)

GFCI-PROTECTED LUMINAIRES (FIXTURES)

GFCI-PROTECTED LUMINAIRES (FIXTURES) AND CEILING PADDLE FANS PERMITTED

Figure 15–25 Indoor pool lighting fixture locations.

Some swimming pools will require the installation of luminaires (lighting fixtures) underwater in the walls of the pool (Figure 15–26). Section 680.23(A)(5) requires lighting fixtures that are installed in the walls of pools to be located so that the top of the luminaire (fixture) lens is at least 18 inches (450 mm) below the pool's normal water level, unless the fixture is listed and identified for use at a depth of not less than 4 inches (100 mm) below the pool's normal water depth. Section 680.23(B)(2)(b) requires that the termination point for the 8 AWG solid copper (or larger) bonding conductor located inside the forming shell for the luminaire (light fixture) must be covered with a listed potting compound or covered in some other way so as to protect the connection from the deteriorating effects of the pool water. The wet-niche, dry-niche, or no-niche luminaires (lighting fixtures) must be connected to a 12 AWG or larger equipment-grounding conductor sized according to Table 250.122 according to Section 680.23(F)(2). Section 680.26(B)(2) and (C) requires that the metal forming shells for the luminaires be bonded to the common bonding grid with a solid copper conductor no smaller than 8 AWG. These lighting fixtures are connected through a junction box, and the connection from the junction box to the forming shell must be made by a conduit, typically rigid nonmetallic conduit, but the conduit could be rigid metal conduit, intermediate metal conduit, or liquid-tight flexible nonmetallic conduit. An insulated copper equipment-grounding conductor must be run along with the circuit conductors in the conduit. Section 680.23(B)(2)(b) requires that when using rigid nonmetallic

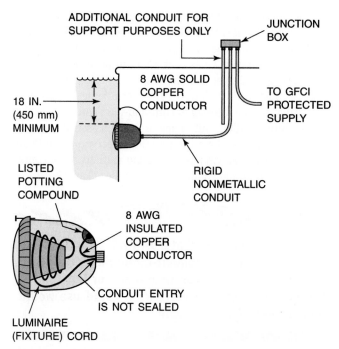

Figure 15-26 **Installation of luminaires (lighting fixtures) underwater in the walls of a permanently installed swimming pool.**

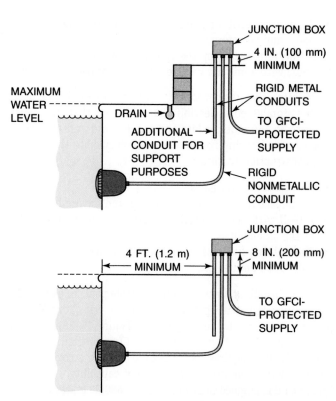

Figure 15-27 **Installation of junction boxes used with underwater lighting.**

conduit, an 8 AWG insulated copper equipment-grounding conductor be installed in the conduit and connected to the forming shell, junction box, or other enclosure in the circuit. This requirement is not necessary if a listed low-voltage lighting system is used that does not require grounding. The supply conductors brought to the junction box by the electrician must be GFCI protected if the voltage is above 15 volts.

There are a few specific rules to follow when installing the junction boxes used with underwater lighting. They are summarized as follows (Figure 15-27):

- Locate the junction box at least 4 feet (1.2 m) from the pool's inside wall, unless separated from the pool by a solid fence, wall, or other permanent barrier—Section 680.21(A)(2)(b)
- The junction box must be made of copper, brass, suitable plastic, or other approved corrosion-resistant material— Section 680.24(A)(1)(2)
- Measured from the inside of the bottom of the junction box, the box must be located at least 8 inches (200 mm) above the maximum water level of the pool—Section 680.24(A)(2)(a)
- The junction box must be equipped with threaded entries or a specifically listed nonmetallic hub—Section 680.24(A)(1)(1)
- Locate the junction box no less than 4 inches (100 mm) above the ground level or pool deck. This is measured from the inside of the bottom of the junction box— Section 680.24(A)(2)(a)

- The junction box support must comply with Section 314.23(E), which requires the box to be supported by at least two or more rigid metal conduit or intermediate metal conduit threaded into the box. Each conduit must be secured within 18 inches (450 mm) of the enclosure. Rigid nonmetallic conduit is not permitted for junction box support, so if it is used, additional support will be required.

Storage Pools

So-called aboveground storage pools are very popular with home owners. The electrical requirements and subsequent circuit installation for a storage pool is not as complicated as it is for a permanently installed swimming pool. In this section, we look at the most common wiring requirements for storage pools.

Most storable pools have no lighting fixtures installed in or on them. However, Section 680.33(A) says that if lighting fixtures are installed, they must be cord-and-plug connected and be a "made at the factory" assembly. The lighting fixture assembly must be properly listed for use with storage pools and must have the following construction features:

- No exposed metal parts
- A lamp that operates at no more than 15 volts

- An impact-resistant polymeric lens, lighting fixture body, and transformer enclosure
- A transformer, to drop the voltage from 120 volts to 15 volts, with a primary voltage rating of not over 150 volts

Lighting fixtures without a transformer can also be used with a storable pool. Section 680.33(B) says that they must operate at 150 volts or less and can be cord-and-plug connected. The lighting fixture assembly must have the following construction features:

- No exposed metal parts
- An impact-resistant polymeric lens and fixture body
- A GFCI with open neutral protection as an integral part of the assembly
- The lighting fixture is permanently connected to a GFCI with open neutral protection

When installing the wiring for a storable pool, an electrician typically installs a receptacle in a location that allows the filter system pump to be plugged in (Figure 15–28). Section 680.32 states that it must be GFCI protected. Section 406.8(B)(2)(a) requires the receptacle to have a cover or enclosure that maintains its weatherproof capability, whether a pump cord is plugged in or not. Section 680.31 states that the cord-and-plug-connected pool filter pump must incorporate an approved double-insulated system and must have a means for grounding the appliance's internal and nonaccessible non-current-carrying metal parts. The grounding means must consist of an equipment-grounding conductor (run with the power supply conductors) in the flexible cord that properly terminates in a grounding-type attachment plug with a fixed grounding contact member.

Spas and Hot Tubs

Electrical installation requirements for spas and hot tubs are found in Part IV of Article 680. A hot tub is made of wood, typically redwood, teak, or oak. A spa is made of plastic, fiberglass, concrete, tile, or some other man-made product. However, even though they are made from different materials and do not look exactly the same, both have certain electrical installation rules that apply to both of them.

Section 680.42 specifies the wiring methods that are allowed for outdoor installations. They include flexible connections using flexible raceway or cord-and-plug connections. For a one-family dwelling unit, a spa or hot tub assembly can be connected using regular wiring methods recognized in Chapter 3 of the *NEC®* as long as the wiring method has a copper equipment-grounding conductor not smaller than 12 AWG.

◤◤◤◤ ⬛ CAUTION ⬛ ◢◢◢◢

CAUTION: Remember that a 14 AWG Type NM cable has a 14 AWG equipment-grounding conductor. This is too small to use with a spa or hot tub, so the minimum size an electrician could use would be a 12 AWG Type NM cable.

Section 680.43 covers indoor installation requirements. The following installation requirements apply to receptacle installation (Figure 15–29):

- At least one 125-volt, 15- or 20-ampere-rated receptacle connected to a general-purpose branch circuit must be installed at least 5 feet (1.5 m) but not more than 10 feet (3 m) from the inside wall of the spa or hot tub.
- All other 125-volt, 15- or 20-ampere-rated receptacles located in the area of the spa or hot tub must be located at least 5 feet (1.5 m) measured horizontally from the inside wall of the spa or hot tub.
- All 125-volt, 15-, 20-, or 30-amp-rated receptacles located within 10 feet (3 m) of the inside walls of the spa or hot tub must be GFCI protected.
- The receptacle that supplied power to the spa or hot tub must be GFCI protected.

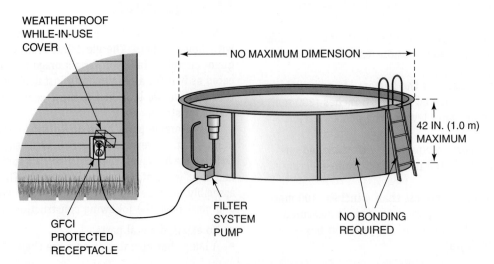

Figure 15–28 Installation of a receptacle for a storable pool. The receptacle will serve the pump motor.

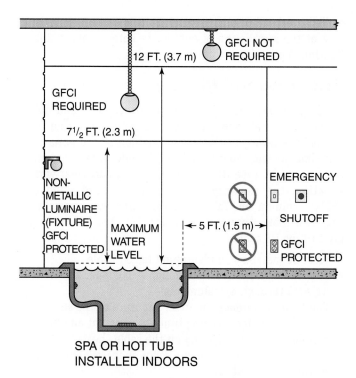

Figure 15–29 Installation requirements that apply to receptacles installed at indoor spa or hot tub locations.

> **CAUTION**
>
> CAUTION: Section 680.43(A)(4) states that a readily accessible, clearly labeled emergency shutoff switch for stopping the motor that provides recirculating and jet system power must be installed at least 5 feet (1.5 m) away, adjacent to, and within sight of the spa or hot tub. This rule does not apply to single-family dwellings, but it is this author's recommendation that it be included with all spa or hot tub installations in a house.

The following installation requirements apply to luminaire (lighting fixture) installations or ceiling-suspended paddle fans installation around a spa or hot tub (Figure 15–29):

- They must not be installed less than 12 feet (3.7 m) above the spa ot hot tub unless GFCI protected.
- They must not be installed less than 7 feet, 6 inches (2.3 m) above the spa or hot tub, even when GFCI protected.
- If located less than 7 feet, 6 inches (2.3 m) above the spa or hot tub, they must be suitable for a damp location and be a recessed luminaire with a glass or plastic lens and a nonmetallic or isolated metal trim and be a surface-mounted luminaire with a glass or plastic globe, a nonmetallic body, or a metal body that is isolated from contact.

- If underwater lighting is installed for a spa or hot tub, the same rules for underwater lighting in a swimming pool apply.
- When installing wall switches for the lighting fixtures or ceiling-suspended paddle fans, they must be kept a minimum of 5 feet (1.5 m) from the inside edge of the spa or hot tub.

Hydromassage Bathtubs

A **hydromassage bathtub** is intended to be filled, used, and then drained after each use. Some people refer to this type of unit as a "whirlpool bath." Part VII of Article 680 covers the installation requirements for hydromassage bathtubs (Figure 15–30). Hydromassage bathtubs are usually installed in a bathroom of a home. Remember that Section 210.8(A)(1) and (B)(1) requires all bathroom receptacles (125-volt, 15- or 20-ampere rated) to be GFCI protected. Additionally, Section 680.71 requires all 125-volt receptacles that are 15-, 20-, or 30-amp rated to be GFCI protected when they are located within 5 feet (1.5 m) measured horizontally of the inside walls of the hydromassage tub. This is true whether the hydromassage tub is installed in a bathroom area or some other area of the house. Section 680.71 also states that all hydromassage bathtubs and any associated electrical equipment must be GFCI protected.

Section 680.74 requires all metal piping, electrical equipment metal parts, and pump motors associated with the hydromassage bathtub to be bonded together using an 8 AWG or larger solid copper conductor. The bonding wire can be bare or insulated.

The hydromassage bathtub electrical equipment must be accessible without causing damage to any part of the building, according to Section 680.73. This means that there will have to be a panel of some type that can be easily removed for access to the electrical equipment.

Electricians usually run a separate circuit to a hydromassge bathtub. It is normally either a 15-amp circuit using a 14 AWG conductor in a cable or raceway or a 20-amp

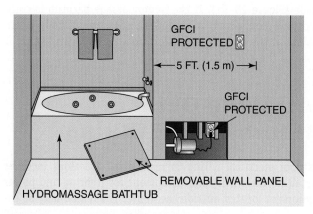

Figure 15–30 The installation requirements for hydromassage bathtubs.

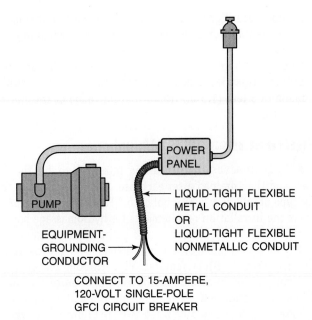

Figure 15–31 Electrical connections for a hydromassage bathtub. The motor, power panel, and the electrical supply leads are shown.

circuit using a 12 AWG conductor. Remember that the circuit must be GFCI protected, which will mean the use of a GFCI circuit breaker at the electrical panel. The circuit wiring is run during the rough-in stage to a location that will be near to where the pump and control/junction box for the hydromassage pump will be located. A junction box is installed as close to this location as possible. Consult the building plans and the plumbing contractor to determine exactly where the hydromassage bathtub will be located, specifically at what spot will the pump and control panel be located. The manufacturers of hydromassage bathtubs usually supply a length of liquid-tight flexible raceway that contains a black ungrounded conductor, a white grounded conductor, and a green equipment-grounding conductor. The length supplied is around 3 feet (900 mm). The length of liquid-tight flex is brought to the junction box, and the proper connections are made (Figure 15–31). If the hydromassage bathtub is cord-and-plug connected, a receptacle (GFCI protected) is installed at the end of the branch circuit at a location that is close to the pump location.

Installing Outdoor Branch-Circuit Wiring

Outdoor electrical wiring in residential situations includes installing the wiring and equipment for lighting and power equipment located outside the house. The wiring may be in-

stalled overhead or underground. In this section, we take a look at the installation of outdoor receptacles and outdoor lighting applications.

Underground Wiring

Since the majority of underground receptacle and lighting circuits installed in residential wiring is done using Type UF cable, a review of this wiring method is appropriate at this time. Article 340 covers Type UF cable. When determining the size of Type UF cable to use for a particular wiring situation, the ampacity of the cable is found using the "60°C" column in Table 310.16, just like nonmetallic sheathed cable. According to Article 340, the following apply to Type UF cable:

- Must be marked as underground feeder cable
- Is available from 14 AWG through 4/0 copper and from 12 AWG through 4/0 aluminum
- Can be used outdoors in direct exposure to the sun only if it is listed as being sunlight resistant and has a sunlight-resistant marking on the cable sheathing
- Can be used with the same fittings as is used with nonmetallic sheathed cable
- Can be buried directly in the ground and installed according to Section 300.5 and Table 300.5

 FROM EXPERIENCE

The outside sheathing of a Type UF cable is much more difficult to remove than the outside sheathing on a nonmetallic sheathed cable. In a Type UF cable, the outer sheathing is molded around each of the circuit conductors. This manufacturing process helps make the cable suitable for underground installation but does make the cable rather difficult to strip. While every electrician eventually finds a way to strip away the UF cable sheathing, the following method has worked well for this author: 1) At the end of the cable you wish to strip, use a knife to cut a 1 inch slit in the cable above the bare grounding conductor. 2) Using your fingers, open up the cable sheathing and grab the end of the bare grounding conductor with your long-nose pliers. 3) While holding on to the cable end with one hand, pull the grounding conductor back about 8 inches. This action will slit the cable sheathing as you are pulling the grounding conductor. 4) Now, using your long-nose pliers, grab the end of the other conductors in the cable and pull them back to the same point as the grounding conductor. 5) Cut off the cable sheathing with a knife or your diagonal cutting pliers.

- When used as an interior wiring method, must be installed according to the same rules as for nonmetallic sheathed cable
- Contains an equipment-grounding conductor (bare) that is used to ground equipment fed by the Type UF cable

Some electricians prefer to use a conduit wiring method when installing underground conductors. All underground installations are considered a wet location by definition and means that only certain types of conduit will be able to be used. Also, any wiring installed in the underground conduits must have a "W" in their insulation designation, such as "THWN" or "XHHW." The "W" means that the conductor insulation is suitable for installation in a wet location (refer to Table 310.13 of the *NEC®*). The conduits installed underground must be of a type that is resistant to corrosion. The manufacturer of the conduit and Underwriters Laboratories (UL) listings will give you information that will let you determine if the conduit type you wish to install underground is suitable for that location. Additional protection to a metal conduit can be provided by applying a protective coating to the conduit (Figure 15–32). Most electricians will use rigid nonmetallic conduit for underground conduit installation. A separate equipment-grounding conductor will have to be installed, and it can either be bare or green insulated. It will be sized according to Table 250.122 of the *NEC®*.

The minimum burial depths for both Type UF cable and for any of the conduit wiring methods are shown in Table 300.5 of the *NEC®*. When installing the underground conductors across a lawn or field, use the minimum depths as shown in Figure 15–33. When installing a cable or conduit system under a residential driveway or parking lot, use the minimum depths as shown in Figure 15–34.

Outdoor Receptacles

Receptacle outlets located outdoors must be installed in weatherproof enclosures. The electrical boxes are usually made of metal and are of the "FS" or "FD" type. This box type is often called a "bell box" by electricians working in the field and has threaded openings or hubs that allow attachment to the box with conduit or a cable connector. Each of these boxes comes from the factory with a few "threaded plugs" that are used in any unused threaded openings to make the box truly weatherproof. These boxes can be mounted on the surface of an outside wall or on some other structural support, such as a wooden post driven into the ground. They are often installed with underground wiring and supported by conduits coming up out of the ground. Section 314.23(E) and (F) covers the support rules for electrical boxes when they are fed using an underground wiring method and supported by conduit. Basically, the rules state that the box must be supported by at least two conduits threaded wrenchtight into the box. For boxes that contain a receptacle device, or if the box is just a junction box with both conduits threaded into the same side, the maximum distance from where the conduit exits the ground to the box is 18 inches (450 mm) (Figure 15–35).

When a receptacle is installed outdoors, the enclosure and cover combination must maintain its weatherproof characteristics whether a cord plug is inserted into the receptacle or not. This is required by Section 406.8(B)(1) of the *NEC®*. This requirement is met by installing a self-closing cover that is deep enough to also cover the attached plug cap on a cord (Figure 15–36).

Outdoor Lighting

Outdoor lighting can be mounted on the side of building structures, on poles, or even on trees. The lighting provides illumination on a home owner's property for activities after dark and for one other very important reason: security. Any luminaire (lighting fixture) installed outdoors and exposed to the weather must be listed as suitable for the location. A label with a marking that states "suitable for wet

Is Supplemental Corrosion Protection Required?

	In Concrete Above Grade?	In Concrete Below Grade?	In Direct Contact with Soil?
Rigid conduit[1]	No	No	No[2]
Intermediate conduit[1]	No	No	No[2]
Electrical metallic tubing[1]	No	Yes	Yes[3]

[1] Severe corrosion can be expected where ferrous metal conduits come out of concrete and enter the soil. Here again, some electrical inspectors and consulting engineers might specify the application of some sort of supplemental nonmetallic corrosion protection.

[2] Unless subject to severe corrosive effects. Different soils have different corrosive characteristics.

[3] In most instances, electrical metallic tubing is not permitted to be installed underground in direct contact with the soil because of corrosion problems.

Figure 15–32 Supplemental corrosion protection may be required for the raceways you are installing underground.

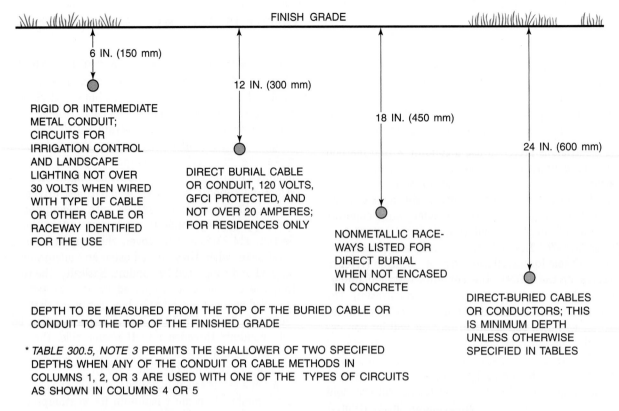

FINISH GRADE

6 IN. (150 mm)

RIGID OR INTERMEDIATE
METAL CONDUIT;
CIRCUITS FOR
IRRIGATION CONTROL
AND LANDSCAPE
LIGHTING NOT OVER
30 VOLTS WHEN WIRED
WITH TYPE UF CABLE
OR OTHER CABLE OR
RACEWAY IDENTIFIED
FOR THE USE

12 IN. (300 mm)

DIRECT BURIAL CABLE
OR CONDUIT, 120 VOLTS,
GFCI PROTECTED, AND
NOT OVER 20 AMPERES;
FOR RESIDENCES ONLY

18 IN. (450 mm)

NONMETALLIC RACE-
WAYS LISTED FOR
DIRECT BURIAL
WHEN NOT ENCASED
IN CONCRETE

24 IN. (600 mm)

DIRECT-BURIED CABLES
OR CONDUCTORS; THIS
IS MINIMUM DEPTH
UNLESS OTHERWISE
SPECIFIED IN TABLES

DEPTH TO BE MEASURED FROM THE TOP OF THE BURIED CABLE OR
CONDUIT TO THE TOP OF THE FINISHED GRADE

TABLE 300.5, NOTE 3 PERMITS THE SHALLOWER OF TWO SPECIFIED
DEPTHS WHEN ANY OF THE CONDUIT OR CABLE METHODS IN
COLUMNS 1, 2, OR 3 ARE USED WITH ONE OF THE TYPES OF CIRCUITS
AS SHOWN IN COLUMNS 4 OR 5

TABLE 300.5 SHOWS LESSER DEPTHS WHEN CABLES OR CONDUITS ARE PROTECTED
BY CONCRETE, BUT IN RESIDENTIAL INSTALLATIONS THIS IS NOT NORMALLY DONE

TABLE 300.5 (NOTE 5): WHERE SOLID ROCK IS ENCOUNTERED, MAKING IT IMPOSSIBLE
TO ATTAIN THE SPECIFIED DEPTHS, ALL WIRING SHALL BE INSTALLED IN A METAL OR
NONMETALLIC RACEWAY LISTED FOR DIRECT BURIAL; THE RACEWAY SHALL BE COVERED
BY A MINIMUM OF 2 IN. (50 mm) OF CONCRETE EXTENDING DOWN TO THE ROCK

Figure 15–33 Minimum burial depths for cables and conduits installed underground.

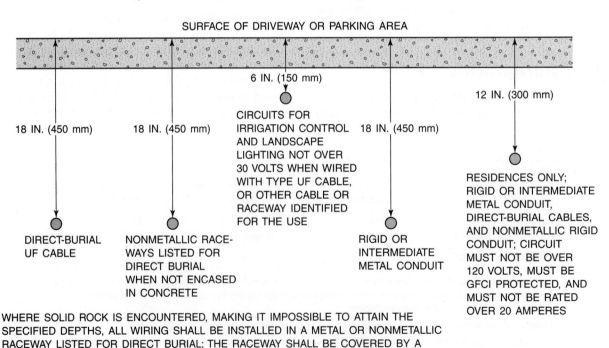

SURFACE OF DRIVEWAY OR PARKING AREA

6 IN. (150 mm)

12 IN. (300 mm)

18 IN. (450 mm)

18 IN. (450 mm)

CIRCUITS FOR
IRRIGATION CONTROL
AND LANDSCAPE
LIGHTING NOT OVER
30 VOLTS WHEN WIRED
WITH TYPE UF CABLE,
OR OTHER CABLE OR
RACEWAY IDENTIFIED
FOR THE USE

18 IN. (450 mm)

DIRECT-BURIAL
UF CABLE

NONMETALLIC RACE-
WAYS LISTED FOR
DIRECT BURIAL
WHEN NOT ENCASED
IN CONCRETE

RIGID OR
INTERMEDIATE
METAL CONDUIT

RESIDENCES ONLY;
RIGID OR INTERMEDIATE
METAL CONDUIT,
DIRECT-BURIAL CABLES,
AND NONMETALLIC RIGID
CONDUIT; CIRCUIT
MUST NOT BE OVER
120 VOLTS, MUST BE
GFCI PROTECTED, AND
MUST NOT BE RATED
OVER 20 AMPERES

WHERE SOLID ROCK IS ENCOUNTERED, MAKING IT IMPOSSIBLE TO ATTAIN THE
SPECIFIED DEPTHS, ALL WIRING SHALL BE INSTALLED IN A METAL OR NONMETALLIC
RACEWAY LISTED FOR DIRECT BURIAL; THE RACEWAY SHALL BE COVERED BY A
MINIMUM OF 2 IN. (50 mm) OF CONCRETE EXTENDING DOWN TO THE ROCK

DEPTH TO BE MEASURED FROM THE TOP OF THE BURIED CABLE OR
CONDUIT TO THE TOP OF THE FINISHED DRIVEWAY

Figure 15–34 Minimum burial depths for cables and raceways under residential driveways or parking lots.

Figure 15–35 Section 314.23(E) and (F) covers the support rules for electrical boxes when they are fed using an underground wiring method and supported by conduit.

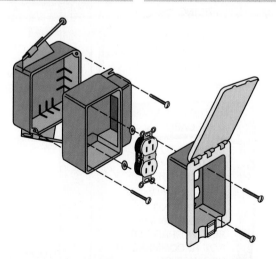

Figure 15–36 **A recessed type of weatherproof receptacle. This style meets the requirements of Section 406.8(B)(1) and (B)(2), which requires that the enclosure be weatherproof while an item is plugged into it and is left unattended.** *Photos courtesy of TayMac Corporation*

Figure 15–37 **Locating lighting fixtures on trees is not prohibited by the *NEC*®. A recognized wiring method that adequately protects the conductors must be used to carry the conductors up the tree trunk to the fixture.**

locations" must be found on the fixture. If a luminaire is to be installed under a canopy or under an open porch, it is considered a "damp" location, and the fixture only needs a label that states "suitable for damp locations." A wet-location luminaire may be used in either a "damp" or a "wet" outdoor location.

Most wiring installed for outdoor lighting is done underground, but overhead wiring from the main house to the location of outdoor lighting may be done. Contrary to popular belief, Section 410.16(H) allows outdoor lighting fixtures to be mounted on trees (Figure 15–37). However, Sections 225.26 and 527.4(J) state that overhead conductor spans cannot be supported by trees or other living or dead vegetation (Figure 15–38). So, if an electrician is installing wiring to a tree-mounted lighting fixture, it will need to be by an underground wiring method. Make sure to use a wiring method that provides adequate protection to the circuit conductors when installing wiring from the point where it emerges from the ground and is run up a tree to the lighting fixture.

A common outdoor wiring installation for electricians in residential wiring is a pole light. Figure 15–39 shows a typical pole light installation.

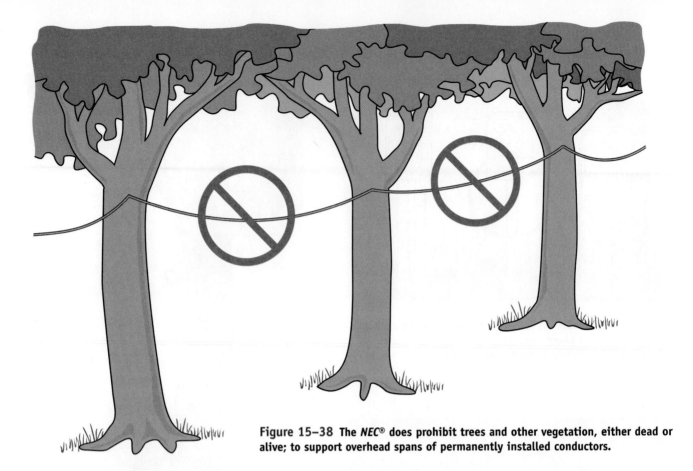

Figure 15–38 The *NEC*® does prohibit trees and other vegetation, either dead or alive; to support overhead spans of permanently installed conductors.

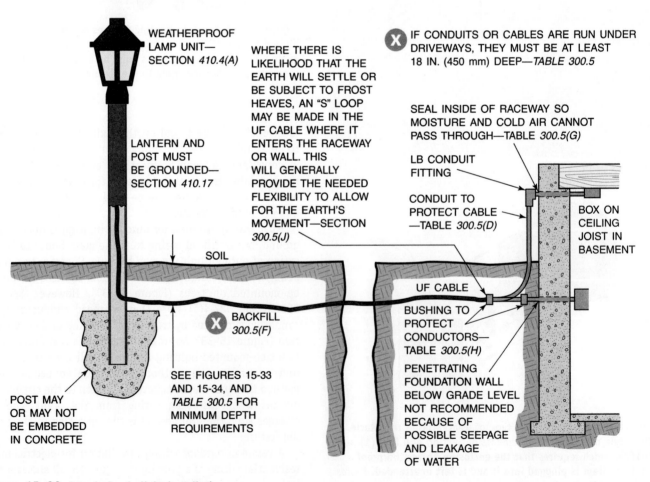

WEATHERPROOF LAMP UNIT— SECTION *410.4(A)*

WHERE THERE IS LIKELIHOOD THAT THE EARTH WILL SETTLE OR BE SUBJECT TO FROST HEAVES, AN "S" LOOP MAY BE MADE IN THE UF CABLE WHERE IT ENTERS THE RACEWAY OR WALL. THIS WILL GENERALLY PROVIDE THE NEEDED FLEXIBILITY TO ALLOW FOR THE EARTH'S MOVEMENT—SECTION *300.5(J)*

X IF CONDUITS OR CABLES ARE RUN UNDER DRIVEWAYS, THEY MUST BE AT LEAST 18 IN. (450 mm) DEEP—*TABLE 300.5*

SEAL INSIDE OF RACEWAY SO MOISTURE AND COLD AIR CANNOT PASS THROUGH—*TABLE 300.5(G)*

LB CONDUIT FITTING

CONDUIT TO PROTECT CABLE —*TABLE 300.5(D)*

BOX ON CEILING JOIST IN BASEMENT

LANTERN AND POST MUST BE GROUNDED— SECTION *410.17*

SOIL

UF CABLE

BACKFILL *300.5(F)*

BUSHING TO PROTECT CONDUCTORS— *TABLE 300.5(H)*

POST MAY OR MAY NOT BE EMBEDDED IN CONCRETE

SEE FIGURES 15-33 AND 15-34, AND *TABLE 300.5* FOR MINIMUM DEPTH REQUIREMENTS

PENETRATING FOUNDATION WALL BELOW GRADE LEVEL NOT RECOMMENDED BECAUSE OF POSSIBLE SEEPAGE AND LEAKAGE OF WATER

Figure 15–39 A typical pole light installation.

Installing the Wiring for a Standby Power System

In many areas, the reliability of electric service to a home is not very good. Ice storms, high winds, heavy snows, and other severe weather conditions can cause power outages. Many home owners have opted to have the electrical contractor install a **standby power system** so they will not need to be without electrical power for prolonged periods of time. In this section, we take a look at a common installation of standby power systems for residential customers. It is beyond the scope of this book to cover all possible standby power installations for houses.

The most common standby power system installation for homes uses a portable **generator** and the associated equipment necessary to safely feed branch circuits already installed as part of the residential wiring system. The generator is a packaged unit that has a gasoline-powered engine turning a generator that produces alternating current electricity. These manufactured generator units are usually started manually, like a lawnmower, but many of the larger units offer electric starting. Portable generators are rated in watts and can range from around 3,000 watts up to 12,000 watts or more. Each generator comes factory equipped with one or more 15- or 20-amp-rated receptacles for direct connection of extension cords for powering small loads. A larger 120/240-volt polarized **twistlock receptacle** is installed on the units so the generator can be connected to power larger loads.

Before installing the standby power system, a home owner must determine which electrical loads are critical and which are not. On the basis of this information, the correct size of generator can be determined. Some generator manufacturers suggest that after all the **critical loads** are added up, another 20% is added for any future loads that may be included. Generators can be purchased that can supply the electrical power needs for a whole house, or, as is more often the case, generators are purchased that can supply only a portion of the house electrical loads.

The installation for the standby power system presented in this section includes a generator that is cord-and-plug connected to a **transfer switch.** There is also wiring installed from the main service panel to the transfer switch and from the transfer switch to the critical load panel. The panel for critical loads has hardwired connections to the branch circuits that are considered to be necessary when the electrical power from the local electric utility is not available. An electrician will install the critical load panel alongside the regular service entrance panel.

The transfer switch is very important. When the transfer switch is in the "normal" position, electrical power from the electric utility is routed to the branch circuits in the critical load panel (Figure 15–40). When the transfer switch is in the "standby generator" position (Figure 15–41), it disconnects the critical load branch circuits from the incoming electric utility power. Now, only the electrical power from the generator can be delivered to the critical load panel branch circuits. The transfer switch eliminates the possibility of the generator voltage being applied to the secondary of the electric utility's supply transformer, resulting in a very high voltage being put back on the "dead" electric utility wiring from the primary of the home supply transformer. This situation is commonly called "backfeeding" and could be a very dangerous situation when utility linemen working on the power lines believe them to be deenergized but they are really "live" with a high voltage developed from the home owner's generator. Many utility linemen have been severely injured or even killed by a home electrical system standby generator not being properly installed with an approved transfer switch. Transfer switches are rated in amperes and for residential applications range from 40 to 200 amps. They are a double-pole, double-throw (DPDT) type of switch. If the transfer switch is connected to the line side of the main service disconnecting means (that is, between the utility meter and the main circuit breaker), the transfer switch must be listed as suitable for use as service entrance equipment. The installation discussed in this section assumes the transfer switch to be located on the load side of the service disconnecting means and therefore does not need to be listed as suitable for service entrance equipment.

The connection of the generator to the transfer switch is generally done by a flexible cord with cord-and-plug connections. There will be a male polarized twistlock receptacle on the generator that is installed by the generator manufacturer. The receptacle installed for the connection to the transfer switch will need to be a polarized twistlock receptacle as well and will need to be sized according to the voltage and current rating of the generator receptacle it will be plugged into. Since the generator is gasoline powered and will need to be located outdoors, the location of the receptacle that the generator is plugged into may be located outdoors on an outside wall or possibly just inside the overhead door of an attached garage. The size of the conductors must be sized according to Section 445.13, which requires them to be at least 115% of the generator's nameplate current rating.

 FROM EXPERIENCE

To save money, some electrical contractors use two two-pole circuit breakers that are mechanically interlocked as the transfer switch. The mechanical interlock keeps them from both being "On" at the same time. One of the two-pole breakers is connected to the incoming service entrance conductors, and the other two-pole breaker is connected to the standby power system generator.

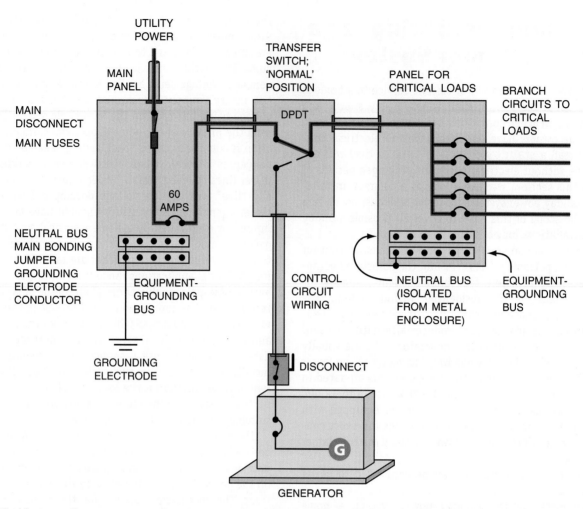

Figure 15–40 A standby power system installation with the transfer switch in the "normal" position. The branch circuits identified as being critical are being fed from the service entrance conductors.

For example, if you were installing an 8-kilowatt, 120/240-volt generator, the minimum conductor size from the generator to the transfer switch would be found as follows:

- 8,000 watts/240 volts = 33.33 amps
- 33.33 amps × 115% = 38.33 amps
- Minimum wire size Type NM cable would be an 8 AWG copper conductor according to Table 310.16 (the "60°C" column)
- Minimum-size Type SO flexible cord would be 8 AWG copper according to Table 400.5(A)

The connection from the transfer switch to the critical load panel is also a 120/240-volt system. It will be sized the same way as for sizing the conductors from the generator to the transfer switch as described previously. A cable wiring method like nonmetallic sheathed cable or a raceway wiring method may be used.

The connection from the main service panel to the transfer switch is also installed as a 120/240-volt system with a two-pole circuit breaker in the main panel sized according to

the number of branch circuits supplied from the critical load panel. Common sizes of the circuit breaker include 30, 50, and 60 amperes. A nonmetallic sheathed cable can be used or the wiring can be installed in a raceway. The size of the conductors is usually the same size as those going from the transfer switch to the critical load panel.

(See page 438 for the recommended procedure for safely connecting a generator's electrical power to the critical load branch circuits.)

Summary

When an electrician installs the electrical wiring system for a house, there are many special wiring applications that may be encountered. Garages, either attached or detached, are often found in residential construction and must be wired as necessary. Many of the common wiring practices that relate to both detached and attached garages were covered in this chapter.

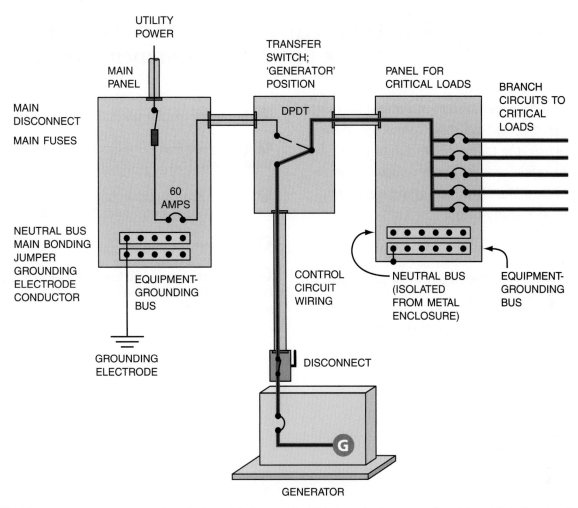

Figure 15–41 A standby power system installation with the transfer switch in the "generator" position. When the transfer switch is in this position, the critical load branch circuits are being fed by the generator power. Notice that with the transfer switch in this position, there is no way that electrical power from the generator can be "backfed" to the electrical utility power lines.

More and more homes have swimming pools, spas or hot tubs, or a hydromassage bathtub installed. All these items require some type of electrical wiring that must be installed by an electrician. This chapter presented the basic installation requirements for these specific installations. Outdoor wiring was also presented with the emphasis being on outdoor receptacles and outdoor luminaires (lighting fixtures). The last special wiring situation discussed in this chapter was standby electrical power systems. Portable generators were described, and the proper installation for a typical standby power system with a portable generator was presented. Not all residential wiring systems will require an electrician to deal with the areas covered in this chapter. However, when these wiring situations are encountered, you will now have a better idea of the electrical wiring requirements and installation techniques needed for these applications.

Procedures

Connecting a Generator's Electrical Power to the Critical Load Branch Circuits

When a standby power system includes a portable generator that is cord-and-plug connected to the transfer switch, the following procedure should be followed to safely connect the generator power to the critical load branch circuits when electrical power from the utility is lost:

1 Plug in the four-wire flexible cord to the polarized twist-lock receptacle on the generator.

2 Plug the other end into the polarized twistlock receptacle that is connected to the transfer switch.

3 Switch the transfer switch to the "generator" position.

4 Making sure that the generator is located outdoors, start the generator.

5 When electrical power from the utility is restored, switch the transfer switch to its "normal" position.

6 Shut down the generator.

7 Unplug the power cord from the generator end and the transfer switch end. Put everything back in its proper storage area.

Review Questions

Directions: Answer the following items with clear and complete responses.

1 Describe what is meant by the phrase "standby power."

2 Describe the function of a transfer switch in a standby power system.

3 A typical transfer switch used in a residential standby power system is a:

a. Double-pole, single-throw switch (DPST)

b. Double-pole, double-throw switch (DPDT)

c. Single-pole, double-throw switch (SPDT)

d. Single-pole, single-throw switch (SPST)

4 The conductors running from a standby system generator to the transfer switch must not be less than _____ % of the generator's output rating. The *NEC®* section that states this is _____.

5 Article _____ covers the requirements for swimming pools, hot tubs, and hydromassage bathtubs.

6 The *NEC®* requires all metal parts of a permanently installed swimming pool to be bonded together. The bonding grid must be connected to a grounding electrode. True or False? Name the *NEC®* section that backs up your answer. Section _____.

7 Receptacles located within 20 feet (6 m) from the inside edge of a swimming pool must be _____ protected.

8 Luminaires (lighting fixtures) not GFCI protected and installed over a swimming pool must be located no less than _____ feet above the maximum water level.

9 All receptacles that are located within 10 feet (3 m) of the inside walls of an indoor hot tub must be GFCI protected. True or False. Name the *NEC®* section that backs up your answer. Section _____.

10 Wall switches must be located at least 5 feet (1.5 m) from the edge of an indoor hot tub. True or False. Name the *NEC®* section that backs up your answer. Section _____.

11 When installing underwater luminaires (light fixtures) in a swimming pool, they must be installed so that the top of the lens is not less than _____ inches below the normal water level, unless they are designed for a lesser depth.

12 Describe the difference between a hot tub and a hydromassage bathtub.

13 All 120-volt receptacles installed within _____ feet of the hydromassage bathtub must be GFCI protected.

14 State the minimum burial depth for a Type UF cable feeding a pole light in a residential application. The circuit is rated 20 amperes and is not GFCI protected.

15 State the minimum burial depth for a Type UF cable feeding a pole light in a residential application. The circuit is rated 20 amperes and is GFCI protected.

16 State the minimum burial depth for a rigid nonmetallic conduit feeding a detached garage and installed across a lawn in a residential application.

17 When receptacles are installed outdoors and are intended to serve cord-and-plug-connected items, the cover must provide a weatherproof environment whether the plug is inserted into the receptacle or not. True or False. Name the *NEC®* section that backs up your answer. Section _____.

18 Describe what kind of electrical hazard can occur when a standby system generator is used, no transfer switch has been installed, and the service entrance main circuit breaker is not turned off.

19 | **Define the following:**

    a. Dry-niche luminaire (light fixture)

    b. Wet-niche luminaire (light fixture)

    c. No-niche luminaire (light fixture)

20 | A _____ or larger solid copper conductor must be used to make the bonding connections in the bonding grid for a permanently installed swimming pool. Name the *NEC*® section that backs up your answer. Section _____.

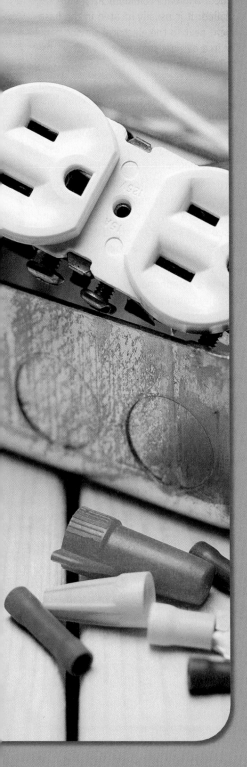

Chapter 16

Video, Voice, and Data Wiring Installation

Today's residential electricians are often required to install video, voice, and data wiring systems in homes. They must be aware of different video, voice, and data wiring materials and installation techniques. This chapter presents material designed to give you a better understanding of the materials, installation methods, and termination techniques as they apply to copper-structured cabling systems.

OBJECTIVES

Upon completion of this chapter, the student should be able to:

- list several common terms and definitions used in video, voice, and data cable installations.

- demonstrate an understanding of EIA/TIA 570 standards for the installation of video, voice, and data wiring in residential applications.

- identify common materials and equipment used in video, voice, and data wiring.

- demonstrate an understanding of the installation of video, voice, and data wiring in residential applications.

Glossary of Terms

bandwidth identifies the amount of data that can be sent on a given cable; it is measured in hertz (Hz) or megahertz (MHz); a higher frequency means higher data-sending capacity

category ratings on the bandwidth performance of UTP cable; categories include 3, 4, 5, 5e, and 6; Category 5 is rated to 100 MHz and is the most widely used

coaxial cable a cable in which the center signal-carrying conductor is centered within an outer shield and separated from the conductor by a dielectric; used to deliver video signals in residential structured cabling installations

EIA/TIA an acronym for the Electronic Industry Association and Telecommunications Industry Association; these organizations create and publish compatibility standards for the products made by member companies

EIA/TIA 570-A the main standard document for structured cabling installations in residential situations

F-type connector a 75-ohm coaxial cable connector that can fit RG-6 and RG-59 cables and is used for terminating video system cables in residential wiring applications

horizontal cabling the connection from the distribution center to the work area outlets

IDC an acronym for "insulation displacement connection"; a type of termination where the wire is "punched down" into a metal holder with a punch-down tool; no prior stripping of the wire is required

jack the receptacle for an RJ-45 plug

jacketed cable a voice/data cable that has a nonmetallic polymeric protective covering placed over the conductors

megabits per second (Mbps) refers to the rate that digital bits (1s and 0s) are sent between two pieces of equipment

megahertz (MHz) refers to the upper frequency band on the ratings of a cabling system

patch cord a short length of cable with an RJ-45 plug on either end; used to connect hardware to the work area outlet or to connect cables in a distribution panel

punch-down block the connecting block that terminates cables directly; 110 blocks are most popular for residential situations

RG-59 a type of coaxial cable typically used for residential video applications; EIA/TIA 570A recommends that RG-6 coaxial cable be used instead of RG-59 because of the better performance characteristics of the RG-6 cable

RG-6 (Series 6) a type of coaxial cable that is "quad shielded" and is used in residential structured cabling systems to carry video signals such as cable and satellite television

ring one of the two wires needed to set up a telephone connection; it is connected to the negative side of a battery at the telephone company; it is the telephone industry's equivalent to an ungrounded "hot" conductor in a normal electrical circuit

RJ-11 a popular name for a six-position UTP connector

RJ-45 the popular name for the modular eight-pin connector used to terminate Category 5 UTP cable

service center the hub of a structured wiring system with telecommunications, video, and data communications installed; it is usually located in the basement next to the electrical service panel or in a garage; sometimes called a "distribution center"

STP an acronym for "shielded twisted pair" cable; it resembles UTP but has a foil shield over all four pairs of copper conductors and is used for better high-frequency performance and less electromagnetic interference

structured cabling an architecture for communications cabling specified by the EIA/TIA TR41.8 committee and used as a voluntary standard by manufacturers to ensure compatibility

tip the first wire in a pair; a conductor in a telephone cable pair that is usually connected to the positive side of a battery at the telephone company's central office; it is the telephone industry's equivalent to a grounded conductor in a normal electrical circuit

UTP an acronym for "unshielded twisted pair" cable; it is comprised of four pairs of copper conductors and graded for bandwidth as "categories" by EIA/TIA 568; each pair of wires is twisted together

work area outlet the jack on the wall that is connected to the desktop computer by a patch cord

Introduction to EIA/TIA 570-A Standards

In today's homes, there is often a need for a "structured cabling" system to be installed. This system is completely separate from the residential electrical power system that we have discussed up to now. The structured cabling system includes the wiring and other necessary components for providing video, voice, and data signals throughout a house. It is installed using special cables and associated equipment by the same electricians who install the regular electrical power system. These systems offer multiple signal outlets for faster Internet connections and the ability to network home computers and peripheral devices like printers. They also provide for a number of conveniently located signal outlets for all the video devices (televisions, VCRs, DVD players, and so on) and telephones that are used in a modern home.

These systems are wired with special cables that include unshielded twisted pair cable (UTP), shielded twisted pair cable (STP), and coaxial cables. Sometimes optical fiber cables (FO) are used in the installation of a structured cabling system. However, since optical fiber cable is normally used only in commercial and industrial structured cabling applications, the installation of optical fiber cable is beyond the scope of this book and is not covered here.

In an effort to standardize the installation of a structured cabling system, the Telecommunications Industry Association (TIA) and the Electronic Industry Association (EIA) standards development committee was formed to develop installation standards for structured cabling systems in residential applications. EIA/TIA 570-A was developed by the committee to establish a standard for a generic cabling system that can accommodate many different applications. It was determined that proper installation and connector termination are critical to a video, voice, and data system's overall performance. By following EIA/TIA 570-A, a structured cabling system will perform properly.

To be able to truly understand the installation requirements as outlined in EIA/TIA 570-A, an electrician needs to become familiar with several terms that are used in the area of structured cabling. These terms are not normally used when referring to the regular electrical power system in a house, and not understanding the meaning of these terms can cause a lot of confusion for an electrician installing a structured cabling system. The following structured cabling system terms and definitions are used often and should become familiar to you:

- **Structured cabling** is a system for video, voice, and data communications cabling specified by EIA/TIA and used as a voluntary standard by manufacturers to ensure compatibility. Notice that the word "voluntary" is used. These installation guidelines are not mandatory. Interestingly enough, the only mandatory installation rules

for installing a structured cabling system are those referenced in the *National Electrical Code®* (*NEC®*). We will look at the *NEC®* rules later in this chapter.

- **EIA/TIA** is the abbreviation for the Electronic Industry Association and the Telecommunications Industry Association. Both organizations create compatibility standards for the products made by their member companies.
- **Bandwidth** is the term that identifies the amount of data that can be sent on a given cable. It is measured in hertz (Hz) or megahertz (MHz). The higher the frequency, the higher the data-sending capacity.
- **Insulation displacement connection (IDC)** is a type of termination where the wire is "punched down" into a metal holder with a punch-down tool (Figure 16–1). No prior stripping of the wire is required.
- **Unshielded twisted pair (UTP)** (Figure 16–2) is the type of cable normally used to install voice and data communication wiring in a house. It is comprised of four pairs of copper conductors and graded for bandwidth as "categories."
- A **category** is the rating, based on the bandwidth performance, of UTP cable.
- **Shielded twisted pair (STP)** is a cable that resembles UTP but has a foil shield over all four pairs of copper conductors and is used for better high-frequency performance and less electromagnetic interference (EMI).
- **Coaxial cable** is a cable in which the center signal-carrying conductor is centered within an outer shield and separated from the conductor by a dielectric. This cable type is used to install video signal wiring in residential applications. It can also be used to carry a high-speed Internet signal.
- **RG-6 (Series 6)** (Figure 16–3) is a type of coaxial cable that is "quad shielded" and is used in residential

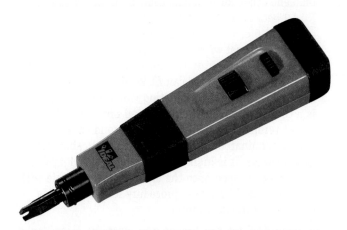

Figure 16–1 A punch-down tool with a 110 blade. This tool terminates voice and data wires to a 110 punch-down block. It also cuts the wires to length. The impact-absorbing cushioned grip reduces operator stress for high-volume users. The actuation mechanism can be adjusted, depending on the operator's preference. *Courtesy of Ideal Industries, Inc.*

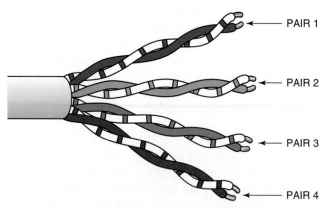

Figure 16–2 An unshielded twisted pair (UTP) cable.

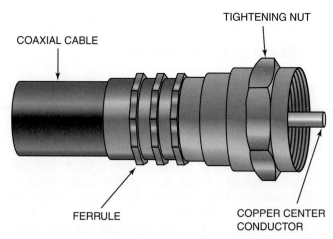

Figure 16–4 An F-type coaxial cable connector. This connector is threaded onto a female adapter at a television outlet.

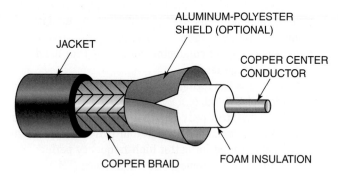

Figure 16–3 An RG-6 coaxial cable.

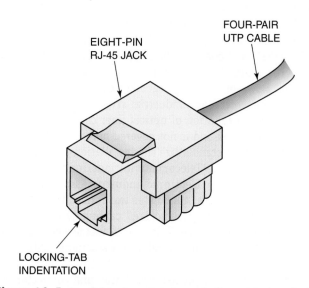

Figure 16–5 An eight-pin connector or jack used to terminate UTP cable. It is commonly called an RJ-45.

structured cabling systems to carry video signals such as cable and satellite television. The letters "RG" stand for "radio guide."

- **RG-59** is a type of coaxial cable typically used for residential video applications. However, EIA/TIA-570A recommends that RG-6 coaxial cable be used instead of RG-59 because of the better performance characteristics of the RG-6 cable. Do not use RG-59 coaxial cable in any new video signal wiring installation.
- An **F-type connector** (Figure 16–4) is a 75-ohm coaxial cable connector that can fit RG-6 and RG-59 cables and is only used for terminating video system cables. It is a connector type that is threaded onto the television outlet termination point for a good, solid connection. It can be terminated on the end of a coaxial cable in a variety of ways. It is highly recommended that F-type connectors be crimped onto the end of a coaxial cable with a proper tool. The proper procedure for doing this is shown later in this chapter.

(See page 456 for the proper procedure for installing an F-type connector on an RG-6 coaxial cable.)

- **RJ-11** is popular name given to a six-position connector or jack. The letters "RJ" stand for "registered jack."
- **RJ-45** is the popular name given to an eight-pin connector or jack used to terminate UTP cable (Figure 16–5).

(See page 457 for the proper procedure for installing an RJ-45 jack on the end of a four-pair UTP Category 5 cable.)

- A **jack** (Figure 16–6) is the term given to the receptacle device that accepts an RJ-11 or RJ-45 plug.
- A **punch-down block** is the connecting block that terminates UTP cables directly; 110 blocks (Figure 16–7) are most popular for residential applications and require a 110-block punch-down tool for making the terminations.
- **Horizontal cabling** is the term used to identify the cables run from a service center that serves as the "hub" for the structured cabling system to the work area outlet.
- The **work area outlet** is the jack on the wall that is connected to a desktop computer by a patch cord (Figure 16–8).

SURFACE-MOUNTED JACK

WALL-MOUNTED JACK

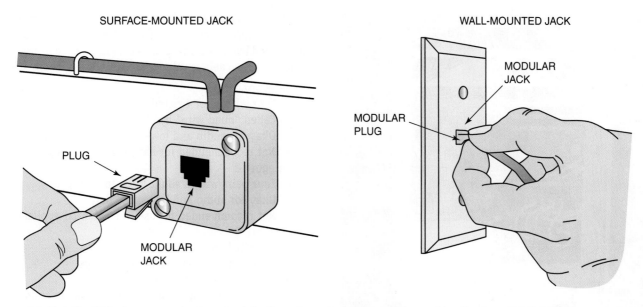

Figure 16–6 A "jack" is the receptacle that the "plug" on the end of voice and data cables fits into. The jack can be surface mounted or flush mounted.

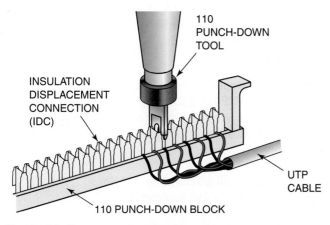

Figure 16–7 A 110-style punch-down block. A punch-down tool with a 110 blade is used to terminate the wires of a UTP cable to the block. *Courtesy of Ideal Industries, Inc.*

- A **patch cord** is the short length of cable with an RJ-45 plug on either end used to connect a home computer to the work area outlet. A patch cord can also be used to interconnect various punch-down blocks in the service center.

(See page 460 for the proper procedure for assembling a patch cord with RJ-45 plugs and Category 5 UTP cable.)

- The **service center** (Figure 16–9) is the hub of a structured wiring system with telecommunications, video, and data communications installed. It is usually located in the basement next to the electrical service panel or in a garage.
- **Megahertz (MHz)** refers to the upper frequency band on the ratings of a cabling system. "Mega" is the prefix

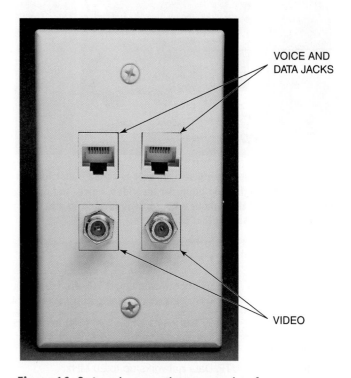

Figure 16–8 A work area outlet may consist of one or more jacks for video, voice, or data connections.

for "one million." "Hertz" is the term used for the number of cycles per second of the specific signal.
- **Megabits per second (Mbps)** refers to the rate that digital bits (1s and 0s) are sent between two pieces of digital electronic equipment. A bit is the smallest unit of measure in the binary system. The expression "bit" comes from the term "binary digit."

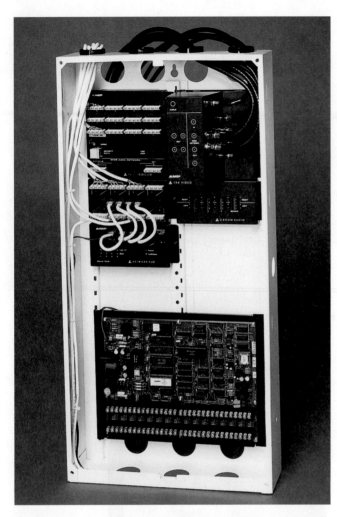

Figure 16–9 A service center, sometimes called a distribution center, is the origination point for the video, voice, and data circuits installed throughout a house. It is located in a basement, garage, or some other utility area.

The cables used for voice and data signal transmission have been categorized by the EIA and TIA. All these cables are tested and listed according to Underwriters Laboratories Standard 444. The cable is categorized as follows (Figure 16–10):

- Category 1
  - Four wires, not twisted
  - Called "quad wire" by electricians
  - Okay for audio and low-speed data transmission
  - Not recommended for use in residential applications
- Category 2
  - Four pairs with a slight twist to each pair
  - Okay for audio and low-speed data
  - Not recommended for use in residential applications
- Category 3
  - 16-MHz bandwidth
  - Supports applications up to 10 Mbps
  - 100-ohm UTP-rated Category 3
  - Declining in popularity but often used in residential telephone applications
- Category 4
  - 20-MHz bandwidth
  - Supports applications up to 16 Mbps
  - 100-ohm UTP-rated Category 4
  - Basically obsolete
- Category 5
  - 100-MHz bandwidth
  - Supports applications up to 100 Mbps
  - 100-ohm UTP-rated Category 5
  - Most often used UTP cable for residential voice and data applications
- Category 5e
  - 100-MHz bandwidth
  - Supports applications up to 100 Mbps
  - 100-ohm UTP-rated Category 5e

Category 1:	Four-pair nontwisted cable. Referred to as "quad wire". Okay for audio and low-speed data. Not suitable for modern audio, data, and imaging applications. This is the old-fashioned plain old telephone service (POTS) cable.
Category 2:	Usually four pair with a slight twist to each pair. OK for audio and low-speed data.
Category 3:	UTP. Data networks up to 16 MHz.
Category 4.	UTP. Data networks up to 20 MHz.
Category 5.	UTP. Data networks up to 100 MHz. By far the most popular, having four pairs with 24 AWG copper connectors, solid (for structured wiring) and stranded (for patch cords). Cable has an outer PVC jacket.
Category 5e:	UTP. Basically an enhanced Category 5 manufactured to tighter tolerances for higher performance. Data networks up to 350 MHz. Will eventually replace Category 5 for new installations.
Category 6:	UTP. Four twisted pairs. Data networks up to 250 MHz. Has a spline between the four pairs to minimize interference between the pairs.
Category 7:	Four twisted pairs surrounded by a metal shield. Data networks up to 750 MHz. Specifications not yet standardized.

Figure 16–10 Cables are rated in "categories." This figure shows the various category types and has a short description for each.

- Higher performance over a minimally compliant Category 5 installation by following Category 5e Technical Specifications
- Rapidly becoming the preferred cable type for both voice and data in residential applications
- Category 6
  - 250-MHz bandwidth
  - Supports applications over 100 Mbps
  - 100-ohm UTP-rated Category 6
  - Current applications designed to run on Category 5 cable, but Category 6 cable sometimes specified in anticipation of future needs
  - The emerging favorite in commercial applications but not used very often in residential applications
- Category 7
  - 750-MHz bandwidth
  - A metal shield around the conductors
  - Not currently used in residential applications

The voice and data cable recommended by EIA/TIA 570-A for use in residential applications has the following characteristics:

- It carries both voice and data.
- It is normally 22 or 24 AWG copper.
- It is always described and connected in pairs.
- The wire pairs should be twisted together to preserve signal quality.

Each pair of wires in a voice/data structured cabling system cable consists of a **tip** and a **ring** wire. This is a carry-over from the old days in the telephone industry. When an operator was asked to connect a call to another party, a cable was plugged into a connection panel in front of the operator. This cable had a connector on the end of it with a tip and a ring that made the proper connection so that the call could go though. Since direct current is used by telephone systems, polarity must be maintained, so the standard wiring practice is to use the tip wire as the positive (+) conductor and the ring as the negative (−) conductor.

Each pair of conductors in a UTP or STP cable is twisted together in a certain way. The reason for the twisted pairs is that it helps prevent interference like induction and 'crosstalk' from other pairs in the same cable and from other sources like electrical power circuits and electric motors. You may have experienced "crosstalk" when you have been using a phone and heard another conversation going on in the background at the same time. **Jacketed cable** with four twisted conductor pairs is recommended for all inside residential voice and data wiring.

```
◄◄◄◄◄◄  ( CAUTION )  ►//◄//◄►
```

CAUTION: The old "quad wire" telephone cable that has been used for years to wire telephone jacks in residential applications is not recommended for use by EIA/TIA 570-A. Use a four-pair UTP cable instead.

Each pair of conductors in voice/data cable has a specific color coding so an electrician will know which pair of conductors connects to the proper terminals at a voice/data jack or in the service center (Figure 16–11). The standard color coding for a four-pair UTP cable is as follows:

- Pair 1: Tip is white/blue; ring is blue.
- Pair 2: Tip is white/orange; ring is orange.
- Pair 3: Tip is white/green; ring is green.
- Pair 4: Tip is white/brown; ring is brown.

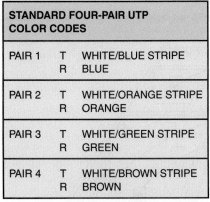

STANDARD FOUR-PAIR UTP COLOR CODES		
PAIR 1	T	WHITE/BLUE STRIPE
	R	BLUE
PAIR 2	T	WHITE/ORANGE STRIPE
	R	ORANGE
PAIR 3	T	WHITE/GREEN STRIPE
	R	GREEN
PAIR 4	T	WHITE/BROWN STRIPE
	R	BROWN

NOTE: FOR SIX-WIRE JACKS USE PAIR 1, 2 AND 3 COLOR CODES. FOR FOUR-WIRE JACKS, USE PAIR 1 AND 2 COLOR CODES.

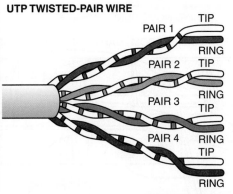

*CAUTION

QUAD WIRE IS NO LONGER ACCEPTABLE FOR INSTALLATION IN MULTILINE ENVIRONMENTS. IF ENCOUNTERED DURING A RETROFIT, QUAD WIRE SHOULD BE REPLACED WITH 100-Ohm UTP IF POSSIBLE. CONNECTING NEW QUAD TO INSTALLED QUAD WILL ONLY AMPLIFY EXISTING PROBLEMS AND LIMITATIONS ASSOCIATED WITH QUAD WIRE; LEAVING EXISTING QUAD IN PLACE AND CONNECTING 100Ω UTP TO IT MAY ALSO BE INEFFECTIVE, AS THE QUAD WIRE MAY NEGATE THE DESIRED EFFECT OF THE UTP.

Figure 16–11 The TIA standard color coding for residential voice and data wiring.

FROM EXPERIENCE

An easy way to remember the color coding for a four-pair UTP cable is to use the acronym "BLOGB." "BL" stands for blue, "O" stands for orange, "G" stands for green, and "B" stands for brown.

The standard color coding for a quad wire (solid-color, non-twisted pair) is as follows:

- Pair 1: Tip is green; ring is red.
- Pair 2: Tip is black; ring is yellow.

Installing Residential Video, Voice, and Data Circuits

In today's residential structured cabling systems, it is common practice to install a service center that serves as the origination point for all video, voice, and data systems in a house. Locate the service center (sometimes called a distribution center) in the basement, garage, or some other utility area of the house. There must be readily available electrical power for the service center, and access to the service-grounding electrode system is required. In new residential construction, the "home runs" from the service center are run in concealed pathways, such as in wall or ceiling cavities. Some electricians choose to install the cable wiring in conduit so that it will be easier to add or take away cables in the future as customer needs change. In existing homes, the cable wiring is concealed as much as possible in attics, basements, and crawl spaces. When the wiring cannot be concealed, it is run on the surface using a surface metal or nonmetallic raceway.

Voice Wiring System

Figure 16–12 shows a suggested wiring layout for the voice system in a typical house. Use four-pair 100-ohm UTP cable. Remember to not use the old quad cable. A Category 3 cable is the absolute minimum performance category that should be installed. It is recommended that at least a Category 5 or 5e be used in all new voice installations. At each wall outlet location, an eight-position RJ-45 jack with T568A wiring should be used. T568B wiring of the jack may also be used (Figure 16–13). Whichever wiring scheme is used, make sure that all jacks in the house are wired the same way. There should be a minimum of one voice jack per outlet location. Each voice outlet location should have a separate home run back to the service center. The home runs are terminated at the service center by punching down the cable to the proper 110 terminal blocks. Once the system has been completely installed, connection of individual telephones to the wall jacks is accomplished by simply plugging the phone into any jack in the house.

The telephone company will bring telephone service to the house by either an overhead wiring method or an underground wiring method. Either way, the telephone company will terminate its wiring at a telephone network interface box mounted on the outside of the house. This enclosure is weatherproof and is suitable for mounting on the

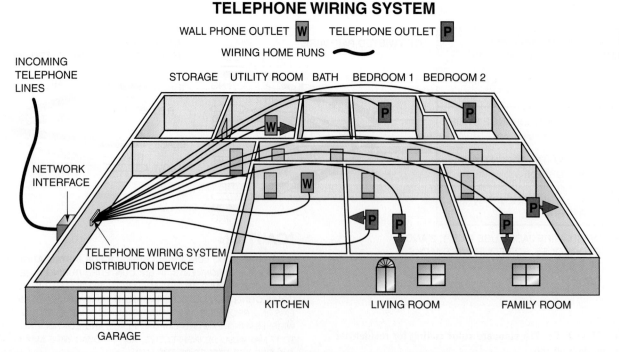

Figure 16–12 A typical residential telephone wiring system installation from a central distribution panel.

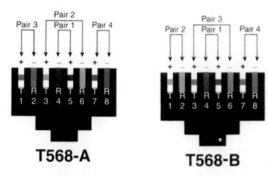

Figure 16–13 **The wiring configurations for a T568-A (A) and a T568-B (B) scheme. Modular plugs are shown in this illustration, but the wiring configuration for the two schemes is the same for jacks and punch-down blocks as well.** *Courtesy of Ideal Industries, Inc.*

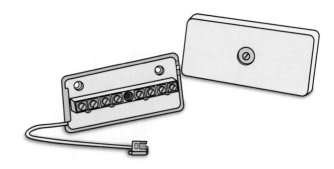

A TELEPHONE JUNCTION BOX FEATURES A SHORT, PREWIRED CORD THAT PLUGS INTO A STANDARD NETWORK INTERFACE, ALLOWING EASY CONNECTION OF ADDITIONAL TELEPHONE CABLES

Figure 16–14 **A telephone junction box.**

outside of the house. If the telephone company terminates its wiring inside a house, a standard network interface box is used. The point where the telephone company's wiring ends and the home owner's interior wiring begins is called the "demarcation point." As an electrician installing the voice system in a home, you will need to install telephone wiring from the demarcation point to the service center.

CAUTION

CAUTION: It is common practice to install a wall-mounted telephone jack in the kitchen of a house. This ensures that a phone will be connected to the system at all times so that in the case of an emergency, portable phones would not have to be found and then plugged into a jack for service. Be sure to install at least one wall-mounted telephone jack in each house.

Even though a service center is recommended to be installed for all new residential telephone installations, many home owners will choose to have the telephone system installed in a more traditional manner. In most cases, when a home owner chooses the traditional method over the newer EIA/TIA method described previously, it is based on the fact that the older method costs less money to install. The older method, even though it is not EIA/TIA recommended, is still the most often used technique for installing a telephone system in a house. It requires an electrician to install a telephone junction box (Figure 16–14) at a convenient location. It is usually located inside a home and is often placed next to the service entrance panel. A connection is made by the electrician from the demarcation point to the junction box with a suitable length of telephone cable. From the telephone junction box, individual runs of telephone cable (Category 3 or higher) are wired to the various telephone outlet locations throughout a house. Figure 16–15 shows a typical telephone system installation using the more traditional method.

Data Wiring System

Figure 16–16 shows a suggested wiring layout for the data wiring system in a typical house. Use a four-pair 100-ohm UTP cable. It is recommended that a Category 5 or 5e be used as the minimum category rated cable for this type of installation. At each wall outlet location, an eight-position RJ-45 jack with T568-A wiring should be used. T568-B wiring of the jack may also be used (see Figure 16–13). Whichever wiring scheme is used, make sure that all jacks in the house are wired the same way. There should be a minimum of one data jack at each wall outlet location. Each data outlet location should have a separate home run back to the service center. The home runs are terminated at the service center by punching down the cable to the proper 110 terminal blocks. The equipment, like home computers, that the data wall outlets are serving are connected to the system with patch cords. One end of the patch cord is plugged into the RJ-45 jack located on a network card that has been installed in the computer. The other end is simply plugged into the wall jack. Once the system has been completely installed and Internet service is brought to the service center, the connection of individual computers and peripheral equipment to the wall jacks is accomplished by simply plugging the equipment into any jack in the house with a patch cord.

Video Wiring System

Figure 16–17 shows a suggested wiring layout for the video wiring system in a typical house. Use a 75-ohm RG-6 coaxial cable. It is recommended that two runs of coaxial cable be installed from the service center to each television outlet location. The two runs of cable will allow for video distribution from any video source as well as distribution to a television at each outlet location. Also, coaxial cable is becoming more popular as a data transmission line for high-speed Internet connections, and the extra cable could be used to serve a computer located close to the television

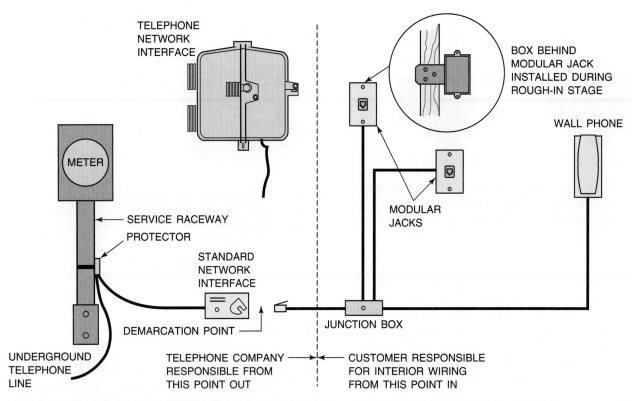

Figure 16–15 A typical residential telephone installation using a more traditional method. The telephone company installs and protects the incoming telephone service cable. It can be installed underground, as shown in this illustration, or overhead. This illustration shows the protector connected to the metal service entrance raceway. This establishes the ground path that helps protect the line from hazardous voltages caused by lightning. This grounding connection is often made to the copper grounding electrode conductor.

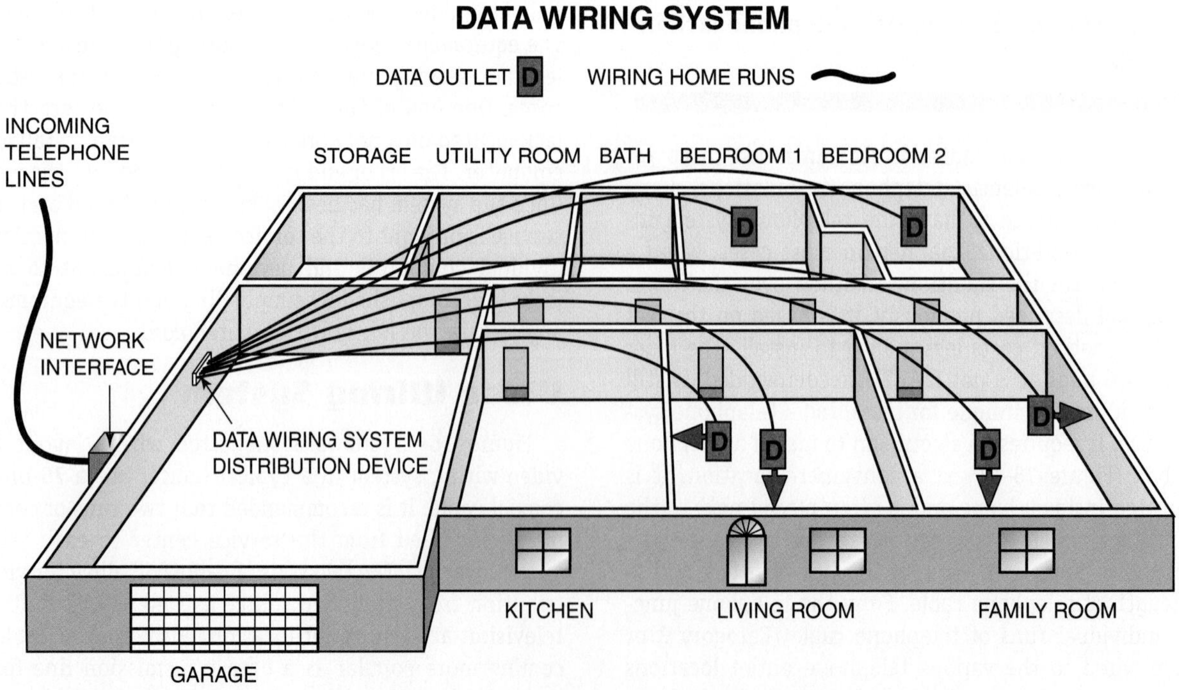

DATA WIRING SYSTEM

DATA OUTLET **D**   WIRING HOME RUNS ~

INCOMING TELEPHONE LINES

STORAGE   UTILITY ROOM   BATH   BEDROOM 1   BEDROOM 2

NETWORK INTERFACE

DATA WIRING SYSTEM DISTRIBUTION DEVICE

KITCHEN   LIVING ROOM   FAMILY ROOM

GARAGE

Figure 16–16 A typical residential data wiring system installation from a central distribution center.

VIDEO WIRING SYSTEM
(HOME RUNS FROM COMMON DISTRIBUTION POINT)

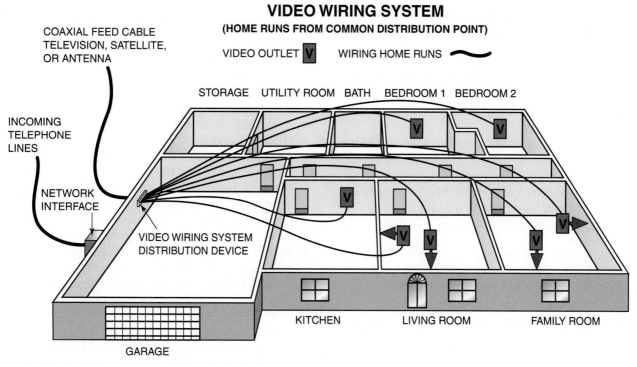

Figure 16–17 A typical residential video wiring system installation from a central distribution center.

outlet location. Two runs are also recommended to be run to a convenient attic or basement location in case a satellite television system will be installed at a later date. At each end of the coaxial cable, male F-type connectors are installed. Threaded F-type connectors are recommended to reduce signal interference. Push-on fittings are not recommended. At the wall-mounted television outlet end, a female-to-female F-type coupler is used. At the service center end, the F-type connector is threaded onto the proper fitting. At the television outlet end, connection to a video device is done using a 75-ohm RG-6 coaxial patch cord. Once the system has been completely installed and cable television or satellite television service is brought to the service center, the connection of individual televisions and other video devices to the wall jacks is accomplished by simply connecting them to any video outlet in the house.

Like the voice (telephone) system described earlier, a more traditional method of installing the coaxial cable throughout a house is often used. Again, it is usually the

higher cost associated with the EIA/TIA recommended method that causes home owners to choose the more traditional coaxial wiring technique. In this method, television service from a cable television company, a television antenna, or a satellite dish is brought to a distribution point in the house. The television service wiring is terminated at a splitter (Figure 16–18). The splitter "splits up" the incoming signal so it can be sent on individual RG-6 coaxial cables from the splitter to different television outlets throughout the house.

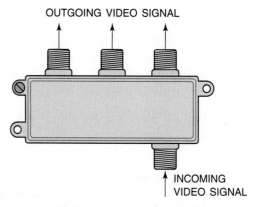

Figure 16–18 A three-way coaxial cable splitter. One signal-carrying cable comes into the splitter. Three signal-carrying cables leave the splitter to go to television outlets located in different areas of a house.

FROM EXPERIENCE

Use 75-ohm termination caps in each unused television outlet throughout the house. This will help keep the signal to the video devices strong and reliable.

Common Video, Voice, and Data Installation Safety Considerations

The following safety items should be considered when installing cabling and jacks for a video, voice, and data application in a house:

- Never install or connect telephone wiring during an electrical storm.
- Jacks should never be installed where a person could use a telephone (hardwired) while in a bathtub, hot tub, or swimming pool.
- Do not run open communications wiring between structures where it may be exposed to lightning.
- Avoid telecommunications wiring in or near damp locations.
- Never place telephone wires near bare power wires or lightning rods.
- Never place voice and data wiring in any conduit, box, or other enclosure that contains power conductors.
- Always maintain adequate separation between voice and data wiring and electrical wiring according to the *NEC®*.
- 50 to 60 volts direct current is normally present on an idle telephone tip and ring pair. An incoming call consists of 90 volts alternating current. This can cause a shock under the right conditions.

CAUTION

CAUTION: Many electricians are under the impression that you cannot get an electrical shock from a telephone line. This is not true. An incoming telephone call consists of 90 volts alternating current between the tip and ring conductors. If an electrician happens to be working on a telephone circuit when a call comes in, a 90-volt shock is possible. This author can tell you from personal experience that a 90-volt telephone circuit shock hurts.

- Always disconnect the dial-tone service from the house when working on an existing phone system. If you cannot disconnect, simply take the receiver off the hook. The direct-current value will drop, and the 90-volt alternating-current ring will not be available.

Common Video, Voice, and Data Installation Practices

The "Star" wiring method is the recommended way to install the wiring. This is an easy method for electricians since it is the same as running home runs in electrical work. The video, voice, and data cables are run from the service center to each outlet location. If it is more convenient, the cables can be run from the various video, voice, and data outlets back to the service center.

The following items should be followed by an electrician when installing the structured wiring cables:

- Keep the cable runs as short as possible.
- Do not splice wires on the cable runs. Run the cables as one continuous length.
- Do not pull the wire with more than 25 pounds of pulling tension (four pair).
- Do not run the wire too close to electrical power wiring (Figure 16–19).
- Do not bend too sharply. It is recommended that UTP cable be bent so that the radius of the bend is no less than four times the diameter of the UTP cable. Coaxial cable should be bent so that the radius of the bend is no less than 10 times the diameter of the coaxial cable.
- Do not install the cable with kinks or knots.
- There are no *NEC®* or EIA/TIA 570-A requirements for the maximum distance between supports for the cables. There is also no requirement for the minimum distance from a box or panel that the cables have to be secured. It is recommended that adequate support be provided so that the cables follow the building framing members closely and that there are no lengths of the cable that sag excessively. A good rule of thumb is to secure all

Purpose	Type of Wire Involved	Minimum Separation
Electric supply	Bare light or power of any voltage	5 ft.
	Open wiring not over 300 volts	2 in.
	Wires in conduit or in armored or nonmetallic sheath cable/power ground wires	None
Radio and television	Antenna lead and ground wires without grounded shield	4 in.
Cable television cables	Community television systems coaxial cables with grounded shield	None
Telephone service drop wire	Aerial or buried	2 in.
Fluorescent lighting	Fluorescent lighting wire	5 in.
Lightning system	Lightning rods and wires	6 ft.

Figure 16–19 This table shows the minimum recommended separations between residential video, voice, and data wiring and the electrical power system conductors.

video, voice, and data cables within 12 inches (300 mm) of an enclosure and then support them so that there is a maximum distance between supports of 4.5 feet (1.4 m).

- Use insulated rounded or depth-stop plastic staples to secure the cables to the building framing members.
- Use tie wraps (secured loosely) when you have a bundle of several cables to support.
- Maintain polarity and match color coding throughout the house.
- To provide compatibility with two-line telephones, which are common in many homes, be sure to wire up the two inner pairs of an RJ-45 jack. Home telephones are typically connected to a telephone jack with a four-wire flat cable that has four-pin plugs on each end. When a four-pin plug is connected to an eight-pin RJ-45 jack, it makes connections with only the four middle pins of the jack. The four middle pins correspond to pair 1 and pair 2 of the voice cabling system.

CAUTION

CAUTION: Always use a T568-A wiring scheme when wiring two-line telephone jacks. This wiring scheme will result in both incoming lines being available at each telephone jack. If a T568-B wiring scheme is used, both pair 1 and pair 3 are used as the two innermost pairs, and the pair 2 telephone line will not be available when a four-pin plug is connected to the RJ-45 jack.

- If conduit is installed, leave a pull string so that the video, voice, and data cable can be easily pulled in later.
- Never run video, voice, or data wiring in the same conduit with power wires.
- Use inner structural walls instead of outer walls for cable runs whenever possible.
- Do not run the cables through bored holes with power wires (Figure 16–20).

FROM EXPERIENCE

A common wiring practice used to support bundles of several video, voice, or data cables is to drive a regular Romex cable staple into a wooden framing member until it is almost all the way in. Then insert a tie wrap through the area between the staple and the framing member, place the tie wrap around the bundle of cables, and loosely tighten the tie wrap. This will hold the bundle of cables in place. Remember to not tighten the tie wrap too tightly.

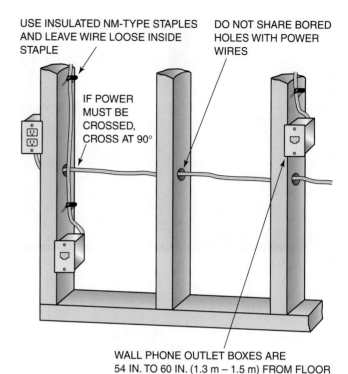

USE INSULATED NM-TYPE STAPLES AND LEAVE WIRE LOOSE INSIDE STAPLE

DO NOT SHARE BORED HOLES WITH POWER WIRES

IF POWER MUST BE CROSSED, CROSS AT 90°

WALL PHONE OUTLET BOXES ARE 54 IN. TO 60 IN. (1.3 m – 1.5 m) FROM FLOOR

Figure 16–20 The suggested installation practice for running video, voice, and data cables through bored holes in building framing members.

- Keep the cables away from heat sources, such as hot water pipes and furnaces.
- Avoid running exposed cables whenever possible.
- Leave about 18 inches of wire at outlets and connection points. This will ensure enough wire to properly make the necessary terminations. Push any extra cabling back into the wall or ceiling in case future work will need to be done on the system.
- Always check for shorts, opens, and grounds when the rough-in is complete.
- It is recommended that a separate 15- or 20-amp branch circuit be run to the service center.

The following items should be considered when terminating the cables:

- Binding posts—most commonly used for residential applications (Figure 16–21):
  - Be careful to not nick the inner conductors when stripping the outer jacket of the cable.
  - Wrap the conductor in a clockwise direction between two washers.
  - Be sure the wire does not get caught in the screw threads—it may break.
  - Trim off any excess exposed bare wire.
  - Use no more than two or three wires under a single screw.
  - Always leave plenty of spare wire at the connection point.

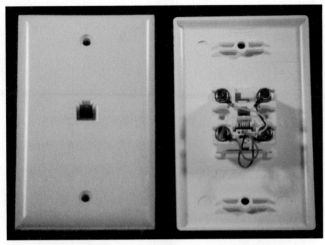

FRONT   BACK

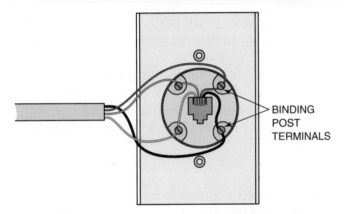

BINDING
POST
TERMINALS

Figure 16–21 Binding post connections are found on the back of some residential telephone wall jacks. This type of jack is fine for voice-only applications. However, it is no longer recommended for use in residential installations.

- Do not overtighten the binding posts with more than 7 inch-pounds of torque.
- Insulation displacement connectors (IDCs) (Figure 16–7):
  - Do not untwist any more of the wire pairs than absolutely necessary to make a termination. Make sure that the twists are no more than 1/2 inch away from where the wires are terminated.
  - IDCs are faster and more reliable than binding post terminations.
  - IDCs require a special punch-down tool.
  - The most common IDC type for residential applications is a 110 clip.
- Wiring jacks for residential applications:
  - EIA/TIA 570-A recommends eight conductor RJ-45 jacks.
  - Six (RJ-11) or four conductor jacks (binding post type) are still widespread but are not recommended to be installed for any new residential installation or system upgrade.

- Category 5 or 5e wiring and devices are recommended for all residential applications.
- Wire one or two jacks per room. Warning: Do not use a screwdriver blade to terminate IDCs. Always use a 110-clip punch-down tool.
- Use standard wiring pattern T568-A or T568-B.
- Install the jacks at the same height as electrical receptacle outlets.
- Cover unused wall boxes with blank wall plates.

Video, Voice, and Data Installations and the NEC®

The installation of a structured cabling system must conform to these *NEC®* rules:

- Sections 800.5 and 820.5: Access to equipment must not be denied by an accumulation of wires and cables or coaxial cables that prevents removal of panels, including suspended ceiling panels. Structured cabling system wiring must always be supported. It cannot just lie on top of panels so that access to other electrical equipment is hindered.
- Sections 800.6 and 820.6: Communications circuits and equipment and coaxial cable must be installed in a neat and workmanlike manner. Cables installed exposed on the outer surface of ceilings and walls must be supported by the structural components of the building structure in such a manner that the cable cannot be damaged by normal building use. The cables must be attached to structural components by straps, staples, hangers, or similar fittings designed and installed so as not to damage the cable. The installation must also conform with Section 300.4(D). Cables must be attached to or supported by the structure with straps, clamps, hangers, and the like. The installation method must not damage the cable. In addition, the location of the cable should be carefully evaluated to ensure that activities and processes within the building do not cause damage to the cable.
- Section 800.50: Communications wires and cables installed as wiring within buildings must be listed as being suitable for the purpose and installed in accordance with Section 800.52. Communications cables must be marked in accordance with Table 800.50 (Figure 16–22). Examples:
  - CMP = communications plenum cable
  - CMR = communications riser cable
  - CMG = communications general-purpose cable
- Section 820.50: Coaxial cables in a building must be listed as being suitable for the purpose, and cables must be marked in accordance to Table 820.50 (Figure 16–23). Examples:
  - CATVP = plenum-rated coaxial
  - CATVR = riser-rated coaxial
  - CATV = general-purpose coaxial
  - CATVX = limited use for dwelling units or in raceways

Table 800.50 Cable Markings

Cable Marking	Type	Reference
MPP	Multipurpose plenum cable	800.51(G) and 800.53(A)
CMP	Communications plenum cable	800.51(A) and 800.53(A)
MPR	Multipurpose riser cable	800.51(G) and 800.53(B)
CMR	Communications riser cable	800.51(B) and 800.53(B)
MPG	Multipurpose general-purpose cable	800.51(G) and 800.53(D) and (E)(1)
CMG	Communications general-purpose cable	800.51(C) and 800.53(D) and (E)(1)
MP	Multipurpose general-purpose cable	800.51(G) and 800.53(D) and (E)(1)
CM	Communications general-purpose cable	800.51(D) and 800.53(D) and (E)(1)
CMX	Communications cable, limited use	800.51(E) and 800.53(C), (D), and (E)
CMUC	Undercarpet communications wire and cable	800.51(F) and 800.53(F)(6)

Figure 16–22 *NEC®* **Table 800.50, Telecommunications Cable Markings.** *Reprinted with permission from NFPA 70-2002, National Electrical Code®, Copyright © 2001, National Fire Protection Association, Quincy, MA 02269. This reprinted material is not the complete and official Position of the NFPA on the referenced subject, which is represented only by the standard in its entirety.*

Table 820.50 Cable Markings

Cable Marking	Type	Reference
CATVP	CATV plenum cable	820.51(A) and 820.53(A)
CATVR	CATV riser cable	820.51(B) and 820.53(B)
CATV	CATV cable	820.51(C) and 820.53(C)
CATVX	CATV cable, limited use	820.51(D) and 820.53(C)

Figure 16–23 *NEC®* **Table 820.50, Coaxial Cable Markings.** *Reprinted with permission from NFPA 70-2002, National Electrical Code®, Copyright © 2001, National Fire Protection Association, Quincy, MA 02269. This reprinted material is not the complete and official Position of the NFPA on the referenced subject, which is represented only by the standard in its entirety.*

- Sections 800.52(A)(2) and 820.52(A)(2): Communications wires and cables and coaxial cable must be separated at least 2 inches (50 mm) from conductors of any electric light, power, Class 1, non–power-limited fire alarm, or medium-power network-powered broadband communications circuits. However, Exception No. 1 says that where either (1) all the conductors of the electric light and power circuits are in a metal-sheathed, metal-clad, nonmetallic-sheathed, Type AC, or Type UF cable or (2) all the conductors of communications circuits are in a raceway, the 2-inch (50-mm) clearance does not have to be adhered to. This exception allows electricians to install video, voice, and data wiring directly alongside of the power and lighting wiring, depending on what wiring method is being used. This is the case in residential wiring since the common wiring methods used to install a residential electrical system are all listed in Exception No. 1.
- Sections 800.52(E) and 820.52(E): Raceways must be used for their intended purpose. Communications cables or coaxial cable must not be strapped, taped, or attached by any means to the exterior of any conduit or raceway as a means of support.

Summary

This chapter gave you a good introduction to the parts and materials, as well as the common installation practices, for a structured cabling system in a residential application. The installation of a structured cabling system by electricians has become quite common throughout the country, and knowing as much as possible about structured cabling is necessary for any electrician doing residential wiring. As these systems get more complex and new materials and installation techniques become available, you will need to consult Web sites, read trade magazines, and attend upgrade classes so that you can stay current in this ever changing area of electrical wiring.

Hardwired structured cabling installation was covered in this chapter, but be aware that wireless technology is available that may someday be the preferred way that the video, audio, voice, and data signals are sent from device to device in a house. However, for the time being and for the foreseeable future, electricians will be asked to install UTP cable, coaxial cable, and the various jacks and connectors so that video, voice, and data signals can be distributed throughout a home.

Procedures

Installing an F-Type Connector on an RG-6 Coaxial Cable

- Put on safety glasses and observe all applicable safety rules.

- Measure and mark 3/4 inch from the end of the cable.

A Using a coaxial cable stripper, remove all 3/4 inch of the outside sheathing of the cable.

B Using the coaxial stripper, remove 1/2 inch of the white dielectric insulation from the center conductor.

C Push the RG-6 F-type connector onto the end of the cable until the white dielectric insulation is even with the internal hole that the center conductor fits through.

D Using a proper-sized crimping tool, crimp the connector ferrule onto the cable. Make sure the crimp is made toward the cable side of the connector. Too close to the ferrule head will pinch the crimp.

- Clean up the work area and put all tools and materials away.

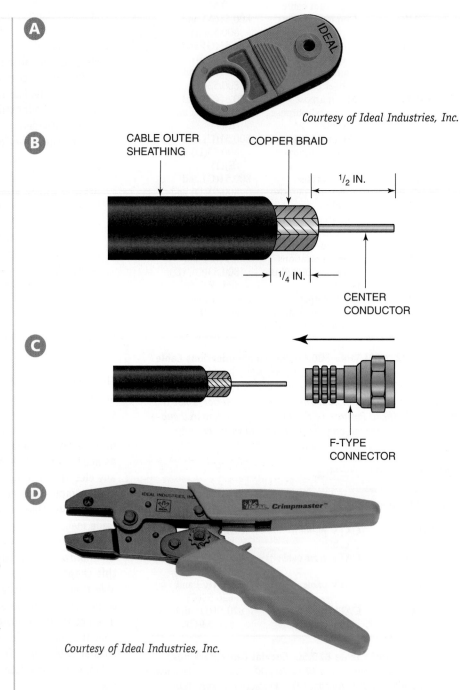

Courtesy of Ideal Industries, Inc.

Courtesy of Ideal Industries, Inc.

Procedures

Installing an RJ-45 Jack on the End of a Four-Pair UTP Category 5 Cable

- Put on safety glasses and observe all applicable safety rules.

A Using a proper UTP stripping tool, remove about 2 inches of jacket from the cable end.

B Determine which wiring scheme you will be using (T568-A or T568-B) and note the color coding and pin numbers on the jack. Note: Most manufacturers of RJ-45 jacks and plugs will have color coding and pin numbering available on their products.

A

Courtesy of Ideal Industries, Inc.

B

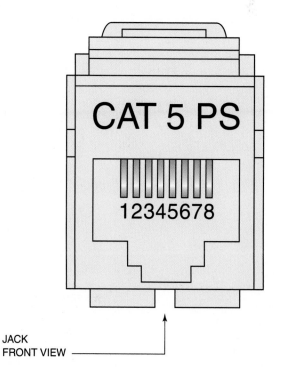

PIN NUMBER

4 6 8 7

B

A

JACK
SIDE VIEW

COLOR CODE FOR T568-A & T568-B WIRING

CAT 5 PS

12345678

JACK
FRONT VIEW

Procedures

Installing an RJ-45 Jack on the End of a Four-Pair UTP Category 5 Cable (continued)

C Route the conductors for termination. Terminate one pair at a time starting from the rear of the jack. Terminating each pair after placement will prevent crushing the inside pairs with the punch-down tool. Note: The cable should be placed so that the cable jacket touches the rear of the jack housing as shown.

C

T568-A			T568-B		
①	5	WHITE/BLUE	①	5	WHITE/BLUE
	4	BLUE		4	BLUE
②	3	WHITE/ORANGE	②	3	WHITE/GREEN
	6	ORANGE		6	GREEN
③	1	WHITE/GREEN	③	1	WHITE/ORANGE
	2	GREEN		2	ORANGE
④	7	WHITE/BROWN	④	7	WHITE/BROWN
	8	BROWN		8	BROWN

PAIR 1

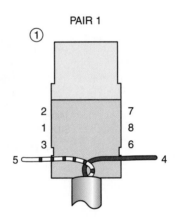

PAIR 2

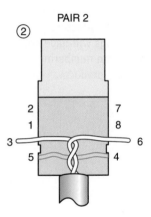

PAIR 3

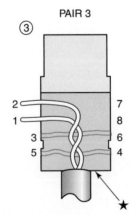

PAIR 4

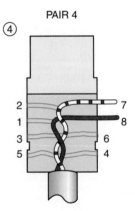

★ KEEP THE CABLE OUTER SHEATHING TIGHT AGAINST THE BODY OF THE RJ-45 JACK

D Using a 110-style punch-down tool, seat each conductor into the proper IDC slot. Be sure to keep the twists within 1/2 inch of the IDC slot. The punch-down tool if used properly will trim any excess wire off flush with the device body.

D

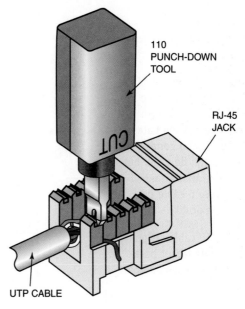

110 PUNCH-DOWN TOOL

CUT

RJ-45 JACK

UTP CABLE

E Place the cap that comes with the RJ-45 device over the terminated wires and press it into place. This cap will provide a more secure connection of the wires to the IDC slots as well as provide some additional strain relief.

- The jack can now be inserted into a wall plate assembly and secured to an electrical box placed at the desired location.

- Clean up the work area and put all tools and materials away.

E

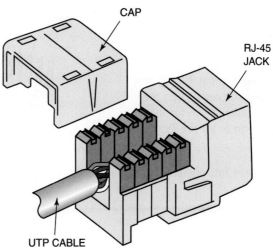

CAP

RJ-45 JACK

UTP CABLE

Procedures

Assembling a Patch Cord with RJ-45 Plugs Using a Length of Category 5 UTP Cable

- Put on safety glasses and observe all applicable safety rules.

- Determine the length of the patch cord and, using a pair of cable cutters, cut the desired length from a spool or box of the Category 5 UTP cable. Solid conductor Category 5 cable can be used, but it is recommended that stranded conductor Category 5 cable be used for added flexibility.

- Using a proper UTP cable stripping tool, strip the cable jacket back about 3/4 inch from each end of the cable.

- Determine whether the connection will be done to the EIA/TIA 568-A color scheme or to the EIA/TIA 568-B color scheme.

A On one end of the patch cord, sort the pairs out so they fit into the plug in the order shown for the T568-A scheme or for the T568-B scheme.

- Insert the pairs into the plug.

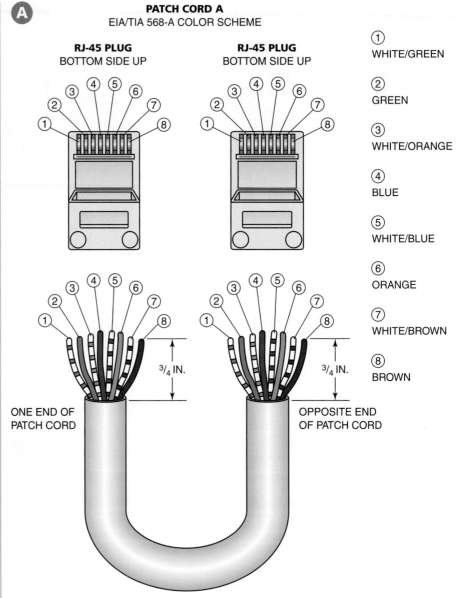

A

PATCH CORD A
EIA/TIA 568-A COLOR SCHEME

RJ-45 PLUG
BOTTOM SIDE UP

RJ-45 PLUG
BOTTOM SIDE UP

ONE END OF PATCH CORD

OPPOSITE END OF PATCH CORD

3/4 IN.

3/4 IN.

① WHITE/GREEN

② GREEN

③ WHITE/ORANGE

④ BLUE

⑤ WHITE/BLUE

⑥ ORANGE

⑦ WHITE/BROWN

⑧ BROWN

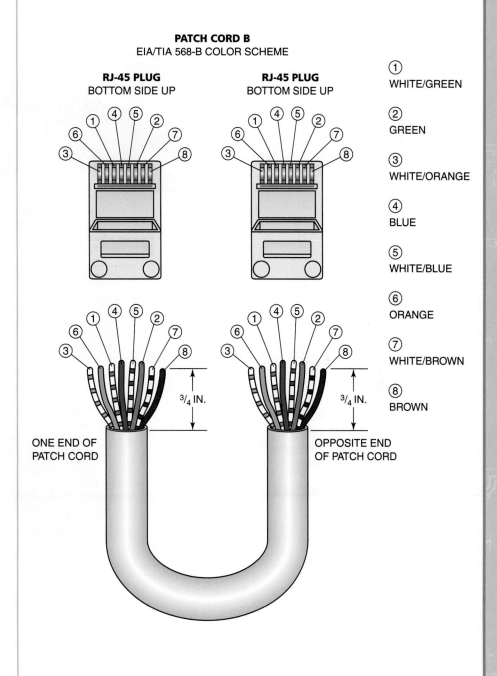

PATCH CORD B
EIA/TIA 568-B COLOR SCHEME

RJ-45 PLUG
BOTTOM SIDE UP

RJ-45 PLUG
BOTTOM SIDE UP

① WHITE/GREEN

② GREEN

③ WHITE/ORANGE

④ BLUE

⑤ WHITE/BLUE

⑥ ORANGE

⑦ WHITE/BROWN

⑧ BROWN

³/₄ IN.

³/₄ IN.

ONE END OF
PATCH CORD

OPPOSITE END
OF PATCH CORD

Procedures

Assembling a Patch Cord with RJ-45 Plugs Using a Length of Category 5 UTP Cable (continued)

B Using a proper crimping tool, crimp the pins with the crimping tool.

- Install a plug on the other end following steps 5, 6, and 7.

B

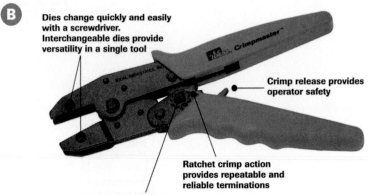

Dies change quickly and easily with a screwdriver. Interchangeable dies provide versatility in a single tool

Crimp release provides operator safety

Ratchet crimp action provides repeatable and reliable terminations

Pressure adjustment for actuation force

Courtesy of Ideal Industries, Inc.

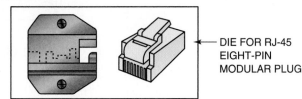

DIE FOR RJ-45 EIGHT-PIN MODULAR PLUG

C Using a tester, test the patch cord to determine that the correct color scheme has been used and that there is continuity between both ends of the patch cord.

- Clean up the work area and put all tools and materials away.

C

Courtesy of Ideal Industries, Inc.

Review Questions

Directions: Answer the following items with clear and complete responses.

1. The most often used category rated UTP cable for wiring from various outlets back to the service center in a residential application is _____.

2. The minimum bending radius for a Category 5–rated cable is recommended to be no less than _____ times the diameter of the cable.

3. The minimum bending radius for a coaxial cable is recommended to be no less than _____ times the diameter of the cable.

4. Define the term "structured cabling."

5. Name the EIA/TIA standard that is used as a guide for the installation of a structured cabling system in a residential application.

6. Describe why each pair of wires in the voice and data cables used in a residential structured cabling system is twisted.

7. Describe the difference between a 568-A and a 568-B connection.

8. What do the letters "IDC" stand for?

9. What is the maximum amount of pulling force that an electrician can exert on a Category 5 cable?

10. Name the document that contains the installation requirements for a structured cabling system that must be followed.

11. A _____ is the rating, based on the bandwidth performance, of UTP cable.

12. A type of coaxial cable that is "quad shielded" and is used in residential structured cabling systems to carry video signals such as cable and satellite television is called _____.

13. The popular name given to an eight-pin connector or jack used to terminate UTP cable is _____.

14. A _____ _____ is the short length of cable with an RJ-45 plug on either end and is used to connect a home computer to the work area outlet.

15. The standard color coding for a four-pair UTP cable is:
    - Pair 1: tip is _____; ring is _____
    - Pair 2: tip is _____; ring is _____
    - Pair 3: tip is _____; ring is _____
    - Pair 4: tip is _____; ring is _____

Residential Electrical System Trim-Out

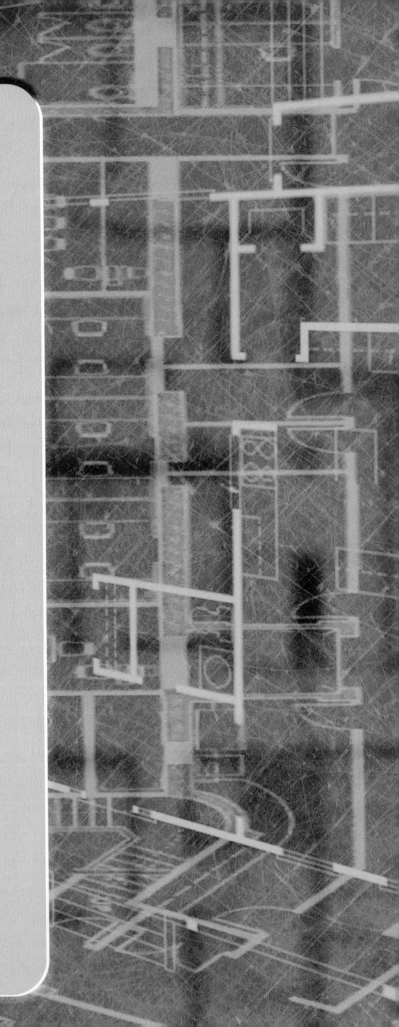

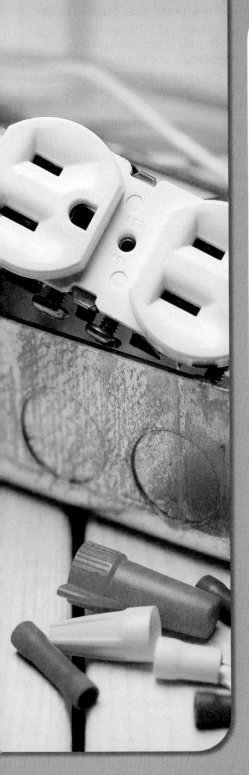

Chapter 17 | Lighting Fixture Installation

nstallation of the lighting fixtures in a house will be one of the last things an electrician does to complete the residential electrical system. The ceilings and walls will have their final coat of paint or other finish already applied. Choosing the lighting fixture types to be used throughout a house is usually done during the initial planning stages for the installation of the electrical system. Planning early allows the electrician to install the proper rough-in wiring and electrical boxes required for a specific lighting fixture type. The electrical connections for most types of lighting fixtures used in residential wiring are very similar. However, the method of mounting the fixtures to a wall or ceiling varies with the type of fixture used. Manufacturers of lighting fixtures enclose installation instructions with each lighting fixture and should always be consulted during the installation. This chapter introduces you to basic lighting fundamentals, lamp types, and common lighting fixture types used in residential lighting systems. The methods most often used for installing lighting fixtures are also covered.

OBJECTIVES

Upon completion of this chapter, the student should be able to:

- ⊗ demonstrate an understanding of lighting basics.
- ⊗ demonstrate an understanding of common lamp and lighting fixture terminology.
- ⊗ demonstrate an understanding of the three different lamp types used in residential wiring applications: incandescent, fluorescent, and high-intensity discharge.
- ⊗ select a lighting fixture for a specific residential living area.
- ⊗ demonstrate an understanding of the installation of common residential lighting fixtures.

Glossary of Terms

ballast a component in a fluorescent lighting fixture that controls the voltage and current flow to the lamp

candlepower a measure of lighting intensity

Class P ballast a ballast with a thermal protection unit built in by the manufacturer; this unit opens the lighting electrical circuit if the ballast temperature exceeds a specified level

efficacy a rating that indicates the efficiency of a light source

fluorescent lamp a gaseous discharge light source; light is produced when the phosphor coating on the inside of a sealed glass tube is struck by energized mercury vapor

foot-candle the unit used to measure how much total light is reaching a surface; 1 lumen falling on 1 square foot of surface produces an illumination of 1 foot-candle

high-intensity discharge (HID) lamp another type of gaseous discharge lamp, except the light is produced without the use of a phosphor coating

incandescent lamp the original electric lamp; light is produced when an electric current is passed through a filament; the filament is usually made of tungsten

lumen the unit of light energy emitted from a light source

luminaire a complete lighting unit consisting of a lamp or lamps together with the parts designed to distribute the light, to position and protect the lamps and ballast (where applicable), and to connect the lamps to the power supply

sconce a wall-mounted lighting fixture

troffer a term commonly used by electricians to refer to a fluorescent lighting fixture installed in the grid of a suspended ceiling

Type IC a light fixture designation that allows the fixture to be completely covered by thermal insulation

Type Non-IC a light fixture that is required to be kept at least 3 inches from thermal insulation

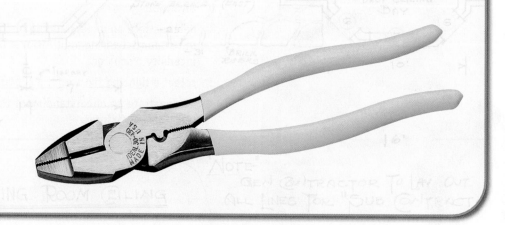

Lighting Basics

Today the lighting industry produces many different styles of lighting fixtures and lamps to provide the home owner with an overwhelming amount of flexibility when planning their lighting system. The *National Electrical Code®* (*NEC®*) uses the term "luminaire" when referring to lighting fixtures. A luminaire is defined as a complete lighting unit consisting of a lamp or lamps together with the parts designed to distribute the light, to position and protect the lamps and ballast (where applicable), and to connect the lamps to the power supply.

The overall performance of a lighting system is a combination of the quantity of light the lamps produce and the quality of that light. Light is the visible portion of the electromagnetic spectrum. Figure 17–1 shows the relationship of the various visible wavelengths. The shortest visible wavelength appears as violet, while the longest visible wavelength appears red. The other colors—blue, green, yellow, and orange—are intermediate wavelengths and are between violet and red.

We tend to think that each object has a fixed color. In reality, an object's appearance results from the way it reflects the light that falls on it. For example, an apple appears red because it reflects light in the red part of the spectrum and absorbs the light of the other wavelengths. The quality of the red wavelength striking the apple determines the shade of red we see.

White light is formed by nearly equal parts of all the visible wavelengths. The balance does not have to be precise for the light to appear white. Red, blue, and green are the primary colors, and they can be combined to make any other color. This means that a light source that contains a good balance of red, blue, and green will provide excellent color appearance for objects it illuminates.

Lamp manufacturers are concerned with three factors: color temperature, color rendering, and lamp efficiency. The color temperature of a light source is a measurement of its color appearance and is measured in degrees Kelvin (K) (Figure 17–2). It is based on the premise that any object heated to a high enough temperature will emit light and that as the temperature rises, the color will shift from red to orange to yellow and then to white. One of the more confusing aspects of lighting is that light from the higher-temperature wavelengths, blue and white, is referred to as "cool," while the light from the lower-temperature wavelengths is referred to as "warm." For example, a lamp with a color temperature rating of 2,000 K is a "warm" yellow light, and a lamp with a color temperature rating of 6,000 K puts off a

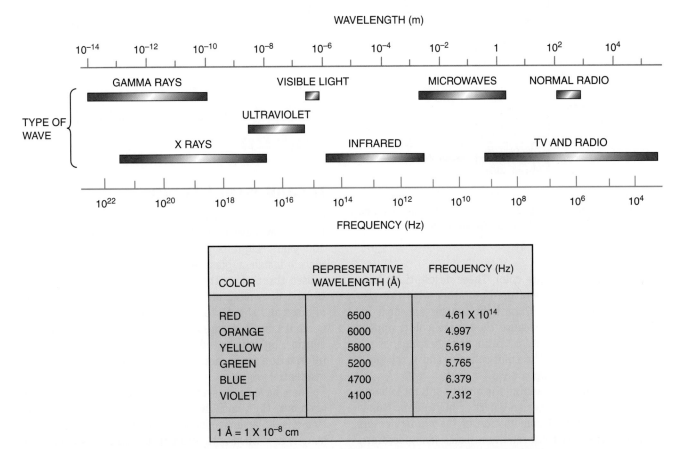

COLOR	REPRESENTATIVE WAVELENGTH (Å)	FREQUENCY (Hz)
RED	6500	4.61 X 10^{14}
ORANGE	6000	4.997
YELLOW	5800	5.619
GREEN	5200	5.765
BLUE	4700	6.379
VIOLET	4100	7.312

1 Å = 1 X 10^{-8} cm

Figure 17–1 The electromagnetic spectrum. Notice that visible light falls between ultraviolet light and infrared light. Red has the longest wavelength and violet the shortest.

combinations produce light with slightly different mixtures of the primary color wavelengths, which causes us to see colors in different shades. This means that the red apple discussed earlier may be seen in different shades of red, depending on the type of lamp that is illuminating it.

Many electrical people think that a higher-wattage lamp will produce more light. They are actually confusing light output with the amount of energy a lamp uses. Light output is measured in "lumens," while the amount of energy used by a lamp type is measured in watts. A lumen is defined as the unit of light that is emitted from a light source. For example, a 20-watt compact fluorescent lamp will produce as much usable light as a 75-watt incandescent lamp and use much less energy. In today's energy-conscious world, the ability to get sufficient light for a lower price is very important. The best indicator of a lamp's performance, called its efficacy, is its LPW (lumens per watt) rating. This rating is a ratio of the number of lumens a lamp produces to each watt of power it uses. The higher the LPW of a light source, the better its efficiency.

A second set of light measurements that are often confused are foot-candles and candlepower. A foot-candle is the measurement of the amount of light that falls on a surface that is exactly 1 foot away from a burning candle. It is a measurement of the light that falls on a surface. Candlepower, on the other hand, measures the intensity of a light source in a specific direction. It has no relationship with any object that is being lit. Figure 17–3 shows the relationship between lumens and foot-candles.

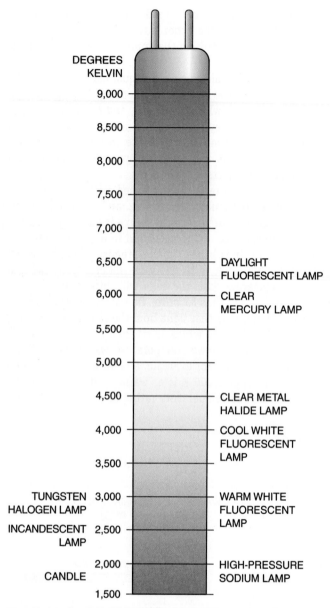

Figure 17–2 Color temperature scale. This scale assigns a numeric value to the color appearance of a light source ranging from orange/red (warm light) to blue/white (cool light).

light that is a "cool" white color. These descriptions have nothing to do with temperature but rather with the way the colors are perceived. Warm light sources are ideal for residential applications because they make colors appear more natural and vibrant. In residential settings, earth tones usually dominate the color scheme, and warm light sources enhance their appearance.

The standard for determining the quality of light is based on the sunlight that strikes an object at high noon. This is when light has the best combination of the primary colors and provides the best color rendition. Lamp manufacturers, attempting to reproduce this effect, have used various combinations of materials with varying degrees of success. These

Overview of Lamp Types Found in Residential Lighting

Incandescent Lamps

Incandescent lamps were the first type of electric lamp. They have used the same basic technology for over 100 years. Light is produced when a tungsten filament, placed inside a glass enclosure, has an electric current passed through it. The resistance of the filament causes it to heat up, giving off light. These lamps come in a variety of sizes and shapes to meet various lighting needs. Incandescent lamp bases also come in several sizes to meet residential lighting needs. Figure 17–4 shows additional information for incandescent lamp and base types used in residential wiring.

There have been several improvements to the incandescent lamp over the years. The straight filament has been replaced by a coiled filament to increase the life of the lamp. The sealed lamp is now filled with an inert gas, such as argon. Early incandescent lamps had the air pumped out to create a vacuum around the filament. This was done to keep oxygen away from

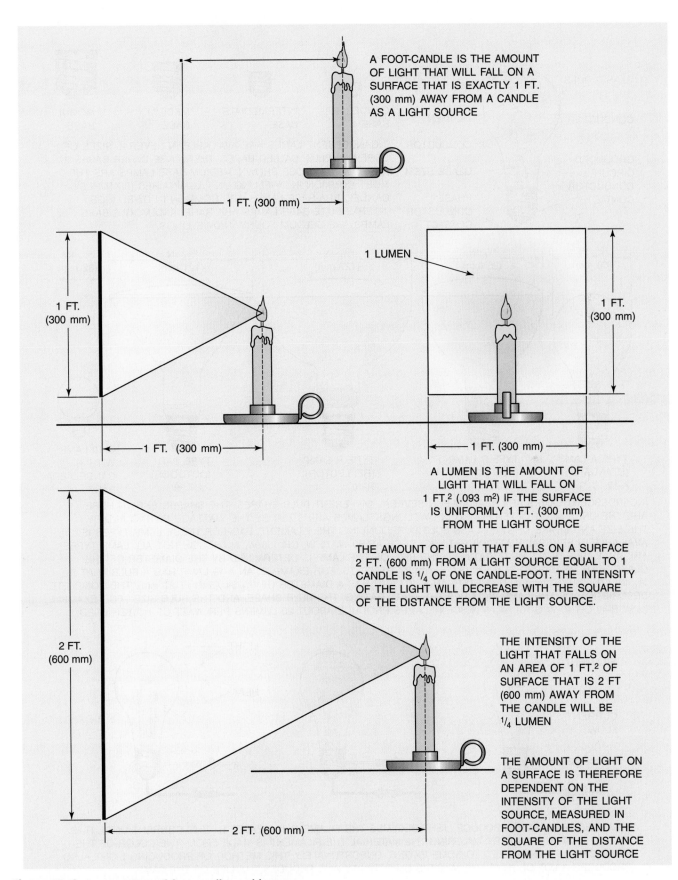

A FOOT-CANDLE IS THE AMOUNT OF LIGHT THAT WILL FALL ON A SURFACE THAT IS EXACTLY 1 FT. (300 mm) AWAY FROM A CANDLE AS A LIGHT SOURCE

1 FT. (300 mm)

1 FT. (300 mm)

1 FT. (300 mm)

1 LUMEN

1 FT. (300 mm)

1 FT. (300 mm)

A LUMEN IS THE AMOUNT OF LIGHT THAT WILL FALL ON 1 FT.^{2} (.093 m^2) IF THE SURFACE IS UNIFORMLY 1 FT. (300 mm) FROM THE LIGHT SOURCE

THE AMOUNT OF LIGHT THAT FALLS ON A SURFACE 2 FT. (600 mm) FROM A LIGHT SOURCE EQUAL TO 1 CANDLE IS $\frac{1}{4}$ OF ONE CANDLE-FOOT. THE INTENSITY OF THE LIGHT WILL DECREASE WITH THE SQUARE OF THE DISTANCE FROM THE LIGHT SOURCE.

THE INTENSITY OF THE LIGHT THAT FALLS ON AN AREA OF 1 FT.^{2} OF SURFACE THAT IS 2 FT (600 mm) AWAY FROM THE CANDLE WILL BE $\frac{1}{4}$ LUMEN

THE AMOUNT OF LIGHT ON A SURFACE IS THEREFORE DEPENDENT ON THE INTENSITY OF THE LIGHT SOURCE, MEASURED IN FOOT-CANDLES, AND THE SQUARE OF THE DISTANCE FROM THE LIGHT SOURCE

2 FT. (600 mm)

2 FT. (600 mm)

Figure 17–3 A comparison of foot-candles and lumens.

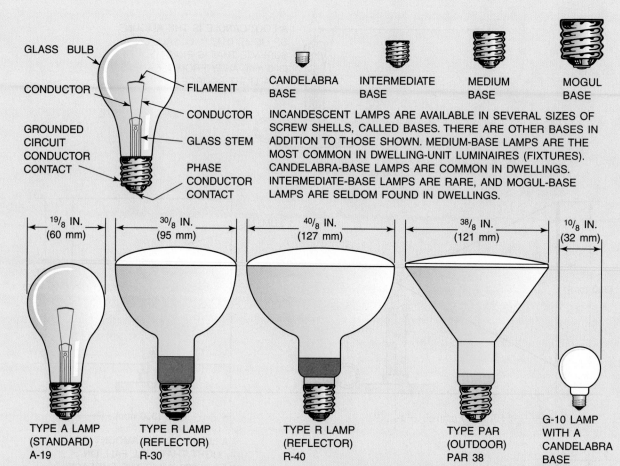

INCANDESCENT LAMPS ARE AVAILABLE IN SEVERAL SIZES OF SCREW SHELLS, CALLED BASES. THERE ARE OTHER BASES IN ADDITION TO THOSE SHOWN. MEDIUM-BASE LAMPS ARE THE MOST COMMON IN DWELLING-UNIT LUMINAIRES (FIXTURES). CANDELABRA-BASE LAMPS ARE COMMON IN DWELLINGS. INTERMEDIATE-BASE LAMPS ARE RARE, AND MOGUL-BASE LAMPS ARE SELDOM FOUND IN DWELLINGS.

INCANDESCENT LAMPS ARE AVAILABLE IN SEVERAL DIFFERENT BULB SHAPES. THE SHAPES SHOWN HERE ARE THE ONES MOST OFTEN FOUND IN DWELLINGS. LAMPS ARE SIZED BY THE WATTAGE RATING AND BY THE SIZE AND SHAPE OF THE GLASS BULB SURROUNDING THE FILAMENT. EACH OF THESE LAMP TYPES IS AVAILABLE IN SEVERAL WATT (W) RATINGS, SUCH AS 60 W, 75 W, OR 150 W, ALTHOUGH NOT ALL LAMP TYPES ARE AVAILABLE IN ALL WATT RATINGS. THE SIZE OF THE LAMP IS DETERMINED BY THE DIAMETER OF THE GLASS BULB MEASURED IN $1/8$ IN. (3.175 mm) INCREMENTS. FOR EXAMPLE, AN A-19 LAMP HAS A BULB THAT IS $19/8$ IN. ($2 3/8$ IN.) (60 mm) IN DIAMETER. AN R-40 LAMP HAS A DIAMETER OF $40/8$ IN. (5 IN.) (127 mm). THE COMPLETE DESCRIPTION OF THE LAMP INCLUDES THE WATT RATING, THE BULB SHAPE, AND THE BULB SIZE—FOR EXAMPLE, 75 W R-40 OR 40 W G-10. INCANDESCENT LAMPS PRODUCE ABOUT 20 LUMENS PER WATT OF POWER USED.

INCANDESCENT LAMPS PRODUCE LIGHT BECAUSE THE FILAMENT IS HEATED BY THE CURRENT UNTIL IT IS SO HOT THAT IT GLOWS. BY TAILORING THE MATERIAL THE FILAMENT IS MADE FROM, THE COLOR OF THE LIGHT CAN BE CONTROLLED TO SOME EXTENT. UNFORTUNATELY, THIS METHOD OF PRODUCING LIGHT ALSO PRODUCES A LOT OF HEAT. THIS WASTES ELECTRICITY AND REDUCES THE LIFE OF THE LAMP.

Figure 17–4 A guide to incandescent lamp types and bases used in residential wiring.

the filament. As the filament heated up, oxygen would cause it to burn out almost immediately. Because a true vacuum could not be created inside the lamp, the remaining oxygen would work to shorten the life of the lamp.

The latest improvement is the tungsten halogen lamp. In standard incandescent lamps, the tungsten from the filament is burned off over time. This causes the filament to become thinner and eventually break. As the tungsten is burned off, it is deposited on the inside of the lamp surface, giving it a black coating, which causes the lamp to dim. By using a thicker filament and filling the lamp with halogen gas, lamp efficiency increases by at least 20%. This occurs because the tungsten that is burned off the filament combines with the halogen gas and is redeposited back onto the filament, extending the life of the filament and the lamp.

CAUTION: Halogen lamps give off extreme amounts of heat and have been associated with several fires. An electrician must carefully follow all manufacturers' installation instructions when installing halogen lamps. These lamps are required to be enclosed by a glass lens to prevent contact with flammable materials.

Incandescent lamps and their associated fixtures are used in every aspect of residential lighting. They are the most common lamp type used in residential systems for several reasons:

- Come in a wide variety of sizes and styles
- Have the lowest initial cost of any lamp type
- Produce a warm light that produces excellent color tones
- Can be easily controlled with dimmers
- Are very inefficient (Because they produce light by heating a solid object, most of the energy they consume is released as heat, not light.)

Fluorescent Lamps and Ballasts

Fluorescent lamps are referred to as "electric discharge lamps" by the NEC®. They have been around for many years and are an excellent light source. Light is produced when an electric current is passed through tungsten cathodes at each end of a sealed glass tube. The tube is filled with an inert gas, such as argon or krypton. A very small amount of mercury is also in the tube. Electrons are emitted from the cathodes and strike particles of mercury vapor. This results in the production of ultraviolet radiation, which causes a phosphor coating on the inside of the glass tube to glow. Figure 17–5 shows additional information about fluorescent lamps used in residential wiring applications.

Today's fluorescent lamps offer more options in light quality than any other type of lamp. Refinements in the composition of phosphor coatings and better control over the generation of primary color light wavelengths allow fluorescent lamps to provide excellent color rendition of virtually all colors. While their light quality is not as good as incandescent lamps, fluorescent lamps offer much better efficacy. Strip luminaires (fixtures without lenses) are usually found in garages, basements, and work areas. Wraparound fixtures (those with lenses) are often used in living areas of a residence. The availability of a wide range of lens types provide for better appearance and better light distribution in living areas that require a large amount of general lighting.

Fluorescent lamps have two electrical requirements. First, a high-voltage source is needed to start the lamp. Second, once the lamp is started, the mercury vapor offers a decreasing amount of resistance to the current. To prevent the lamp from drawing more and more current and quickly burning itself out, the current flow must be regulated. The **ballast** was developed to fill these needs (Figure 17–6).

A ballast is a device that provides both the voltage surge needed to start the lamp and the current control that allows the lamp to operate efficiently. There are two types available today: the magnetic ballast and the electronic ballast. While both perform the same functions, the electronic ballast offers several advantages over the magnetic ballast. When compared to a magnetic ballast, electronic ballasts produce their full light output using 25% to 40% less energy, are more reliable, last longer, produce a constant flicker-free light, and are more expensive.

Section 410.73(E) of the NEC® requires that all fluorescent ballasts installed indoors have thermal protection built into the unit by the manufacturer for both new and replacement installations. This protection causes the electrical circuit to the fixture to open when the temperature level inside the unit exceeds a preset level. Heat, caused by electrical shorts, grounds, or a lack of air circulation, could cause a fire. Ballasts with built-in thermal protection are called **Class P ballasts.**

Another factor to consider with fluorescent ballasts is their location. Standard ballasts, either magnetic or electronic, do not perform well in cold temperatures. A special ballast has been developed to overcome this problem. This unit is expensive and has had some reliability problems. Most electricians recommend that incandescent luminaires be used in these situations.

Fluorescent ballasts have three types of circuitry. The first is called "preheat" (Figure 17–7). They are easily identified because they have a "starter." The most common starter looks like a silver button on the fixture. When the ballast receives power, the starter causes the cathodes to glow, or preheat, for a few seconds before the arc is established to produce light. Lamps used with preheat ballasts have two pins (bipin) on each end of the lamp. Preheat lamps and ballasts cannot be used with dimmers.

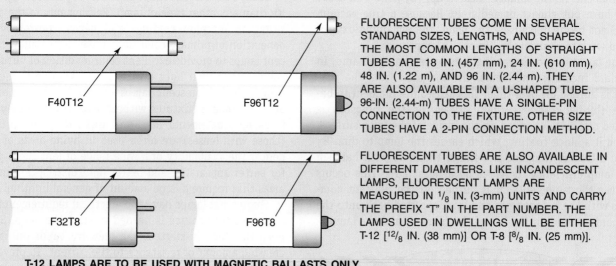

FLUORESCENT TUBES COME IN SEVERAL STANDARD SIZES, LENGTHS, AND SHAPES. THE MOST COMMON LENGTHS OF STRAIGHT TUBES ARE 18 IN. (457 mm), 24 IN. (610 mm), 48 IN. (1.22 m), AND 96 IN. (2.44 m). THEY ARE ALSO AVAILABLE IN A U-SHAPED TUBE. 96-IN. (2.44-m) TUBES HAVE A SINGLE-PIN CONNECTION TO THE FIXTURE. OTHER SIZE TUBES HAVE A 2-PIN CONNECTION METHOD.

FLUORESCENT TUBES ARE ALSO AVAILABLE IN DIFFERENT DIAMETERS. LIKE INCANDESCENT LAMPS, FLUORESCENT LAMPS ARE MEASURED IN $1/8$ IN. (3-mm) UNITS AND CARRY THE PREFIX "T" IN THE PART NUMBER. THE LAMPS USED IN DWELLINGS WILL BE EITHER T-12 [$12/8$ IN. (38 mm)] OR T-8 [$8/8$ IN. (25 mm)].

T-12 LAMPS ARE TO BE USED WITH MAGNETIC BALLASTS ONLY.
T-8 LAMPS ARE TO BE USED WITH ELECTRONIC BALLASTS ONLY.
USING THE WRONG LAMP OR THE WRONG BALLAST WILL SHORTEN THE LIFE OF BOTH COMPONENTS

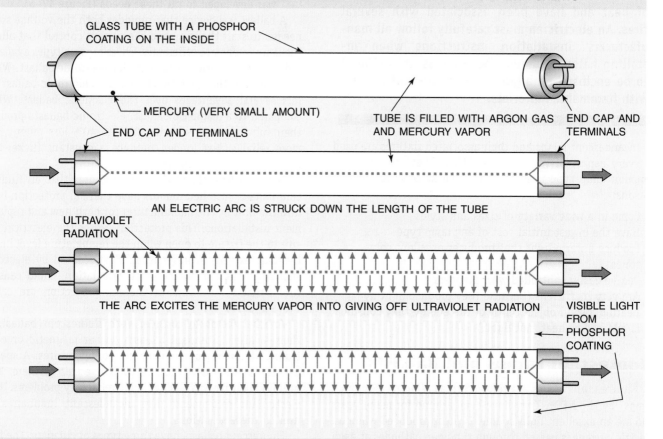

LIGHT IS PRODUCED BY AN ELECTRIC ARC THAT PRODUCES ULTRAVIOLET RADIATION. THE ULTRAVIOLET RADIATION CAUSES THE PHOSPHOR MATERIAL TO PRODUCE VISIBLE LIGHT. BY CONTROLLING THE MATERIAL COMPOSITION OF THE COATING, THE COLOR OF THE LIGHT CAN BE CONTROLLED TO SOME DEGREE. THIS METHOD OF LIGHTING PRODUCES MORE LIGHT AND LESS HEAT FOR EACH WATT OF ENERGY USED. WHEREAS AN INCANDESCENT LAMP WILL PRODUCE ABOUT 14 LUMENS PER WATT OF POWER USED, A FLUORESCENT LAMP PROVIDES ABOUT 80 TO 100 LUMENS PER WATT OF POWER USED.

Figure 17–5 A guide to fluorescent lamps used in residential wiring.

Figure 17–6 An electronic Class P ballast that is thermally protected as required in Section 410.73(E) of the *NEC*®. *Courtesy of Motorola Lighting.*

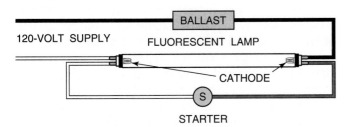

Figure 17–7 A preheat ballast circuit.

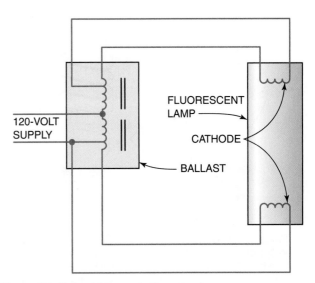

Figure 17–8 A rapid-start ballast circuit.

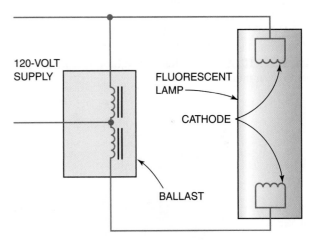

Figure 17–9 An instant-start ballast circuit.

The most common type of ballast circuitry is the "rapid start" (Figure 17–8). This type of starting circuitry does not require a starter. Instead, the filaments remain energized by a low-voltage circuit from the ballast. This allows the lamps to come to full power in less than 1 second. To ensure the lamps start properly, ballast manufacturers recommend the ballast and the fluorescent fixture case be grounded to the supply circuit grounding conductor. Lamps used with rapid-start ballasts are bipin and may be dimmed with a special dimming ballast.

The third type of ballast circuitry is the "instant start" (Figure 17–9). These ballasts provide a high-voltage surge to start the lamp instantly. They require special lamps that do not require preheating of the cathodes. The main disadvantage of this system is that lamp life is shortened by 20% to 40%. Most instant-start lamps have a single pin on each end of the lamp and cannot be used with dimmers.

The fastest-growing application of fluorescent lighting is in compact fluorescent lamps. As shown in Figure 17–10, these lamps are designed to replace the more popular incandescent lamps. Basically, it is a narrow fluorescent tube that is doubled back on itself with both ends terminated in a plastic base. The base also contains the ballast. While compact fluorescent lamps cost more to purchase than incandescent lamps, in the long term they are a better value because of their longer lamp life and more efficient operation.

High-Intensity Discharge Lamps

High-intensity discharge (HID) lamps are another type of gaseous discharge lamp. The method used to produce light is similar to fluorescent lighting. An arc is established between two cathodes in a glass tube filled with a metallic gas, such as mercury, halide, or sodium. The arc causes the metallic gas to produce radiant energy, resulting in light.

The electrodes are only a few inches (or less) apart in the lamp arc tube, not at opposite ends of a glass tube like a fluorescent lamp. This causes the arc to generate extremely high temperatures, allowing the metallic elements of the gas to release large amounts of visible energy. Because the energy is released as light energy, no phosphors are needed in HID lighting (Figure 17–11).

Like other electric discharge lighting, HID lamps require the use of a ballast to operate properly. It is extremely important that the ballast be designed for the lamp type and wattage being used. All HID lamps require a warm-up period before they are able to provide full light output. The main disadvantage of HID lighting is that even a very brief power interruption can cause the lamp to restart its arc and have to warm up again—a process that usually takes several minutes. For this reason, HID lighting is not generally used inside residences. It is used for outdoor security and area lighting applications.

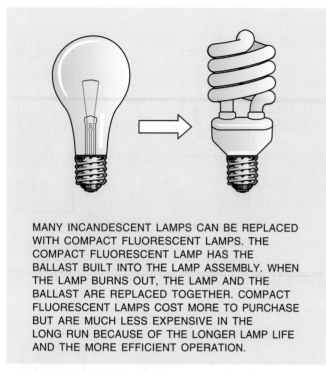

MANY INCANDESCENT LAMPS CAN BE REPLACED WITH COMPACT FLUORESCENT LAMPS. THE COMPACT FLUORESCENT LAMP HAS THE BALLAST BUILT INTO THE LAMP ASSEMBLY. WHEN THE LAMP BURNS OUT, THE LAMP AND THE BALLAST ARE REPLACED TOGETHER. COMPACT FLUORESCENT LAMPS COST MORE TO PURCHASE BUT ARE MUCH LESS EXPENSIVE IN THE LONG RUN BECAUSE OF THE LONGER LAMP LIFE AND THE MORE EFFICIENT OPERATION.

Figure 17–10 A compact fluorescent lamp is a great replacement for an incandescent lamp.

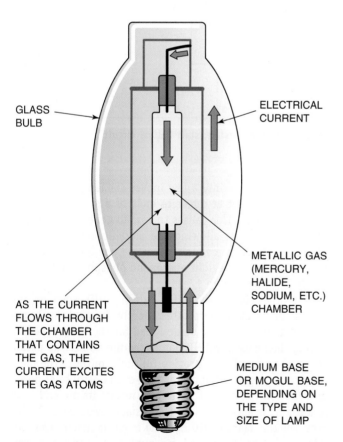

GLASS BULB

ELECTRICAL CURRENT

METALLIC GAS (MERCURY, HALIDE, SODIUM, ETC.) CHAMBER

AS THE CURRENT FLOWS THROUGH THE CHAMBER THAT CONTAINS THE GAS, THE CURRENT EXCITES THE GAS ATOMS

MEDIUM BASE OR MOGUL BASE, DEPENDING ON THE TYPE AND SIZE OF LAMP

Figure 17–11 A typical HID lamp and an explanation of how it produces light.

Mercury vapor, metal halide, and sodium vapor are the three types of HID lamps available today. Their names reflect the material inside the arc chamber. Mercury vapor is the oldest HID lamp and is virtually obsolete as a light source. Even though this lamp is very efficient and has an extremely long life, it is rarely used in new residential applications but may be encountered in existing dwellings as a light source for outside area lighting. Metal halide lamps are the most energy efficient source of white light on the market today. These lamps offer excellent color rendition and long life. They are widely used for large areas that require good illumination, such as highways, parking lots, and the interiors of large commercial buildings. In residential systems, they are limited to large outdoor applications. Sodium vapor lamps provide the highest efficacy of any lamp type. These lamps are produced in both high- and low-pressure versions. While both are very efficient, their poor color rendition limits their use to areas where high-quality lighting is not an important factor. These lamps are seldom found in any residential lighting system but can be used to illuminate residential driveways and other outdoor areas.

Table 17–1 lists the different types of lamps discussed in this section and lists each lamp's characteristics.

Selecting the Appropriate Lighting Fixture

Today, the lighting industry produces many different types of lighting fixtures to satisfy almost every need. Residential luminaires are readily available and easy to install. They are usually either incandescent or fluorescent and can be surface mounted, recessed, or mounted in suspended ceilings.

Installation requirements for light fixtures are contained in Article 410 of the *NEC*®. This article also covers fixture mounting, supporting, clearances, and construction requirements. Each fixture comes with specific installation instructions provided by the manufacturer. It is very important for the electrician to read, understand, and follow these instructions. Each fixture also comes with labeling that informs the installer of any installation restrictions pertaining to location and wiring methods. Some common information found on a label includes the following:

- For wall mount only
- Ceiling mount only
- Maximum lamp wattage: _____ watts
- Lamp type
- Suitable for operation in an ambient temperature not exceeding _____°F (°C)
- Suitable for use in suspended ceilings
- Suitable for damp locations
- Suitable for wet locations
- Suitable for mounting on low-density cellulose fiberboard
- For supply connections, use wire rated at least _____°F (°C)

Table 17-1   Characteristics of Lamp Types Used in Residential Wiring

	Incandescent	Fluorescent	Mercury	Metal Halide	HPS	LPS
Lumens per watt	6–23	25–100	30–65	65–120	75–140	130–180
Wattage range	40–1,500	4–215	40–1000	175–1,500	35–1,000	35–180
Life (hours)	750–8,000	9,000–20,000	16,000–24,000	5,000–15,000	20,000–24,000	18,000
Color temperature (K)	2,400–3,100	2,700–7,500	3,000–6,000	3,000–5,000	2,000	1,700
Color Rendition Index	90–100	50–110	25–55	60–70	20–25	0
Potential for good color rendition	High	Highest	Fair	Good	Color discrimination	No color discrimination
Lamp cost	Low	Moderate	Moderate	High	High	Moderate
Operational cost	High	Good	Moderate	Moderate	Low	Low

Note. The values given have are generic for general service lamps. A survey of lamp manufacturers' catalogs should be made before specifying or purchasing any lamp.

- Thermally protected
- Type IC
- Type Non-IC

Each light fixture and lamp installed in any residential lighting system should carry the label of a nationally recognized testing laboratory. This ensures the installer that the luminaire has met a minimum set of performance standards. The three major testing laboratories are Underwriters Laboratories (UL), Canadian Standards Association (CSA), and Electrical Testing Laboratories (ETL). These testing labs were discussed in Chapter 2. If an electrician finds a fixture that does not carry one of these labels, he or she should check with the local inspector to be sure the luminaire is approved for installation.

Because of the very personal nature of lighting and the vast number of choices available, selecting the "perfect" light fixture for a particular location can be a very daunting task. There is no set rule explaining who will choose the luminaires for a residence. If a building contractor builds a house to sell (a "spec house"), the contractor will usually choose the fixtures. If a building contractor builds a house for a specific individual, the customer will have the final say on the type of fixtures installed.

Residential lighting can be divided into four separate groups: general lighting, accent lighting, task lighting, and security lighting. The requirements of each group are unique and require different approaches. The electrician must work closely with the home owner or the builder to ensure that the lighting they install produces the desired effect. A description of each lighting group follows:

- General lighting provides a room or other area with adequate overall lighting. It can range from very basic, like a plastic keyless fixture in a garage, to very ornate, such as a chandelier hanging over the home foyer. General lighting is often used with either accent lighting or task lighting in bedrooms, kitchens, living rooms, and garages or workshops.

- Accent lighting is used to focus your attention on a particular area or object, such as a fireplace or work of art. Recessed spotlights and track lighting are two popular fixture types used for accent lighting. To be effective, accent lighting should be at least five times brighter than the room general lighting.
- Task lighting provides an adequate light source for tasks to be performed without distracting glare or shadows. It is important that the task lighting complement the general lighting of the room. To avoid too much contrast, task lighting should be no more than three times brighter than the general room lighting.
- Security lighting is designed to meet safety and security concerns around the exterior of the house. When properly designed and installed, security lighting will enhance the appearance of the residence and increase its value.

There are four factors that should be considered when selecting light fixtures for a particular location:

- Match both the lamp and the fixture to the desired application. Is the light source appropriate for the type of lighting required? For example, you would not want to use floodlights to accent a wall-hung painting. Does the light fixture complement the light fixtures in the rest of the room? Lighting manufacturers will provide a wealth of information to simplify this decision. They have pamphlets and booklets that give information on the types of fixtures that will work in most residential situations.
- How important is color rendition to the lighting application? Color rendition is a term used to describe how well a particular light source allows us to see an object's true color. All lamps are assigned a number reflecting how their light compares to sunlight at noon. The higher the assigned number, the better the color rendition of the lamp. Most incandescent lamps have a color rendition number of 95, and fluorescent lamps have a color rendition of 80 to 85. In residential lighting, color rendition

is very important for indoor applications. For unfinished basements, garages, and outdoor locations, the color rendition may not be as important.

- The energy efficiency of the lamp and fixture must be considered. While the most light output for the lowest cost (purchase price plus operating cost) is certainly important, in some cases a less efficient luminaire will provide a better lighting result and must be considered. The goal is find the right light source with the highest efficiency for the lowest total cost.
- How long will a particular lamp last? In residential lighting, this is usually not a high priority because most lamps are relatively inexpensive and are easy to replace. Lamp life becomes an issue in areas where lamps are difficult to reach. Lamps located in cathedral ceilings or in chandelier fixtures hung in open foyers are examples of areas where lamp life is important.

Recessed Luminaires

Recessed luminaires (Figure 17–12) are fixtures installed above the ceiling. They are designed so that little or none of the fixture, lamp, lens, or trim extends below the level of the sheetrock or drop ceiling tile. Both fluorescent and incandescent recessed fixtures are available.

In residential lighting systems, recessed fluorescent lighting is not used that often. When it is used, it is usually in areas of a house such as kitchens, laundry rooms, finished basements, or recreation rooms. This type of fixture produces a softer and more even light distribution. These fixtures are available in two-, three-, and four-lamp models. The two-lamp fixtures require one ballast, while the three- and four-lamp fixtures usually require two ballasts. Recessed fluorescent fixtures have lenses to protect the lamps and to disperse the light in a particular pattern. Each lens is designed to provide maximum light output for that particular light fixture.

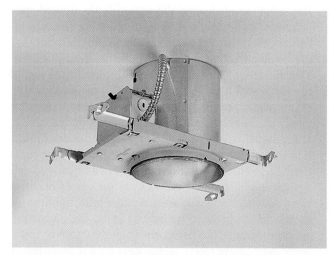

Figure 17–12 A recessed luminaire (lighting fixture). *Photo courtesy of Progress Lighting.*

Recessed incandescent lighting is a very versatile type of lighting available for residential lighting and, as a result, is used often in residential lighting systems. The light fixture comes in two separate pieces: the recessed fixture rough-in frame (Figure 17–13) and the trim. Each recessed incandescent fixture, commonly called a "can," is available with trim rings that will fit only that can. The variety of trim rings available allow the fixture to be used as general lighting, accent lighting, task lighting, or security lighting.

The rough-in frame is made of metal and holds the light fixture and a junction box. This framework is installed in the ceiling joists or trusses, and the fixture is connected to the electrical system by means of the junction box before the ceiling is installed. The trim rings and lamp are installed after the ceiling is painted. It is important the electrician know the trim ring to be installed so that the proper fixture may be installed during the rough-in phase.

The electrician must also know the insulation rating of the fixture housing. Recessed incandescent fixtures are rated either **Type IC** (insulated ceiling) or **Type Non-IC.** Type IC fixtures are designed to be in direct contact with thermal insulation. Type Non-IC fixtures must be installed so that the insulation is no closer than 3 inches (75 mm) to any part of the fixture. Figure 17–14 shows both types of recessed incandescent lighting fixtures. This allows the heat generated by the lamp to dissipate away from the fixture, preventing damage to the fixture and possibly fire. This is covered in Section 410.66 of the *NEC®*. Each recessed incandescent fixture is equipped with a built-in thermal protection unit, as stated in Section 410.65(C) of the *NEC®*. This is to lessen the threat of fire from heat buildup.

There are four main types of trim rings available for recessed incandescent light fixtures (Figure 17–15). They are held in place with springs or clips. The open baffle is used to provide general lighting. This trim type allows for a softer illumination of a wide area. This trim is available in various colors, with either smooth or stepped sides. The down-light trim ring is a good choice for task lighting. This trim provides a more focused beam of light for the task at hand. The trim rings come with openings of different sizes to accommodate lamps of different size and wattage. The fish-eye trim ring is used mainly for accent lighting but also works well in task lighting and general lighting situations. This trim can be rotated along two axes, allowing the light to be aimed to best fit the desired application. The fourth trim type is the waterproof trim.

 FROM EXPERIENCE

Recessed incandescent light fixture rough-in frames and trim rings are sold separately. Be sure to get both parts when purchasing the light fixtures.

Figure 17–13 An incandescent recessed lighting fixture rough-in frame. The adjustable mounting brackets allow the frame to be easily installed between two ceiling joists. The cable feeding the fixture is terminated in the attached junction box. The *NEC®*'s installation requirements for recessed fixtures are found in Sections 410.64 through 410.72.

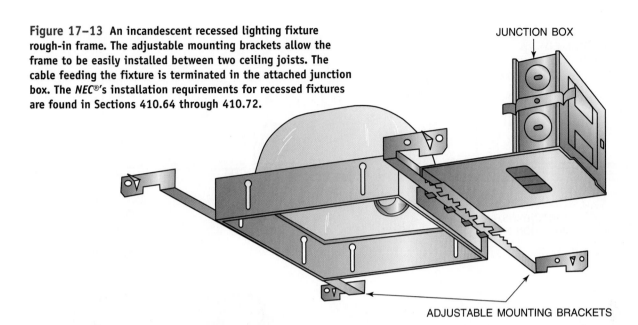

JUNCTION BOX

ADJUSTABLE MOUNTING BRACKETS

Figure 17–14 (A) A Type Non-IC recessed fixture installed with the thermal insulation kept away from the fixture. (B) A Type IC fixture that may be completely covered with thermal insulation. (C) The required clearances from thermal insulation for a Type Non-IC recessed fixture according to Section 410.66 of the *NEC®*.

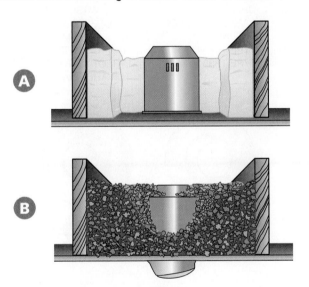

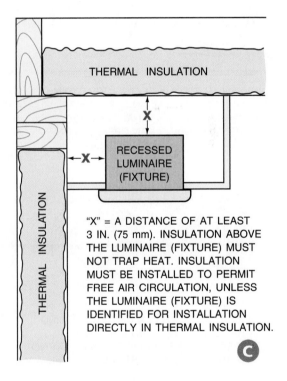

THERMAL INSULATION

X

THERMAL INSULATION

RECESSED LUMINAIRE (FIXTURE)

X

"X" = A DISTANCE OF AT LEAST 3 IN. (75 mm). INSULATION ABOVE THE LUMINAIRE (FIXTURE) MUST NOT TRAP HEAT. INSULATION MUST BE INSTALLED TO PERMIT FREE AIR CIRCULATION, UNLESS THE LUMINAIRE (FIXTURE) IS IDENTIFIED FOR INSTALLATION DIRECTLY IN THERMAL INSULATION.

C

 FROM EXPERIENCE

The fish-eye trim is not a good choice for high ceilings or cathedral ceilings. Changing lamps can be extremely difficult unless you have the proper ladder. Telescoping tools designed for changing lamps will not work with this style of trim ring. They simply push the fish-eye part of the trim ring back up into the can.

Basically, a down-light trim ring with a glass or acrylic lens, it is designed for use in wet locations. Outdoors, this trim ring is used to provide general lighting on decks and porches. Indoors, this trim is used to light bathroom shower and tub units.

Surface-mounted Luminaires

Surface-mounted luminaires make up the majority of light fixtures installed in a residence. There are a multitude of styles and sizes available, so you are limited only

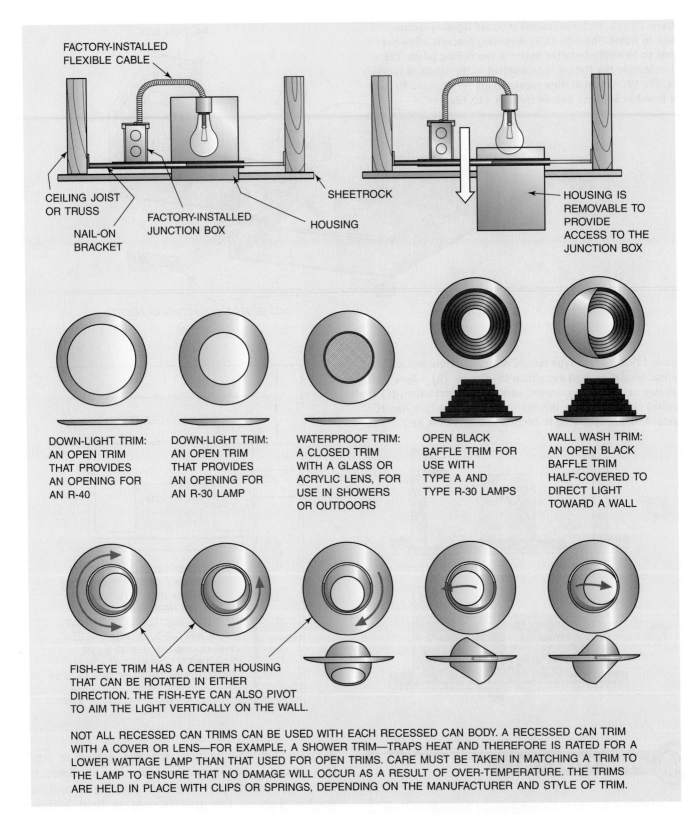

FACTORY-INSTALLED
FLEXIBLE CABLE

CEILING JOIST
OR TRUSS

NAIL-ON
BRACKET

FACTORY-INSTALLED
JUNCTION BOX

SHEETROCK

HOUSING

HOUSING IS
REMOVABLE TO
PROVIDE
ACCESS TO THE
JUNCTION BOX

DOWN-LIGHT TRIM:
AN OPEN TRIM
THAT PROVIDES
AN OPENING FOR
AN R-40

DOWN-LIGHT TRIM:
AN OPEN TRIM
THAT PROVIDES
AN OPENING FOR
AN R-30 LAMP

WATERPROOF TRIM:
A CLOSED TRIM
WITH A GLASS OR
ACRYLIC LENS, FOR
USE IN SHOWERS
OR OUTDOORS

OPEN BLACK
BAFFLE TRIM FOR
USE WITH
TYPE A AND
TYPE R-30 LAMPS

WALL WASH TRIM:
AN OPEN BLACK
BAFFLE TRIM
HALF-COVERED TO
DIRECT LIGHT
TOWARD A WALL

FISH-EYE TRIM HAS A CENTER HOUSING
THAT CAN BE ROTATED IN EITHER
DIRECTION. THE FISH-EYE CAN ALSO PIVOT
TO AIM THE LIGHT VERTICALLY ON THE WALL.

NOT ALL RECESSED CAN TRIMS CAN BE USED WITH EACH RECESSED CAN BODY. A RECESSED CAN TRIM
WITH A COVER OR LENS—FOR EXAMPLE, A SHOWER TRIM—TRAPS HEAT AND THEREFORE IS RATED FOR A
LOWER WATTAGE LAMP THAN THAT USED FOR OPEN TRIMS. CARE MUST BE TAKEN IN MATCHING A TRIM TO
THE LAMP TO ENSURE THAT NO DAMAGE WILL OCCUR AS A RESULT OF OVER-TEMPERATURE. THE TRIMS
ARE HELD IN PLACE WITH CLIPS OR SPRINGS, DEPENDING ON THE MANUFACTURER AND STYLE OF TRIM.

Figure 17–15 A guide to incandescent recessed lighting fixture trims used in residential applications.

by your imagination. Surface-mounted luminaires can be either incandescent or fluorescent and are designed for either wall or ceiling mounting. However, you should always check the manufacturer's instructions because some fixtures are designed as "wall mount only" or "ceiling mount only." Improper mounting can cause heat buildup, resulting in damage to the fixture or fire.

Wall-mounted lighting fixtures (Figure 17–16), also called **sconces,** are usually connected to either metallic or nonmetallic lighting outlet boxes after the sheetrock is installed and finished. Sconces can be used in any room in the house and are used for general lighting or accent lighting. There are many styles and shapes of sconces to meet almost any lighting need. Units generally have one or two lamps with either a candelabra or a medium base.

Ceiling-mounted fixtures (Figure 17–17) are also connected to metallic or nonmetallic lighting outlet boxes after the ceiling has been installed and finished. Ceiling-mounted fixtures are designed to use either incandescent or fluorescent lamps and may be installed on the bottom of dropped ceilings if listed for that purpose. For this type of mounting, special brackets, available in most electrical supply stores, must be used, and the fixture must be properly supported to the ceiling framework. Installation requirements for this type of installation are found in Section 410.16(C) of the *NEC®* and are discussed in greater detail later in this chapter.

There are three basic types of ceiling mounted fixtures: direct mount, pendant, and chandelier. Direct-mount fixtures (Figure 17–18) are the most common type used in residential lighting applications. Some manufacturers refer to this type of mounting as "close to ceiling." They can be used to provide general lighting in any room in a house. Incandescent models usually have one, two, or three lamps, while fluorescent models may have two, three, or four lamps.

Chandelier fixtures (Figure 17–19) are multiple lamp fixtures that are suspended from an electrical lighting outlet box. A special bracket is attached to the light box, and the chain is attached to this bracket. A small, electrical cable runs along the chain between the light box and the fixture providing electrical power. Chandeliers are found in dining rooms, foyers, and other more formal areas of a residence.

Pendant fixtures (Figure 17–20) are similar to chandeliers in that they are also suspended from a lighting outlet box. However, pendant fixtures differ in two ways. First, they are one-lamp fixtures and are smaller than chandeliers. Second, they may be suspended from the lighting outlet box by means of a chain, a hollow metal rod, or a special electrical cord. If a cord is used, it must be listed for use with a pendant fixture since it must provide power to the fixture and also support its weight. Pendant fixtures are used to provide general, task, and accent lighting in many areas of a house.

Figure 17–16 A typical wall-mounted luminaire (lighting fixture). *Photo courtesy of Progress Lighting.*

Figure 17–17 A typical ceiling-mounted luminaire (lighting fixture). *Photo courtesy of Progress Lighting.*

Figure 17–18 A direct-mount lighting fixture. The lighting fixture shown is mounted directly to the ceiling in a room like a kitchen. This fixture is designed to take two 40-watt, U-shaped fluorescent tubes. *Photo courtesy of Progress Lighting.*

Figure 17–19 Several examples of chandelier-type lighting fixtures. They are commonly installed over a dining room table. *Photo courtesy of Progress Lighting.*

Figure 17–20 An example of a pendant-type lighting fixture that could be installed over a small table in the eating area of a kitchen. Other pendant-type lighting fixture styles are popular for installation in areas of a house with high ceilings. *Photo courtesy of Progress Lighting.*

Installing Common Residential Lighting Fixtures

Installing lighting fixtures in a house is one of the last things an electrician does during the installation of the electrical system. Before an electrician actually installs the lighting fixtures, there are three factors that need to be considered:

- You must remember that light fixtures are very fragile and are easily damaged or broken. It is important to handle and store these fixtures with care.

> **CAUTION**
>
> **CAUTION: Make sure the power to the circuit is disconnected before starting to install a lighting fixture. Many electricians are needlessly shocked because they *assumed* the power had been turned off. Always check the circuit with a voltage tester before beginning and make sure the circuit is locked out.**

- Read the manufacturer's instructions that come with each lighting fixture. The instructions can make the installation process much easier if they are read *before* you start the installation. Nothing is more frustrating than to get three-fourths of the installation complete, only to realize you did not install a piece in the proper location—now you have to take the fixture completely apart and start over. This happens many times and can be avoided by simply reading the instructions. Remember, instructions are not a means of last resort.

 FROM EXPERIENCE

On some lighting fixtures, especially the ones hung with chain, the fixture wiring is not color coded, and at first glance the grounded wire and the ungrounded wire are not distinguishable from each other. However, you will find that one of the wires has a smooth surface and that one has a ribbed surface. The wire with the ribbed surface is the grounded conductor, and the wire with the smooth surface is the ungrounded conductor.

- When installing incandescent lighting fixtures, an electrician must ensure that they are connected to the electrical system with the proper polarity. Section 410.23 of the *NEC®* states that the grounded conductor must be connected to the silver screw shell. The ungrounded, or "hot," conductor is connected to the brass contact in the bottom center of the screw shell. If the fixture has both a black and a white conductor, and the polarity has been determined, you may proceed with the installation. If the fixture has two wires of the same color, usually black, a continuity tester can be used to determine the proper polarity. One test lead of the tester is placed against the silver screw shell, and the other test lead is placed against the bare end of one of the fixture conductors. If continuity is indicated, that wire is the grounded conductor and should be marked with white tape. If continuity is not indicated, the wire must be the ungrounded conductor and should be marked accordingly.

Direct Connection to a Lighting Outlet Box

There are not many incandescent fixtures that are mounted directly to the outlet box. The two main types of fixtures in this category are keyless fixtures and pull-chain fixtures (Figure 17–21). Keyless fixtures are typically made of porcelain or plastic and are often used in

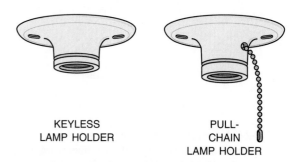

KEYLESS LAMP HOLDER PULL-CHAIN LAMP HOLDER

Figure 17–21 Keyless and pull-chain direct-connection fixtures. They may be made of porcelain, plastic, or fiberglass.

basements, crawl spaces, attics, and garages. Because the fixture is made from nonmetallic materials, it has no provisions for grounding.

(See page 488 for installation steps for installing a light fixture directly to an outlet box.)

Direct Connection to the Ceiling

Surface-mounted fluorescent fixtures (Figure 17–22) are installed and connected to the electrical system in a slightly different manner. Attaching the fixture to the ceiling requires a different approach because the fixture is not attached directly to the lighting outlet box. Instead, the fixture is attached directly to the ceiling, and the electrical wiring is brought into the fixture from an electrical lighting outlet box. Surface-mounted fluorescent fixtures have wiring compartments where the lighting branch-circuit wiring is connected to the fixture wiring. The electrical connections are very basic: white wire to white wire, black wire to black wire; and the grounding wire to the fixture's metal body.

There are two ways that a surface-mounted fluorescent fixture can be connected to the lighting branch-circuit wiring. The first way is to have the wiring method used, such as a Type NM cable, connected directly to the lighting fixture. In this scenario, a length of cable is installed at the location of the fixture(s) during the rough-in stage and left long enough to hang down from the ceiling a short distance. It is then coiled up out of the way and placed between the ceiling joists. When the ceiling workers put up the ceiling, they will put a small hole in the ceiling material and bring the cable down through the hole so that a short length hangs down below the ceiling. During the installation of the surface-mounted fixture, the electrician attaches this cable to the lighting fixture with a cable clamp, secures the fixture to the ceiling, and makes the

necessary electrical connections. Manufacturers provide knockouts on their surface-mounted fluorescent fixtures for this installation method.

The second wiring method is similar to the one where the lighting fixture is attached directly to a lighting outlet box. During the rough-in stage, a lighting outlet box is installed at the location for the surface-mounted fluorescent fixture, and the lighting circuit wiring is run to the box. When a surface-mounted fluorescent fixture is installed, it will cover the lighting outlet box. The wiring is brought from the lighting outlet box down and into the wiring compartment of the surface-mounted fluorescent fixture. When using this method, make sure to follow Section 410.14(B) of the *NEC®*, which requires surface-mounted fixtures to allow access to the wiring in the lighting outlet box *after* the surface-mounted fixture has been installed. This means that a large opening will have to be knocked out in the surface-mounted fluorescent fixture. The hole is then aligned with the lighting outlet box that the fixture is covering (Figure 17–23). The large hole allows easy access to the wiring in the lighting outlet box.

(See page 489 for installation steps for installing a cable-connected surface-mounted fluorescent lighting fixture directly to a ceiling.)

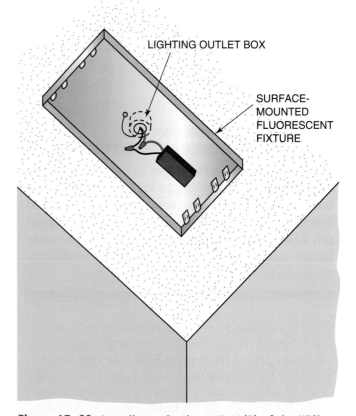

Figure 17–23 According to Section 410.14(B) of the *NEC®*, there must be access to the lighting outlet box even when a surface-mounted fluorescent lighting fixture is installed over the box location.

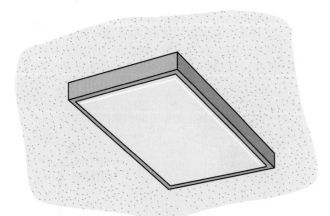

Figure 17–22 A surface-mounted fluorescent fixture. If they are to be mounted on low-density cellulose fiberboard, they must be marked with a label that says, "Suitable for Surface Mounting on Low-Density Cellulose Fiberboard."

Figure 17–24 Examples of fixtures that use the strap to outlet box mounting method. *Photo courtesy of Progress Lighting.*

Strap to Lighting Outlet Box

This light fixture installation process is very similar to the direct connection method discussed previously. The main difference is that a metal strap is connected to the lighting outlet box and the fixture is attached to the metal strap with headless bolts and decorative nuts. The metal strap must be connected to the system-grounding conductor. For this type of installation, it is important to read, understand, and follow the manufacturer's instructions. Figure 17–24 shows an example of a lighting fixture that is installed using the strap to lighting outlet box method.

(See page 490 for installation steps for the installation of a strap to lighting outlet box lighting fixture.)

Stud and Strap Connection to a Lighting Outlet Box

There are larger and heavier types of light fixtures that use the stud and strap method of installation. Hanging fixtures, like a chandelier or pendant fixture, often require extra mounting support when compared to smaller light fixtures. The electrical connections for each installation are the same as discussed previously, so we will not spend a lot of time on them here. It is extremely important to read and follow the manufacturer's instructions because there are some slight variations with this type of installation.

(See page 491 for installation steps for installing a chandelier-type light fixture using the stud and strap connection method to a lighting outlet box.)

Recessed Luminaire Installation

At this stage of the process, recessed fixtures are very easy to finish out. The recessed fixture rough-in frame has already been installed and connected to the electrical sys-

tem (Figure 17–25). All that remains to be done is to install the lamps and the trim rings or lenses.

For recessed incandescent lighting fixtures, make sure the power to the circuit is turned off, review the instructions, and wear the proper safety gear:

1. Install the appropriate trim rings with the provided springs or clips. Make sure the trim ring is listed for use with the installed can light.
2. Install the recommended lamp.
3. Test the fixture for proper operation.

For recessed fluorescent lighting fixtures, keep the following items in mind when you are completing the installation:

1. Make sure the power to the circuit is turned off, review the instructions, and wear the proper protective gear.
2. Install the recommended lamps.
3. Test the fixture for proper operation.
4. Install the fixture lens.

Luminaire Installation in Dropped Ceilings

When installing luminaires in dropped ceilings, also called "suspended ceilings," it is important to make sure the fixture is listed for that type of installation. The two listings to look for are "suitable for use in suspended ceilings" and "suitable for mounting on low-density cellulose fiberboard." The second listing is important because most drop ceiling tiles are made of that material. These fixtures are not considered recessed because there is usually a great deal of room above them and easy access to the electrical outlet box on each fixture.

There are several types of light fixtures that may be used in a suspended ceiling. They include recessed incandescent and fluorescent fixtures and any surface-mounted fixture that can be connected to a metallic or nonmetallic outlet box and is listed as suitable for installation in a dropped ceiling. It is important to note that the lighting outlet boxes must be connected to the ceiling support grid by means of an approved mounting bar. Fluorescent lighting fixtures installed in a dropped ceiling grid are often referred to by electricians as a **troffer.**

(See page 492 for installation steps for installing a fluorescent fixture in a dropped ceiling.)

Outside Luminaire Installation

When installing light fixtures outside a residence, you can use any of the methods that have been discussed previously in this chapter. Review the method that is appropriate for the fixture you are installing.

There are several factors that must be considered when installing light fixtures outdoors. One is whether the fixture is listed for the desired application. Outdoor fixtures are listed for use in either damp or wet locations. A damp location is any area that is subject to some degree of moisture

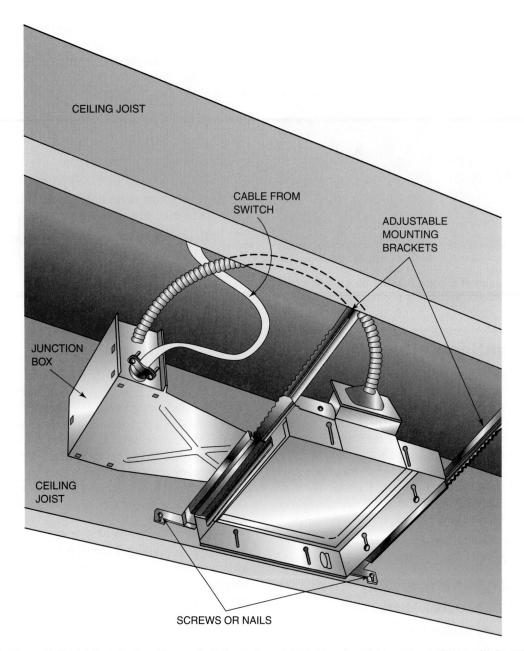

CEILING JOIST

CABLE FROM
SWITCH

ADJUSTABLE
MOUNTING
BRACKETS

JUNCTION
BOX

CEILING
JOIST

SCREWS OR NAILS

Figure 17–25 Recessed fixture installation. The rough-in frame is prewired. The electrician runs a cable to the junction box and connects the fixture to the lighting branch circuit by making the necessary electrical connections in the junction box. The fixture is secured to the ceiling joists with screws or nails.

but not saturation from water or other liquids. Examples of a damp location are fixtures installed in the eaves of a house or in the ceiling of an open porch. A wet location is any area that is subject to saturation from water or other liquids. Wet location fixtures may be installed in damp locations, but damp location fixtures cannot be installed in wet locations. Figure 17–26 shows several examples of outdoor lighting fixtures that are suitable for installation in a wet (or damp) location.

Always check with the local code enforcement office for regulations concerning the placement of outdoor fixtures. Many communities have restrictions limiting the amount of light

that may be seen from neighboring properties. There may also be regulations that limit the brightness of outdoor fixtures and the amount of light that is directed up into the sky.

Summary

There is much more to a good lighting system than making good electrical connections. The electrician must have the knowledge to select the type of fixture and lamp that will provide the desired lighting effect in each location of the resi-

Figure 17–26 Examples of outdoor lighting fixtures. These fixtures are designed for wall mounting. *Photo courtesy of Progress Lighting.*

dence. There are three main types of lighting used in residential lighting. Incandescent lighting, where light is produced by heating a filament, is the most common. Incandescent lamps provide excellent color rendition in all phases of home lighting. They are the most versatile lamps because they are compatible with dimmers to vary the light output. Incandescent lamps are also the least efficient lamp used in residential lighting. It is also the least expensive lamp type to purchase.

Fluorescent lighting gives the best result when used for general lighting in kitchens, laundry rooms, basements, and work areas. It is a very efficient light source with long life. The major drawbacks to fluorescent lighting are cold temperature operations and incompatibility with dimmers.

HID lighting produces the whitest light and is the most efficient lamp used in residential lighting. The major drawback to HID lighting is the warm-up period before full light output is achieved. This warm-up period can last several minutes and

occurs every time the lamp is lit and after power interruptions. This limits HID lighting to outdoor applications.

It is extremely important to install luminaires according to the manufacturer's instructions. Even similar-looking fixtures from different manufacturers can have different installation procedures. Reading the instructions before you start the installation will result in a smoother, faster, and correct installation. The installation steps provided in this chapter are meant to be a guide to the learner. They are not meant to replace the manufacturer's instructions.

Above all, be safe. Always make sure the power to the circuit is off before you start the installation, and test the circuit before you touch the conductors. Many electricians have been shocked because they thought that simply having the switch controlling the fixture turned off was adequate protection. Also, always use the proper safety equipment to ensure your safety.

Procedures

Installation Steps for Installing a Light Fixture Directly to an Outlet Box

- Put on safety glasses and observe all applicable safety rules.

- Using a voltage tester, verify that there is no electrical power at the lighting outlet where the fixture will be installed. If electrical power is present, turn off the power and lock out the circuit.

- Locate and identify the un-grounded, grounded, and grounding conductors in the lighting outlet box.

A The grounding conductor will not be connected to this fixture. If there is a grounding conductor in a nonmetallic box, simply coil it up and push it to the back or bottom of the electrical box. Do not cut it off, as it may be needed if another type of light fixture is installed at that location. If there are two or more grounding conductors in a nonmetallic box, connect them together with a wirenut and push them to the back or bottom of the lighting outlet box. If there is a grounding conductor in a metal outlet box, it must be connected to the outlet box by means of a listed grounding screw or clip. If there are two or more grounding conductors in a metal box, use a wirenut to connect them together along with a grounding pigtail. Attach the grounding pigtail to the metal box with a listed grounding screw.

- Connect the white grounded conductor(s) to the silver screw or white fixture pigtail. If there is one grounded conductor in the box, strip approximately 3/4 inch (19 mm) of insulation from the end of the conductor and form a loop at the end of the conductor using an approved tool, such as a

A

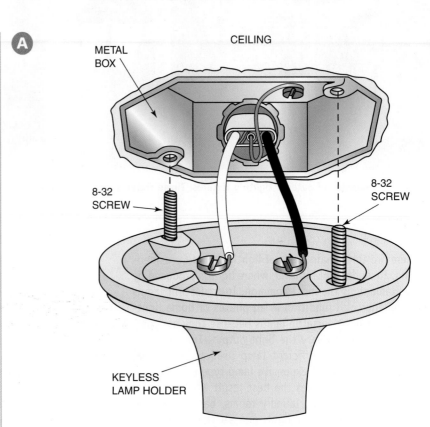

CEILING

METAL BOX

8-32 SCREW

8-32 SCREW

KEYLESS LAMP HOLDER

T-stripper. Once the loop is made, slide it around the silver terminal screw on the fixture so the end is pointing in a clockwise direction. Hold the conductor in place and tighten the screw. If there are two or more grounded conductors in the box, strip the ends as described previously and use a wirenut to connect them and a white pigtail together. Attach the pigtail to the silver grounded screw as described previously.

- The black ungrounded conductor(s) is connected to the brass-colored terminal screw on the fixture. The connection procedure is the same as for the grounded conductor(s).

- Now the fixture is ready to be attached to the lighting outlet box. Make sure the grounding conductors are positioned so they will not come in contact with the grounded or ungrounded screw terminals. Align the mounting holes on the fixture with the mounting holes on the lighting outlet box. Insert the 8-32 screws that are usually equipped with the fixture through the fixture holes. Then thread them into the mounting holes in the outlet box.

- Tighten the screws until the fixture makes contact with the ceiling or wall. Be careful not to overtighten the screws, as you may damage the fixture.

- Install the proper lamp, remembering not to exceed the recommended wattage.

- Turn on the power and test the light fixture.

Procedures | Installation Steps for Installing a Cable-Connected Fluorescent Lighting Fixture Directly to the Ceiling

- Put on safety glasses and observe all applicable safety rules.

- Using a voltage tester, verify that there is no electrical power at the lighting outlet where the fixture will be installed. If electrical power is present, turn off the power and lock out the circuit.

- Place the fixture on the ceiling in the correct position, making sure it is aligned and the electrical conductors have a clear path into the fixture.

- Mark on the ceiling the location of the mounting holes.

- Use a stud finder to determine if the mounting holes line up with the ceiling trusses. If they do, screws will be used to mount the fixture. If they do not, toggle bolts will be necessary. For some installations, a combination of screws and toggle bolts will be required.

- If screws are used, drill holes into the ceiling using a drill bit that has a smaller diameter than the screws to be used. This will make installing the screws easier. If toggle bolts are to be used, use a flat-bladed screwdriver to punch a hole in the sheetrock only large enough for the toggle to fit through.

- Remove a knockout from the fixture where you wish the conductors to come through. Install a cable connector in the knockout hole.

- Place the fixture in its correct position and pull the cable through the connector and into the fixture. Tighten the cable connector to secure the cable to the fixture.

- This part of the process may require the assistance of a coworker. If using toggle bolts, put the bolt through the mounting hole and start the toggle on the end of the bolt.

- With a coworker holding the fixture, install the mounting screws or push the toggle through the hole until the wings spring open. This will hold the fixture in place until the fixture is secured to the ceiling.

- Make the necessary electrical connections. The grounding conductor should be properly wrapped around the fixture grounding screw and the screw tightened. The white grounded conductor is connected to the white conductor lead, then the black ungrounded conductor is connected to the black fixture conductor.

- Install the wiring cover by placing one side in the mounting clips, squeezing it, and then snapping the other side into its mounting clips.

- Install the recommended lamps. Usually, they have two contact pins on each end of the lamp. Align the pins vertically, slide them up into the lamp holders at each end of the fixture, and rotate the lamp until it snaps into place.

- Test the fixture and lamps for proper operation.

- Install the fixture lens cover.

Procedures

Installation Steps for the Installation of a Strap to Lighting Outlet Box Lighting Fixture

- Put on safety glasses and observe all applicable safety rules.

- Using a voltage tester, verify that there is no electrical power at the lighting outlet where the fixture will be installed. If electrical power is present, turn off the power and lock out the circuit.

- Before starting the installation process, read and understand the manufacturer's instructions.

Ⓐ Mount the strap to the outlet box using the slots in the strap. With metal boxes, the screws are provided with the box. With nonmetallic boxes, you must provide your own 8-32 mounting screws. Put the 8-32 screws through the slot and thread them into the mounting holes on the outlet box. Tighten the screws to secure the strap to the box.

- Identify the proper threaded holes on the strap and install the fixture-mounting headless bolts in the holes so the end of the screw will point down.

- Make the necessary electrical connections. Make sure that *all* metal parts (including the outlet box), the strap, and the fixture are properly connected to the grounding conductor in the power feed cable.

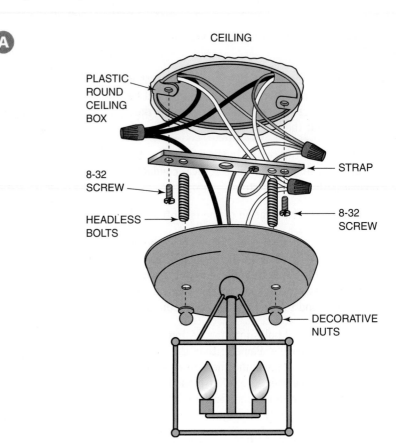

- Neatly fold the conductors into the outlet box. Align the headless bolts with the mounting holes on the fixture. Slide the fixture over the headless bolts until the screws stick out through the holes. Do not be alarmed if the mounting screws seem to be too long. Thread the provided decorative nuts onto the headless bolts. Keep turning the nuts until the fixture is secure to the ceiling or wall.

- Install the recommended lamp and test the fixture operation.

- Install any provided lens or globe. They are usually held in place by three screws that thread into the fixture. Start the screws into the threaded holes, position the lens or globe so it touches the fixture, and tighten the screws until the globe or lens is snug. Do *not* over tighten the screws. You may return the next day and find the globe or lens cracked or broken.

Procedures

Installation Steps for Installing a Chandelier-Type Light Fixture Using the Stud and Strap Connection to a Lighting Outlet Box

- Put on safety glasses and observe all applicable safety rules.

- Using a voltage tester, verify that there is no electrical power at the lighting outlet where the fixture will be installed. If electrical power is present, turn off the power and lock out the circuit.

- Before starting the installation process, read and understand the manufacturer's instructions.

A Install the mounting strap to the outlet box using 8-32 screws.

- Thread the stud into the threaded hole in the center of the mounting strap. Make sure that enough of the stud is screwed into the strap to make a good secure connection.

- Measure the chandelier chain for the proper length, remove any un-needed links and install one end to the light fixture.

- Thread the light fixture's chain-mounting bracket on to the stud. Remove the holding nut and slide it over the chain.

- Slide the canopy over the chain.

- Attach the free end of the chain to the chain-mounting bracket.

- Weave the fixture wires and the grounding conductor up through the chain links, being careful to keep the chain links straight. Section 410.28(F) of the *NEC*® states

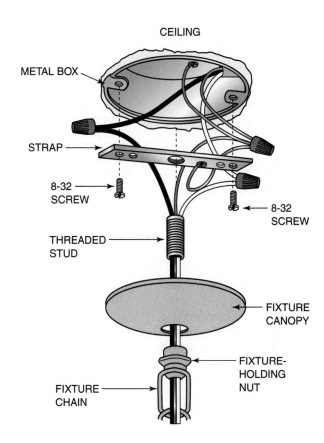

that the conductors must not bear the weight of the fixture. As long as the chain is straight and the conductors make all the bends, the chain will support the fixture properly.

- Now run the fixture wires up through the fixture stud and into the lighting outlet box.

- Make all necessary electrical connections.

- Slide the canopy up the chain until it is in the proper position. Slide the nut up the chain and thread it on to the chain-mounting bracket until the canopy is secure.

- Install the recommended lamp and test the fixture for proper operation.

Procedures

Installation Steps for Installing a Fluorescent Fixture (Troffer) in a Dropped Ceiling

- Put on safety glasses and observe all applicable safety rules.

- Before starting the installation process, read and understand the manufacturer's instructions.

- During the rough-in stage, mark the location of the fixtures on the ceiling.

- Using standard wiring methods, place lighting outlet boxes on the ceiling near the marked fixture locations and connect them to the lighting branch circuit.

- Once the dropped ceiling grid has been installed by the ceiling contractor, install the fluorescent light fixtures in the ceiling grid at the proper locations. Some electricians refer to this action as "laying in" the fixture. Once the fixture is installed, some electricians refer to the fixtures as being "laid in."

 Support the fixture according to *NEC®* requirements. Section 410.16(C) requires that all framing members used to support the ceiling grid be securely fastened to each other and to the building itself. The fixtures themselves must be securely fastened to the grid by an approved means, such as bolts, screws, rivets, or clips. This is to prevent the fixture from falling and injuring someone.

- Using a voltage tester, verify that there is no electrical power at the lighting outlet where the fixture will be installed. If electrical power is present, turn off the power and lock out the circuit.

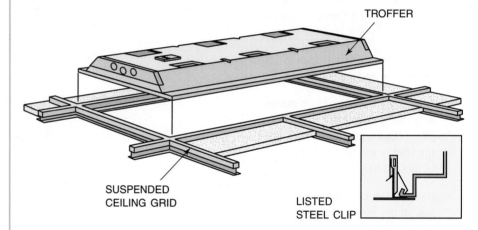

A

IMPORTANT: TO PREVENT THE LUMINAIRE (FIXTURE) FROM INADVERTENTLY FALLING, *410.16(C)* OF THE CODE REQUIRES THAT (1) SUSPENDED CEILING FRAMING MEMBERS THAT SUPPORT RECESSED LUMINAIRES (FIXTURES) MUST BE SECURELY FASTENED TO EACH OTHER, AND MUST BE SECURELY ATTACHED TO THE BUILDING STRUCTURE AT APPPROPRIATE INTERVALS, AND (2) RECESSED LUMINAIRES (FIXTURES) MUST BE SECURELY FASTENED TO THE SUSPENDED CEILING FRAMING MEMBERS BY BOLTS, SCREWS, RIVETS, OR SPECIAL LISTED CLIPS PROVIDED BY THE MANUFACTURER OF THE LUMINAIRE (FIXTURE) FOR THE PURPOSE OF ATTACHING THE LUMINAIRE (FIXTURE) TO THE FRAMING MEMBER.

TROFFER

SUSPENDED CEILING GRID

LISTED STEEL CLIP

B Connect the fixture to the electrical system. This is done by means of a "fixture whip." A fixture whip is often a length of Type NM, Type AC, or Type MC cable. It can also be a raceway with approved conductors, such as flexible metal conduit or electrical nonmetallic tubing. The fixture whip must be at least 18 inches (450 mm) long and no longer than 6 feet (1.8 m).

• Make all necessary electrical connections. The fixture whip should already be connected to the outlet box mounted in the ceiling. Using an approved connector, connect the cable or raceway to the fixture outlet box and run the conductors into the outlet box. Make sure that all metal parts are properly connected to the grounding system. Connect the white grounded conductors together and then the black ungrounded conductors together. Close the connection box.

• Install the recommended lamps and test the fixture for proper operation.

• Install the lens on the fixture.

B

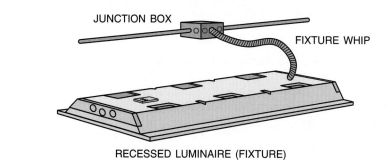

JUNCTION BOX

FIXTURE WHIP

RECESSED LUMINAIRE (FIXTURE)

Review Questions

Select the most appropriate answer.

1. The *NEC®* refers to a fluorescent lamp as an _____ _____ lamp.

2. The two types of lamps used most often in residential lighting are _____ and _____.

3. The shortest visible wavelength in the magnetic spectrum appears _____ in color.

4. What causes an apple to appear red?

5. Why are "warm" light sources preferred for residential lighting?

6. Why are halogen lamps considered dangerous?

7. The most inefficient type of residential lighting is the _____ lamp.

    a. fluorescent
    b. incandescent
    c. halogen
    d. HID

8. What causes a fluorescent lamp to give off light?

9. What is the purpose of a ballast?

10. What is a Class P ballast?

11. HID stands for _____ _____ _____.

12. List the four groups of residential lighting.

    a.
    b.
    c.
    d.

13. What is a Type Non-IC fixture, and where is it covered in the *NEC®*?

14. Another name for a wall-mounted light fixture is a _____.

15. The requirements for dropped ceiling light fixture installations are covered in Section _____ of the *NEC®*.

16. When connecting the grounded conductor to an incandescent light fixture, it is always connected to the silver _____ _____.

17. What size screws are used to attach the mounting strap to a lighting outlet box? _____.

18. The two listings you should look for before installing a fixture in a dropped ceiling are:

    a.
    b.

19. What is a fixture whip?

20. Fixtures rated for damp locations can be used in wet locations. True False

21. A T-8 fluorescent lamp is _____ inch(es) in diameter.

22. List the four types of incandescent lamp bases presented in this chapter.

    a.
    b.
    c.
    d.

23. List three shapes that fluorescent lamps are available in.

    a.
    b.
    c.

24. Define the term "luminaire."

25. A fluorescent lighting fixture that is "laid in" to the grid of a suspended ceiling is commonly called a _____ by many electricians.

Chapter 18 | Device Installation

The trim-out stage of the residential wiring system installation is when the required receptacles and switches are installed by the electrician. In order to have the branch circuits installed during the rough-in stage work properly, the conductor connections to the various receptacles and switches on the circuits must be made properly. This means that all conductor terminations to the devices must be done in such a way that they function properly for many years of trouble-free service. It is imperative that the devices be installed in a professional, high-quality manner. At best, a device not installed properly simply does not work correctly. At worst, a device not installed properly could cause a fire. The fire could result in serious damage to the home or, worse yet, serious personal injury to the home's occupants. Good workmanship needs to be practiced at all times but is especially important when connecting devices to a branch circuit. This chapter presents the proper selection and installation procedures for the receptacle and switch types required on the branch circuits that make up a residential electrical system. It also presents a discussion on installing ground fault circuit interrupter receptacles, arc fault circuit interrupter receptacles, and transient voltage surge suppressors.

OBJECTIVES

Upon completion of this chapter, the student should be able to:

- demonstrate an understanding of the proper way to splice wires together using a wirenut.
- demonstrate an understanding of the proper way to terminate circuit conductors to a switch or receptacle device.
- select the proper receptacle for a specific residential application.
- demonstrate an understanding of the proper installation techniques for receptacles.
- select the proper switch type for a specific residential application.
- demonstrate an understanding of the proper installation techniques for switches.
- demonstrate an understanding of GFCI and AFCI receptacle installation.
- demonstrate an understanding of TVSS devices.

Glossary of Terms

arc fault circuit interrupter (AFCI) receptacle a receptacle device designed to control dangerous arcs that occur at receptacles and other outlets in residential bedrooms

duplex receptacle the most common receptacle type used in residential wiring; it has two receptacles on one strap; each receptacle is capable of providing power to a cord-and-plug-connected electrical load

impulse a type of transient voltage that originates outside the home and is usually caused by utility company switching or lightning strikes

receptacle a device installed in an electrical box for the connection of an attachment plug

ring wave a type of transient voltage that originates inside the home and is usually caused by home office photocopiers, computer printers, the cycling on and off of heating, ventilating, and air-conditioning equipment, and spark igniters on gas appliances like furnaces and ranges

single receptacle a single contact device with no other contact device on the same strap (yoke)

split-wired receptacle a duplex receptacle wired so that the top outlet is "hot" all the time and the bottom outlet is switch controlled

strap (yoke) the metal frame that a receptacle or switch is built around; it is also used to mount a switch or receptacle to a device box

transient voltage surge suppressor (TVSS) (receptacle or strip) an electrical device designed to protect sensitive electronic circuit boards from voltage surges

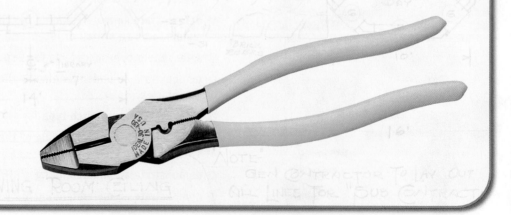

Selecting the Appropriate Receptacle Type

hapter 2 presented an overview of **receptacle** devices. It is appropriate at this time for you to review that information. During the rough-in stage, the electrical boxes and a wiring method with the required number of conductors are installed at each receptacle outlet location. Surface-mounted receptacles, like those sometimes used for an electric range or electric clothes dryer, will not need separate electrical boxes. During the trim-out stage, an electrician selects the proper receptacle type for a specific location and installs it. The installation for regular flush-mount wall receptacles will consist of making the proper electrical connections to the terminal screws on the receptacle and then securely mounting the device to the electrical box. A plastic or metal cover is then attached to the receptacle, which finishes off the installation. A surface-mounted receptacle installation involves attaching the receptacle base to a wall or floor surface with screws, making the proper terminations to the terminal screws (or lugs), and securing the surface mount cover on the receptacle to complete the installation. A short review of receptacle devices is presented in the following paragraphs.

A receptacle is an electrical device that allows the home owner to access the electrical system with cord-and-plug-connected equipment, such as lamps, computers, kitchen appliances, and tools. The *National Electrical Code®* (*NEC®*) defines a receptacle as a contact device installed at an outlet for the connection of an attachment plug. A **single receptacle** is a single contact device with no other contact device on the same strap (yoke). A **duplex receptacle** has two contact devices on the same strap (yoke). A multiple receptacle has more than two contact devices on the same strap (yoke) (Figure 18–1). In residential applications, general-use receptacles are usually 125-volt, 15- or 20-amp devices and have specific slot configurations (Figure 18–2) and are found in regular and decorator styles (Figure 18–3). Standard colors available from the factory are brown, ivory,

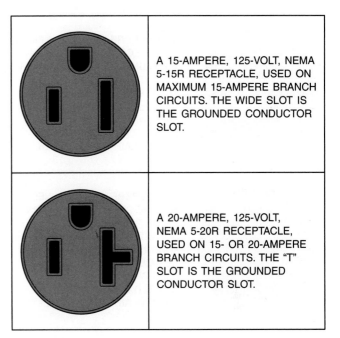

A 15-AMPERE, 125-VOLT, NEMA 5-15R RECEPTACLE, USED ON MAXIMUM 15-AMPERE BRANCH CIRCUITS. THE WIDE SLOT IS THE GROUNDED CONDUCTOR SLOT.

A 20-AMPERE, 125-VOLT, NEMA 5-20R RECEPTACLE, USED ON 15- OR 20-AMPERE BRANCH CIRCUITS. THE "T" SLOT IS THE GROUNDED CONDUCTOR SLOT.

Figure 18–2 125-volt, 15- or 20-amp devices have specific slot configurations.

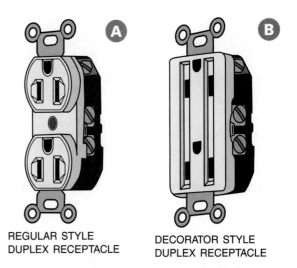

REGULAR STYLE DUPLEX RECEPTACLE

DECORATOR STYLE DUPLEX RECEPTACLE

Figure 18–3 Regular (A) and decorator-style (B) duplex receptacles.

and white, but receptacles in other colors are available by special order. For appliances that require just 240 volts, like a room air conditioner, receptacles are made with special blade configurations to meet the requirements of the appliance (Figure 18–4). For appliances that require 120/240 volts, such as electric clothes dryers and electric ranges, special blade configurations to meet the requirements of the appliance are also available (Figure 18–5).

General-use receptacles are made from high-strength plastic with a metal **strap (yoke)** running through it or

SINGLE RECEPTACLE

MULTIPLE RECEPTACLE (DUPLEX)

MULTIPLE RECEPTACLE (TRIPLEX)

Figure 18–1 Single, duplex, and triplex receptacle devices.

A 15-AMPERE, 250-VOLT, NEMA 6-15R RECEPTACLE, USED ON MAXIMUM 15-AMPERE BRANCH CIRCUITS. NOTE THE SLOT ARRANGEMENT SO STANDARD 125-VOLT ATTACHMENT PLUG CAPS WILL NOT FIT; COULD BE USED FOR A 240-VOLT WINDOW AIR CONDITIONER.

A 20-AMPERE, 250-VOLT, NEMA 6-20R RECEPTACLE, USED ON 15- OR 20-AMPERE BRANCH CIRCUITS. NOTE THE SLOT ARRANGEMENT SO STANDARD 125-VOLT ATTACHMENT PLUG CAPS WILL NOT FIT; COULD BE USED FOR A 240-VOLT WINDOW AIR CONDITIONER.

Figure 18–4 240-volt receptacles with special blade configurations.

behind it. The metal strap provides a means for the grounding blade of a male plug to be attached to the grounding system and for the receptacle to be attached to the device box. Section 406.3 of the *NEC®* requires each receptacle used in new construction to be grounded to the electrical system. This is accomplished by attaching a branch-circuit grounding conductor to a green grounding screw, which is connected to the receptacle strap.

Receptacles rated 125 volts and 15 or 20 amperes come with a long slot for the grounded plug blade of the attachment plug, a narrow slot for the ungrounded plug blade of the attachment plug, and a U-shaped slot for the attachment plug grounding prong (Figure 18–6). There are terminal screws on each side of the receptacle, brass colored for the ungrounded conductors and silver colored for the grounded (neutral) conductors. In between the screw terminals of a 125-volt duplex receptacle is a removable tab used for **split-wired receptacles.** There is also a green grounding screw on each receptacle for grounding.

Many receptacles also have holes in the back, which are used for "quick wiring" or "back wiring." Back wiring can be done in one of two ways: screw connected or push-in. Screw

50-AMPERE
3-POLE, 3-WIRE
125/250-VOLT
NEMA 10-50R
PERMITTED
PRIOR TO 1996 *NEC®*

50-AMPERE
4-POLE, 4-WIRE
125/250-VOLT
NEMA 14-50R
REQUIRED BY 1996 *NEC®*
FOR NEW INSTALLATIONS

30-AMPERE
3-POLE, 3-WIRE
125/250-VOLT
NEMA 10-30R
PERMITTED
PRIOR TO 1996 *NEC®*

30-AMPERE
4-POLE, 4-WIRE
125/250-VOLT
NEMA 14-30R
REQUIRED BY 1996 *NEC®*
FOR NEW INSTALLATIONS

Figure 18–5 120/240-volt range (50-amp) and dryer (30-amp) receptacles have special blade configurations to meet the requirements of the appliance.

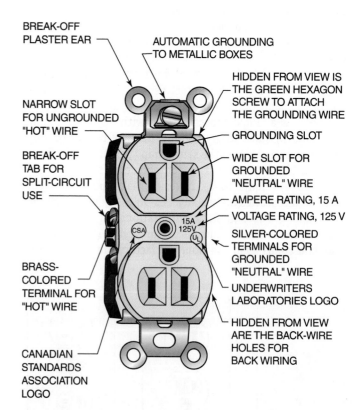

BREAK-OFF PLASTER EAR

AUTOMATIC GROUNDING TO METALLIC BOXES

HIDDEN FROM VIEW IS THE GREEN HEXAGON SCREW TO ATTACH THE GROUNDING WIRE

NARROW SLOT FOR UNGROUNDED "HOT" WIRE

GROUNDING SLOT

BREAK-OFF TAB FOR SPLIT-CIRCUIT USE

WIDE SLOT FOR GROUNDED "NEUTRAL" WIRE

AMPERE RATING, 15 A

VOLTAGE RATING, 125 V

SILVER-COLORED TERMINALS FOR GROUNDED "NEUTRAL" WIRE

BRASS-COLORED TERMINAL FOR "HOT" WIRE

UNDERWRITERS LABORATORIES LOGO

HIDDEN FROM VIEW ARE THE BACK-WIRE HOLES FOR BACK WIRING

CANADIAN STANDARDS ASSOCIATION LOGO

Figure 18–6 125-volt, 15- or 20-amp receptacles used in residential wiring come with a long slot for the grounded plug blade of the attachment plug, a short slot for the ungrounded plug blade of the attachment plug, and a U-shaped slot for the attachment plug grounding prong.

Some wiring situations call for a split receptacle. This wiring technique will allow the bottom half of a duplex receptacle to be controlled by a switch, and the top half will be "hot" at all times. A load like a lamp can be plugged into the bottom half of the receptacle and be controlled by a single-pole switch. An electrician creates a split-wired receptacle by removing the connecting tab between the brass terminal screws on the "hot" side of a duplex receptacle (Figure 18–7).

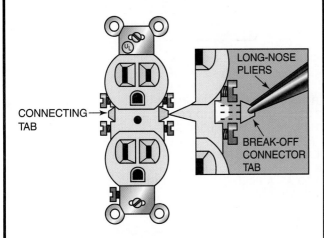

CONNECTING TAB

LONG-NOSE PLIERS

BREAK-OFF CONNECTOR TAB

Figure 18–7 An electrician creates a split-wired receptacle by removing the connecting tab between the brass terminal screws on the "hot" side of a duplex receptacle.

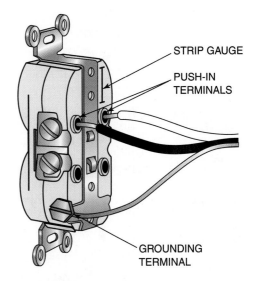

STRIP GAUGE

PUSH-IN TERMINALS

GROUNDING TERMINAL

Figure 18–8 A duplex receptacle may be back wired by using the "push-in terminals" on the back of the device.

CAUTION

CAUTION: Conductors attached to a receptacle using the screwless push-in method often come loose when the receptacle is being secured to the electrical outlet box. For this reason, many electrical contractors will not allow their employees to use the screwless push-in connections for receptacles or switches.

connected is by far the most recommended method. This connection feature is found on higher-quality (and higher-priced) receptacles. However, the time saved in making this type of connection can more than pay for the extra cost of the receptacle. In this method, the conductor is stripped back according to a strip gauge located on the receptacle and inserted into the proper hole. The terminal screw is then tightened, securing the conductor to the receptacle. In the push-in method, the wire is inserted into the proper hole (called a "push-in terminal") and held in place by a thin copper strip that pushes against the conductor (Figure 18–8). The drawback to this method is that as current flows through the copper strip, the strip can weaken, allowing the conductor to pull free. This can cause open circuits, short circuits, ground faults, or fires. The push-in method is UL listed for use with only 14 AWG solid copper conductors. Aluminum conductors, stranded conductors, or conductors sized 12 AWG and larger may not be used. The holes on the back of push-in receptacles are sized so that 12 AWG conductors will not fit into them.

In most cases, the amperage rating of a receptacle must match or be higher than the amperage rating of the circuit. This is true for 15-, 30-, 40-, and 50-amp receptacles. On 20-amp circuits, either 20-amp single receptacles, 20-amp duplex receptacles, or 15-amp duplex receptacles may be used; 15-amp-rated single receptacles may not be used on a 20-amp-rated circuit, but 20-amp-rated receptacles may be used on 15-amp circuits.

Ground Fault Circuit Interrupter Receptacles

Ground fault circuit interrupter receptacles (Figure 18–9) were developed to protect individuals from electric shock hazards. GFCI receptacles protect the locations where they are installed and can also protect regular duplex receptacles installed farther down the circuit. Ground-fault protection is also available with a GFCI circuit breaker. The breaker provides ground-fault protection for the entire circuit. GFCI breakers are covered in

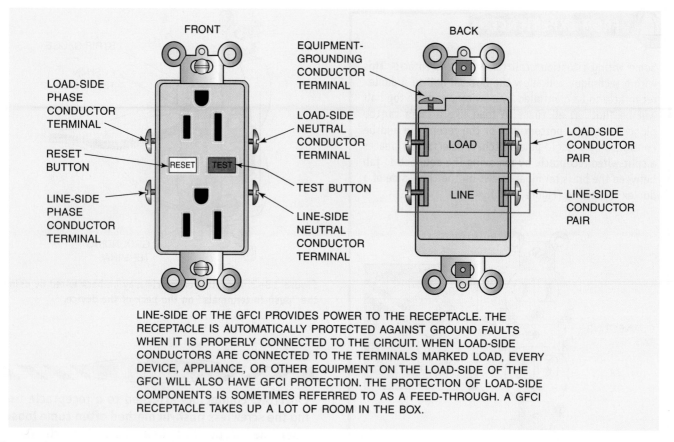

FRONT   BACK

LOAD-SIDE PHASE CONDUCTOR TERMINAL

EQUIPMENT-GROUNDING CONDUCTOR TERMINAL

LOAD-SIDE NEUTRAL CONDUCTOR TERMINAL

RESET BUTTON

LINE-SIDE PHASE CONDUCTOR TERMINAL

TEST BUTTON

LINE-SIDE NEUTRAL CONDUCTOR TERMINAL

LOAD

LINE

LOAD-SIDE CONDUCTOR PAIR

LINE-SIDE CONDUCTOR PAIR

LINE-SIDE OF THE GFCI PROVIDES POWER TO THE RECEPTACLE. THE RECEPTACLE IS AUTOMATICALLY PROTECTED AGAINST GROUND FAULTS WHEN IT IS PROPERLY CONNECTED TO THE CIRCUIT. WHEN LOAD-SIDE CONDUCTORS ARE CONNECTED TO THE TERMINALS MARKED LOAD, EVERY DEVICE, APPLIANCE, OR OTHER EQUIPMENT ON THE LOAD-SIDE OF THE GFCI WILL ALSO HAVE GFCI PROTECTION. THE PROTECTION OF LOAD-SIDE COMPONENTS IS SOMETIMES REFERRED TO AS A FEED-THROUGH. A GFCI RECEPTACLE TAKES UP A LOT OF ROOM IN THE BOX.

Figure 18–9 Ground fault circuit interrupter receptacles.

greater detail in Chapter 19. Shock hazards exist because of wear and tear on conductor insulation, defective materials, or misuse of equipment. This can result in contact with "hot" conductors or the metal parts of equipment that are in contact with "hot" conductors. Now would be a good time to review the shock hazards discussed in Chapter 1. Remember that the effect of the electric shock on the body is determined by the amount of current flowing through the body, the path the current takes through the body, and the length of time the current flows through the body.

A GFCI operates by monitoring the current flow on both the "hot" conductor and the grounded conductor. Whenever the GFCI senses a current on the "hot" conductor that is 4 to 6 milliamps greater than the current flowing on the grounded conductor, it trips the circuit off. This imbalance means that some of the current is taking a path other than the normal return path on the grounded conductor (Figure 18–10).

A GFCI is designed to protect against ground faults only. It does not provide protection against short circuits, which occur when two "hot" conductors touch each other or when a "hot" conductor comes in contact with a grounded conductor. The circuit overcurrent protection device (fuse or circuit breaker) provides this type of protection. A GFCI will not provide overcurrent protection for the circuit.

CAUTION

CAUTION: A GFCI will not protect against a person touching both a "hot" conductor and a grounded conductor at the same time. When this occurs, the current flowing through the conductors is the same, so there is no current imbalance for the GFCI to detect.

Ground-fault protection is required in many areas, both inside and outside the house. These requirements are listed in Section 210.8 of the *NEC®* and were covered in Chapter 14. Now is a good time to review these requirements.

There are several factors to consider when purchasing a GFCI receptacle:

- Look for a lockout feature that does not allow the GFCI to be reset if it is not functioning properly.
- Make sure the GFCI cannot be reset if the line and load connections are reversed.
- Some GFCI receptacles have an indicator light that will glow when the test button is pushed and the connections are wired properly. This same light will also glow when the GFCI has been tripped.

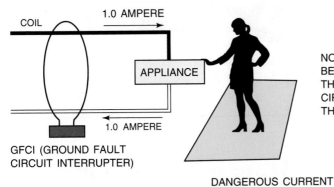

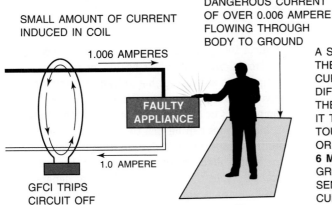

NO CURRENT IS INDUCED IN THE COIL BECAUSE BOTH WIRES ARE CARRYING THE SAME CURRENT. THE GROUND FAULT CIRCUIT INTERRUPTER DOES NOT TRIP THE CIRCUIT OFF.

A SMALL AMOUNT OF CURRENT IS INDUCED IN THE COIL BECAUSE OF THE UNBALANCE OF CURRENT IN THE CONDUCTORS. THIS CURRENT DIFFERENCE IS AMPLIFIED SUFFICIENTLY BY THE GROUND FAULT INTERRUPTER TO CAUSE IT TO TRIP THE CIRCUIT OFF BEFORE THE PERSON TOUCHING THE FAULTY APPLIANCE IS INJURED OR KILLED. **NOTE: CURRENT VALUES ABOVE 6 MILLIAMPERES ARE CONSIDERED DANGEROUS.** GROUND FAULT CIRCUIT INTERRUPTERS MUST SENSE AND OPERATE WHEN THE GROUND CURRENT EXCEEDS 6 MILLIAMPERES.

Figure 18–10 GFCI operation.

Because of the electronic nature of GFCI circuits, they should be tested at least once a month. Testing is done by simply pushing the "Push to Test" button on the face of the device. If the GFCI fails to trip, install a new GFCI device. If the GFCI trips, push the "Reset" button, and the GFCI will be reset and working properly. This test is to ensure that the tripping circuit and mechanism will operate properly if a ground fault does occur. GFCI receptacles should also be checked often in areas prone to lightning strikes. GFCI circuitry may be damaged by voltage surges from nearby lightning strikes.

Arc Fault Circuit Interrupter Receptacles

According to Section 210.12 of the 2002 *NEC®*, *all* 120-volt, 15- and 20-amp outlets in residential bedrooms must be **arc fault circuit interrupter (AFCI)** protected. This comes as the result of a series of studies by the Consumer Product Safety Commission that indicated that a high percentage of the 150,000 residential electrical fires that occur annually in the United States start in bedrooms.

Arcing in residential wiring applications typically occurs for two reasons. The first is an overloaded extension cord or a broken wire in an extension cord or power cord. The over-

load condition can cause the insulation to break down, creating conditions where arcing between the conductors can occur. When a conductor is broken, the ends can be close enough that the current can bridge the gap by arcing. In either case, the resulting arc can be at a low enough level that it would not be detected by the circuit overload device. The arc will, however, be able to ignite flammable materials that it contacts.

The second reason for arcing occurs when a plug is knocked out of a receptacle when furniture or other similar objects are being moved. As the plug blades are pulled from the receptacle slot, arcing occurs. The more load on the circuit, the greater the arc. Again, this is a low-level arcing that will not necessarily trip the overcurrent device but can cause a fire.

AFCI receptacles (Figure 18–11) are designed to trip when they sense rapid fluctuations in the current flow that is typical of arcing conditions. They are set up to recognize the "signature" of dangerous arcs and trip the circuit off when one occurs. AFCIs can distinguish between dangerous arcs and the operational arcs that occur when a plug is inserted or removed from a receptacle or a switch is turned on or off.

AFCIs come in two styles to meet the requirements of Section 210.12: circuit breakers and receptacles. An AFCI circuit breaker is installed in the service panel to protect the entire circuit. The breaker also provides some protection to extension cords plugged into circuit receptacles. AFCI circuit

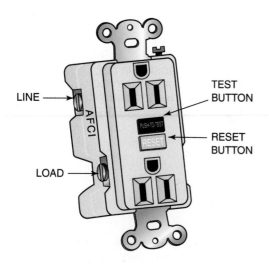

LINE

AFCI

LOAD

TEST BUTTON

RESET BUTTON

Figure 18–11 An AFCI receptacle.

FROM EXPERIENCE

Several manufacturers are working on units that combine GFCI and AFCI protection. These combination devices may be available by the time this book is published.

Transient Voltage Surge Supressors

Many of the appliances used in homes today contain sensitive electronic circuitry, circuit boards, or microprocessors. These components are susceptible to transient voltages, also known as voltage surges or voltage spikes. Voltage spikes can destroy components or cause microprocessors to malfunction.

Line surges can be line to line, line to neutral, or line to ground. Transient voltages are grouped into two categories. The first, called an **impulse,** is when the transient voltage starts outside the residence. They are generally caused by power company equipment. The second category is the **ring wave.** This line surge begins inside the residence and is caused by inductive loads, such as the spark igniters on gas clothes dryers, gas ranges, or gas water heaters, electric motors, and computers.

Transient voltage surge suppressors (TVSS) are used to minimize voltage surges and spikes. A TVSS device works by clamping the high-surge voltage to a level that a piece of equipment like a computer can withstand (Figure 18–12). A TVSS reacts to a voltage surge in less than 1 nanosecond (one-billionth of a second) and permits only a safe amount of energy to enter the connected load.

breakers are covered in greater detail in Chapter 19. AFCI receptacles provide maximum protection to extension and power cords that are plugged into them. Many of the AFCI receptacles are feed-through receptacles that, when properly wired, can protect both light fixtures and receptacles that are farther down the circuit. To meet *NEC®* requirements that the entire branch circuit be protected, some AFCI receptacles can sense arcing that occurs on the conductors between the receptacle and the circuit panel as well as arcing that occurs downstream from the receptacle.

At present, AFCI protection is required for new bedroom installations only. Be sure to check with local code enforcement officials, as several areas are now requiring AFCI protection for all living areas in a residence. As the cost of AFCI protection decreases and its reliability is proven, look for AFCI protection to be required in more areas of the residence.

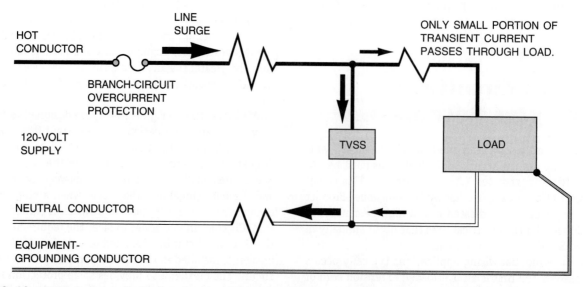

HOT CONDUCTOR

LINE SURGE

ONLY SMALL PORTION OF TRANSIENT CURRENT PASSES THROUGH LOAD.

BRANCH-CIRCUIT OVERCURRENT PROTECTION

120-VOLT SUPPLY

TVSS

LOAD

NEUTRAL CONDUCTOR

EQUIPMENT-GROUNDING CONDUCTOR

Figure 18–12 A TVSS device works by clamping the high-surge voltage to a level that a piece of equipment like a computer can withstand.

TVSS devices are available in several different styles. One style is a device that looks very much like a GFCI receptacle and mounts in a device box (Figure 18–13). Class A surge suppressors provide branch-circuit protection for all receptacles on the circuit, just like a feed-through GFCI receptacle. Another surge suppressor style is one unit that protects the entire house for transient voltages. This unit is hard-wired directly into the main electrical panel.

The type of surge suppressor you are probably most familiar with is the plug-in strip. (Figure 18–14). These are most often used to provide surge protection to computer equipment and audiovisual equipment. Several electrical components can be protected by one surge strip. The better strips will also provide filtering of the electricity being delivered to sensitive electronic equipment. Filtering will help eliminate electrical "noise" that may be present on the incoming electrical conductors.

Figure 18–13 A receptacle model TVSS device.

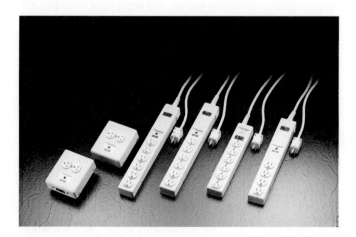

Figure 18–14 Plug-in TVSS strips. The strips allow multiple loads to be protected from voltage surges at the same time. *Courtesy of Hubbell, Incorporated.*

Installing Receptacles

The amount of work required to install the receptacles is directly proportional to the amount of work that is put into the rough-in phase of the installation process. The more circuit testing and wire marking that occurs during the rough-in stage, the easier receptacle installation will be during the trim-out stage. Circuits should always be checked before the sheetrock is installed and finished. Circuit tracing and, if necessary, cable or raceway replacement is much easier during the rough-in stage.

Before the actual installation begins, check to make sure that no receptacles are buried under the sheetrock. Drywall installers are very conscientious about exposing all the receptacles, but on occasion they will miss one. Marking receptacle and switch locations on the floor and having an accurate circuit layout will make finding buried receptacle boxes much easier. Once a buried box location is determined, a keyhole saw is used to carefully cut the drywall to expose the device box.

Pigtails

The listing instructions for receptacles and switches normally allow only one wire to be terminated to each terminal screw. However, many electrical boxes containing receptacles (or switches) will have many circuit conductors requiring connections to the device terminal screws. The best way to make the necessary connections so that only one conductor gets connected to a terminal screw is to use a "pigtail." Figure 18–15 shows the correct method for connecting circuit conductors to a device with a pigtail splice. The usual method of making pigtail splices in an electrical box involves using a

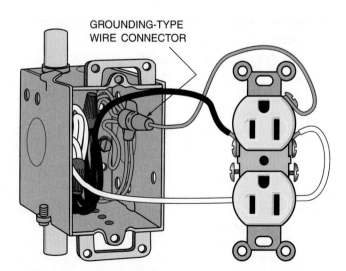

Figure 18–15 This connection shows the use of a white and a black pigtail. Notice the use of a green grounding wirenut used to provide a grounding pigtail that is terminated to the green grounding screw on the receptacle.

wirenut to connect the necessary wires together. Wirenuts were first introduced in Chapter 2. Other connector types were also discussed. Conductor splices made by following the proper installation procedures for wirenuts will result in a device installation that is safe and trouble free.

(See page 510 for the correct procedure for connecting wires together with a wirenut.)

Terminal Loops

The preferred method for connecting circuit conductors to a receptacle (or switch) is to form terminal loops in the wire, put the loop under a terminal screw, and tighten the screw the proper amount. This sounds like a simple wiring practice, but a terminal screw that is not tightened properly or a wire that is not looped properly around a screw will typically be the cause of future problems. An electrician must practice good workmanship when connecting wires to any device so that the installation will be trouble free.

(See page 511 for the correct procedure for using terminal loops to connect circuit conductors to terminal screws on a receptacle or switch.)

Receptacle Installation

The first step in connecting a receptacle to the branch-circuit wiring is to locate the proper conductors and pull them out from the device box. Once all the circuit conductors have been properly identified, the conductors are ready to be connected to the receptacle. For regular duplex receptacles, there should be three wires: an ungrounded black conductor, a grounded white conductor, and a bare grounding conductor. Remember that the hot conductor is connected to the brass-colored terminal, the grounded conductor is connected to the silver screw terminal, and the grounding conductor is connected to the green grounding screw.

(See page 512 for the correct procedure for installing duplex receptacles in a nonmetallic electrical outlet box.)

(See page 514 for the correct procedure for installing duplex receptacles in a metal electrical outlet box.)

A split-wired receptacle has a switch controlling half the receptacle, and the other half is "hot" all the time (Figure 18–16). This method of wiring is generally used when a lamp is providing the general room lighting instead of an overhead light fixture. Before installing this type of receptacle, the tab connecting the terminal screws on the "hot" side of the receptacle must be removed. Do not remove the tab connecting the silver terminal screws. The connections for this wiring situation were also presented in Chapter 13 (see Figure 13–18).

GFCI and AFCI receptacles are connected to the electrical system in much the same manner as with regular duplex receptacles. However, on the back of the GFCI and AFCI receptacles, one of the brass screw terminals and one of the silver screw terminals are marked for the "Line," or incoming power conductors. The other set of screw terminals are

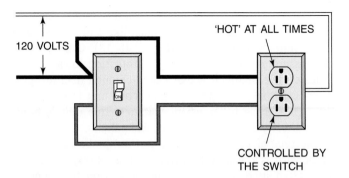

Figure 18–16 A split-wired duplex receptacle has a switch controlling the bottom half of the receptacle. The other half is "hot" at all times.

marked as "Load" terminals, or the outgoing power conductors. On new GFCI and AFCI receptacles, the "Load" terminals are covered with a yellow tape. If there is only one cable in the device box, the connection is quite simple. The grounding conductor is connected to the green grounding screw. The white grounded conductor is connected to the silver line terminal screw, and the black ungrounded conductor is connected to the brass line terminal screw. The conductors can either be wrapped around the screw terminals or back wired as previously discussed.

If the GFCI or AFCI is a feed-through receptacle, the connection will be slightly different. The grounding conductors need to be connected together with a grounding pigtail, and the pigtail will be connected to the green grounding screw. The conductors from the cable supplying power to the receptacle will be connected to the "Line" terminals, while the cable running to the receptacles downstream from the receptacle will be connected to the "Load" terminals. Remember, the "Load" terminals will have the yellow tape over them that will need to be removed. Now all the receptacles connected to the load side of the receptacle will be protected by the feed-through GFCI or AFCI receptacle. If a ground fault or higher level arc occurs, the receptacle will trip off, stopping the current flow to the receptacle and all the other receptacles connected to it.

(See page 515 for the correct procedure for installing feed-through GFCI and AFCI duplex receptacles in nonmetallic electrical outlet boxes.)

Once the particular receptacle you are installing is connected to the electrical system, it is time to secure it to the device box. Make sure the conductors inside the box are pushed to the back of the device box, leaving enough room to install the receptacle. Next, position the receptacle so the grounding prong is on the top (Figure 18–17). While this is not an *NEC®* requirement, it is the recommended positioning of a receptacle. This will lessen the chances of a piece of metal coming in contact with both the "hot" and grounded conductors, causing arcing and a short circuit. Carefully push the receptacle into the device box, checking that the ears on

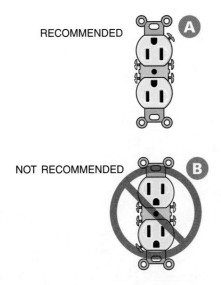

RECOMMENDED

NOT RECOMMENDED

Figure 18–17 Position the receptacle so the grounding slot is on the top (A) Although the *NEC*® does not prohibit the grounding slot from being on the bottom (B), this practice is not recommended.

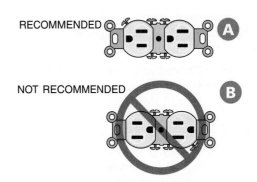

RECOMMENDED

NOT RECOMMENDED

Figure 18–18 When a receptacle is installed in a horizontal position (A), the grounded slot should be on the top. Although the *NEC*® does not prohibit the grounded slot from being on the bottom (B), this practice is not recommended.

the top and bottom of the receptacle yoke will rest against the sheetrock when the receptacle is installed. This is a requirement of Section 406.4 of the *NEC*®. Once you have determined that the receptacle ears will rest against the drywall properly, the receptacle can be attached to the device box using the provided 6-32 machine screws.

On occasion, a receptacle will need to be installed in the horizontal position. In this situation, the recommended receptacle position is to have the grounded conductor slot, the long slot, on the top (Figure 18–18). Again, this is to help keep metal objects away from the "hot" conductor plug blade.

If the receptacle ears do not rest against the drywall, use spacers to properly position the receptacle. Spacers are needed when the device box was installed so that the front

FROM EXPERIENCE

There are many common items used as spacers by electricians. For example, some electricians use a length of bare 14 AWG grounding conductor wrapped several times around the shank of a no. 2 Phillips screwdriver. The coils of wire can be cut off to provide the needed spacers. Many electricians use small flat washers, and some electricians use plastic tubing cut to the desired length. Still other electricians remove and save the mounting ears from receptacles and switches installed in previous installations that do not use the mounting ears.

FROM EXPERIENCE

Be careful not to strip out the threads in the device box mounting holes when installing the receptacles, especially when you are working with nonmetallic boxes. This creates many problems trying to get the receptacle to mount securely to the device box. When installing the receptacle, turn it approximately 30 degrees in both directions and push it into the box. Pull it back out, align the receptacle with the mounting holes, and then install the receptacle. This takes all the pressure off the threads while you tighten the screws. If you do encounter a stripped mounting hole, use your triple-tap tool described in Chapter 3 to rethread the hole.

edge of the box is not perfectly flush with the wall covering. There are several commercial products available for this application. It is important that the receptacle is positioned properly with respect to the drywall. This will allow the receptacle cover to fit well over the receptacle.

Now the receptacle cover, or plate, can be installed. Receptacle covers come in two types. The most common is the regular receptacle cover. This cover comes in the same colors that receptacles do. It is important to purchase the covers from the same company as the receptacles to ensure a proper color match. The second cover has a rectangular hole in it to fit GFCI, AFCI, TVSS, and decorator-style receptacles. In most cases, the cover is provided with the GFCI, the AFCI, and the TVSS receptacles. For the decorator-style receptacles and regular receptacles, the covers are purchased separately.

The cover is attached to a regular receptacle with a single screw in the center of the cover plate. For cover plates with rectangular openings, two screws are needed. It is important to make sure the receptacle is flush with the wall and straight. This allows the cover to be straight and flush with the receptacle. The device and its cover are the part of the installation the home owner and everyone who enters the

FROM EXPERIENCE

Keeping receptacles and covers straight is not a difficult task. When installing receptacles, push them as far right or left as you can before tightening the screws. This will align the receptacles in a vertical position. When installing the cover, use a torpedo level to help you keep it straight. When you are done, step back about 6 or 8 feet and look at it. If the cover plate looks crooked, re-align it until it looks straight.

house will see. The quality of this part of the installation will determine, to a large extent, the impression that others will have concerning your work. Another nice touch is to align the cover screw slots to face in the same direction.

When connecting a receptacle to a multiwire circuit, it is important that the conductors be pigtailed when there are two cables in the device box. Section 300.13 (B) of the *NEC®* requires that the continuity of the grounded conductor not depend on device connections to receptacles, lamp holders, and so on. This means that you cannot connect one of the grounded conductors to one of the silver terminal screws and the second grounded conductor to the other silver terminal screw (Figure 18–19). If a problem developed with the receptacle, the "hot" conductor could be unaffected while the grounded conductor would be open. Any receptacle downstream would have power to the "hot" side of the receptacle but there would be no return path through the electrical system. This would create a hazard for anyone using any of those receptacles. This is one reason it is a very good wiring practice to pigtail all conductors whenever possible. In this way, you ensure a complete circuit no matter what happens to one receptacle.

Selecting the Appropriate Switch

The primary factor in selecting the appropriate switch is having the proper switch, or combination of switches, control the desired luminaire (light fixture) or receptacle. Chapter 2 presented a very detailed look at the different switch types used in residential wiring. It also showed how each switch actually works. Chapter 13 presented several different switching circuits and how the various switch types are connected into those circuits. A review of the information presented in Chapters 2 and 13 is appropriate at this time.

Single-pole, three-way, and four-way switches are the main switch types used in residential wiring. Single-pole switches are used to control one or more light fixtures or receptacles from a single location. Three-way switches are used

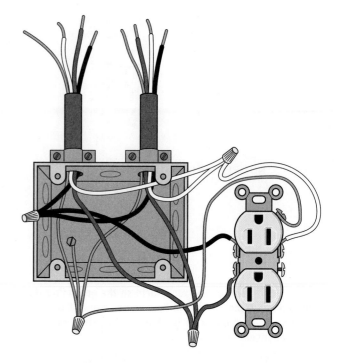

A PROPER WAY TO CONNECT GROUNDED NEUTRAL CONDUCTORS IN A MULTIWIRE BRANCH CIRCUIT.

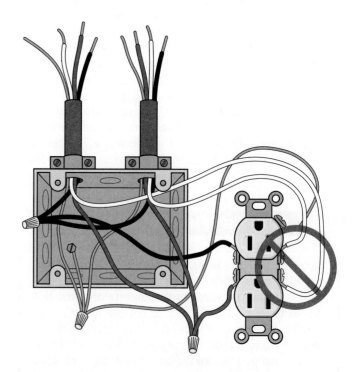

B IMPROPER WAY TO CONNECT GROUNDED NEUTRAL CONDUCTORS IN A MULTIWIRE BRANCH CIRCUIT. NOT PERMITTED.

Figure 18–19 Connecting the grounded neutral conductors to a split-wired receptacle on a multiwire branch circuit.

FROM EXPERIENCE

When choosing the color for switches or receptacles, be sure to purchase all the devices and their covers from the same manufacturer. While many companies may list the color you want, the shade of that color may vary. Ivory is the color with the most variation from different manufacturers. It is very difficult to obtain a professional finish if the colors of the devices and covers do not match.

FROM EXPERIENCE

It is extremely important to read the installation instructions before you begin installing switches. On one house wiring project, the home owner provided the three-way dimmer switches. We could not get one set to operate correctly until we read the instructions. To our surprise, this particular type of three-way dimmer required a dimmer switch be placed in *both* locations. As soon as we installed the second dimmer, the circuit worked fine.

in pairs to control one or more light fixtures or receptacles from two locations. Four-way switches are used with three-way switches to control one or more light fixtures from three or more locations. Both single-pole and three-way switches are available as dimmer switches. Double-pole switches are used to control 240-volt loads, such as electric water heaters. The last type of switch you may have to use in a residence is a low-voltage switch, such as the type needed for controlling a door chime. Residential switches are generally rated at 15 amps, 120/277 volts. Higher-ampere-rated switches are available. They can be wired to screw terminals or may be back wired. Remember that push-in back wiring is not a recommended method of installation.

Once the appropriate switch has been determined for a particular location on the wiring system, the color and style of the switch must be decided. Switch color is usually determined by the room color and varies throughout the house. Ivory and white are the two most popular colors installed in residences. Brown is the color that is preferred in rooms with darker-colored walls. The two styles of switches used most are

the toggle switch and the rocker switch (Figure 18–20). Toggle switches are the standard switch used. Rocker switches present a cleaner line, and some people find them easier to operate than toggle switches. Toggle switches are available as snap switches or quiet switches. The designation refers to the sound the switch makes when it is operated. Snap switches have a definite "snap" sound when operated. Most home owners prefer the quiet switch. Switch manufacturers offer a complete line of switches in a wide variety of colors.

Installing Switches

Installing switches is very similar to installing receptacles. The more effort put into the rough-in phase, the easier the installation process will be. If the conductors were properly marked and grouped after the circuit was tested during the rough-in stage, connecting and installing the switches will be very easy. If, however, the conductors are simply pushed back into the box as a group, then the installation process will be much more time consuming because many of the conductors in multigang boxes will have to be traced with a continuity tester to determine the proper switch connections. Connecting the conductors to the switch is done basically in the same manner as connecting conductors to receptacles. The insulation is stripped from the conductor, and a loop is made in the end of the conductor. The loop is placed around the proper terminal in a clockwise manner and the terminal screw tightened. If the switch is back wired, the insulation is stripped from the conductor, the end of the conductor is inserted into the proper hole, and the terminal screw is tightened.

For single-pole switches, two conductors (usually black) will be connected to the switch. One is the incoming power wire, and the other wire runs to the light fixture or receptacle. Be sure to ground the switch as per Section 404.9(B) (see Figure 13–8). Also be sure the switch is set up so it will read "Off" when the toggle is in the down position. If the two conductors to be connected are black and white, you will be

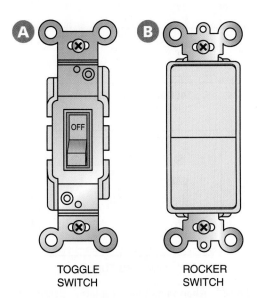

Figure 18–20 The two styles of switches used in residential wiring are the toggle switch (A) and the rocker switch (B).

TOGGLE SWITCH

ROCKER SWITCH

using a connection called a "switch loop" (see Figure 13–9). On this connection, the white wire is used as an ungrounded conductor and must be identified as such. Section 200.7(C) requires that the white conductor be identified at both ends as a "hot" conductor. Most electricians use black electrical tape to mark the conductor.

For three-way switches, three conductors will be connected to the switch. The two traveler conductors are connected to the two brass-colored terminals, and the conductor that provides power or the conductor that takes power to the light fixture is connected to the third, dark-colored terminal. If the conductors are not marked, looking into the device box can be an easy way to identify the conductors. If the conductors come from two different cables, the black and red wires in the three-wire cable are the travelers. The black conductor in the two-wire cable will be connected to the common terminal. If all three conductors come from the same cable, a continuity tester is used to identify the conductors. Section 200.7(C) requires the white conductor to be reidentified if it is used as an ungrounded conductor. Remember to ground the switch as per Section 404.9(B) (see Figures 13–11, 13–12, and 13–13).

Four-way switches have four conductors connected to them. Two conductors will come from one three-wire cable, and two conductors will come from a second three-wire cable. The conductors that come from one cable will be connected to the two screw terminals that are the same color. The remaining two conductors will be connected to the two screw terminals that are a different color. Remember to properly ground the switch (see Figures 13–14 and 13–15).

Once the switches are connected to the electrical system, they are ready to install in the device box. Again, this procedure is very similar to the receptacle installation procedure. The main difference is the number of switches that are installed in multigang boxes. In residential installations, two- and three-gang switch boxes are common. In these installations, particular care must be taken to ensure there is enough room in the device box for all the conductors and switches. Installing a switch in the device box is easily done by tilting the switch and pushing it into the device box, pulling it back out and aligning the mounting holes, and then installing the 6-32 mounting screws and tightening them. Make sure the ears on the yoke of the switch are mounted flush with the drywall. Use the spacer method described in the receptacle installation section of this chapter to obtain the desired result.

Installing faceplates on switches is done in just about the same manner as installing cover plates on receptacles. However, instead of one screw, switch plates require two screws to attach them to a switch. Because there are so many multigang switch boxes, it is important to make sure all the switches are level so the faceplate will be level when it is installed. The mounting holes are large enough to allow enough movement to keep the switches level. Using the short, colored 6-32 screws that come with a faceplate, attach the faceplate to the switches and the wall surface. Do not overtighten the screws on a plastic faceplate, or you may crack it and have to replace it. Align the screw slots in one direction, usually vertical, to provide a more finished look (Figure 18–21).

If metal faceplates are installed, be sure to provide a means to ground them to the grounding system, as required in Section 404.9(B). The metal screws provided with the switch cover are an acceptable means of grounding these faceplates. If installing decorator switch covers, make sure the provided screws are metal and not nylon. Many manufacturers of decorator switch covers provide nylon screws to keep from damaging the faceplate finish. These screws do not meet the requirements of Section 404.9(B). You will need to install metal screws to properly ground these faceplates.

FROM EXPERIENCE

When installing switches in multigang device boxes, push all the switches either all the way to the left or all the way to the right, then tighten the mounting screws. This will space the switches so that the faceplate will fit the first time.

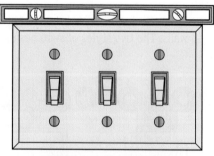

ALIGN SCREW SLOTS
LIKE THIS,

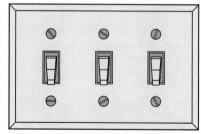

NOT LIKE THIS.

Figure 18–21 When installing faceplates, always align the screw slots in one direction, usually vertical, to provide a more finished look.

Summary

In this chapter, we have looked at how to install the many different types of switches and receptacles used in residential wiring. Grounding-type duplex receptacles are the primary means by which a home owner accesses the home electrical system. GFCI receptacles are specialized devices that help prevent people from receiving electrical shock. The *NEC*® requires that GFCIs be installed in specific locations, such as kitchens, bathrooms, garages, and unfinished basements. AFCIs are installed to control dangerous arc faults that can cause fires. AFCIs are required to be installed in all residential bedrooms. TVSSs are devices designed to protect sensitive electronic circuits found in many of today's appliances and audiovisual equipment from voltage surges.

Switches are devices used to control light fixtures from one, two, or more locations. Switches can also control all or part of receptacles. Dimming switches are used on incandescent and special fluorescent light fixtures to control the light output. Dimmers can be used to control light fixtures from one location or in circuits that control light fixtures from two or more locations.

When installing receptacles or switches, it is important that the conductors be securely attached to the device. If the side screw terminals are used, be sure the conductor loop is wrapped around the screw terminal in a clockwise direction and then properly tightened. If the device is back wired, make sure the conductor is inserted into the proper hole and then secured to the device. All devices installed in new construction are required by the *NEC*® to be grounded to the house grounding system.

The installation of devices, either receptacles or switches, and their cover plates are a very important part of the electrical wiring process. This part of the installation is the most visible part of the electrical system, and many people form their impression of the quality of your work by how well the switches, receptacles, and their covers look. Switches and receptacles installed so they are straight and flush with the wall surface, along with covers that are level and flush with the wall surface, demonstrate your commitment to quality workmanship.

Procedures

Connecting Wires Together with a Wirenut

- Wear safety glasses and observe all applicable safety rules.

- Using a wire stripper, remove approximately 3/4 inch of insulation from the end of the wires.

 Place the ends of the wires together and then place the wirenut over the ends of the conductors. If splicing two wires together, there is no need to twist the wires together before putting on the wirenut. If splicing three or more wires together, it is a good wiring practice to twist the wires together before putting on the wirenut.

 While exerting a slight inward pressure on the wirenut, twist the wirenut in a clockwise direction. Continue to twist the wirenut until it is tight on the wires.

 Test the splice by pulling on the conductors. If any of the wires come loose, redo the splice. The completed splice should be secure and have no bare wire showing below the bottom of the wirenut.

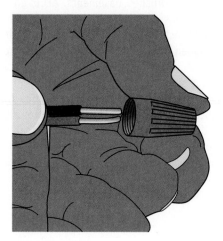

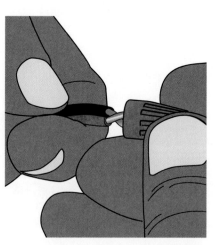

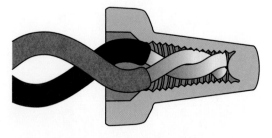

Procedures

Using Terminal Loops to Connect Circuit Conductors to Terminal Screws on a Receptacle or Switch

- Wear safety glasses and observe all applicable safety rules.

- Using a wire stripper, remove approximately 3/4 inch of insulation from the end of the wire.

Ⓐ Using long-nose pliers or the wire strippers, make a loop at the end of the wire.

Ⓑ Place the loop around the terminal screw so that the loop is going in the clockwise direction. This will cause the loop to close around the screw post. If the loop is put on so it is going in the counterclockwise direction, the loop will open up and actually become loose as the terminal screw is turned clockwise to tighten it.

Ⓒ Using a screwdriver, tighten the terminal screw until it is snug. Then tighten it approximately one-quarter turn more. Note: Electricians should use a torque screwdriver, set to the proper torque for the screw you are tightening, to tighten the terminal screws the proper amount.

Ⓐ

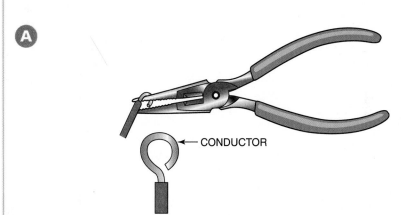

CONDUCTOR

Ⓑ

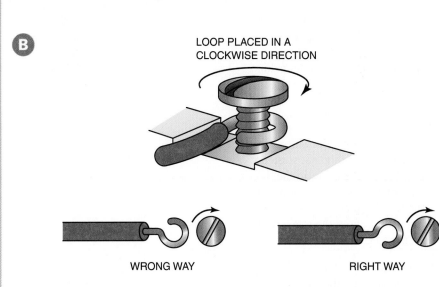

LOOP PLACED IN A CLOCKWISE DIRECTION

WRONG WAY   RIGHT WAY

Ⓒ

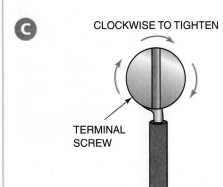

CLOCKWISE TO TIGHTEN

TERMINAL SCREW

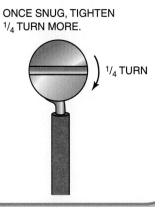

ONCE SNUG, TIGHTEN 1/4 TURN MORE.

1/4 TURN

Procedures

Installing Duplex Receptacles in a Nonmetallic Electrical Outlet Box

- Wear safety glasses and observe all applicable safety rules.

A Using a wire stripper, remove approximately 3/4 inch of insulation from the end of the insulated wires.

B Using long-nose pliers or wire strippers, make a loop at the end of each of the wires.

- Place the loop on the black wire around a brass terminal screw so that the loop is going in the clockwise direction. While pulling the loop snug around the screw terminal, tighten the screw to the proper amount with a screwdriver.

- Place the loop on the white wire around a silver terminal screw so that the loop is going in the clockwise direction. While pulling the loop snug around the screw terminal, tighten the screw to the proper amount with a screwdriver.

C Complete the installation by placing the loop on the bare grounding wire around the green terminal screw so that the loop is going in the clockwise direction. While pulling the loop snug around the screw terminal, tighten the screw to the proper amount with a screwdriver.

A

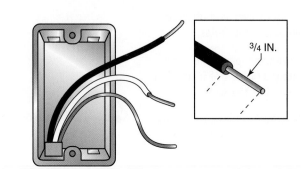

3/4 IN.

B

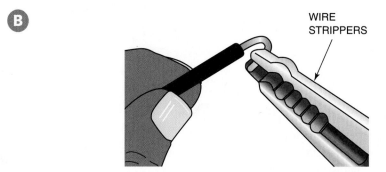

WIRE STRIPPERS

C

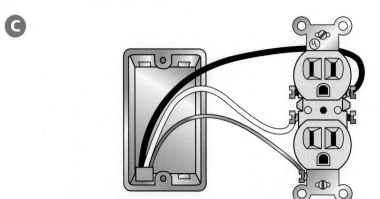

D Place the receptacle into the out-let box by carefully folding the conductors back into the device box.

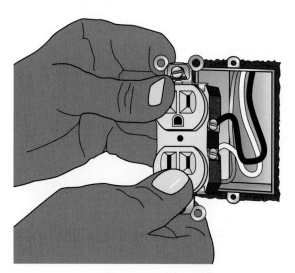

E Secure the receptacle to the de-vice box using the 6-32 screws. Mount the receptacle so it is verti-cally aligned.

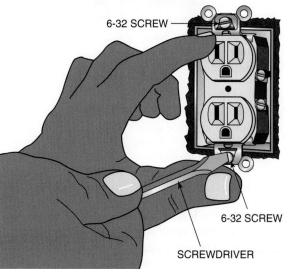

6-32 SCREW

6-32 SCREW

SCREWDRIVER

F Attach the receptacle cover plate to the receptacle. Be careful to not tighten the mounting screw(s) too much. Plastic faceplates tend to crack very easily.

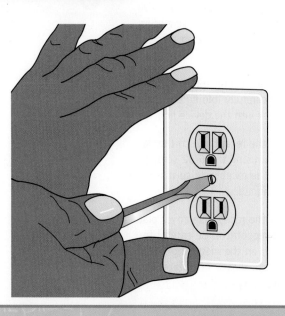

Procedures

Installing Duplex Receptacles in a Metal Electrical Outlet Box

- Wear safety glasses and observe all applicable safety rules.

A Attach a 6- to 8-inch-long grounding pigtail to the metal electrical outlet box with a 10-32 green grounding screw. The pigtail can be a bare or green insulated copper conductor.

B Attach another 6- to 8-inch-long grounding pigtail to the green screw on the receptacle.

C Using a wirenut, connect the branch-circuit grounding conductor(s), the grounding pigtail attached to the box, and the grounding pigtail attached to the receptacle together.

- Using a wire stripper, remove approximately 3/4 inch of insulation from the end of the insulated wires.

- Using long-nose pliers or wire strippers, make a loop at the end of each of the wires.

- Place the loop on the black wire around a brass terminal screw and the loop on the white wire around a silver terminal screw so that the loops are going in the clockwise direction. Tighten the screws to the proper amount with a screwdriver.

- Place the receptacle into the outlet box by carefully folding the conductors back into the device box.

- Secure the receptacle to the device box using the 6-32 screws. Mount the receptacle so it is vertically aligned.

- Attach the receptacle cover plate to the receptacle. Be careful to not tighten the mounting screw(s) too much. Plastic faceplates tend to crack very easily.

A

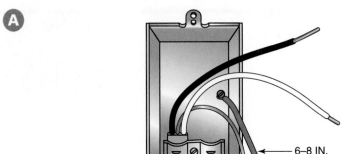

6–8 IN. PIGTAIL JUMPER

B

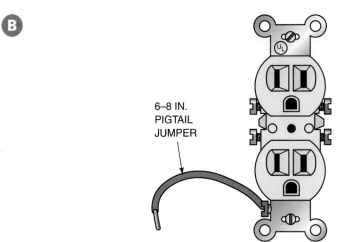

6–8 IN. PIGTAIL JUMPER

C

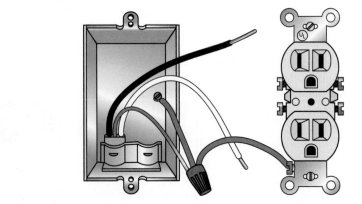

Procedures

Installing Feed-Through GFCI and AFCI Duplex Receptacles in Nonmetallic Electrical Outlet Boxes

- Wear safety glasses and observe all applicable safety rules.

A At the electrical outlet box containing the GFCI or AFCI feed-through receptacle, use a wirenut to connect the branch-circuit grounding conductors and a grounding pigtail together. Connect the grounding pigtail to the receptacle's green grounding screw.

- At the electrical outlet box containing the GFCI or AFCI feed-through receptacle, identify the incoming power conductors and connect the white grounded wire to the line-side silver screw and the incoming black ungrounded wire to the line-side brass screw.

- At the electrical outlet box containing the GFCI or AFCI feed-through receptacle, identify the outgoing conductors and connect the white grounded wire to the load-side silver screw and the outgoing black ungrounded wire to the load-side brass screw.

- Secure the GFCI or AFCI receptacle to the electrical box with the 6-32 screws provided by the manufacturer.

- A proper GFCI or AFCI cover is provided by the device manufacturer; attach it to the receptacle with the short 6-32 screws also provided.

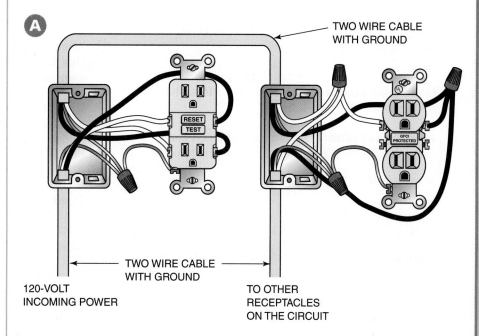

A

TWO WIRE CABLE WITH GROUND

RESET TEST

GFCI PROTECTED

TWO WIRE CABLE WITH GROUND

120-VOLT INCOMING POWER

TO OTHER RECEPTACLES ON THE CIRCUIT

- At the next "downstream" electrical outlet box containing a regular duplex receptacle, connect the white grounded wire(s) to the silver screw(s) and the black ungrounded wire(s) to the brass screw(s) in the usual way. Place a label on the receptacle that states "GFCI Protected." These labels are provided by the manufacturer.

- Continue to connect and label any other "downstream" duplex receptacles as outlined in the previous step.

Review Questions

Directions: Answer the following items with clear and complete responses.

1 Section _____ of the *NEC*® covers the required locations for GFCIs in residential applications.

2 The _____ conductor is connected to the silver screw terminal on a duplex receptacle.

3 _____ _____ _____ _____ are required by the *NEC*® for all residential bedroom electrical outlets. The *NEC*® section that states this is _____.

4 Which type of switch has four screw terminals and does not have "ON/OFF" marked on the switch toggle?

a. Three-way switch
b. Four-way switch
c. Double-pole switch
d. Single-pole switch

5 When a receptacle is installed horizontally in a device box, it is recommended that the _____ prong be on the top.

6 To GFCI protect receptacles that are downstream from a GFCI receptacle, the wiring going to those receptacles is connected to the _____ terminals on the GFCI receptacle.

7 A GFCI is designed to protect you if you touch both the "hot" and the grounded conductors of a circuit at the same time. True or False.

8 The two categories of transients that a TVSS device can protect sensitive electronic equipment from are:

1.
2.

9 A GFCI is designed to protect you if you come in contact with a circuit "hot" conductor and the metal frame of an appliance. True or False.

10 When installed in noncombustible materials, a device box cannot be set back from the wall surface more than _____.

11 When wiring a split-circuit receptacle, it is recommended that the part of the duplex receptacle controlled by a switch be the _____ half and the part that is "hot" at all times is the _____ half.

12 Three-way switches are used to control a luminaire (light fixture) from three locations. True or False.

13 Section _____ of the *NEC*® states that all switches must be effectively grounded, except when _____.

14 If nonmetallic device boxes are used, how are metal cover plates grounded?

15 Conductor reidentification, required when connecting a switch loop, is covered in Section _____ of the *NEC*®.

16 A GFCI will trip when it senses a current imbalance of _____ or more.

17 A 15-amp duplex receptacle may be installed on a 20-amp-rated circuit. True or False.

18 A 15-amp single receptacle may be installed on a 20-amp-rated branch circuit. True or False.

19 The metal frame of a switch or receptacle is called a(n) _____ or _____.

20 The push-in method of connecting conductors to a switch or receptacle is UL listed for use with only _____ solid copper conductors. Aluminum conductors, stranded conductors, or conductors sized _____ and larger may not be used.

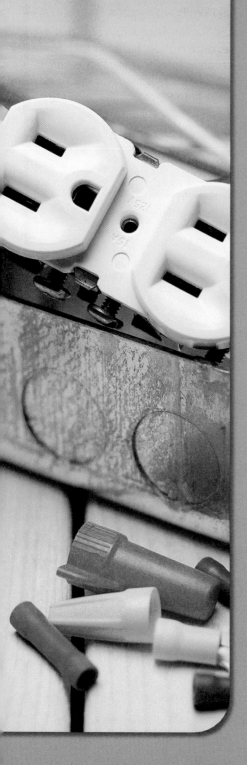

Chapter 19

Service Panel Trim-Out

An electrical panel serves as the origination point of the branch-circuit wiring. It contains the fuses or circuit breakers that provide overcurrent protection to the circuits that make up a residential wiring system. During the rough-in stage, the various cable or raceway wiring methods used to wire the system's branch circuits were installed. Enough conductor length is left at each electrical panel so that at some future time an electrician can connect the branch-circuit wiring to the overcurrent protection devices in the panel. Some electricians make these connections at the same time they rough-in the branch-circuit wiring, but most electricians make the connections during the trim-out stage. They refer to this process as "trimming out the panel."

In this chapter, we look at trimming out the panel. An overview of the types of overcurrent protection devices used in electrical panels will be presented. Both ground fault circuit interrupter (GFCI) circuit breakers and arc fault circuit interrupter (AFCI) circuit breakers are discussed. Since most electrical panels in today's residential electrical systems use circuit breakers, an overview of common methods of circuit breaker installation and panel trim-out procedures is also presented.

OBJECTIVES

Upon completion of this chapter, the student should be able to:

- ⊗ select the proper overcurrent protection device for a specific residential branch circuit.
- ⊗ demonstrate an understanding of common fuses and circuit breakers used in residential wiring.
- ⊗ demonstrate an understanding of installing circuit breakers or fuses in a panel.
- ⊗ demonstrate an understanding of the common techniques for trimming out a residential panel.

Glossary of Terms

arc fault circuit interrupter (AFCI) circuit breaker a device intended to provide protection from the effects of arc faults by recognizing characteristics unique to arcing and by functioning to deenergize the circuit when an arc fault is detected

cartridge fuse a fuse enclosed in an insulating tube that confines the arc when the fuse blows; this fuse may be either a ferrule or a blade type

Edison-base plug fuse a fuse type that uses the same standard screw base as an ordinary lightbulb; different fuse sizes are interchangeable

with each other; this fuse type may only be used as a replacement for an existing Edison-base plug fuse

interrupting rating the highest current at rated voltage that a device is intended to interrupt under standard test conditions

overcurrent protection device (OCPD) a fuse or circuit breaker used to protect an electrical circuit from an overload, a short circuit, or a ground fault

stab a term used to identify the location on a loadcenter's ungrounded bus bar where a circuit breaker is snapped on

Type S plug fuse a fuse type that uses different fuse bases for each fuse size; an adapter that matches each Type S fuse size is required; this fuse type is not interchangeable with other Type S fuse sizes

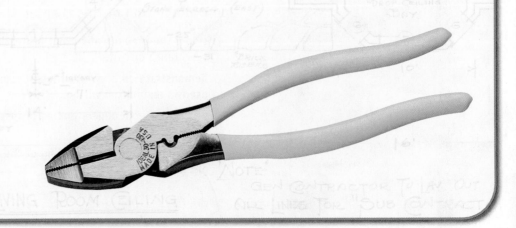

Understanding Residential Overcurrent Protection Devices

Chapter 2 presented a good overview of the different types of **overcurrent protection devices (OCPDs)** used in residential wiring. Both circuit breakers and fuses are used to provide overcurrent protection in residential wiring, but there is no question that between the two, circuit breakers are used the most. Circuit breakers are available as a single-pole device for 120-volt applications and as a two-pole device for 240-volt applications (Figure 19–1). Ground fault circuit interrupter (GFCI) and **arc fault circuit interrupter (AFCI)** circuit

breakers are also available (Figure 19–2). Fuses are available in two basic styles: plug and cartridge. The plug-type fuses are available as an **Edison-base** model and as a **Type S** model (Figure 19–3). **Cartridge fuses** are available as a ferrule model or a blade-type model (Figure 19–4). Fuses may be found with time-delay or non-time-delay characteristics. A review of the OCPD information in Chapter 2 is appropriate at this time.

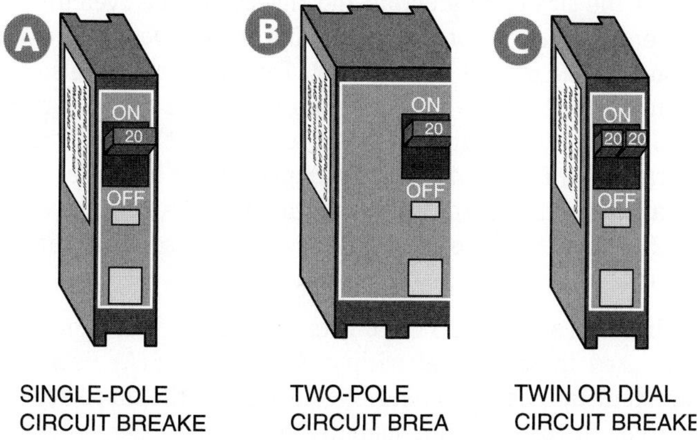

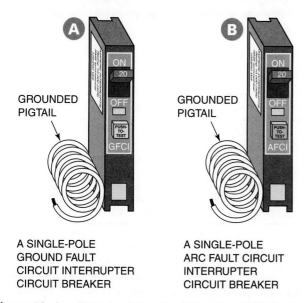

SINGLE-POLE CIRCUIT BREAKE   TWO-POLE CIRCUIT BREA   TWIN OR DUAL CIRCUIT BREAKE

Figure 19–1 Circuit breakers are available as a single-pole device for 120-volt applications (A) and as a two-pole device for 240-volt applications (B) Twin or Dual Circuit breakers can be used for two 120-volt branch circuits (C).

A SINGLE-POLE GROUND FAULT CIRCUIT INTERRUPTER CIRCUIT BREAKER   A SINGLE-POLE ARC FAULT CIRCUIT INTERRUPTER CIRCUIT BREAKER

GROUNDED PIGTAIL   GROUNDED PIGTAIL

Figure 19–2 A GFCI circuit breaker (A) and an AFCI circuit breaker (B).

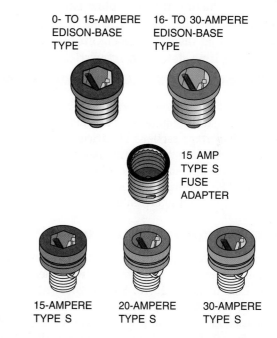

0- TO 15-AMPERE EDISON-BASE TYPE   16- TO 30-AMPERE EDISON-BASE TYPE

15 AMP TYPE S FUSE ADAPTER

15-AMPERE TYPE S   20-AMPERE TYPE S   30-AMPERE TYPE S

Figure 19–3 Plug fuses are available in an Edison-base style and a Type S style.

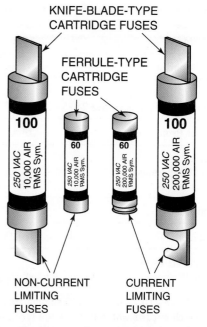

KNIFE-BLADE-TYPE CARTRIDGE FUSES

FERRULE-TYPE CARTRIDGE FUSES

NON-CURRENT LIMITING FUSES   CURRENT LIMITING FUSES

Figure 19–4 Cartridge fuses are available as a ferrule model or a blade-type model.

CAUTION: Use time-delay fuses to protect circuits that supply motor loads. Remember that electric motors draw approximately six to eight times their normal running current at start-up. A time-delay fuse will allow the large inrush of current to get the motor started. Using a non-time-delay fuse will result in having the fuse blow each time the motor starts. Circuit breakers used in residential wiring are called inverse time circuit breakers and have built-in time-delay characteristics.

Article 240 of the *National Electrical Code®* (*NEC®*) provides many installation requirements for OCPDs. A discussion of the more important sections of Article 240 that apply specifically to overcurrent protection devices used in residential wiring is presented in the next few paragraphs. A discussion on OCPD interrupting ratings versus equipment short-circuit ratings is also presented.

General Requirements for OCPDs

Section 240.4 states that circuit conductors must be protected against overcurrent in accordance with their ampacity as specified in Table 310.16. Section 240.6(A) lists the standard sizes of fuses and circuit breakers. Common standard sizes used in residential wiring include 15, 20, 25, 30, 35, 40, 45, 50, 60, 70, 80, 90, and 100 amperes. Other standard sizes are available for larger electrical loads. Because all

the ampacities listed in Table 310.16 do not correspond with a standard fuse or circuit breaker size, Section 240.4 (B) allows the next-higher standard overcurrent device rating to be used, provided that all the following conditions are met:

- The conductors being protected are not part of a multi-outlet branch circuit supplying receptacles for cord-and-plug-connected portable loads.
- The ampacity of the conductors does not correspond with the standard ampere rating of a fuse or a circuit breaker.
- The next-higher standard rating selected does not exceed 800 amperes.

Section 240.4(B) applies in most residential wiring situations, but there are a couple of specific requirements that must be observed first:

- Table 210.24 (Figure 19–5) summarizes the requirements for the size of conductors and the size of the overcurrent protection for branch circuits where two or more receptacle outlets are required. The first footnote also indicates that the wire sizes are for copper conductors.
- Section 210.3 indicates that branch-circuit conductors rated 15, 20, 30, 40, and 50 amperes must be protected at their ratings. For example, a branch circuit wired with 20-amp-rated wire must have a 20-amp fuse or circuit breaker.

In summary, Table 310.16 lists the ampacities of conductors used in residential wiring. Section 240.6(A) lists the standard ratings of overcurrent devices. If the ampacity of the conductor does not match the rating of the standard overcurrent device, Section 240.4(B) permits the use of the

Table 210.24 Summary of Branch-Circuit Requirements

Circuit Rating	15 A	20 A	30 A	40 A	50 A
Conductors (min. size):					
Circuit wires[1]	14	12	10	8	6
Taps	14	14	14	12	12
Fixture wires and cords — See 240.5					
Overcurrent Protection	**15 A**	**20 A**	**30 A**	**40 A**	**50 A**
Outlet devices:					
Lampholders permitted	Any type	Any type	Heavy duty	Heavy duty	Heavy duty
Receptacle rating[2]	15 max. A	15 or 20 A	30 A	40 or 50 A	50 A
Maximum Load	**15 A**	**20 A**	**30 A**	**40 A**	**50 A**
Permissible load	See 210.23(A)	See 210.23(A)	See 210.23(B)	See 210.23(C)	See 210.23(C)

[1]These gauges are for copper conductors.
[2]For receptacle rating of cord-connected electric-discharge luminaires (lighting fixtures), see 410.30(C).

Figure 19–5 Table 210.24 of the *NEC®* summarizes the requirements for the size of conductors and the size of the overcurrent protection for branch circuits where two or more receptacle outlets are required. *Reprinted with permission from NFPA 70-2002, National Electrical Code®, Copyright © 2001, National Fire Protection Association, Quincy, MA 02269. This reprinted material is not the complete and official Position of the NFPA on the referenced subject, which is represented only by the standard in its entirety.*

next-larger standard overcurrent device. However, if the ampacity of a conductor matches the standard rating of Section 240.6(A), that conductor must be protected by the standard-size device. For example, in Table 310.16, a 3 AWG, 75°C copper, Type THWN is listed as having an ampacity of 100 amperes. That conductor would be protected by a 100-ampere overcurrent device. Remember that Section 310.15(B)(6) allows an increase in the ampacity of the conductor and the overcurrent device for a single-phase, 120/240-volt residential service entrance.

Section 240.4(D) requires the OCPD to not exceed 15 amperes for 14 AWG, 20 amperes for 12 AWG, and 30 amperes for 10 AWG copper, or 15 amperes for 12 AWG and 25 amperes for 10 AWG aluminum and copper-clad aluminum after any correction factors for ambient temperature and number of conductors have been applied. Since these wire sizes are common in residential wiring, circuits installed with 14 AWG copper conductors will have a 15-amp OCPD, circuits installed using 12 AWG copper conductors will have a 20-amp OCPD, and circuits installed with 10 AWG copper conductors will have 30-amp OCPDs. This rule has also been explained in previous chapters of this book.

Section 240.24 addresses the location of the OCPDs in the house. Section 240.24(A) requires that OCPDs be readily accessible. Section 240.24(C) requires the OCPDs to be located where they will not be exposed to physical damage. Part (D) prohibits the OCPDs from being located in the vicinity of easily ignitable material. Examples of locations where combustible materials may be stored are linen closets, paper storage closets, and clothes closets. Finally, Section 240.24(E) says that in dwelling units, OCPDs cannot be located in bathrooms.

No discussion of the general requirements for OCPDs is complete without a discussion of interrupting ratings. Section 110.9 states that all fuses and circuit breakers intended to interrupt the circuit at fault levels must have an adequate interrupting rating wherever they are used in the electrical system. Fuses or circuit breakers that do not have adequate interrupting ratings could rupture while attempting to clear a short circuit. The *NEC®* defines the term **interrupting rating** as the highest current at rated voltage that a device is intended to interrupt under standard test conditions. It is important that the test conditions match the actual installation needs. Interrupting ratings should not be confused with short-circuit current ratings.

Short-circuit current ratings are marked on the equipment. This marking appears on many pieces of equipment, such as panelboards. Remember, the basic purpose of overcurrent protection is to open the circuit before conductors or conductor insulation is damaged when an overcurrent condition occurs. An overcurrent condition can be the result of an overload, a ground fault, or a short circuit and must be eliminated before the conductor insulation damage point is reached. Fuses and circuit breakers must be se-

lected to ensure that the short-circuit current rating of any electrical system component is not exceeded should a short circuit or a high-level ground fault occur. Electrical system components include wire, bus structures in loadcenters, switches, disconnect switches, and other electrical distribution equipment, all of which have limited short-circuit ratings and would be damaged or destroyed if those short-circuit ratings were exceeded. Merely providing OCPDs with sufficient interrupting ratings does not ensure adequate short-circuit protection for the system components. When the available short-circuit current exceeds the short-circuit current rating of an electrical component, the OCPD must limit the let-through current to within the rating of that electrical component. Utility companies can determine and provide information on available short-circuit current levels at the service equipment. Calculating the available short-circuit current is beyond the scope of this book.

> **CAUTION**
>
> **CAUTION: Fuses and circuit breakers that are subjected to fault currents that exceed their interrupting ratings may actually explode like a bomb. Damage to people and equipment can result. It is extremely important for overcurrent protection devices to have an interrupting rating high enough to handle any available fault current.**

The interrupting rating of most circuit breakers is either 5,000 or 10,000 amps. The interrupting rating of fuses can be as high as 200,000 amps. For circuit breakers, if the interrupting rating is other than 5,000 amps, it must be marked on the breaker. The interrupting rating for all fuses must be marked on the fuse. The interrupting rating of a fuse is defined as the maximum amount of current the device can handle without failing. The interrupting rating of a device must be at least as high as the available fault current. The available fault current is the maximum amount of current that can flow through the circuit if a fault occurs. It depends on the size of the utility company transformer that is supplying the electrical power to the residential wiring system.

> **CAUTION**
>
> **CAUTION: When replacing fuses, never use a fuse that has a lower interrupting rating than the one that is being replaced. This could cause severe damage to the circuit components.**

General Requirements for Plug Fuses

Section 240.50 applies to plug fuses of both the Edison-base type and Type S type. This section limits the voltage on the circuits using this fuse type to the following:

- Circuits not exceeding 125 volts between conductors. All 120-volt residential branch circuits meet this requirement.
- Circuits supplied by a system having a grounded neutral where the line-to-neutral voltage does not exceed 150 volts. All residential 120/240-volt circuits meet this requirement. An example would be a 120/240-volt electric clothes dryer branch circuit. A fusible disconnect that uses plug fuses could be used with a 120/240-volt dryer branch circuit.

Part (B) of Section 240.50 requires each fuse, fuseholder, and adapter to be marked with its ampere rating. Section 240.50(C) requires plug fuses of 15-ampere and lower rating to be identified by a hexagonal configuration of the window, cap, or other prominent part to distinguish them from fuses of higher ampere ratings. Figure 19–6 shows some examples of Edison-base plug fuses and Type S fuses. Note the hexagonal feature on the 10- and 15-ampere fuses.

Specific Requirements for Edison-Base Plug Fuses

Section 240.51 applies specifically to Edison-base plug fuses and requires the following:

- Plug fuses of the Edison-base type must be classified at not over 125 volts and 30 amperes or below.

Figure 19–6 Some examples of Edison-base plug fuses and Type S fuses. Note the hexagonal feature on the 10- and 15-ampere fuses.

- Plug fuses of the Edison-base type can be used only for replacements in existing installations where there is no evidence of overfusing or tampering.

CAUTION

CAUTION: Edison-base plug fuses may *not* be used as OCPDs in any new electrical installation. If an electrician chooses to use plug fuses in a new electrical installation, they must be Type S plug fuses.

Specific Requirements for Type S Plug Fuses

Section 240.53 applies specifically to Type S plug fuses and requires the following:

- Type S fuses are classified at not over 125 volts and 0 to 15 amperes, 16 to 20 amperes, and 21 to 30 amperes. Type S plug fuses are found in three sizes for residential work: 15-amp rated, which is blue in color; 20-amp rated, which is orange in color; and 30 amp-rated, which is green in color.
- Type S fuses of a higher ampere rating cannot be interchangeable with a lower-ampere-rated fuse. For example, a 20-amp Type S plug fuse will not work if inserted into a 15-amp adapter. This feature prevents overfusing of branch-circuit conductors.
- They must be designed so that they cannot be used in any fuseholder other than a Type S fuseholder or a fuseholder with a Type S adapter inserted.

Section 240.54 applies specifically to Type S fuses, adapters, and fuseholders, and requires the following:

- Type S adapters must fit Edison-base fuseholders.
- Type S fuseholders and adapters must be designed so that either the fuseholder itself or the fuseholder with a

FROM EXPERIENCE

Once a Type S fuse adapter has been installed in an Edison-base screw shell, it is virtually impossible to remove. Therefore, make sure that you put the correct adapter for the size of Type S fuse you wish to use. Color coding makes this choice fairly simple. The bottom inside of the Type S fuse adapter is blue to match the 15-amp Type S fuse, orange to match the 20-amp Type S fuse, and green to match the 20-amp Type S fuse. Anytime this author has tried to remove a Type S fuse adapter once it has been completely installed always resulted in the destruction of the adapter.

Type S adapter inserted cannot be used for any fuse other than a Type S fuse.
- Type S adapters must be designed so that once inserted in a fuseholder, they cannot be removed.
- Type S fuses, fuseholders, and adapters must be designed so that tampering or shunting (bridging) would be difficult.
- Dimensions of Type S fuses, fuseholders, and adapters must be standardized to permit interchangeability regardless of the manufacturer.

Figure 19–7 shows various Type S fuse and adapter ratings and the range of fuse ratings that fit into a specific adapter.

Specific Requirements for Cartridge Fuses

Sections 240.60 and 240.61 apply specifically to cartridge fuses and state the following:

- Cartridge fuses and fuseholders of the 300-volt type are permitted to be used in circuits not exceeding 300 volts between conductors.
- Fuses rated 600 volts or less are permitted to be used for voltages at or below their ratings.
- Fuseholders must be designed so that it will be difficult to put a fuse of any given class into a fuseholder that is designed for a current lower or a voltage higher than that of the class to which the fuse belongs. Fuseholders for current-limiting fuses cannot permit insertion of fuses that are not current limiting.
- Fuses must be plainly marked, either by printing on the fuse barrel or by a label attached to the barrel showing the following: ampere rating, voltage rating, interrupting rating where other than 10,000 amperes, current limiting where applicable, and the name or trademark of the manufacturer.
- Cartridge fuses and fuseholders are to be classified according to their voltage and amperage ranges.

Specific Requirements for Circuit Breakers

Section 240.81 states that circuit breakers must clearly indicate whether they are in the open "Off" or closed "On" position. Where circuit breaker handles are operated vertically, the "Up" position of the handle must be the "On" position. Also, circuit breakers must be designed so that any fault must be cleared before the circuit breaker can be reset. This means that even if the handle is held in the "On" position, the circuit breaker will remain tripped as long as there is a trip-rated fault on the circuit.

Section 240.83 requires that circuit breakers be marked with their ampere rating in a manner that will be durable and visible after installation. This marking is permitted to be made visible by removal of a trim or cover. Part (B) says that circuit breakers rated at 100 amperes or less and 600 volts or less must have the ampere rating molded, stamped, etched, or similarly marked on their handles. Part (C) requires every circuit breaker having an interrupting rating other than

TYPE S FUSE INFORMATION

Type S Fuse Ampere Ratings	Type S Adapter Rating	Type S Fuse Ampere Ratings That Fit into This Adapter
3/10, 1/2, 8/10, 4/10, 6/10, 1	1	1 ampere and smaller
1⅛, 1¼	1¼	All smaller
1 4/10, 1 6/10	1 6/10	All smaller
1 8/10, 2	2	1 8/10, 2
2¼, 3½	2½	1 8/10, 2, 2¼, 2½
2 8/10, 3 2/10	3 2/10	1 8/10, 2, 2¼, 2½, 2 8/10, 3 2/10
3½, 4	4	3½, 4
4½, 5	5	3½, 4, 4½, 5
5 6/10, 6¼	6¼	3½, 4, 4½, 5, 5 6/10, 6¼
7, 8	8	7, 8
9, 10	10	7, 8, 9, 10
12, 14	14	7, 8, 9, 10, 12, 14
15	15	15
20	20	20
25	30	20, 25, 30
30	30	20, 25, 30

Figure 19–7 This table shows various Type S fuse and adapter ratings and the range of fuse ratings that fit into a specific adapter.

5,000 amperes to have its interrupting rating shown on the circuit breaker. Part (D) requires circuit breakers used as switches in 120- and 277-volt fluorescent lighting circuits to be listed and marked SWD or HID. Circuit breakers used as switches in high-intensity discharge lighting circuits must be listed and marked as HID. Circuit breakers marked SWD are 15- or 20-ampere breakers that have been subjected to additional endurance and temperature testing. If high-intensity discharge (HID) lighting such as high-pressure sodium or metal halide lighting is used, the breaker used for switching must be marked HID. Part (E) says that circuit breakers must be marked with a voltage rating not less than the nominal system voltage that is indicative of their capability to interrupt fault currents between phases or phase to ground.

A circuit breaker with a straight voltage rating, such as 240 volts, is permitted to be applied in a circuit in which the nominal voltage between any two conductors does not exceed the circuit breaker's voltage rating. A circuit breaker with a slash rating, such as 120/240 volts, is permitted to be applied in a solidly grounded circuit where the voltage of any conductor to ground does not exceed the lower of the two values of the circuit breaker's voltage rating and the voltage between any two conductors does not exceed the higher value of the circuit breaker's voltage rating. The slash (/) between the lower and higher voltage ratings in the marking indicates that the circuit breaker has been tested for use on a circuit with the higher voltage between phases and with the lower voltage to ground.

GFCI and AFCI Circuit Breakers

GFCI Circuit Breakers

You know that some receptacles in a home require GFCI protection. There are two ways that an electrician can provide GFCI protection to the receptacles that require it. One way is to use GFCI receptacles. This method was presented in an earlier chapter. The other way is to use a GFCI circuit breaker. An advantage to the GFCI circuit breaker is that the whole circuit becomes protected. However, with the GFCI receptacle, only that receptacle location and any "downstream" receptacles are GFCI protected. A disadvantage is the relatively high cost of the GFCI breaker as compared to the GFCI receptacle.

The GFCI circuit breaker looks very similar to a "regular" circuit breaker (Figure 19–8). However, there are two differences that you should be aware of. The first is that the GFCI breaker has a white pigtail attached to it. It is all curled up (like a pig's tail) when the GFCI breaker is first taken out of the box that it comes in. The second difference is the "Push to Test" button that is located on the front of the breaker. This allows the breaker to be tested for correct operation once it has been installed and energized.

CAUTION: A GFCI circuit breaker should be tested at least once a month, just like a GFCI receptacle.

AFCI Circuit Breakers

An AFCI is a device intended to provide protection from the effects of arc faults by recognizing characteristics unique to arcing and by functioning to deenergize the circuit when an arc fault is detected. Section 210.12(B) requires all branch circuits that supply 125-volt, single-phase, 15- and 20-ampere-rated outlets installed in dwelling unit bedrooms to be protected by an AFCI listed to provide protection of the entire branch circuit. The basic objective of an AFCI device is to deenergize the branch circuit when an arc fault is detected.

Section 210.12 requires that AFCI protection be provided on branch circuits that supply *all* outlets in dwelling unit bedrooms. This means that *all* receptacle outlets, lighting outlets, or any other power outlet in a residential bedroom (like a smoke detector outlet) must be AFCI protected. The requirement is limited to 15- and 20-ampere 125-volt circuits. There is no rule against providing AFCI protection on other circuits or in locations other than bedrooms. Because circuits are often shared between a bedroom and other areas, such as closets and hallways, providing AFCI protection on the complete circuit would comply with Section 210.12.

AFCIs are evaluated to UL 1699, Safety Standard for Arc-Fault Circuit Interrupters, using testing methods that create or simulate arcing conditions to determine the product's ability to detect and interrupt arcing faults. These devices are also tested

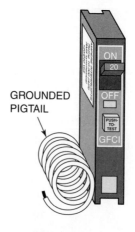

GROUNDED PIGTAIL

A SINGLE-POLE GROUND FAULT CIRCUIT INTERRUPTER CIRCUIT BREAKER

A REGULAR SINGLE-POLE CURCUIT BREAKER

Figure 19–8 A GFCI circuit breaker has a white pigtail attached to it. It also has a "Push to Test" button for testing the ground-fault interrupting capabilities of the breaker.

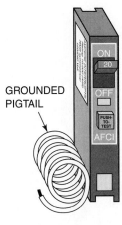

GROUNDED
PIGTAIL

A SINGLE-POLE
ARC FAULT CIRCUIT
INTERRUPTER
CIRCUIT BREAKER

A REGULAR
SINGLE-POLE
CIRCUIT BREAKER

Figure 19–9 An AFCI circuit breaker looks very similar to a GFCI circuit breaker.

to verify that arc detection is not any less effective by the presence of loads and circuit characteristics that may mask the hazardous arcing condition. In addition, these devices are evaluated to determine resistance to unwanted tripping due to the presence of arcing that occurs in control and utilization equipment under normal operating conditions or to a loading condition that closely mimics an arcing fault, such as a solid-state electronic ballast or a dimmed lighting load.

The *NEC®* is clear that the objective is to provide protection of the entire branch circuit. For instance, a cord AFCI could not be used to comply with the requirement of Section 210.12 to protect the entire branch circuit. This means that for an electrician to comply with Section 210.12, an AFCI circuit breaker will need to be used so that the entire branch circuit, from the overcurrent protective device to the last outlet installed on the circuit, is AFCI protected.

AFCI circuit breakers look very similar to GFCI circuit breakers (Figure 19–9). The AFCI breaker has a "Push to Test" button, but it is typically a different color than the "Push to Test" button on a GFCI breaker.

Installing Circuit Breakers in a Panel

At this point in the installation of the electrical system, the branch wiring has been installed into the service entrance panel or subpanels, and it is time to install the OCPD for each branch circuit.

It is important for an electrician to trim out the panel in a neat and professional manner. The service panel or subpanel will not only look better but also allow better air circulation around the wires, circuit breakers, or fuses. This will

FROM EXPERIENCE

It is a good idea to provide the home owner with at least two extra circuit breakers to be used as spares. This allows the home owner to add to the electrical system without having to replace the existing panel or add a subpanel.

result in the reduction of heat buildup. Trimming out a panel in a neat manner will also allow for easier troubleshooting of branch-circuit problems should they arise.

In this section, we look at safely installing circuit breakers in electrical panels. We do not cover fuse installation since the majority of branch-circuit OCPDs used in today's homes are circuit breakers.

Before installing circuit breakers, some factors must be considered. First, make sure that the circuit breakers being used are compatible with the loadcenter already installed. Circuit breakers are not always interchangeable with different manufacturers. Second, remember that Section 408.15 of the *NEC®* states that no more than 42 overcurrent devices may be installed in any lighting and appliance panelboard. Most 100-amp-rated electrical panels used in residential wiring are designed for 20 or 24 overcurrent devices. Most 200-amp-rated electrical panels are designed for 40 circuits. Be sure that your panel will handle the number of circuits the electrical system requires.

Safety Rules

When installing or removing circuit breakers from an electrical panel in a house where the service entrance conductors have been connected to the local utility company system and a kilowatt-hour meter has been installed in the meter socket, there a couple of safety rules to follow:

- Always turn off the electrical power at the main circuit breaker when working in an energized main breaker panel. This will disconnect the power from the load side of the panel, but be aware that the line side of the panel will still be energized (Figure 19–10). If you are working on an energized subpanel, find the circuit breaker in the service panel, turn it off, and lock it in the "Off" position.
- Test the panel you are working on with a voltage tester to verify that the electrical power is indeed off. Never assume that the panel is deenergized.

CAUTION

CAUTION: When installing or removing circuit breakers from an electrical panel, make sure that the panel has been deenergized and the electrical power supplying the panel has been locked out.

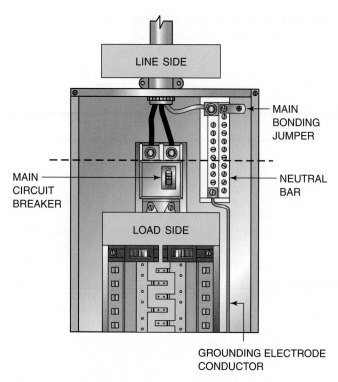

Figure 19–10 The line side versus the load side in an electrical panel. Remember that the line side will still be "hot" even when the main circuit breaker is off. Only the load side will be deenergized.

Circuit Breaker Installation

Circuit breakers are installed in an electrical panel by attaching them to the bus bar assembly in the panel. The bus bar assembly is connected to the incoming service entrance conductors or, in the case of a subpanel, to the incoming feeder conductors. The bus bar distributes the electrical power to each of the circuit breakers located in the panel.

The circuit breakers are attached to the bus bar by contacts in the breakers that are snapped onto the bus bar at specific locations, commonly called stabs. A single-pole circuit breaker has one stab contact, and a two-pole circuit breaker has two stab contacts (Figure 19–11). Circuit breakers are constructed in such a way that the end opposite the stab contacts has a slot that fits onto a retainer clip in the panel. Between the retainer clip and the circuit breaker contacts being snapped onto the bus bar stabs, the breaker is held in place. In panels typically used in commercial and industrial locations, the circuit breakers are more likely to be bolted onto the bus bar rather than just snapping onto them. Once all the breakers have been installed and the wiring has been correctly attached to them, a cover is put on the panel. The cover also helps hold the circuit breakers in place. Figure 19–12 shows how both single-pole circuit breakers and double-pole circuit breakers are attached to a bus bar in a main breaker panel.

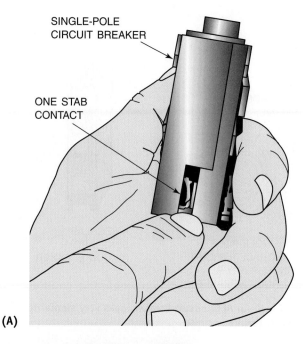

(A)

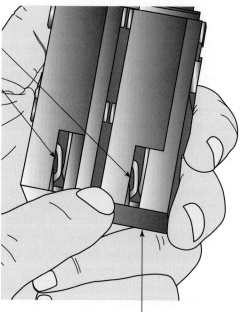

(B)

Figure 19–11 A single-pole circuit breaker (A) has one stab contact, and a two-pole circuit breaker (B) has two stab contacts.

CAUTION

CAUTION: Only use approved handle ties to connect two single-pole circuit breakers together as a double-pole breaker.

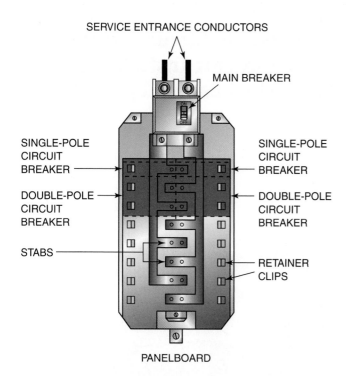

SERVICE ENTRANCE CONDUCTORS

MAIN BREAKER

SINGLE-POLE CIRCUIT BREAKER

SINGLE-POLE CIRCUIT BREAKER

DOUBLE-POLE CIRCUIT BREAKER

DOUBLE-POLE CIRCUIT BREAKER

STABS

RETAINER CLIPS

PANELBOARD

Figure 19–12 A single-pole circuit breaker and a double-pole circuit breaker attached to a bus bar in a main breaker panel.

FROM EXPERIENCE

The covers used for electrical panels come from the factory with several partially punched-out rectangular pieces of metal that will need to be removed by an electrician. The number and location of the pieces to be removed is determined by the number of circuit breakers used in the panel and their actual location. Once the correct number of pieces are removed, the cover will fit correctly onto the panel and over the installed circuit breakers. Be careful to not take out too many of the pieces. If you do, a plug must be inserted in the spot to prevent the "hot" ungrounded panel bus bar from being exposed.

Most branch circuits in a residential wiring system are 120-volt circuits. They will be wired with 14 AWG or 12 AWG copper conductors. Remember that 14 AWG conductors require a single-pole circuit breaker that is 15-amp rated and that 12 AWG conductors require a single-pole circuit breaker rated at 20 amps.

(See page 528 for the correct procedure for installing a single-pole circuit breaker for a 120-volt branch circuit installed with nonmetallic sheathed cable [Type NM].)

(See page 530 for the correct procedure for installing a single-pole GFCI circuit breaker for a 120-volt branch circuit installed with nonmetallic sheathed cable [Type NM].)

(See page 532 for the correct procedure for installing a single-pole AFCI circuit breaker for a 120-volt bedroom branch circuit installed with nonmetallic sheathed cable [Type NM].)

Many residential branch circuits serve appliances like electric water heaters, air conditioners, and electric heating units. These loads require 240 volts to operate properly. Like the 120-volt branch circuits, the 240-volt branch circuits require a 15-amp circuit breaker when wired with 14 AWG wire, a 20-amp circuit breaker when wired with 12 AWG wire, and a 30-amp circuit breaker when wired with 10 AWG wire. However, since it is a 240-volt circuit, a two-pole circuit breaker will be needed.

(See page 534 for the correct procedure for installing a two-pole circuit breaker for a 240-volt branch circuit installed with nonmetallic sheathed cable [Type NM].)

Branch circuits in residential wiring can also supply 120/240 volts to appliances such as electric clothes dryers and electric ranges. This branch-circuit installation requires a two-pole circuit breaker, just like the 240-volt-only application described in the preceding Paragraph. The big difference with this type of circuit is that a three-wire cable with a grounding conductor is used.

(See page 536 for the correct procedure for installing a two-pole circuit breaker for a 120/240-volt branch circuit installed with nonmetallic sheathed cable [Type NM].)

Summary

The main service panel and/or subpanels contain the OCPDs for the various branch circuits that make up a residential wiring system. The OCPDs may consist of fuses or circuit breakers. It is during the trim-out stage that an electrician will "trim out" the electrical panel(s) used in a house. OCPDs are installed in the panel(s), and the proper wiring terminations are made to them one branch circuit at a time.

This chapter presented an overview of the types of OCPDs used in residential wiring. Edison-base plug fuses, Type S plug fuses, and cartridge fuses were presented. Since circuit breakers are used the most as overcurrent protection in residential installations, a thorough discussion of the different types of circuit breakers was also presented. This included GFCI and AFCI circuit breakers. Circuit breaker installation was covered in detail, and several actual installation procedures were presented.

Trimming out the panel for a residential wiring system is a very important part of the trim-out stage. It must be done in a neat and professional manner. Making sure that the OCPD is installed properly and that the correct size has been chosen will allow the branch circuits to supply electricity to all the loads located throughout a house in a safe and trouble-free manner.

Procedures

Installing a Single-Pole Circuit Breaker for a 120-Volt Branch Circuit Installed with Nonmetallic Sheathed Cable (Type NM)

- Put on safety glasses and observe all applicable safety rules.

- Using a voltage tester, make sure that the panel is deenergized and that the electrical power source for the panel has been locked out.

A Locate a free space in the panel for the circuit breaker. Insert the slot on the back of the circuit breaker onto the retainer clip. Then, using firm pressure, snap the single-pole circuit breaker onto the bus bar stab.

- Determine the location on the equipment-grounding bar where you want to terminate the bare equipment-grounding conductor. Loosen the termination screw. (Note: In a main service panel the equipment-grounding conductor is terminated on the grounded neutral terminal bar.)

A

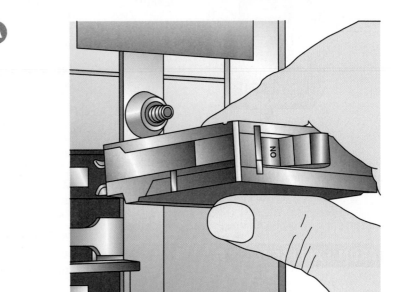

B Insert the end of the bare equipment-grounding conductor into the termination hole on the equipment-grounding bar and tighten the screw according to the torque requirements listed on the panel enclosure.

- Determine the location on the grounded neutral terminal bar where you want to terminate the white grounded circuit conductor. Loosen the termination screw.

B

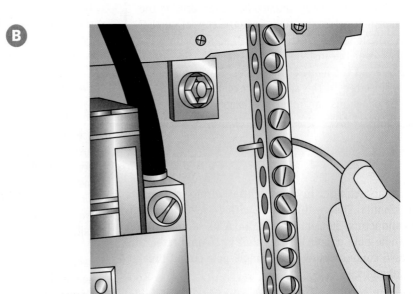

C Using a wire stripper, remove approximately 1/2 inch (13 mm) from the end of the white grounded circuit conductor and insert the tip of the white conductor into the termination hole on the grounded terminal bar. Tighten the screw according to the torque requirements listed on the panel enclosure.

C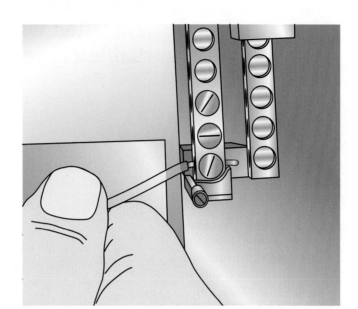

D Using a wire stripper, remove approximately 1/2 inch (13 mm) from the end of the black ungrounded circuit conductor and insert the tip of the black wire under the terminal screw of the single-pole circuit breaker. (Note: You may have to loosen the terminal screw on the circuit breaker.) Tighten the screw according to the torque requirements listed for the breaker.

D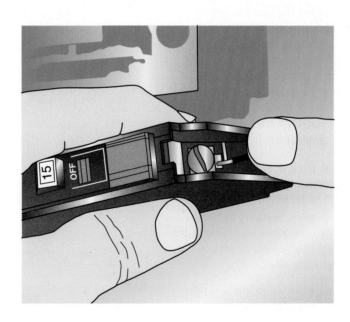

Procedures

Installing a Single-Pole GFCI Circuit Breaker for a 120-Volt Branch Circuit Installed with Nonmetallic Sheathed Cable (Type NM)

- Put on safety glasses and observe all applicable safety rules.

- Using a voltage tester, make sure that the panel is deenergized and that the electrical power source for the panel has been locked out.

- Locate a free space in the panel for the circuit breaker. Insert the slot on the back of the circuit breaker onto the retainer clip. Then, using firm pressure, snap the single-pole GFCI circuit breaker onto the bus bar stab.

- Determine the location on the equipment-grounding bar where you want to terminate the bare equipment-grounding conductor. Loosen the termination screw. (Note: In a main service panel the equipment-grounding conductor is terminated on the grounded neutral terminal bar.)

- Insert the end of the bare equipment-grounding conductor into the termination hole on the equipment-grounding bar and tighten the screw according to the torque requirements listed on the panel enclosure.

 Connect the white grounded pigtail of the GFCI circuit breaker to the grounded terminal bar in the panel. The pigtail usually comes with its end stripped back at the factory.

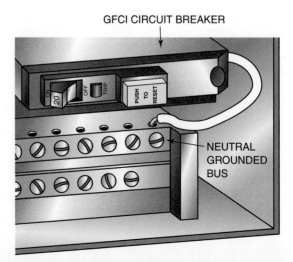

A

GFCI CIRCUIT BREAKER

NEUTRAL GROUNDED BUS

B Using a wire stripper, remove approximately 1/2 inch (13 mm) from the end of the white grounded branch-circuit conductor and insert the tip of the white conductor into the terminal on the GFCI breaker marked "load neutral." Tighten the screw according to the torque requirements listed for the breaker.

B

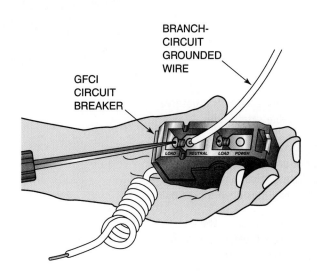

C Using a wire stripper, remove approximately 1/2 inch (13 mm) from the end of the black ungrounded circuit conductor and insert the tip of the black conductor into the terminal on the GFCI breaker marked "load." Tighten the screw according to the torque requirements listed for the breaker.

C

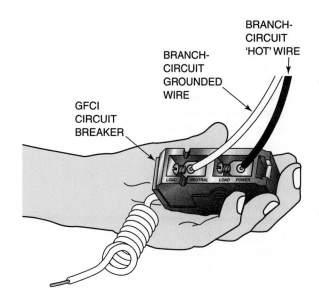

Procedures

Installing a Single-Pole AFCI Circuit Breaker for a 120-Volt Bedroom Branch Circuit Installed with Nonmetallic Sheathed Cable (Type NM)

- Put on safety glasses and observe all applicable safety rules.

- Using a voltage tester, make sure that the panel is deenergized and that the electrical power source for the panel has been locked out.

- Locate a free space in the panel for the circuit breaker. Insert the slot on the back of the circuit breaker onto the retainer clip. Then, using firm pressure, snap the single-pole AFCI circuit breaker onto the bus bar stab.

- Determine the location on the equipment-grounding bar where you want to terminate the bare equipment-grounding conductor. Loosen the termination screw. (Note: In a main service panel the equipment-grounding conductor is terminated on the grounded neutral terminal bar.)

- Insert the end of the bare equipment-grounding conductor into the termination hole on the equipment-grounding bar and tighten the screw according to the torque requirements listed on the panel enclosure.

 Connect the white grounded pigtail of the AFCI circuit breaker to the grounded terminal bar in the panel. The pigtail usually comes with its end stripped back at the factory.

A

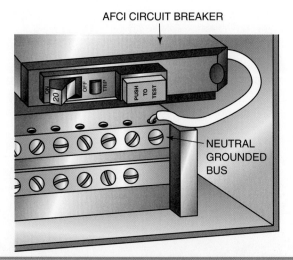

AFCI CIRCUIT BREAKER

NEUTRAL GROUNDED BUS

B Using a wire stripper, remove approximately 1/2 inch (13 mm) from the end of the white grounded branch-circuit conductor and insert the tip of the white conductor into the terminal on the AFCI breaker marked "load neutral." Tighten the screw according to the torque requirements listed for the breaker.

B

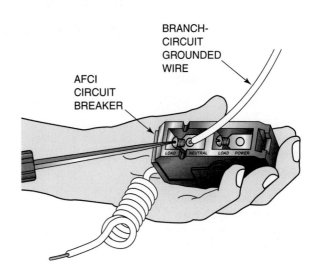

BRANCH-CIRCUIT GROUNDED WIRE

AFCI CIRCUIT BREAKER

C Using a wire stripper, remove approximately 1/2 inch (13 mm) from the end of the black ungrounded branch-circuit conductor and insert the tip of the black conductor into the terminal on the AFCI breaker marked "load." Tighten the screw according to the torque requirements listed for the breaker.

C

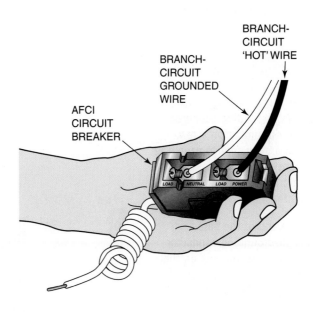

BRANCH-CIRCUIT 'HOT' WIRE

BRANCH-CIRCUIT GROUNDED WIRE

AFCI CIRCUIT BREAKER

Procedures

Installing a Two-Pole Circuit Breaker for a 240-Volt Branch Circuit Installed with Nonmetallic Sheathed Cable (Type NM)

- Put on safety glasses and observe all applicable safety rules.

- Using a voltage tester, make sure that the panel is deenergized and that the electrical power source for the panel has been locked out.

- Locate a free space in the panel for the two-pole circuit breaker. Remember that this breaker takes up two regular spaces. Insert the slots on the back of the circuit breaker onto the retainer clips. Then, using firm pressure, snap the double-pole circuit breaker onto the bus bar stabs.

- Determine the location on the equipment-grounding bar where you want to terminate the bare equipment-grounding conductor. Loosen the termination screw. (Note: In a main service panel the equipment-grounding conductor is terminated on the grounded neutral terminal bar.)

A Insert the end of the bare equipment-grounding conductor into the termination hole on the equipment-grounding bar and tighten the screw according to the torque requirements listed on the panel enclosure.

A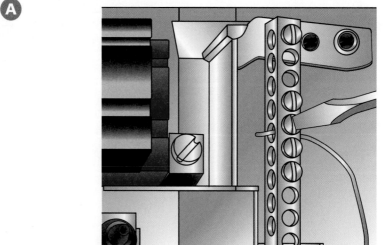

Using a wire stripper, remove approximately 1/2 inch (13 mm) from the end of the black ungrounded branch-circuit conductor and insert the tip of the black wire under one of the terminal screws of the two-pole circuit breaker. (Note: You may have to loosen the terminal screws on the circuit breaker.) Tighten the screw according to the torque requirements listed for the breaker.

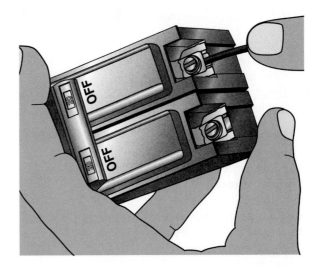

 Using a wire stripper, remove approximately 1/2 inch (13 mm) from the end of the white ungrounded branch-circuit conductor. Reidentify this white wire as a "hot" conductor by wrapping it with some black electrical tape. Insert the tip of the reidentified white wire under the other terminal screw of the two-pole circuit breaker. Tighten the screw according to the torque requirements listed for the breaker.

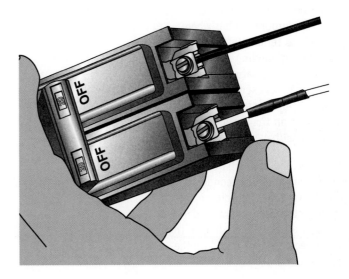

Procedures

- Put on safety glasses and observe all applicable safety rules.

- Using a voltage tester, make sure that the panel is deenergized and that the electrical power source for the panel has been locked out.

- Locate a free space in the panel for the two-pole circuit breaker. Remember that this breaker takes up two regular spaces. Insert the slots on the back of the circuit breaker onto the retainer clips. Then, using firm pressure, snap the double-pole circuit breaker onto the bus bar stabs.

- Determine the location on the equipment-grounding bar where you want to terminate the bare equipment-grounding conductor. Loosen the termination screw. (Note: In a main service panel the equipment-grounding conductor is terminated on the grounded neutral terminal bar.)

- Insert the end of the bare equipment-grounding conductor into the termination hole on the equipment-grounding bar and tighten the screw according to the torque requirements listed on the panel enclosure.

- Determine the location on the grounded terminal bar where you want to terminate the white grounded circuit conductor. Loosen the termination screw.

 Using a wire stripper, remove approximately 1/2 inch (13 mm) from the end of the white grounded branch-circuit conductor and insert the tip of the white conductor into the termination hole on the grounded terminal bar. Tighten the screw according to the torque requirements listed on the panel enclosure.

Installing a Two-Pole Circuit Breaker for a 120/240-Volt Branch Circuit Installed with Nonmetallic Sheathed Cable (Type NM)

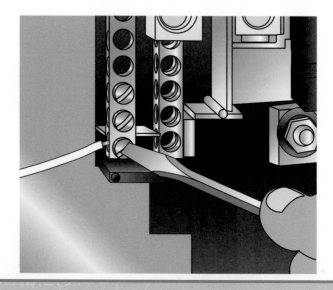

B Using a wire stripper, remove approximately 1/2 inch (13 mm) from the end of the black ungrounded branch-circuit conductor and insert the tip of the black wire under one of the terminal screws of the two-pole circuit breaker. Tighten the screw according to the torque requirements listed for the breaker.

B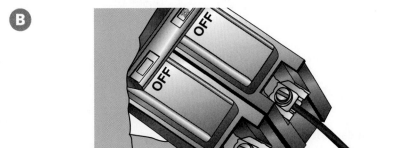

C Using a wire stripper, remove approximately 1/2 inch (13 mm) from the end of the red ungrounded branch-circuit conductor. Insert the tip of the red wire under the other terminal screw of the two-pole circuit breaker. Tighten the screw according to the torque requirements listed for the breaker.

C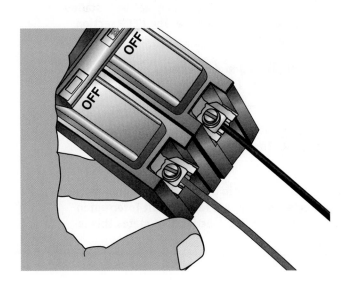

Review Questions

Directions: Answer the following items with clear and complete responses.

1. Describe a Type S fuse.

2. The maximum voltage between conductors on circuits protected with plug fuses is _____ volts.

3. List the standard sizes of fuses and circuit breakers up to and including 100 amperes.

4. A 240-volt electric water heater branch circuit will require a _____ pole circuit breaker.

5. A 120-volt dishwasher branch circuit will require a _____ pole circuit breaker.

6. Overcurrent protection devices may not be installed in clothes closets. True or False. The *NEC®* section that states this is _____.

7. Overcurrent protection devices may be installed in bathrooms. True or False. The *NEC®* section that states this is _____.

8. Overcurrent protection devices must be accessible. True or False. The *NEC®* section that states this is _____.

9. All electrical outlets in dwelling unit bedrooms must be protected with an Arc Fault Circuit Interrupter. True or False. The *NEC®* section that states this is _____.

10. Explain why it is important for an electrician to trim out the panel in a neat and professional manner.

11. Name the two types of cartridge fuses found in residential wiring.

12. Name the maximum overcurrent protective devices allowed for the following residential branch-circuit conductor sizes:

    1) 14 AWG copper _____
    2) 12 AWG copper _____
    3) 10 AWG copper _____

13. What does the hexagonal configuration on a plug fuse indicate?

14. A circuit breaker used to switch on or off fluorescent lighting must have the letters _____ written on the breaker. The *NEC®* section that states this is _____.

15. Name an advantage to using a GFCI circuit breaker instead of a GFCI receptacle to provide the *NEC®*-required ground-fault protection.

Maintaining and Trouble-shooting a Residential Electrical Wiring System

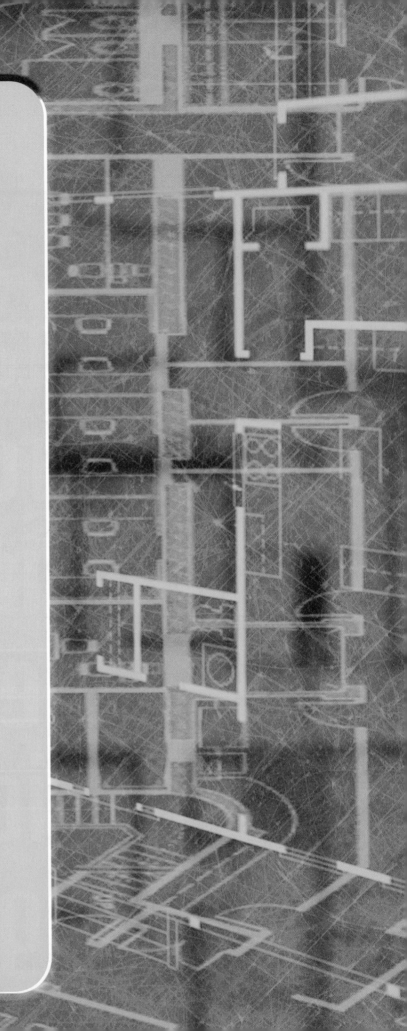

SECTION FIVE
MAINTAINING AND TROUBLESHOOTING A RESIDENTIAL ELECTRICAL WIRING SYSTEM

- Chapter 20:
  Checking Out and Troubleshooting Electrical Wiring Systems

Chapter 20

Checking Out and Troubleshooting Electrical Wiring Systems

T*he installation of a residential electrical system must meet the minimum installation requirements of the* National Electrical Code® *(NEC®). An electrician who follows the installation information presented in the previous chapters of this book should have an electrical installation that is essentially safe and free from defects. The installed electrical system should provide home owners with the electrical service they desire. However, during the installation process, mistakes can happen. Some electrical equipment may be faulty. Other tradespeople, such as carpenters and plumbers working in the house, may inadvertently damage a part of the electrical system. Electricians can mix up traveler wires on three-way and four-way switching circuits. Wirenut connections may not be properly tightened. In other words, a just completed residential electrical installation may have problems that are not readily apparent without a thorough* checkout *of all parts of the system. When problems are found,* troubleshooting *the problem will have to be done, and then the identified problem will need to be fixed. This chapter looks at common checkout, troubleshooting, and repair procedures that an electrician should be familiar with.*

OBJECTIVES

Upon completion of this chapter, the student should be able to:

- follow a checklist to determine if the basic requirements of the *NEC®* were met in the electrical system installation.
- demonstrate an understanding of how to test for current and voltage in an energized circuit.
- demonstrate an understanding of how to test for continuity in existing branch-circuit wiring and wiring devices.
- demonstrate an understanding of the common testing techniques to determine whether a circuit has a short circuit, a ground fault, or an open circuit.
- troubleshoot common residential electrical system problems.

Glossary of Terms

buried box an electrical box that has been covered over with sheetrock or some other building material

checkout the process of determining if all parts of a recently installed electrical system are functioning properly

ground fault an unintended low-resistance path in an electrical circuit through which some current flows to ground using a pathway other than the intended pathway; it results when an ungrounded "hot" conductor unintentionally touches a grounded surface or grounded conductor

open circuit a circuit that is energized but does not allow useful current to flow on the circuit because of a break in the current path

receptacle polarity when the white-colored grounded conductor is attached to the silver screw of the receptacle and the "hot" ungrounded circuit conductor is attached to the brass screw on the receptacle; this results in having the long slot always being the grounded slot and the short slot always being the ungrounded slot on 125-volt, 15- or 20-amp-rated receptacles used in residential wiring

short circuit an unintended low-resistance path through which current flows around rather than along a circuit's intended current path; it results when two circuit conductors come in contact with each other unintentionally

troubleshooting the process of determining the cause of a malfunctioning part of a residential electrical system

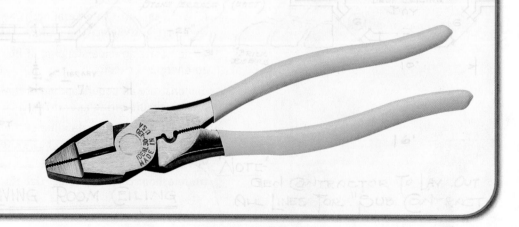

Determining if All Applicable NEC® Installation Requirements Are Met

It is common practice for an electrical inspector to inspect the electrical installation at the end of the electrical system rough-in and at the end of the electrical system trim-out to make sure that the system has been installed according to the requirements of national, state, and local electrical codes. At the end of the rough-in stage and on completion of the residential electrical system trim-out stage, an electrician should go over the following list to make sure the basic installation requirements of the *NEC®* have been met. State and local electrical codes are beyond the scope of this book but should always be considered by the electrician doing the residential electrical system installation. Making sure that these guidelines are met will result in an electrical system installation that will very likely pass an inspection by the electrical inspector. As an electrician goes over this list, make note of those areas that were not installed according to the *NEC®*. Any areas of the installation that are found to be deficient will need to be updated to meet the minimum *NEC®* standards.

General Circuitry

NEC® Sections 210.11 and 422.12

In addition to the branch circuits installed to supply general illumination and receptacle outlets in dwelling units, the following *minimum* requirements also apply:

- Two 20-amp circuits for the kitchen receptacles
- One 20-amp circuit for the laundry receptacles
- One 20-amp circuit for the bathroom receptacles
- One separate, individual branch circuit for central heating equipment

NEC® Section 210.52

Receptacles installed in the kitchen to serve countertop surfaces must be supplied by at least two separate small-appliance branch circuits.

NEC® Section 300.3

All conductors of the same circuit, including grounding and bonding conductors, must be contained in the same raceway, cable, or trench.

NEC® Section 408.4

All circuits and circuit modifications must be legibly identified as to purpose or use on a directory located on the face or inside of the electrical panel doors.

NEC® Section 240.3

The rating of the fuse or circuit breaker determines the minimum size of the circuit conductor, per the following table:

Fuse or Circuit Breaker Size	Minimum-Size Wire (CU)	Minimum-Size Wire (AL)
15 amp	14 AWG	12 AWG
20 amp	12 AWG	10 AWG
30 amp	10 AWG	8 AWG
40 amp	8 AWG	6 AWG
50 amp	6 AWG	4 AWG

Note: Conductors that supply motors, air-conditioning units, and other special equipment may have overcurrent protection that exceeds the general limitations in this chart. Also, only use aluminum conductors when they are allowed in your area by the authority having jurisdiction.

NEC® Section 406.3

Receptacle outlets must be of the grounding type, be effectively grounded, and be wired to have the proper polarity.

NEC® Section 210.52

Receptacle outlets in habitable rooms must be installed so that no point measured horizontally along the floor line in any wall space is more than 6 feet (1.8 m) from a receptacle outlet. A receptacle must be installed in each wall space 2 feet (600 mm) or more in width.

At kitchen countertops, receptacle outlets must be installed so that no point along the wall line is more than 24 inches (600 mm) measured horizontally from a receptacle outlet in that space.

A receptacle outlet must be installed at each counter space that is 12 inches (300 mm) or wider and at each island counter or peninsular space 24 inches (600 mm) by 12 inches (300 mm) or larger. Countertop spaces separated by range tops, sinks, or refrigerators are separate spaces.

Outdoor receptacles, accessible at grade level and no more than 6.5 feet (2 m) above grade, must be installed at the front and back of a dwelling.

NEC® Section 210.12

All branch circuits supplying 125-volt, 15- and 20-ampere outlets in dwelling unit bedrooms must be protected by a listed arc fault circuit interrupter device (AFCI). Generally, these are circuit breaker–type devices and cannot be installed on multiwire branch circuits.

Required Ground Fault Circuit Interrupter Protection

NEC® Section 210.8

In dwellings, ground fault circuit interrupter (GFCI) protection must be provided for all receptacle outlets installed in bathrooms, garages, crawl spaces, unfinished basements, kitchen countertops, locations within 6 feet (1.8 m) of wet-bar sinks that serve the countertop, boathouses, and outdoors. Receptacles that are not readily accessible may be exempt from the GFCI requirement.

NEC® Section 680.71

A hydromassage bathtub must have GFCI protection.

All 125-volt receptacles not exceeding 30 amperes installed within 5 feet (1.5 m) of the inside walls of the hydromassage bathtub must be GFCI protected.

All equipment associated with a hydromassage bathtub must be accessible without damaging the building structure or finish.

Wiring Methods

NEC® Section 314.23

All electrical boxes must be securely supported by the building structure.

NEC® Section 314.27

When boxes are used as the sole support for a ceiling-suspended paddle fan, they must be listed and labeled for such use.

NEC® Section 334.30

Type NM cable must be secured at intervals not exceeding 4½ feet (1.4 m) and within 12 inches (300 mm) of each electrical box and within 8 inches (200 mm) of each single-gang nonmetallic electrical box.

NEC® Section 314.17

The outer jacket of Type NM cable must extend into a single-gang nonmetallic electrical box a minimum of 1/4 inch (6 mm).

NEC® Section 300.14

The minimum length of conductors at all boxes must be at least 6 inches (150 mm). At least 3 inches (75 mm) must extend outside the box. This includes grounding conductors.

NEC® Section 300.4

Where cables are installed through bored holes in joists, rafters, or wood framing members, the holes must be bored so that the edge of the hole is not less than 1.25 inches (8 mm) from the nearest edge of the wood member. Where this distance cannot be maintained or where screws or nails are likely to penetrate the cable, they must be protected by a steel plate at least 1/16 inch (1.6 mm) thick and of appropriate length and width.

NEC® Section 300.22

Type NM cable must not be installed in spaces used for environmental air; however, Type NM cable is permitted to pass through perpendicular to the long dimension of such spaces.

NEC® Sections 250.134, 314.4, and 404.9

All metal electrical equipment including boxes, cover plates, and plaster rings must be grounded. All switches, including dimmer switches, must be grounded.

NEC® Sections 110.12 and 314.17

Unused openings in boxes must be effectively closed. When openings in nonmetallic boxes are broken out and not used, the entire box must be replaced.

NEC® Section 110.14

Only one conductor can be installed under a terminal screw. In boxes with more than one ground wire, the ground wires must be spliced with an approved mechanical connector and then, using a "jumper" or a "pigtail," be attached to the grounding terminal screw of the device. In metal boxes, the equipment-grounding wires must connect to the box with a green grounding screw or other approved method.

NEC® Sections 110.14 and 300.15

Splices must be made with an approved method, like a "wirenut," and must be made in listed electrical boxes or enclosures. When splicing underground conductors, the method and items used must be identified for such use.

NEC® Sections 314.25 and 410.12

In a completed installation, all outlet boxes must have a cover, lamp holder, canopy for a luminaire (light fixture), and an appropriate cover plate for switches and receptacles.

NEC® Section 314.19

Junction boxes must be installed so that the wiring contained in them can be rendered accessible without removing any part of the building.

NEC® Section 314.16

The volume of electrical boxes must be sufficient for the number of conductors, devices, and cable clamps contained within the box. Nonmetallic boxes are marked with their cubic inch capacity.

NEC® Section 410.8

Storage space, as applied to an electrical installation in a closet, is the volume bounded by the sides and back closet walls and planes extending from the closet floor vertically to a height of 6 feet (1.8 m) or the highest clothes hanging rod and parallel to the walls at a horizontal distance of 24 inches (600 mm) from the sides and back of the closet walls, respectively, and continuing vertically to the closet ceiling parallel to the walls at a horizontal distance of 12 inches (300 mm) or the shelf width, whichever is greater.

Luminaires (lighting fixtures) installed in clothes closets must have the following minimum clearances from the defined storage area:

- 12 inches (300 mm) for surface incandescent fixtures
- 6 inches (150 mm) for recessed incandescent fixtures
- 6 inches (150 mm) for fluorescent fixtures

Incandescent luminaires with open or partially enclosed lamps and pendant fixtures or lamp holders are not permitted in clothes closets.

NEC® *Section 410.66*

Recessed lighting fixtures installed in insulated ceilings or installed within 1/2 inch (13 mm) of combustible material must be approved for insulation contact and labeled Type IC.

Equipment Listing and Labeling

NEC® *Section 110.3*

All electrical equipment must be installed and used in accordance with the listing requirements and manufacturer's instructions. All electrical equipment, including luminaires, devices, and appliances, are *listed* and *labeled* by a nationally recognized testing laboratory (NRTL) as having been tested and found suitable for a specific purpose. Underwriters Laboratories (UL) and the Canadian Standards Association (CSA) are two of the recognized agencies.

Service Entrances

NEC® *Section 310.15*

Service entrance conductor sizes for 120/240-volt residential services must not be smaller than those given in Table 310.15(B)(6):

Service Size in Amps	Minimum CU Size	Minimum AL Size
100	4 AWG	2 AWG
150	1 AWG	2/0 AWG
200	2/0 AWG	4/0 AWG
250	4/0 AWG	300 Kcmil
300	250 Kcmil	350 Kcmil
350	350 Kcmil	500 Kcmil
400	400 Kcmil	600 Kcmil

NEC® *Section 110.14*

Conductors of dissimilar metals must not be intermixed in a terminal or splicing device unless the device is listed for the purpose. Listed antioxidant compound must be used on all aluminum conductor terminations, unless information from the device manufacturer specifically states that it is not required.

NEC® *Section 300.7*

Portions of raceways and sleeves subject to different temperatures (where passing from the interior to the exterior of a building) must be sealed with an approved material to prevent condensation from entering the service equipment.

NEC® *Section 230.54*

Where exposed to weather, service entrance conductors must be enclosed in rain-tight enclosures and arranged in a drip loop to drain.

NEC® *Section 300.4*

Where raceways containing ungrounded conductors 4 AWG or larger enter a cabinet, box, or electrical enclosure, the conductors must be protected by an insulated bushing providing a smoothly rounded insulating surface.

NEC® *Section 230.70*

The electrical service disconnecting means must be installed at a readily accessible location either outside a house or inside at a location that is nearest to the point of entrance of the service entrance conductors. No excess "inside run" is allowed.

NEC® *Sections 230.70 and 240.24*

Electrical panels containing fuses or circuit breakers must be readily accessible and must not be located in bathrooms or in the vicinity of easily ignitable materials such as clothes closets.

NEC® *Section 110.26*

Sufficient working space must be provided around electrical equipment. When the voltage to ground does not exceed 150 volts, the depth of that space in the direction of access to live parts must be a minimum of 3 feet (900 mm). The minimum width of that space in front of electrical equipment must be the width of the equipment or 30 inches (750 mm), whichever is greater. This work space must be clear and extend from the floor to a height of 6.5 feet (2 m). This space cannot be used for storage.

All work spaces must be provided with illumination.

Grounding

NEC® *Section 250.50*

The house electrical service must be connected to a grounding electrode system consisting of a metal underground water pipe in direct contact with earth for 10 feet

(3.0 m) or more. If a metal water pipe is not available as the grounding electrode, any other electrode as specified in Section 250.52 is allowed. An additional electrode must supplement the water pipe electrode. If the metal water pipe is used as part of the grounding system, a bonding jumper must be placed around the water meter.

NEC® Sections 250.64 and 250.66

The grounding electrode conductor must be unspliced and its size is determined, using the size of the service entrance conductors, by Table 250.66. The conductor that is the sole connection to a rod, pipe, or plate electrode is not required to be larger than 6 AWG copper:

Size of Service Entrance Conductor		Size of the Grounding Electrode Conductor	
Copper	Aluminum	Copper	Aluminum
4 AWG	2 AWG	8 AWG	6 AWG
1 AWG	2/0 AWG	6 AWG	4 AWG
2/0 or 3/0	4/0 or 250 Kcmil	4 AWG	2 AWG
4/0–350 Kcmil	300 Kcmil–500 Kcmil	2 AWG	1/0 AWG
400 Kcmil	600 Kcmil	1/0	3/0

NEC® Section 250.28

A main bonding jumper or the green bonding screw provided by the panel manufacturer must be installed in the service panel to electrically bond the grounded service conductor and the equipment-grounding conductors to the service enclosure.

NEC® Section 250.104

The interior metal water piping and other metal piping that may become energized must be bonded to the service equipment with a bonding jumper sized the same as the grounding electrode conductor.

Underground Wiring

NEC® Section 300.5

Direct-buried cable or conduit or other raceways must meet the following minimum cover requirements:

Direct-Buried Cable	Rigid Metal Conduit or Intermediate Metal Conduit	Rigid Nonmetallic Conduit (PVC)
24 inches (600 mm)	6 inches (150 mm)	18 inches (450 mm)

Residential branch circuits rated 20 amps or 15 amps at 120 volts and that have GFCI protection at their source are allowed a minimum cover of 12 inches (300 mm).

Underground service laterals must have their location identified by a warning ribbon placed in the trench at least 12 inches (300 mm) above the underground conductors.

Where subject to movement, direct-buried cables or raceways must be arranged to prevent damage to the enclosed conductors or connected equipment.

Conductors emerging from underground must be installed in rigid metal conduit, intermediate metal conduit, or Schedule 80 rigid nonmetallic conduit to provide protection from physical damage. This protection must extend from 18 inches (450 mm) below grade or the minimum cover distance to a height of 8 feet (2.4 m) above finished grade or to the point of termination aboveground.

Determining if the Electrical System Is Working Properly

Once an electrician has determined that all applicable *NEC®* installation requirements have been met, a check of the electrical system must be done to determine if everything is working properly. The check of the system is usually done after an electrical inspector has okayed the installation and the local electric utility has connected the house service entrance to the utility electrical system. At this time, the main service disconnecting means is turned on, which energizes the fuse or circuit breaker panel. Each circuit is then energized one by one, and each receptacle and lighting outlet on the circuit is checked for proper voltage, proper polarity of the connections, and proper switch control.

The electrical plans used as a guide for the initial installation of the electrical system can be used during the checkout. As discussed earlier in this book, it is a good idea to draw out a cabling diagram on the building electrical plan so that an electrician will know exactly which receptacle and lighting outlets, as well as which switch locations, are on a particular circuit. By comparing the outlets that are included on the electrical plan cabling diagram to the actual outlets being checked, an electrician is able to determine which outlets are working properly and which outlets are not. If a lighting outlet or power outlet is found to not be working as intended, a troubleshooting procedure will need to be initiated to find the problem. Once the cause of the problem is found, the electrician will need to correct the problem and then verify that the correction has caused the circuit and its components to operate properly.

Testing Receptacle Outlets

In Chapter 4, you were introduced to the various types of test and measurement instruments commonly used in residential wiring. You were also instructed on how to properly take a voltage reading using either a voltage tester or a voltmeter. One of the most important parts of a residential electrical system to check is the various receptacle outlets. An electrician will need to check each receptacle outlet to determine if the proper voltage is available and that the wiring connections have resulted in the proper receptacle polarity. Proper **receptacle polarity** simply means that the white-colored grounded conductor is attached to the silver screw of the receptacle and the "hot" ungrounded circuit conductor is attached to the brass screw on the receptacle. Receptacles wired with the proper polarity will ensure that electrical equipment that is cord-and-plug connected will have the "hot" ungrounded conductor going where it is supposed to go and the grounded conductor going where it is supposed to go.

CAUTION: Remember to always wear safety glasses and observe proper safety procedures when using measurement instruments to test a circuit.

Receptacle outlets can be checked with a voltage tester, a voltmeter, or a plug tester, such as the one shown in Figure 20–1. As mentioned in Chapter 4, a solenoid voltage tester is probably the most often used voltage tester used in residential wiring. It can be used to check receptacles to determine if they are energized with the correct voltage, to determine if the receptacles are wired for the proper polarity, and to determine if the receptacle is properly grounded.

(See page 553 for the correct procedure for testing 120-volt receptacles with a voltage tester.)

(See page 555 for the correct procedure for testing 240-volt receptacles with a voltage tester.)

Testing Lighting Outlets

Luminaires (lighting fixtures) are not usually taken apart and tested individually. Electricians usually test a luminaire

Figure 20–1 A plug-in receptacle tester. This device tests for correct wiring, an open ground, reversed polarity, an open "hot," the "hot" on the grounded side, and the "hot" and ground reversed. Neon lamps glow in a certain configuration to indicate the result of the test. *Courtesy of Ideal Industries, Inc.*

(light fixture) by energizing the lighting circuit that the luminaire is on and activating the switch or switches that are supposed to control the lighting outlet. It is assumed that the luminaire (lighting fixture) has been properly wired if the lamp(s) light in the luminaire when the proper switches are activated.

Since the trim-out of lighting fixtures is very straightforward, there are not a lot of things that can go wrong during their installation. It is basically a "white to white" and a "black to black" wiring connection at each fixture. The grounding connection is also very straightforward and is done by wire-nutting the circuit-grounding conductor to the fixture's grounding conductor or attaching the circuit grounding conductor to the metal body of the lighting fixture with a proper screw or clip. Thus, if the lamps in the lighting fixture do not come on when the switches are activated, troubleshooting the problem is usually limited to the following:

- Check to see if the circuit is turned off. If it is, turn it on.
- Make sure that the lamps installed in the lighting fixture are not "burned out."
- Check to see that all circuit conductor connections are tight and that they all have continuity.
- Check for damage to the circuit wiring.
- Check to see if the lighting outlet is being fed from a **buried box** that has been sheetrocked over and never trimmed out.
- Make sure the luminaire is properly grounded, especially if it is a fluorescent fixture.
- Check out the switching scheme for the proper connections. See the next section for ideas about switch checking.

Testing for Proper Switch Connections

There are a few procedures to follow when checking out the various switches used to control luminaires (lighting fixtures) in a house. The easiest switch type to check out is a single-pole switch. Remember that they are used to control a lighting load from one location. When a lighting circuit is energized, a single-pole switch can be checked out by simply opening and closing the switch and observing if the lighting fixture(s) it is controlling also turns on and off. If it does, an electrician can rightly assume that the switch is connected properly.

Three-way and four-way switching arrangements are not as simple to check out since there are several different switching combinations. The most often found problem with three-way and four-way switches is having an electrician mix up the traveler wires when the switches were installed as part of the trim-out process. This can be as simple as having a traveler wire attached to a common (black) terminal on a three-way switch or having all traveler wires mixed up with each other on a four-way switch.

(See page 559 for the correct procedure for testing a standard three-way switching arrangement.)

The following example shows what can happen when a traveler conductor and a common conductor in a three-way switching circuit are mixed up (Figure 20–2).

• When three-way switch A is toggled, the lighting load is turned on and then off (Figure 20–3).
• When three-way switch A is left in the down position, toggling three-way switch B will also turn the lighting load on and off (Figure 20–4).

• However, with three-way switch A in the up position, the lighting load cannot be turned on regardless of the position of three-way switch B (Figure 20–5).

CAUTION: In order to ensure that three-way and four-way switching systems are working properly, it is necessary to test the switches in all possible configurations.

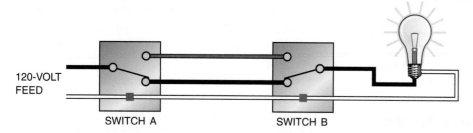

AN EXAMPLE OF THE WIRING FOR A NORMAL THREE-WAY SWITCHING SYSTEM. NOTICE THAT THE TOGGLING OF EITHER SWITCH WILL REVERSE THE ON-OFF STATUS OF THE LAMP.

Figure 20–2 A three-way switching connection that is properly wired. Toggling of either switch will turn the lamp on or off.

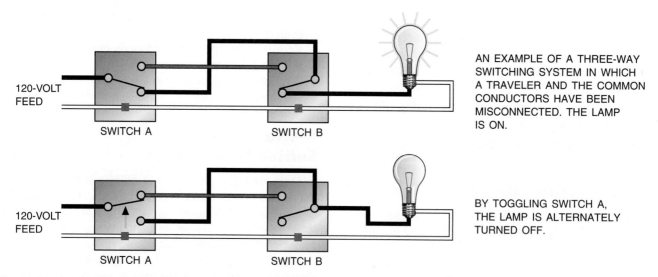

AN EXAMPLE OF A THREE-WAY SWITCHING SYSTEM IN WHICH A TRAVELER AND THE COMMON CONDUCTORS HAVE BEEN MISCONNECTED. THE LAMP IS ON.

BY TOGGLING SWITCH A, THE LAMP IS ALTERNATELY TURNED OFF.

Figure 20–3 A three-way switching connection in which a traveler wire and the common wire have been misconnected. Toggling switch A will cause the lamp to go on or off.

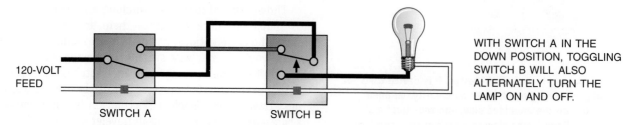

WITH SWITCH A IN THE DOWN POSITION, TOGGLING SWITCH B WILL ALSO ALTERNATELY TURN THE LAMP ON AND OFF.

Figure 20–4 When switch A is left in the down position, toggling switch B will also cause the lamp to go on or off.

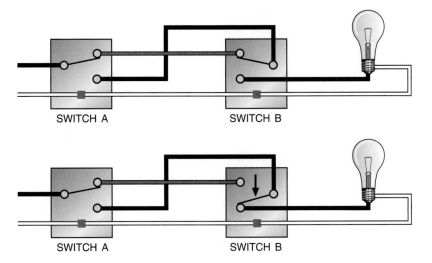

WITH SWITCH A IN THE UP POSITION, THE LAMP IS OFF AND WILL REMAIN OFF REGARDLESS OF THE POSITION OF SWITCH B.

SWITCH A   SWITCH B

SWITCH A   SWITCH B

IN ORDER TO ENSURE THAT THE THREE-WAY AND FOUR-WAY SWITCHING SYSTEMS ARE WORKING PROPERLY, IT IS NECESSARY TO TEST THE SYSTEM IN ALL POSSIBLE SWITCH POSITION CONFIGURATIONS.

Figure 20–5 With switch A in the up position, the lamp is off and will stay off no matter which position switch B is in.

Troubleshooting Common Residential Electrical Circuit Problems

The previous sections in this chapter presented some common checkout procedures for receptacles, luminaires (lighting fixtures), and switches. The testing procedures are relatively straightforward, and if something is found to not work properly, troubleshooting procedures are initiated to find the cause of the problem, and the problem is fixed. This is great if you are checking out and testing receptacles, lighting fixtures, or switches. But what happens when you turn on a circuit breaker to energize a circuit for testing and the circuit breaker immediately trips off. What if a circuit breaker is turned on to energize a circuit for testing, and even though you have turned the breaker on, there is no electrical power on the circuit? These situations are common and require good troubleshooting techniques to locate the problem and fix it.

In the first scenario described in the previous paragraph, the likely cause of the circuit breaker tripping (or fuse blowing) as soon as the circuit is energized is a **short-circuit** condition. This condition is an unintended low-resistance path through which current flows around rather than along a circuit's intended current path. It results when two circuit conductors come in contact with each other unintentionally. Because the short-circuit current path has a very low resistance, the actual current flow on a 120- or 240-volt residential circuit becomes very high and causes a circuit breaker to trip the instant the circuit is energized (Figure 20–6). Until the short circuit is fixed, the circuit breaker will not stay on.

Another possible cause for having a circuit breaker trip (or fuse blow) as soon as a circuit is energized is a **ground fault** condition. A ground fault is an unintended low-resistance path in an electrical circuit through which some current flows to ground using a pathway other than the intended pathway. It results when an ungrounded "hot" conductor unintentionally touches a grounded surface or grounded conductor (Figure 20–7). In the second scenario, the likely cause of no electrical power in a circuit (or parts of a circuit) is an **open-circuit** condition. This condition results in a circuit that is energized but does not allow useful current to flow on the circuit because of a break in the current path (Figure 20–8). Knowing the likely cause of a problem in a circuit is great, but finding the location of the problem in the circuit will require troubleshooting the circuit.

(See page 561 for the correct procedure for determining which receptacle outlet on a circuit has a ground fault using a continuity tester.)

 FROM EXPERIENCE

There are many times when a ground-fault condition is caused by a bare grounding conductor making contact with one of the screw terminals in the box. Removing the switch cover will often reveal the problem. Moving the bare grounding conductor away from the terminal screw that it was touching will take care of the problem.

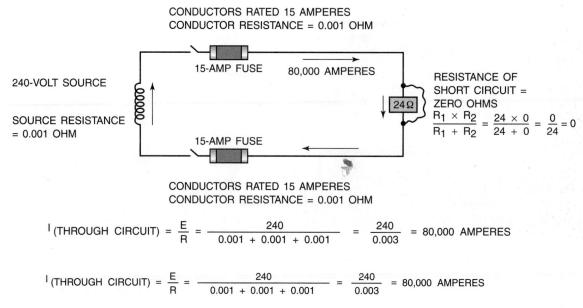

$$I \text{ (THROUGH CIRCUIT)} = \frac{E}{R} = \frac{240}{0.001 + 0.001 + 0.001} = \frac{240}{0.003} = 80,000 \text{ AMPERES}$$

$$I \text{ (THROUGH CIRCUIT)} = \frac{E}{R} = \frac{240}{0.001 + 0.001 + 0.001} = \frac{240}{0.003} = 80,000 \text{ AMPERES}$$

Figure 20–6 An example of a short circuit. The circuit conductors will get extremely hot because of the high current flow. The conductor insulation will melt off, and the conductors themselves will melt unless the fuses open the circuit in a very short amount of time.

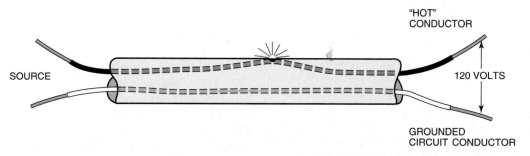

"HOT" CONDUCTOR COMES IN CONTACT WITH METAL RACEWAY OR OTHER METAL OBJECT. IF THE RETURN GROUND PATH HAS LOW RESISTANCE (IMPEDANCE), THE OVERCURRENT DEVICE PROTECTING THE CIRCUIT WILL CLEAR THE FAULT. IF THE RETURN GROUND PATH HAS HIGH RESISTANCE (IMPEDANCE), THE OVERCURRENT DEVICE WILL NOT CLEAR THE FAULT. THE METAL OBJECT WILL THEN HAVE A VOLTAGE TO GROUND THE SAME AS THE "HOT" CONDUCTOR HAS TO GROUND. IN HOUSE WIRING, THIS VOLTAGE TO GROUND IS 120 VOLTS. PROPER GROUNDING AND GROUND-FAULT CIRCUIT INTERRUPTER PROTECTION IS DISCUSSED ELSEWHERE IN THIS TEXT. THE CALCULATION PROCEDURE FOR A GROUND FAULT IS THE SAME AS FOR A SHORT CIRCUIT; HOWEVER, THE VALUES OF "R" CAN VARY GREATLY BECAUSE OF THE UNKNOWN IMPEDANCE OF THE GROUND RETURN PATH. LOOSE LOCKNUTS, BUSHINGS, SET SCREWS ON CONNECTORS AND COUPLINGS, POOR TERMINATIONS, RUST, AND SO ON. ALL CONTRIBUTE TO THE RESISTANCE OF THE RETURN GROUND PATH, MAKING IT EXTREMELY DIFFICULT TO DETERMINE THE ACTUAL GROUND-FAULT CURRENT VALUES.

Figure 20–7 An example of a ground fault. A section of the "hot" circuit conductor that has had its insulation scraped off during the installation process is touching a grounded surface.

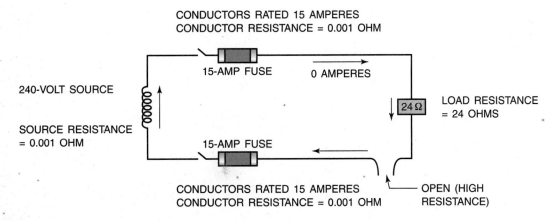

Figure 20-8 An example of an open circuit. There is no current flow through the circuit when there is an "open" in the circuit. Notice that even without current flow, a voltage can be present on the circuit.

When performing troubleshooting tests with a continuity tester, electricians need to be aware of "feed-through," which occurs when a load, like a lightbulb, is still connected in a branch circuit that is being tested. The current from the continuity tester can flow through the load and cause a false reading. When testing lighting branch circuits with a continuity tester, remove all lamps from lighting fixtures. When using a continuity tester to test a receptacle circuit, make sure there are no portable items plugged into any of the receptacles on the circuit. Figure 20-9 shows some testing situations on a lighting circuit with the load still connected in the circuit.

Summary

An important stage in the installation of a residential electrical system is making sure that all *NEC*®, local, and state electrical requirements have been met. This chapter presented a checklist of common *NEC*® requirements that need to be included in virtually all residential wiring installations. If an electrician finds that any of the items listed are not included in the electrical system, he or she must bring the installation up to code.

The other very important part of this last stage is checking out each circuit to verify that everything is working the way it is supposed to before you leave this job and go on to the next one. Several testing procedures were presented in this chapter. If an electrical circuit is found to be working improperly or not at all, troubleshooting procedures will need to be started so that the reason for the problem can be found and fixed. This chapter presented some common troubleshooting procedures, including the "halving the circuit" method for finding problems in a circuit.

As we learned earlier in this book, the *NEC*® requires electricians to install a residential electrical system in a neat and workmanlike manner. While many people interpret this to mean different things, there is one interpretation that all electrical people agree on: The electrical installation must be installed in a professional manner and must meet the minimum standards of the *NEC*®. This will result in an electrical system that should give reliable and safe service to the home owner for a long, long time.

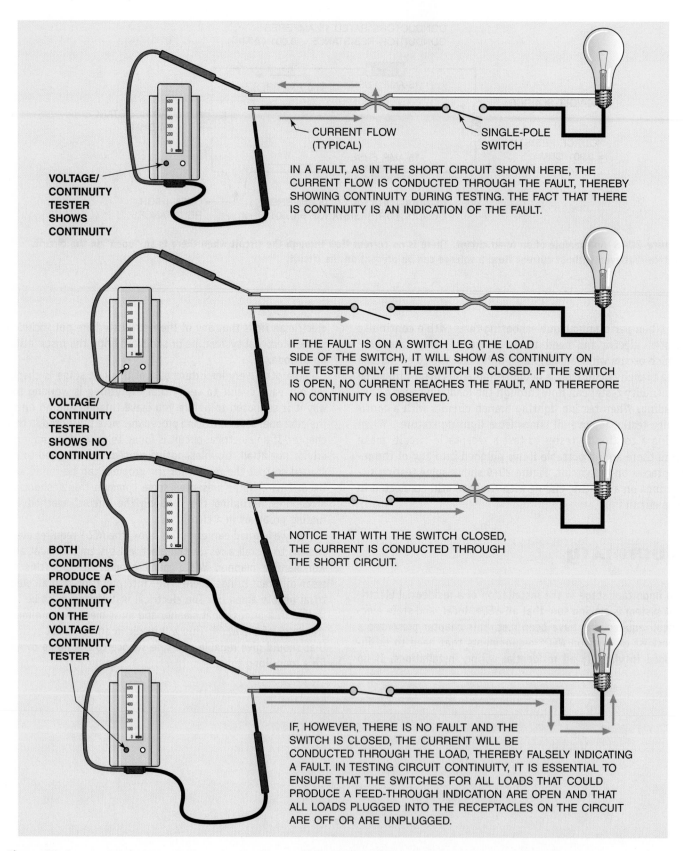

VOLTAGE/
CONTINUITY
TESTER
SHOWS
CONTINUITY

CURRENT FLOW
(TYPICAL)

SINGLE-POLE
SWITCH

IN A FAULT, AS IN THE SHORT CIRCUIT SHOWN HERE, THE
CURRENT FLOW IS CONDUCTED THROUGH THE FAULT, THEREBY
SHOWING CONTINUITY DURING TESTING. THE FACT THAT THERE
IS CONTINUITY IS AN INDICATION OF THE FAULT.

VOLTAGE/
CONTINUITY
TESTER
SHOWS NO
CONTINUITY

IF THE FAULT IS ON A SWITCH LEG (THE LOAD
SIDE OF THE SWITCH), IT WILL SHOW AS CONTINUITY ON
THE TESTER ONLY IF THE SWITCH IS CLOSED. IF THE SWITCH
IS OPEN, NO CURRENT REACHES THE FAULT, AND THEREFORE
NO CONTINUITY IS OBSERVED.

BOTH
CONDITIONS
PRODUCE A
READING OF
CONTINUITY
ON THE
VOLTAGE/
CONTINUITY
TESTER

NOTICE THAT WITH THE SWITCH CLOSED,
THE CURRENT IS CONDUCTED THROUGH
THE SHORT CIRCUIT.

IF, HOWEVER, THERE IS NO FAULT AND THE
SWITCH IS CLOSED, THE CURRENT WILL BE
CONDUCTED THROUGH THE LOAD, THEREBY FALSELY INDICATING
A FAULT. IN TESTING CIRCUIT CONTINUITY, IT IS ESSENTIAL TO
ENSURE THAT THE SWITCHES FOR ALL LOADS THAT COULD
PRODUCE A FEED-THROUGH INDICATION ARE OPEN AND THAT
ALL LOADS PLUGGED INTO THE RECEPTACLES ON THE CIRCUIT
ARE OFF OR ARE UNPLUGGED.

Figure 20–9 A feed-through condition may cause false readings during testing. When a load is connected in the circuit under test, the current from the continuity tester may flow through the load and indicate a fault between the "hot" ungrounded conductor and the grounded conductor. When using a continuity tester to troubleshoot a circuit, always disconnect the load from the circuit.

Procedures

Testing 120-Volt Receptacles with a Voltage Tester to Determine Proper Voltage, Polarity, and Grounding

- Wear safety glasses and observe all applicable safety rules.

(A) Test between the ungrounded conductor connection (short slot) and the grounded conductor connection (long slot). The meter should indicate 120 volts.

(A)

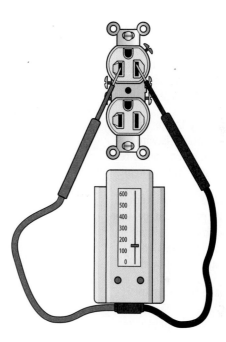

(B) Test between the ungrounded conductor connection (short slot) and the grounding conductor connection (U-shaped grounding slot). The meter should indicate 120 volts.

(B)

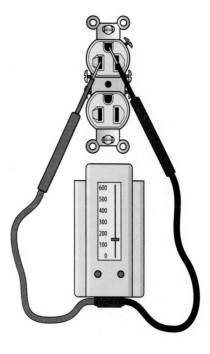

Procedures

Testing 120-Volt Receptacles with a Voltage Tester to Determine Proper Voltage, Polarity, and Grounding (continued)

C Test between the grounded conductor connection (long slot) and the grounding conductor connection (U-shaped grounding slot). The meter should indicate 0 volts.

- If the tests produce the proper voltage, the receptacle is wired properly, and you can go on to test the next receptacle on the circuit.

- If the tests in the previous steps do not produce the proper voltage, there is a problem with the circuit wiring, and the following troubleshooting procedures should be followed:

  - Check to see if the circuit is turned off.
  - Check to see that all circuit conductor connections are tight and that they all have continuity.
  - Check for damage to the circuit wiring.
  - Check to see if the receptacle is being fed from a "buried box" that has been sheetrocked over and never trimmed out.
  - Correct the problem(s) found during troubleshooting and retest the receptacle for proper voltage and grounding.

C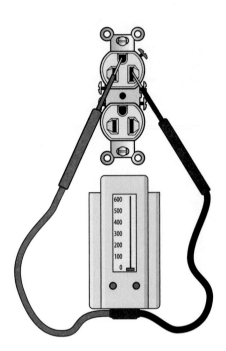

- If the test in step 'C' produces a voltage of 120 volts, receptacle polarity is reversed, and the following troubleshooting procedures should be followed:

  - Turn off the circuit power.
  - Remove the receptacle from the outlet box.
  - Reverse the wire connections so that the "white" grounded conductor is attached to a silver screw and the "hot" ungrounded conductor is attached to a brass screw.
  - Turn on the power and retest the receptacle for proper polarity.

Procedures

Testing 120/240-Volt Range and Dryer Receptacles with a Voltage Tester to Determine Proper Voltage, Polarity, and Grounding

- Wear safety glasses and observe all applicable safety rules.

A Test between the ungrounded conductor connections. The meter should indicate 240 volts.

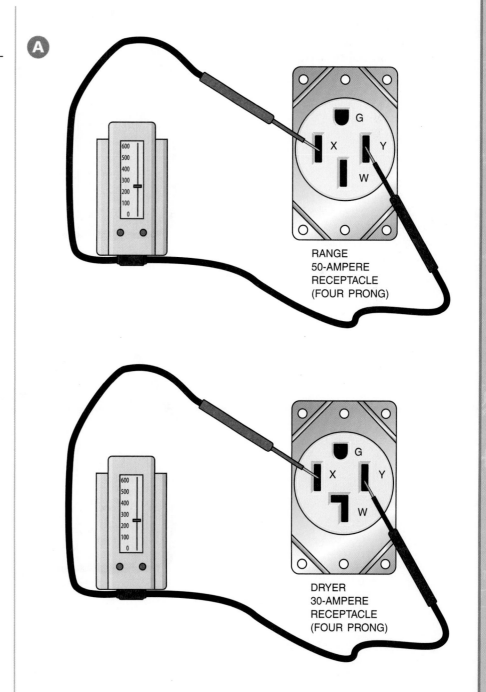

RANGE
50-AMPERE
RECEPTACLE
(FOUR PRONG)

DRYER
30-AMPERE
RECEPTACLE
(FOUR PRONG)

Testing 120/240-Volt Range and Dryer Receptacles with a Voltage Tester to Determine Proper Voltage, Polarity, and Grounding (continued)

B Test between one of the ungrounded conductor connections and the grounded conductor connection. The meter should indicate 120 volts.

C Test between the other ungrounded conductor connection and the grounded conductor connection. The meter should indicate 120 volts.

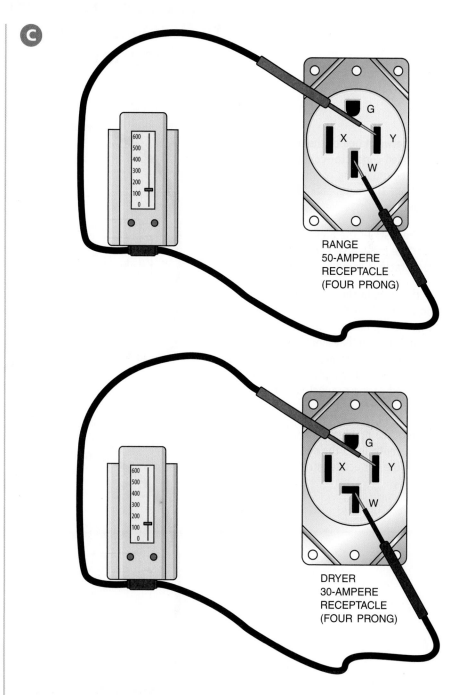

RANGE
50-AMPERE
RECEPTACLE
(FOUR PRONG)

DRYER
30-AMPERE
RECEPTACLE
(FOUR PRONG)

Procedures

Testing 120/240-Volt Range and Dryer Receptacles with a Voltage Tester to Determine Proper Voltage, Polarity, and Grounding (continued)

D Test between one of the ungrounded conductor connections and the grounding conductor connection. The meter should indicate 120 volts.

- If the tests in the previous steps produce the proper voltage, the receptacle is wired properly.

- If the tests in the previous steps do not produce the proper voltage, there is a problem with the circuit wiring, and troubleshooting procedures will need to begin. More than likely, it is simply a matter of the feed wiring connected to the wrong terminals.

- Correct the problem by rewiring the receptacle so the correct wires go to the correct terminals.

D

RANGE
50-AMPERE
RECEPTACLE
(FOUR PRONG)

DRYER
30-AMPERE
RECEPTACLE
(FOUR PRONG)

Procedures

Testing a Standard Three-Way Switching Arrangement

- Wear safety glasses and observe all applicable safety rules.

- Energize the three-way switching circuit.

A Position the toggle on three-way switch A so that the lighting load is energized.

- At three-way switch A, toggle it on and off and observe that the lighting load also goes on and off. Leave the switch in a position that keeps the lighting load on.

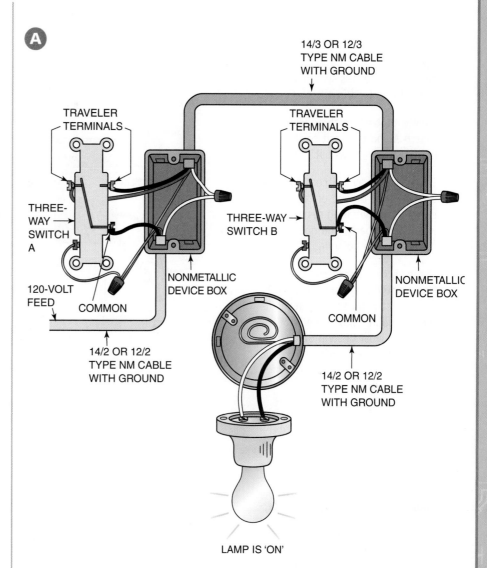

A

14/3 OR 12/3 TYPE NM CABLE WITH GROUND

TRAVELER TERMINALS

TRAVELER TERMINALS

THREE-WAY SWITCH A

THREE-WAY SWITCH B

120-VOLT FEED

COMMON

COMMON

NONMETALLIC DEVICE BOX

NONMETALLIC DEVICE BOX

14/2 OR 12/2 TYPE NM CABLE WITH GROUND

14/2 OR 12/2 TYPE NM CABLE WITH GROUND

LAMP IS 'ON'

Procedures

Testing a Standard Three-Way Switching Arrangement (continued)

B Go to three-way switch B and toggle it so the lighting load goes on and off. Leave this switch so that the lighting load is off.

- Finally, go back to three-way switch A and toggle it so the lighting load comes back on.

- If the three-way switches turn the lighting load on and off as required, the switching system was installed properly, and nothing more needs to be done.

- If the three-way switches do not turn the lighting load on and off as the test requires, troubleshooting procedures should be initiated. The most likely problem will be an improper connection of the traveler wires on one or both of the three-way switches.

- Turn off the electrical power to the circuit before initiating troubleshooting procedures. Lock out the power.

- Verify that the traveler wires are correctly wired to the traveler terminals of the three-way switch.

- If the traveler wires are incorrectly connected, rewire the circuit, turn the electrical power on, and retest the circuit. The switches should now work properly.

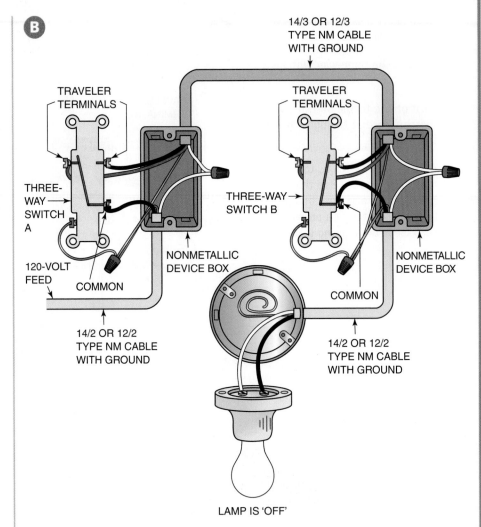

B

14/3 OR 12/3 TYPE NM CABLE WITH GROUND

TRAVELER TERMINALS

TRAVELER TERMINALS

THREE-WAY SWITCH A

THREE-WAY SWITCH B

120-VOLT FEED

COMMON

NONMETALLIC DEVICE BOX

NONMETALLIC DEVICE BOX

COMMON

14/2 OR 12/2 TYPE NM CABLE WITH GROUND

14/2 OR 12/2 TYPE NM CABLE WITH GROUND

LAMP IS 'OFF'

Procedures

Determining Which Receptacle Outlet Box Has a Ground Fault Using a Continuity Tester

- Wear safety glasses and observe all applicable safety rules.

A In the electrical panel where the circuit originates, turn off the circuit breaker and disconnect the circuit ungrounded conductor (the black wire) from the breaker, the circuit grounded conductor (the white wire) from the neutral terminal bar, and the circuit equipment-grounding conductor (the bare or green wire) from the grounding terminal.

B At the electrical panel, use a continuity tester and test between the ungrounded circuit conductor (black) of the circuit and the grounded (white) circuit conductor. If there is continuity, then the problem is a short circuit. Then test between the ungrounded circuit conductor (black) and the circuit-grounding conductor (bare or green). If there is continuity, then the problem is a ground fault.

- To find out where either the short circuit or ground fault is located on the circuit, locate a receptacle outlet box that is approximately midway on the circuit. Consulting a cabling diagram of the branch circuit will help with this determination.

- At the midpoint receptacle outlet box, remove the circuit conductors. This will separate the circuit into two halves: the line side and the load side.

A
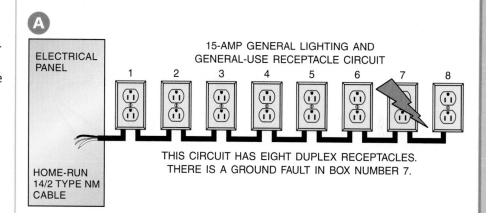
THIS CIRCUIT HAS EIGHT DUPLEX RECEPTACLES. THERE IS A GROUND FAULT IN BOX NUMBER 7.

B
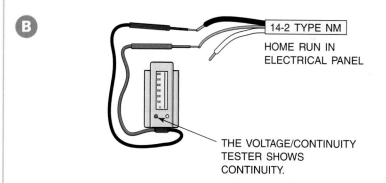

ocedures

Determining Which Receptacle Outlet Box Has a Ground Fault Using a Continuity Tester (continued)

C At the receptacle outlet box, perform the same tests with the continuity tester as described previously. This will allow you to determine if the problem is in the wiring from the electrical panel to your location (the line side) or in the wiring from your location to the end of the circuit (the load side).

C

NO CONTINUITY INDICATED

CONTINUITY IS INDICATED

TEST THE LOAD-SIDE PHASE AND GROUNDING CONDUCTORS FOR CONTINUITY. IF THERE IS CONTINUITY, THE FAULT IS SOMEWHERE ON THAT HALF OF THE CIRCUIT.

D Once you determine which half of the branch circuit contains the problem, split the part of the circuit you are working with in half and repeat the continuity testing.

D

NO CONTINUITY INDICATED

CONTINUITY IS INDICATED

E Continue to split the part of the circuit you are working with in half and test for continuity until you locate the problem.

- Once the problem is located, make the necessary repairs, reconnect the circuit components, reconnect the wiring in the electrical panel, and retest the circuit.

E

PROBLEM IS LOCATED HERE

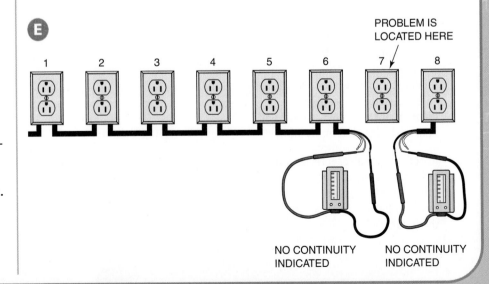

NO CONTINUITY INDICATED

NO CONTINUITY INDICATED

Review Questions

Directions: Answer the following items with clear and complete responses.

1 An electrician tests a 15-amp, 125-volt-rated duplex receptacle with a voltage tester. The tester indicates 120 volts between the ungrounded (short) slot and the grounded (long) slot. The tester also indicates 120 volts between the grounded (long) slot and the grounding slot (U-shaped). The likely cause is

a. an open equipment-grounding path.

b. reversed polarity at the receptacle.

c. an open grounded circuit conductor.

d. a short circuit in the cable feeding the receptacle.

2 An electrician tests a 15-amp, 125-volt-rated duplex receptacle with a voltage tester. The tester indicates 120 volts between the ungrounded (short) slot and the grounded (long) slot. The tester also indicates 0 volts between the ungrounded (short) slot and the grounding slot (U-shaped). The likely cause is

a. an open equipment-grounding path.

b. reversed polarity at the receptacle.

c. an open grounded circuit conductor.

d. a short circuit in the cable feeding the receptacle.

3 When checking out the electrical circuits for correct operation, an electrician should turn them all on at the same time. True or False.

4 When troubleshooting for short circuits or ground faults, the circuit being tested should be energized with electrical power. True or False.

5 A continuity tester can be used to find short circuits, ground faults, and open circuits in deenergized circuits. True or False.

Glossary

accessible (as applied to wiring methods) capable of being removed or exposed without damaging the building structure or finish or not permanently closed in by the structure or finish of the building

accessible, readily (readily accessible) capable of being reached quickly for operation, renewal, or inspections without requiring a person to climb over or remove obstacles or to use portable ladders

ambient temperature the temperature of the air that surrounds an object on all sides

American Wire Gauge (AWG) a scale of specified diameters and cross sections for wire sizing that is the standard wire-sizing scale in the United States

ammeter (clamp-on) a measuring instrument that has a movable jaw that is opened and then clamped around a current-carrying conductor to measure current flow

ammeter (in-line) a measuring instrument that is connected in series with a load and measures the amount of current flow in the circuit

ampacity the current in amperes that a conductor can carry continuously under the conditions of use without exceeding its temperature rating

ampere the unit of measure for electrical current flow

analog meter a meter that uses a moving pointer (needle) to indicate a value on a scale

approved when a piece of electrical equipment is approved, it means that it is acceptable to the authority having jurisdiction (AHJ)

arc the flow of a high amount of current across an insulating medium, like air

arc blast a violent electrical condition that causes molten metal to be thrown through the air

arc fault circuit interrupter (AFCI) circuit breaker a device intended to provide protection from the effects of arc faults by recognizing characteristics unique to arcing and by functioning to deenergize the circuit when an arc fault is detected

arc fault circuit interrupter (AFCI) receptacle a receptacle device designed to control dangerous arcs that occur at receptacles and other outlets in residential bedrooms

architect a qualified person who creates and designs drawings for a residential construction project

auger a drill bit type with a spiral cutting edge used to bore holes in wood

auto-ranging meter a meter feature that automatically selects the range with the best resolution and accuracy

backboard the surface on which a service panel or subpanel is mounted; it is usually made of plywood and is painted a flat black color

backfeeding a wiring technique that allows electrical power from an existing electrical panel to be fed to a new electrical panel by a short length of cable; this technique is commonly used when an electrician is upgrading an existing service entrance

back-to-back bend a type of conduit bend that is formed by two 90-degree bends with a straight length of conduit between the two bends

ballast a component in a fluorescent lighting fixture that controls the voltage and current flow to the lamp

balloon frame a type of frame in which studs are continuous from the foundation sill to the roof; this type of framing is found mostly in older homes

band joist the framing member used to stiffen the ends of the floor joists where they rest on the sill

bandwidth identifies the amount of data that can be sent on a given cable; it is measured in hertz (Hz) or megahertz (MHz); a higher frequency means higher data-sending capacity

bathroom branch circuit a branch circuit that supplies electrical power to receptacle outlets in a bathroom; lighting outlets may also be served by the circuit as long as other receptacle or lighting outlets outside the bathroom are not connected to the circuit; it is rated at 20 amperes

bender a tool used to make various bends in electrical conduit raceway

bimetallic strip a part of a circuit breaker that is made from two different metals with unequal thermal expansion rates; as the strip heats up, it will tend to bend

blueprint architectural drawings used to represent a residential building; it is a copy of the original drawings of the building

bonding the permanent joining of metal parts to form an electrically conductive path that ensures electrical continuity and the capacity to conduct safely any current likely to be imposed on the metal object; the purpose of bonding is to establish an effective path for fault current that facilitates the operation of the overcurrent protective device

bonding jumper a conductor used to ensure electrical conductivity between metal parts that are required to be electrically connected

bottom plate the lowest horizontal part of a wall frame which rests on the subfloor

box fill the total space taken up in an electrical box by devices, conductors, and fittings; box fill is measured in cubic inches

box offset bend a type of conduit bend that uses two equal bends to cause a slight change of direction for a conduit at the point where it is attached to an electrical box

branch circuit the circuit conductors between the final overcurrent device (fuse or circuit breaker) and the outlets

break lines lines used to show that part of the actual object is longer than what the drawing is depicting

buried box an electrical box that has been covered over with sheetrock or some other building material

bushing (insulated) a fiber or plastic fitting designed to screw onto the ends of conduit or a cable connector to provide protection to the conductors

cabinet an enclosure for a panel board that is designed for either flush or surface mounting; a swinging door is provided

cable a factory assembly of two or more insulated conductors that have an outer sheathing that holds everything together; the outside sheathing can be metallic or nonmetallic

cable hook also called an "eyebolt"; it is the part used to attach the service drop cable to the side of a house in an overhead service entrance installation

candlepower a measure of lighting intensity

cartridge fuse a fuse enclosed in an insulating tube that confines the arc when the fuse blows; this fuse may be either a ferrule or a blade type

category ratings on the bandwidth performance of UTP cable; categories include 3, 4, 5, 5e, and 6; Category 5 is rated to 100 MHz and is the most widely used

ceiling joists the horizontal framing members that rest on top of the wall framework and form the ceiling

centerline a series of short and long dashes used to designate the center of items, such as windows and doors

checkout the process of determining if all parts of a recently installed electrical system are functioning properly

chuck key a small wrench, usually in a T shape, used to open or close a chuck on a power drill

circuit breaker a device designed to open and close a circuit manually and to open the circuit automatically on a predetermined overcurrent without damage to itself when properly applied within its rating

circuit (electrical) an arrangement consisting of a power source, conductors, and a load

circular mils the diameter of a conductor in mils (thousandths of inches) times itself; the number of circular mils is the cross-sectional area of a conductor

Class P ballast a ballast with a thermal protection unit built in by the manufacturer; this unit opens the lighting electrical circuit if the ballast temperature exceeds a specified level

coaxial cable a cable in which the center signal-carrying conductor is centered within an outer shield and separated from the conductor by a dielectric; used to deliver video signals in residential structured cabling installations

combination switch a device with more than one switch type on the same strap or yoke

computer-aided drafting (CAD) the making of building drawings using a computer

concentric knockout a series of removable metal rings that allow the knockout size to vary according to how many of the metal rings are removed; the center of the knockout hole stays the same as more rings are removed; some standard residential wiring sizes are 1/2, 3/4, 1, 1¼, 1½, 2, and 2½ inches

conductor a material that allows electrical current to flow through it; examples are copper, aluminum, and silver

conduit a raceway with a circular cross section, such as electrical metallic tubing, rigid metal conduit, or intermediate metal conduit;

conduit "LB" a piece of electrical equipment that is connected in-line with electrical conduit to provide for a 90° change of direction

connector a fitting that is designed to secure a cable or length of conduit to an electrical box

continuity tester a testing device used to indicate whether there is a continuous path for current flow through an electrical circuit or circuit component

copper-clad aluminum an aluminum conductor with an outer coating of copper that is bonded to the aluminum core

cord-and-plug connection an installation technique where electrical appliances are connected to a branch circuit with a flexible cord with an attachment plug; the attachment plug end is plugged into a receptacle of the proper size and type

counter-mounted cooktop a cooking appliance that is installed in the top of a kitchen countertop; it contains surface cooking elements, usually two large elements and two small elements

crimp a process used to squeeze a solderless connector with a tool so that it will stay on a conductor

critical loads the electrical loads that are determined to require electrical power from a standby power generator when electrical power from the local electric utility company is interrupted

current the intensity of electron flow in a conductor

cutter a hardened steel device used to cut holes in metal electrical boxes

detail drawing a part of the building plan that shows an enlarged view of a specific area

deteriorating agents a gas, fume, vapor, liquid, or any other item that can cause damage to electrical equipment

device a piece of electrical equipment that is intended to carry but not use electrical energy; examples include switches, lamp holders, and receptacles

device box an electrical device that is designed to hold devices such as switches and receptacles

die the component of a knockout punch that works in conjunction with the cutter and is placed on the opposite side of the metal box or enclosure

digital meter a meter where the indication of the measured value will be given as an actual number in a liquid crystal display (LCD)

dimension a measurement of length, width, or height shown on a building plan

dimension line a line on a building plan with a measurement that indicates the dimension of a particular object

dimmer switch a switch type that raises or lowers the lamp brightness of a lighting fixture

disconnecting means a term used to describe a switch that is able to deenergize an electrical circuit or piece of electrical equipment; sometimes referred to as the "disconnect"

DMM a series of letters that stands for "digital multimeter"

double insulated an electrical power tool type constructed so the case is isolated from electrical energy and is made of a nonconductive material

double-pole switch a switch type used to control two separate 120-volt circuits or one 240-volt circuit from one location

draft-stops also called "fire-stops"; the material used to reduce the size of framing cavities in order to slow the spread of fire; in wood frame construction, it consists of full-width dimension lumber placed between studs or joists

drip loop an intentional loop put in service entrance conductors at the point where they extend from a weatherhead; the drip loop conducts rainwater to a lower point than the weatherhead, helping to ensure that no water will drip down the service entrance conductors and into the meter enclosure

dry-niche luminaire a lighting fixture intended for installation in the wall of a pool that goes in a niche; it has a fixed lens that seals against water entering the niche and surrounding the lighting fixture

dual-element time-delay fuses a fuse type that has a time-delay feature built into it; this fuse type is used most often as an overcurrent protection device for motor circuits

duplex receptacle the most common receptacle type used in residential wiring; it has two receptacles on one strap; each receptacle is capable of providing power to a cord-and-plug-connected electrical load

dwelling unit one or more rooms for the use of one or more persons as a housekeeping unit with space for eating, living, and sleeping and permanent provisions for cooking and sanitation

eccentric knockout a series of removable metal rings that allow a knockout size to vary according to how many of the metal rings are removed; the center of the knockout hole changes as more metal rings are removed; common sizes are the same as for concentric knockouts

Edison-base plug fuse a fuse type that uses the same standard screw base as an ordinary lightbulb; different fuse sizes are interchangeable with each other; this fuse type may only be used as a replacement for an existing Edison-base plug fuse

efficacy a rating that indicates the efficiency of a light source

EIA/TIA an acronym for the Electronic Industry Association and Telecommunications Industry Association; these organizations create and publish compatibility standards for the products made by member companies

EIA/TIA 570-A the main standard document for structured cabling installations in residential situations

electric range a stand-alone electric cooking appliance that typically has four cooking elements on top and an oven in the bottom of the appliance

electrical drawings a part of the building plan that shows the electrical supply and distribution for the building electrical system

electrical shock the sudden stimulation of nerves and muscle caused by electricity flowing through the body

elevation drawing a drawing that shows the side of the house that faces in a particular direction; for example, the north elevation drawing shows the side of the house that is facing north

enclosure the case or housing of apparatus or the fence or walls surrounding an installation to prevent personnel from accidentally contacting energized parts or to protect the equipment from physical damage

equipment a general term including material, fittings, devices, appliances, luminaires (lighting fixtures), apparatus, and other parts used in connection with an electrical installation

equipment-grounding conductor the conductor used to connect the non-current-carrying metal parts of equipment, raceways, and other enclosures to the system-grounded conductor, the grounding electrode conductor, or both at the service equipment

exothermic welding a process for making bonding connections on the bonding grid for a permanently installed swimming pool using specially designed connectors, a form, a metal disk, and explosive powder; this process is sometimes called "Cad-Weld"

extension lines lines used to extend but not actually touch object lines and have the dimension lines drawn between them

feeder the circuit conductors between the service equipment and the final branch-circuit overcurrent protection device

field bend any bend or offset made by installers, using proper tools and equipment, during the installation of conduit systems

fish the process of installing cables in an existing wall or ceiling

fitting an electrical accessory, like a locknut, that is used to perform a mechanical rather than an electrical function

floor joists horizontal framing members that attach to the sill plate and form the structural support for the floor and walls

floor plan a part of the building plan that shows a bird's-eye view of the layout of each room

fluorescent lamp a gaseous discharge light source; light is produced when the phosphor coating on the inside of a sealed glass tube is struck by energized mercury vapor

foot-candle the unit used to measure how much total light is reaching a surface; 1 lumen falling on 1 square foot of surface produces an illumination of 1 foot-candle

footing the concrete base on which a dwelling foundation is constructed; it is located below grade

forming shell the support structure designed and used with a wet-niche lighting fixture; it is installed in the wall of a pool

foundation the base of the structure, usually poured concrete or concrete block, on which the framework of the house is built; it sits on the footing

four-way switch a switch type that, when used in conjunction with two three-way switches, will allow control of a 120-volt lighting load from more than two locations

framing the building "skeleton" that provides the structural framework of the house

F-type connector a 75-ohm coaxial cable connector that can fit RG-6 and RG-59 cables and is used for terminating video system cables in residential wiring applications

fuse that opens a circuit when the fusible link is melted away by the extreme heat caused by an overcurrent

ganging the joining together of two or more device boxes for the purpose of holding more than one device

general lighting circuit a branch-circuit type used in residential wiring that has both lighting and receptacle loads connected to it; a good example of this circuit type is a bedroom branch circuit that has both receptacles and lighting outlets connected to it

generator a rotating machine used to convert mechanical energy into electrical energy

girders heavy beams that support the inner ends of floor joists

ground a conducting connection, whether intentional or accidental, between an electrical circuit or equipment and the earth or to some conducting body that serves in place of the earth

ground fault an unintended low-resistance path in an electrical circuit through which some current flows to ground using a pathway other than the intended pathway; it results when an ungrounded "hot" conductor unintentionally touches a grounded surface or grounded conductor

ground fault circuit interrupter (GFCI) a device that protects people from dangerous levels of electrical current by measuring the current difference between two conductors of an electrical circuit and tripping to an open position if the measured value exceeds 6 milliamperes

grounded connected to earth or to some conducting body that serves in place of the earth

grounded conductor a system or circuit conductor that is intentionally grounded

grounding an electrical connection to an object that conducts electrical current to the earth

grounding conductor a conductor used to connect equipment or the grounded conductor of a wiring system to a grounding electrode or electrodes

grounding electrode a part of the building service entrance that connects the grounded service (neutral) conductor to the earth

grounding electrode conductor the conductor used to connect the grounding electrode to the equipment-grounding conductor, to the grounded conductor, or to both at the service

handy box a type of metal device box used to hold only one device; it is surface mounted

hardwired an installation technique where the circuit conductors are brought directly to an electrical appliance and terminated at the appliance

harmonics a frequency that is a multiple of the 60 Hz fundamental; harmonics cause distortion of the voltage and current AC waveforms

hazard a potential source of danger

heat pump a reversible air-conditioning system that will heat a house in cool weather and cool a house in warm weather

hidden line a line on a building plan that shows an object hidden by another object on the plan; hidden lines are drawn using a dashed line

high-intensity discharge (HID) lamp another type of gaseous discharge lamp, except the light is produced without the use of a phosphor coating

home run the part of the branch-circuit wiring that originates in the loadcenter and provides electrical power to the first electrical box in the circuit

horizontal cabling the connection from the distribution center to the work area outlets

hydraulic a term used to describe tools that use a pressurized fluid, like oil, to accomplish work

hydromassage bathtub a permanently installed bathtub with recirculating piping, pump, and associated equipment; it is designed to accept, circulate, and discharge water each use

hydronic system a term used when referring to a hot water heating system

IDC an acronym for "insulation displacement connection"; a type of termination where the wire is "punched down" into a metal holder with a punch-down tool; no prior stripping of the wire is required

impulse a type of transient voltage that originates outside the home and is usually caused by utility company switching or lightning strikes

incandescent lamp the original electric lamp; light is produced when an electric current is passed through a filament; the filament is usually made of tungsten

individual branch circuit a circuit that supplies only one piece of electrical equipment; examples are one range, one space heater, or one motor

inductive heating the heating of a conducting material in an expanding and collapsing magnetic field; inductive heating will occur when current-carrying conductors of a circuit are brought through separate holes in a metal electrical box or enclosure

insulated refers to a conductor that is covered by a material that is recognized by the *National Electrical Code*® as electrical insulation

insulator a material that does not allow electrical current to flow through it; examples are rubber, plastic, and glass

interconnecting the process of connecting together smoke detectors so that if one is activated, they all will be activated

interrupting rating the highest current at rated voltage that a device is intended to interrupt under standard test conditions

inverse time circuit breaker a type of circuit breaker that has a trip time that gets faster as the fault current flowing through it gets larger; this is the circuit breaker type used in house wiring

jack he receptacle for an RJ-45 plug

jacketed cable a voice/data cable that has a nonmetallic polymeric protective covering placed over the conductors

jet pump a type of water pump used in home water systems; the pump and electric motor are separate items that are located away from the well in a basement, garage, crawl space, or other similar area

jug handles a term used to describe the type of bend that must be made with certain cable types; bending cable too tightly will result in damage to the cable and conductor insulation; bending them in the shape similar to a "handle" on a "jug" will help satisfy *NEC*® bending requirements

junction box a box whose purpose is to provide a protected place for splicing electrical conductors

kilowatt-hour meter an instrument that measures the amount of electrical energy supplied by the electric utility company to a dwelling unit

knockout (KO) a part of an electrical box that is designed to be removed, or "knocked out," so that a cable or raceway can be connected to the box

knockout plug a piece of electrical equipment used to fill unused openings in boxes, cabinets, or other electrical equipment

knockout punch a tool used to cut holes in electrical boxes for the attachment of cables and conduits

laundry branch circuit a type of branch circuit found in residential wiring that supplies electrical power to laundry areas; no lighting outlets or other receptacles may be connected to this circuit

leader a solid line that may or may not be drawn at an angle and has an arrow on the end of it; it is used to connect a note or dimension to a part of the building

legend a part of a building plan that describes the various symbols and abbreviations used on the plan

level perfectly horizontal; completely flat; a tool used to determine if an object is level

lighting outlet an outlet intended for the direct connection of a lamp holder, a luminaire (lighting fixture), or a pendant cord terminating in a lamp holder

line side the location in electrical equipment where the incoming electrical power is connected; an example is the line-side lugs in a meter socket where the incoming electrical power conductors are connected

load (electrical) a part of an electrical circuit that uses electrical current to perform some function; an example would be a lightbulb (produces light) or electric motor (produces mechanical energy)

load side the location in electrical equipment where the outgoing electrical power is connected; an example is the load-side lugs in a meter socket where the outgoing electrical power conductors to the service equipment are connected

loadcenter a type of panelboard normally located at the service entrance in a residential installation and usually containing the main service disconnect switch

lug a device commonly used in electrical equipment used for terminating a conductor

lumen the unit of light energy emitted from a light source

luminaire a complete lighting unit consisting of a lamp or lamps together with the parts designed to distribute the light, to position and protect the lamps and ballast (where applicable), and to connect the lamps to the power supply

Madison hold-its thin metal straps that are used to hold "old-work" electrical boxes securely to an existing wall or ceiling

main bonding jumper a jumper used to provide the connection between the grounded service conductor and the equipment-grounding conductor at the service

manual ranging meter a meter feature that requires the user to manually select the proper range

mast kit a package of additional equipment that is required for the installation of a mast-type service entrance; it can be purchased from an electrical distributor

Material Safety Data Sheet (MSDS) a form that lists and explains each of the hazardous materials that electricians may work with so they can safely use the material and respond to an emergency situation

megabits per second (Mbps) refers to the rate that digital bits (1s and 0s) are sent between two pieces of equipment

megahertz (MHz) refers to the upper frequency band on the ratings of a cabling system

megohmmeter a measuring instrument that measures large amounts of resistance and is used to test electrical conductor insulation

meter enclosure the weatherproof electrical enclosure that houses the kilowatt-hour meter; also called the "meter socket" or "meter trim"

mil 1 mil is equal to .001 inches; this is the unit of measure for the diameter of a conductor

multimeter a measuring instrument that is capable of measuring many different electrical values, such as voltage, current, resistance, and frequency, all in one meter

multiwire circuit a circuit in residential wiring that consists of two ungrounded conductors that have 240 volts between them and a grounded conductor that has 120 volts between it and each ungrounded conductor

nameplate the label located on an appliance that contains information such as amperage and voltage ratings, wattage ratings, frequency, and other information needed for the correct installation of the appliance

National Electrical Code® (*NEC*®) a document that establishes minimum safety rules for an electrician to follow when performing electrical installations; it is published by the National Fire Protection Association (NFPA)

National Electrical Manufacturers Association (NEMA) an organization that establishes certain construction standards for the manufacture of electrical equipment; for example, a NEMA Type 1 box purchased from Company X will meet the same construction standards as a NEMA Type 1 box from Company Y

new work box an electrical box without mounting ears; this style of electrical box is used to install electrical wiring in a new installation

nipple an electrical conduit of less than 2 feet in length used to connect two electrical enclosures

noncontact voltage tester a tester that indicates if a voltage is present by lighting up, making a noise, or vibrating; the tester is not actually connected into the electrical circuit but is simply brought into close proximity of the energized conductors or other system parts

no-niche luminaire a lighting fixture intended for above or below the water-level installation; it does not have a forming shell that it fits into but rather sits on the surface of the pool wall; it can be located above or below the waterline

nonlinear loads a load where the load impedance is not constant, resulting in harmonics being present on the electrical circuit

object line a solid dark line that is used to show the main outline of the building

Occupational Safety and Health Administration (OSHA) since 1971, OSHA's job has been to establish and enforce workplace safety rules

offset bend a type of conduit bend that is made with two equal-degree bends in such a way that the conduit changes elevation and avoids an obstruction

ohm the unit of measure for electrical resistance

ohmmeter a measuring instrument that measures values of resistance

Ohm's law the mathematical relationship between current, voltage, and resistance in an electrical circuit

old work box an electrical box with mounting ears; this style of electrical box is used to install electrical wiring in existing installations

open circuit a circuit that is energized but does not allow useful current to flow on the circuit because of a break in the current path

outlet a point on the wiring system at which current is taken to supply electrical equipment; an example is a lighting outlet or a receptacle outlet

outlet box a box that is designed for the mounting of a receptacle or a lighting fixture

overcurrent any current in excess of the rated current of equipment or the ampacity of a conductor; it may result from an overload, a short circuit, or a ground fault

overcurrent protection device (OCPD) a fuse or circuit breaker used to protect an electrical circuit from an overload, a short circuit, or a ground fault

overload a larger-than-normal current amount flowing in the normal current path

panelboard a panel designed to accept fuses or circuit breakers used for the protection and control of lighting, heating, and power circuits; it is designed to be placed in a cabinet and placed in or on a wall; it is accessible only from the front

patch cord a short length of cable with an RJ-45 plug on either end; used to connect hardware to the work area outlet or to connect cables in a distribution panel

permanently installed swimming pool a swimming pool constructed totally or partially in the ground with a water depth capacity of greater than 42 inches (1 m); all pools, regardless of depth, installed in a building are considered permanent

pigtail a short length of wire used in an electrical box to make connections to device terminals

platform frame a method of wood frame construction in which the walls are erected on a previously constructed floor deck or platform

plumb perfectly vertical; the surface of the item you are leveling is at a right angle to the floor or platform you are working from

polarity the positive or negative direction of DC voltage or current

polarized plug a two-prong plug that distinguishes between the grounded conductor and the "hot" conductor by having the grounded conductor prong wider than the hot conductor prong; this plug will fit into a receptacle only one way

pool cover, electrically operated a motor-driven piece of equipment designed to cover and uncover the water surface of a pool

porcelain standoff a fitting that is attached to a service entrance mast, which provides a location for the attachment of the service drop conductors in an overhead service entrance

power source a part of an electrical circuit that produces the voltage required by the circuit

pryout (PO) small parts of electrical boxes that can be "pried" open with a screwdriver and twisted off so that a cable can be secured to the box

pulling in the process of installing cables through the framework of a house

punch-down block the connecting block that terminates cables directly; 110 blocks are most popular for residential situations

raceway an enclosed channel of metal or nonmetallic materials designed expressly for holding wires or cables; raceways used in residential wiring include rigid metal conduit, rigid nonmetallic conduit, intermediate metal conduit, liquid-tight flexible conduit, flexible metal conduit, electrical nonmetallic tubing, and electrical metallic tubing

rafters part of the roof structure that is supported by the top plate of the wall sections; the roof sheathing is secured to the rafters and then covered with shingles or other roofing material to form the roof

rain-tight constructed or protected so that exposure to a beating rain will not result in the entrance of water under specified test conditions

receptacle a device installed in an electrical box for the connection of an attachment plug

receptacle polarity when the white-colored grounded conductor is attached to the silver screw of the receptacle and the "hot" ungrounded circuit conductor is attached to the brass screw on the receptacle; this results in having the long slot always being the grounded slot and the short slot always being the ungrounded slot on 125-volt, 15- or 20-amp-rated receptacles used in residential wiring

reciprocating to move back and forth

redhead sometimes called a "red devil"; an insulating fitting required to be installed in the ends of a Type AC cable to protect the wires from abrasion

reel a drum having flanges on each end; reels are used for wire or cable storage

resistance the opposition to current flow

RG-59 a type of coaxial cable typically used for residential video applications; EIA/TIA 570A recommends that RG-6 coaxial cable be used instead of RG-59 because of the better performance characteristics of the RG-6 cable

RG-6 (Series 6) a type of coaxial cable that is "quad shielded" and is used in residential structured cabling systems to carry video signals such as cable and satellite television

ribbon a narrow board placed flush in wooden studs of a balloon frame to support floor joists

ring one of the two wires needed to set up a telephone connection; it is connected to the negative side of a battery at the telephone company; it is the telephone industry's equivalent to an ungrounded "hot" conductor in a normal electrical circuit

ring wave a type of transient voltage that originates inside the home and is usually caused by home office photocopiers, computer printers, the cycling on and off of heating, ventilating, and air-conditioning equipment, and spark igniters on gas appliances like furnaces and ranges

riser a length of raceway that extends up a utility pole and encloses the service entrance conductors in an underground service entrance

RJ-11 a popular name for a six-position UTP connector

RJ-45 the popular name for the modular eight-pin connector used to terminate Category 5 UTP cable

Rome Xa trade name for nonmetallic sheathed cable (NMSC); this is the term most electricians use to refer to NMSC

roof flashing/weather collar two parts of a mast-type service entrance that, when used together, will not allow water to drip down and into a house through the hole in the roof that the service mast extends through

rough-in the stage in an electrical installation when the raceways, cable, boxes, and other electrical equipment are installed; this electrical work must be completed before any construction work can be done that covers wall and ceiling surfaces

running boards pieces of board lumber nailed or screwed to the joists in an attic or basement; the purpose of using running boards is to have a place to secure cables during the rough-in stage of installing a residential electrical system

saddle bend a type of conduit bend that results in a conduit run going over an object that is blocking the path of the run; there are two styles: a three-point saddle and a four-point saddle

safety switch a term used sometimes to refer to a disconnect switch; a safety switch may use fuses or a circuit breaker to provide overcurrent protection

scaffolding also referred to as staging; a piece of equipment that provides a platform for working in high places; the parts are put together at the job site and then taken apart and reconstructed when needed at another location

scale the ratio of the size of a drawn object and the object's actual size

schedule a table used on building plans to provide information about specific equipment or materials used in the construction of the house

sconce a wall-mounted lighting fixture

sectional drawing a part of the building plan that shows a cross-sectional view of a specific part of the dwelling

secured (as applied to electrical cables) fastened in place so the cable cannot move; common securing methods include staples, tie wraps, and straps

self-contained spa or hot tub a factory-fabricated unit consisting of a spa or hot tub vessel having integrated water-circulating, heating, and control equipment

service the conductors and equipment for delivering electric energy from the serving utility to the wiring system of the premises served

service center the hub of a structured wiring system with telecommunications, video, and data communications installed; it is usually located in the basement next to the electrical service panel or in a garage; sometimes called a "distribution center"

service conductors the conductors from the service point to the service disconnecting means

service disconnect a piece of electrical equipment installed as part of the service entrance that is used to disconnect the house electrical system from the electric utility's system

service drop the overhead service conductors from the last pole to the point connecting them to the service entrance conductors at the building

service entrance conductors, overhead system the service conductors between the terminals of the service equipment and a point usually outside the building where they are joined by tap or splice to the service drop

service entrance conductors, underground system the service conductors between the terminals of the service equipment and the point of connection to the service lateral

service entrance the part of the wiring system where electrical power is supplied to the residential wiring system from the electric utility; it includes the main panelboard, the electric meter, overcurrent protection devices, and service conductors

service equipment the necessary equipment connected to the load end of the service conductors supplying a building and intended to be the main control and cutoff of the supply

service head the fitting that is placed on the service drop end of service entrance cable or service entrance raceway and is designed to minimize the amount of moisture that can enter the cable or raceway; the service head is commonly referred to as a "weatherhead"

service lateral the underground service conductors between the electric utility transformer, including any risers at a pole or other structure, and the first point of connection to the service entrance conductors in a meter enclosure

service mast a piece of rigid metal conduit or intermediate metal conduit, usually 2 or 2½ inches in diameter, that provides service conductor protection and the proper height requirements for service drops

service point the point of connection between the wiring of the electric utility and the premises wiring

service raceway the rigid metal conduit, intermediate metal conduit, electrical metallic tubing, rigid nonmetallic conduit, or any other approved raceway that encloses the service entrance conductors

setout the distance that the face of an electrical box protrudes out from the face of a building framing member; this distance is dependent on the thickness of the wall or ceiling finish surface material

shall a term used in the *National Electrical Code®* that means that the rule must be followed

sheath the outer covering of a cable that is used to provide protection and to hold everything together as a single unit

sheathing boards sheet material like plywood that is fastened to studs and rafters; the wall or roofing finish material will be attached to the sheathing

sheetrock a popular building material used to finish off walls and ceilings in residential and commercial construction; it is available in standard sizes, such as 4 by 8 feet, and is constructed of gypsum sandwiched between a paper front and back

short circuit an unintended low-resistance path through which current flows around rather than along a circuit's intended current path; it results when two circuit conductors come in contact with each other unintentionally

sill a length of wood that sets on top of the foundation and provides a place to attach the floor joists

sill plate a piece of equipment that, when installed correctly, will help keep water from entering the hole in the side of a house that the service entrance cable from the meter socket to the service panel goes through

single receptacle a single contact device with no other contact device on the same strap (yoke)

single-pole switch a switch type used to control a 120-volt lighting load from one location

small-appliance branch circuit a type of branch circuit found in residential wiring that supplies electrical power to receptacles located in kitchens and dining rooms; no lighting outlets are allowed to be connected to this circuit type

spa or hot tub a hydromassage pool or tub designed for immersion of users; usually has a filter, heater, and motor-driven pump; it can be installed indoors or outdoors, on the ground, or in a supporting structure; a spa or hot tub is not designed to be drained after each use

specifications a part of the building plan that provides more specific details about the construction of the building

spliced connecting two or more conductors with a piece of approved equipment like a wirenut; splices must be done in approved electrical boxes

split-wired receptacle a duplex receptacle wired so that the top outlet is "hot" all the time and the bottom outlet is switch controlled

stab a term used to identify the location on a loadcenter's ungrounded bus bar where a circuit breaker is snapped on

standby power system a backup electrical power system that consists of a generator, transfer switch, and associated electrical equipment; its purpose is to provide electrical power to critical branch circuits when the electrical power from the utility company is not available

storable swimming pool a swimming pool constructed on or above the ground with a maximum water depth capacity of 42 inches (1 m) or a pool with nonmetallic, molded polymeric walls (or inflatable fabric walls) regardless of size or water depth capacity

STP an acronym for "shielded twisted pair" cable; it resembles UTP but has a foil shield over all four pairs of copper conductors and is used for better high-frequency performance and less electromagnetic interference

strap (yoke) the metal frame that a receptacle or switch is built around; it is also used to mount a switch or receptacle to a device box

strip to damage the threads of the head of a bolt or screw

structure in a two-story house, the first-floor ceiling joists are the second floor's floor joists

structured cabling an architecture for communications cabling specified by the EIA/TIA TR41.8 committee and used as a voluntary standard by manufacturers to ensure compatibility

stub-up a type of conduit bend that results in a 90-degree change of direction

subfloor the first layer of floor material that covers the floor joists; usually 4-by-8-plywood or particleboard

submersible pump a type of water pump used in home water systems; the pump and electric motor are enclosed in the same housing and are lowered down a well casing to a level that is below the water line

supplemental grounding electrode a grounding electrode that is used to "back up" a metal water pipe grounding electrode

supported (as applied to electrical cables) held in place so the cable is not easily moved; common supporting methods include running cables horizontally through holes or notches in framing members or using staples, tie wraps, or straps after a cable has been properly secured close to a box according to the *NEC*®

switch box a name used to refer to a box that just contains switches

switch loop a switching arrangement where the feed is brought to the lighting outlet first and a two-wire loop run from the lighting outlet to the switch

symbol a standardized drawing on the building plan that shows the location and type of a particular material or component

take-up the amount that must be subtracted from a desired stub-up height so the bend will come out right; the take-up is different for each conduit size

tempered treated with heat to maximize the metal hardness

thermostat a device used with a heating or cooling system to establish a set temperature for the system to achieve; they are available as line voltage models or low-voltage models; some thermostats can also be programmed to keep a home at a specific temperature during the day and another temperature during the night

thinwall a trade name often used for electrical metallic tubing

threaded hub the piece of equipment that must be attached to the top of a meter socket so that a raceway or a cable connector can be attached to the meter socket

three-way switch a switch type used to control a 120-volt lighting load from two locations

tip the first wire in a pair; a conductor in a telephone cable pair that is usually connected to the positive side of a battery at the telephone company's central office; it is the telephone industry's equivalent to a grounded conductor in a normal electrical circuit

top plate the top horizontal part of a wall framework

torque the turning or twisting force applied to an object when using a torque tool; it is measured in inch-pounds or foot-pounds

transfer switch a switching device for transferring one or more load conductor connections from one power source to another

transformer the electrical device that steps down the 120-volt house electrical system voltage to the 16 volts a chime system needs to operate correctly

transient voltage surge suppressor (TVSS) (receptacle or strip) an electrical device designed to protect sensitive electronic circuit boards from voltage surges

troffer a term commonly used by electricians to refer to a fluorescent lighting fixture installed in the grid of a suspended ceiling

troubleshooting the process of determining the cause of a malfunctioning part of a residential electrical system

true RMS meter a type of meter that allows accurate measurement of AC values in harmonic environments

twistlock receptacle a type of receptacle that requires the attachment plug to be inserted and then turned slightly in a clockwise direction to lock the plug in place; the attachment plug must be turned slightly counterclockwise to release the plug so it can be removed from the receptacle

Type IC a light fixture designation that allows the fixture to be completely covered by thermal insulation

Type Non-IC a light fixture that is required to be kept at least 3 inches from thermal insulation

Type S plug fuse a fuse type that uses different fuse bases for each fuse size; an adapter that matches each Type S fuse size is required; this fuse type is not interchangeable with other Type S fuse sizes

utility box a name used to refer to a metal single gang, surface mounted device box; also called a handy box

utility pole a wooden circular column used to support electrical, video, and telecommunications utility wiring; it may also support the transformer used to transform the high utility company voltage down to the lower voltage used in a residential electrical system

UTP an acronym for "unshielded twisted pair" cable; it is comprised of four pairs of copper conductors and graded for bandwidth as "categories" by EIA/TIA 568; each pair of wires is twisted together

ventricular fibrillation very rapid irregular contractions of the heart that result in the heartbeat and pulse going out of rhythm with each other

volt the unit of measure for voltage

voltage the force that causes electrons to move from atom to atom in a conductor

voltage tester a device designed to indicate approximate values of voltage or to simply indicate if a voltage is present

volt-ampere a unit of measure for alternating current electrical power; for branch-circuit, feeder, and service calculation purposes, a watt and a volt-ampere are considered the same

voltmeter a measuring instrument that measures a precise amount of voltage

VOM a name sometimes used in reference to a "multimeter";, the letters stand for "volt-ohm milliammeter"

wall studs the parts that form the vertical framework of a wall section

wallboard a thin board formed from gypsum and layers of paper that is used often as the interior wall sheathing in residential applications; commonly called "sheetrock"

wall-mounted oven a cooking appliance that is installed in a cabinet or wall and is separated from the counter-mounted cooktop

wet location installations underground or in concrete slabs or masonry in direct contact with the earth in locations subject to saturation with water or other liquids, such as in unprotected areas exposed to the weather

wet-niche luminaire a type of lighting fixture intended for installation in a wall of a pool; it is accessible by removing the lens from the forming shell; this luminaire type is designed so that water completely surrounds the fixture inside the forming shell

Wiggy a trade name for a solenoid type of voltage tester

wirenut a piece of electrical equipment used to mechanically connect two or more conductors together

wiring a term used by electricians to describe the process of installing a residential electrical system

work area outlet the jack on the wall that is connected to the desktop computer by a patch cord

Index

Note: Items in **bold** indicate table or figure entry.